AGROCHEMICALS
Desk Reference

2nd EDITION

John H. Montgomery

LEWIS PUBLISHERS

Boca Raton   New York

Library of Congress Cataloging-in-Publication Data

Montgomery, John H. (John Harold), 1955–
    Agrochemicals desk reference / by John H. Montgomery. — 2nd ed.
       p. cm.
    Includes bibliographical references and index.
    ISBN 1-56670-167-8 (alk. paper)
    1. Agricultural chemicals—Environmental aspects—Handbooks,
manuals, etc. 2. Agricultural chemicals—Handbooks, manuals, etc.
I. Title.
TD196.A34M66   1997
628.5′2—dc21
                                            97-77
                                          CIP

© 1997 by CRC Press LLC
Lewis Publishers is an imprint of CRC Press LLC

No claim to original U.S. Government works
International Standard Book Number 1-56670-167-8
Library of Congress Card Number 97-77
Printed in the United States of America 1 2 3 4 5 6 7 8 9 0
Printed on acid-free paper

PREFACE TO SECOND EDITION

The continuation of research publications dealing with the fate, transport, and reme-diation of hazardous substances in the environment has stimulated this new edition. More than 500 additional references are cited in this book. Most of these citations deal with the fate and transport of agrochemicals in various environmental compartments (e.g., air, soil, groundwater, surface water).

The sections on Environmental Fate have been expanded and include much more information than the previous edition. These sections are subdivided into the following categories: Biological, Soil, Plant, Surface Water, Groundwater, Photolytic, and Chemi-cal/Physical.

During the research process, sufficient information on the fate of additional agrochem-icals necessitated their inclusion in this edition. These additional compounds are Benflu-ralin, Captan, Dicofol, Dienoclor, Fenoxaprop-ethyl, Metaldehyde, Pentachloronitroben-zene and Triclopyr.

When available, odor thresholds in air, taste thresholds in water, experimentally deter-mined bioconcentration factors and half-lives of chemicals in various media are also included. The latter includes photolytic, hydrolysis, biodegradation, and volatilization half-lives. Moreover, four appendices have been added. These include conversion factors between different concentration units, bulk density values of soils and selected rocks, porosity values of selected rocks and unconsolidated sediments, and solubility of miscel-laneous organic compounds not profiled in the text.

PREFACE

The use of pesticides, herbicides, fungicides and other agricultural chemicals has increased significantly over the last several decades. These "agrochemicals" have varying degrees of stability in the environment. Despite their degradation in the environment by a variety of mechanisms, the parent compounds and their degradates persist long enough to adversely impact soils and/or groundwater. As a consequence, the contamination of groundwater by agrochemicals has gained considerable and more serious attention in recent years. The sources of agrochemicals adversely impacting the groundwater environment include, but are not limited to, hazardous waste sites, municipal landfills, forested land areas, agricultural operations (e.g., farmlands, pesticide-treated forests), domestic practices (e.g., fertilizers, herbicides, etc.) and non-point source areas.

The protection of groundwater from agrochemicals and other pollutants requires that regulators and public and private interests cooperate. To this end, professionals from environmental consultants to the local fire officers need reliable, accurate and readily accessible data to accomplish these tasks. Unfortunately, the necessary data are scattered throughout many reference books, papers and journals. This book is designed to include, in one reference, all the information on widely used agrochemicals needed by those involved in the protection and remediation of the groundwater environment. Its format is easy to use by a wide spectrum of professionals. The data fields have been selected to fulfill the minimum technical requirements of the user based on the extensive experience of the author, his colleagues and others.

This book should be useful to government agencies, environmental scientists, emergency response teams and cleanup contractors (for its physicochemical properties, health and exposure data and uses category); environmental personnel from consulting and industrial firms (for its health and exposure data and physicochemical data); chemical engineers, soil scientists and industrial hygienists (for its synonym index, uses category, symptoms of exposure and physicochemical data); real estate developers, insurance underwriters and environmental attorneys (for its uses category and synonym index).

The book is based on more than 1,200 references. Most of the citations reviewed from the documented literature and included in this book pertained to the fate and transport of agrochemicals in the subsurface environment. This book is broad enough and comprehensive enough to serve its purpose as a desk reference, but small enough to be taken out into the field. Every effort has been made to select the most accurate information and present it factually and accurately. The publisher and author would appreciate hearing from readers regarding corrections and suggestions for material that might be included for use in future editions.

The author is grateful to the staff of CRC Press, in particular, Mr. Neil Levine and Ms. Evelyn Krasnow, for their invaluable contributions, suggestions and assistance during the preparation of this book. The author also extends thanks to the many anonymous reviewers for their comments and suggestions on draft proofs.

CONTENTS

INTRODUCTION

The compounds profiled include those herbicides, insecticides and fumigants most commonly found in the groundwater environment. The number of compounds included in this book was narrowed down based on information available in the documented literature. Compounds profiled in this book include pesticides having the potential for contaminating groundwater in New Jersey (Murphy and Fenske, 1987), compounds for which physicochemical data are available from the U.S. Department of Agriculture (1990) and those compounds which have been identified as organic Priority Pollutants promulgated by the U.S. Environmental Protection Agency (U.S. EPA) under the Clean Water Act of 1977 (40 CFR 136, 1977).

The compound headings are those commonly used by the U.S. EPA and many agricultural organizations. Positional and/or structural prefixes set in italic type are not an integral part of the chemical name and are disregarded in alphabetizing. These include *asym-, sym-, n-, sec-, cis-, trans-, α-, β-, γ-, o-, m-, p-, N-*, etc.

Synonyms: These are listed alphabetically following the convention used for the compound headings. Compounds in boldface type are the Chemical Abstracts Service (CAS) names listed in the eighth or ninth Collective Index. If no synonym appears in boldface type, then the compound heading is the CAS assigned name. Synonyms include chemical names, common or generic names, trade names, registered trademarks, government codes and acronyms. All synonyms found in the literature are listed.

Although synonyms were retrieved from several references, most of them were retrieved from the Registry of Toxic Effects of Chemical Substances (RTECS, 1985).

Structure: This is given for every compound regardless of its complexity. The structural formula is a graphic representation of atoms or group(s) of atoms relative to each other. Clearly, the limitation of structural formulas is that they depict these relationships in two dimensions.

DESIGNATIONS

Chemical Abstracts Service (CAS) Registry Number: This is a unique identifier assigned by the American Chemical Society to chemicals recorded in the CAS Registry System. This number is used to access various chemical databases such as the Hazardous Substances Data Bank (HSDB), CAS Online, Chemical Substances Information Network and many others. This entry is also useful to conclusively identify a substance regardless of the assigned name.

Department of Transportation (DOT) Designation: This is a four-digit number assigned by the U.S. Department of Transportation (DOT) for hazardous materials and is identical to the United Nations identification number (which is preceded by the letters UN). This number is required on shipping papers, on placards or orange panels on tanks and on a label or package containing the material. These numbers are widely used for personnel responding to emergency situations, e.g., overturned tractor trailers, in which the identification of the transported material is quickly and easily determined. Additional information may be obtained through the U.S. Department of Transportation, Research and Special Programs Administration, Materials Transportation Bureau, Washington, DC 20590.

Molecular Formula (mf): This is arranged by carbon, hydrogen and remaining elements in alphabetical order in accordance with the system developed by Hill (1900). Molecular formulas are useful in identifying isomers (i.e., compounds with identical molecular formulas) and are required if one wishes to calculate the formula weight of a substance.

Formula Weight (fw): This is calculated to the nearest hundredth using the empirical formula and the 1981 Table of Standard Atomic Weights as reported in Weast (1986). Formula weights are required for many calculations, such as converting weight/volume units, e.g., mg/L or g/L, to molar units (mol/L); with density for calculating molar volumes; and for estimating Henry's law constants.

Registry of Toxic Effects of Chemical Substances (RTECS) Number: Many compounds are assigned a unique accession number consisting of two letters followed by seven numerals. This number is needed to quickly and easily locate additional toxicity and health-based data which are cross-referenced in the RTECS (1985). Contact the National Institute for Occupational Safety and Health (NIOSH), U.S. Department of Health and Human Services, 4676 Columbia Parkway, Cincinnati, OH 45226 for additional information.

PHYSICAL AND CHEMICAL PROPERTIES

Appearance and Odor: The appearance, including the physical state (solid, liquid, or gas) of a chemical at room temperature (20–25°C) is provided. If the compound can be detected by the olfactory sense, the odor is noted. Unless noted otherwise, the information provided in this category is for the pure substance and was obtained from many sources (Hawley, 1981; Verschueren, 1983; Windholz et al., 1983; CHRIS Hazardous Chemical Data, 1984; Sax, 1984; Sittig, 1985; Toxic and Hazardous Industrial Chemicals Safety Manual for Handling and Disposal with Toxicity and Hazard Data, 1986; Sax and Lewis, 1987; Hazardous Substances Data Bank, 1989).

Bioconcentration Factor (BCF): The bioconcentration factor is defined as the ratio of the chemical accumulated in tissue to the concentration in water. Generally, high bioconcentration factors tend to be associated with very lipophilic compounds. Conversely, low bioconcentration factors are associated with compounds having high aqueous solubilities.

Bioconcentration factors have been shown to be correlated with the octanol/water partition coefficient in aquatic organisms (Davies and Dobbs, 1984; de Wolf et al., 1992; Isnard and Lambert, 1988) and fish (Davies and Dobbs, 1984; Kenaga, 1980; Isnard and Lambert, 1988; Neely et al., 1974; Ogata et al., 1984; Oliver and Niimi, 1985).

Boiling Point (bp): This is defined as the temperature at which the vapor pressure of a liquid equals the atmospheric pressure. Unless noted otherwise, all boiling points are reported at one atmosphere pressure (760 mmHg). Although not used in environmental assessments, boiling points for aromatic compounds have been found to be linearly correlated with aqueous solubility (Almgren et al., 1979). Boiling points are also useful in assessing entry of toxic substances into the body. Body contact with high-boiling liquids is the most common means of entry into the body, whereas the inhalation route is the most common for low-boiling liquids (Shafer, 1987).

Dissociation Constant (pK$_a$): In an aqueous solution, an acid (HA) will dissociate into the carboxylate anion (A$^-$) and hydrogen ion (H$^+$) and may be represented by the general equation:

$$HA_{(aq)} \leftrightarrow H^+ + A^- \qquad [1]$$

At equilibrium, the ratio of the products (ions) to the reactant (non-ionized electrolyte) is related by the equation:

$$K_a = ([H^+][A^-]/[HA]) \qquad [2]$$

where K$_a$ is the dissociation constant. This expression shows that K$_a$ increases if there is increased ionization and vice versa. A strong acid (weak base) such as hydrochloric acid ionizes readily and has a large K$_a$, whereas a weak acid (or stronger base) such as benzoic acid ionizes to a lesser extent and has a lower K$_a$. The dissociation constants for weak acids are sometimes expressed as K$_b$, the dissociation constant for the base, and both are related to the dissociation constant for water by the expression:

$$K_w = K_a + K_b \qquad [3]$$

where K$_w$ is the dissociation constant for water (10^{-14} at 25°C), K$_a$ is the acid dissociation constant, K$_b$ is the base dissociation constant.

The dissociation constant is usually expressed as pK$_a$ = $-\log_{10}$K$_a$. Equation [3] becomes:

$$pK_w = pK_a + pK_b \qquad [4]$$

When the pH of the solution and the pK$_a$ are equal, 50% of the acid will have dissociated into ions. The percent dissociation of an acid or base can be calculated if the pH of the solution and the pK$_a$ of the compound are known (Guswa et al., 1984):

For organic acids: $\qquad \alpha_a = [100/(1 + 10^{(pH-pKa)})] \qquad [5]$

For organic bases: $\qquad \alpha_b = [100/(1 + 10^{(pKw-pKb-pH)})] \qquad [6]$

where α_a is the percent of the organic acid that is nondissociated, α_b is the percent of the organic base that is nondissociated, pK$_a$ is the $-\log_{10}$ dissociation constant for an acid, pK$_w$ is the $-\log_{10}$ dissociation constant for water (14.00 at 25°C), pK$_b$ is the $-\log_{10}$ dissociation constant for base (pK$_b$ = pK$_w$–pK$_a$) and pH is the $-\log_{10}$ hydrogen ion activity (concentration) of the solution.

Since ions tend to remain in solution, the degree of dissociation will affect processes such as volatilization, photolysis, adsorption and bioconcentration (Howard, 1989).

Henry's Law Constant (K$_H$): Sometimes referred to as the air-water partition coefficient, Henry's law constant is defined as the ratio of the partial pressure of a compound in air to the concentration of the compound in water at a given temperature under equilibrium conditions. If the vapor pressure and solubility of a compound are known, this parameter can be calculated at 1.0 atm (760 mmHg) as follows:

$$K_H = Pfw/760S \qquad [7]$$

where K_H is Henry's law constant (atm · m³/mol), P is the vapor pressure (mmHg), S is the solubility in water (mg/L) and fw is the formula weight (g/mol).

Henry's law constant can also be expressed in dimensionless form and may be calculated using one of the following equations:

$$K_{H'} = K_H/RK \quad \text{or} \quad K_{H'} = S_a/S \qquad [8]$$

where $K_{H'}$ is Henry's law constant (dimensionless), R is the ideal gas constant (8.20575×10^{-5} atm · m³/mol · K), K is the temperature of water (degrees Kelvin), S_a is the solute concentration in air (mol/L), and S is the aqueous solute concentration (mol/L).

It should be noted that estimating Henry's law constant assumes that the gas obeys the ideal gas law and the aqueous solution behaves as an ideally dilute solution. The solubility and vapor pressure data inputted into the equations are valid only for the pure compound and must be in the same standard state at the same temperature.

The major drawback in estimating Henry's law constant is that both the solubility and the vapor pressure of the compound are needed in equation [7]. If one or both of these parameters is unknown, an empirical equation based on quantitative structure-activity relationships (QSAR) may be used to estimate Henry's law constants (Nirmalakhandan and Speece, 1988). In this QSAR model, only the structure of the compound is needed. From this, connectivity indexes (based on molecular topology), polarizability (based on atomic contributions) and the propensity of the compound to form hydrogen bonds can easily be determined. These parameters, when regressed against known Henry's law constants for 180 organic compounds, yielded an empirical equation that explained more than 98% of the variance in the data set having an average standard error of only 0.262 logarithm units.

Henry's law constant may also be estimated using the bond or group contribution method developed by Hine and Mookerjee (1975). The constants for the bond and group contributions were determined using experimentally determined Henry's law constants for 292 compounds. The authors found that those estimated values significantly deviating from observed values (particularly for compounds containing halogen, nitrogen, oxygen and sulfur substituents) could be explained by "distant polar interactions", i.e., interactions between polar bonds or structural groups.

A more recent study for estimating Henry's law constants using the bond contribution method was provided by Meylan and Howard (1991). In this study, the authors updated and revised the method developed by Hine and Mookerjee (1975) due to new experimental data that have become available since 1975. Bond contribution values were determined for 59 chemical bonds based on known Henry's law constants for 345 organic compounds. A good statistical fit [correlation coefficient (r^2) = 0.94] was obtained when the bond contribution values were regressed against known Henry's law constants for all compounds. For selected chemicals classes, r^2 increased slightly to 0.97.

Russell et al. (1992) conducted a similar study using the same data set from Hine and Mookerjee (1975). They developed a computer-assisted model based on five molecular descriptors which were related to the compound's bulk, lipo-philicity and polarity. They found that 63 molecular structures were highly correlative with the log of Henry's law constants ($r^2 = 0.96$).

Henry's law constants provided an indication of the relative volatility of a substance. According to Lyman et al. (1982), if $K_H < 10^{-7}$ atm $\cdot$ m³/mol, the substance has a low volatility. If K_H is $> 10^{-7}$ but $< 10^{-5}$ atm $\cdot$ m³/mol, the substance will volatilize slowly. Volatilization becomes an important transfer mechanism in the range $10^{-5} < H < 10^{-3}$ atm $\cdot$ m³/mol. Values of $K_H > 10^{-3}$ atm $\cdot$ m³/mol indicate volatilization will proceed rapidly.

The rate of volatilization will also increase with an increase in temperature. ten Hulscher et al. (1992) studied the temperature dependence of Henry's law constants for three chlorobenzenes, three chlorinated biphenyls and six polynuclear aromatic hydrocarbons. They observed that over the temperature range of 10 to 55°C, Henry's law constant was doubled for every 10°C increase in temperature. This temperature relationship should be considered when assessing the role of chemical volatilization from large surface water bodies whose temperatures are generally higher than those typically observed in ground-water.

Hydrolysis Half-Life (H-$t_{1/2}$): The hydrolysis half-life of a chemical is the time that it takes to reach one-half or 50% of its original concentration. The rate of chemical hydrolysis is highly dependent upon the compound's solubility, tempera-ture and pH. Since other environmental factors such as photolysis, volatility (i.e., Henry's law constants) and adsorption can affect the rate of hydrolysis, these factors are virtually eliminated by performing hydrolysis experiments under carefully controlled laboratory conditions. The hydrolysis half-lives reported in the literature were calculated using experimentally deter-mined hydrolysis rate constants.

Ionization Potential (IP): The ionization potential of a compound is defined as the energy required to remove a given electron from the molecule's atomic orbit (outermost shell) and is expressed in electron volts (eV). One electron volt is equivalent to 23,053 cal/mol.

Knowing the ionization potential of a contaminant is required in determining the appropriate photoionization lamp for detecting that contaminant or family of contaminants. Photoionization instruments are equipped with a radiation source (ultra-violet lamp), pump, ionization chamber, an amplifier and a recorder (either digital or meter). Generally, compounds with ionization potentials smaller than the radiation source (UV lamp rating) being used will readily ionize and will be detected by the instrument. Conversely, com-pounds with ionization potentials higher than the lamp rating will not ionize and will not be detected by the instrument.

Soil/Sediment Partition Coefficient (K_{oc}): The soil/sediment partition or sorption coef-ficient is defined as the ratio of adsorbed chemical per unit weight of organic carbon to the aqueous solute concentration. This value provides an indication of the tendency of a chemical to partition between particles containing organic carbon and water. Compounds that bind strongly to organic carbon have characteristically low solubilities, whereas compounds with low tendencies to adsorb onto organic particles have high solubilities.

Non-ionizable chemicals that sorb onto organic materials in an aquifer (i.e., organic carbon) are retarded in their movement in groundwater. The sorbing solute travels at a linear velocity that is lower than the groundwater flow velocity by a factor of R_d, the retardation factor. If the K_{oc} of a compound is known, the retardation factor may be calculated using the following equation from Freeze and Cherry (1974) for unconsolidated sediments:

$$R_d = V_w/V_c = [1 + (BK_d/n_e)]$$ [9]

where R_d is the retardation factor (unitless), V_w is the average linear velocity of groundwater (e.g., ft/day), V_c is the average linear velocity of contaminant (e.g., ft/day), B is the average soil bulk density (g/cm^3), n_e is the effective porosity (unitless), and K_d is the distribution (sorption) coefficient (cm^3/g).

By definition, K_d is defined as the ratio of the concentration of the solute on the solid to the concentration of the solute in solution. This can be represented by the Freundlich equation:

$$K_d = VM_S/MM_L = C_S/C_L{}^n$$ [10]

where V is the volume of the solution (cm^3), M_S is the mass of the sorbed solute (g), M is the mass of the porous medium (g), M_L is the mass of the solute in solution (g), C_S is the concentration of the sorbed solute (g/cm^3), C_L is the concentration of the solute in the solution (g/cm^3) and n is a constant.

Values of n are normally between 0.7 and 1.1 although values of 1.6 have been reported (Lyman et al., 1982). If n is unknown, it is assumed to be unity and a plot of C_S versus C_L will be linear. The distribution coefficient is related to K_{oc} by the equation:

$$K_{oc} = K_d/f_{oc}$$ [11]

where f_{oc} is the fraction of naturally occurring organic carbon in soil.
Sometimes K_d is expressed on an organic-matter basis and is defined as:

$$K_{om} = K_d/f_{om}$$ [12]

where f_{om} is the fraction of naturally occurring organic matter in soil. The relationship between K_{oc} and K_{om} is defined as:

$$K_{om} = 0.58K_{oc}$$ [13]

where the constant 0.58 is assumed to represent the fraction of carbon present in the soil or sediment organic matter (Allison, 1965).
For fractured rock aquifers in which the porosity of the solid mass between fractures is insignificant, Freeze and Cherry (1974) report the retardation equation as:

$$R_d = V_w/V_c = [1 + (2K_A/b)]$$ [14]

where K_A is the distribution coefficient (cm) and b is the aperture of fracture (cm).
To calculate the retardation factors for ionizable compounds such as acids and bases, the fraction of un-ionized acid (α_a) or base (α_b) needs to be determined (see **Dissociation Constant**). According to Guswa et al. (1984), if it is assumed only the un-ionized portion of the acid is adsorbed onto the soil, the retardation factor for the acid becomes:

$$R_a = [1 + (\alpha_a BK_d/n_e)]$$ [15]

However, for a base they assume that the ionized portion is exchanged with a monovalent ion and the un-ionized portion of the base is adsorbed hydrophobically. Therefore, the retardation factor for the base is:

$$R_b = \{1 + [(\alpha_b B K_d)/n_e] + [CECB(1 - \alpha_b)]/(100\Sigma z^+ n_e)\} \qquad [16]$$

where CEC is the cation exchange capacity of the soil (cm^3/g) and Σz^+ is the sum of all positively charged particles in the soil (milliequivalents/cm^3). Guswa et al. (1984) report that the term Σz^+ is approximately 0.001 for most agricultural soils.

Correlations between K_{oc} and bioconcentration factors in fish and beef have shown a log-log linear relationship (Kenaga, 1980) as well as solubility of organic compounds in water (Means et al., 1980; Abdul et al., 1987). Moreover, the log K_{oc} has been shown to be related to molecular connectivity indices (Sabljić and Protić, 1982; Koch, 1983; Govers et al., 1984; Sabljić, 1984; Gerstl and Helling, 1987; Sabljić, 1987; Meylan et al., 1992) and high performance liquid chromatography (HPLC) capacity factors (Haky and Young, 1984; Hodson and Williams, 1988; Szabo et al., 1990, 1990a).

In instances where experimentally determined K_{oc} values are not available, they can be estimated using recommended regression equations as cited in Lyman et al. (1982) or Meylan et al. (1992). All the K_{oc} estimations are based on regression equations in which the aqueous solubility or the K_{ow} of the substance is known.

Melting Point (mp): The melting point of a substance is defined as the temperature at which a solid substance undergoes a phase change to a liquid. The reverse process, the temperature at which a liquid freezes to a solid, is called the freezing point. For a given substance, the melting point is identical to the freezing point.

Unless noted otherwise, all melting points are reported at the standard pressure of 1.0 atmosphere (760 mmHg). Although the melting point of a substance is not directly used in predicting its behavior in the environment, it is useful in determining the phase in which the substance would be found under typical conditions.

Octanol/Water Partition Coefficient (K_{ow}): The K_{ow} of a substance is the *n*-octanol/water partition coefficient and is defined as the ratio of the solute concentration in the water-saturated *n*-octanol phase to the solute concentration in the *n*-octanol-saturated water phase. Values of K_{ow} are therefore unitless.

The partition coefficient has been recognized as a key parameter in predicting the environmental fate of organic compounds. The log K_{ow} has been shown to be linearly correlated with log bioconcentration factors (BCF) in aquatic organisms (Davies and Dobbs, 1984; Isnard and Lambert, 1988; de Wolf et al., 1992), in fish (Neely et al., 1974; Kenaga, 1980; Davies and Dobbs, 1984; Ogata et al., 1984; Oliver and Niimi, 1985; Isnard and Lambert, 1988), log soil/sediment partition coefficients (K_{oc}) (Chiou et al., 1979; Kenaga and Goring, 1980), log of the solubility of organic compounds in water (Hansch et al., 1968; Chiou et al., 1977; Yalkowsky and Valvani, 1979, 1980; Banerjee et al., 1980; Chiou et al., 1982; Tewari et al., 1982; Miller et al., 1984, 1985; Isnard and Lambert, 1988), molecular surface area (Yalkowsky and Valvani, 1979, 1980; Miller et al., 1984; Funasaki et al., 1985; Camilleri et al., 1988; Woodburn et al., 1992), molar refraction (Yoshida et al., 1983), molecular connectivity indices (Govers et al., 1984; Woodburn et al., 1992), reversed-phase liquid chromatography (RPLC) retention factors (Khaledi and Breyer, 1989; Woodburn et al., 1992), RPLC capacity factors (Braumann, 1986; Minick et al., 1989), isocratic RPLC capacity factors (Hafkenscheid and Tomlinson, 1983); HPLC capacity factors (Carlson et al., 1975; Harnisch et al., 1983; Brooke et al., 1986; Eadsforth, 1986; Kraak et al., 1986; DeKock and Lord, 1987; Hammers et al., 1982; Miyake et al., 1982, 1987, 1988; Szabo et al., 1990), HPLC retention times (Mirrlees et al., 1976; Veith and Morris, 1978; Sarna et al., 1984; Webster et al., 1985; Burkhard and Kuehl, 1986), reversed-phase thin layer chromatography retention parameters (Bruggeman et al., 1982), gas chromato-graphy retention indices (Valkó et al., 1984), distribution coefficients (Campbell et al., 1983), solvatochromic parameters (Sadek et al., 1985), biological responses (Schultz et al., 1989), log of the n-hexane/water and L-α-phosphatidylcholine dimyristol partitioning coefficients (Gobas et al., 1988), molecular descriptors and physicochemical properties (Bodor et al., 1989; Warne et al., 1990), molecular structure (Suzuki, 1991), linear solvation energy relationships (LSER) (Kamlet et al., 1988) and substituent constants which are based on empirically derived atomic or group constants and structural factors (Hansch and Anderson, 1967; Hansch et al., 1972). Variables needed for employing the LSER method have recently been presented by Hickey and Passino-Reader (1991).

For ionizable compounds (e.g., acids, amines and phenols), K_{ow} values are a function of pH. Unfortunately, many investigators have neglected to report the pH of the solution at which the K_{ow} was determined. If a K_{ow} value is used for an ionizable compound for which the pH is known, both values should be noted.

Photolysis Half-Life (P-$t_{1/2}$): The photolysis half-life of a chemical is the time required for the parent chemical to reach one-half or 50% of its original concentration. Chemicals will undergo photolysis if they can absorb sunlight. Photolysis can occur in air, soil, water and plants. The rate of photolysis is dependent upon the pH, temperature, presence of sensitizers, sorption to soil, and depth of the compound in soil and water. Lyman et al. (1982) present an excellent overview of the photolysis process.

Solubility in Organics (S$_o$): The presence of small quantities of solvents can enhance a compound's solubility in water (Nyssen et al., 1987). Consequently, its fate and transport in soils, sediments and groundwater will be changed due to the presence of these cosolvents. For example, soils contaminated with compounds having low water solubilities tend to remain bound to the soil by adsorbing onto organic carbon and/or by interfacial tension with water. A solvent introduced to an unsaturated soil environment (e.g., a surface spill, leaking aboveground tank, etc.) may come in contact with existing soil contaminants. As the solvent interacts with the existing contamination, it may mobilize it, thereby facilitating its migration. Consequently, the organic solvent can facilitate the leaching of contaminants from the soil to the water table. Therefore, the presence of cosolutes must be considered when predicting the fate and transport of contaminants in the unsaturated zone, the water table and surface water bodies.

Solubility in Water (S$_w$): The water solubility of a compound is defined as the saturated concentration of the compound in water at a given temperature and pressure. This parameter is perhaps the most important factor in estimating a chemical's fate and transport in the aquatic environment. Compounds with high water solubilities tend to desorb from soils and sediments (i.e., they have low K_{oc} values), are less likely to volatilize from water and are susceptible to biodegradation. Conversely, compounds with low solubilities tend to adsorb onto soils and sediments (have high K_{oc}), volatilize more readily from water and bioconcentrate in aquatic organisms. The more soluble compounds commonly enter the water table more readily than their less soluble counterparts.

The water solubility of a compound varies with temperature, pH (particularly, ionizable compounds such as acids and bases) and other dissolved constituents, e.g., inorganic salts (electrolytes) and organic chemicals including naturally occurring organic carbon, such as humic and fulvic acids. At a given temperature, the variability/discrepancy of water-solubility measurements documented by investigators may be attributed to one or more of the following: (1) purity of the compound, (2) analytical method employed, (3) particle size (for solid solubility determinations only), (4) adsorption onto the container and/or suspended solids, (5) time allowed for equilibrium conditions to be reached, (6) losses due to volatilization and (7) chemical transformations (e.g., hydrolysis).

The water solubility of chemical substances has been related to bioconcentration factors (BCF), soil/sediment partition coefficients (K_{oc}) (Means et al., 1980; Abdul et al., 1987), n-octanol/water partition coefficients (K_{ow}) (Hansch et al., 1968; Chiou et al., 1977; Yalkowsky and Valvani, 1979, 1980; Chiou et al., 1982; Miller et al., 1984, 1985), HPLC capacity factors (Hafkenscheid and Tomlinson, 1981; Whitehouse and Cooke, 1982), molecular descriptors and physicochemical properties (Warne et al., 1990), soil organic matter K_{om} (Chiou et al., 1983), total molecular surface area (Hermann, 1972; Amidon et al., 1975; Valvani et al., 1976; Lande and Banerjee, 1981; Lande et al., 1985), the compound's molecular structure (Nirmalakhandan and Speece, 1988a, 1989; Suzuki, 1989), boiling points (Almgren et al., 1979) and for homologous series of hydrocarbons or classes of organic compounds-carbon number (Bell, 1973; Robb, 1976; Mitra et al., 1977; Krzyzanowska and Szeliga, 1978) and molar volumes (Lande and Banerjee, 1981). With the exception of the molecular structure-solubility relationship, regression equations generated from the other relationships have demonstrated a log-log linear relationship for these properties. The reported regression equations are useful in estimating the solubility of a compound in water if experimental values are not available. In addition, the solubility of a compound may be estimated from experimentally determined Henry's law constants (Kamlet et al., 1987) or from measured infinite dilution activity coefficients (Wright et al., 1992).

Unless noted otherwise, all reported solubilities were determined using distilled and/or deionized water. For some compounds, solubilities were determined using groundwater, natural seawater or artificial seawater.

Specific Density (ρ): The specific density, also known as relative density, is defined as:

$$\rho = d_s/d_w \qquad [17]$$

where d_s is the density of a substance (g/mL or g/cm^3) and d_w is the density of distilled water (g/mL or g/cm^3). Values of specific density are unitless and are reported in the form ρ at T_s/T_w where ρ is the specific density of the substance, T_s is the temperature of substance at the time of measurement (°C) and T_w is the water temperature (°C).

For example, the value 1.1750 at 20/4°C indicates a specific density of 1.1750 for the substance at 20°C with respect to water at 4°C. At 4°C, the density of water is exactly 1.0000 g/mL (g/cm^3). Therefore, the specific density of a substance is equivalent to the density of the substance relative to the density of water at 4°C.

The density of a hydrophobic substance enables it to sink or float in water. Density values are especially important for liquids migrating through the un-saturated zone and encountering the water table as "free product." Generally, liquids that are less dense than water "float" on the water table. Conversely, organic liquids that are more dense than water commonly "sink" through the water table, e.g., dense nonaqueous phase liquids (DNAPLs) such as chloroform, dichloroethane and tetrachloroethylene.

Hydrophilic substances, on the other hand, behave differently. Acetone, which is less dense than water, does not float on water because it is freely miscible with it in all proportions. Therefore, the solubility of a substance must be considered in assessing its behavior in the subsurface.

Environmental Fate: Chemicals released in the environment are susceptible to several degradation pathways. These include chemical (i.e., hydrolysis, oxidation, reduction, dealkylation, dealkoxylation, decarboxylation, methylation, isomerization and conjugation), photolysis or photooxidation and biodegradation. Compounds transformed by one or more of these processes may result in the formation of more toxic or less toxic substances. In addition, the transformed product(s) will behave differently than the parent compound due to changes in their physicochemical properties. Many researchers focus their attention on transformation rates rather than the transformation products. Consequently, only limited data exist on the transitional and resultant end products. Where available, compounds that are transformed into identified products as well as environmental fate rate constants and/or half-lives are listed.

In addition to chemical transformations occurring under normal environmental conditions, abiotic degradation products are also included. Types of abiotic transformation processes or treatment technologies fall into two categories — physical and chemical. Types of physical processes used in removing or eliminating hazardous wastes include sedimentation, centrifugation, flocculation, oil/water separation, dissolved air flotation, heavy media separation, evaporation, air stripping, steam stripping, soil vapor extraction with or without air sparging, distillation, soil flushing, chelation, liquid-liquid extraction, supercritical extraction, filtration, carbon adsorption, reverse osmosis, ion exchange, electrodialysis and vitrification. This information can be useful in evaluating abiotic degradation as a possible remedial measure. Chemical processes include neutralization, precipitation, hydrolysis (acid or base catalyzed), photolysis, oxidation-reduction, catalytic oxidation by hydrogen peroxide, alkaline chlorination, electrolytic oxidation, ozonolysis, catalytic dehydrochlorination and alkali metal dechlorination.

Most of the abiotic chemical transformation products reported in this book are limited to only three processes: hydrolysis, photooxidation and chemical oxidation-reduction. These processes are the most widely studied and reported in the literature. Detailed information describing the above technologies, their availability/limitation and company sources is available (U.S. Environmental Protection Agency, 1987).

Vapor Density (vap d): The vapor density of a substance is defined as the ratio of the mass of vapor per unit volume. An equation for estimating vapor density is readily derived from a varied form of the ideal gas law:

$$PV = MRK/fw \qquad [18]$$

where P is the vapor pressure (atm), V is the volume (L), M is the mass (g), R is the ideal gas constant (8.20575×10^{-2} atm · L/mol · K) and K is the temperature (degrees Kelvin). Recognizing that the density of any substance is defined as:

$$d = M/V \qquad [19]$$

Substituting this equation into equation [18] and rearranging and simplifying results in an expression to determine the vapor density (g/L):

$$p = Pfw/RK \qquad [20]$$

At standard temperature (293.15K) and pressure (1 atm), equation [20] simplifies to:

$$p = fw/24.47 \qquad [21]$$

The specific vapor density of a substance relative to air is determined using:

$$p_v = fw/24.47p_{air} \qquad [22]$$

where p_v is the specific vapor density of a substance (unitless) and p_{air} is the vapor density of air (g/L).

The specific vapor density, pv, is simply the ratio of the vapor density of the substance to that of air under the same pressure and temperature. According to Weast (1986), the vapor density of dry air at 20°C and 760 mmHg is 1.204 g/L. At 25°C, the vapor density of air decreases slightly to 1.184 g/L. Calculated specific vapor densities are reported relative to air (set equal to 1) only for compounds which are liquids at room temperature (i.e., 25°C). These are reported in addition to the calculated vapor densities.

Vapor Pressure (vp): The vapor pressure of a substance is defined as the pressure exerted by the vapor (gas) of a substance when it is under equilibrium conditions. It provides a semi-quantitative rate at which it will volatilize from soil and/or water. The vapor pressure of a substance is a required input parameter for calculating the air-water partition coefficient (see **Henry's Law Constant**), which in turn is used in estimating the volatilization rate of compounds from groundwater to the unsaturated zone and from surface water bodies to the atmosphere.

FIRE HAZARDS

Flash Point (fl p): The flash point is defined as the minimum temperature at which a substance releases ignitable flammable vapors in the presence of an ignition source. e.g., spark or flame. Flash points may be determined by two methods — Tag closed cup (ASTM method D56) or Cleveland open cup (ASTM method D93). Unless noted otherwise, all flash point values represent closed cup method determinations. Flash point values determined by the open cup method are slightly higher than those determined by the closed cup method; however, the open cup method is more representative of actual conditions.

According to Sax (1984), a material with a flash point of 100°F or less is considered dangerous, whereas a material having a flash point greater than 200°F is considered to have a low flammability. Substances with flash points within this temperature range are considered to have moderate flammabilities.

Lower Explosive Limit (LEL): The minimum concentration (vol % in air) of a flammable gas or vapor required for ignition or explosion to occur in the presence of an ignition source (see also Flash Point).

Upper Explosive Limit (UEL): The maximum concentration (vol % in air) of a flammable gas or vapor required for ignition or explosion to occur in the presence of an ignition source (see also Flash Point).

HEALTH HAZARD DATA

Immediately Dangerous to Life or Health (IDLH): According to the National Institute of Occupational Safety and Health (1987), the IDLH level ". . . for the purpose of respirator selection represents a maximum concentration from which, in the event of respirator failure, one could escape within 30 minutes without experiencing any escape-impairing or irreversible health effects." Concentrations are reported in parts per million (ppm) or milligrams per cubic meter (mg/m^3).

Exposure Limits: The permissible exposure limits (PEL) in air, set by the Occupational Health and Safety Administration (OSHA), can be found in the Code of Federal Regulations (General Industry Standards for Toxic and Hazardous Substances, 1977). Unless noted otherwise, the PEL are 8-hour time-weighted average (TWA) concentrations. If NIOSH (1994) and/or the American Conference of Governmental Industrial Hygienists (ACGIH) has published recommended exposure limits, these are also included. The ACGIH's recommended exposure limits, commonly known as threshold limit values (TLV), are subdivided into three exposure classes (Threshold Limit Values and Biological Exposure Indices for 1987–1988). The TLVs, which are updated annually, are defined as follows:

Threshold Limit Value-Time Weighted Average (TLV-TWA) — the time-weighted average (TWA) concentration for a normal 8-hour workday and a 40-hour workweek, to which nearly all workers may be repeatedly exposed, day after day, without adverse effect.

Threshold Limit Value-Short Term Exposure Limit (TLV-STEL) — the concentration to which workers can be exposed continuously for a short period of time without suffering from 1) irritation, 2) chronic or irreversible tissue damage, or 3) narcosis of sufficient degree to increase the likelihood of accidental injury, impair self-rescue or materially reduce work efficiency and provided that the daily TLV-TWA is not exceeded. It is not a separate independent exposure limit; rather, it supplements the TWA limit where there are recognized acute toxic effects from a substance whose toxic effects are primarily of a chronic nature. STELs are recommended only where toxic effects have been reported from high short-term exposures in either humans or animals.

STEL is defined as a 15-minute time-weighted average exposure which should not be exceeded at any time during a workday even if the 8-hour TWA is within the TLV. Exposures at the STEL should not be longer than 15 minutes and should not be repeated more than four times per day. There should be at least 60 minutes between successive exposures at the STEL. An averaging period other than 15 minutes may be recommended when this is warranted by observed biological effects.

Threshold Limit Value-Ceiling (TLV-C) — the concentration that should not be exceeded during any part of the working exposure.

For additional information from OSHA, write to Technical Data Center, U.S. Department of Labor, Washington, DC 20210. The ACGIH's address is 6500 Glenway Ave., Building D-7, Cincinnati, OH 45211-7881.

Formulation Types: Types of formulations, e.g., emulsifiable concentrates, water or oil solubles, water dispersible liquids and granules, are provided. These were obtained from several sources (Hartley and Kidd, 1987; Ashton and Monaco, 1991; Worthing and Walker, 1991).

Toxicology: Information on toxicity to aquatic life was obtained primarily from the Royal Society of Chemistry (Hartley and Kidd, 1987) and the Chemical Hazard Response Information System (CHRIS) Manual (1984). Information on toxicity to rats and/or mice were obtained from Ashton and Monaco (1991), Hartley and Kidd (1987) and RTECS (1985). The absence of toxicity data does not imply that toxic effects do not exist.

Uses: Descriptions of specific agricultural uses are based on one or more of the following sources — HSDB (1989), CHRIS Manual (1984), Sittig (1985) and Verschueren (1983). This information is useful in attempting to identify potential sources of the industrial and environmental contamination.

REFERENCES

Abdul, S.A., T.L. Gibson, and D.N. Rai. "Statistical Correlations for Predicting the Partition Coefficient for Nonpolar Organic Contaminants between Aquifer Organic Carbon and Water," *Haz. Waste Haz. Mater.,* 4(3):211–222 (1987).

Allison, L.E. "Organic Carbon" in *Methods of Soil Analysis, Part 2.,* Black, C., Evans, D., White, J., Ensminger, L., and F. Clark, eds. (Madison, WI: American Society of Agronomy, 1965), pp. 1367–1378.

Almgren, M., F. Grieser, J.R. Powell, and J.K. Thomas. "A Correlation between the Solubility of Aromatic Hydrocarbons in Water and Micellar Solutions, with Their Normal Boiling Points," *J. Chem. Eng. Data,* 24(4):285–287 (1979).

Amidon, G.L., S.H. Yalkowsky, S.T. Anik, and S.C. Valvani. "Solubility of Nonelectrolytes in Polar Solvents. V. Estimation of the Solubility of Aliphatic Monofunctional Compounds in Water Using a Molecular Surface Area Approach," *J. Phys. Chem.,* 79(21):2239–2246 (1975).

Ashton, F.M. and T.J. Monaco. *Weed Science* (New York: John Wiley & Sons, Inc., 1991), 466 p.

Banerjee, S., S.H. Yalkowsky, and S.C. Valvani. "Water Solubility and Octanol/Water Partition Coefficients of Organics. Limitations of the Solubility-Partition Coefficient Correlation," *Environ. Sci. Technol.,* 14(10):1227–1229 (1980).

Bell, G.H. "Solubilities of Normal Aliphatic Acids, Alcohols and Alkanes in Water," *Chem. Phys. Lipids,* 10:1–10 (1973).

Bodor, N., Z. Gabanyi, and C.-K. Wong. "A New Method for the Estimation of Partition Coefficient," *J. Am. Chem. Soc.,* 111(11):3783–3786 (1989).

Braumann, T. "Determination of Hydrophobic Parameters By Reversed-Phase Liquid Chromatography: Theory, Experimental Techniques, and Application in Studies on Quantitative Structure-Activity Relationships," *J. Chromatogr.*, 373:191–225 (1986).

Brooke, D.N., A.J. Dobbs, and N. Williams. "Octanol:Water Partition Coefficients (P): Measurement, Estimation, and Interpretation, Particularly for Chemicals with P >10^5," *Ecotox. Environ. Safety*, 11(3):251–260 (1986).

Bruggeman, W.A., J. Van Der Steen, and O. Hutzinger. "Reversed-Phase Thin-Layer Chromatography of Polynuclear Aromatic Hydrocarbons and Chlorinated Biphenyls. Relationship with Hydrophobicity as Measured by Aqueous Solubility and Octanol-Water Partition Coefficient," *J. Chromatogr.*, 238:335–346 (1982).

Burkhard, L.P. and D.W. Kuehl. "*n*-Octanol/Water Partition Coefficients by Reverse Phase Liquid Chromatography/Mass Spectrometry for Eight Tetrachlorinated Planar Molecules," *Chemosphere*, 15(2):163–167 (1986).

Camilleri, P., S.A. Watts, and J.A. Boraston. "A Surface Area Approach to Determination of Partition Coefficients," *J. Chem. Soc., Perkin Trans. 2*, (September 1988), pp. 1699–1707.

Campbell, J.R., R.G. Luthy, and M.J.T. Carrondo. "Measurement and Prediction of Distribution Coefficients for Wastewater Aromatic Solutes," *Environ. Sci. Technol.*, 17(10):582–590 (1983).

Carlson, R.M., R.E. Carlson, and H.L. Kopperman. "Determination of Partition Coefficients by Liquid Chromatography," *J. Chromatogr.*, 107:219–223 (1975).

Chiou, C.T., V.H. Freed, D.W. Schmedding, and R.L. Kohnert. "Partition Coefficients and Bioaccumulation of Selected Organic Chemicals," *Environ. Sci. Technol.*, 11(5):475–478 (1977).

Chiou, C.T., L.J. Peters, and V.H. Freed. "A Physical Concept of Soil-Water Equilibria for Nonionic Organic Compounds," *Science (Washington, DC)*, 206(4420):831–832 (1979).

Chiou, C.T., P.E. Porter, and D.W. Schmedding. "Partition Equilibria of Nonionic Organic Compounds between Organic Matter and Water," *Environ. Sci. Technol.*, 17(4):227–231 (1983).

Chiou, C.T., D.W. Schmedding, and M. Manes. "Partitioning of Organic Compounds in Octanol-Water Systems," *Environ. Sci. Technol.*, 16(1):4–10 (1982).

"CHRIS Hazardous Chemical Data, Vol. 2," U.S. Department of Transportation, U.S. Coast Guard, U.S. Government Printing Office (November, 1984).

Davies, R.P. and A.J. Dobbs. "The Prediction of Bioconcentration in Fish," *Water Res.*, 18(10):1253–1262 (1984).

DeKock, A.C. and D.A. Lord. "A Simple Procedure for Determining Octanol-Water Partition Coefficients using Reverse Phase High Performance Liquid Chromatography (RPHPLC)," *Chemosphere,* 16(1):133–142 (1987).

de Wolf, W., J.H.M. de Bruijn, W. Sienen, and J.L.M. Hermens. "Influence of Biotransformation on the Relationship between Bioconcentration Factors and Octanol-Water Partition Coefficients," *Environ. Sci. Technol.*, 26(6):1197–1201 (1992).

Eadsforth, C.V. "Application of Reverse-Phase H.P.L.C. for the Determination of Partition Coefficients," *Pestic. Sci.,* 17(3):311–325 (1986).

Freeze, R.A. and J.A. Cherry. *Groundwater* (Englewood Cliffs, NJ: Prentice-Hall, Inc., 1974), 604 p.

Funasaki, N., S. Hada, and S. Neya. "Partition Coefficients of Aliphatic Ethers — Molecular Surface Area Approach," *J. Phys. Chem.,* 89(14):3046–3049 (1985).

"General Industry Standards for Toxic and Hazardous Substances," U.S. Code of Federal Regulations, 29 CFR 1910.1000, Subpart Z (January 1977).

Gerstl, Z. and C.S. Helling. "Evaluation of Molecular Connectivity as a Predictive Method for the Adsorption of Pesticides in Soils," *J. Environ. Sci. Health,* B22(1):55–69 (1987).

Gobas, F.A.P.C., J.M. Lahittete, G. Garofalo, W.Y. Shiu, and D. Mackay. "A Novel Method for Measuring Membrane-Water Partition Coefficients of Hydrophobic Organic Chemicals: Comparison with 1-Octanol-Water Partitioning," *J. Pharm. Sci.,* 77(3):265–272 (1988).

Govers, H., C. Ruepert, and H. Aiking. "Quantitative Structure-Activity Relation-ships for Polycyclic Aromatic Hydrocarbons: Correlation between Molecular Connectivity, Physico-Chemical Properties, Bioconcentration and Toxicity in *Daphnia Pulex,*" *Chemosphere,* 13(2):227–236 (1984).

"Guidelines Establishing Test Procedures for the Analysis of Pollutants," U.S. Code of Federal Regulations, 40 CFR 136, 5–285 (1996).

Guswa, J.H., W.J. Lyman, A.S. Donigan, Jr., T.Y.R. Lo, and E.W. Shanahan. *Groundwater Contamination and Emergency Response Guide* (Park Ridge, NJ: Noyes Publications, 1984), 490 p.

Hafkenscheid, T.L. and E. Tomlinson. "Estimation of Aqueous Solubilities of Organic Non-Electrolytes Using Liquid Chromatographic Retention Data," *J. Chromatogr.,* 218:409–425 (1981).

Hafkenscheid, T.L. and E. Tomlinson. "Correlations Between Alkane/Water and Octan-1-ol/Water Distribution Coefficients and Isocratic Reversed-Phase Liquid Chromato-graphic Capacity Factors of Acids, Bases and Neutrals," *Int. J. Pharm.,* 16:225–239 (1983).

Haky, J.E. and A.M. Young. "Evaluation of a Simple HPLC Correlation Method for the Estimation of the Octanol-Water Partition Coefficients of Organic Compounds," *J. Liq. Chromatogr.,* 7(4):675–689 (1984).

Hammers, W.E., G.J. Meurs, and C.L. De Ligny. "Correlations between Chromato-graphic Capacity Ratio Data on Lichrosorb RP-18 and Partition Coefficients in the Octanol-Water System," *J. Chromatogr.,* 247:1–13 (1982).

Hansch, C. and S.M. Anderson. "The Effect of Intramolecular Hydrophobic Bonding on Partition Coefficients," *J. Org. Chem.,* 32:2853–2586 (1967).

Hansch, C., A. Leo, and D. Nikaitani. "On the Additive-Constitutive Character of Partition Coefficients," J. Org. Chem., 37(20):3090–3092 (1972).

Hansch, C., J.E. Quinlan, and G.L. Lawrence. "The Linear Free-Energy Relation-ship between Partition Coefficients and Aqueous Solubility of Organic Liquids," *J. Org. Chem.,* 33(1):347–350 (1968).

Harnisch, M., H.J. Mockel, and G. Schulze. "Relationship between Log P_{ow} Shake-Flask Values and Capacity Factors Derived from Reversed-Phase High Performance Liquid Chromatography for *n*-Alkylbenzene and Some OECD Reference Substances," *J. Chromatogr.,* 282:315–332 (1983).

Hartley, D. and H. Kidd, Eds. The Agrochemicals Handbook, 2nd ed. (England: Royal Society of Chemistry, 1987).

Hawley, G.G. The Condensed Chemical Dictionary (New York: Van Nostrand Reinhold Co., 1981), 1135 p.

Hazardous Substances Data Bank. National Library of Medicine, Toxicology Information Program (1989).

Hermann, R.B. "Theory of Hydrophobic Bonding. II. The Correlation of Hydro-carbon Solubility in Water with Solvent Cavity Surface Area," *J. Phys. Chem.*, 76(19):2754–2759 (1972).

Hickey, J.P. and D.R. Passino-Reader. "Linear Solvation Energy Relationships: Rules of Thumb" for Estimation of Variable Values," *Environ. Sci. Technol.*, 25(10):1753–1760 (1991).

Hill, E.A. "On a System of Indexing Chemical Literature; Adopted by the Classification Division of the U.S. Patent Office," *J. Am. Chem. Soc.*, 22(8):478–494 (1900).

Hine, J. and P.K. Mookerjee. "The Intrinsic Hydrophobic Character of Organic Compounds. Correlations in Terms of Structural Contributions," *J. Org. Chem.*, 40(3):292–298 (1975).

Hodson, J. and N.A. Williams. "The Estimation of the Adsorption Coefficient (K_{oc}) for Soils by High Performance Liquid Chromatography," *Chemosphere,* 19(1):67–77 (1988).

Howard, P.H. Handbook of Environmental Fate and Exposure Data for Organic Chemicalsρ Volume I. Large Production and Priority Pollutants (Chelsea, MI: Lewis Publishers, Inc., 1989), 574 p.

Isnard, S. and S. Lambert. "Estimating Bioconcentration Factors from Octanol-Water Partition Coefficient and Aqueous Solubility," *Chemosphere,* 17(1):21–34 (1988).

Kamlet, M.J., R.M. Doherty, M.H. Abraham, P.W. Carr, R.F. Doherty, and R.W. Taft. "Linear Solvation Energy Relationships. 41. Important Differences between Aqueous Solubility Relationships for Aliphatic and Aromatic Solutes," *J. Phys. Chem.*, 91(7):1996–2004 (1987).

Kamlet, M.J., R.M. Doherty, P.W. Carr, D. Mackay, M.H. Abraham, and R.W. Taft. "Linear Solvation Energy Relationships. 44. Parameter Estimation Rules That Allow Accurate Prediction of Octanol/Water Partition Coefficients and Other Solubility and Toxicity Properties of Polychlorinated Biphenyls and Polycyclic Aromatic Hydrocarbons," *Environ. Sci. Technol.*, 22(5):503–509 (1988).

Kenaga, E.E. "Correlation of Bioconcentration Factors of Chemicals in Aquatic and Terrestrial Organisms with Their Physical and Chemical Properties," *Environ. Sci. Technol.*, 14(5):553–556 (1980).

Kenaga, E.E. and C.A.I. Goring. "Relationship between Water Solubility, Soil Sorption, Octanol-Water Partitioning and Concentration of Chemicals in Biota," in *Aquatic Toxicology, ASTM STP 707,* Eaton, J.G., P.R. Parrish, and A.C. Hendricks, Eds. (Philadelphia, PA: American Society for Testing and Materials, 1980), pp. 78–115.

Khaledi, M.G. and E.D. Breyer. "Quantitation of Hydrophobicity with Micellar Liquid Chromatography," *Anal. Chem.*, 61(9):1040–1047 (1989).

Koch, R. "Molecular Connectivity Index for Assessing Ecotoxicological Behaviour of Organic Compounds," *Toxicol. Environ. Chem.*, 6(2):87–96 (1983).

Kraak, J.C., H.H. Van Rooij, and J.L.G. Thus. "Reversed-Phase Ion-Pair Systems for the Prediction of *n*-Octanol-Water Partition Coefficients of Basic Compounds by High-Performance Liquid Chromatography," *J. Chromatogr.*, 352:455–463 (1986).

Krzyzanowska, T. and J. Szeliga. "A Method for Determining the Solubility of Individual Hydrocarbons," *Nafta,* 28:414–417 (1978).

Lande, S.S. and S. Banerjee. "Predicting Aqueous Solubility of Organic Non-electrolytes from Molar Volume," *Chemosphere,* 10(7):751–759 (1981).

Lande, S.S., D.F. Hagen, and A.E. Seaver. "Computation of Total Molecular Surface Area from Gas Phase Ion Mobility Data and its Correlation with Aqueous Solubilities of Hydrocarbons," *Environ. Toxicol. Chem.*, 4(3):325–334 (1985).

Lyman, W.J., W.F. Reehl, and D.H. Rosenblatt. *Handbook of Chemical Property Estimation Methods: Environmental Behavior of Organic Compounds* (New York: McGraw-Hill, Inc., 1982).

Means, J.C., S.G. Wood, J.J. Hassett, and W.L. Banwart. "Sorption of Polynuclear Aromatic Hydrocarbons by Sediments and Soils," *Environ. Sci. Technol.,* 14(2):1524–1528 (1980).

Meylan, W. and P.H. Howard. "Bond Contribution Method for Estimating Henry's Law Constants," *Environ. Toxicol. Chem.,* 10(10):1283–1293 (1991).

Meylan, W., P.H. Howard, and R.S. Boethling. "Molecular Topology/Fragment Contribution Method for Predicting Soil Sorption Coefficients," *Environ. Sci. Technol.,* 26(8):1560–1567 (1992).

Miller, M.M., S. Ghodbane, S.P. Wasik, Y.B. Tewari, and D.E. Martire. "Aqueous Solubilities, Octanol/Water Partition Coefficients, and Entropies of Melting of Chlorinated Benzenes and Biphenyls," *J. Chem. Eng. Data,* 29(2):184–190 (1984).

Miller, M.M., S.P. Wasik, G.-L. Huang, W.-Y. Shiu, and D. Mackay. "Relationships between Octanol-Water Partition Coefficient and Aqueous Solubility," *Environ. Sci. Technol.,* 19(6):522–529 (1985).

Minick, D.J., D.A. Brent, and J. Frenz. "Modeling Octanol-Water Partition Coefficients by Reversed-Phase Liquid Chromatography," *J. Chromatogr.,* 461:177–191 (1989).

Mirrlees, M.S., S.J. Moulton, C.T. Murphy, and P.J. Taylor. "Direct Measurement of Octanol-Water Partition Coefficient by High-Pressure Liquid Chromatography," *J. Med. Chem.,* 19(5):615–619 (1976).

Mitra, A., R.K. Saksena, and C.R. Mitra. "A Prediction Plot for Unknown Water Solubilities of Some Hydrocarbons and Their Mixtures," *Chem. Petro-Chem. J.,* 8:16–17 (1977).

Miyake, K., F. Kitaura, N. Mizuno, and H. Terada. "Phosphatidylcholine-Coated Silica as a Useful Stationary Phase for High-Performance Liquid Chromatographic Determination of Partition Coefficients between Octanol and Water," *J. Chromatogr.,* 389(1):47–56 (1987).

Miyake, K., N. Mizuno, and H. Terada. "Effect of Hydrogen Bonding on the High-Performance Liquid Chromatographic Behaviour of Organic Compounds. Relationship between Capacity Factors and Partition Coefficients," *J. Chromatogr.,* 439:227–235 (1988).

Miyake, K. and H. Terada. "Determination of Partition Coefficients of Very Hydrophobic Compounds by High-Performance Liquid Chromatography on Glyceryl-Coated Controlled-Pore Glass," *J. Chromatogr.,* 240(1):9–20 (1982).

Murphy, E. and R. Fenske. "Pesticide Use in New Jersey: Implications for Groundwater Quality," Office of Science and Research, New Jersey Department of Environmental Protection, 1987), 214 p.

Neely, W.B., D.R. Branson, and G.E. Blau. "Partition Coefficient to Measure Bioconcentration Potential of Organic Chemicals in Fish," *Environ. Sci. Technol.,* 8(13):1113–1115 (1974).

"NIOSH Pocket Guide to Chemical Hazards," U.S. Department of Health and Human Services, U.S. Government Printing Office (1994), 398 p.

Nirmalakhandan, N.N. and R.E. Speece. "QSAR Model for Predicting Henry's Constant," *Environ. Sci. Technol.,* 22(11):1349–1357 (1988).

Nirmalakhandan, N.N. and R.E. Speece. "Prediction of Aqueous Solubility of Organic Compounds Based on Molecular Structure," *Environ. Sci. Technol.,* 22(3):328–338 (1988a).

Nirmalakhandan, N.N. and R.E. Speece. "Prediction of Aqueous Solubility of Organic Compounds Based on Molecular Structure. 2. Application to PNAs, PCBs, PCDDs, etc.," *Environ. Sci. Technol.*, 23(6):708–713 (1989).

Nyssen, G.A., E.T. Miller, T.F. Glass, C.R. Quinn II, J. Underwood, J., and D.J. Wilson. "Solubilities of Hydrophobic Compounds in Aqueous-Organic Solvent Mixtures," *Environ. Monit. Assess.*, 9(1):1–11 (1987).

Ogata, M., K. Fujisawa, Y. Ogino, and E. Mano. "Partition Coefficients as a Measure of Bioconcentration Potential of Crude Oil in Fish and Sunfish," *Bull. Environ. Contam. Toxicol.*, 33(5):561–567 (1984).

Oliver, B.G. and A.J. Niimi. "Bioconcentration Factors of Some Halogenated Organics for Rainbow Trout: Limitations in Their Use for Prediction of Environmental Residues," *Environ. Sci. Technol.*, 19(9):842–849 (1985).

"Registry of Toxic Effects of Chemical Substances," U.S. Department of Health and Human Services, National Institute for Occupational Safety and Health (1985), 2050 p.

Robb, I.D. "Determination of the Aqueous Solubility of Fatty Acids and Alcohols," Aust. *J. Chem.*, 18:2281–2285 (1966).

Russell, C.J., S.L. Dixon, and P.C. Jurs. "Computer-Assisted Study of the Relationship between Molecular Structure and Henry's Law Constant," *Anal. Chem.*, 64(13):1350–1355 (1992).

Sabljić, A. "On the Prediction of Soil Sorption Coefficients of Organic Pollutants from Molecular Structure: Application of Molecular Topology Model," *Environ. Sci. Technol.*, 21(4):358–366 (1987).

Sabljić, A. "Predictions of the Nature and Strength of Soil Sorption of Organic Pollutants by Molecular Topology," *J. Agric. Food Chem.*, 32(2):243–246 (1984).

Sabljić, A. and M. Protić. "Relationship between Molecular Connectivity Indices and Soil Sorption Coefficients of Polycyclic Aromatic Hydrocarbons," *Bull. Environ. Contam. Toxicol.*, 28(2):162–165 (1982).

Sadek, P.C., P.W. Carr, R.M. Doherty, M.J. Kamlet, R.W. Taft, and M.H. Abraham. "Study of Retention Processes in Reversed-Phase High-Performance Liquid Chromatography by the Use of the Solvatochromic Comparison Method," *Anal. Chem.*, 57(14):2971–2978 (1985).

Sarna, L.P., P.E. Hodge, and G.R.B. Webster. "Octanol-Water Partition Coefficients of Chlorinated Dioxins and Dibenzofurans by Reversed-Phase HPLC Using Several C_{18} Columns," *Chemosphere*, 13(9):975–983 (1984).

Sax, N.I. *Dangerous Properties of Industrial Materials* (New York: Van Nostrand Reinhold Co., 1984), 3124 p.

Sax, N.I. and R.J. Lewis, Sr. *Hazardous Chemicals Desk Reference* (New York: Van Nostrand Reinhold Co., 1987), 1084 p.

Schultz, T.W., S.K. Wesley, and L.L. Baker. "Structure-Activity Relationships for Di and Tri Alkyl and/or Halogen Substituted Phenols," *Bull. Environ. Contam. Toxicol.*, 43(2):192–198 (1989).

Shafer, D. *Hazardous Materials Training Handbook* (Madison, CT: Bureau of Law and Business, Inc., 1987), 206 p.

Sittig, M. *Handbook of Toxic and Hazardous Chemicals and Carcinogens* (Park Ridge, NJ: Noyes Publications, 1985), 950 p.

Suzuki, T. "Development of an Automatic Estimation System for both the Partition Coefficient and Aqueous Solubility," *J. Comput.-Aided Mol. Des.*, 5:149–166 (1991).

Szabo, G., S.L. Prosser, and R.A. Bulman. "Adsorption Coefficient (K_{oc}) and HPLC Retention Factors of Aromatic Hydrocarbons," *Chemosphere,* 21(4/5):495–505 (1990).

Szabo, G., S.L. Prosser, and R.A. Bulman. "Determination of the Adsorption Coefficient (K_{oc}) of Some Aromatics for Soil by RP-HPLC on Two Immobilized Humic Acid Phases," *Chemosphere,* 21(6):777–788 (1990a).

Szabo, G., S.L. Prosser, and R.A. Bulman. "Prediction of the Adsorption Co-efficient (K_{oc}) for Soil by a Chemically Immobilized Humic Acid Column using RP-HPLC," *Chemosphere,* 21(6):729–739 (1990).

ten Hulscher, Th.E.M., L.E. van der Velde, and W.A. Bruggeman. "Temperature Dependence of Henry's Law Constants for Selected Chlorobenzenes, Polychlorinated Biphenyls and Polycyclic Aromatic Hydrocarbons," *Environ. Toxicol. Chem.,* 11(11):1595–1603 (1992).

Tewari, Y.B., M.M. Miller, S.P. Wasik, and D.E. Martire. "Aqueous Solubility of Octanol/Water Partition Coefficient of Organic Compounds at 25.0°C," *J. Chem. Eng. Data,* 27(4):451–454 (1982).

Threshold Limit Values and Biological Exposure Indices for 1987–1988 (Cincinnati, OH: American Conference of Governmental Industrial Hygienists, 1987), 114 p.

Toxic and Hazardous Industrial Chemicals Safety Manual for Handling and Disposal with Toxicity and Hazard Data (Tokyo, Japan: International Technical Information Institute, 1986), 700 p.

U.S. EPA. "A Compendium of Technologies Used in the Treatment of Hazardous Wastes," Office of Research and Development, U.S. EPA Report-625/8-87-014 (1987), 49 p.

U.S. Department of Agriculture. Agricultural Research Service Pesticide Properties Database. Systems Research Laboratory, Beltsville, MD (1990).

Valkó, K., O. Papp, and F. Darvas. "Selection of Gas Chromatographic Stationary Phase Pairs for Characterization of the 1-Octanol-Water Partition Coefficient," *J. Chromatogr.,* 301:355–364 (1984).

Valvani, S.C., S.H. Yalkowsky, and G.L. Amidon. "Solubility of Nonelectrolytes in Polar Solvents. VI. Refinements in Molecular Surface Area Computations," *J. Phys. Chem.,* 80(8):829–835 (1976).

Veith, G.D. and R.T. Morris. "A Rapid Method for Estimating Log P for Organic Chemicals," U.S. EPA Report-600/3-78-049 (1978), 15 p.

Verschueren, K. *Handbook of Environmental Data on Organic Chemicals* (New York: Van Nostrand Reinhold Co., 1983), 1310 p.

Warne, M. St.J., D.W. Connell, D.W. Hawker, and G. Schüürmann. "Prediction of Aqueous Solubility and the Octanol-Water Partition Coefficient for Lipophilic Organic Compounds Using Molecular Descriptors and Physicochemical Properties," *Chemosphere,* 21(7):877–888 (1990).

Weast, R.C., Ed. *CRC Handbook of Chemistry and Physics,* 67th ed. (Boca Raton, FL: CRC Press, Inc., 1986), 2406 p.

Webster, G.R.B., K.J. Friesen, L.P. Sarna, and D.C.G. Muir. "Environmental Fate Modeling of Chlorodioxins: Determination of Physical Constants," *Chemosphere,* 14(6/7):609–622 (1985).

Whitehouse, B.G. and R.C. Cooke. "Estimating the Aqueous Solubility of Aromatic Hydrocarbons by High Performance Liquid Chromatography," *Chemosphere,* 11(8):689–699 (1982).

Windholz, M., S. Budavari, R.F. Blumetti, and E.S. Otterbein, Eds., *The Merck Index,* 10th ed. (Rahway, NJ: Merck and Co., 1983), 1463 p.

Woodburn, K.B., J.J. Delfino, and Rao, P.S.C. "Retention of Hydrophobic Solutes on Reversed-Phase Liquid Chromatography Supports: Correlation with Solute Topology and Hydrophobicity Indices," *Chemosphere,* 24(8):1037–1046 (1992).

Worthing, C.R. and S.B. Walker, Eds. *The Pesticide Manual — A World Compendium,* 9th ed. (Great Britain: British Crop Protection Council, 1991), 1141 p.

Wright, D.A., S.I. Sandler, and D. DeVoll. "Infinite Dilution Activity Coefficients and Solubilities of Halogenated Hydrocarbons in Water at Ambient Temperature," *Environ. Sci. Technol.,* 26(9):1828–1831 (1992).

Yalkowsky, S.H. and S.C. Valvani. "Solubilities and Partitioning 2. Relationships between Aqueous Solubilities, Partition Coefficients, and Molecular Surface Areas of Rigid Aromatic Hydrocarbons," *J. Chem. Eng. Data,* 24(2):127–129 (1979).

Yalkowsky, S.H. and S.C. Valvani. "Solubility and Partitioning I: Solubility of Nonelectrolytes in Water," *J. Pharm. Sci.,* 69(8):912–922 (1980).

Yoshida, K., S. Tadayoshi, and F. Yamauchi. "Relationship between Molar Refraction and *n*-Octanol/Water Partition Coefficient," *Ecotox. Environ. Safety,* 7(6):558–565 (1983).

BIBLIOGRAPHY

The books listed below were used in the preparation of the Agrochemicals Desk Reference. Most of the physicochemical properties included for each compound can be readily obtained from these and many other sources. Individual citations were not provided; however, information on transformation products, soil sorption and aquatic toxicity data for the majority of the compounds is widely scattered in many journal articles throughout the literature. For this reason, individual citations have been provided.

Ashton, F.M. and T.J. Monaco. *Weed Science* (New York: John Wiley & Sons, Inc., 1991), 466 p.

"CHRIS Hazardous Chemical Data, Vol. 2," U.S. Department of Transportation, U.S. Coast Guard, U.S. Government Printing Office (November, 1984).

Cremlyn, R.J. *Agrochemicals — Preparation and Mode of Action* (New York: John Wiley & Sons, Inc., 1991), 396 p.

Dean, J.A. *Handbook of Organic Chemistry* (New York: McGraw-Hill, Inc., 1987), 957 p.

Hartley, D. and H. Kidd, Eds. *The Agrochemicals Handbook,* 2nd ed. (England: Royal Society of Chemistry, 1987).

Hawley, G.G. *The Condensed Chemical Dictionary* (New York: Van Nostrand Reinhold Co., 1981), 1135 p.

Keith, L.H. and D.B. Walters. *National Toxicology Program's Chemical Solubility Compendium,* (Chelsea, MI: Lewis Publishers, Inc., 1992), 437 p.

Keith, L.H. and D.B. Walters. *National Toxicology Program's Chemical Data Compendium — Volume II, Chemical and Physical Properties,* (Chelsea, MI: Lewis Publishers, Inc., 1992), 1642 p.

Montgomery, J.H. *Groundwater Chemicals Desk Reference — 2nd Edition* (Boca Raton, FL: CRC Press, Inc., 1996), 1345 p.

Murphy, E. and R. Penske. *Pesticides in New Jersey: Implications in Groundwater Quality* (Trenton, NJ: New Jersey Department of Environmental Protection, Office of Science and Research, 1987), 214 p.

"NIOSH Pocket Guide to Chemical Hazards," U.S. Department of Health and Human Services, U.S. Government Printing Office (1994), 398 p.

Que Hee, S.S. and R.G. Sutherland. *The Phenoxyalkanoic Herbicides. Vol. 1. Chemistry, Analysis, and Environmental Pollution* (Boca Raton, FL: CRC Press, Inc, 1981), 321 p.

"Registry of Toxic Effects of Chemical Substances," U.S. Department of Health and Human Services, National Institute for Occupational Safety and Health (1985), 2050 p.

Sax, N.I. and R.J. Lewis, Sr. *Hazardous Chemicals Desk Reference* (New York: Van Nostrand Reinhold Co., 1987), 1084 p.

Sittig, M. *Handbook of Toxic and Hazardous Chemicals and Carcinogens* (Park Ridge, NJ: Noyes Publications, 1985), 950 p.

Threshold Limit Values and Biological Exposure Indices for 1987–1988 (Cincinnati, OH: American Conference of Governmental Industrial Hygienists, 1987), 114 p.

Verschueren, K. *Handbook of Environmental Data on Organic Chemicals* (New York: Van Nostrand Reinhold Co., 1983), 1310 p.

Windholz, M., S. Budavari, R.F. Blumetti, and E.S. Otterbein, Eds. *The Merck Index,* 10th ed. (Rahway, NJ: Merck & Co., 1983), 1463 p.

Worthing, C.R., and S.B. Walker, Eds. *The Pesticide Manual — A World Compendium,* 9th ed. (Great Britain: British Crop Protection Council, 1991), 1141 p.

ABBREVIATIONS AND SYMBOLS

Å	angstrom
α	alpha
α_a	percent of acid that is nondissociated
α_b	percent of base that is nondissociated
$\approx$	approximately equal to
ACGIH	American Conference of Governmental Industrial Hygienists
ASTM	American Society for Testing and Materials
asym	asymmetric
atm	atmosphere
b	aperture of fracture
B	average soil bulk density (g/cm^3)
β	beta
BCF	bioconcentration factor
bp	boiling point
C	ceiling
°C	degrees Centigrade (Celsius)
cal	calorie
CAS	Chemical Abstracts Service
CEC	cation exchange capacity (meq/L unless noted otherwise)
CERCLA	Comprehensive Environmental Response, Compensation and Liability Act
CHRIS	Chemical Hazard Response Information System
cm	centimeter
C_L	concentration of solute in solution
C_S	concentration of sorbed solute
DOT	Department of Transportation (U.S.)
d_s	density of a substance
d_w	density of water
δ	delta
EC_{50}	concentration necessary for 50% of the aquatic species tested showing abnormal behavior
et al.	and others
eV	electron volts
°F	degrees Fahrenheit
fl p	flash point
f_{oc}	fraction of organic carbon
fw	formula weight
γ	gamma
g	gram
gal	gallon
GC/MS	gas chromatography/mass spectrometry
>	greater than
$\geq$	greater than or equal to
ha	hectacre
HPLC	high performance liquid chromatography
HSDB	Hazardous Substances Data Bank

H-$t_{1/2}$	hydrolysis half-life
IDLH	immediately dangerous to life or health
IP	ionization potential
K	kelvin (°C + 273.15)
K_a	acid dissociation constant
K_A	distribution coefficient (cm)
K_b	base dissociation constant
K_d	distribution coefficient (cm^3/g)
kg	kilogram
K_H	Henry's law constant (atm · m^3/mol · K)
$K_{H'}$	Henry's law constant (dimensionless)
K_{oc}	soil/sediment partition coefficient (organic carbon basis)
K_{om}	soil/sediment partition coefficient (organic matter basis)
K_{ow}	*n*-octanol/water partition coefficient
kHz	kilohertz
kPa	kilopascal
K_w	dissociation constant for water (10^{-14} at 25°C)
<	less than
≤	less than or equal to
L	liter
lb	pound
LC_{50}	lethal concentration necessary to kill 50% of the aquatic species tested
LC_{100}	lethal concentration necessary to kill 100% of the aquatic species tested
LD_{50}	lethal dose necessary to kill 50% of the mammals tested
lel	lower explosive limit
m	meter
m-	meta (as in *m*-dichlorobenzene)
M	molarity (moles/liter)
M	mass
meq	milliequivalents
mg	milligram
min	minute(s)
mL	milliliter
M_L	mass of sorbed solute
mmHg	millimeters of mercury
mmol	millimole
mol	mole
mp	melting point
M_S	mass of solute in solution
mV	millivolt
N	normality (equivalents/liter)
n-, *N*-	normal (as in *n*-propyl, *N*-nitroso)
n_e	effective porosity
ng	nanogram
NIOSH	National Institute for Occupational Safety and Health
nm	nanometer
o-	ortho (as in *o*-dichlorobenzene)
OSHA	Occupational Safety and Health Administration
ρ	specific density (unitless)

p-	para (as in p-dichlorobenzene)
P	pressure
Pa	pascal
p_{air}	vapor density of air
PEL	permissible exposure limit
pH	$-\log_{10}$ hydrogen ion activity (concentration)
pK_a	$-\log_{10}$ dissociation constant of an acid
pK_b	$-\log_{10}$ dissociation constant of a base
pK_w	$-\log_{10}$ dissociation constant of water
ppb	parts per billion (μg/L)
ppm	parts per million (mg/L)
$P\text{-}t_{1/2}$	photolysis half-life
p_v	specific vapor density
QSAR	quantitative structure-activity relationships
R	ideal gas constant (8.20575×10^{-5} atm $\cdot$ m^3/mol)
R_a	retardation factor for an acid
R_b	retardation factor for a base
RCRA	Resource Conservation and Recovery Act
R_d	retardation factor
R_f	retention factor
RTECS	Registry of Toxic Effects of Chemical Substances
S	solubility
S_a	solute concentration in air (mol/L)
SARA	Superfund Amendments and Reauthorization Act
sec-	secondary (as in sec-butyl)
S_o	solubility in organics
S_w	solubility in water
sp.	species
spp.	species (plural)
STEL	short-term exposure limit
sym	symmetric
t-	tertiary (as in t-butyl; but $tert$-butyl)
$t_{1/2}$	half-life
TLC	thin layer chromatography
TLV	threshold limit value
TOC	total organic carbon (mg/L)
T_s	temperature of a substance
T_w	temperature of water
TWA	time-weighted average
μ	micro (10^{-6})
μg	microgram
uel	upper explosive limit
$unsym$	unsymmetric
U.S. EPA	U.S. Environmental Protection Agency
UV	ultraviolet
V, vol	volume
vap d	vapor density
Vc	average linear velocity of contaminant (e.g., ft/day)
vp	vapor pressure

V_w	average linear velocity of groundwater (e.g., ft/day)
W	watt
λ	wavelength
wt	weight
z^+	positively charged species (milliequivalents/cm^3)

ACEPHATE

Synonyms: Acetylphosphoramidothioic acid *O,S*-dimethyl ester; *N*-Acetylphosphora-midothioic acid *O,S*-dimethyl ester; Chevron RE 12420; *O,S*-Dimethylacetylphosphoroa-midothioate; ENT 27822; *N*-(Methoxy(methylthio)phosphinoyl)acetamide; Orthene; Orthene 755; Ortho 12420; Ortran; Ortril; RE 12420; 75 SP.

$$CH_3CNH - P \overset{\overset{O}{\|}}{\underset{}{}} \begin{matrix} \overset{O}{\|} \\ \end{matrix}$$

$$
\begin{array}{cc}
O & O \\
\| & \| \quad OCH_3 \\
CH_3CNH & -P \\
& \quad \backslash SCH_3
\end{array}
$$

Designations: CAS Registry Number: 30560-19-1; mf: $C_4H_{10}NO_3PS$; fw: 183.16; RTECS: TB4760000.

Properties: Colorless to white solid. Mp: 64–68°C (impure), 93°C; ρ: 1.35 at 20/4°C; H-$t_{1/2}$ at 40°C (hours): 60 (pH 9), 710 (pH 3); K_H: 5.2 × 10^{-13} atm · m³/mol at 20–24°C (approximate — calculated from water solubility and vapor pressure); log K_{oc}: 0.48; log K_{ow}: −1.87 (calculated); S_o (g/L): acetone (>100), ethanol (<50); S_w: 790 g/L at 20°C; vp: 1.7 × 10^{-6} mmHg at 24°C.

Soil properties and adsorption data

Soil	K_d (mL/g)	f_{oc} (%)	K_{oc} (mL/g)	pH
Soil 1	1.04	0.35	297	6.3
Soil 2	0.65	5.05	13	5.1
Soil 3	0.54	2.73	20	5.1
Soil 4	3.03	0.46	659	7.5

Source: Arienzo et al., 1993.

Environmental Fate

Soil. In aerobic and anaerobic soils, methamidophos and carbon dioxide were identified as the major soil metabolites (Hartley and Kidd, 1987). The estimated half-life in soil is 3 days (Wauchope, 1988).

Plant. Acephate is quickly absorbed, translocated and transformed in pine seedlings (Werner, 1974) and cotton plants (Bull, 1979). The chemical was metabolized via cleavage of the amide bond to form methamidophos (*O,S*-dimethyl phosphoramidothioate) and an unknown, but insecticidally active compound, which were identified in the roots, stems and leaves (Werner, 1974). Methamidophos was also found in cotton leaves following a single application of acephate. Four additional degradation products were formed — two of which were tentatively identified as *O,S*-dimethyl phosphorothioate and S-methyl acetylphosphoramidothioate. The amount of methamidophos and the four products repre-sented about 9 and 5% of the applied amount, respectively (Bull, 1979).

1

Chemical/Physical. Emits toxic fumes of phosphorus, nitrogen and sulfur oxides when heated to decomposition (Sax and Lewis, 1987).

Symptoms of Exposure: Slight irritation of eyes and skin.

Formulation Types: Wettable powder; water-soluble powder; encapsulated granules.

Toxicity: LC_{50} (96-hour) for rainbow trout >1 g/L, channel catfish 2.23 g/L, largemouth black bass 1.725 g/L, bluegill sunfish 2.05 g/L and goldfish 9.55 g/L (Hartley and Kidd, 1987); acute oral LD_{50} for male and female rats is 945 and 866 mg/kg, respectively (Hartley and Kidd, 1987), 700 mg/kg (RTECS, 1985).

Uses: Contact and systemic insecticide for control of sucking and chewing insects in cotton, ornamentals, forestry, tobacco, fruits, vegetables and other crops.

ACROLEIN

Synonyms: Acraldehyde; Acrylaldehyde; Acrylic aldehyde; Allyl aldehyde; Aqualin; Aqualine; Biocide; Crolean; Ethylene aldehyde; Magnacide; NSC 8819; Propenal; **2-Propenal**; Prop-2-en-1-al; 2-Propen-1-one; RCRA waste number P003; Slimicide; UN 1092.

$$CH_2=CHCHO$$

Designations: CAS Registry Number: 107-02-8; DOT: 1092 (inhibited); mf: C_3H_4O; fw: 56.06; RTECS: AS1050000.

Properties: Colorless to yellow, watery liquid with a very sharp, pungent, irritating odor. Polymerizes readily if not inhibited. Mp: $-86.9°C$; bp: $52.7°C$; ρ: 0.847 at $15.6°C$, 0.8410 at $20/4°C$; fl p: $-25°C$ (closed cup), $-18°C$ (open cup); lel: 2.8%; uel: 31%; H-$t_{1/2}$: 3.5 days (pH 5), 1.5 days (pH 7), 4 hours (pH 10); K_H: 4.4×10^{-6} atm · m^3/mol at $25°C$; IP: 10.10 eV; log K_{oc}: -0.31; log K_{ow}: -0.10; S_o: completely miscible with lower alcohols, acetone, benzene, ethyl ether and saturated hydro-carbons; S_w: 200 g/L at $25°C$; vap d: 2.29 g/L at $25°C$, 1.94 (air = 1); vp: 135.7, 220 and 330 mmHg at 10, 20 and $30°C$, respectively.

Environmental Fate

Biological. Microbes in site water degraded acrolein to β-hydroxypropionaldehyde (Kobayashi and Rittman, 1982). This product also forms when acrolein is hydrated in distilled water (Burczyk et al., 1968). When 5 and 10 mg/L of acrolein were statically incubated in the dark at $25°C$ with yeast extract and settled domestic wastewater inoculum, complete degradation was observed after 7 days (Tabak et al., 1981). Activated sludge was capable of degrading acrolein at concentrations of 2,300 ppm but no other information was provided (Wierzbicki and Wojcik, 1965).

Photolytic. Photolysis products include carbon monoxide, ethylene, free radicals and a polymer (Calvert and Pitts, 1966). Anticipated products from the reaction of acrylonitrile with ozone or hydroxyl radicals in the atmosphere are glyoxal, formaldehyde, formic acid and carbon dioxide (Cupitt, 1980). The major product reported from the photooxidation of acrolein with nitrogen oxides is formaldehyde with a trace of glyoxal (Altshuller, 1983). Osborne et al. (1962) reported that acrolein was stable at $30°C$ and UV light (λ = 313 nm) in the presence and absence of oxygen.

Groundwater. The half-life for acrolein in groundwater was estimated to range from 14 days to 8 weeks (Howard et al., 1991).

Chemical/Physical. Wet oxidation of acrolein at $320°C$ yielded formic and acetic acids (Randall and Knopp, 1980). May polymerize in the presence of light and explosively in the presence of concentrated acids (Worthing and Hance, 1991) forming disacryl, a white plastic solid (Windholz et al., 1983; Humburg et al., 1989). In distilled water, acrolein

3

was hydrolyzed to β-hydroxypropionaldehyde (Burczyk et al., 1968; Reinert and Rodgers, 1987; Kollig, 1993). The reported hydrolysis rate constant at pH 7 is 6.68×10^8/year (Kollig, 1993). The estimated hydrolysis half-life in water is 22 days (Burczyk et al., 1968).

Exposure Limits: NIOSH REL: TWA 0.1 ppm, STEL 0.3 ppm, IDLH 2 ppm; OSHA PEL: TWA 0.1 ppm; ACGIH TLV: TWA 0.1 ppm, STEL 0.3 ppm.

Symptoms of Exposure: Severe irritation of eyes, skin, mucous membranes; abnormal pulmonary function; delayed pulmonary edema, chronic respiratory disease.

Formulation Types: Liquid (includes an inhibitor such as hydroquinone to prevent polymerization).

Toxicity: EC50 (96-hour) for oysters 55 μg/L (salt water); EC50 (24-hour) for salmon 80 μg/L (fresh water); LC_{50} (24-hour) for bluegill sunfish 0.079 mg/L, mosquito fish 0.39 mg/L, rainbow trout 0.15 mg/L and shiners 0.04 mg/L; LC_{50} (48-hour) for oysters 0.56 mg/L and shrimps 0.10 mg/L (Worthing and Hance, 1991); acute oral LD_{50} for rats 46 mg/kg (Ashton and Monaco, 1991), 25,100 μg/kg (RTECS, 1985).

Uses: Contact herbicide and algicide; injected in water for the control of submerged and floating weeds in irrigation ditches and canals.

ACRYLONITRIL

Synonyms: Acritet; Acrylon; Acrylonitrile monomer; An; Carbacryl; Cyanoethylene; ENT 54; Fumigrain; Miller's fumigrain; Nitrile; Propenenitrile; **2-Propenenitrile**; RCRA waste number U009; TL 314; UN 1093; VCN; Ventox; Vinyl cyanide.

$$CH_2=CHCN$$

Designations: CAS Registry Number: 107-13-1; DOT: 1093; mf: C_3H_3N; fw: 53.06; RTECS: AT5250000.

Properties: Clear, colorless, watery liquid with a sweet irritating odor resembling peach pits. Slowly turns yellow on exposure to visible light. Mp: $-83°C$; bp: $77.5–79°C$; ρ: 0.8060 at 20/4°C; fl p: $-1°C$; lel: 3.05%; uel: $17.0 ± 0.5\%$; H-$t_{1/2}$: 1,220 years at 25°C and pH 7; K_H: $1.10 × 10^{-4}$ atm · m³/mol at 25°C; IP: 10.91 eV; log K_{oc}: -1.13 (calculated); log K_{ow}: -0.92 to 1.20; S_o: soluble in ethanol, ethyl ether, acetone, benzene, carbon tetrachloride, toluene; miscible with alcohol and chloroform; S_w: 80 g/L at 25°C; vap d: 2.17 g/L at 25°C, 1.83 (air = 1); vp: 83 mmHg at 20°C.

Environmental Fate

Biological. Degradation by the microorganism *Nocardia rhodochrous* yielded ammonium ion and propionic acid, the latter being oxidized to carbon dioxide and water (DiGeronimo and Antoine, 1976). When 5 and 10 mg/L of acrylonitrile were statically incubated in the dark at 25°C with yeast extract and settled domestic wastewater inoculum, complete degradation was observed after 7 days (Tabak et al., 1981).

Photolytic. In an aqueous solution at 50°C, UV light photooxidized acrylonitrile to carbon dioxide. After 24 hours, the concentration of acrylonitrile was reduced 24.2% (Knoevenagel and Himmelreich, 1976).

Chemical/Physical. Ozonolysis of acrylonitrile in the liquid phase yielded formaldehyde and the tentatively identified compounds glyoxal, an epoxide of acrylonitrile and acetamide (Munshi et al., 1989). In the gas phase, cyanoethylene oxide was reported as an ozonolysis product (Munshi et al., 1989a). Anticipated products from the reaction of acrylonitrile with ozone or hydroxyl radicals in air included formaldehyde, formic acid, HC(*O*)CN and cyanide ions (Cupitt, 1980). Wet oxidation of acrylonitrile at 320°C yielded formic and acetic acids (Randall and Knopp, 1980). Incineration or heating to decomposition releases toxic nitrogen oxides (Sittig, 1985) and cyanides (Lewis, 1990). Polymerizes readily in the absence of oxygen or on exposure to visible light (Windholz et al., 1983).

The hydrolysis rate constant for acrylonitrile at pH 2.87 and 68°C was determined to be $6.4 × 10^{-3}$/hour, resulting in a half-life of 4.5 days. At 68.0°C and pH 7.19, no hydrolysis/disappearance was observed after 2 days. However, when the pH was raised to 10.76, the hydrolysis half-life was calculated to be 1.7 hours (Ellington et al., 1986).

Acrylonitrile hydrolyzes to acrylamide which undergoes further hydrolysis forming acrylic acid and ammonia (Kollig, 1993).

Exposure Limits: NIOSH REL: TWA 1 ppm, 15-min C 1 ppm, IDLH 85 ppm; OSHA PEL: TWA 2 ppm, 15-min C 10 ppm; ACGIH TLV: TWA 2 ppm.

Symptoms of Exposure: Asphyxia, eye irritation, headache, sneezing, nausea, vomiting, weakness, light-headedness, skin vesiculation, scaling dermatitis.

Formulation Types: Liquid.

Toxicity: LC100 (24-hour) all fish 100 mg/L (fresh water); acute oral LD_{50} for rats 78 mg/kg (Verschueren, 1983).

Uses: Grain fumigant.

ALACHLOR

Synonyms: Alanex; Alochlor; Bronco; Bullet; Cannon; **2-Chloro-N-(2,6-diethylphenyl)-N-(methoxymethyl)acetamide**; 2-Chloro-2′,6′-diethyl-N-(methoxymethyl)acetanilide; CP 50144; Lariat; Lasso; Lasso II; Lasso EC; Lazo; Metachlor; Methachlor; Pillarzo.

Designations: CAS Registry Number: 15972-60-8; mf: $C_{14}H_{20}ClNO_2$; fw: 269.77; RTECS: AE1225000.

Properties: Odorless, cream-colored solid or crystals. Mp: 39.5–41.5°C; bp: 100°C at 0.02 mmHg (decomposes at 105°C); ρ: 1.133 at 25/15.6°C; K_H: 8.26×10^{-9} atm · m³/mol at 23°C (Fendinger and Glotfelty, 1988); log K_{oc}: 1.63–2.28; log K_{ow}: 2.64, 2.90; S_o: soluble in acetone, benzene, chloroform, ethanol, ethyl ether, ethyl acetate; S_w: 242 mg/L at 25°C; vp: 3.10×10^{-5} mmHg at 25°C.

Soil properties and adsorption data

Soil	K_d (mL/g)	f_{oc} (%)	K_{oc} (mL/g)	pH
Drummer silty clay	3.70	1.97	188	—
Dundee silt loam (conventional tillage)	3.49	1.02	342	5.8
Dundee silt loam (no tillage)	5.39	1.67	323	5.5
Dupo silt loam	0.88	0.70	126	—
Lintonia loam	0.35	0.41	86	—
Sand	0.30	0.41	73	6.5
Silt	0.90	0.70	129	8.1
Spinks sand loam	1.30	1.39	94	—

Source: U.S. Department of Agriculture, 1990; Locke, 1992.

Environmental Fate

Soil. Degradation products identified in an upland soil after 80 days of incubation included 7-ethyl-1-hydroxyacetyl-2,3-dihydroindole, 8-ethyl-2-hydroxy-1-(methyl-methoxy)-1,2,3,4-tetrahydroquinone, 2′,6′-diethyl-2-hydroxy-N-(methoxymethyl)aceta-nilide and 9-ethyl-1,5-dihydro-1-(methoxymethyl)-5-methyl-1,4-benzoxazepin-2-(3H)-one (Chou, 1977). In an upland soil, the microorganism *Rhizoctonia solani* degraded alachlor to unidentified water soluble products (Lee, 1986). Degradation of alachlor by soil fungi gave 2-chloro-2′,6′-diethylacetanilide, 2,6-diethylaniline, 1-chloroacetyl-2,3-dihydro-7-ethylindole, 2′,6′-diethyl-N-methoxymethylaniline and chloride ions (Tiedje and

Hagedorn, 1975). Novick et al. (1986) reported that alachlor and its metabolites may not be mineralized in soils pretreated with the herbicide but may persist for long periods of time. They concluded that leaching of these metabolites to groundwater will be degraded but very slowly. At concentrations of 0.073 and 10 µg/mL, <8% of ^{14}C-ring-labeled alachlor was mineralized in 30 days (Novick et al., 1986). 2-Chloro-2',6'-diethylacetanilide was the major metabolite that formed following the incubation of alachlor in a Sawyer fine sandy loam at 0% relative humidity (Hargrove and Meikle, 1971).

In soil, alachlor is photolytically unstable. Fang (1979) reported that after an 8-hour exposure to sunlight, degradation yields in a sandy loam, loam, silt loam and clay were 22, 27, 36 and 39%, respectively. Degradation appeared to be enhanced in soils having low organic carbon content and low pH.

Persistence in soil is approximately 6 to 10 weeks (Hartley and Kidd, 1987) but is moderately slower in sandy soils low in organic matter (Ashton and Monaco, 1991) and much slower under sterilized conditions. In nonsterile soils, 39–54% of the applied amount degraded after 28 days (Fang, 1983). The half-lives for alachlor in soil containing 6 and 15% moisture were 23 and 5.7 days, respectively (Walker and Brown, 1985). In a Ustollic Haplargid clay with 1.1% organic matter, 50% soil moisture and a pH of 7.8, the dissipation rates of alachlor at 10 and 20°C were 0.046 and 0.053/day, respectively. Similarly, the dissipation rates in an Aridic Argiustoll soil with 25% organic matter, 50% soil moisture and a pH of 8.0 at 10 and 20°C were 0.018 and 0.028/day, respectively (Zimdahl and Clark, 1982).

Groundwater. According to the U.S. EPA (1986) alachlor has a high potential to leach to groundwater.

Plant. In plants, alachlor is absorbed, translocated and transformed into glutathione (GSH) conjugates (Breaux et al., 1987; *O'*Connell, 1988). Four hours after treating corn with ^{14}C-labeled alachlor, 46% was converted to GSH conjugates and 42% was unreacted alachlor (*O'*Connell, 1988).

Surface Water. 2,6-Dichloroaniline, 2-chloro-2',6'-diethylacetanilide and 2-hydroxy-2',6'-diethylacetanilide were reported as possible degradation products of alachlor that were identified in the Mississippi River and its tributaries (Pereira and Rostad, 1990).

Photolytic. Alachlor, applied as a thin film on borosilicate glass, underwent photodegradation by sunlight via four major photodegradative pathways. These included dechlorination, *N*-dealkylation, *N*-deacylation and cyclized *N*-dealkylated products (Cessna and Muir, 1991). Photolysis of alachlor in aqueous solutions was studied by Somich et al. (1988) and reported by Hapeman-Somich (1991) and Somich et al. (1988). The photolysis study was performed using a 220-mL reactor equipped with a medium pressure lamp (λ ≤240 nm). Two major photoproducts included hydroxyalachlor and an unreported lactam. Other products identified were norchloralachlor, 2',6'-diethylacetanilide and 2-hydroxy-2',6'-diethyl-*N*-methylacetanilide. Degradation was rapid (half-life 1.6 minutes) and appeared to follow first-order kinetics. The decrease in pH during the reaction, from 6.6 to 2.5, indicated the formation of an acid, possibly hydrochloric (Somich et al., 1988).

Major photoproducts reported in soil treated with alachlor were 2',6'-diethylacetanilide, 1-chloro-2',6'-diethylacetanilide, 2,6-diethylaniline, chloroacetic acid, 2',6'-diethyl-*N*-methoxymethylaniline and 1-chloroacetyl-2,3-dihydro-7-ethylindole. Organic matter in soil enhanced the rate of photolysis (Chesters et al., 1989).

Chiron et al. (1995) investigated the photodegradation of alachlor (20 µg/L) in distilled water and Ebro River water using an xenon arc irradiation. Photolytic half-lives of alachlor in distilled water and river water were 140 and 84 minutes, respectively. In distilled water, alachlor completely disappeared after 10 hours of irradiation. The major photoproducts were hydroxyalachlor and 8-ethyl-1-methoxymethyl-4-methyl-2-oxo-1,2,3,4-tetrahydro-quinone.

Chemical/Physical. Stable to light (Hartley and Kidd, 1987) but is hydrolyzed in strongly acidic or alkaline solutions (Windholz et al., 1983) to give methanol, chloroacetic acid, formaldehyde and 2,6-diethylaniline (Sanborn et al., 1977; Sittig, 1985). Alachlor decomposed in 5 M hydrochloric acid at 46°C. After 72 hours, 2-chloro-2',6'-diethylace-tanilide was the major decomposition product identified (Hargrove and Merkle, 1971).

Tirey et al. (1993) evaluated the degradation of alachlor at elevated temperatures (275–700°C) in the presence of oxygen and identified a large number of reaction products. Of the 27 products reported, six could not be tentatively identified by GC/MS. Reaction products identified at 10 different temperatures include 2-chloro-2',6'-diethylacetanilide, methoxydimethylindole, 2,6-diethylaniline, ethylquinoline, chloromethyl acetate, 2,6-diethylbenzonitrile, methylethylquinoline, dihydrodimethyl-1(2H)-naphthaleneone, 4,6-dimethoxy-1,2-naphthoquinone, carbon dioxide, nitric oxide, hydrogen chloride, ethyl-methylquinoline, phenylethenylbenzonitrile, isocyanatonaphthalene, methoxyacetaldehyde, carbon monoxide, hydrogen cyanide, hydrogen azide, 2-propenenitrile and benzonitrile.

Koskinen et al. (1994) studied the ultrasonic decomposition of alachlor (3.1 nM) in pure water using a sonicator operating in the continuous mode at a maximum output of 20 kHz. Decomposition followed first-order kinetics and the rate increased with increasing power and temperature. The disappearance half-lives at 24, 27, and 30°C were 375, 180 and 86 minutes, respectively. The authors suggested that decomposition of alachlor was probably the result of the herbicide reacting with highly reactive hydroxyl free radicals.

Formulation Types: Emulsifiable concentrate (4 lb/gal); microscopic capsules (4 lb/gal); granules (15%).

Toxicity: LC_{50} (96-hour) for technical grade — rainbow trout 1.8 mg/L, bluegill sunfish 2.8 mg/L (Hartley and Kidd, 1987); for Lasso EC formulation the LC_{50} (48-hour) for *Daphnia magna* is 35 mg/L and the LC_{50} (96-hour) for rainbow trout and bluegill sunfish are 4.2 and 6.4 mg/L, respectively (Humburg et al., 1989); acute oral LD_{50} of technical alachlor for rats 930 mg/kg (Ashton and Monaco, 1991), Lasso 2,416 mg/kg, Lasso EC 1,000 mg/kg, Lasso II >5,010 mg/kg, Bronco 3,152 mg/kg (Humburg et al., 1989), 1,200 mg/kg (RTECS, 1985).

Uses: Preemergence, early postemergence or soil-incorporated herbicide used to control most annual grasses and many annual broad-leaved weeds in beans, corn, cotton, milo, peanuts, peas, soybeans, sunflower and certain woody ornamentals.

ALDICARB

Synonyms: Aldecarb; Ambush; Carbanolate; ENT 27093; **2-Methyl-2-(methylthio)pro-panal *O*-((methylamino)carbonyl)oxime**; 2-Methyl-2-(methylthio)propionaldehyde *O*-(methylcarbamoyl)oxime; NCI-C08640; OMS 771; RCRA waste number P070; Temic; Temik; Temik G10; Temik 10 G; UC 21149; Union Carbide 21149; Union Carbide UC-21149.

$$CH_3SC(CH_3)_2CH=NOCONHCH_3$$

Designations: CAS Registry Number: 116-06-3; DOT: 2757; mf: $C_7H_{14}N_2O_2S$; fw: 190.25; RTECS: UE2275000.

Properties: Colorless crystals with a faint sulfurous odor. Mp: 99–100°C; bp: decomposes; ρ: 1.195 at 25/4°C; H-$t_{1/2}$ (pH-buffered distilled water at 20°C): 131 days (pH 3.95), 559 days (pH 6.02), 324 days (pH 7.96), 55 days (pH 8.85), 6 days (pH 9.85); K_H: 1.45×10^{-9} atm · m³/mol at 20–25°C (approximate — calculated from water solubility and vapor pressure); log K_{oc}: 0.85–1.67; log K_{ow}: 0.70–1.13; P-$t_{1/2}$: 0.23 days (Atkinson, 1987); S_o (wt %): acetone (35), benzene (15), chlorobenzene (15), chloroform (35), ethyl ether (20), isopropane (20), methylene chloride (30), toluene (10), xylene (5); S_w: 5.73 g/L at 25°C (Ferreira and Seiber, 1981); vp: 3.47×10^{-5} mmHg at 25°C (Ferreira and Seiber, 1981).

Soil properties and adsorption data

Soil	K_d (mL/g)	f_{oc} (%)	K_{oc} (mL/g)	pH (meq/100 g)	CEC
Arredondo sand	0.20	0.80	25	6.80	—
Astatula sand	0.08	0.17	48	6.29	—
Astatula sand	0.03	0.12	25	5.89	—
Batcombe silt loam	0.87	2.05	42	6.10	—
Cecil sandy loam	0.18	0.90	20	5.60	—
Clarion soil	0.78	2.64	29	5.00	21.02
Eustis fine sand	0.17	0.69	25	5.40	—
Fine sand (0–15 cm)	0.39	1.04	38	6.15	—
Fine sand (15–30 cm)	0.11	0.29	38	5.65	—
Fine sand (90–105 cm)	0.03	0.52	6	5.80	—
Fine sand (105–120 cm)	0.03	0.52	6	5.95	—
Harps soil	1.13	3.80	29	7.30	37.83
Peat soil	4.16	18.36	22	6.98	77.34
Rothamsted Farm	0.60	1.51	40	5.10	—
Sarpy fine sandy loam	0.19	0.51	37	7.30	5.71
Thurman loamy fine sand	0.22	1.07	21	6.83	6.10

Soil properties and adsorption data *(continued)*

Soil	K_d (mL/g)	f_{oc} (%)	K_{oc} (mL/g)	pH (meq/100 g)	CEC
Webster silty clay loam	0.76	3.97	20	7.30	—
Woburn sandy loam	0.06	0.78	8	7.00	—
Woburn sandy loam	0.30	3.43	9	6.34	—

Source: Felsot and Dahm, 1979; Bromilow et al., 1980; Lord et al., 1980; Briggs, 1981; Bilkert and Rao, 1985; U.S. Department of Agriculture, 1990.

Environmental Fate

Biological. Jones (1976) reported several fungi degraded aldicarb to water-soluble constituents, namely, methyl(methylsulfonyl)propionamide and methyl(methylsulfonyl)propanol. The fungi tested, in order of effectiveness of degrading aldicarb, were *Gliocladium catenulatum, Penicillium multicolor = Cunninghamella elegans, Rhizoctonia* sp. and *Trichoderma harzianum* (Jones, 1976).

Soil. In soils, aldicarb quickly degrades via oxidation to 2-methyl-2-(methylsulfonyl)propionaldehyde-*O*-(methylcarbamoyl)oxime (aldicarb sulfoxide), 2-methyl-2-(methylsulfonyl)propionaldehyde-*O*-(methylcarbamoyl)oxime (aldicarb sulfone) (Coppedge et al., 1967; Bull et al., 1968; Andrawes et al., 1971; Smelt et al., 1978; Malik and Yadav, 1979; Ou et al., 1985; Macalady et al., 1986; Zhong et al., 1986; Smelt et al., 1987; Lemley et al., 1988; Day, 1991; Miles, 1991), aldicarb sulfoxide oxime, aldicarb sulfone oxime and aldicarb sulfoxide nitrile, TLC polar products and two unidentified compounds (Ou et al., 1985). Aldicarb sulfoxide, aldicarb sulfone and water soluble noncarbamate compounds were found in field soils 2 years after aldicarb application (Andrawes et al., 1971). Malik and Yadav (1979) found that the highest and lowest levels of residuals remained in fine sand and clay, respectively. Twelve weeks after applying aldicarb (20 ppm) to soil, the amount of aldicarb and aldicarb sulfoxide remaining were 0.7 and 9.5%, respectively.

In both soil and water, chemical and biological mediated reactions can transform aldicarb to the corresponding sulfoxide and sulfone via oxidation (Alexander, 1981). Reduction of aldicarb in natural waters and sediments yields 2-methyl-2-methyl thiopropionaldehyde and 2-methyl-2-methyl thiopropionitrile (Wolfe, 1992). The rate of microbial degradation of aldicarb to its metabolites, aldicarb sulfoxide and aldicarb sulfone in soils, was essentially the same in soils showing enhanced carbofuran degradation. In addition, the persistence of these compounds was not dramatically altered under these conditions (Racke and Coats, 1988). Metabolites identified in fallow, sandy loam soils include the oxidation products aldicarb sulfone and aldicarb sulfoxide (Bromilow and Leistra, 1980). The primary degradative pathway of aldicarb in surface soils is oxidation by microorganisms (Zhong et al., 1986). Aldicarb is rapidly converted to the sulfoxide in the presence of oxidizing agents (Hartley and Kidd, 1987) and microorganisms (Zhong et al., 1986). Further oxidation to the sulfone by microorganisms occurs at a much slower rate (Zhong et al., 1986).

Rajagopal et al. (1989) used numerous compounds to develop a proposed pathway of degradation of aldicarb in soil. These compounds included aldicarb oxime, *N*-hydroxymethyl aldicarb, *N*-hydroxymethyl aldicarb sulfoxide, *N*-demethyl aldicarb sulfoxide, *N*-demethyl aldicarb sulfone, aldicarb sulfoxide, aldicarb sulfone, *N*-hydroxymethyl aldicarb sulfone, aldicarb oxime sulfone, aldicarb sulfone aldehyde, aldicarb sulfone alcohol, aldicarb nitrile sulfone, aldicarb sulfone amide, aldicarb sulfone acid, aldicarb oxime sulfoxide, aldicarb sulfoxide aldehyde, aldicarb sulfoxide alcohol, aldicarb nitrile sulfoxide, aldicarb sulfoxide amide, aldicarb sulfoxide acid, elemental sulfur, carbon dioxide and water. Mineralization was more rapid in aerobic surface soils than in either aerobic or anaerobic subsurface soils. In surface soils (30 cm depth) under aerobic conditions, half-lives ranged from 20 to 361 days. In subsurface soils (20 and 183 cm depths), half-lives under aerobic and anaerobic conditions were 131–233 and 223–1,130 days, respectively (Ou et al., 1985). The reported half-lives in soil ranged from approximately 70 days (Jury et al., 1987) to several months (Jones et al., 1986). Bromilow et al. (1980) reported the half-life for aldicarb in soil to be 9.9 days at 15°C and pH 6.34–7.0.

In aerobic soils, aldicarb degraded rapidly (half-life 7 days) releasing carbon dioxide. Mineralization half-lives for the incubation of aldicarb in aerobic and anaerobic soils were 20–361 and 223–1,130 days, respectively. At an application rate of 20 ppm, the half-lives for aldicarb in clay, silty clay loam and fine sandy loam were 9, 7 and 12 days, respectively (Coppedge et al., 1967). Other soil metabolites may include acids, amides and alcohols (Hartley and Kidd, 1987).

Groundwater. In Florida groundwater, aldicarb was converted to aldicarb sulfoxide under aerobic conditions. Conversely, under anaerobic conditions (ph 7.7), oxidative metabolites (aldicarb sulfoxide and aldicarb sulfone) reverted back to the parent compound (aldicarb). Half-lives in unfiltered and filtered groundwater were 635 and 62 days, respectively (Miles and Delfino, 1985). In sterile anaerobic groundwater at pH 8.2, aldicarb slowly hydrolyzed to the aldicarb oxime. In a microorganism-enriched groundwater at pH 6.8, aldicarb rapidly degraded to aldicarb nitrile (Trehy et al., 1984). At 20°C and pH values of 7.96 and 8.5, hydrolysis half-lives of 324 and 170 days were reported, respectively (Given and Dierberg, 1985).

According to the U.S. EPA (1986) aldicarb has a high potential to leach to groundwater.

Plant. In plants, aldicarb is rapidly metabolized to the corresponding sulfoxide, sulfone (Andrawes et al., 1973; Cremlyn, 1991) and water soluble noncarbamate compounds (Andrawes et al., 1973). In cotton, however, Coppedge et al. (1967) found that aldicarb was rapidly metabolized to aldicarb sulfoxide but further oxidation to aldicarb sulfone was much slower. At moderate temperatures, aldicarb is completely oxidized to aldicarb sulfoxide in cotton plants within 4 to 9 days. This metabolite is hydrolyzed to the corresponding oxime which was reported as the principal metabolite in the cotton plant (Metcalf et al., 1966). In a later study, Bartley et al. (1970) identified up to 10 metabolites in cotton plants at harvest time. Aldicarb degraded primarily via a reductive pathway forming conjugates of 2-methyl-2-(methylsulfinyl)propanol and lesser quantities of conjugated 2-methyl-2-(methylsulfonyl)propanol, 2-methyl-2-(methylsulfonyl)propionaldehyde oxime and 2-methyl-2-(methylsulfinyl)propionaldehyde oxime. Oxidation also occurred and this led to the formation of nonconjugated 2-methyl-2-(methylsulfinyl)propionamide, 2-methyl-2-(methylsulfinyl)propionic acid and 2-methyl-2-(methylsulfonyl)propionic acid (Bartley et al., 1970).

Photolytic. Though no products were identified, aldicarb vapor may react with hydroxyl radicals in the atmosphere. A half-life of 0.24 days was reported (Atkinson, 1987).

Chemical/Physical. The hydrolysis of aldicarb in water is both acid and base catalyzed (Banks and Tyrell, 1984). Hansen and Spiegel (1983) studied the hydrolysis rate of aldicarb, aldicarb sulfoxide and aldicarb sulfone in aqueous buffer solutions. At a given temperature, the rate of hydrolysis increases rapidly above pH 7.5. The reported hydrolysis half-lives for aldicarb are as follows: 4,580 days at pH 5.5 and 5°C, 3,240 days at pH 5.5 and 15°C, 1,950 days at pH 7.5 and 5°C, 1,900 days at pH 7.5 and 15°C, 1,380 days at pH 8.5 and 5°C and 170 days at pH 8.5 and 15°C (Hansen and Spiegel, 1983). The hydrolysis half-lives for aldicarb in a sterile 1% ethanol/water solution at 25°C and pH values of 4.5, 6.0, 7.0 and 8.0, were 25, 38, 35 and 38 weeks, respectively (Chapman and Cole, 1982). In a sterilized buffer solution, an hydrolysis half-life of 1,170 hours was observed (Ferreira and Seiber, 1981). Aldicarb degrades rapidly in the chlorination of drinking water forming aldicarb sulfoxide which subsequently degrades to aldicarb sulfone, (chloromethyl)sulfonyl species and N-chloroaldicarb sulfoxide (Miles, 1991a).

Banks and Tyrrell (1985) studied the Cu^{2+}-promoted decomposition of aldicarb in aqueous solution over the pH region 2.91–5.51. 2-Methyl-2-(methylthio)propionitrile and 2-methyl-2-(methylthio)propanal formed at yields of 82 and 18%, respectively.

When aldicarb in acetonitrile was irradiated by UV light ($\lambda = 254$ nm), methyl-amine, dimethyl disulfide, tetramethylsuccinonitrile and 1-(methylthio)-2,3-dicyano-2,3-dimethylbutane were produced. Minor amounts of N,N'-dimethylurea also formed. The same products were formed when the photolyses were carried out in the presence of acetophenone, benzophenone, or benzonitrile (Freeman and McCarthy, 1984).

Tirey et al. (1993) evaluated the degradation of aldicarb at nine different temperatures. When aldicarb was oxidized at a temperature range of 275–650°C, the following reaction products were identified by GC/MS: acetone, 2-methyl-2-propenenitrile, methylthiopropene, dimethyldisulfide, carbon dioxide, carbon monoxide, nitrous oxide, isocyanatomethane, hydrogen azide, sulfur dioxide, hydrogen cyanide and 2-propenenitrile.

Formulation Types: Granules.

Toxicity: LC_{50} (96-hour) for rainbow trout 0.88 mg/L and bluegill sunfish 1.5 mg/L (Hartley and Kidd, 1987); LC_{50} (72-hour) for bluegill sunfish 100 µg/L (Day, 1991); acute oral LD_{50} for rats 930 µg/kg (Hartley and Kidd, 1987), 650 µg/kg (RTECS, 1985).

Uses: Systemic insecticide, acaricide, nematocide.

ALDRIN

Synonyms: Aldrec; Aldrex; Aldrex 30; Aldrite; Aldrosol; Altox; Compound 118; Drinox; ENT 15949; Hexachlorohexahydro-*endo,exo*-dimethanonaphthalene; **1,2,3,4,10,10-Hexachloro-1,4,4a,5,8,8a-hexahydro1,4:5,8-dimethanonaphthalene**; 1,2,3,4,10,10-Hexachloro-1,4,4a,5,8,8a-hexahydro-1,4-*endo,exo*-5,8-dimethanonaphthalene; 1,2,3,4,10,10-Hexachloro-1,4,4a,5,8,8a-hexahydro-*exo*-1,4-endo-5,8-dimethanonaphthalene; 1,4,4a,5,8,8a-Hexahydro-1,4-*endo,exo*-5,8-dimethanonaphthalene; HHDN; NA 2761; NA 2762; NCI-C00044; Octalene; RCRA waste number P004; Seedrin; Seedrin liquid.

Designations: CAS Registry Number: 309-00-2; DOT: 2761; mf: $C_{12}H_8Cl_6$; fw: 364.92; RTECS: IO2100000.

Properties: White, odorless crystals when pure; technical grades are tan to dark brown with a mild chemical odor. Mp: 104°C (pure), 49–60°C (technical); bp: 145°C at 2 mmHg; ρ: 1.70 at 20/4°C; fl p: nonflammable; H-$t_{1/2}$: 760 days at 25°C and pH 7 (Ellington et al., 1987); K_H: 4.96 × 10^{-4} atm · m^3/mol at 25°C; log BCF: freshwater clam (*Corbicula manilensis*) 4.13 (Hartley and Johnston, 1983); log K_{oc}: 2.61, 4.69; log K_{ow}: 5.17–7.4; P-$t_{1/2}$: 113.49 hours (absorbance λ = 227.0 nm, concentration on glass plates = 6.7 μg/cm^2); S_o (g/L): acetone (5–10), benzene (>600), dimethyl sulfoxide (1–5), 95% ethanol (<1), xylene (>600); S_w: 17–180 μg/L at 25°C; vp: 2.31 × 10^{-5} mmHg at 20°C, 1.52 × 10^{-4} mmHg at 25°C.

Soil properties and adsorption data

Soil	K_d (mL/g)	f_{oc} (%)	K_{oc} (mL/g)	pH
Batcombe silt loam	996	2.05	48,585	6.1
Rothamsted Farm	730	1.51	48,344	5.1

Source: Lord et al., 1980; Briggs, 1981.p

Environmental Fate

Biological. Dieldrin is the major metabolite formed from the microbial degradation of aldrin via epoxidation (Lichtenstein and Schulz, 1959; Korte et al., 1962; Kearney and Kaufman, 1976). Microorganisms responsible for this reaction were identified as *Aspergillus niger, Aspergillus flavus, Penicillium chrysogenum* and *Penicillium notatum* (Korte et al., 1962). Dieldrin may further degrade to photodieldrin (Kearney and Kaufman, 1976). A pure culture of the marine alga namely *Dunaliella* sp. degraded aldrin to dieldrin and

14

the diol at yields of 23.2 and 5.2%, respectively (Patil et al., 1972). In four successive 7-day incubation periods, aldrin (5 and 10 mg/L) was recalcitrant to degradation in a settled domestic wastewater inoculum (Tabak et al., 1981). In a mixed microbial population under anaerobic conditions, nearly all aldrin (87%) degraded to two unidentified products in 4 days (Maule et al., 1987).

Soil. Patil and Matsumura (1970) reported 13 of 20 soil microorganisms were able to degrade aldrin to dieldrin under laboratory conditions. Harris and Lichtenstein (1961) studied the volatilization of aldrin (4 ppm) in Plainfield sand and quartz sands. Air was passed over the soil at a wind speed of 1 L/min for 6 hours at 22°C and 38% relative humidity. No volatilization was observed when both soils were dry. When the soils were wet, however, the amount of aldrin that volatilized from wet Plainfield and quartz sands was 4 and 2.19%, respectively. When the relative humidity was increased to 100%, 4.52 and 7.33% of the aldrin volatilized from Plainfield and quartz sands.

Aldrin was found to be very persistent in an agricultural soil. Fifteen years after application of aldrin (20 lb/acre), 5.8% of the applied dosage was recovered as dieldrin and 0.2% was recovered as photodieldrin (Lichtenstein et al., 1971).

Plant. Photoaldrin and photodieldrin formed when aldrin was codeposited on bean leaves and exposed to sunlight (Ivie and Casida, 1971). Dieldrin and 1,2,3,4,7,8-hexa-chloro-1,4,4a,6,7,7a-hexahydro-1,4-*endo*-methyleneindene-5,7-dicarboxylic acid were identified in aldrin-treated soil on which potatoes were grown (Klein et al., 1973).

Surface Water. Under oceanic conditions, aldrin may undergo dihydroxylation at the chlorine free double bond to produce aldrin diol (Verschueren, 1983). When raw water obtained from the Little Miami River in Ohio containing aldrin (10 μg/L) was placed in a sealed glass jar and exposed to sunlight or artificial fluorescent light for 8 weeks, about 80% was converted to dieldrin (Eichelberger and Lichtenberg, 1971).

Photolytic. Aldrin exhibits low absorption at wavelengths greater than 290 nm (Gore et al., 1971). Photolysis of 0.33 ppb aldrin in San Francisco Bay water by sunlight produced photodieldrin. The half-life was 1.1 days. (Singmaster, 1975). When an aqueous solution containing aldrin was photooxidized by UV light at 90–95°C, 25, 50 and 75% degraded to carbon dioxide after 14.1, 28.2 and 109.7 hours, respectively (Knoevenagel and Him-melreich, 1976). Aldrin in a hydrogen peroxide solution (5 μM) was irradiated by UV light ($\lambda = 290$ nm). After 12 hours, the aldrin concentration was reduced 79.5%. Dieldrin, photoaldrin and an unidentified compound were reported as metabolites (Draper and Crosby, 1984).

After a short-term (<1 hour) exposure to sunlight, aldrin on silica gel chromatoplates was converted to photoaldrin. Photodecomposition was accelerated by several photosen-sitizing agents (Ivie and Casida, 1971a). Photoaldrin was formed when a benzene solution containing aldrin and benzophenone as a sensitizer was exposed to UV light ($\lambda = 268–356$ nm) (Rosen and Carey, 1968). Photodegradation of aldrin by sunlight for 1 month yielded the following products: dieldrin, photodieldrin, photoaldrin and a polymeric substance (Rosen and Sutherland, 1967).

Photolysis of solid aldrin using a high pressure mercury lamp with a pyrex filter (λ >300 nm) yielded a polymeric substance with small amounts of photoaldrin, dieldrin, hydrochloric acid and carbon dioxide (Gäb et al., 1974). When aldrin (166.8 mg) adsorbed on silica gel surface was irradiated by a mercury high pressure lamp (λ >230 nm), however, high yields of epoxides were formed. Products identified and their respective conversion yields were dieldrin (63.9%), photoaldrin (5.1%), photodieldrin (2.7%), photoketoaldrin (4.7%), polymer and polar products (23.6%) (Gäb et al., 1975).

Sunlight and UV light can convert aldrin to photoaldrin (Georgacakis and Khan, 1971). Oxygen atoms can also convert aldrin to dieldrin (Saravanja-Bozanic et al., 1977). When aldrin vapor (5 mg) in a reaction vessel was irradiated by a sunlamp for 45 hours, 14–34% degraded to dieldrin (50–60 µg) and photodieldrin (20–30 µg). However, when the aldrin vapor concentration was reduced to 1 µg and irradiation time extended to 14 days, 60% degraded to dieldrin (0.63 µg), photo-dieldrin (0.02 µg) and photoaldrin (0.02 µg) (Crosby and Moilanen, 1974). Aldrin in a hydrogen peroxide solution (5 µM) was irradiated by UV light λ = 290 nm). After 12 hours, the aldrin concentration was reduced 79.5%. Dieldrin, photoaldrin and an unidentified compound were reported as the end products (Draper and Crosby, 1984).

When an aqueous solution of aldrin (0.07 µM) in natural water samples collected from California and Hawaii was irradiated (λ <220 nm) for 36 hours, 25% was photooxidized to dieldrin (Ross and Crosby, 1985). Dieldrin also formed as the major product when a film of aldrin was irradiated at 254 nm (Roburn, 1963).

Chemical/Physical. In an aqueous solution containing peracetic acid, aldrin was transformed to dieldrin in the dark (Ross and Crosby, 1975). Aldrin is oxidized in the presence of oxygen forming dieldrin (Saravanja-Bozanic et al., 1977). During storage aldrin slowly decomposes, releasing hydrogen chloride (Hartley and Kidd, 1987). The hydrolysis rate constant for aldrin at pH 7 and 25°C was determined to be 3.8×10^{-5}/hr, resulting in a half-life of 760 days. The hydrolysis half-lives are reduced significantly at varying pHs and temperature. At a temperature of 68.0°C and pH values of 3.03, 6.99 and 10.70, the half-lives were 1.9, 2.9 and 2.5 days, respectively (Ellington et al., 1986). No disappearance of aldrin was observed in water after 2 weeks at pH 11 and 85°C (Kollig, 1993).

In hexane and 10% acetone water solvents, ozone readily oxidized aldrin to dieldrin (Hoffman and Eichelsdoerfer, 1971).

When heated to decomposition, toxic chlorides are released (Lewis, 1990).

Exposure Limits: NIOSH REL: 0.25 mg/m^3, IDLH 25 mg/m^3; OSHA PEL: TWA 0.25 mg/m^3; ACGIH TLV: TWA 0.25 mg/m^3.

Symptoms of Exposure: Headache, dizziness; nausea, vomiting, malaise; myoclonic jerks of limbs; clonic, tonic convulsions; coma; hematuria, azotemia.

Formulation Types: Wettable powder; emulsifiable concentrate; dustable powder; granules.

Toxicity: LC$_{50}$ (96-hour) for American eel 5 ppb, mummichog 4–8 ppb, striped killifish 17 ppb, Atlantic silverside 13 ppb, striped mullet 100 ppb, bluehead 12 ppb, northern puffer 36 ppb, fathead minnow 28 µg/L, bluegill sunfish 13 µg/L, rainbow trout 17.7 µg/L, coho salmon 45.9 µg/L, chinook 7.5 µg/L, striped bass 10 µg/L, pumpkinseed 20 µg/L and white perch 42 µg/L; LC$_{50}$ (48-hour) for mosquito fish 36 ppb; LC$_{50}$ (24-hour) for bluegill sunfish 260 ppb (Verschueren, 1983); acute oral LD$_{50}$ for rats is 38–67 mg/kg (Hartley and Kidd, 1987).

Uses: Formerly as insecticide and fumigant; manufacture and use has been discontinued in the U.S.

ALLIDOCHLOR

Synonyms: Alidochlor; CDAA; CDAAT; 2-Chloro-*N*,*N*-diallylacetamide; **2-Chloro-*N*,*N*-di-2-propenylacetamide**; α-Chloro-*N*,*N*-diallylacetamide; CP 6343; Diallylchloroacetamide; *N*,*N*-Diallylchloroacetamide; *N*,*N*-Diallyl-2-chloroacetamide; *N*,*N*-Diallyl-α-chloroacetamide; NCI-C04035; Radox; Randox; Randox T.

$$CH_2 = CHCH_2 \diagdown \quad \overset{O}{\overset{\|}{NCCH_2Cl}}$$
$$CH_2 = CHCH_2 \diagup$$

Designations: CAS Registry Number: 93-71-0; mf: $C_8H_{12}ClNO$; fw: 173.65; RTECS: AB5250000.

Properties: Amber liquid. Mp: <25°C; bp: 92°C at 2 mmHg; log K_{oc}: 1.83 (calculated); log K_{ow}: 0.97 (calculated); S_o: soluble in ethanol, hexane, xylene; S_w: 1.97 g/L at 25°C; vap d: 7.10 g/L at 25°C, 6.02 (air = 1).

Environmental Fate

Plant. Allidochlor is translocated in plants to chloroacetic acid and diallylamine. The diallylamine is further transformed to carbon dioxide. The acid undergoes further degradation to glycollic acid which breaks down to glyoxalic acid. Glyoxalic acid undergoes further degradation to give formic acid, glycine and carbon dioxide (Cremlyn, 1991).

Chemical/Physical. Emits very toxic fumes of phosphorus oxides and chlorine when heated to decomposition (Sax and Lewis, 1987).

Symptoms of Exposure: Strong irritant.

Toxicity: Acute oral LD_{50} for rats 700 mg/kg (RTECS, 1985).

Uses: Selective preemergence herbicide used to control annual grass weeds and some broad-leaved weeds in maize, millet, soybeans, sorghum, sugarcane, vegetables and ornamentals.

AMETRYN

Synonyms: A 1093; Amtrex; Ametryne; Crisatine; 2-Ethylamino-4-isopropylamino-6-methylmercapto-*s*-triazine; 2-Ethylamino-4-isopropylamino-6-methylthio-*s*-triazine; 2-Ethylamino-4-isopropylamino-6-methylthio-1,3,5-triazine; *N*-Ethyl-*N'*-isopropyl-6-methylthio-1,3,5-triazine-2,4-diyldiamine; **N-Ethyl-*N'*-(1-methylethyl)-6-(methylthio)-1,3,5-triazine-2,4-diamine**; Evik; Evik 80W; G 34,162; Gesapax; 2-Methylmercapto-4-ethylamino-*s*-triazine; 2-Methylthio-4-ethylamino-6-isopropylamino-*s*-triazine.

$$CH_3S \diagdown \quad N \diagdown \quad NHCH_2CH_3$$

NHCH(CH_3)_2

Designations: CAS Registry Number: 834-12-8; mf: $C_9H_{17}N_5S$; fw: 227.35; RTECS: XY9100000.

Properties: White crystalline powder. Mp: 84–86°C; ρ: 1.19 at 20/4°C; fl p: nonflammable; pK_a: 4.1; H-t_{1/2}: 32 days (pH 1), >200 days (pH 13); K_H: 1.36×10^{-9} atm · m³/mol at 20°C (approximate — calculated from water solubility and vapor pressure); log K_{oc}: 2.23–2.44; log K_{ow}: 2.61 (Liu and Qian, 1995); S_o (g/L at 20°C): acetone (500), methylene chloride (600), hexane (14), methanol (450), toluene (400); S_w at 26°C: 17.7, 1.78, 0.849, 0.857 and 0.846 mM at pH values 2.0, 3.0, 5.0, 7.0 and 10.0, respectively (Ward and Weber, 1968); vp: 8.4×10^{-7} mmHg at 20°C.

Soil properties and adsorption data

Soil	K_d (mL/g)	f_{oc} (%)	K_{oc} (mL/g)	pH (meq/100 g)	CEC
Aguadilla loamy sand	3.08	1.44	214	7.4	10.0
Aguirre clay loam	2.76	0.75	368	9.0	14.3
Alonso clay	7.11	1.84	386	5.1	13.8
Altura loam	2.55	2.13	120	8.0	27.6
Bayamón sandy clay loam	3.20	0.98	326	4.7	5.0
Catalina clay	2.78	1.09	255	4.7	11.8
Cataño sand	2.08	1.21	172	7.9	6.9
Cayaguá sandy loam	3.97	1.15	345	5.2	7.3
Cialitos clay loam	13.32	2.82	472	5.4	18.6
Coloso clay loam	9.16	2.13	430	5.7	23.0
Coto clay	2.51	1.84	136	7.7	14.0
Fe clay loam	3.72	1.96	190	7.5	27.6
Fortuna silty clay loam	18.23	1.90	959	5.4	23.3
Fraternidad clay	3.34	1.21	276	6.3	36.0
Fraternidad clay	8.14	2.42	336	5.9	58.0

Soil properties and adsorption data *(continued)*

Soil	K_d (mL/g)	f_{oc} (%)	K_{oc} (mL/g)	pH (meq/100 g)	CEC
Guanicá clay	4.04	2.77	146	8.1	52.1
Humata silty clay loam	4.70	0.98	480	4.5	10.1
Josefa silt loam	7.16	1.90	376	6.0	16.8
Juncos silty clay	10.90	1.55	703	6.2	13.4
Mabí clay	6.49	2.26	287	7.0	55.2
Mabí clay loam	5.58	2.82	198	5.7	31.0
Mercedita silty clay	2.38	1.38	173	8.1	19.9
Moca clay	13.61	2.19	621	5.8	31.0
Múcara loam	1.05	1.90	550	5.8	19.6
Nipe clay loam	9.52	3.06	311	5.7	11.9
Pandura sandy loam	3.63	1.15	316	5.7	7.7
Río Pedras silty clay	3.20	2.02	158	4.9	11.5
San Anton loam	4.33	1.55	279	6.7	26.1
Sand	0.88	0.35	253	5.6	—
Sandy loam	4.84	1.74	278	6.1	—
Silt loam	3.78	1.68	225	6.9	—
Silt loam	4.97	2.90	171	7.0	—
Silt loam	2.81	1.22	230	7.0	—
Toa loam	9.72	1.16	837	5.3	13.0
Toa sandy loam	2.08	0.34	612	6.0	8.0
Talante sandy loam	6.19	0.80	774	5.1	4.0
Vega Alta sandy loam	5.14	2.02	254	5.0	5.6
Via loam	8.49	1.32	643	5.1	39.9

Source: Liu et al., 1970; U.S. Department of Agriculture, 1990.

Environmental Fate

Biological. Cook and Hütter (1982) reported that bacterial cultures were capable of degrading ametryne forming the corresponding hydroxy derivative (hydroxyametryne).

Soil. Although no products were reported, the half-life in soil is 70–120 days (Worthing and Hance, 1991).

Groundwater. According to the U.S. EPA (1986) ametryn has a high potential to leach to groundwater.

Plant. Ametryn is metabolized by tolerant plants into nontoxic hydroxy and dealkylated derivatives (Humburg et al., 1989).

Photolytic. The dye-sensitized photodecomposition of ametryn was studied in aqueous, aerated solutions (Rejto et al., 1983). When an aqueous ametryn solution was irradiated in sunlight for several hours, 2-(methylthio)-4-(isopropylamino)-6-amino-*s*-triazine and 2-(methylthio)-4-(isopropylamino)-6-acetamido-*s*-triazine formed in yields of 55 and 2.6%, respectively (Rejto et al., 1983). Further irradiation of the solution led to the formation of 2-(methylthio)-4,6-diamino-*s*-triazine which eventually decomposed to unidentified products (Rejto et al., 1983). The UV ($\lambda = 253.7$ nm) photolysis of ametryn in water, methanol, ethanol, *n*-butanol and benzene yielded the 2-H analog 4-ethylamino-6-isopropylamino-*s*-triazine. Photodegradation was not observed at wavelengths >300 nm (Pape and Zabik, 1970).

Chemical/Physical. Hydrolyzes to the 6-hydroxy analog, especially in the presence of strong acids and alkalies (Hartley and Kidd, 1987).

Symptoms of Exposure: Eye and skin irritant.

Formulation Types: Wettable concentrate (80%); suspension concentrate.

Toxicity: LC_{50} (96-hour) for rainbow trout 8.8 mg/L, bluegill sunfish 4.1 mg/L, goldfish 14.1 mg/L, carp <1.0 mg/L (Hartley and Kidd, 1987), oyster >1.0 ppm (Humburg et al., 1989); acute oral LD_{50} for rats 1,110 mg/kg (Hartley and Kidd, 1987), 508 mg/kg (RTECS, 1985).

Uses: Herbicide used to control broad-leaved and grass weeds in corn, sugarcane, certain citrus subtropical fruits (bananas, pineapple) and in noncropland. Preharvest and postharvest desiccant used in potatoes to control both crop and weeds.

AMINOCARB

Synonyms: A 363; Bay 44646; Bayer 5080; Bayer 44646; 4-Dimethylamine-*m*-cresyl methylcarbamate; 4-Dimethylamino-3-cresyl methylcarbamate; **4-(Dimethyl-amino)-3-methylphenol methylcarbamate**; 4-(Dimethylamino)-*m*-tolyl methylcarbamate; ENT 25784; Matacil; Mitacil.

$$(CH_3)_2N-\underset{\substack{\\ CH_3}}{\overset{\substack{CH_3 \\ }}{\bigcirc}}-O-\overset{\overset{\textstyle O}{\|}}{C}NHCH_3$$

Designations: CAS Registry Number: 2032-59-9; mf: $C_{11}H_{16}N_2O_2$; fw: 208.26; RTECS: FC0175000.

Properties: Colorless to white crystals. Mp: 93–94°C; log K_{oc}: 1.92 (calculated); log K_{ow}: 1.73; S_o: soluble in most polar organic solvents; S_w: 872, 915 and 1,360 mg/L at 10, 20 and 30°C, respectively.

Environmental Fate

Plant/Surface Water. Several transformation products reported by Day (1991) include 4-amino-*m*-tolyl-*N*-methylcarbamate (AA), 4-amino-3-methylphenol (AC), 4-formamido-*m*-tolyl-*N*-methylcarbamate (FA), *N*-(4-hydroxy-2-methylphenyl)-*N*-methylformamide (FC), 4-methylformamido-*m*-tolyl-*N*-methylcarbamate (MFA), 4-methylamino-*m*-tolyl-*N*-methylcarbamate (MAA), 3-methyl-4-(methylamino)phenyl-*N*-methylcarbamate (MAC), phenol, methylamine and carbon dioxide. MAA was not detected in natural water but was detected in fish tissues following exposure to aminocarb-treated water in the laboratory. The metabolites FA, AC and MAC were detected in Canadian forests treated with aminocarb but the metabolites AA, MAA and FC were not detected (Day, 1991).

On and/or in bean plants, aminocarb degrades with the carbamate moiety remaining intact. Methylcarbamate derivatives identified include the 4-methylamino, 4-amino, 4-methylformamido and 4-formamido analogs (Abdel-Wahab et al., 1966).

Photolytic. When aminocarb in ethanol was irradiated by UV light, extensive degradation was observed. No degradation products were identified; however, two unidentified cholinesterase inhibitors were reported (Crosby et al., 1965).

Chemical/Physical. Aminocarb is hydrolyzed in purified water to 4-(dimethylamino)-3-methylphenol which is then converted to 2-methyl-1,4-benzoquinone. This compound was then oxidized to form the following compounds: 6-(dimethylamino)-2-methyl-1,4-benzoquinone, 6-(methylamino)-2-methyl-1,4-benzoquinone, 5-(dimethylamino)-2-methyl-1,4-benzoquinone and 5-(methylamino)-2-methyl-1,4-benzoquinone (Leger and Mallet, 1988). When aminocarb was irradiated by a high pressure xenon-mercury lamp (λ = 253.7 nm) in aerated and degassed ethyl alcohol and cyclohexene solutions, 4-dimethylamino-3-methyl phenol formed as the major product. A duplicate run using an excitation wavelength of >300 nm yielded that same phenol as the major product. Since aminocarb absorbs radiation in the solar region (at λ >300 nm), this compound would be expected to undergo photochemical degradation in the environment (Addison et al., 1974).

Emits toxic fumes of nitrogen oxides when heated to decomposition (Sax and Lewis, 1987).

Formulation Types: Wettable powder.

Toxicity: Acute oral LD_{50} for rats 30–50 mg/kg (Hartley and Kidd, 1987), for male and female rats, 40 and 38 mg/kg, respectively (Windholz et al., 1983).

Uses: Nonsystemic, broad-spectrum insecticide used to control the spruce budworm in forests; molluscicide.

AMITROLE

Synonyms: Amazol; Amerol; Aminotriazole; 2-Aminotriazole; 3-Aminotriazole; 3-Amino-*s*-triazole; 2-Amino-1,3,4-triazole; 3-Amino-1,2,4-triazole; **3-Amino-1*H*-1,2,4-triazole**; Amino triazole weed killer 90; Aminotriazole-spritzpulver; Amitol; Amitril; Amitril T.L.; Amitrol; Amitrol 90; Amitrol T; Amizol; Amizol D; Amizol F; AT; AT-90; ATA; AT liquid; Azaplant; Azolan; Azole; Campaprim A 1544; Cytrol; Cytrol Amitrole-T; Cytrole; Diurol; Diurol 5030; Domatol; Domatol 88; Elmasil; Emisol; Emisol 50; Emisol F; ENT 25455; Fenamine; Fenavar; Herbidal total; Herbizole; Kleer-lot; Orga 414; Radoxone TL; Ramizol; RCRA waste number U011; Simazol; Triazolamine; 1*H*-1,2,4-Triazol-3-amine; UN 2588; USAF XR-22; Vorox; Vorox AA; Vorox SS; Weedar ADS; Weedar AT; Weedazin; Weedazin arginit; Weedazol; Weedazol GP2; Weedazol super; Weedazol T; Weedazol TL; Weedex granulat; Weedoclor; X-all liquid.

Designations: CAS Registry Number: 61-82-5; mf: $C_2H_4N_4$; fw: 84.08; RTECS: XZ3850000.

Properties: Colorless to white, odorless crystalline solid. Mp: 157–159°C; ρ: 1.138 at 20/4°C; fl p: nonflammable; $H_{-1/2}$: estimated to be 40 days (Reinert and Rodgers, 1987); K_H: 1.63×10^{-15} atm · m³/mol at 20°C (approximate — calculated from water solubility and vapor pressure); log K_{oc}: 1.73–2.31; log K_{ow}: -0.15; S_o: moderately soluble in acetonitrile, chloroform, ethanol, methanol and methylene chloride but only sparingly soluble in ethyl acetate; S_w: 280 g/L at 20°C (pH 7.0); vp: 4.13×10^{-9} mmHg at 20°C.

Soil properties and adsorption data

Soil	K_d (mL/g)	f_{oc} (%)	K_{oc} (mL/g)	pH
Sand	0.68	0.46	147	5.6
Sandy loam	3.52	1.74	202	6.1
Silt loam	3.79	3.42	111	6.8
Silt loam	1.57	2.90	54	7.0

Source: U.S. Department of Agriculture, 1990.

Environmental Fate

Soil. When radiolabeled amitrole-5-^{14}C was incubated in a Hagerstown silty clay loam, 50 and 70% of the applied amount evolved as $^{14}CO_2$ after 3 and 20 days, respectively. In autoclaved soil, however, no $^{14}CO_2$ was detected. The chemical degradation in soil was probably via hydroxyl radicals (Kaufman et al., 1968). The average persistence in soils is 2–4 weeks (Hartley and Kidd, 1987).

Plant. Amitrole is transformed in plants to form the conjugate β-(3-amino-1,2,4-triazol-1-yl)-α-alanine (Humburg et al., 1989) and/or 3-(3-amino-*s*-triazole-1-yl)-2-aminopropionic acid (Duke et al., 1991). Amitrole is metabolized in Canada thistle (*Cirsium arvense* L.) to three unknown compounds which were more phytotoxic than amitrole (Herrett and Bagley, 1964). In field horsetail, amitrole was absorbed and translocated but degradation products were not reported (Coupland and Peabody, 1981).

Surface Water. In pond water, adsorption to suspended sediments was an important process. The initial half-life was reported to be no more than 68 days. After 120 days, 20% of the applied amount remained (Grzenda et al., 1966). The biodegradation half-life of amitrole in water is approximately 40 days (Reinert and Rodgers, 1987).

Photolytic. Direct photolysis of amitrole is not expected to occur since the herbicide shows little or no absorption greater than 295 nm (Gore et al., 1971).

Chemical/Physical. Reacts with acids and bases forming soluble salts (Hartley and Kidd, 1987). Emits toxic fumes of nitrogen oxides when heated to decomposition (Sax and Lewis, 1987); however, incineration with polyethylene results in more than 99% decomposition (Sittig, 1985).

An aqueous solution of amitrole has been shown to decompose in the following free radical systems: Fenton's reagent, UV irradiation and riboflavin-sensitized photodecomposition (Plimmer et al., 1967). Amitrole-5-^{14}C reacted with Fenton's reagent to give radiolabeled carbon dioxide, unlabeled urea and unlabeled cyanamide. Significant degradation of amitrole was observed when an aqueous solution was irradiated by a sunlamp (λ = 280–310 nm). In addition to ring compounds, it was postulated that other products may have formed from the polymerization of amitrole free radicals (Plimmer et al., 1967).

When an aqueous solution containing amitrole was subjected to UV light for 8 hours, no degradation was observed. However, when humic acid (100 mg) was added, the compound degraded with a photolysis half-life of 7.5 hours (Jensen-Korte et al., 1987).

Exposure Limits: NIOSH REL: TWA 0.2 mg/m^3; ACGIH TLV: TWA 0.2 mg/m^3.

Formulation Types: Water soluble powder or concentrate.

Toxicity: LC$_{50}$ (48-hour) for bluegill sunfish 100 ppm and coho salmon 325 mg/L (Verschueren, 1983); acute oral LD$_{50}$ for rats 1,100–2,500 mg/kg (Verschueren, 1983).

Uses: Nonselective, foliage-applied, systemic, triazole herbicide used in uncropped land and orchards to control certain grasses and to kill annual and perennial grasses and weeds. It is also effective on poison ivy, poison oak and aquatic weeds.

ANILAZINE

Synonyms: Anilazin; B 622; Bortrysan; (o-Chloroanilo)dichlorotriazine; 2-Chloro-N-(4,6-dichloro-1,3,5-triazin-2-yl)aniline; 2,4-Dichloro-6-o-chloroanilo-s-triazine; 2,4-Dichloro-6-(2-chloroanilo)-1,3,5-triazine; **4,6-Dichloro-N-(2-chlorophenyl)-1,3,5-triazin-2-amine**; Direz; Dyrene; Dyrene 50W; ENT 26058; Kemate; NCI-C08684; Triasyn; Triazin; Triazine; Zinochlor.

Designations: CAS Registry Number: 101-05-3; mf: $C_9H_5C_{13}N_4$; fw: 275.54; RTECS: XY7175000.

Properties: White to tan crystals. Mp: 159–160°C; ρ: 1.8 at 20/4°C; H-$t_{1/2}$ (hours): 730 (pH 4), 790 (pH 7), 22 (pH 9); K_H: 1.12×10^{-6} atm · m³/mol at 20°C (approxi-mate — calculated from water solubility and vapor pressure); log K_{oc}: 3.48; log K_{ow}: 3.01; So (g/L at 20°C): hexane (1.7), methylene chloride (90), 2-propanol (8), at 30°C: acetone (100), chlorobenzene (60), toluene (50), xylene (40); S_w: 8 mg/L at 20°C; vp: 2.48×10^{-5} mmHg at 20°C.

Environmental Fate

Soil. Anilazine is readily degraded by soil bacteria (Harris et al., 1968). The reported half-life of anilazine in soil is approximately 12 hours (Hartley and Kidd, 1987).

Plant. In plants, one or both of the chlorine atoms on the triazine ring may be replaced by thio or amino groups (Hartley and Kidd, 1987).

Chemical/Physical. Anilazine is subject to hydrolysis (Windholz et al., 1983) releasing chlorine gas (Hartley and Kidd, 1987).

Exposure Limits: The German Research Society's recommended Maximum Allowable Concentration is 0.2 mg/m³.

Formulation Types: Suspension concentrate; wettable powder.

Toxicity: LC_{50} (96-hour) for bluegill sunfish, goldfish and carp <1.0 mg/L; LC_{50} (48-hour) for rainbow trout 150 µg/L (Hartley and Kidd, 1987); acute oral LD_{50} for rats >5,000 mg/kg (Hartley and Kidd, 1987), 2,700 mg/kg (RTECS, 1985).

Uses: Nonsystemic, foliar fungicide used in potatoes, tomatoes, wheat, barley and ornamentals.

ANTU

Synonyms: Anturat; Bantu; Chemical 109; Krysid; **1-Naphthalenylthiourea**; 1-(1-Naphthyl)-2-thiourea; α-Naphthylthiourea; *N*-1-Naphthylthiourea; α-Naphthylthiocarbamide; Rattrack.

$$NHCSNH_2$$

Designations: CAS Registry Number: 86-88-4; DOT: 1651; mf: $C_{11}H_{10}N_2S$; fw: 202.27; RTECS: YT9275000.

Properties: Colorless to gray, odorless solid. Bitter taste. Mp: 198°C; bp: decomposes; ρ: 1.895 (calculated); H-$t_{1/2}$: 361 days at 25°C and pH 7; So (g/L at 25°C): acetone (24.3), triethylene glycol (86); S_w: 600 mg/L at 25°C; vp: $\approx$ 0 mmHg at 20°C.

Environmental Fate

Chemical/Physical. The hydrolysis rate constant for ANTU at pH 7 and 25°C was determined to be 8×10^{-5}/hour, resulting in a half-life of 361 days (Ellington et al., 1988).

Emits very toxic fumes of nitrogen and sulfur oxides when heated to decomposition (Lewis, 1990).

Exposure Limits: NIOSH REL: TWA 0.3 mg/m³, IDLH 100 mg/m³; OSHA PEL: TWA 0.3 mg/m³.

Symptoms of Exposure: Vomiting; dyspnea; cyanosis; coarse pulmonary rales after ingestion of large doses.

Formulation Types: Tracking powder.

Toxicity: Acute oral LD_{50} for Norwegian rats 6–8 mg/kg (Hartley and Kidd, 1987).

Use: Rodenticide.

ASPON

Synonyms: A 42; ASP 51; Bis-*O,O*-di-*n*-propylphosphorothionic anhydride; E 8573; ENT 16894; NPD; Propylthiopyrophosphate; Stauffer ASP-51; Tetra-*n*-propyl dithionopyrophosphate; **Tetrapropyl dithiopyrophosphate**; Tetra-*n*-propyl dithiopyrophosphate.

$$CH_3CH_2CH_2O \diagdown \overset{\overset{S}{\parallel}}{P} \diagup \overset{\overset{S}{\parallel}}{O-P} \diagup OCH_2CH_2CH_3$$

Designations: CAS Registry Number: 3244-90-4; mf: $C_{12}H_{28}O_5P_2S_3$; fw: 378.46; RTECS: XN4550000.

Properties: Liquid. Mp: <25°C; H-$t_{1/2}$: 32 days at 40°C and pH 7; log K_{oc}: 3.80–3.88; log K_{ow}: 2.15 (calculated); S_w: 160 mg/L at room temperature; vap d: 15.47 g/L at 25°C, 13.11 (air = 1).

Toxicity: Acute oral LD_{50} for rats 450 mg/kg (RTECS, 1985).

Uses: An organophosphorus, nonsystemic insecticide used to control chinch bugs in lawns.

ASULAM

Synonyms: Asilan; Asulfox F; Asulox; Asulox 40; Jonnix; MB 9057; Methyl *N*-(4-aminobenzenesulfonyl)carbamate; **Methyl((4-aminophenyl)sulfonyl)carbamate**; Methyl sulfanilylcarbamate.

$$SO_2NHCOOH$$

$$NH_2$$

Designations: CAS Registry Number: 3337-71-1; mf: $C_8H_{10}N_2O_4S$; fw: 230.24; RTECS: FD1190000.

Properties: Odorless, colorless crystals or white crystalline powder. Mp: 142–144°C (decomposes); fl p: nonflammable; pK_a: 4.82 (imido group); K_H: 1.1×10^{-12} atm · m³/mol at 25°C (approximate — calculated from water solubility and vapor pressure); log K_{oc}: 1.80–2.16; log K_{ow}: 0.763; So (g/L): acetone (300), *N,N*-dimethylformamide (>800), ethanol (180), methanol (290), 2-butanone (280); S_w: 5 g/L at 20–25°C; vp: 10^{-8} mmHg at 25°C.

Soil properties and adsorption data

Soil	K_d (mL/g)	f_{oc} (%)	K_{oc} (mL/g)	pH
Sand	0.80	0.58	138	6.8
Sand	1.00	3.19	31	6.9
Sandy loam	1.35	2.15	63	7.0
Sandy loam	2.72	1.86	146	7.2

Source: U.S. Department of Agriculture, 1990.

Environmental Fate

Soil. It is not persistent in soils since its half-life is approximately 6–14 days (Hartley and Kidd, 1987). The short persistence time is affected by soil temperature and moisture content. The half-life of asulam in a heavy clay soil having a moisture content of 34% and maintained at 20°C was 7 days (Smith and Walker, 1977). In soil, sulfanilamide was reported as a product of hydrolysis. In non-sterile soils, this compound further degraded to unidentifiable products (Smith, 1988) which may include substituted anilines (Bartha, 1971).

Photolytic. The reported photolytic half-lives for asulam in water at pH 3 and 9 were 2.5 and 9 days, respectively (Humburg et al., 1989).

Chemical/Physical. Forms water-soluble salts (Hartley and Kidd, 1987). When heated to 75°C, asulam decomposed to sulfanilic acid, carbamic acid and sulfanilamide. At 90°C, 4-nitro- and 4-nitrosobenzene sulfonic acids were released (Rajagopal et al., 1984).

Formulation Types: Aqueous solution of the sodium salt (3.34 lb/gal).

Toxicity: LC_{50} (96-hour) for goldfish, rainbow trout, channel catfish >5 g/L, harlequin fish >1.7 g/L and bluegill sunfish >3 g/L (Worthing and Hance, 1991); acute oral LD_{50} for rats is >8,000 mg/kg (Ashton and Monaco, 1991), 2,000 mg/kg (RTECS, 1985), >5,000 mg/kg (potassium salt) (Verschueren, 1983) and mice 5,000 mg/kg (potassium salt) (Verschueren, 1983).

Uses: Systemic, pre- and postemergence herbicide used to control several perennial grasses and certain broad-leaved weeds such as brackenfern, crabgrass, itchgrass, paragrass, tansy ragwort, and wild mustard, in alfalfa, uncropped land, certain ornamentals and turf.

ATRAZINE

Synonyms: A 361; Aatrex; Aatrex 4L; Aatrex 4LC; Aatrex nine-*o*; Aatrex 80W; Aktikon; Aktikon PK; Aktinit A; Aktinit PK; Argezin; Atazinax; Atranex; Atrasine; Atratol A; Atrazin; Atred; Atrex; Candex; Cekuzina-T; 2-Chloro-4-ethylamineisopropylamine-*s*-triazine; 1-Chloro-3-ethylamino-5-isopropylamino-*s*-triazine; 1-Chloro-3-ethylamino-5-isopropylamino-2,4,6-triazine; 2-Chloro-4-ethylamino-6-isopropylamino-*s*-triazine; 2-Chloro-4-ethylamino-6-isopropylamino-1,3,5-triazine; **6-Chloro-*N*²-ethyl-*N*⁴-isopropyl-1,3,5-triazine-2,4-diamine**; 2-Chloro-4-(2-propylamino)-6-ethylamino-*s*-triazine; Crisatrina; Crisazine; Cyazin; Farmco atrazine; Fenamin; Fenamine; Fenatrol; G 30027; Geigy 30027; Gesaprim; Gesoprim; Griffex; Hungazin; Hungazin PK; Inakor; Oleogesaprim; Primatol; Primatol A; Primaze; Radazin; Radizine; Shell atrazine herbicide; Strazine; Triazine A; Triazine A 1294; Vectal; Vectal SC; Weedex A; Wonuk; Zeazin; Zeazine.

Designations: CAS Registry Number: 1912-24-9; DOT: 1609; mf: $C_8H_{14}ClN_5$; fw: 215.68; RTECS: XY5600000.

Properties: Colorless powder or white crystalline solid. The reported odor threshold concentration in air and the taste threshold concentration in water are 9.2 ppm and 20 ppb, respectively (Young et al., 1996). Mp: 171–174°C; bp: decomposes; ρ: 1.187 at 20/4°C; fl p: nonflammable; pK_a: 1.62 at 20°C, 1.70 at 21°C; H-$t_{1/2}$: 1,771 years at 25°C and pH 7; K_H: 3.04 × 10⁻⁹ atm · m³/mol at 20°C (approximate — calculated from water solubility and vapor pressure); log K_{oc}: 1.95–2.71; log K_{ow}: 2.27 (Liu and Qian, 1995); P-$t_{1/2}$: 58.67 hours (absorbance λ = 240.0 nm, concentration on glass plates = 6.7 μg/cm²); So (g/kg at 20°C): ethyl acetate (18), *n*-octanol (10), *n*-pentane (0.36), at 27°C: chloroform (52), dimethyl sulfoxide (183), ethyl acetate (28), ethyl ether (12), methanol (18); Sw (mg/L): 33.8 at 22°C (Mills and Thurman, 1994), at 26°C: 0.144, 0.161 and 0.170 M at pH values of 3.0, 7.0 and 10.0, respectively (Ward and Weber, 1968); vp: 3 × 10⁻⁷ mmHg at 20°C, 1.4 × 10⁻⁶ mmHg at 30°C.

Soil properties and adsorption data

Soil	K_d (mL/g)	f_{oc} (%)	K_{oc} (mL/g)	pH	Salinity	TOC (mg/L)	CEC[a]
Bates silt loam	0.80	0.80	100	6.5	—	—	9.3
Alluvium 1	0.20	0.09	224	7.6	—	—	—
Alluvium 2	0.19	0.15	123	7.6	—	—	—
Baxter silty clay loam	2.30	1.21	190	6.0	—	—	11.2
Begbroke silt loam	1.00	1.90	53	7.1	—	—	—

Soil properties and adsorption data *(continued)*

Soil	K_d (mL/g)	f_{oc} (%)	K_{oc} (mL/g)	pH	Salinity	TOC (mg/L)	CEC[a]
Bentonite	1.30	—	—	—	—	80.0	—
Boyce loam	0.78	1.27	61	8.0	—	—	20.0
Cecil loamy sand	0.89	0.90	99	5.6	—	—	6.8
Chehalis sandy loam	0.83	0.98	84	5.2	—	—	16.9
Chillum silt loam	4.00	2.54	157	4.6	—	—	7.6
Choptank River, MD	—	—	4,860	—	9.92	108.5	—
Choptank River, MD	—	—	13,600	—	1.24	98.6	—
Choptank River, MD	—	—	8,540	—	1.50	65.5	—
Choptank River, MD	—	—	6,990	—	14.20	104.6	—
Choptank River, MD	—	—	4,840	—	17.00	59.3	—
Choptank River, MD	—	—	7,930	—	5.71	74.2	—
Clarksville silty clay loam	1.70	0.80	212	5.7	—	—	5.7
Collombey	0.86	1.28	67	7.8	—	—	—
Cumberland silt loam	1.40	0.69	203	6.4	—	—	6.5
Dark sandy loam	12.30	12.00	102	6.3	—	—	18.0
Deschutes sandy loam	0.68	0.51	132	5.9	—	—	12.9
Drummer clay loam	6.30	3.63	174	8.0	—	—	40.0
Drummer clay loam	12.60	3.63	347	3.9	—	—	40.0
Drummer clay loam	6.50	3.63	179	6.0	—	—	40.0
Drummer clay loam	8.20	3.63	226	5.3	—	—	40.0
Drummer clay loam	10.70	3.63	295	4.7	—	—	40.0
Eldon silt loam	2.50	1.73	144	5.9	—	—	12.9
Eudora silt loam	1.10	1.00	110	—	—	—	—
Eustis fine sand	0.62	0.56	111	5.6	—	—	5.2
Evouettes	1.98	2.09	95	6.1	—	—	—
Gerald silt loam	3.20	1.55	206	4.7	—	—	11.0
Gila silty clay	0.65	0.63	103	8.0	—	—	29.1
Glendale sandy clay loam	0.62	0.50	124	7.4	—	—	—
Grundy silty clay loam	4.80	2.07	232	5.6	—	—	13.5
Hagerstown silty clay loam	3.70	2.48	149	5.5	—	—	12.5
Handford sandy loam	0.08	0.43	19	6.1	—	—	6.0[b]
Handford sandy loam	0.27	0.43	63	6.1	—	—	6.0[b]
Handford sandy loam	0.61	0.43	142	6.1	—	—	6.0[b]
Hickory Hill silt	7.07	3.27	216	—	—	—	—
Illite	6.20	—	—	—	—	—	25.0
Kaipoioi LBF	9.30	16.70	56	5.4	—	—	—
Kaolinite	0.10	—	—	—	—	—	5.0
Kapaa HFL	2.90	5.70	51	4.4	—	—	—
Keith fine sandy loam	1.49	1.67	89	6.3	—	—	—
Knox silt loam	3.60	1.67	216	5.4	—	—	18.8
Kula RP	6.00	12.70	47	5.9	—	—	—
Lakeland sandy loam	1.00	1.90	53	6.2	—	—	2.9

Soil properties and adsorption data *(continued)*

Soil	K_d (mL/g)	f_{oc} (%)	K_{oc} (mL/g)	pH	Salinity	TOC (mg/L)	CEC[a]
Lebanon silt loam	2.20	1.04	212	4.9	—	—	7.7
Lindley loam	2.60	0.86	302	4.7	—	—	6.9
Lintonia loamy sand	0.60	0.34	176	5.3	—	—	3.2
Loam	2.03	1.45	140	6.6	—	—	—
Marian silt loam	2.20	0.80	275	4.6	—	—	9.9
Marshall silty clay loam	4.50	2.42	186	5.4	—	—	21.3
Menfro silt loam	1.70	1.38	123	5.3	—	—	9.1
Metolius sandy loam	0.62	0.80	77	7.1	—	—	18.5
Molokai LHL	2.30	2.30	100	6.3	—	—	—
Monona silty clay loam	1.92	1.67	115	5.8	—	—	—
Newtonia silt loam	1.80	0.92	196	5.2	—	—	8.8
Oswego silty clay loam	2.70	1.67	162	6.4	—	—	21.0
Powder silt loam	1.20	1.67	72	7.7	—	—	31.9
Putnam silt loam	1.90	1.09	174	5.3	—	—	12.3
Quincy sandy loam	0.26	0.28	93	8.0	—	—	11.0
Rhinebeck silty clay loam	1.98	3.13	63	6.7	—	—	—
Salix loam	2.30	1.21	190	6.3	—	—	17.9
Sand	0.42	0.46	91	5.6	—	—	—
Sand 1	0.20	0.18	112	7.5	—	—	—
Sand 2	0.31	0.12	263	7.1	—	—	—
Sand 3	0.73	0.14	519	7.2	—	—	—
Sand 4	1.68	0.10	1,680	5.6	—	—	—
Sand 5	0.90	0.10	890	6.1	—	—	—
Sand 6	0.35	0.09	387	6.9	—	—	—
Sandy loam	0.99	1.74	57	6.1	—	—	—
Sandy loam	1.00	1.93	52	7.1	—	—	11.0
Sarpy loam	2.20	0.75	293	7.1	—	—	14.3
Sharkey clay	3.10	1.44	215	5.0	—	—	28.2
Sharpsburg silty clay loam	3.89	2.19	178	5.2	—	—	—
Shelby loam	3.20	2.07	154	4.3	—	—	20.1
Silt loam	1.46	1.22	120	7.0	—	—	—
Summit silty clay	5.60	2.82	198	4.8	—	—	35.1
Till 1	2.58	0.32	806	8.2	—	—	—
Till 2	2.83	0.33	858	7.7	—	—	—
Tujunga loamy sand	0.09	0.33	27	6.3	—	—	0.5[b]
Tujunga loamy sand	0.15	0.33	45	6.3	—	—	0.5[b]
Tujunga loamy sand	0.37	0.33	112	6.3	—	—	0.5[b]
Union silt loam	4.10	1.04	394	5.4	—	—	6.8
Valentine loamy fine sand	0.77	0.80	96	5.9	—	—	—
Valois silt loam	1.98	1.64	121	5.9	—	—	—
Vermiculite	0.10	—	—	—	—	—	130.0

Soil properties and adsorption data *(continued)*

Soil	K_d (mL/g)	f_{oc} (%)	K_{oc} (mL/g)	pH	Salinity	TOC (mg/L)	CEC[a]
Vetroz	2.88	3.25	89	6.7	—	—	—
Wabash clay	3.70	1.27	291	5.7	—	—	40.3
Waverley silt loam	3.00	1.15	261	6.4	—	—	12.8
Webster silty clay loam	6.03	3.87	156	7.3	—	—	54.7
Wehadkee silt loam	1.80	1.09	165	5.6	—	—	10.2
Woodburn silt loam	1.45	1.32	110	5.2	—	—	12.8

Source: Talbert and Fletchall, 1965; Harris, 1966; McGlamery and Slife, 1966; Hance, 1967; Green and Obien, 1969; Grover and Hance, 1969; Obien and Green, 1969; Leistra, 1970; Colbert et al., 1975; Dao and Lavy, 1978; Rao and Davidson, 1979; Brown and Flagg, 1981; Burkhard and Guth, 1981; Singh et al., 1990; U.S. Department of Agriculture, 1990; Gamerdinger et al., 1991; Mills and Thurman, 1994; Roy and Krapac, 1994. a) meq/100 g. b) cmol/kg.

Environmental Fate

Biological. Twelve fungi in a basal salts medium and supplemented with sucrose degraded atrazine via monodealkylation. Degradation by *Aspergillus fumigatus* and *Penicillium janthinellum* yielded 2-chloro-4-amino-6-isopropylamino-*s*-triazine and *Rhizopus stolonifer* yielded 2-chloro-4-ethylamino-6-amino-*s*-triazine (Paris and Lewis, 1973). Atrazine was transformed by the culture *Nocardia* forming 2-chloro-4-amino-*s*-triazine (Giardina et al., 1980, 1982).

Deethylatrazine was the major metabolite found in an alluvial aquifer in south central Ohio. Ring-labeled atrazine was recalcitrant to degradation by soil and aquifer microorganisms under aerobic conditions. Aerobic incubations of sediments mineralized <0.1–1.5% [ethyl-2-^{14}C]atrazine to carbon dioxide. The rate of mineralization was higher in shallow sediments (5.2–5.6 m) than in deep sediments (17.7–18.1 m). This suggests that microorganisms in the shallow sediments were better adapted at degrading atrazine than microorganisms in the deeper sediments (McMahon et al., 1992).

Soil. Atrazine undergoes chemical hydrolysis in soil forming hydroxyatrazine, a nonphytotoxic hydroxy analog (Armstrong et al., 1967; Harris, 1967; Skipper et al., 1967; Knuelsi et al., 1969; Obien and Green, 1969; Goswami and Green, 1971; Helling, 1971; Esser et al., 1975; Khan, 1978; Cremlyn, 1991; Somasundaram et al., 1991; Nair and Schnoor, 1992). The rate of hydrolysis, which followed first-order kinetics, was higher in low pH soils (Harris et al., 1969; Obien and Green, 1969; Best and Weber, 1974; Hiltbold and Buchanan, 1977; Lowder and Weber, 1982) and is favorable in soils containing high organic matter (Jordan et al., 1970). Atrazine hydrolysis is also catalyzed with additions of small amounts of sterilized soil (Armstrong et al., 1967), by the presence of fulvic and humic acids (Armstrong et al., 1967; Li and Felbeck, 1972; Khan, 1978; Junk et al., 1980; Lowder and Weber, 1982), or by Al^{+3}- and H^+-montmorillonite clays (Russell et al., 1968; Skipper et al., 1978). In addition, the rate of hydrolysis in soils was more a function of the organic carbon content (f_{oc}) rather than the amount of clay present (Armstrong et al., 1967; Anderson et al., 1980a). The catalytic action of fulvic acids is reduced if Cu(II) is present (Haniff et al., 1985).

Li and Felbeck (1972) reported that the half-lives for atrazine at 25°C and pH 4 with and without fulvic acid (2%) were 1.73 and 244 days, respectively. The hydrolysis half-lives in a 5 mg/L fulvic acid solution and 25°C at pH values of 2.9, 4.5, 6.0 and 7.0 were

34.8, 174, 398 and 742 days, respectively. The only product identified was 2-ethylamino-4-hydroxy-6-isopropylamino-1,3,5-triazine (Khan, 1978). The primary degradative pathway appears to be chemical (i.e., hydrolysis) rather than microbial (Armstrong et al., 1967; Skipper et al., 1967; Best et al., 1974; Gormly and Spalding, 1979; Geller, 1980; Gaynor et al., 1981; Lowder and Weber, 1982).

In soil-water suspensions under both aerobic and anaerobic conditions, ^{14}C-ring labeled atrazine degraded very slowly (half-life >90 days). In aerobic soils, only 0.59% evolved as $^{14}CO_2$ (Goswami and Green, 1971). Geller (1980) reported that atrazine degradation in soil and water was probably due to chemical rather than biological processes. Muir and Baker (1978) reported, however, that dechlorination was a major metabolic process in soils and that 2-chloro-4-ethylamino-6-amino-s-triazine and 2-chloro-4-amino-6-isopropylamino-s-triazine were the major metabolites formed. Though atrazine is very persistent in soils, mineralization of ^{14}C-ring labeled atrazine to $^{14}CO_2$ was detected by Skipper and Volk (1972).

Nine years after applying uniformly ^{14}C-ring labeled atrazine to a mineral soil in Germany, 50% of the ^{14}C residues were in the bound (nonextractable) form. Residues detected included atrazine (0.11 ppm), deethylatrazine (trace), deisopropylatrazine (trace), hydroxyatrazine (0.10 ppm), deethylhydroxyatrazine (0.13 ppm) and deisopropylhydroxyatrazine (0.07 ppm). Except for deethylatrazine, these metabolites were found in higher concentrations in humic soils (Capriel et al., 1985). The half-life in soil ranged from 6 to 10 weeks (Hartley and Kidd, 1987). In biologically active soils, the chlorine atom at the C-6 position is replaced by a hydroxyl group and hydroxyatrazine is formed (Beynon et al., 1972). Pure cultures of soil fungi were capable of degrading atrazine to two or more metabolites. These included 2-chloro-4-amino-6-isopropylamino-s-triazine and 2-chloro-4-ethylamino-6-amino-s-triazine plus the suspected intermediates 2-chloro-4-hydroxy-6-ethylamino-s-triazine, 2-hydroxy-4-amino-6-ethylamino-s-triazine and 2,4-dihydroxy-6-ethylamino-s-triazine (Kaufman and Kearney, 1970).

In a soil-core microcosm study, Winkelmann and Klaine (1991) observed that the concentration of atrazine decreased exponentially over a 6-month period. Metabolites identified in soil included DEA, deisopropylatrazine, DAA and hydroxyatrazine. The half-life in soil is 71 days (Jury et al., 1987). Under laboratory conditions, the half-lives for atrazine in a Hatzenbühl soil (pH 4.8) and Neuhofen soil (pH 6.5) at 22°C were 53 and 113 days, respectively (Burkhard and Guth, 1981). Atrazine degradation products identified in soil were deethylatrazine, deisopropylatrazine, deethyldeisopropylatrazine and hydroxyatrazine (Patumi et al., 1981). Microbial attack of atrazine gave deethylated atrazine and deisopropyl atrazine as major and minor metabolites, respectively (Sirons et al., 1973).

A facultative anaerobic bacterium isolated from a stream sediment utilized atrazine as a carbon and nutrient source. Microbial growth was observed but no degradation products were isolated. At 30°C, the half-life was estimated to be 7 days (Jessee et al., 1983).

The half-lives for atrazine in soil incubated in the laboratory under aerobic conditions ranged from 24 to 109 days (Obien and Green, 1969; Zimdahl et al., 1970; Beynon et al., 1972; Hance, 1974). In field soils, the half-lives for atrazine ranged from 12 to 53 days with an average half-life of 20 days (Sikka and Davis, 1966; Marriage et al., 1975; Khan and Marriage, 1977; Hall and Hartwig, 1978). The mineralization half-lives for ^{14}C-ring labeled atrazine in soil ranged from 9.5 to 28.3 years (Goswami and Green, 1971; Skipper and Volk, 1972).

The disappearance half-lives of atrazine in soil ranged from 12 days when applied as an emulsion to 28 days when applied as granules (Schulz et al., 1970). The half-lives of atrazine in a sandy loam, clay loam and an organic amended soil under non-sterile conditions were 315-1,947, 162–5,824 and 54–934 days, respect-ively, while under sterile conditions the half-lives were 324–1,042, 116–3,000 and 53–517 days, respectively (Schoen and Winterlin, 1987). Atrazine has a low volatility. In the laboratory, 0.5, 0.8 and 0.8% of ^{14}C-ring labeled atrazine was lost to volatilization from soils planted with maize at a concentration of 1 ppm and barley at soil concentrations of 1 and 6 ppm, respectively (Scheunert et al., 1986).

Groundwater. According to the U.S. EPA (1986) atrazine has a high potential to leach to groundwater.

Plant. In tolerant plants, atrazine is readily transformed to hydroxyatrazine which may further degrade via dealkylation of the side chains and subsequent hydrolysis of the amino groups with some evolution of carbon dioxide (Castelfranco et al., 1961; Roth and Knuesli, 1961; Humburg et al., 1989). In corn juice, atrazine was converted to hydroxyatrazine (Montgomery and Freed, 1964). In both roots and shoots of young bean plants, atrazine underwent monodealkylation forming 2-chloro-4-amino-6-isopropylamino-*s*-triazine. This metabolite was found to be less phytotoxic than atrazine (Shimabukuro, 1967).

Roots of the marsh grass *Spartina alterniflora* were placed in an aqueous atrazine solution (5 × 10⁻⁵ M) for 2 days. The atrazine was readily absorbed by the roots and translocated to the shoots. Metabolites identified by TLC were 2-chloro-4-amino-6-ethyl-amino-*s*-triazine, 2-chloro-4-amino-6-isopropylamino-*s*-triazine and 2-chloro-4,6-diamino-*s*-triazine (Pillai et al., 1977). Atrazine was found to be persistent in a peach orchard soil several years after nine consecutive annual applications of the herbicide. After 2 and 3.5 years following the last application of atrazine, metabolites identified in the soil included deethylatrazine, hydroxyatrazine, deethylhydroxyatrazine and deisopropylhy-droxyatrazine. These metabolites were also found in oat plants grown in the same soil, but they were translocated into the plant where they underwent detoxification via conju-gation (Khan and Marriage, 1977). Lamoureux et al. (1970) studied the metabolism and detoxification of atrazine in treated sorghum leaf sections. They found a third major pathway for the metabolism of atrazine which led to the formation of *S*-(4-ethylamino-6-isopropylamino-2-*s*-triazino)glutathione (III) and γ-L-glutamyl-*S*-(4-ethylamino-6-isopro-pylamino-2-*s*-triazino)-L-cysteine (IV). It was suggested that compound III was formed by the enzymatic catalyzation of glutathione with atrazine. Compound IV formed from the reaction of compound III with a carboxypeptidase or by the enzymatically catalyzed condensation of γ-L-glutamyl-L-cysteine with atrazine.

Surface Water. Desethyl- and desisopropylatrazine were degradation products of atra-zine identified in the Mississippi River and its tributaries (Pereira and Rostad, 1990). Under laboratory conditions, atrazine in distilled water and river water was completely degraded after 21.3 and 7.3 hours, respectively (Mansour et al., 1989).

Photolytic. In aqueous solutions, atrazine is converted exclusively to hydroxyatrazine by UV light (λ = 253.7 nm) (Pape and Zabik, 1970; Khan and Schnitzer, 1978) and natural sunlight (Pape and Zabik, 1972). Irradiation of atrazine in methanol, ethanol and *n*-butanol afforded Atratone (2-methoxy-4-ethylamino-6-isopropylamino-*s*-triazine), 2-ethoxy-*s*-tri-azine analog and 2-*n*-butoxy-*s*-triazine analog, respectively (Pape and Zabik, 1970). Atra-zine did not photodegrade when irradiated in methanol, ethanol, or in water at wavelengths >300 nm (Pape and Zabik, 1970). Hydroxyatrazine, two de-*N*-alkyl and the de-*N,N'*-dialkyl analogs of atrazine were produced in the presence of fulvic acid (0.01%).

The photolysis half-life of atrazine in aqueous solution containing fulvic acid remained essentially unchanged but increased at lower pHs (Khan and Schnitzer, 1978). Similar results were achieved in the work by Minero et al. (1992) who reported that the photolysis

(λ >340 nm) of aqueous solution of atrazine containing dissolved humic acid (10 ppm organic carbon) was three times higher in the presence of humic acid. Dehalogenation was the significant degradative pathway but some dealkylation and deamination also occurred (Minero et al., 1992).

An aqueous solution (15°C) of atrazine (10 mg/L) containing acetone (1% by volume) as a photosensitizer was exposed to UV light (λ $\geq$290 nm). The reported photolysis half-lives with and without the sensitizer were 25 and 4.9 hours, respectively (Burkhard and Guth, 1976). Photoproducts formed were hydroxytriazines, two de-N-alkyl and the de-N,N'-dialkyl analogs (Burkhard and Guth, 1976).

Pelizzetti et al. (1990) studied the aqueous photocatalytic degradation of atrazine (ppb level) using simulated sunlight (λ >340 nm) and titanium dioxide as a photocatalyst. Atrazine rapidly degraded from 2 ppb to <0.1 ppb in a few minutes. The following intermediates were identified via HPLC and/or GC/MS: 2,4-diamino-6-hydroxy-N-ethyl-N'-(1-methylethyl)-1,3,5-triazine, ammeline (2,4-diamino-6-hydroxy-1,3,5-triazine), ammelide (2-diamino-4,6-dihydroxy-1,3,5-triazine), 2,4-diamino-6-chloro-1,3,5-triazine, 2,4-diamino-6-chloro-N-(1-methylethyl)-1,3,5-triazine, 2,4-diamino-6-chloro-N-ethyl-1,3,5-triazine, 2-amino-4-chloro-6-hydroxy-1,3,5-triazine, 2-chloro-4,6-dihydroxy-1,3,5-triazine, cyanuric acid (2,4,6-trihydroxy-1,3,5-triazine) and 2-acetylamino-4-amino-6-chloro-N-(1-methylethyl)-1,3,5-triazine. Complete degradation of atrazine gave cyanuric acid (2,4,6-trihydroxy-1,3,5-triazine), chloride and nitrate ions. Mineralization of cyanuric acid to carbon dioxide was not observed (Pelizzetti et al., 1990). Nearly identical photodegradation products were reported by Pelizzetti et al. (1992). They studied the photocatalytic degradation of atrazine in water containing a suspension of titanium dioxide (0.5 g/L) as the catalyst. Irradiation was carried out in Pyrex glass cells using an 1500-W Xenon lamp (λ cutoff = 340 nm). The major photoprocesses of degradation were alkyl chain oxidation and subsequent oxidation leading to the formation of hydroxy derivatives. Dehalogenation was also observed but this was considered a minor degradative pathway. Cyanuric acid was the major photoproduct formed. There was only a partial conversion or mineralization to carbon dioxide (Pelizzetti et al., 1992).

The dye-sensitized photodecomposition of atrazine was studied in aqueous, aerated solutions. When the solution was irradiated in sunlight for several hours, 2-chloro-4-(isopropylamino)-6-amino-s-triazine and 2-chloro-4-(isopropylamino)-6-acetamido-s-triazine formed in yields of 70 and 7%, respectively (Rejto et al., 1983). Further irradiation of the solution led to the formation of 2-chloro-4,6-diamino-s-triazine which eventually degraded to unidentified products (Rejto et al., 1983). Hydroxyatrazine was the major intermediate compound formed when atrazine (100 μg/L) in both oxygenated estuarine water (Jones, 1982; Mansour et al., 1989) and estuarine sediments were exposed to sunlight. The rate of degradation was slightly higher in water (half-life 3–12 days) than in sediments (half-life 1–4 weeks) (Jones, 1982).

In soil, the photolytic half-life of atrazine (10 ppm) in two agricultural soils was 330 and 385 days (Jones et al., 1982).

Chemical/Physical. The hydrolysis half-lives of atrazine in aqueous buffered solutions at 25°C and pH values of 1, 2, 3, 4, 11, 12 and 13 were reported to be 3.3, 14, 58, 240, 100, 12.5 and 1.5 days, respectively (Armstrong et al., 1967). Atrazine does not hydrolyze in uncatalyzed solutions, even under elevated temperatures. The estimated half-life of atrazine in neutral, uncatalyzed water at pH 6.97 and 25°C is 1,800 years. Under acidic conditions, hydrolysis proceeds via mono- and di-protonated forms (Plust et al., 1981). Atrazine is stable in slightly acidic or basic media, but is hydrolyzed to hydroxy derivatives by alkalies and strong mineral acids (Windholz et al., 1983). Atrazine reacts with strong mineral acids forming hydroxyatrazine (Montgomery and Freed, 1964).

In the presence of hydroxy or perhydroxy radicals generated from Fenton's reagent, atrazine undergoes oxidative dealkylation in aqueous solutions (Kaufman and Kearney, 1970). Major products identified by mass spectrometry included deisopropylatrazine (2-chloro-4-ethylamino-6-amino-s-triazine), 2-chloro-4-amino-6-isopropylamino-s-triazine (DEA) and the dealkylated dealkylatrazine (DAA, 2-chloro-4,6-diamino-s-triazine) (Kaufman and Kearney, 1970).

The primary ozonation by-products of atrazine (15 µg/L) in natural surface water and synthetic water were deethylatrazine, deisopropylatrazine, a didealkylated atrazine (2-chloro-4,6-diamino-s-triazine), a deisopropylatrazine amide (4-acetamido-4-amino-6-chloro-s-triazine), 2-amino-4-hydroxy-6-isopropylamino-s-triazine and an unknown compound. The types of compounds formed were pH dependent. At high pH, low alkalinity, or in the presence of hydrogen peroxide, hydroxyl radicals formed from ozone yielded s-triazine hydroxy analogs via hydrolysis of the Cl-Cl bond. At low pH and low alkalinity, which minimized the production of hydroxy radicals, dealkylated atrazine and an amide were the primary byproducts formed (Adams and Randtke, 1992).

Arnold et al. (1995) investigated the degradation of atrazine by Fenton's reagent at 25°C. Complete degradation of atrazine was achieved <30 seconds using a 2.69 mM solution of (1:1) ferrous sulfate:hydrogen peroxide. Degradation decreased from 99% at pH 3 to 37% at pH 9. During the course of the reaction (pH 3), 10 intermediate compounds were detected but they completely disappeared after 11.5 hours. Two end products remained: 2-chloro-4,6-diamino-s-triazine and 2-acet-amido-4-amino-6-chloro-s-triazine.

Products formed from the combustion of atrazine at 900°C included carbon monoxide, carbon dioxide, hydrochloric acid and ammonia (Kennedy et al., 1972, 1972a). At 250°C, however, atrazine decomposes to yellow flakes which were tentatively identified as primary or secondary amines (Stojanovic et al., 1972). Tirey et al. (1993) evaluated the degradation of atrazine at 12 different temperatures. When atrazine was oxidized at the temperature range of 300–800°C, numerous reaction products were identified by GC/MS. These include, but are not limited to, 2-chloro-4,6-bis(isopropylamino)-s-triazine, 2,6-bis(1-methylethyl)benzenamine, 4-chloro-2,5-dimethoxybenzenamine, 1-propene, ethanedinitrile, hydrogen cyanide, hydrogen azide, hydrogen chloride, acrolein, acetonitrile, acetone, propenenitrile, 2-isocyanatopropane, cyclobutanone, 1-cyclopropylethanone, nitrous oxide, 6-methyl-1H-purine, carbon monoxide, carbon dioxide, 3-ethyl-2,5-dimethylpyrazine, 4-methoxy-N-methylbenzenamine.

Koskinen et al. (1994) studied the ultrasonic decomposition of atrazine (3.1 nM) in pure (distilled/deionized) water using a sonicator operating in the continuous mode at a maximum output of 20 kHz. Decomposition followed first-order kinetics and the rate increased with increasing power and temperature. The disappearance half-lives of analytical and formulated grades of atrazine were 330 and 2,772 minutes, respectively. The authors suggested that decomposition of atrazine was probably the result of the herbicide reacting with highly reactive hydroxyl free radicals.

Exposure Limits: OSHA PEL: TWA 5 mg/m³; ACGIH TLV: TWA 5 mg/m³.

Symptoms of Exposure: Dermatitis; severely irritates skin, eyes, nose and throat.

Formulation Types: Liquid concentrate (4 lb/gal); wettable powder (80%); dispersible granules (85.5%).

Toxicity: LC_{50} (96-hour) for rainbow trout 8.8 mg/L, bluegill sunfish 16 mg/L, carp 76 mg/L, perch 16 mg/L, catfish 7.6 mg/L and guppies 4.3 mg/L (Hartley and Kidd, 1987); LC_{50} (48-hour) for rainbow trout 12.6 ppm (Sanborn et al., 1977); acute oral LD_{50} of the 80% formulation for rats is 5,100 mg/kg (Ashton and Monaco, 1991), 750 mg/kg (RTECS, 1985).

Uses: Preemergence and postemergence herbicide for control of some annual grasses and broad-leaved weeds in corn, fallow land, rangeland, sorghum, non-cropland, certain tropical plantations, evergreen nurseries, fruit crops and lawns.

AZINPHOS-METHYL

Synonyms: Bay 9027; Bay 17147; Bayer 9027; Bayer 17147; Benzotriazinedithio-phosphoric acid dimethoxy ester; Carfene; Cotneon; Cotnion methyl; Crysthion 2L; Crysthyon; DBD; *S*-(3,4-Dihydro-4-oxobenzo(α)(1,2,3)triazin-3-ylmethyl)-*O,O*-dimethyl phosphorodithioate; *S*-(3,4-Dihydro-4-oxo-1,2,3-benzotriazin-3-yl-methyl)-*O,O*-dimethyl phosphorodithioate; *O,O*-Dimethyl-*S*-(benzaziminomethyl) dithiophosphate; *O,O*-Dimethyl-*S*-(1,2,3-benzotriazinyl-4-keto)methyl phosphorodithioate; *O,O*-Di-methyl-*S*-(3,4-dihydro-4-keto-1,2,3-benzotriazinyl-3-methyl) dithiophosphate; Dimethyldithiophosphoric acid *N*-methylbenzazimide ester; *O,O*-Dimethyl-S-(4-oxo-3*H*-1,2,3-benzotriazine-3-methyl) phosphorodithioate; *O,O*-Dimethyl-*S*-(4-oxobenzotriazino-3-methyl) phosphorodithioate; *O,O*-Dimethyl-*S*-(4-oxo-1,2,3-benzotriazino-3-methyl) thiothionophosphate; *O,O*-Dimethyl *S*-((4-oxo-1,2,3-benzotriazin-3-yl)methyl) phosphorodithioate; *O,O*-Dimethyl *S*-((4-oxo-1,2,3-benzotriazin-3(4*H*)yl)methyl) phosphorodithioate; ENT 23233; Gothnion; Gusathion; Gusathion 20; Gusathion 25; Gusathion K; Gusathion M; Gusathion methyl; Guthion; 3-(Mercaptomethyl)-1,2,3-benzotriazin-4(3*H*)-one-*O,O*-dimethyl phosphorodithioate-*S*-ester; Methylazinphos; *N*-Methylbenzazimide, dimethyldithiophosphoric acid ester; Methyl guthion; Metiltriazotion; NA 2783; NCI-C00066; **Phosphorodithioic acid *O,O*-dimethyl ((4-oxo-1,2,3-benzotriazin-3(4*H*)-yl)methyl) ester**; Phosphorodithioic acid *O,O*-dimethyl ester, *S*-ester with 3-mercaptomethyl-1,2,3-benzotriazin-4(3*H*)-one; R 1582.

Designations: CAS Registry Number: 86-50-0; DOT: 2783; mf: $C_{10}H_{12}N_3O_3PS_2$; fw: 317.33; RTECS: TE1925000.

Properties: Colorless crystals or brown waxy solid having a low odor threshold in water (0.2 μg/kg). Mp: 72.4°C; bp: decomposes >200°C; ρ: 1.44 at 20/4°C; fl p: nonflammable; H-$t_{1/2}$ (pH 8.6): 36.4, 27.9 and 7.2 days at 6, 25 and 40°C, respectively; K_H: 7.50×10^{-9} atm · m³/mol at 20°C; log K_{oc}: 2.47–3.53; log K_{ow}: 2.69, 2.75; S_o: soluble in benzene, carbon tetrachloride, chloroform, ethylbenzene, methylene chloride, 2-propanol, toluene, xylene and many other organic solvents; S_w: 9.5, 20.9, 29 and 43.6 mg/L at 10, 20, 25 and 30°C, respectively; vp: 1.6×10^{-6} mmHg at 20°C.

Soil properties and adsorption data

Soil	K_d (mL/g)	f_{oc} (%)	K_{oc} (mL/g)	pH
Ca-hectorite (30°C)	1,418.0	—	—	—
Ca-hectorite (45°C)	1,036.0	—	—	—

Soil properties and adsorption data *(continued)*

Soil	K_d (mL/g)	f_{oc} (%)	K_{oc} (mL/g)	pH
Ca-montmorillonite (30°C)	92.3	—	—	—
Ca-montmorillonite (45°C)	158.6	—	—	—
Clayey loam	9.9	0.29	3,414	6.0
Cohansey sand	96.7	2.55	3,789	—
Cu-montmorillonite (30°C)	2,072.0	—	—	—
Cu-montmorillonite (45°C)	2,598.0	—	—	—
Loamy sand	7.6	1.62	469	6.6
Sandy loam	3.3	1.10	300	6.4
Silt loam	16.8	2.90	579	7.9
Silt loam	11.0	1.04	1,058	5.5
Silt loam	28.5	2.67	1,067	5.4

Source: Sánchez Martin and Sánchez Camazano; 1984; Reduker et al., 1988; U.S. Department of Agriculture, 1990.

Environmental Fate

Biological. Mixed cultures of microorganisms obtained from soil, raw sewage, activated sludge, and settled sludge were all able to degrade azinphosmethyl. When this dithioate pesticide was incubated in a stirred flask containing a mixed culture for 4 days, the concentration decreased from 99 to 40 mg/L (Barik et al., 1984).

Soil. The principal degradation products in soil and by selected soil microorganisms are benzazimide, thiomethylbenzazimide, bis(benzazimidylmethyl)disulfide and anthranilic acid. Benzazimide is further transformed only by *Pseudomonas* sp. DSM 5030 at a sufficient rate to 5-hydroxybenzazimide (Engelhardt and Wallnöfer, 1983).

When radiolabeled azinphos-methyl was incubated in soil, 50 and 93% of the applied amount degraded to carbon dioxide after 44 and 197 days, respectively (Engelhardt et al., 1984). The presence of benzamide, salicylic acid and $^{14}CO_2$ from [carbonyl-^{14}C]- and [ring-U-^{14}C]azinphos-methyl indicated that the 1,2,3-benzotriazinone ring is cleaved. The key intermediates in the degradation of azinphosmethyl in soil were 3-(mercaptomethyl)-1,2,3-benzotriazin-4(3*H*)-one, 3-((methylthio)methyl)-1,2,3-benzotriazin-4(3*H*)-one, 3-((methylsulfonyl)methyl)-1,2,3-benzotriazin-4(3*H*)-one and 3-((methylsulfinyl)methyl)-1,2,3-benzotriazin-4(3*H*)-one (Engelhardt et al., 1984). Other metabolites identified included *O,O*-dimethyl *S*-((4-oxo-1,2,3-benzotriazin-4(3*H*)-yl)methyl) phosphorodithioate, bis(3-(thiomethyl)-1,2,3-benzotriazin-4(3*H*)-one), 3,3′-thiobis(methylene)-1,2,3-benzotriazin-4(3*H*)-one, 3-methyl-1,2,3-benzotriazin-4(3*H*)-one, 3,3′-oxybis(methylene)bis(1,2,3-benzotriazin-4(3*H*)-one), 3-(hydroxymethyl)-1,2,3-benzotriazin-4(3*H*)-one, 1,2,3-benzotriazin-4(3*H*)-one, *S*-methyl *S*-((oxo-3*H*-1,2,3-benzotriazin-3-yl)methyl) dithiophosphate, *O*-methyl *S*-((oxo-3*H*-1,2,3-benzotriazin-3-yl)methyl) dithiophosphate, (4-oxo-3*H*-1,2,3-benzotriazin-3-yl)methanesulfonic acid and anthranilic acid (Engelhardt et al., 1984).

In a dry soil (2–3% moisture content), the half-lives were 484, 88 and 32 days at 6, 25 and 40°C, respectively. In a soil containing 50% moisture, the half-lives were much shorter, i.e., 64, 13 and 5 days at 6, 25 and 40°C, respectively (Yaron et al., 1974).

Azinphos-methyl will not leach to any great extent in soil (Helling, 1971). Staiff et al. (1975) studied the persistence of azinphos-methyl in a test plot over an 8-year period. At the end of the eighth year, virtually no azinphos-methyl was detected 30 cm below the surface. In a sugarcane runoff plot, azinphos-methyl was applied at a rate of 0.84 kg/ha

4 times each year in 1980 and 1981. Runoff losses in 1980 and 1981 were 0.08 and 0.55 of the applied amount, respectively (Smith et al., 1983).

Photolytic. Azinphosphos-methyl on glass plates was decomposed by UV light (λ = 2537 Å) to anthranilic acid, benzazimide, methyl benzazimide sulfide, *N*-methyl benzazimide and an unidentified water-soluble compound (Liang and Lichtenstein, 1972).

Chemical/Physical. At temperatures >200°C, azinphos-methyl decomposes (Windholz et al., 1983) emitting toxic fumes of phosphorus, nitrogen and sulfur oxides (Sax and Lewis, 1987; Lewis, 1990).

Decomposed by alkaline solutions to give anthranilic acid and other products (Sittig, 1985).

Exposure Limits: NIOSH REL: 0.2 mg/m³; OSHA PEL: TWA 0.2 mg/m³; ACGIH TLV: TWA 0.2 mg/m³.

Symptoms of Exposure: Miosis, eye ache, chest wheezing, rhinorrhea of the forehead, cyanosis, salivation, laryngeal spasm, nausea, vomiting, anorexia, diarrhea, sweating, paralysis, convulsions and low blood pressure.

Formulation Types: Emulsifiable concentrate; wettable powder; dustable powder.

Toxicity: LC$_{50}$ (96-hour) for rainbow trout 20 µg/L, bluegill sunfish 4.6 µg/L, guppies 100 µg/L (Hartley and Kidd, 1987), goldfish >1,000 µg/L (Worthing and Hance, 1991), fathead minnow 93 µg/L, minnow 240 µg/L, goldfish 4.3 mg/L, bluegill sunfish 5.2 µg/L (Katz, 1961), largemouth bass 52 µg/L, rainbow trout 14 µg/L, brown trout 4 µg/L, coho salmon 17 µg/L, perch 13 µg/L, channel catfish 3.29 mg/L and black bullhead 3.5 mg/L (Macek and McAllister, 1970); acute oral LD$_{50}$ for rats 4–20 mg/kg (Hartley and Kidd, 1987), 7 mg/kg (RTECS, 1985).

Uses: Nonsystemic insecticide and acaricide for control of insects pests in blueberry, grape, maize, vegetable, cotton and citrus crops.

BENDIOCARB

Synonyms: Bencarbate; **2,2-Dimethyl-1,3-benzodioxol-4-ol methylcarbamate**; 2,2-Dimethyl-1,3-benzodioxol-4-ol *N*-methylcarbamate; 2,2-Dimethylbenzo-1,3-dioxol-4-ol methylcarbamate; Dycarb; Ficam; Ficam D; Ficam ULV; Ficam W; Garvox; 2,3-Isopropylidenedioxyphenyl methylcarbamate; Multamat; Multimet; NC 6897; Niomil; Rotate; Tattoo; Turcam.

Designations: CAS Registry Number: 22781-23-3; mf: $C_{11}H_{13}NO_4$; fw: 223.23; RTECS: FC1140000.

Properties: Colorless to white crystals. Mp: 129–130°C; ρ: 1.25 at 20/4°C; pK_a: 8.8; H-$t_{1/2}$: 4 days at 25°C and pH 7; K_H: 3.6 × 10⁻⁶ atm · m³/mol at 20°C (approximate — calculated from water solubility and vapor pressure); log K_{oc}: 2.76; log K_{ow}: 1.70; So (g/kg at 25°C): acetone (200), benzene (40), chloroform (200), 1,4-dioxane (200), hexane (0.35), methylene chloride (200); S_w: 40 mg/L at 20°C; vp: 4.95 × 10⁻⁶ mmHg at 20°C.

Environmental Fate

Soil. Though no products were identified, the reported half-life in soil is several days to a few weeks (Hartley and Kidd, 1987). When soil containing bendiocarb was incubated for 4 weeks, the amount remaining was 30.4–47.2% of the applied amount. After one application, the amount of bendiocarb recovered was only 0.9–1.3% (Racke and Coates, 1988). Biodegradation of bendiocarb in soil was enhanced if the soil was pretreated with carbofuran or trimethacarb (Dzantor and Felsot, 1989).

Plant. At planting, bendiocarb was applied at a rate of 1.7 kg/ha. During harvest (110 days following planting), the residues of bendiocarb found in corn (leaves and stems), husks, cobs and kernels were 12, 5, 4 and 3 ppb, respectively. At an application rate of 2.5 kg/ha, residues found in leaves and stems, husks, cobs and kernels were 24, 18, 11 and 1 ppb, respectively (Szeto et al., 1984).

Chemical/Physical. Bendiocarb is stable to light but in water hydrolyzes to 2,3-isopropylidenedioxyphenol, methylamine and carbon dioxide. The hydrolysis half-life at pH 7 and 25°C is 4 days (Worthing and Hance, 1991).

Formulation Types: Wettable powder; dustable powder; granules; suspension concentrate; seed treatment; aerosol.

Toxicity: LC_{50} (96-hour) for rainbow trout 1.55 mg/L (Hartley and Kidd, 1987); acute oral LD_{50} for rats 40–156 mg/kg (Hartley and Kidd, 1987).

Uses: Contact insecticide used to control beetles, wireworms, flies, wasps and mosquitoes in beets and maize.

BENFLURALIN

Synonyms: Balan; Balfin; Benefex; Benefin; Benfluraline; Bethrodine; Bonalan; *N*-Butyl-*N*-ethyl-2,6-dinitro-4-trifluoromethylaniline; **N-Butyl-N-ethyl-2,6-dinitro-4-(trifluorom-ethyl)benzenamine**; *N*-Butyl-*N*-ethyl-α,α,α-trifluoro-2,6-dinitro-*p*-toluidine; EL 110; Quilan.

$$CH_3CH_2 \diagdown \quad \diagup CH_2CH_2CH_2CH_3$$

(chemical structure of benfluralin with N, two NO₂ groups, and CF₃)

Designations: CAS Registry Number: 1861-41-1; mf: $C_{13}H_{16}F_3N_3O_4$; fw: 335.29.

Properties: Yellow to orange crystalline solid. Mp: 65–66.5°C; bp: 148–149 at 7 mmHg; fl p: nonflammable; K_H: 3.5×10^{-4} atm · m³/mol at 25°C (approximate — calculated from water solubility and vapor pressure); log Kd: 1.43 (sandy soil at pH 7.7), 2.07 (clay loam at pH 6.9); log K_{ow}: 5.29; So (g/L at 25°C): acetone (650), acetonitrile (>250), 2-butanone (580), chloroform (>500), *N,N*-dimethylformamide (450), 1,4-dioxane (600), ethanol (24), methanol (40), xylene (420–450); S_w: 100 μg/L at 25°C; vp: 7.8×10^{-5} mmHg at 25°C, 3.89×10^{-4} mmHg at 30°C.

Environmental Fate

Soil. Benfluralin degraded faster in flooded soils under anaerobic conditions than when oxygen was present. The major route of benfluralin degradation in flooded soil is the formation of polar products (Golab et al., 1970) including *N*-butyl-*N*-ethyl-α,α,α-trifluoro-5-nitrotoluene-3,4-diamine, *N*-butyl-*N*-ethyl-α,α,α-trifluorotoluene-3,4,5-triamine, *N*-butyl-*N*-ethyl-α,α,α-trifluorotoluene-3,4,5-triamine, *N*-butyl-α,α,α-trifluoro-2,6-dinitro-*p*-toluidine, *N*-ethyl-α,α,α-trifluoro-2,6-dinitro-*p*-toluidine, *N*-butyl-α,α,α-trifluoro-5-nitrotoluene-3,4-diamine, *N*-butyl-α,α,α-trifluorotoluene-3,4,5-triamine, *N*-ethyl-α,α,α-trifluoro-5-nitrotoluene-3,4-diamine, α,α,α-trifluoro-5-nitrotoluene-3,4-diamine, and α,α,α-trifluorotoluene-3,4,5-triamine (Williams, 1977). The rate of degradation was highest under anaerobic conditions.

Nash (1988) reported a dissipation half-life of 2 days for benfluralin in soil.

Photolytic. Though no products were identified, benfluralin is subject to photodegradation by UV light (Worthing and Hance, 1991). A photodegradation yield of 79% was achieved after the herbicide in dry soil was subjected to sunlight for 7 days (Parochetti and Dec, 1978).

Chemical/Physical. In aqueous solutions (pH 5–9 and 26°C), benfluralin is stable up to 30 days (Worthing and Hance, 1991).

Symptoms of Exposure: Eye and skin irritant.

Formulation Types: Wettable powder; granules; emulsifiable concentrate.

Toxicity: Acute oral LD_{50} for rats >10,000 mg/kg, mice >5,000 mg/kg, dogs and rabbits >2,000 mg/kg (Worthing and Hance, 1991).

Uses: A trifluoromethyl dinitroaniline selective preemergence herbicide for control of annual grasses and some broad-leaved weeds in vegetables and turf.

BENOMYL

Synonyms: Arilate; BBC; Benlat; Benlate; Benlate 50; Benlate 50W; Benomyl 50W; BNM; **(1-((Butylamino)carbonyl)-1*H*-benzimidazol-2-yl)carbamic acid methyl ester**; 1-(Butylcarbamoyl)-2-benzimidazolecarbamic acid methyl ester; D 1991; Du Pont 1991; F 1991; Fungicide 1991; Fundasol; Fundazol; MBC; Methyl (1-((butyl-amino)carbonyl)-1*H*-benzimidazol-2-yl)carbamate; Methyl 1-(butylcarbamoyl)-2-benzimidazolylcarbamate; Tersan 1991.

Designations: CAS Registry Number: 17804-35-2; mf: $C_{14}H_{18}N_4O_3$; fw: 290.62; RTECS: DD6475000.

Properties: Colorless to white crystalline solid with a faint, acrid odor. Mp: decomposes before melting; bp: decomposes; log K_{oc}: 3.28 (calculated); log K_{ow}: 1.40–3.11; So (g/kg at 25°C): acetone (18), chloroform (94), *N,N*-dimethylform-amide (53), ethanol (4), *n*-heptane (10), xylene (10) and many other organic solvents except oils; S_w: 2.8 mg/L at 25°C (stable only at pH 7); vp: negligible at room temperature.

Environmental Fate

Biological. Mixed cultures can grow on benomyl as the sole carbon source. It was proposed that benomyl degraded to butylamine and methyl 2-benzimidazole-carbamate (MBC), the latter undergoing further degradation to 2-aminobenzimidazole then to carbon dioxide and other products (Fuchs and de Vries, 1978).

Soil. In soil and water, benomyl is transformed to methyl-2-benzimidazole and 2-aminobenzimidazole (Rhodes and Long, 1974; Ramakrishna et al., 1979; Rajagopal et al., 1984). Benomyl is easily hydrolyzed in soil to methyl-2-benzimidazole carbamate (Li and Nelson, 1985). On bare soil and turf, benomyl degraded to methyl 2-benzimidazolecar-bamate and 2-aminobenzimidazole. Most of the residues were immobile in soils containing organic matter and were limited to the top four inches of soil (Baude et al., 1974). In a loamy garden soil, benomyl was degraded by two fungi and four strains of bacteria to nonfungistatic compounds (Helweg, 1972). The reported half-lives in bare soil and turf are 3–6 and 6–12 months, respectively (Baude et al., 1974).

Plant. On apple foliage treated with a Benlate formulation, benomyl was transformed to MBC. Benomyl dissipated quickly and the reported half-life on foliage was 3–7 days (Chiba and Veres, 1981).

Chemical/Physical. In aqueous solutions, especially in the presence of acids, benomyl hydrolyzes to the strongly fungicidal methyl-2-benzimidazolecarbamate (carbendazim) (Clemons and Sisler, 1969; Peterson and Edgington, 1969; Zbozinek, 1984; Cremlyn, 1991; Worthing and Hance, 1991) and butyl isocyanate (Zbozinek, 1984; Worthing and Hance, 1991). The latter is unstable in water and decomposes to butylamine and carbon dioxide (Zbozinek, 1984). In highly acidic and alkaline aqueous solutions (pH <1 and pH >11), benomyl is completely converted to 3-butyl-2,4-dioxo-[1,2-a]-s-triazinobenzimidazole (STB) with smaller quantities of methyl 2-benzimidazolecarbamate (MBC). In addition 1-(2-benzimidazolyl)-3-n-butylurea was identified but only under highly alkaline conditions (Singh and Chiba, 1985).

Emits toxic fumes of nitrogen oxides when heated to decomposition (Lewis, 1990).

Exposure Limits: OSHA PEL: TWA 15 mg/m^3 (total dust), 5 mg/m^3 (respirable fraction); ACGIH TLV: TWA 10 mg/m^3.

Symptoms of Exposure: Eye and skin irritant.

Formulation Types: Wettable powder.

Toxicity: LC$_{50}$ (96-hour) for rainbow trout 0.17 mg/L and goldfish 4.2 mg/L (Hartley and Kidd, 1987); acute oral LD$_{50}$ for rats >10,000 mg/kg (Hartley and Kidd, 1987), 10 mg/kg (RTECS, 1985).

Uses: Post-harvest systemic fungicide used to control fungi and mildew on cotton, roses, soft fruits, tomatoes, cucumbers and other vegetables.

BENTAZONE

Synonyms: BAS 351-H; Basagran; Bendioxide; Bentazon; 3-Isopropyl-1*H*-2,1,3-benzothi-adiazin-4(3*H*)-one-2,2-dioxide; **3-(1-Methylethyl)-1*H*-2,1,3-benzothiadiazin-4(3*H*)-one-2,2-dioxide.**

Designations: CAS Registry Number: 25057-89-0; mf: $C_{10}H_{12}N_2O_3S$; fw: 240.28; RTECS: DK9900000.

Properties: Odorless, colorless crystals to white crystalline powder. Mp: 137–139°C; bp: decomposes at 200°C; ρ: 1.47; fl p: nonflammable (>100°C); pK_a: 3.4; K_H: <4.7 × 10^{-11} atm · m³/mol at 20°C (approximate — calculated from water solubility and vapor pressure); log K_{oc}: 3.82; log K_{ow}: -0.46; So (g/kg at 20°C): acetone (1,507), benzene (33), chloroform (180), cyclohexane (0.2), ethanol (861), ethyl acetate (650), ethyl ether (616), olive oil (27); S_w: 500 mg/L at 20°C; vp: <7.5 × 10^{-8} mmHg at 20°C.

Soil properties and adsorption data

Soil	K_d (mL/g)	f_{oc} (%)	K_{oc} (mL/g)	pH
Dundee silt loam	0.039	1.02	3.82	5.60

Source: Gaston et al., 1996.

Environmental Fate

Soil. Under aerobic conditions, bentazone was reported to degrade to 6- and 8-hydrox-ybentazone compounds. In addition, anthranilic acid and isopropylamide were reported as soil hydrolysis products (Otto et al., 1978). Persistence in soil is less than 6 weeks (Hartley and Kidd, 1987). Bentazone is readily adsorbed onto organic carbon and therefore, is not expected to leach to groundwater (Abernathy and Wax, 1973). The dissipation half-life of bentazone in field soil is 5 days (Ross et al., 1989).

Plant. Undergoes hydroxylation of the aromatic ring and subsequent conju-gation in plants (Otto et al., 1978; Hartley and Kidd, 1987) forming 6- and 8-hydroxybentazone compounds (Otto et al., 1978). The half-life in and/or on plants is 2–3 days (Ross et al., 1989).

Photolytic. Humburg et al. (1989) reported that 30% degradation of bentazone on glass plates occurred when exposed to UV light (λ = 200–400 nm); however, no photoproduct(s) were reported. The natural sunlight and simulated sunlight irradiation of bentazone was studied in aqueous solution (500 ppm), on soil and as thin films on Pyrex plates (Nilles and Zabik, 1975). The natural sunlight irradiation of a saturated aqueous solution of bentazone for 115 hours yielded six major photoproducts including *o*-amino-*N*-isopropylbenzamide, *o*-nitro-*N*-isopropylbenzamide and *o*-nitroso-*N*-isopropylbenzamide. These three compounds were also observed following the simulated irradiation of bentazone as thin films of glass. The simulated sunlight irradiation of bentazone on soil produced six major photoproducts including *o*-nitro-*N*-isopropylbenzamide and *o*-nitroso-*N*-iso-propy-lbenzamide (Nilles and Zabik, 1975).

Chiron et al. (1995) investigated the photodegradation of bentazone (20 µg/L) in distilled water and Ebro River water using an xenon arc irradiation. In distilled water, bentazone completely disappeared after 16 hours of irradiation. Photodegradation appeared to follow pseudo-first-order kinetics with a half-life of about 2.5 hours. The presence of humic substances (4 mg/L) increased the rate of photodegradation and the disappearance of bentazone was achieved in 8 hours. No significant breakdown photoproducts were identified.

Symptoms of Exposure: Ingestion causes apathy, ataxia, anorexia, prostration, tremors, occasional vomiting and diarrhea.

Formulation Types: Soluble concentrate; emulsifiable concentrate (4 lb/gal).

Toxicity: LC_{50} (96-hour) for rainbow trout 190 mg/L and bluegill sunfish 616 mg/L (Hartley and Kidd, 1987); acute oral LD_{50} for rats is 1,100 mg/kg (RTECS, 1985).

Uses: Selective, contact, postemergence herbicide used to control a variety of annual and perennial broad-leaved weeds in most grass and legume crops.

α-BHC

Synonyms: Benzenehexachloride-α-isomer; α-Benzenehexachloride; ENT 9232; α-HCH; α-Hexachloran; α-Hexachlorane; α-Hexachlorcyclohexane; α-Hexachlorocyclohexane; 1,2,3,4,5,6-Hexachloro-α-cyclohexane; **1α,2α,3β,4α,5β,6β-Hexachlorocyclohexane**; α-1,2,3,4,5,6-Hexachlorocyclohexane; α-Lindane; TBH.

Designations: CAS Registry Number: 319-84-6; DOT: 2761; mf: $C_6H_6Cl_6$; fw: 290.83; RTECS: GV3500000.

Properties: Brownish-to-white crystalline solid with a phosgene-like odor (technical grade). Mp: 159.1°C; bp: 288°C; ρ: ≈ 1.87; fl p: nonflammable; K_H: 5.3 × 10^{-6} atm · m^3/mol at 20°C (approximate — calculated from water solubility and vapor pressure); log K_{oc}: 3.279; log K_{ow}: 3.46–3.89; P-$t_{1/2}$: 91.09 hours (absorbance λ ≤300 nm, concentration on glass plates = 6.7 μg/cm^2); S_o: soluble in ethanol, benzene, chloroform, cod liver oil (22.1, 23.6, 31.2 g/L at 4, 12 and 20°C, respectively), n-octanol (6.4, 9.4 and 14.9 g/L at 4, 12 and 20°C, respectively); S_w: 1.63 mg/L at 20°C; vp: 2.5 × 10^{-5} mmHg at 20°C, 1.73 × 10^{-3} mmHg at 25°C.

Environmental Fate

Biological. Clostridium sphenoides degraded α-BHC to δ-3,4,5,6-tetrachloro-1-cyclohexane (Heritage and MacRae, 1977a). In four successive 7-day incubation periods, α-BHC (5 and 10 mg/L) was recalcitrant to degradation in a settled domestic wastewater inoculum (Tabak et al., 1981).

Soil. Under aerobic conditions, indigenous microbes in contaminated soil produced pentachlorocyclohexane. However, under methanogenic conditions, α-BHC was converted to chlorobenzene, 3,5-dichlorophenol and the tentatively identified compound 2,4,5-trichlorophenol (Bachmann et al., 1988).

Photolytic. When an aqueous solution containing α-BHC was photooxidized by UV light at 90–95°C, 25, 50 and 75% degraded to carbon dioxide after 4.2, 24.2 and 40.0 hours, respectively (Knoevenagel and Himmelreich, 1976). In basic, aqueous solutions, α-BHC dehydrochlorinates forming pentachlorocyclohexene, which is further transformed to trichlorobenzenes. In a buffered aqueous solution at pH 8 and 5°C, the calculated hydrolysis half-life is 26 years (Ngabe et al., 1993).

Chemical/Physical. Emits very toxic chloride fumes when heated to decomposition (Lewis, 1990). α-BHC will hydrolyze via trans-dehydrochlorination of the axial chlorines resulting in the formation of hydrochloric acid and the intermediate 1,3,4,5,6-pentachlorocyclo-hexene. The intermediate will undergo further hydrolysis resulting in the formation of 1,2,3-trichlorobenzene, 1,2,4-trichlorobenzene and hydrochloric acid (Kollig, 1993).

Formulation Types: Emulsifiable concentrate; wettable powder; fumigant; suspension concentrate; granules; dustable powder.

Toxicity: LC_{50} (96-hour) for guppies >1.4 mg/L (Verschueren, 1983); acute oral LD_{50} for rats 177 mg/kg (RTECS, 1985).

Use: Not produced commercially in the U.S. and its sale is prohibited by the U.S. EPA.

β-BHC

Synonyms: *trans*-α-Benzenehexachloride; β-Benzenehexachloride; Benzene-*cis*-hexachloride; ENT 9233; β-HCH; β-Hexachlorobenzene; **1α,2β,3α,4β,5α,6β-Hexachlorocyclohexane**; β-Hexachlorocyclohexane; 1,2,3,4,5,6-Hexachloro-β-cyclohexane; 1,2,3,4,5,6-Hexachloro-*trans*-cyclohexane; β-1,2,3,4,5,6-Hexachlorocyclohexane; β-Isomer; β-Lindane; TBH.

Designations: CAS Registry Number: 319-85-7; DOT: 2761; mf: $C_6H_6Cl_6$; fw: 290.83; RTECS: GV4375000.

Properties: Solid. Mp: 311.7°C; bp: sublimes; ρ: 1.89 at 19/4°C; fl p: nonflammable; K_H: 2.3×10^{-7} atm · m³/mol at 20–25°C (approximate — calculated from water solubility and vapor pressure); log K_{oc}: 3.322–3.553; log K_{ow}: 3.80–4.50; P-$t_{1/2}$: 151.80 hours (absorbance λ ≤300 nm, concentration on glass plates = 6.7 μg/cm²); S_o: soluble in ethanol, benzene and chloroform; S_w: 240 ppb at 25°C; vp: 2.8×10^{-7} mmHg at 20°C.

Soil properties and adsorption data

Soil	K_d (mL/g)	f_{oc} (%)	K_{oc} (mL/g)	pH
Ca-Bentonite (19.8°C)	4.0	—	—	—
Ca-Bentonite (30°C)	4.5	—	—	—
Ca-Staten peaty muck (20°C)	456	12.76	3,574	—
Ca-Staten peaty muck (30°C)	437	12.76	3,425	—
Ca-Venado clay (20°C)	63	—	—	—
Ca-Venado clay (30°C)	60	—	—	—
Silica gel (19.8°C)	2.1	—	—	—
Silica gel (30°C)	1.6	—	—	—

Source: Mills and Biggar, 1969.

Environmental Fate

Soil. No biodegradation of β-BHC was observed under denitrifying and sulfate-reducing conditions in a contaminated soil collected from The Netherlands (Bachmann et al., 1988). In four successive 7-day incubation periods, β-BHC (5 and 10 mg/L) was recalcitrant to degradation in a settled domestic wastewater inoculum (Tabak et al., 1981).

Chemical/Physical. Emits very toxic fumes of chlorides, hydrochloric acid and phosgene when heated to decomposition (Lewis, 1990). β-BHC will not hydrolyze to any reasonable extent (Kollig, 1993).

Formulation Types: Emulsifiable concentrate; wettable powder; fumigant; suspension concentrate; granules; dustable powder.

Toxicity: Acute oral LD_{50} for rats 6,000 mg/kg (RTECS, 1985).

Use: Insecticide.

δ-BHC

Synonyms: δ-Benzenehexachloride; ENT 9234; δ-HCH; δ-Hexachlorocyclohexane; δ-1,2,3,4,5,6-Hexachlorocyclohexane; δ-(aeeeee)-1,2,3,4,5,6-Hexachlorocyclohexane; 1α,2α,3α,4β,5β,6β-**Hexachlorocyclohexane**; 1,2,3,4,5,6-Hexachloro-δ-cyclohexane; δ-Lindane; TBH.

Designations: CAS Registry Number: 319-86-8; DOT: 2761; mf: $C_6H_6Cl_6$; fw: 290.83; RTECS: GV4550000.

Properties: Solid with a musty-like odor. Mp: 140.8°C; bp: 60°C at 0.36 mmHg; ρ: ≈ 1.87; fl p: nonflammable; H-$t_{1/2}$: 191 days (pH 7), 11 hours (pH 9); K_H: 2.5 × 10^{-7} atm · m^3/mol at 20–25°C (approximate — calculated from water solubility and vapor pressure); log K_{oc}: 3.279; log K_{ow}: 2.80, 4.14; P-$t_{1/2}$: 153.84 hours (absorbance λ ≤300 nm, concentration on glass plates = 6.7 μg/cm^2); S_o: soluble in acetone, ethanol, benzene and chloroform; S_w: 21.3 ppm at 25°C; vp: 1.7 × 10^{-5} mmHg at 20°C.

Environmental Fate

Biological. Dehydrochlorination of δ-BHC by a *Pseudomonas* sp. under aerobic conditions was reported by Sahu et al. (1992). They also reported that when deionized water containing δ-BHC was inoculated with *Pseudomonas* sp., the concentration of δ-BHC decreased to undetectable levels after 8 days with concomitant formation of chloride ions and δ-pentachlorocyclohexane. In four successive 7-day incubation periods, δ-BHC (5 and 10 mg/L) was recalcitrant to degradation in a settled domestic wastewater inoculum (Tabak et al., 1981).

Chemical/Physical. δ-BHC dehydrochlorinates in the presence of alkalies. The hydrolysis half-lives at pH values of 7 and 9 are 191 days and 11 hours, respectively (Worthing and Hance, 1991).

Formulation Types: Emulsifiable concentrate; wettable powder; fumigant; suspension concentrate; granules; dustable powder.

Toxicity: Acute oral LD_{50} for rats 1,000 mg/kg (RTECS, 1985).

Use: Insecticide.

BIFENOX

Synonyms: MC-4379; **Methyl 5-(2,4-dichlorophenoxy)-2-nitrobenzoate**; Modown.

Designations: CAS Registry Number: 42576-02-3; mf: $C_{14}H_9Cl_2NO_5$; fw: 342.14; RTECS: DG7890000.

Properties: Yellow-tan crystalline solid. Mp: 84–86°C; bp: decomposes; ρ: 1.155; fl p: nonflammable; K_H: 1.09×10^{-7} atm · m³/mol at 25°C; log K_{oc}: 2.24–4.39; log K_{ow}: 4.48; So (g/kg at 25°C): acetone (400), ethanol (<50), chlorobenzene (400), xylene (300), kerosene and other aliphatic solvents (<10); S_w: 350 µg/L at 25°C; vp: 2.40×10^{-6} mmHg at 30°C.

Soil properties and adsorption data

Soil	K_d (mL/g)	f_{oc} (%)	K_{oc} (mL/g)	pH
Loam	36.0	2.26	1,593	6.7
Loamy sand	79.0	0.81	9,729	6.9
Sand	2.6	0.17	1,529	7.5
Sandy clay loam	156.0	0.64	24,451	7.3
Silt loam	408.0	1.16	35,172	7.4

Source: U.S. Department of Agriculture, 1990.

Environmental Fate

Soil. Bifenox degrades in soil forming 5-(2,4-dichlorophenoxy)-2-nitrobenzoic acid and methyl 5-(2,4-dichlorophenoxy)anthranilate (Hartley and Kidd, 1987; Smith 1988). The average half-life in soils is 7–14 days (Hartley and Kidd, 1987; Humburg et al., 1989).

Plant. Rapidly undergoes ring hydroxylation and subsequent conjugation in rice plants (Ashton and Monaco, 1991).

Photolytic. The UV photolysis (λ = 300 nm) of bifenox in various solvents was studied by Ruzo et al. (1980). In water, 2,4-dichloro-3'-(carboxymethyl)-4'-hydroxydiphenyl ether and 2,4-dichloro-3'-(carboxymethyl)-4'-aminodiphenyl ether were identified. In cyclohexane, 2,4-dichloro-4'-nitrodiphenyl ether and methyl formate were the major products. In methanol, a dichloromethoxy phenol was identified. Photodegradation occurred via reductive dechlorination, decarboxy-methylation, nitro group reduction and cleavage of the ether linkage (Ruzo et al., 1980).

Formulation Types: Emulsifiable concentrate; wettable powder; granules; suspension concentrate; flowable powder (4 lb/gal).

54

Toxicity: LC_{50} (96-hour) for rainbow trout 0.87 mg/L and bluegill sunfish 0.64 mg/L (Hartley and Kidd, 1987); acute oral LD_{50} for rats 6,400 mg/kg (RTECS, 1985).

Uses: Selective preemergence or postemergence herbicide used to effectively control a wide variety of broad-leaved weeds (such as bindweed, jimsonweed, kochia, mustards, pigweeds, sesbania, smartweed and velvet-leaf) in tolerant crops (corn, grain sorghum, maize, rice and soybeans).

BIPHENYL

Synonyms: Bibenzene; **1,1′-Biphenyl**; Diphenyl; Lemonene; Phenylbenzene.

Designations: CAS Registry Number: 92-52-4; mf: $C_{12}H_{10}$; fw: 154.21; RTECS: NU8050000.

Properties: Colorless to white scales or crystals with a pleasant but peculiar odor. Mp: 71°C; bp: 255.9°C; ρ: 0.8660 at 20/4°C; fl p: 112.8°C; lel: 0.6% at 100°C; uel: 5.8% at 155°C; K_H: 1.93–4.15 × 10^{-4} atm · m^3/mol at 25°C; IP: 8.27 eV; log K_{oc}: 3.04–3.32; log K_{ow}: 3.16–4.09; S_o: soluble in alcohol, benzene, ether (199.3 g/kg at 23°C) and triolein (415 g/kg at 37°C; S_w: 5.94–7.48 ppm at 25°C; vp: 5.84 × 10^{-4} mmHg at 20.70°C.

Soil properties and adsorption data

Soil	K_d (mL/g)	f_{oc} (%)	K_{oc} (mL/g)	pH (meq/100 g)	CEC
Clay loam	15.6	1.42	1,100	5.91	12.4
Clay loam	124.0	10.40	1,200	4.89	35.0
Light clay	31.8	1.51	2,100	5.18	13.2
Light clay	56.8	3.23	1,800	5.26	28.3
Sandy loam	90.0	7.91	1,100	5.51	26.3

Source: Kishi et al., 1990.

Environmental Fate

Biological. Reported biodegradation products include 2,3-dihydro-2,3-dihydroxybiphenyl, 2,3-dihydroxybiphenyl, 2-hydroxy-6-oxo-6-phenylhexa-2,4-dienoate, 2-hydroxy-3-phenyl-6-oxohexa-2,4-dienoate, 2-oxopenta-4-enoate, phenylpyruvic acid (Verschueren, 1983), 2-hydroxybiphenyl, 4-hydroxybiphenyl and 4,4′-hydroxybiphenyl (Smith and Rosazza, 1974). Under aerobic conditions, *Beijerinckia* sp. degraded biphenyl to *cis*-2,3-dihydro-2,3-dihydroxybiphenyl. In addition, *Oscillatoria* sp. and *Pseudomonas putida* degraded biphenyl to 4-hydroxybiphenyl and benzoic acid, respectively (Kobayashi and Rittman, 1982). The microbe *Candida lipolytica* degraded biphenyl into the following products: 2-, 3- and 4-hydroxybiphenyl, 4,4′-dihydroxybiphenyl and 3-methoxy-4-hydroxybiphenyl (Cerniglia and Crow, 1981). With the exception of 3-methoxy-4-hydroxybiphenyl, these products were also identified as metabolites by *Cunninghanella elegans* (Dodge et al., 1979). In activated sludge, 15.2% mineralized to carbon dioxide after 5 days (Freitag et al., 1985).

Photolytic. A carbon dioxide yield of 9.5% was achieved when biphenyl adsorbed on silica gel was irradiated with light (λ >290 nm) for 17 hours. Irradiation of biphenyl (λ

>300 nm) in the presence of nitrogen monoxide resulted in the formation of 2- and 4-nitrobiphenyl (Fukui et al., 1980).

Chemical/Physical. The aqueous chlorination of biphenyl at 40°C over a pH range of 6.2 to 9.0 yielded *o*-chlorobiphenyl and *m*-chlorobiphenyl (Snider and Alley, 1979). In an acidic aqueous solution (pH 4.5) containing bromide ions and a chlorinating agent (sodium hypochlorite), 4-bromobiphenyl formed as the major product. Minor products identified include 2-bromobiphenyl, 2,4- and 4,4'-di-bromobiphenyl (Lin et al., 1984).

Biphenyl will not hydrolyze since it has no hydrolyzable functional group.

Exposure Limits: NIOSH REL: TWA 0.2 ppm, IDLH 100 mg/m^3; OSHA PEL: TWA 0.2 ppm; ACGIH TLV: TWA 0.2 ppm.

Symptoms of Exposure: Irritation of throat, eyes; headache; nausea; fatigue; numbness in limbs, liver damage (Humburg et al., 1989).

Formulation Types: Solution.

Toxicity: Acute oral LD_{50} for rats 3,280 mg/kg (RTECS, 1985).

Uses: Fungistat for oranges; plant disease control.

BIS(2-CHLOROETHYL)ETHER

Synonyms: Bis(β-chloroethyl)ether; Chlorex; 1-Chloro-2-(β-chloroethoxy)ethane; Chloroethyl ether; 2-Chloroethyl ether; (β-Chloroethyl)ether; DCEE; Dichlorodi-ethyl ether; 2,2′-Dichlorodiethyl ether; β,β′-Dichlorodiethyl ether; Dichloroether; Dichloroethyl ether; α,α′-Dichloroethyl ether; Di(β-chloroethyl)ether; Di(2-chloro-ethyl)ether; *sym*-Dichloroethyl ether; 2,2′-Dichloroethyl ether; Dichloroethyl oxide; ENT 4504; **1,1′-Oxybis(2-chloroethane)**; RCRA waste number U025; UN 1916.

$$ClCH_2CH_2-O-CH_2CH_2Cl$$

Designations: CAS Registry Number: 111-44-4; DOT: 1916; mf: $C_4H_8Cl_2O$; fw: 143.01; RTECS: KN0875000.

Properties: Colorless, clear liquid with a strong fruity odor. Mp: −52.2 to −24.5°C; bp: 178.5°C; ρ: 1.2199 at 20/4°C; fl p: 55°C; H-$t_{1/2}$: 2.5 years at 25°C and pH 7; K_H: 1.3 × 10^{-5} atm · m³/mol; IP: 9.85 eV; log K_{oc}: 1.15; log K_{ow}: 1.12, 1.58; S_o: soluble in acetone, ethanol, benzene and ether; S_w: 10.7 g/L at 20°C; vap d: 5.84 g/L at 25°C, 4.94 (air = 1); vp: 0.71 mmHg at 20°C.

Environmental Fate

Biological. When 5 and 10 mg/L of bis(2-chloroethyl)ether were statically incubated in the dark at 25°C with yeast extract and settled domestic wastewater inoculum, complete degradation was observed after 7 days (Tabak et al., 1981).

Chemical/Physical. Bis(2-chloroethyl)ether is subject to hydrolysis (Enfield and Yates, 1990; NIOSH, 1994). The hydrolysis rate constant for bis(2-chloroethyl)ether at pH 7 and 25°C was determined to be 2.6 × 10^{-5}/hour, resulting in a half-life of 3.0 years. Products of hydrolysis include 2-(2-chloroethoxy)ethanol, bis(2-hydroxyethyl)ether, hydrochloric acid and/or 1,4-dioxane (Ellington et al., 1988; Kollig, 1993).

Emits chlorinated acids when incinerated (Sittig, 1985).

Exposure Limits: NIOSH REL: TWA 5 ppm (30 mg/m³), STEL 10 ppm, IDLH 100 ppm; OSHA PEL: C 15 ppm (90 mg/m³); ACGIH TLV: TWA 30 mg/m³, STEL 60 mg/m³.

Symptoms of Exposure: Inhalation may cause irritation of eyes and respiratory system.

Formulation Types: Fumigant.

Toxicity: Acute oral LD_{50} for rats 75–105 mg/kg (Verschueren, 1983).

Uses: Soil fumigant; acaricide.

BIS(2-CHLOROISOPROPYL)ETHER

Synonyms: BCIE; BCMEE; Bis(β-chloroisopropyl)ether; Bis(2-chloro-1-methyl-ethyl)ether; 1-Chloro-2-(β-chloroisopropoxy)propane; 2-Chloroisopropyl ether; β-Chloroisopropyl ether; (2-Chloro-1-methylethyl)ether; Dichlorodiisopropyl ether; Dichloroisopropyl ether; 2,2′-Dichloroisopropyl ether; NCI-C50044; **2,2′-Oxybis-(1-chloropropane)**; RCRA waste number U027; UN 2490.

$$\begin{array}{ccc} CH_2Cl & & CH_2Cl \\ \diagdown & & \diagup \\ & CH-O-CH & \\ \diagup & & \diagdown \\ CH_3 & & CH_3 \end{array}$$

Designations: CAS Registry Number: 108-60-1; mf: $C_6H_{12}Cl_2O$; fw: 171.07; RTECS: KN1750000.

Properties: Colorless to brown oily liquid. Mp: –20°C; bp: 187°C; ρ: 1.103 at 20/4°C; fl p: 85°C; K_H: 1.1×10^{-4} atm · m^3/mol; log K_{oc}: 1.79; log K_{ow}: 2.58; S_o: soluble in acetone, ethanol, benzene and ether; S_w: 1,700 mg/L at 20°C; vap d: 6.99 g/L at 25°C, 5.91 (air = 1); vp: 0.85 mmHg at 20°C.

Environmental Fate

Biological. When bis(2-chloroisopropyl)ether (5 and 10 mg/L) was statically incubated in the dark at 25°C with yeast extract and settled domestic wastewater inoculum, complete biodegradation was achieved after 14 days (Tabak et al., 1981).

Chemical/Physical. Kollig (1993) reported that bis(2-chloroisopropyl)ether is subject to hydrolysis forming hydrochloric acid and the intermediate (2-hydroxy-isopropyl-2-chloroisopropyl)ether. The latter undergoes further hydrolysis yielding bis(2-hydroxyiso-propyl)ether.

Formulation Types: Emulsifiable concentrate; granules.

Toxicity: LC_{50} (48-hour) for carp >40 mg/L (Hartley and Kidd, 1987); acute oral LD_{50} for rats 503.6 mg/kg (Hartley and Kidd, 1987), 240 mg/kg (RTECS, 1985).

Uses: Apparently used as a nematocide in Japan but is not registered in the U.S. for use as a pesticide.

BROMACIL

Synonyms: Borea; Bromax; Bromax 4G; Bromax 4L; Bromazil; 5-Bromo-3-*sec*-butyl-6-methyluracil; **5-Bromo-6-methyl-3-(1-methylpropyl)-2,4(1***H***3***H***)-pyrimidinedione**; 5-Bromo-6-methyl-3-(1-methylpropyl)uracil; Cynogan; Du Pont herbicide 976; Eerex granular weed killer; Eerex water soluble concentrate weed killer; Herbicide 976; Hyvar; Hyvarex; Hyvar X; Hyvar × Bromacil; Hyvar XL; Hyvar × weed killer; Hyvar X-WS; Krovar I; Krovar II; Nalkil; Uragan; Uragon; Urox B; Urox B water soluble concentrate weed killer; Uron HX; Urox HX granular weed killer; Weed Blast.

Designations: CAS Registry Number: 314-40-9; mf: $C_9H_{13}BrN_2O_2$; fw: 261.12; RTECS: YQ9100000.

Properties: Colorless to white, odorless, crystalline solid. Mp: 157.5–160.0°C; ρ: 1.55 at 25/4°C; fl p: active ingredient and dry formulations are nonflammable; pK_a: <7; K_H: 1.05 × 10^{-10} atm · m^3/mol at 25°C (approximate — calculated from water solubility and vapor pressure); log K_{oc}: ≈ 1.51; log K_{ow}: 1.84–2.04; So (g/L at 25°C): acetone (167), acetonitrile (71), ethanol (134), xylene (32); S_w: 815 mg/L at 25°C; vp: 2.48 × 10^{-7} mmHg at 25°C.

Soil properties and adsorption data

Soil	K_d (mL/g)	f_{oc} (%)	K_{oc} (mL/g)	pH
Adkins loamy sand	0.30	0.40	76	7.3
Baldwin Lake	2.89	1.00	289	8.0–9.0
Basinger fine sand	0.57	0.61	93	5.8
Bet Dagan	0.11	0.84	13	—
Bet Dagan I	0.10	0.40	25	7.9
Bet Dagan II	0.32	1.01	32	7.8
Big Bear Lake	3.93	7.90	50	6.7–7.2
Boca fine sand	0.76	1.67	46	7.1
Castle Lake	6.35	14.50	44	4.8–5.8
Chobee fine sandy loam	0.76	1.39	55	7.2
Clear Lake	5.26	13.30	40	5.7–6.8
Delta (Hill Slough)	2.72	4.40	62	6.6–7.2
Gilat	0.16	0.55	29	7.8
Gilat	0.13	0.73	18	—
Golan	0.16	0.68	28	—
Holopaw fine sand	0.33	0.50	66	6.1
Hula-1	19.75	29.88	65	6.3

Soil properties and adsorption data *(continued)*

Soil	K_d (mL/g)	f_{oc} (%)	K_{oc} (mL/g)	pH
Hula-2	5.05	7.85	64	6.9
Jenks Lake	1.20	3.00	40	5.3–6.2
Keyport silt loam	1.50	1.20	125	5.4
Kinneret	0.41	1.79	23	—
Kinneret-A	1.30	4.55	29	—
Kinneret-F	1.10	4.31	26	—
Kinneret-G	0.79	2.55	31	—
Malkiya	0.66	3.38	20	—
Mivtahim	0.03	0.06	50	8.5
Mivtahim	0.03	0.26	12	—
Mockingbird Canyon	0.56	2.10	27	6.5–7.4
Netanya	0.03	0.10	30	—
Neve Yaar	0.40	1.64	24	—
Neve Yaar	0.39	1.18	33	7.7
Neve Yaar	1.12	1.22	89	7.3
Oxidized Hula-2	1.71	3.08	55	6.9
Oxidized Neve Ya'ar	0.60	0.47	126	7.3
Pineda fine sand	0.57	1.22	93	7.1
Riviera fine sand	0.47	0.94	50	6.2
Sa'ad	0.63	0.56	112	7.6
San Joaquin Marsh	0.68	2.60	26	7.3–8.1
Semiahmoo mucky peat	21.33	27.80	77	5.4
Shefer	0.24	0.72	33	7.2
Shefer	0.21	0.82	70	—
Wabasso sand	0.57	0.61	93	6.6

Source: Rhodes et al., 1970; Angemar et al., 1984; Corwin and Farmer, 1984; Gerstl and Mingelgrin, 1984; Madhun et al., 1986; Reddy et al., 1992.

Environmental Fate

Soil. Metabolites tentatively identified in soil were 5-bromo-3-(3-hydroxy-1-methylpropyl)-6-methyluracil, 5-bromo-3-*sec*-butyl-6-hydroxymethyluracil, 5-bromo-3-(2-hydroxy-1-methylpropyl)-6-methyluracil and carbon dioxide. The presence of uracil products suggests that bromacil was degraded via hydroxylation of the side chain alkyl groups. In the laboratory, 25.3% of ^{14}C-bromacil degraded in soil to carbon dioxide after 9 weeks but mineralization in the field was not observed. The half-life of bromacil in a silt loam was 5–6 months (Gardiner et al., 1969).

To a neutral sandy loam soil maintained at a soil water holding capacity of 60%, 2.88 ppm of 2-^{14}C-bromacil was applied. After 600 days, 22.1% (0.64 ppm) of the applied amount was converted to $^{14}CO_2$ (Wolf and Martin, 1974). The evolution of $^{14}CO_2$ was significantly reduced when the soil water holding capacity was maintained at 100%, i.e., <0.5% $^{14}CO_2$ after 145 days (Wolf and Martin, 1974).

Residual activity in soil is limited to approximately 7 months (Hartley and Kidd, 1987). In California soils, bromacil was persistent for 30 months (Lange et al., 1968; Weber and Best, 1972). The reported half-life in soil is 60 days (Alva and Singh, 1991); however,

Gardiner et al. (1969) reported a half-life of 5–6 months in a silt loam when applied at a rate of 4 lbs/acre.

The average half-life for bromacil in soil incubated in the laboratory under aerobic conditions was 132 days (Zimdahl et al., 1970). In field soils, the average disappearance half-life was 349 days (Gardiner et al., 1969; Leistra et al., 1975). Under aerobic conditions, the mineralization half-lives for bromacil in soil ranged from 151 days to 4.5 years (Gardiner et al., 1969; Wolf and Martin, 1974).

Groundwater. According to the U.S. EPA (1986) bromacil has a high potential to leach to groundwater.

Plant. Bromacil is slowly absorbed and translocated in plants. Gardiner et al. (1969) reported 17% of the herbicide was translocated to the stems and leaves and 83% remained in the roots. A 6-hydroxymethyl metabolite was reported as the major metabolite (Gardiner et al., 1969). In orange seedlings, 5-bromo-3-*sec*-butyl-6-hydroxymethyluracil was reported as a major metabolite (Humburg et al., 1989).

Photolytic. When a dilute aqueous solution (1–10 mg/L) of bromacil was exposed to sunlight for 4 months, the *N*-dealkylated photoproduct, 5-bromo-6-methyluracil, formed in small quantities. This compound is less stable than bromacil and upon further irradiation, the debrominated product, 6-methyluracil, was formed (Moilanen and Crosby, 1974). Acher and Dunkelblum (1979) studied the dye-sensitized photolysis of aerated aqueous solutions of bromacil using sunlight as the irradiation source. After one hour, a mixture of diastereoisomers of 3-*sec*-butyl-5-acetyl-5-hydroxyhydantoin formed in an 83% yield. In a subsequent study another minor intermediate was identified as a 5,5′-photoproduct of 3-*sec*-butyl-6-methyluracil. In this study, the rate of photooxidation increased with pH. The most effective sensitizers were riboflavin (10 ppm) and methylene blue (2–5 ppm) (Acher and Saltzman, 1980). Direct photodegradation of bromacil is not significant (Ishihara, 1963; Acher and Dunkelblum, 1979).

Chemical/Physical. When bromacil was heated to 900°C, carbon monoxide, carbon dioxide, chlorine, hydrochloric acid and ammonia were produced (Kennedy et al., 1972). At 250°C, bromacil changed from a white crystalline solid to a black solid which was tentatively identified as 3-*sec*-butyl methyluracil (Stojanovic et al., 1972).

Emits very toxic fumes of bromides and nitrogen oxides when heated to decomposition (Lewis, 1990).

Exposure Limits: NIOSH REL: 1 ppm; ACGIH TLV: TWA 1 ppm.

Symptoms of Exposure: May irritate eyes, nose, throat and skin.

Formulation Types: Soluble concentrate (2 and 4 lb/gal); wettable powder (80%); granules (4%).

Toxicity: LC_{50} (48-hour) for rainbow trout 75 mg/L, bluegill sunfish 71 mg/L and carp 164 mg/L (Hartley and Kidd, 1987); acute oral LD_{50} for rats 5,200 mg/kg (Ashton and Monaco, 1991), 641 mg/kg (RTECS, 1985).

Uses: Uracil herbicide applied to soil to control a wide variety of annual and perennial grasses, broad-leaved weeds and general vegetation on uncropped land. It is also used for selective weed control in apple, asparagus, cane fruit, hops and citrus crops.

BROMOXYNIL

Synonyms: Brittox; Brominal; Brominex; Brominil; Broxynil; Buctril; Buctril 20; Buctril 21; Buctril industrial; Butilchlorofos; Chipco Buctril; Chipco crabkleen; 2,6-Dibromo-4-cyanophenol; 3,5-Dibromo-4-hydroxybenzonitrile; **3,5-Dibromo-4-hydroxyphenylcyanide**; ENT 20852; 4-Hydroxy-3,5-dibromobenzonitrile; MB 10064; M&B 10064; ME4 Brominal; Nu-lawn weeder; Oxytril M; Partner.

Designations: CAS Registry Number: 1689-84-5; mf: $C_7H_3Br_2NO$; fw: 276.93; RTECS: DI3150000.

Properties: Odorless, colorless crystals or a light tan to creamy powder. Technical grades may impart a slight odor. The reported odor threshold concentration in air is >11 ppm (Young et al., 1996). Mp: 190°C, 194–195°C; bp: sublimes at 135°C and 0.15 mmHg; fl p: >82°C (ester formulation); pK_a: 4.06; K_H: 1.4×10^{-6} atm · m³/mol at 20–25°C (approximate — calculated from water solubility and vapor pressure); log K_{oc}: 2.48 (calculated); log K_{ow}: <2; So (g/L at 20°C): acetone (170); benzene (10), *N,N*-dimethylformamide (610), ethanol (70), methanol (90), cyclohexanone (170), tetrahydrofuran (410); S_w: 130 mg/L at 25°C; vp: 4.8×10^{-6} mmHg at 20°C.

Environmental Fate

Biological. Duke et al. (1991) reported that bromoxynil can be converted to 3,5-dibromo-4-hydroxybenzoic acid by a microbial nitrolase.

Soil. In soils, *Klebsiella pneumoniae* metabolized bromoxynil to 3,5-dibromo-4-hydroxybenzoic acid and ammonia (McBride et al., 1986). In soil, bromoxynil undergoes nitrile and then amide hydrolysis yielding 3,5-dibromo-4-hydroxybenzoic acid and 3,5-dibromo-4-hydroxybenzamide (Smith, 1988). Degradation was rapid in a heavy clay soil, sandy loam and clay loam. After 1 week, only 10% of the applied dosage was recovered.

Nitrification in soils is inhibited when bromacil is applied at a concentration of <50 ppm (Debona and Audus, 1970). The half-life in soil is approximately 10 days (Hartley and Kidd, 1987).

Plant. In plants, the cyano group is hydrolyzed to an amido group which is subsequently oxidized to a carboxylic acid. Hydrolyzes to hydroxybenzoic acid (Hartley and Kidd, 1987). In plants, bromoxynil may hydrolyze to a benzoic acid (Humburg et al., 1989). Bromoxynil-resistant cotton was recently developed by inserting a *bxn* gene cloned from the soil bacterium *Klebsiella ozaenae*. This gene, which encodes a specific nitrolase, converted bromoxynil to its primary metabolite 3,5-dibromo-4-hydroxybenzoic acid (Stalker et al., 1988).

Chemical/Physical. Emits toxic fumes of nitrogen oxides and bromine when heated to decomposition (Sax and Lewis, 1987). Reacts with bases forming water-soluble salts (Worthing and Hance, 1991). Bromacil is stable to UV light (Hartley and Kidd, 1987).

Formulation Types: Suspension concentrate; emulsifiable concentrate.

Toxicity: LC_{50} (48-hour) for harlequin fish 5.0 mg/L (potassium salt) (Hartley and Kidd, 1987); LC_{50} for rainbow trout 100 ppb, goldfish 170 ppb and catfish 23 ppb (Humburg et al., 1989); acute oral LD_{50} for rats 190 mg/kg (RTECS, 1985).

Uses: Selective contact foliage-applied herbicide used to control many broad-leaved weeds in cereals.

BROMOXYNIL OCTANOATE

Synonyms: Bronate; Buctril; **2,6-Dibromo-4-cyanophenyl octanoate**; 3,5-Dibromo-4-octanoyloxybenzonitrile; M&B 10731; RP 16272.

Designations: CAS Registry Number: 1689-99-2; mf: $C_{15}H_{17}Br_2NO_2$; fw: 403.13; RTECS: DI3325000.

Properties: Cream-colored, waxy solid. Mp: 45–46°C; bp: sublimes at 90°C and 0.1 mmHg; K_H: 3.16 × 10^{-5} atm · m^3/mol at 25°C; log K_{oc}: 4.25 (calculated); log K_{ow}: 5.06; So (g/L at 25°C): acetone (100–170), benzene (700), chloroform (800), cyclohexanone (550), methanol (740), methylene chloride (800), toluene (760); S_w: 80 µg/L at 25°C; vp: 4.80 × 10^{-6} mmHg at 20°C.

Environmental Fate

Chemical/Physical. Emits toxic fumes of nitrogen oxides and bromine when heated to decomposition (Sax and Lewis, 1987). Hydrolyzes, especially at pH >9, to bromoxynil and octanoic acid (Hartley and Kidd, 1987; Worthing and Hance, 1991).

Formulation Types: Emulsifiable concentrate; suspension concentrate.

Toxicity: LC_{50} (48-hour) for rainbow trout 150 µg/L (Hartley and Kidd, 1987), goldfish 460 µg/L and catfish 63 µg/L (Worthing and Hance, 1991); acute oral LD_{50} for rats 365 mg/kg (Hartley and Kidd, 1987).

Uses: Selective contact foliage-applied herbicide used to control many broad-leaved weeds such as bindweed, chickweed and *Veronica* spp. in cereals.

BUTIFOS

Synonyms: B 1776; Butiphos; Butyl phosphorotrithioate; Chemagro 1,776; Chemagro B1776; Def; Def defoliant; Degreen; E-Z-Off; Fos-fall 'A'; Ortho phosphate defoliant; *S,S,S*-**Tributyl phosphorotrithioate**; *S,S,S*-Tributyl trithiophosphate.

$$(CH_3CH_2CH_2CH_2S)_3P$$

Designations: CAS Registry Number: 78-48-8; mf: $C_{12}H_{27}OPS_3$; fw: 314.52; RTECS: TG5425000.

Properties: Pale yellow liquid with a mercaptan-like odor. Mp: $<-25°C$; bp: 150°C at 0.3 mmHg; ρ: 1.057 at 20/4°C; K_H: 2.88×10^{-4} atm · m^3/mol at 20°C (approximate — calculated from water solubility and vapor pressure); log K_{oc}: 3.44 (calculated); log K_{ow}: 3.23; S_o: soluble in most organic solvents; S_w: 2.3 mg/L at 20°C; vap d: 12.85 g/L at 25°C, 10.89 (air = 1); vp: 1.60×10^{-3} mmHg at 20°C.

Formulation Types: Emulsifiable concentrate.

Toxicity: LC_{50} (96-hour) for rainbow trout <5.0 mg/L and bluegill sunfish 1.0 mg/L (Hartley and Kidd, 1987); acute oral LD_{50} for male and female rats 233 and 200 mg/kg, respectively (Hartley and Kidd, 1987), 150 mg/kg (RTECS, 1985).

Uses: Defoliant for cotton.

BUTYLATE

Synonyms: **Bis(2-methylpropyl)carbamothioic acid *S*-ethyl ester**; Butilate; Diisobutylthiocarbamic acid *S*-ethyl ester; Diisocarb; *S*-Ethyl bis(2-methylpropyl)-carbamothioate; Ethyl *N,N*-diisobutylthiocarbamate; *S*-Ethyl diisobutylthiocarbamate; *S*-Ethyl *N,N*-diisobutylthiocarbamate; Ethyl-*N,N*-diisobutylthiolcarbamate; Genate Plus 6.7EC; R 1910; Stauffer R 1910; Sutan.

$$CH_3CH_2SC \overset{\overset{\displaystyle O}{\|}}{\underset{}{}} -N \overset{\diagup CH_2CHCH_3 \; \overset{\displaystyle CH_3}{|}}{\diagdown CH_2CHCH_3 \; \underset{\displaystyle CH_3}{|}}$$

Designations: CAS Registry Number: 2008-41-5; mf: $C_{11}H_{23}NOS$; fw: 217.37; RTECS: EZ7525000.

Properties: Clear, colorless to amber, semi-volatile liquid with an aromatic odor. Mp: <25°C; bp: 138°C at 21 mmHg; ρ: 0.9402 and 0.9417 at 25/4 and 20/4°C, respectively; fl p: 110°C (open cup); K_H: 5.5×10^{-6} atm · m³/mol at 20–25°C (approximate — calculated from water solubility and vapor pressure); log K_{oc}: 2.73 (calculated); log K_{ow}: 4.15; S_o: miscible with acetone, ethanol, kerosene, methyl isobutyl ketone, xylene and many other organic solvents at 20°C; S_w: 45 mg/L at 22–25°C; vap d: 8.88 g/L at 25°C, 7.53 (air = 1); vp: 1.3×10^{-2} mmHg at 20°C.

Environmental Fate

Soil. Hydrolyzes in soil to ethyl mercaptan, carbon dioxide and diisobutylamine (Hartley and Kidd, 1987). Butylate is probably subject to degradation of soil microorganisms. It was reported that butylate may degrade via hydrolysis of the ester linkage forming the corresponding mercaptan (ethyl mercaptan), alkylamine (diisobutylamine) and carbon dioxide. Transthiolation and oxidation of the mercaptan forms the alcohol which may further oxidize to afford a metabolic pool (Kaufman, 1967). Somasundaram and Coats (1991) reported butylate in soils is oxidized to the corresponding sulfoxide. The reported half-life in soil is approximately 1.5–10 weeks (Worthing and Hance, 1991). The reported half-life of butylate in a loam soil at 21–27°C was 3 weeks (Humburg et al., 1989). Residual activity in soil is limited to approximately 4 months (Hartley and Kidd, 1987).

Groundwater. According to the U.S. EPA (1986) butylate has a high potential to leach to groundwater.

Plant. In plants, butylate is metabolized to carbon dioxide, diisobutylamine, fatty acids, conjugates of amines and other compounds (Hartley and Kidd, 1987; Humburg et al., 1989).

Formulation Types: Granules (10%) with an inert safener; emulsifiable concentrate (6.7 lb/gal).

Toxicity: LC$_{50}$ (96-hour) for rainbow trout 4.2 mg/L and bluegill sunfish 6.9 mg/L (Hartley and Kidd, 1987); acute oral LD$_{50}$ for male and female rats 4,659 and 5,431 mg/kg, respectively (Hartley and Kidd, 1987), 4,000 mg/kg (RTECS, 1985).

Uses: Selective, systemic, herbicide for the preemergent control of annual grasses and broad-leaved weeds in corn.

CAPTAN

Synonyms: Aacaptan; Agrosol S; Agrox 2-way and 3-way; Amercide; Bangton; Bean seed protectant; Captaf; Captaf 85W; Captan 50W; Captane; Captanstreptomycin 7.5-0.1 potato seed piece protectant; Captex; ENT 26538; Essofungicide 406; Flit 406; Fungus ban type II; Glyodex 3722; Granox PFM; Gustafson captan 30-DD; Hexacap; Kaptan; Le captane; Malipur; Merpan; Microcheck 12; NA 9099; NCI-C00077; Neracid; Orthocide; Orthocide 7.5; Orthocide 50; Orthocide 406; Osocide; SR 406; Stauffer captan; 3a,4,7,7a-Tetrahydro-*N*-(trichloro-methanesulphenyl)phthalimide; **3a,4,7,7a-Tetrahydro-2-((trichlorome-thyl)thio)-1*H*-isoindole-1,3(2*H*)-dione**; 1,2,3,6-Tetrahydro-*N*-trichloromethylthio)-phthalimide; *N*-Trichloromethylmercapto-4-cyclohexene-1,2-dicarboximide; *N*-Trichloromethylthiocyclohex-4-ene-1,2-dicarboximide; *N*-((Trichloromethylthio)-4-cyclohexene-1,2-dicarboximide; *N*-((Trichloromethyl)thio)tetrahydrophthalamide; Trichlorometh-ylthio-1,2,5,6-tetrahydrophthalamide; *N*-Trichloromethylthio-3a,4,7,7a-tetrahydro-phthalamide; Vancide 89; Vancide 89RE; Vancide P-75; Vangard K; Vanguard K; Vanicide; Vondcaptan.

Designations: CAS Registry Number: 133-06-2; DOT 9099; mf: $C_9H_8Cl_3NO_2S$; fw: 300.57; RTECS: GW5075000.

Properties: Odorless crystals from carbon tetrachloride. Technical grades have a pungent odor. Mp: 178°C; ρ: 1.74 at room temperature; log K_{oc}: 2.08–2.13; log K_{ow}: 2.35, 2.79; So (g/kg at 25°C): acetone (21), chloroform (70), cyclohexanone (23), 2-propanol (1.7), xylene (20); S_w: 3.3 mg/L at 25°C; vp: <9.8 × 10^{-6} mmHg at 20°C.

Environmental Fate

Biological. In water, captan reacted with the fungicide L-cysteine forming a compound with an absorption maximum of 272 μm which was identified as 2-thiazolidinethione-4-carboxylic acid. In addition, tetrahydrophthalimide also formed (Lukens and Sisler, 1958).

Soil. The half-lives of captan in soil ranged from 3.5 days (pH 6.4, moisture content 17.5%) to >50 days (pH 6.2, moisture content 1.6%) (Burchfield, 1959). The half-lives of captan in a sandy loam, clay loam and an organic amended soil under non-sterile conditions were 10–354, 3.4–587 and 7.0–213 days, respectively, while under sterile conditions the half-lives were 9.4–373, 2.8–303 and 4.3–191 days, respectively (Schoen and Winterlin, 1987).

Chemical/Physical. Captan undergoes hydrolysis under neutral or alkaline conditions yielding 4-cyclohexene-1,2-dicarboximide, carbon dioxide, hydrogen chloride and sulfur (Wolfe et al., 1976a). The estimated hydrolysis half-life of captan in a phosphate buffer solution at pH 7.07 and 28–29°C is 2.96 hours. Between the pH range 2 to 6, the estimated hydrolysis half-life is 10.70 hours (Wolfe et al., 1976a). Worthing and Hance (1991) reported hydrolysis half-lives of 8.3–32.4 hours and <2 minutes at pH values 7 and 10, respectively.

Decomposes at or near the melting point (Hartley and Kidd, 1987).

Exposure Limits: ACGIH TLV: TWA 5 mg/m^3.

Symptoms of Exposure: May cause skin irritation. Ingestion may cause vomiting and/or diarrhea.

Formulation Types: Wettable powder; dustable powder; seed treatment.

Toxicity: Acute inhalation LC_{50} (4-hour) for mice 4 mg/L; LC_{50} (96-hour) for bluegill sunfish 72 µg/L, brook trout 34 µg/L, harlequin fish 300 µg/L; acute oral LD_{50} for rats 9,000 mg/kg (Worthing and Hance, 1991), mallard ducks and pheasants >5,000 mg/kg, bobwhite quail 2,000–4,000 mg/kg (Hartley and Kidd, 1987).

Uses: Fungicide used to control diseases on many fruit, ornamental and vegetable crops. Bacteriostat in soaps.

CARBARYL

Synonyms: Arylam; Carbamine; Carbatox; Carbatox 60; Carbatox 75; Carpolin; Caryl-derm; Cekubaryl; Crag sevin; Denapon; Devicarb; Dicarbam; ENT 23969; Experimental insecticide 7744; Gamonil; Germain's; Hexavin; Karbaspray; Karbatox; Karbosep; Methylcarbamate-1-naphthalenol; Methylcarbamate-1-naphthol; Methyl carbamic acid 1-naphthyl ester; *N*-Methyl-1-naphthylcarbamate; *N*-Methyl-α-naphthylcarbamate; *N*-Methyl-α-naphthylurethan; NA 2757; NAC; **1-Naphthalenol methylcarbamate**; 1-Naphthol-*N*-methylcarbamate; 1-Naphthyl methylcarbamate; 1-Naphthyl-*N*-methylcarbamate; α-Naphthyl-*N*-methylcarbamate; OMS 29; Panam; Ravyon; Rylam; Seffein; Septene; Sevimol; Sevin; Sok; Tercyl; Toxan; Tricarnam; UC 7744; Union Carbide 7744.

Designations: CAS Registry Number: 63-25-2; DOT: 2757; mf: $C_{12}H_{11}NO_2$; fw: 201.22; RTECS: FC5950000.

Properties: White crystals. The reported odor threshold concentration in air and taste threshold concentration in water are 37 and 44 ppb, respectively (Young et al., 1996). Mp: 142.2°C; ρ: 1.232 at 20/20°C; fl p: 195°C; H-$t_{1/2}$ (days at 27°C): 1,500 (pH 5), 15 (pH 7), 0.15 (pH 9) (Wolfe et al., 1978); K_H: 1.27×10^{-5} atm · m^3/mol at 20°C (approximate — calculated from water solubility and vapor pressure); log K_{oc}: 2.02–2.59; log K_{ow}: 2.31–2.81; P-$t_{1/2}$: 51.66 hours (absorbance λ = 257.5 nm, concentration on glass plates = 6.7 μg/cm^2) and approximately 45 hours (buffered solution, λ >280 nm); S_o: moderately soluble in acetone, cyclohexanone, *N,N*-dimethylformamide and isophorone; S_w: 72.4, 104 and 130 mg/L at 10, 20 and 30°C, respectively; vp: 1.36×10^{-6} mmHg at 25°C (Ferreira and Seiber, 1981).

Soil properties and adsorption data

Soil	K_d (mL/g)	f_{oc} (%)	K_{oc} (mL/g)	pH
Batcombe silt loam	2.13	2.05	104	6.1
Catlin	5.56	2.01	280	6.2
Commerce	1.81	0.68	265	6.7
Loamy sand	2.93	0.58	505	5.3
Rothamsted Farm	1.56	1.51	103	5.1
Sand	2.45	0.23	1,056	7.7
Sand	3.89	2.38	163	5.4
Sandy clay loam	0.44	1.71	26	8.1
Silt loam	3.29	2.09	157	6.3

Soil properties and adsorption data *(continued)*

Soil	K_d (mL/g)	f_{oc} (%)	K_{oc} (mL/g)	pH
Silt loam	4.69	3.07	153	5.0
Tracy	3.96	1.12	353	6.2

Source: Lord et al., 1980; Briggs, 1981; McCall et al., 1981; U.S. Department of Agriculture, 1990.

Environmental Fate

Biological. Fourteen soil fungi metabolized methyl-^{14}C-labeled carbaryl via hydroxylation to 1-naphthyl-*N*-hydroxymethylcarbamate, 4-hydroxy-1-naphthylmethylcarbamate and 5-hydroxy-1-naphthylmethylcarbamate (Bollag and Liu, 1972). Carbaryl was degraded by a culture of *Aspergillus terreus* to 1-naphthylcarbamate. The half-life was 8 days (Liu and Bollag, 1971a).

Various microorganisms isolated from soil hydrolyzed carbaryl to 1-napthol. For example, *Fusarium solani* degraded carbaryl 82% after 12 days at a temperature of 26–28°C (Bollag and Liu, 1971).

In a small watershed, carbaryl was applied to corn seed farrows at a rate of 5.03 kg/ha active ingredient. Carbaryl was stable up to 166 days, but after 135 days, 95% had disappeared. The long lag time suggests that carbaryl degradation was primarily due to microbial degradation (Caro et al., 1974).

Soil. The rate of hydrolysis of carbaryl in flooded soil increased when the soil was pretreated with the hydrolysis product, 1-naphthol (Rajagopal et al., 1986). Carbaryl is hydrolyzed in both flooded and nonflooded soils but the rate is slightly higher under flooded conditions (Rajagopal et al., 1983). When ^{14}C-carbonyl-labeled carbaryl (200 ppm) was added to five different soils and incubated at 25°C for 32 days, evolution of ^{14}C-carbon dioxide varied from 2.2–37.4% (Kazano et al., 1972). Metabolites identified in soil included 1-naphthol (hydrolysis product) (Sud et al., 1972; Ramanand et al., 1988a), hydroquinone, catechol, pyruvate (Sud et al., 1972), coumarin, carbon dioxide (Kazano et al., 1972), 1-naphthylcarbamate, 1-naphthyl *N*-hydroxymethylcarbamate, 5-hydroxy-1-naphthylmethylcarbamate, 4-hydroxy-1-naphthylmethylcarbamate and 1-naphthyl hydroxymethylcarbamate (Liu and Bollag, 1971, 1971a). 1-Naphthol was readily degraded by soil microorganisms (Sanborn et al., 1977).

When carbaryl was applied to soil at a rate of 1,000 L/ha, more than 50% remained in the upper 5 cm (Meyers et al., 1970). The half-lives of carbaryl in a sandy loam, clay loam and an organic amended soil under non-sterile conditions were 96–1,462, 211–2,139 and 51–4,846 days, respectively, while under sterile conditions the half-lives were 67–5,923, 84–9,704 and 126–4,836, respectively (Schoen and Winterlin, 1987).

Liu and Bollag (1971) reported that the fungus *Gliocladium roseum* degraded carbaryl to 1-naphthyl *N*-hydroxymethylcarbamate, 4-hydroxy-1-naphthylmethylcarbamate and 1-naphthylhydroxymethylcarbamate.

Sud et al. (1972) discovered that a strain of *Achromobacter* sp. utilized carbaryl as the sole source of carbon in a salt medium. The organism grew on the degradation products 1-naphthol, hydroquinone and catechol. 1-Naphthol, a metabolite of carbaryl in soil, was recalcitrant to further degradation by a bacterium tentatively identified as an *Arthrobacter* sp. under anaerobic conditions (Ramanand et al., 1988a). Carbaryl or its metabolite 1-naphthol at normal and ten times the field application rate had no effect on the growth of *Rhizobium* sp. or *Azotobacter chroococcum* (Kale et al., 1989). The half-lives for carbaryl under flooded and nonflooded conditions were 13–14 and 23–28 days, respectively (Venkateswarlu et al., 1980).

Rajagopal et al. (1984) identified the following degradates of carbaryl in soil and in microbial cultures: 5,6-dihydrodihydroxy carbaryl, 2-hydroxy carbaryl, 4-hydroxy carbaryl, 5-hydroxy carbaryl, 1-naphthol, N-hydroxymethyl carbaryl, 1-naphthyl carbamate, 1,2-dihydroxynaph-thalene, 1,4-dihydroxynaphthalene, o-coumaric acid, o-hydroxybenzalpyruvate, 1,4-naphthoquinone, 2-hydroxy-1,4-naphthoquinone, coumarin, γ-hydroxy-γ-o-hydroxyphenyl-α-oxobutyrate, 4-hydroxy-1-tetralone, 3,4-di-hydroxy-1-tetralone, pyruvic acid, salicylaldehyde, salicylic acid, phenol, hydroquinone, catechol, carbon dioxide and water. When carbaryl was incubated at room temperature in a mineral salts medium by soil-enrichment cultures for 30 days, 26.8 and 31.5% of the applied insecticide remained in flooded and nonflooded soils, respectively (Rajagopal et al., 1984a). A *Bacillus* sp. and the enrichment cultures both degraded carbaryl to 1-naphthol. Mineralization to carbon dioxide was negligible (Rajagopal et al., 1984a).

The half-lives for carbaryl in soil incubated in the laboratory under aerobic conditions ranged from 9 to 34 days with an average half-life of 22 days (Kazano et al., 1972). In field soils, the half-lives for carbaryl ranged from 4 to 25 days with an average half-life of 12 days (Caro et al., 1974; Kuhr et al., 1974; Johnson and Stansbury, 1975).

Groundwater. According to the U.S. EPA (1986) butylate has a high potential to leach to groundwater.

Plant. In plants, the N-methyl group may be subject to oxidation or hydroxylation (Kuhr, 1968). The presence of pinolene (β-pinene polymer) in carbaryl formulations increases the amount of time carbaryl residues remain on tomato leaves and decreases the rate of decomposition. The half-life in plants ranges from 1.3 to 29.5 days (Blazquez et al., 1970).

Surface Water. In a laboratory aquaria containing estuarine water, 43% of dissolved carbaryl was converted to 1-naphthol in 17 days at 20°C (pH = 7.5–8.1). The half-life of carbaryl in estuarine water without mud at 8°C was 38 days. When mud was present, both carbaryl and 1-naphthol decreased to less than 10% in the estuarine water after 10 days. Based on a total recovery of only 40%, it was postulated that the remainder was evolved as methane (Karinen et al., 1967). The rate of hydrolysis of carbaryl increased with an increase in temperature (Karinen et al., 1967) and in increases of pH above 7.0 (Rajagopal et al., 1984). The presence of a micelle [hexadecyltrimethylammonium bromide (HDATB), 3×10^{-3} M] in natural waters greatly enhanced the hydrolysis rate. The hydrolysis half-lives in natural water samples with and without HDATB were 0.12-0.67 and 9.7–138.6 hours, respectively (González et al., 1992). In a sterilized buffer solution, an hydrolysis half-life of 87 hours was observed (Ferreira and Seiber, 1981). In the dark, carbaryl was incubated in 21°C water obtained from the Holland Marsh drainage canal. Degradation was complete after 4 weeks (Sharom et al., 1980).

In pond water, carbaryl rapidly degraded to 1-naphthol. The latter was further degraded, presumably by *Flavobacterium* sp., into hydroxycinnamic acid, salicylic acid and an unidentified compound (Hazardous Substances Data Bank, 1989). Four days after carbaryl (30 mg/L and 300 μg/L) was added to Fall Creek water, >60% was mineralized to carbon dioxide. At pH 3, however, <10% was converted to carbon dioxide (Boethling and Alexander, 1979). Under these conditions, hydrolysis of carbaryl to 1-naphthol was rapid. The authors could not determine how much carbon dioxide was attributed to biodegradation of carbaryl and how much was due to the biodegradation of 1-naphthol (Boethling and Alexander, 1979). Hydrolysis half-lives for carbaryl in filtered and sterilized Hickory Hills (pH 6.7) and U.S. Department of Agriculture Number 1 pond water (pH 7.2) were 30 and 12 days, respectively (Wolfe et al., 1978).

Six brooks and three rivers in Maine contaminated by carbaryl originated from forest spraying. The disappearance half-lives of carbaryl in Clayton, Carry, Squirrel, Burntland, Mud and Wyman brooks were 15, 19, 77, 111, 42 and 33–83 hours, respectively. Disappearance half-lives of 58–200, 21–71 and 24 hours were reported for Presque Isle, Machia and Penobscot rivers, respectively (Stanley and Trial (1980).

Photolytic. Based on data for phenol, a structurally related compound, an aqueous solution containing the 1-naphthoxide ion (3×10^{-4} M) in room light would be expected to photooxidize to give 2-hydroxy-1,4-naphthoquinone (Tomkiewicz et al., 1971). 1-Naphthol, methyl isocyanate and other unidentified cholinesterase inhibitors were reported as products formed from the direct photolysis of carbaryl by sunlight (Wolfe et al., 1976).

In an aqueous solution at 25°C, the photolysis half-life of carbaryl by natural sunlight or UV light ($\lambda = 313$ nm) is 6.6 days (Wolfe et al., 1978a). A photolysis half-life of 1.88 days was found when carbaryl in an acidic (pH 5.5), buffered aqueous solution was irradiated with UV light ($\lambda > 290$ nm) (Wolfe et al., 1976).

When carbaryl in ethanol was irradiated by UV light, degradation was observed but no products were identified (Crosby et al., 1965).

Peris-Cardells et al. (1993) studied the photocatalytic degradation of carbaryl (200 μg/L) in aqueous solutions containing a suspended catalyst (titantium dioxide) in a continuous flow system equipped with a UV lamp. The investigators concluded that the degradation yield depends on the ratio of the insecticide and catalyst. The rate of degradation was independent of pH and temperature. Under optimal experimental conditions, >99% degradation was achieved in less than 1 minute of irradiation.

Chemical/Physical. Ozonation of carbaryl in water yielded 1-naphthol, naphthoquinone, phthalic anhydride, *N*-formylcarbamate of 1-naphthol (Martin et al., 1983), naphthoquinones and acidic compounds (Shevchenko et al., 1982). Hydrolysis and photolysis of carbaryl forms 1-naphthol (Wauchope and Haque, 1973; Rajagopal et al., 1984, 1986; Miles et al., 1988; Ramanand et al., 1988a; MacRae, 1989; Somasundaram et al., 1991) and 2-hydroxy-1,4-naphthoquinone (Wauchope and Haque, 1973), respectively.

In aqueous solutions, carbaryl hydrolyzes to 1-naphthol (Vontor et al., 1972; Boethling and Alexander, 1979), methylamine and carbon dioxide (Vontor et al., 1972) especially under alkaline conditions (Wolfe et al., 1978).

Miles et al. (1988) studied the rate of hydrolysis of carbaryl in phosphate-buffered water (0.01 M) at 26°C with and without a chlorinating agent (10 mg/L hypochlorite solution). The hydrolysis half-lives at pH 7 and 8 with and without chlorine were 3.5 and 10.3 days and 0.05 and 1.2 days, respectively (Miles et al., 1988). The reported hydrolysis half-lives for carbaryl in water at pH 7, 8, 9 and 10 were 10.5 days, 1.3 days, 2.5 hours and 15.0 minutes, respectively (Aly and El-Dib, 1971). 1-Naphthol, the major product of carbaryl hydrolysis (Kazano et al., 1972), dissociates in water to the 1-naphthoxide ion (Wauchope and Haque, 1973). The hydrolysis half-lives for carbaryl in a sterile 1% ethanol/water solution at 25°C and pH values of 4.5, 6.0, 7.0 and 8.0, were 300, 58, 2.0 and 0.27 weeks, respectively (Chapman and Cole, 1982).

Products reported from the combustion of carbaryl at 900°C include carbon monoxide, carbon dioxide, ammonia and oxygen (Kennedy et al., 1972a).

Carbaryl is stable to light and heat and in slightly acidic and alkaline solutions (Hartley and Kidd, 1987).

Exposure Limits: NIOSH REL: TWA 5 mg/m³, IDLH 100 mg/m³; OSHA PEL: TWA 5 mg/m³; ACGIH TLV: TWA 5 mg/m³.

Symptoms of Exposure: Miosis, blurred vision, tearing, nasal discharge, salivation, sweating, abdominal cramps, nausea, vomiting, diarrhea, tremor, cyanosis, convulsions, severe eye and skin irritation.

Formulation Types: Emulsifiable concentrate; wettable powder; granules; suspension concentrate; dustable powder; oil-miscible flowable concentrate.

Toxicity: LC_{50} (96-hour) for rainbow trout 1.3 mg/L, sheepshead minnow 2.2 mg/L (Worthing and Hance, 1991), bluegill sunfish 6.76 mg/L, goldfish 13.2 mg/L (Hartley and Kidd, 1987), catfish 15.8 mg/L, bullhead 20.0 mg/L, carp 5.3 mg/L, bass 6.4, brown trout 2.0 mg/L, coho salmon 0.76 mg/L (Verschueren, 1983); LC_{50} (24-hour) for bluegill sunfish 3.4 ppm and rainbow trout 3.5 ppm (Verschueren, 1983); acute oral LD_{50} for male and female rats 850 and 500 mg/kg, respectively (Hartley and Kidd, 1987).

Uses: Contact insecticide used to control most insects on fruits, vegetables and ornamentals.

CARBOFURAN

Synonyms: Bay 70143; **2,3-Dihydro-2,2-dimethyl-7-benzofuranol methyl-carbamate**; 2,2-Dimethyl-7-coumaranyl-*N*-methylcarbamate; 2,2-Dimethyl-2,3-dihydro-7-benzofuranyl-*N*-methylcarbamate; ENT 27164; Furadan; Methyl carbamic acid 2,3-dihydro-2,2-dimethyl-7-benzofuranyl ester; NIA 10242; Niagara 10242.

Designations: CAS Registry Number: 1563-66-2; DOT: 2757; mf: $C_{12}H_{15}NO$; fw: 221.26; RTECS: FB9450000.

Properties: Odorless, white crystalline solid. Mp: 153–154°C; bp: decomposes >150°C; ρ: 1.18 at 20/20°C; fl p: nonflammable; H-$t_{1/2}$ (25°C): 170, 690, 690, 8.2 and 1.0 weeks at pH values of 4.5, 5.0, 6.0, 7.0 and 8.0, respectively; K_H: 3.88×10^{-8} atm · m³/mol at 30–33°C (approximate — calculated from water solubility and vapor pressure); log K_{oc}: 1.98–2.32; log K_{ow}: 1.60–2.32; P-$t_{1/2}$: 70.64 hours (absorbance λ = 258.0 nm, concentration on glass plates = 6.7 µg/cm²); S_o: acetone (15%), acetonitrile (14%), benzene (4%), cyclohexane (9%), *N,N*-dimethylformamide (27%), di-methylsulfoxide (25%), methylene chloride (>200 g/L), 2-propanol (20–50 g/L); S_w: 291, 320 and 375 mg/L at 10, 20 and 30°C, respectively; vp: 4.85×10^{-6} mmHg at 25°C (Ferreira and Seiber, 1981).

Soil properties and adsorption data

Soil	K_d (mL/g)	f_{oc} (%)	K_{oc} (mL/g)	pH (meq/100 g)	CEC
Bryce silty clay loam	2.22	7.50	30	—	55.5
Bryce-Swygert silty clay	1.40	2.70	52	—	34.4
Catlin	1.20	2.01	60	6.2	—
Chaslcus	0.61	1.62	38	8.1	—
Commerce	1.12	0.68	160	6.7	—
Drummer silty clay loam	1.13	3.10	36	—	24.8
Earth/Humus	1.65	4.16	40	6.5	—
Flanagan silt loam	1.39	3.50	40	—	27.7
Gilford-Hoopeston-Ade sandy loam	0.74	1.20	62	—	7.5
Houghton muck	8.74	16.80	52	—	72.4
Humus	3.48	10.97	32	6.8	—
Kari soil	2.38	5.10	47	2.7	32.2
Limagne	0.72	2.08	35	8.0	—
Loam	1.08	3.02	34	7.1	—

Soil properties and adsorption data *(continued)*

Soil	K_d (mL/g)	f_{oc} (%)	K_{oc} (mL/g)	pH (meq/100 g)	CEC
Loamy sand	0.56	1.80	31	6.1	—
Plainfield-Bloomfield sand	0.25	0.40	63	—	1.7
Pokkali soil	2.08	4.12	50	3.6	31.2
Sand	0.10	0.99	10	6.9	—
Sandy loam	1.25	4.23	30	5.6	—
Sediment A	0.33	1.50	22	7.7	—
Sediment B	0.30	1.85	16	7.1	—
Silt loam	0.30	2.26	13	4.9	—
Soil #1 (0–30 cm)	0.48	1.46	33	6.0	—
Soil #1 (30–50 cm)	0.49	0.93	53	6.1	—
Soil #1 (50–91 cm)	0.52	1.24	42	6.1	—
Soil #1 (91–135 cm)	0.18	0.70	26	6.4	—
Soil #1 (135–165 cm)	0.09	0.17	53	6.5	—
Soil #1 (>165 cm)	0.08	0.07	117	6.7	—
Soil #2 (0–27 cm)	0.38	1.50	25	5.9	—
Soil #2 (27–53 cm)	0.15	0.16	95	6.3	—
Soil #2 (53–61 cm)	0.07	0.14	48	6.4	—
Soil #2 (61–81 cm)	0.15	0.28	55	6.4	—
Soil #2 (81–157 cm)	0.29	0.50	58	6.0	—
Soil #2 (>157 cm)	0.51	1.30	39	5.9	—
Tracy	1.07	1.12	95	6.2	—
Versailles	0.26	1.10	24	6.4	—

Source: Jamet and Piedallu, 1975; Felsot and Wilson, 1980; McCall et al., 1981; Sukop and Cogger, 1992; Singh et al., 1990a; U.S. Department of Agriculture, 1990.

Environmental Fate

Biological. Carbofuran or their metabolites (3-hydroxycarbofuran and 3-ketocarbofuran) at normal and ten times the field application rate had no effect on *Rhizobium* sp. However, in a nitrogen-free culture medium, *Azotobacter chroococcum* growth was inhibited by carbofuran, 3-hydroxycarbofuran and 3-ketocarbofuran (Kale et al., 1989).

Under *in vitro* conditions, 15 of 20 soil fungi degraded carbofuran to one or more of the following compounds: 3-hydroxycarbofuran, 3-ketocarbofuran, carbofuran phenol and 3-hydroxyphenol (Arunachalam and Lakshmanan, 1988).

Derbyshire et al. (1987) reported that an enzyme, isolated from the microorganism *Achromobacter* sp., hydrolyzed carbofuran via the carbamate linkage forming 2,3-dihydro-2,2-dimethyl-7-benzofuranol. The optimum pH and temperature for the degradation of carbofuran and two other pesticides (aldicarb and carbaryl) were 9.0–10.5 and 45 and 53°C, respectively. Degradation of carbofuran in poorly drained clay-muck, well drained clay-muck and poorly drained clay field test plots from British Columbia was very low. After 2 successive years of treatment, carbofuran concentrations began to increase (Williams et al., 1976).

Soil. Carbofuran is relatively persistent in soil, especially in dry, acidic and low temperature soils (Ahmad et al., 1979; Caro et al., 1973; Fuhremann and Lichtenstein,

1980; Gorder et al., 1982; Greenhalgh and Belanger, 1981; Ou et al., 1982). In alkaline soil with high moisture content, microbial degradation of carbofuran was more important than leaching and chemical degradation (Gorder et al., 1982; Greenhalgh and Belanger, 1981).

Carbofuran phenol is formed from the hydrolysis of carbofuran at pH 7.0. Carbofuran phenol was also found to be the major biodegradation product by *Azospirillum lipoferum* and *Streptomyces* spp. isolated from a flooded alluvial soil (Venkateswarlu and Sethunathan, 1984). The hydrolysis of carbofuran to carbofuran phenol was catalyzed by the addition of rice straw in an anaerobic flooded soil where it accumulated (Venkateswarlu and Sethunathan, 1979). The rate of transformation of carbofuran in soil increased with repeated applications (Harris et al., 1984). In an alluvial soil, carbaryl and its analog carbosulfan 2,3-dihydro-2,2-dimethyl-7-benzofuranyl [(di-*n*-butyl)aminosulfenyl methylcarbamate] both degraded faster at 35°C than at 25°C with carbosulfan degrading to carbofuran (Sahoo et al., 1990). An enrichment culture isolated from a flooded alluvial soil (Ramanand et al., 1988) and a bacterium tentatively identified as an *Arthrobacter* sp. (Ramanand et al., 1988a) readily mineralized carbofuran to carbon dioxide at 35°C. Mineralization was slower at lower temperatures (20–28°C). Under anaerobic conditions, carbofuran did not degrade (Ramanand et al., 1988a). The reported half-lives in soil are 1–2 months (Hartley and Kidd, 1987); 11–13 days at pH 6.5 and 60–75 days for a granular formulation (Ahmad et al., 1979).

In soils, microorganisms degraded carbofuran to carbofuran phenol (Ou et al., 1982), 3-hydroxycarbofuran then to 3-ketocarbofuran (Ou et al., 1982; Kale et al., 1989).

Carbofuran hydrolyzes in soil and water forming carbofuran phenol, carbon dioxide and methylamine (Seiber et al., 1978; Rajagopal et al., 1986; Somasundaram et al., 1989, 1991). Hydrolysis of carbofuran occurs in both flooded and nonflooded soils, but the rate is slightly higher under flooded conditions (Venkateswarlu et al., 1977) especially when the soil is pretreated with the hydrolysis product, carbofuran phenol (Rajagopal et al., 1986). In addition, the hydrolysis of carbofuran was found to be pH dependent in both deionized water and rice paddy water. At pH values of 7.0, 8.7 and 10.0, the hydrolysis half-lives in deionized water were 864, 19.4 and 1.2, hours, respectively. In paddy water, the hydrolysis half-lives at pH values of 7.0, 8.7 and 10.0 were and 240, 13.9 and 1.3, respectively (Seiber et al., 1978).

The half-lives for carbofuran in soil incubated in the laboratory under aerobic conditions ranged from 6 to 93 days with an average half-life of 37 days (Getzin, 1973; Venkateswarlu et al., 1977). In flooded soils, Venkateswarlu (1977) reported a half-life of 44 days. In field soils, half-lives ranging from 31 to 117 days with an average half-life of 68 days were reported (Caro et al., 1973; Mathur et al., 1976; Talekar et al., 1977). The average mineralization half-life for carbofuran in soil was 535 days (Getzin, 1973).

Gorder et al. (1982) studied the persistence of carbofuran in an alkaline cornfield soil having a high moisture content. They concluded that microbial degradation was more important than leaching in the dissipation from soil. Similarly, Greenhalgh and Belanger (1981) reported that microbial degradation was more important than chemical attack in a humic soil and a sandy loam soil.

Plant. Carbofuran is rapidly metabolized in plants to nontoxic products (Cremlyn, 1991). Metcalf et al. (1968) reported that carbofuran undergoes hydroxylation and hydrolysis in plants, insects and mice. Hydroxylation of the benzylic carbon gives 3-hydroxycarbofuran which is subsequently oxidized to 3-ketocarbofuran. In carrots, carbofuran initially degraded to 3-hydroxycarbofuran. This compound reacted with naturally occurring angelic acid in carrots forming a conjugated metabolite identified as 2,3-dihydro-2,2-

dimethyl-7-(((methylamino)-carbonyl)oxy)-3-benzofuranyl (Z)-2-methyl-2-butenoic acid (Sonobe et al., 1981). Metabolites identified in three types of strawberries (Day-Neutral, Tioga and Tufts) were 2,3-dihydro-2,2-dimethyl-3-hydroxy-7-benzofuranyl-N-methylcarbamate, 2,3-dihydro-2,2-dimethyl-3-oxo-7-benzofuranyl-N-methylcarbamate, 2,3-dihydro-2,2-dimethyl-3-benzofuranol, 2,3-dihydro-2,2-dimethyl-3,7-benzofuranol and 2,3-dihydro-2,2-dimethyl-3-oxo-7-benzofuranol (Archer et al., 1977). Oat plants were grown in two soils treated with [^{14}C]carbofuran. Most of the residues recovered in oat leaves were in the form of carbofuran and 3-hydroxycarbofuran. Other metabolites identified were 3-ketocarbofuran, a 3-keto-7-phenol and a 3-hydroxy-7-phenol (Fuhremann and Lichtenstein, 1980).

In cornfield soils, carbofuran losses were high but in autoclaved soils, virtually no losses were observed. This insecticide appears to be poorly absorbed onto the soil matrix and therefore, is readily leachable (Felsot et al., 1981).

Photolytic. 2,3-Dihydro-2,2-dimethylbenzofuran-4,7-diol and 2,3-dihydro-3-keto-2,2-dimethylbenzofuran-7-ylcarbamate were formed when carbofuran dissolved in water was irradiated by sunlight for 5 days (Raha and Das, 1990).

Surface Water. Sharom et al. (1980) reported that the half-lives for carbofuran in sterilized and non-sterilized water collected from the Holly Marsh in Ontario, Canada were 2.5 to 3 weeks at pH values of 7.8–8.0 and 8.0, respectively. The half-lives observed in distilled water were 2 and 3.8 weeks at pH values of 7.0–7.2 and 6.8, respectively. They reported that chemical degradation of dissolved carbofuran was more significant than microbial degradation.

Chemical/Physical. Releases toxic nitrogen oxides when heated to decomposition (Sax and Lewis, 1987; Lewis, 1990).

The hydrolysis half-lives for carbofuran in a sterile 1% ethanol/water solution at 25°C and pH values of 4.5, 5.0, 6.0, 7.0 and 8.0 were 170, 690, 690, 8.2 and 1.0 week, respectively (Chapman and Cole, 1982).

Exposure Limits: OSHA PEL: TWA 0.1 mg/m^3; ACGIH TLV: TWA 0.1 mg/m^3.

Symptoms of Exposure: Early symptoms include burning eyes, dimming of vision, miosis, loss of accommodation, headache, light-headedness, weakness and nausea. Later symptoms include blurred vision, constriction of pupils, abdominal cramps, excessive salivation, perspiration and vomiting.

Formulation Types: Wettable powder; granules; suspension concentrate; flowable concentrate.

Toxicity: LC$_{50}$ (96-hour) for rainbow trout 280 µg/L, bluegill sunfish 240 µg/L and channel catfish 210 µg/L; acute oral LD$_{50}$ for rats 8.2–14.1 mg/kg (corn oil) (Hartley and Kidd, 1987), 5,300 µg/kg (RTECS, 1985), 11 mg/kg (Verschueren, 1983).

Uses: Broad-spectrum, systemic insecticide, nematocide and acaricide applied in soil to control soil insects and nematodes or on foliage to control insects and mites.

CARBON DISULFIDE

Synonyms: Carbon bisulfide; Carbon bisulphide; Carbon disulphide; Carbon sulfide; Carbon sulphide; Dithiocarbonic anhydride; NCI-C04591; RCRA waste number P022; Sulphocarbonic anhydride; UN 1131; Weeviltox.

$$S=C=S$$

Designations: CAS Registry Number: 75-15-0; DOT: 1131; mf: CS_2; fw: 76.13; RTECS: FF6650000.

Properties: Highly refractive, mobile, clear, colorless to pale yellow liquid; sweet, pleasing, ethereal odor when pure. Technical grades have foul odors. Mp: $-111.5°C$; bp: $46.2°C$; ρ: 1.27055 at 15/4°C, 1.2632 at 20/4°C; fl p: $-30°C$; lel: 1.3%; uel: 50%; K_H: 1.33×10^{-2} atm · m³/mol at 20–22°C (approximate — calculated from water solubility and vapor pressure); IP: 10.06 eV, 10.080 eV; log K_{oc}: 2.38–2.55 (calculated); log K_{ow}: 1.84, 2.16 (calculated); S_o: miscible with benzene, ethanol, carbon tetrachloride, chloroform, ethyl ether, propanol and many other organic solvents; S_w: 2.3 g/L at 22°C; vap d: 3.11 g/L at 25°C, 2.63 (air = 1); vp: 297.5 mmHg at 20°C.

Environmental Fate

Chemical/Physical. Carbon disulfide hydrolyzes in alkaline solutions to carbon dioxide and hydrogen disulfide (Peyton et al., 1976). In an aqueous alkaline solution containing hydrogen peroxide, dithiopercarbonate, sulfide, elemental sulfur and polysulfides may be expected to form (Elliott, 1990). In an aqueous alkaline solution (pH ≥8), carbon disulfide reacted with hydrogen peroxide forming sulfate and carbonate ions. However, when the pH is lowered to 7–7.4, colloidal sulfur is formed (Adewuyi and Carmichael, 1987). An aqueous solution containing carbon disulfide reacts with sodium hypochlorite forming carbon dioxide, sulfuric acid and sodium chloride (Patnaik, 1992). Forms a hemihydrate which decomposes at $-3°C$ (Keith and Walters, 1992).

Burns with a blue flame releasing carbon dioxide and sulfur dioxide (Windholz et al., 1983). Emits very toxic sulfur oxides when heated to decomposition (Lewis, 1990).

Carbon disulfide oxidizes in the troposphere producing carbonyl sulfide. The atmospheric half-lives for carbon disulfide and carbonyl sulfide were estimated to be approximately 2 years and 13 days, respectively (Khalil and Rasmussen, 1984).

Exposure Limits: NIOSH REL: TWA 1 ppm, STEL 10 ppm, IDLH 500 ppm; OSHA PEL: TWA 20 ppm, C 30 ppm; ACGIH TLV: TWA 10 ppm.

Symptoms of Exposure: Dizziness, headache, poor sleep, fatigue, nervousness; anorexia, weight loss; psychosis; polyneuropathy; Parkinson-like symptoms; ocular changes; cardiovascular system irregularity; gastrointestinal discomfort; burns, dermatitis. Contact with skin causes burning pain, erythema and exfoliation.

Formulation Types: Dilute solution.

Toxicity: Acute oral LD_{50} for rats 3,188 mg/kg (RTECS, 1985).

Uses: Formerly used as an insecticide.

CARBON TETRACHLORIDE

Synonyms: Benzinoform; Carbona; Carbon chloride; Carbon tet; ENT 4705; Fasciolin; Flukoids; Freon 10; Halon 104; Methane tetrachloride; Necatorina; Necatorine; Perchloromethane; R 10; RCRA waste number U211; Tetrachloormetaan; Tetrachlorocarbon; **Tetrachloromethane**; Tetrafinol; Tetraform; Tetrasol; UN 1846; Univerm; Vermoestricid.

$$CCl_4$$

Designations: CAS Registry Number: 56-23-5; DOT: 1846; mf: CCl_4; fw: 153.82; RTECS: FG4900000.

Properties: Clear, colorless, heavy, liquid with a strong sweetish, ether-like odor and burning taste. Mp: $-22.96°C$; bp: $76.5°C$; ρ: 1.5940 at 20/4°C, 1.585 at 25/4°C; H-$t_{1/2}$: 40.5 years at 25°C and pH 7; K_H: 3.02×10^{-2} atm $\cdot$ m³/mol at 25°C; IP: 11.47 eV; log K_{oc}: 2.35–2.64; log K_{ow}: 2.73, 2.83; S_o: miscible with acetone, ethanol, benzene, chloroform, carbon disulfide, dimethylsulfoxide, ethyl ether, light petroleum, petroleum ether, solvent naphtha, volatile oils, dehydrated alcohol, fat solvents and most organic solvents; S_w: 800 mg/L at 20°C; vap d: 6.29 g/L at 25°C, 5.31 (air = 1); vp: 90 mmHg at 20°C, 137 mmHg at 30°C.

Environmental Fate

Biological. Carbon tetrachloride was degraded by denitrifying bacteria forming chloroform (Smith and Dragun, 1984). An anaerobic species of *Clostridium* biodegraded carbon tetrachloride by reductive dechlorination yielding trichloromethane, dichloromethane and unidentified products (Gälli and McCarty, 1989). Chloroform also formed by microbial degradation of carbon tetrachloride using denitrifying bacteria (Smith and Dragun, 1984).

Carbon tetrachloride (5 and 10 mg/L) showed significant degradation with rapid adaptation in a static-culture flask-screening test (settled domestic wastewater inoculum) conducted at 25°C. Complete degradation was observed after 14 days of incubation (Tabak et al., 1981).

Chemical/Physical. Under laboratory conditions, carbon tetrachloride partially hydrolyzed to chloroform and carbon dioxide (Smith and Dragun, 1984). Complete hydrolysis yielded carbon dioxide and hydrochloric acid (Kollig, 1993). Carbon tetrachloride slowly reacts with hydrogen sulfide in aqueous solution yielding carbon dioxide via the intermediate carbon disulfide. However, in the presence of two micaceous minerals (biotite and vermiculite) and amorphous silica, the rate of transformation increased. At 25°C and a hydrogen sulfide concentration of 1 mM, the half-lives for carbon tetrachloride were calculated to be 2,600, 160 and 50 days for the silica, vermiculite and biotite studies, respectively. In all three studies, the major transformation pathway is the formation of carbon disulfide which undergoes hydrolysis yielding carbon dioxide (81–86% yield) and hydrogen sulfide ions. Minor intermediates detected include chloroform (5–15% yield), carbon monoxide (1–2% yield) and a nonvolatile compound tentatively identified as formic acid (3–6% yield) (Kriegman-King and Reinhard, 1992).

Anticipated products from the reaction of carbon tetrachloride with ozone or hydroxyl radicals in the atmosphere are phosgene and chloride radicals (Cupitt, 1980). Phosgene is hydrolyzed readily to hydrochloric acid and carbon dioxide (Morrison and Boyd, 1971).

Matheson and Tratnyek (1994) studied the reaction of fine-grained iron metal in an anaerobic aqueous solution (15°C) containing carbon tetrachloride (151 μM). Initially, carbon tetrachloride underwent rapid dehydrochlorination forming chloroform, which further degraded to methylene chloride and chloride ions. The rate of reaction decreased with each dehydrochlorination step. However, after 1 hour of mixing, the concentration of carbon tetrachloride decreased from 151 to approximately 15 μM. No additional products were identified although the authors concluded that environmental circumstances may exist where degradation of methylene chloride may occur. They also reported that reductive dehalogenation of carbon tetrachloride and other chlorinated hydrocarbons used in this study appears to take place in conjunction with the oxidative dissolution or corrosion of the iron metal through a diffusion-limited surface reaction.

The evaporation half-life of carbon tetrachloride (1 mg/L) from water at 25°C using a shallow-pitch propeller stirrer at 200 rpm at an average depth of 6.5 cm is 29 minutes (Dilling, 1977).

Exposure Limits: NIOSH REL: STEL 1 hour 2 ppm, IDLH 200 ppm; OSHA PEL: TWA 10 ppm, C 25 ppm, 5-minute/4-hour peak 200 ppm; ACGIH TLV: TWA 5 ppm.

Symptoms of Exposure: Central nervous system depression; nausea, vomiting; liver, kidney damage; skin irritation.

Formulation Types: Fumigant (mixture containing 1,2-dichloroethane or 1,2-dibromoethane and 1,2-dichloroethane).

Toxicity: EC50 (24-hour) for *Daphnia magna* 74.5 mg/L (Lilius et al., 1995); LC_{50} (14-day) for guppies 67 ppm; acute oral LD_{50} for rats 2,920 mg/kg (Verschueren, 1983), 2,800 mg/kg (RTECS, 1985).

Uses: Agricultural fumigant.

CARBOPHENOTHION

Synonyms: Acarithion; Akarithion; *S*-((4-Chlorophenyl)thio)methyl *O,O*-diethyl phos-phorodithioate; *S*-((*p*-Chlorophenyl)thio)methyl *O,O*-diethyl phosphorodithioate; *S*-(4-Chlorophenylthiomethyl)diethyl phosphorothiolothionate; Dagadip; *O,O*-Diethyl *p*-chlo-rophenylmercaptomethyl dithiophosphate; *O,O*-Diethyl *S*-(4-chlorophenylthio)methyl dithiophosphate; *O,O*-Diethyl *S*-(*p*-chlorophenylthio)methyl phosphorodithioate; Endyl; ENT 23708; Garrathion; Lethox; Nephocarp; Oleoakarithion; R 1303; **Phosphorodithioic acid *S*-(((4-chlorophenyl)thio)methyl) *O,O*-diethyl ester**; Stauffer R 1303; Trithion; Trithion miticide.

Designations: CAS Registry Number: 786-19-6; mf: $C_{11}H_{16}ClO_2PS_3$; fw: 342.96; RTECS: TD5250000.

Properties: Colorless to pale amber liquid with a mercaptan-like odor. Mp: <25°C; bp: 82°C at 0.01 mmHg; ρ: 1.271 at 25/4°C; K_H: 4.5 × 10^{-7} atm · m³/mol at 20°C; log K_{oc}: 3.92–4.98; log K_{ow}: 4.75 (calculated); S_o: miscible with most organic solvents and vege-table oils; S_w: 610, 630 and 730 µg/L at 10, 20 and 30°C, respectively; vap d: 14.02 g/L at 25°C, 11.88 (air = 1); vp: 8.03 × 10^{-6} mmHg at 20°C.

Soil properties and adsorption data

Soil	K_d (mL/g)	f_{oc} (%)	K_{oc} (mL/g)	pH
Elkhorn sandy loam	82.3	0.09	95,698	6.0
Hugo gravelly sandy loam	62.8	0.12	52,333	5.5
Sweeney sandy clay loam	54.6	0.65	8,394	6.3
Tierra clay loam	98.6	0.33	29,894	6.2

Source: Rao and Davidson, 1982.

Environmental Fate

Soil. Though no products were reported, the half-life was reported to be ≥100 days (Verschueren, 1983).

Chemical/Physical. Oxidizes to sulfoxide, sulfone, thiol, thiosulfone and thiosulfoxide (Hartley and Kidd, 1987). Emits toxic fumes of chlorine, phosphorus and sulfur oxides when heated to decomposition (Sax and Lewis, 1987).

Symptoms of Exposure: Headache, weakness and dizziness.

Formulation Types: Wettable powder; dustable powder; granules; emulsifiable concentrate; seed treatment.

Toxicity: Very toxic to fish (Hartley and Kidd, 1987); acute oral LD_{50} for male and female rats 30 and 10 mg/kg, respectively (Hartley and Kidd, 1987), 6,800 mg/kg (RTECS, 1985).

Uses: Nonsystemic insecticide and acaricide for controlling mites, aphids and other insects on deciduous fruit trees.

CARBOXIN

Synonyms: 5-Carboxanilido-2,3-dihydro-6-methyl-1,4-oxathiin; Carboxine; D 735; DCMO; 2,3-Dihydro-5-carboxanilido-6-methyl-1,4-oxatiin; 2,3-Dihydro-6-methyl-1,4-oxathiin-5-carboxanilide; 5,6-Dihydro-2-methyl-1,4-oxathiin-3-carboxanilide; **5,6-Dihydro-2-methyl-N-phenyl-1,4-oxathiin-3-carboxamide**; F 735; Flo pro V seed protectant; Vitavax.

Designations: CAS Registry Number: 5234-68-4; mf: $C_{12}H_{13}NO_2S$; fw: 235.31; RTECS: RP4550000.

Properties: Colorless crystals. Mp: 93–95°C; ρ: 1.36; H-$t_{1/2}$: <3 days when exposed to light; K_H: 3.4 × 10^{-10} atm · m^3/mol at 20–25°C (approximate — calculated from water solubility and vapor pressure); log K_{oc}: 2.41; log K_{ow}: 2.17; So (g/kg at 25°C): acetone (600), benzene (150), dimethyl sulfoxide (1,500), methanol (210); S_w: 170 mg/L at 25°C; vp: 1.88 × 10^{-7} mmHg at 20°C.

Environmental Fate

Biological. The sulfoxidation of carboxin to carboxin sulfoxide by the fungus *Ustilago maydis* was reported by Bollag and Liu (1990).

Soil. Carboxin oxidized in soil forming carboxin sulfoxide. The half-life in soil was reported to be 24 hours (Worthing and Hance, 1991).

Plant. In plants (barley, cotton and wheat) and water, carboxin oxidizes to the corresponding sulfoxide (Worthing and Hance, 1991).

Formulation Types: Wettable powder; seed treatment; suspension concentrate.

Toxicity: LC$_{50}$ (96-hour) for rainbow trout 2 mg/L and bluegill sunfish 1.2 mg/L (Hartley and Kidd, 1987); acute oral LD$_{50}$ for rats 3,820 mg/kg (Hartley and Kidd, 1987), 430 mg/kg (RTECS, 1985).

Uses: Systemic plant fungicide.

CHLORAMBEN

Synonyms: ACP-M-728; Ambiben; Amiben; Amiben DS; Amibin; **3-Amino-2,5-dichlorobenzoic acid**; NCI-C00055; Ornamental weeder; Vegaben; Vegiben.

Designations: CAS Registry Number: 133-90-4; mf: $C_7H_5Cl_2NO_2$; fw: 206.02; RTECS: DG1925000.

Properties: Odorless, colorless to white crystalline solid or amorphous powder. Mp: 200–201°C; fl p: nonflammable; log K_{oc}: 2.28; log K_{ow}: 1.11; So (g/L): acetone (233 at 29°C), benzene (0.2 at 24°C), chloroform (0.9 at 25°C); N,N-dimethylformamide (1,206 at 20–25°C), ethanol (173 at 25°C), ethyl ether (70 at 20–25°C), methanol (223 at 20–25°C), 2-propanol (113 at 25°C); S_w: 700 mg/L at 25°C; vp: 7×10^{-3} mmHg at 100°C.

Soil properties and adsorption data

Soil	K_d (mL/g)	f_{oc} (%)	K_{oc} (mL/g)	pH (meq/100 g)	CEC
Anselmo sandy loam	0.10	0.98	10.20	7.0	7.0
Crowley sandy loam[a]	4.50	0.92	48.91	6.5	12.6
Ella loamy sand	0.24	0.92	26.09	3.8	—
Fargo clay	0.30	4.56	6.58	7.9	30.1
Gallion fine sandy loam[a]	1.10	0.35	31.43	6.2	2.8
Illite	0.30	—	—	—	25.0
Kaolinite	3.70	—	—	—	5.0
Keith silt loam	0.30	1.67	17.96	6.2	11.6
Kewaunee clay	0.11	2.19	5.02	6.4	—
Monona silt loam	0.50	2.42	20.66	5.8	17.5
Muck	4.10	23.61	17.37	6.9	65.6
Poygan silty clay	0.24	5.70	4.21	7.2	--
Sharkey clay[a]	6.50	0.75	86.67	6.7	33.4
Sharpsburg silty clay loam	1.00	2.54	39.37	5.8	24.5
Taloka sandy loam[a]	1.90	0.40	47.50	6.4	4.6

Source: Schliebe et al., 1965; Wildung et al., 1968; Talbert et al., 1970. a) Adsorption data for the methyl ester.

Environmental Fate

Soil. In soils, chloramben was degraded by microorganisms but no products were identified (Humburg et al., 1991). The main degradative pathway of chloramben in soil is decarboxylation and subsequent mineralization to carbon dioxide. The calculated half-lives in Ella loamy sand, Kewaunee clay and Poygan silty clay were 120–201, 182–286 and 176–314 days, respectively (Wildung et al., 1968). Persistence in soil is 6–8 weeks (Hartley and Kidd, 1987).

Groundwater. According to the U.S. EPA (1986) and Ashton and Monaco (1991), chloramben has a high potential to leach to groundwater, especially in sandy soils during heavy rains.

Plant. Degrades in plants to *N*-glucoside, glucose ester, conjugates and insoluble residues (Ashton and Monaco, 1991).

Photolytic. Plimmer and Hummer (1969) studied the irradiation of chloramben in water (2–4 mg/L) under a 450-W mercury vapor lamp (λ >2,800 Å) for periods of 2 to 20 hours. Chloride ion was released and a complex mixture of colored products was observed. It was postulated that amino free radicals reacted with each other via polymerization and oxidation processes. The experiment was repeated except the solution contained sodium bisulfite as an inhibitor under a nitrogen atmosphere. Oxidation did not occur and loss of the 2-chloro substituent gave 3-amino-5-chlorobenzoic acid (Plimmer and Hummer, 1969).

Chloramben (sodium salt) in aqueous solutions (100 mg/L) was rapidly photodegraded in outdoor sunlight and under a 360-W mercury arc lamp (Crosby and Leitis, 1969). In sunlight, the solution became yellow-brown. Subsequent analysis by gas-liquid chromatography did not resolve any compounds other than chloramben. However, analysis by TLC indicated at least 12 unidentified products. These products were reportedly formed via replacement of chlorine by a hydroxy group, reductive dechlorination and abstraction of hydrogen from the amine group (oxidation). No photodegradation products could be identified in the solutions irradiated with the mercury arc lamp (Crosby and Leitis, 1969).

Chemical/Physical. Emits toxic fumes of nitrogen oxides and chlorine when heated to decomposition (Sax and Lewis, 1987).

Forms water-soluble salts with alkalies.

Exposure Limits: An experimental carcinogen.

Formulation Types: Soluble concentrate (1.8 lb/gal); granules (10%); water-soluble powder or granules (75%).

Toxicity: Nontoxic to fish (Hartley and Kidd, 1987); acute oral LD_{50} for rats 5,620 mg/kg (Hartley and Kidd, 1987), 3,500 mg/kg (RTECS, 1985).

Uses: Preemergence or preplant herbicide used in many vegetable and field crops to control annual broad-leaved weeds and grasses. Also for postemergent control of common ragweed, redroot pigweed, smartweed and velvet-leaf.

CHLORDANE

Synonyms: A 1068; Aspon-chlordane; Belt; CD-68; Chlordan; γ-Chlordan; Chlori-dan; Chlorindan; Chlor kil; Chlorodane; Chlortox; Corodane; Cortilan-neu; Di-chlorochlordene; Dowklor; ENT 9932; ENT 25552; HCS 3260; Kypchlor; M 140; M 410; NA 2762; NCI-C00099; Niran; Octachlor; 1,2,4,5,6,7,8,8-Octachlor-2,3,3a,4,7,7a-hexahydro-4,7-methanoindane; Octachlorodihydrodicyclopentadiene; 1,2,4,5,6,7,8,8-Octachloro-2,3,3a,4,7,7a-hexahydro-4,7-methanoindene; **1,2,4,5,6,7,8,8-Octachloro-2,3,3a,4,7,7a-hexahydro-4,7-methano-1*H*-indene**; 1,2,4,5,6,7,8,8-Octachloro-3a,4,7,7a-hexahydro-4,7-methyleneindane; Octachloro-4,7-methanohydroindane; Octachloro-4,7-methanotetrahydroindane; 1,2,4,5,6,7,8,8-Octachloro-4,7-methano-3a,4,7,7a-tetrahydroindane; 1,2,4,5,6,7,8,8-Octachloro-3a,4,7,7a-tetrahydro-4,7-methanoindan; 1,2,4,5,6,7,8,8-Octachloro-3a,4,7,7a-tetrahydro-4,7-methanoindane; Octaklor; Octaterr; Orthoklor; RCRA waste number U036; SD 5532; Shell SD 5532; Synklor; Tat chlor 4; Topichlor 20; Topiclor; Topiclor 20; Toxichlor; Velsicol 1068.

cis   trans

Note: Chlordane is a mixture of *cis*- and *trans*-chlordane and other complex chlor-inated hydrocarbons including heptachlor and nonachlor. According to Brooks (1974), technical chlordane has the approximate composition: *trans*-chlordane (24%), four chlordene isomers ($C_{10}H_6Cl_8$) (21.5%), *cis*-chlordane (19%), heptachlor (10%), nonachlor (7%), pentachlorocyclopentadiene (2%), hexachlorocyclopenta-diene (>1%), octachlorocyclopentene (1%), $C_{10}H_{7-8}Cl_{6-7}$ (8.5%) and other unidentified compounds (6%).

Designations: CAS Registry Number: 57–74-9; DOT: 2762; mf: $C_{10}H_6Cl_8$; fw: 409.78; RTECS: PB9800000.

Properties: Colorless to amber to yellowish-brown viscous liquid with an aromatic, slight pungent odor similar to chlorine. Mp: <25°C; bp: 175°C at 2 mmHg; ρ: 1.59–1.63 at 20/4°C; fl p (kerosene solution): 55.6°C, 107.2°C (open cup); lel: 0.7% (kerosene solution); uel: 5% (kerosene solution); K_H: 4.8 × 10^{-5} atm · m^3/mol at 25°C; log BCF: white suckers 3.74 (Roberts et al., 1977), green alga (*Scenedesmus quadricauda*) 3.38–4.18 (Glooschenko et al., 1979); log K_{oc}: 4.58–5.57; log K_{ow}: 6.00; S_o: miscible with cyclohexanone, deodorized kerosene, petroleum solvents, 2-propanol, trichloroethylene, aliphatic and aromatic solvents; S_w: 56 ppb at 25°C; vap d: 16.75 g/L at 25°C, 14.15 (air = 1); vp: 9.75 × 10^{-6} mmHg at 30°C (Nash, 1983).

Soil properties and adsorption data

Soil	K_d (mL/g)	f_{oc} (%)	K_{oc} (mL/g)	pH
Sand	28	0.04	70,000	—
Sand	30	0.04	75,000	—
Silt	190	0.36	52,788	—
Silt	220	0.36	61,111	—

Source: Johnson-Logan et al., 1992.

Environmental Fate

Biological. In four successive 7-day incubation periods, chlordane (5 and 10 mg/L) was recalcitrant to degradation in a settled domestic wastewater inoculum (Tabak et al., 1981).

Soil. The actinomycete, *Nocardiopsis* sp., isolated from soil extensively degraded pure *cis-* and *trans-*chlordane to dichlorochlordene, oxy-chlordane, heptachlor, heptachlor *endo-*epoxide, chlordene chlorohydrin and 3-hydroxy-*trans*-chlordane. Oxychlordane is slowly degraded to 1-hydroxy-2-chlorochlordene (Beeman and Matsumura, 1981). The reported half-life in soil is approximately one year (Hartley and Kidd, 1987).

The percentage of chlordane remaining in a Congaree sandy loam soil after 14 years was 40% (Nash and Woolson, 1967).

Chlordane did not degrade in settled domestic wastewater after 28 days (Tabak et al., 1981).

Plant. Alfalfa plants were sprayed with chlordane at a rate of 1 lb/acre. After 21 days, 95% of the residues had volatilized (Dorough et al., 1972).

Photolytic. Chlordane should not undergo direct photolysis since it does not absorb UV light at wavelengths greater than 280 nm (Gore et al., 1971).

Chemical/Physical. In an alkaline medium or solvent, carrier, diluent or emulsifier having an alkaline reaction, chlorine will be released (Windholz et al., 1983). Technical grade chlordane passed over a 5% platinum catalyst at 200°C resulted in the formation of tetrahydrodicyclopentadiene (Musoke et al., 1982).

Chlordane (1 mM) in methyl alcohol (30 mL) underwent dechlorination in the presence of nickel boride (generated by the reaction of nickel chloride and sodium borohydride). The catalytic dechlorination of chlordane by this method yielded a pentachloro derivative as the major product having the empirical formula $C_{10}H_9C_{15}$ (Dennis and Cooper, 1976).

Chlordane is subject to hydrolysis via the nucleophilic substitution of chlorine by hydroxyl ions to yield 2,4,5,6,7,8,8-heptachloro-3a,4,7,7a-tetrahydro-4,7-methano-1*H*-indene which is resistant to further hydrolysis (Kollig, 1993). The hydrolysis half-life at pH 7 and 25°C was estimated to be >197,000 years (Ellington et al., 1988).

Emits very toxic fumes of chlorides when heated to decomposition (Lewis, 1990).

Exposure Limits: NIOSH REL: IDLH 0.5 mg/m³, IDLH 100 mg/m³; OSHA PEL: TWA 0.5 mg/m³; ACGIH TLV: TWA 0.5 mg/m³, STEL 2 mg/m³.

Symptoms of Exposure: Blurred vision, confusion, ataxia, delirium, cough, abdominal pain, nausea, vomiting, diarrhea; irritability, tremor, convulsions, anuria.

Formulation Types: Emulsifiable concentrate; wettable powder; granules; suspension concentrate; dustable powder; oil.

Toxicity: LC_{50} (96-hour) for rainbow trout 90 µg/L and bluegill sunfish 70 µg/L (Hartley and Kidd, 1987); EC50 (96-hour) for goldfish 0.5 mg/L; acute oral LD_{50} for rats 365–590 mg/kg (Hartley and Kidd, 1987), 343 mg/kg (RTECS, 1985).

Uses: Insecticide and fumigant.

cis-CHLORDANE

Synonyms: α-Chlordane; β-Chlordane; α-1,2,4,5,6,7,8,8-Octachloro-3a,4,7,7a-tetrahydro-4,7-methanoindan.

Designations: CAS Registry Number: 5103-74-2; DOT: 2762; mf: $C_{10}H_6Cl_8$; fw: 409.78; RTECS: PC0175000.

Properties: Solid with a chlorine-like odor. Mp: 107.0–108.8°C; bp: 175°C (isomeric mixture); H-t$_{1/2}$: >197,000 years at 25°C and pH 7; K$_H$: 8.75 × 10^{-4} atm · m^3/mol at 23°C; log BCF: freshwater clam (*Corbicula manilensis*) 3.68 (Hartley and Johnston, 1983); log K$_{oc}$: 4.55–4.94; log K$_{ow}$: 5.93 (calculated); S$_o$: cod liver oil (89.6 g/L at 4°C), *n*-octanol (56.6, 62.0 and 85.6 g/L at 4, 12 and 20°C, respectively), miscible with aliphatic and aromatic solvents; S$_w$: 51 μg/L at 20–25°C; vp: 3.6 × 10^{-5} mmHg at 25°C.

Soil properties and adsorption data

Soil	K$_d$ (mL/g)	f$_{oc}$ (%)	K$_{oc}$ (mL/g)	pH
Clay	162	0.30	54,000	—
Loam	1,380	2.03	68,000	—
Loess	166	0.47	35,310	—
Muck	24,915	24.57	87,096	—
Sand	91	0.30	30,400	—

Source: Erstfeld et al., 1996.

Environmental Fate

Groundwater. According to the U.S. EPA (1986) *cis*-chlordane has a high potential to leach to groundwater.

Photolytic. Irradiation of *cis*-chlordane by a 450-W high-pressure mercury lamp gave photo-*cis*-chlordane (Ivie et al., 1972).

Chemical/Physical. In an alkaline medium or solvent, carrier, diluent or emulsifier having an alkaline reaction, chlorine will be released (Windholz et al., 1983). Based on a reported hydrolysis half-life of >197,000 years at 25°C and pH 7 (Ellington et al., 1988), chemical hydrolysis is not expected to be an environmentally relevant fate process (Lyman et al., 1982).

Emits very toxic fumes of chlorides when heated to decomposition (Lewis, 1990).

Symptoms of Exposure: Blurred vision; confusion; ataxia; delirium; cough; abdominal pain, nausea, vomiting, diarrhea; irritability; tremor, convulsions, anuria.

Formulation Types: See Chlordane.

Toxicity: See Chlordane.

Use: Insecticide.

trans-CHLORDANE

Synonyms: α-Chlordan; *cis*-Chlordan; α-Chlordane; α(*cis*)-Chlordane; γ-Chlordane; 1,2,4,5,6,7,8,8-Octachloro-3a,4,7,7a-tetrahydro-4,7-methanoindan.

Designations: CAS Registry Number: 5103-71-9; DOT: 2762; mf: $C_{10}H_6Cl_8$; fw: 409.78; RTECS: PB9705000.

Properties: Solid with a chlorine-like odor. Mp: 103.0–105.0°C; bp: 175°C (isomeric mixture); K_H: 1.34×10^{-3} atm · m³/mol at 23°C; log BCF: freshwater clam (*Corbicula manilensis*) 3.64 (Hartley and Johnston, 1983); log K_{oc}: 4.67–5.04; log K_{ow}: 8.69, 9.65 (calculated); S_o: *n*-octanol (83.7, 112.9 and 142.1 g/L at 4, 12 and 20°C, respectively), miscible with aliphatic and aromatic solvents; vp: 2.78×10^{-5} mmHg at 25°C.

Soil properties and adsorption data

Soil	K_d (mL/g)	f_{oc} (%)	K_{oc} (mL/g)	pH
Muck	26,303	24.57	107,152	—
Loam	2,239	2.03	110,282	—
Loess	219	0.47	46,509	—
Clay	251	0.30	83,730	—
Sand	141	0.30	47,083	—

Source: Erstfeld et al., 1996.

Environmental Fate

Photolytic. Irradiation of *trans*-chlordane by a 450-W high-pressure mercury lamp gave photo-*trans*-chlordane (Ivie et al., 1972).

Chemical/Physical. In an alkaline medium or solvent, carrier, diluent or emulsifier having an alkaline reaction, chlorine will be released (Windholz et al., 1983).

Emits very toxic fumes of chlorides when heated to decomposition (Lewis, 1990).

Symptoms of Exposure: Blurred vision; confusion; ataxia; delirium; cough; abdominal pain, nausea, vomiting, diarrhea; irritability; tremor, convulsions, anuria.

Formulation Types: See Chlordane.

Toxicity: See Chlordane.

Use: Insecticide.

CHLORDIMEFORM

Synonyms: Acaron; Bermat; C 8514; Carzol; CDM; Chlorfenamidine; *N'*-(**4-Chloro-2-methylphenyl**)-*N,N*-**dimethylmethanimidamide**; Chlorophenamidin; Chlorophenamidine; *N'*-(4-Chloro-*o*-tolyl)-*N,N*-dimethylformamidine; Chlorphenamidine; CIBA 8514; CIBA C8514; *N,N*-Dimethyl-*N'*-(2-methyl-4-chlorophenyl)-formamidine; ENT 27335; ENT 27567; EP 333; Fundal; Fundal 500; Fundex; Galecron; NSC 190935; RS 141; Schering 36268; SN 36268; Spanon; Spanone.

$$Cl-\underset{CH_3}{\underset{|}{\bigcirc}}-N=CHN\overset{CH_3}{\underset{CH_3}{<}}$$

Designations: CAS Registry Number: 6164-98-3; mf: $C_{10}H_{13}ClN_2$; fw: 196.68; RTECS: LQ4375000.

Properties: Colorless to buff-colored crystals. Mp: 32.0°C; bp: 156–157°C at 0.4 mmHg; ρ: 1.105 at 25/4°C; K_H: 3.38 × 10⁻⁴ atm · m³/mol at 20°C (approximate — calculated from water solubility and vapor pressure); log K_{oc}: 2.30 (calculated); log K_{ow}: 1.80, 2.89; S_o: very soluble in acetone, benzene, chloroform, ethyl acetate, hexane, methanol; S_w: 203 and 270 mg/L at 10 and 20°C, respectively; vp: 3.5 × 10⁻⁴ mmHg at 20°C.

Environmental Fate

Plant. Principal soil and/or plants metabolites are *p*-chloro-*o*-toluidine, *N'*-(4-chloro-*o*-tolyl)-*N*-methylformamidine (desmethylchlorphenamidine) and *N*-formyl-*p*-chloro-*o*-toluidine (Verschueren, 1983). *p*-Chloro-*o*-toluidine, *N'*-(4-chloro-*o*-tolyl)-*N*-methylform-amidine (desmethylchlor phenamidine) and *N*-formyl-*p*-chloro-*o*-toluidine were identified in rice grains and straws at concentrations of 3–61, 0.2–1, 10–38 and 80–6,900, 10–180, 67–500 ppb, respectively (Iizuka and Masuda, 1979).

Witkonton and Ercegovich (1972) studied the transformation of chlordimeform in six different fruits following foliar spray application. They found 4'-chloro-*o*-formotoluidide was the only major metabolite identified in apples, pears, cherries, plums, strawberries and peaches.

Chemical/Physical. Reacts with acids forming soluble salts (Hartley and Kidd, 1987).

Emits toxic fumes of nitrogen oxides and chlorine when heated to decomposition (Sax and Lewis, 1987).

Symptoms of Exposure: Eye and skin irritant.

Formulation Types: Emulsifiable concentrate; water soluble powder.

Toxicity: LC$_{50}$ (96-hour) for rainbow trout 7.1 mg/L, bluegill sunfish 1.0 mg/L and barbel 4.5 mg/L (Hartley and Kidd, 1987); acute oral LD$_{50}$ for rats 340 mg/kg (Hartley and Kidd, 1987), 160 mg/kg (RTECS, 1985); 238 mg/kg (Windholz et al., 1983).

Uses: Broad-spectrum acaricide used against adult mites, eggs and larvae. Also used as an insecticide against cockroaches, cotton bollworm, cotton budworm, and other pests.

CHLOROBENZILATE

Synonyms: Acar; Acaraben; Acaraben 4E; Akar; Akar 50; Akar 338; Benzilan; Benz-*o*-chlor; Chlorbenzilat; Chlorbenzilate; Chlorobenzylate; Compound 338; 4,4'-Dichlorobenzilate; 4,4'-Dichlorobenzilic acid ethyl ester; ENT 18596; **Ethyl 4-chloro-α-(4-chlorophenyl)-α-hydroxybenzene acetate**; Ethyl-4,4-dichlorobenzilate; Ethyl-*p,p*'-dichlorobenzilate; Ethyl-4,4'-dichlorodiphenyl glycollate; Ethyl-*p,p*'-dichlorodiphenyl glycollate; Ethyl-2-hydroxy-2,2-bis(4-chlorophenyl)acetate; Folbex; Folbex smoke-strips; G 338; G 23,992; Geigy 338; Kop-mite; NCI-C00408; NCI-C60413; RCRA waste number U038.

Designations: CAS Registry Number: 510–15–6; mf: $C_{16}H_{14}Cl_2O_3$; fw: 325.21; RTECS: DD2275000.

Properties: Colorless to yellow, viscous liquid or crystals. Mp: 36–37.5°C; bp: 415°C; ρ: 1.2816 at 20/4°C; K_H: 3.85×10^{-8} atm · m³/mol at 20°C (approximate — calculated from water solubility and vapor pressure); log K_{oc}: 3.24–3.67; log K_{ow}: 4.58; So (kg/kg at 20°C): acetone (1.0), hexane (0.6), methanol (1.0), methylene chloride (1.0), octanol (0.70), toluene (1.0); S_w: 10 mg/L at 20°C; vp: 9×10^{-7} mmHg at 20°C.

Soil properties and adsorption data

Soil	K_d (mL/g)	f_{oc} (%)	K_{oc} (mL/g)	pH
Loam	48.4	2.78	1,741	5.9
Sand	16.4	0.35	4,713	7.7
Sandy loam	20.5	0.75	2,719	7.9
Sandy loam	48.6	2.32	2,095	7.8

Source: U.S. Department of Agriculture, 1990.

Environmental Fate

Biological. Rhodotorula gracilis, a yeast isolated from an insecticide-treated soil, degraded chlorobenzilate in a basal medium supplemented by sucrose. Metabolites identified by this decarboxylation process were 4,4'-dichlorobenzilic acid, 4,4'-dichlorobenzophenone and carbon dioxide (Miyazaki et al., 1969, 1970).

Soil. Though no products were identified, the half-life of chlorobenzilate in two fine sandy soils was estimated to be 1.5–5 weeks (Wheeler, 1973).

Photolytic. Chlorobenzilate should not undergo direct photolysis since it does not absorb UV light at wavelengths greater than 290 nm (Gore et al., 1971).

Chemical/Physical. Emits toxic fumes of chlorine when heated to decomposition (Sax and Lewis, 1987).

Symptoms of Exposure: Irritation of eyes and skin.

Formulation Types: Emulsifiable concentrate; wettable powder; fumigant.

Toxicity: LC_{50} (96-hour) for rainbow trout 0.6 mg/L and bluegill sunfish 1.8 mg/L (Hartley and Kidd, 1987); acute oral LD_{50} for rats 2,784–3,880 mg/kg (Hartley and Kidd, 1987), 700 mg/kg (RTECS, 1985).

Uses: Nonsystemic pesticide and acaricide.

CHLOROPICRIN

Synonyms: Acquinite; Chlor-o-pic; Dolochlor; G 25; Larvacide 100; Microlysin; NA 1583; NCI-C00533; Nitrochloroform; Nitrotrichloromethane; Pic-clor; Picfume; Picride; Profume A; PS; S 1; **Trichloronitromethane**; Tri-clor; UN 1580.

$$Cl_3CNO_2$$

Designations: CAS Registry Number: 76-06-2; DOT: 1580; mf: CCl_3NO_2; fw: 164.38; RTECS: PB6300000.

Properties: Colorless, slightly oily liquid with a sharp, penetrating, tear gas-like odor. Mp: −64.5°C and −69.2°C (corrected); bp: 111.8°C; ρ: 1.6558 at 20/4°C, 1.6483 at 25/4°C; fl p: detonates; K_H: 2.05×10^{-3} atm · m³/mol; log K_{oc}: 0.82 (calculated); log K_{ow}: 1.03, 2.09; P-$t_{1/2}$: 20 days (simulated atmosphere), 3 days in aqueous solution (sunlight irradiation); S_o: soluble in acetic acid and acetone; miscible with benzene, carbon disulfide, carbon tetrachloride, ethyl ether, methanol and ethanol; S_w: 2,270 mg/L at 25°C; vp: 16.9, 23.8 and 33 mmHg at 20, 25 and 30°C, respectively.

Environmental Fate

Biological. Four *Pseudomonas* sp., including *Pseudomonas putida* (ATCC culture 29607) isolated from soil, degraded chloropicrin by sequential reductive dechlori-nation. The proposed degradative pathway is chloropicrin → nitrodichloromethane → nitrochloromethane → nitromethane + small amounts of carbon dioxide. In addi-tion, a highly water soluble substance tentatively identified as a peptide, was produced by a nonenzymatic mechanism (Castro et al., 1983).

Photolytic. Photodegrades under simulated atmospheric conditions to phosgene and nitrosyl chloride. Photolysis of nitrosyl chloride yields chlorine and nitrous oxide (Moilanen et al., 1978; Woodrow et al., 1983). When aqueous solution of chloropicrin (1 mM) is exposed to artificial UV light (λ <300 nm), protons, carbon dioxide, hydrochloric and nitric acids are formed (Castro and Belser, 1981).

Chemical/Physical. Releases very toxic fumes of nitrogen oxides and chlorine when heated to decomposition (Sax and Lewis, 1987; Lewis, 1990).

Reacts with alcoholic sodium sulfite solutions and ammonia forming methane-trisulfonic acid and guanidine, respectively (Sittig, 1985).

Exposure Limits: NIOSH REL: TWA 0.1 ppm, IDLH 2 ppm; OSHA PEL: TWA 0.1 ppm; ACGIH TLV: TWA 0.1 ppm, STEL 0.3 ppm.

Symptoms of Exposure: Eye and skin irritation, lacrimation; cough, pulmonary edema; nausea, vomiting.

Formulation Types: Fumigant (mixture containing methyl bromide, methyl isothiocyanate, 1,2-dichloropropane and 1,3-dichloro-1-propene).

Toxicity: Acute oral LD_{50} for rats 250 mg/kg (RTECS, 1985).

Uses: Disinfecting cereals and grains; fumigant and soil insecticide; fungicide; rat exterminator.

CHLOROTHALONIL

Synonyms: Bravo; Bravo 6F; Bravo-W-75; Chloroalonil; DAC 2787; Daconil; Daco-nil 2787; Daconil 2787 flowable fungicide; Dacosoil; 1,3-Dicyanotetrachlorobenzene; Exotherm; Exotherm termil; Forturf; NCI-C00102; Nopcocide; Nopcocide *N*-96; Nopcocide N40D & N96; Sweep; TCIN; *m*-TCPN; Termil; **2,4,5,6-Tetra-chloro-1,3-benzenedicarbonitrile**; 2,4,5,6-Tetrachloro-3-cyanobenzonitrile; Tetrachloroisophthalonitrile; *m*-Tetrachlorophthalonitrile; TPN.

Designations: CAS Registry Number: 1897-45-6; mf: $C_8Cl_4N_2$; fw: 265.89; RTECS: NT2600000.

Properties: Colorless to white, odorless crystals. Mp: 250–251°C; bp: >350°C; ρ: 1.8 (pure), 1.7 at 25/4°C; K_H: 1.96×10^{-7} atm · m³/mol at 25°C; log K_{oc}: 2.76, 3.14; log K_{ow}: 2.64; So (g/kg at 25°C): acetone (20), cyclohexanone (30), *N,N*-dimethylformamide (30), dimethylsulfoxide (20), kerosene (<10), xylene (80); S_w: 600 µg/L at 25°C; vp: 5.70×10^{-7} mmHg at 25°C, 9.75×10^{-6} mmHg at 40°C.

Soil properties and adsorption data

Soil	K_d (mL/g)	f_{oc} (%)	K_{oc} (mL/g)	pH
Cohansey sand	25	2.55	980	—

Source: Reduker et al., 1988.

Environmental Fate

Biological. From the first-order biotic and abiotic rate constants of chlorothalonil in estuarine water and sediment/water systems, the estimated biodegradation half-lives were 8.1–10 and 1.8–5 days, respectively (Walker et al., 1988).

Soil. Metabolites identified in soil were 1,3-dicyano-4-hydroxy-2,5,6-trichlorobenzene, 1,3-dicarbamoyl-2,4,5,6-tetrachlorobenzene and 1-carbamoyl-3-cyano-4-hydroxy-2,5,6-trichlorobenzene (Rouchaud et al., 1988). The half-life was reported as 4.1 days (Gilbert, 1976) and 1.5-3 months (Hartley and Kidd, 1987).

Groundwater. According to the U.S. EPA (1986) chlorothalonil has a high potential to leach to groundwater.

Plant. Degrades in plants to 4-hydroxy-2,5,6-trichloroisophthalonitrile (Hartley and Kidd, 1987), 1,3-dicyano-4-hydroxy-2,5,6-trichlorobenzene and 1,3-dicarbamoyl-2,4,5,6-tetrachlorobenzene (Rouchaud et al., 1988). No evidence of degradation products were reported in apple foliage 15 days after application. The half-life of chlorothalonil was 4.1 days (Gilbert, 1976).

Chemical/Physical. Emits toxic fumes of nitrogen oxides, cyanides and chlorine when heated to decomposition (Sax and Lewis, 1987). Chlorothalonil is resistant to hydrolysis under acidic conditions. At pH 9, chlorothalonil (0.5 ppm) hydrolyzed to 4-hydroxy-2,5,6-trichloroisophthalonitrile and 3-cyano-2,4,5,6-tetrachlorobenzamide. Degradation followed first-order kinetics at a rate of 1.8% per day (Szalkowski and Stallard, 1977).

Exposure Limits: An experimental carcinogen.

Formulation Types: Wettable powder; suspension concentrate; fogging concentrate.

Toxicity: LC_{50} (96-hour) for bluegill sunfish 62 µg/L, rainbow trout 49 µg/L and channel catfish 430 µg/L (Hartley and Kidd, 1987); acute oral LD_{50} for albino rats >10,000 mg/kg (Verschueren, 1983).

Uses: Fungicide; bactericide; nematocide.

CHLOROXURON

Synonyms: C 1983; *N'*-**(4-(4-Chlorophenoxy)phenyl)-*N*,*N*-dimethylurea**; 3-(*p*-(*p*-Chlorophenoxy)phenyl)-1,1-dimethylurea; Chloroxifenidim; CIBA 1983; Norex; Tenoran 50W.

$$Cl-\langle\bigcirc\rangle-O-\langle\bigcirc\rangle-NHCON(CH_3)_2$$

Designations: CAS Registry Number: 1982-47-4; mf: $C_{15}H_{15}ClN_2O_2$; fw: 290.75; RTECS: UG1490000.

Properties: Odorless, colorless powder or white crystals. Mp: 151–152°C; ρ: 1.34 at 20/4°C; fl p: nonflammable; log K_{oc}: 3.12–3.27; log K_{ow}: 3.20; So (g/kg): acetone (44), methanol (35), methylene chloride (106), toluene (4); S_w: 3.7 mg/L at 20°C; vp: 1.79 × 10^{-11} mmHg at 20°C.

Soil properties and adsorption data

Soil	K_d (mL/g)	f_{oc} (%)	K_{oc} (mL/g)	pH (meq/100 g)	CEC
Great House SP	475	12.00	3,958	6.3	18.0
Loye clay loam	40	2.55	1,569	7.5	—
Rosemaunde sandy clay loam	70	1.76	3,977	6.7	14.0
Toll Farm HP	330	11.70	2,820	7.4	41.0
Transcosed silty clay loam	175	3.69	4,742	6.2	—
Uvrier sandy loam	14	0.75	1,857	7.8	—
Vétroz humus soil	110	8.29	1,327	—	—
Weed Res. sandy loam	120	1.93	6,218	7.1	11.0

Source: Geissbühler et al., 1963; Hance, 1965.

Environmental Fate

Soil. Hartley and Kidd (1987) reported 4-(4-chlorophenoxy)aniline as a soil metabolite. Chloroxuron was degraded by microorganisms in humus soil and a sandy loam to form *N'*-(4-chlorophenoxy)phenyl-*N*-methylurea, *N'*-(4-chloro-phenoxy)phenylurea and (4-chlorophenoxy)aniline and two minor unidentified compounds (Geissbühler et al., 1963a). Residual activity in soil is limited to approximately 4 months (Hartley and Kidd, 1987).

Plant. In plants, chloroxuron is degraded to monomethylated and demethylated derivatives followed by decarboxylation forming 4-(4-chlorophenoxy)aniline (Humburg et al., 1989).

Photolytic. The UV irradiation of an aqueous solution of chloroxuron for 13 hours resulted in 90% decomposition of the herbicide. Products identified (% yield) were mono- (2.2%) and didemethylated (4.2%) products and carbon dioxide (64%) (Plimmer, 1970).

Chemical/Physical. Hydrolyzes in strong acids and bases forming 4-(4-chloro-phe-noxy)aniline (Hartley and Kidd, 1987). Emits toxic fumes of nitrogen oxides, cyanides and chlorine when heated to decomposition (Sax and Lewis, 1987).

Symptoms of Exposure: Ingestion may cause depression, locomotion, hyperpnea, gasping, coma, death.

Formulation Types: Wettable powder.

Toxicity: LC_{50} (96-hour) for rainbow trout >100 mg/L, bluegill sunfish 28 mg/L and crucian carp >150 mg/L (Hartley and Kidd, 1987); LC_{50} (48-hour) for bluegill sunfish 25.0 mg/L (Verschueren, 1983); acute oral LD_{50} for male and female rats 3,700 and 5,400 mg/kg, respectively (Hartley and Kidd, 1987).

Uses: Postemergence herbicide used to control most annual grasses and broad-leaved weeds.

CHLORPROPHAM

Synonyms: Beet-Kleen; Bud-nip; Chlor-IFC; Chlor-IPC; Chloro-IFK; Chloro-IPC; **(3-Chlorophenyl)carbamic acid 1-methylethyl ester**; Chloropropham; CICP; CI-IPC; CIPC; Ebanil; ENT 18060; Fasco Wy-hoe; Furloe; Furloe 4EC; Isopropyl *m*-chlorocarbanilate; Isopropyl-3-chlorocarbanilate; Isopropyl-3-chlorophenylcarbamate; Isopropyl-*N*-(3-chlorophenyl)carbamate; Isopropyl-*N*-(*m*-chlorophenyl)carbamate; *O*-Isopropyl-*N*-(3-chlorophenyl)carbamate; Liro CIPC; Metoxon; 1-Methylethyl-3-chlorophenylcarbamate; Nexoval; Prevenol; Prevenol 56; Preventol; Preventol 56; Preweed; Sprout nip; Sprout-nip EC; Spud-nic; Spud-nie; Stopgerme-S; Taterpex; Triherbicide CIPC; Unicrop CIPC; Y 3.

$$NHCOOCH(CH_3)_2$$

Designations: CAS Registry Number: 101-21-3; mf: $C_{10}H_{12}ClNO_2$; fw: 213.67; RTECS: FD8050000.

Properties: Colorless to pale brown crystals with a faint characteristic odor. Mp: 40.7–41.4°C; bp: 247°C (decomposes); ρ: 1.180 at 30/4°C; fl p: high; H-$t_{1/2}$: >10,000 days (calculated assuming pseudo-first-order kinetics); K_H: 2.1 × 10^{-8} atm · m^3/mol at 20–25°C (approximate — calculated from water solubility and vapor pressure); log K_{oc}: 1.65–2.98; P-$t_{1/2}$: 121 days at pH 5-9 (calculated); S_o: very soluble in ethanol, 2-propanol and ketones but miscible with acetone and carbon disulfide; S_w: 89 mg/L at 25°C; vp: 10^{-5} mmHg at 25°C (estimated).

Soil properties and adsorption data

Soil	K_d (mL/g)	f_{oc} (%)	K_{oc} (mL/g)	pH (meq/100 g)	CEC
Ascalon sandy clay loam	1.9	0.86	221	7.3	12.7
Barnes clay loam	8.9	4.00	223	7.4	33.8
Beltsville silt loam	6.6	1.40	471	4.3	4.2
Benevola clay (subsoil)	2.3	1.31	176	7.6	20.1
Benevola silty clay (topsoil)	5.6	2.71	206	7.7	19.5
Berkley clay (subsoil)	2.1	4.65	45	7.3	34.4
Berkley silty clay (topsoil)	11.7	4.65	252	7.1	33.7
Bosket silt loam	1.4	0.57	246	5.8	8.4
Cecil sandy clay loam	3.6	1.10	327	5.3	3.6
Chester loam	6.9	1.68	411	4.9	5.2
Chillum silt loam	9.7	2.55	380	4.6	7.6
Christiana loam	1.4	0.57	246	4.4	5.6
Crosby silt loam	7.5	1.91	393	4.8	11.5
Dundee silty clay loam	3.1	0.97	320	5.0	18.1

Soil properties and adsorption data *(continued)*

Soil	K_d (mL/g)	f_{oc} (%)	K_{oc} (mL/g)	pH (meq/100 g)	CEC
Garland clay	1.4	0.66	212	7.7	23.2
Hagerstown silty clay loam	7.8	2.50	312	5.5	12.5
Hagerstown silty clay loam	3.1	1.31	237	7.5	8.8
Iredell clay (subsoil)	1.6	0.62	258	5.6	20.9
Iredell silt loam (topsoil)	12.9	3.06	422	5.4	17.0
Lakeland sandy loam	2.1	1.89	111	6.2	2.9
Montalto clay (subsoil)	0.3	0.87	345	5.9	8.4
Norfolk sandy loam	0.4	0.08	500	5.1	0.2
Ruston sandy loam	2.8	1.06	264	5.1	3.4
Sharkey clay	6.6	2.26	292	6.2	40.2
Soil (0–6 inches)	37.0	4.90	755	—	—
Soil (6–12 inches)	26.0	2.70	962	—	—
Soil (12–18 inches)	12.0	1.30	923	—	—
Soil (18–24 inches)	5.0	0.80	625	—	—
Sterling clay loam	2.1	0.95	221	7.7	22.5
Thurlow clay loam	4.3	1.26	341	7.7	21.6
Tifton loamy sand	1.8	0.57	316	4.9	2.4
Toledo silty clay	8.9	2.81	317	5.5	29.8
Tripp loam	2.2	0.86	256	7.6	14.7
Truckton sandy loam	0.8	0.26	308	7.0	4.4
Wehadkee silt loam	2.5	1.12	223	5.6	10.2
Wooster silt loam	3.3	1.32	250	4.7	6.8

Source: Harris and Sheets, 1965; Roberts and Wilson, 1965.

Environmental Fate

Soil. Hydrolyzes in soil forming 3-chloroaniline (Bartha, 1971; Hartley and Kidd, 1987; Smith, 1988; Rajagopal et al., 1989). In soil, *Pseudomonas striata* Chester, a *Flavobacterium* sp., an *Agrobacterium* sp. and an Achromobacter sp. readily degraded chlorpropham to 3-chloroaniline and 2-propanol. Subsequent degradation by enzymatic hydrolysis yielded carbon dioxide, chloride ions and unidentified compounds (Kaufman, 1967; Rajagopal et al., 1989). Hydrolysis products that may form in soil and in microbial cultures include N-phenyl-3-chlorocarbamic acid, 3-chloroaniline, 2-amino-4-chlorophenol, monoisopropyl carbonate, 2-propanol, carbon dioxide and condensation products (Rajagopal et al., 1989). The reported half-lives in soil at 15 and 29°C are 65 and 30 days, respectively (Hartley and Kidd, 1987).

Plant. Chlorpropham is rapidly metabolized in plants (Ashton and Monaco, 1991). Metabolites identified in soybean plants include isopropyl-N-4-hydroxy-3-chlorophenyl-carbamate, 1-hydroxy-2-propyl-3'-chlorocarbanilate and isopropyl-N-5-chloro-2-hydrox-yphenylcarbamate (Humburg et al., 1989). Isopropyl-N-4-hydroxy-3-chlorophenylcarbam-ate was the only metabolite identified in cucumber plants (Humburg et al., 1989).

Photolytic. The photodegradation rate of chlorpropham in aqueous solution was enhanced in the presence of a surfactant (TMN-10) (Tanaka et al., 1981). In a later study, Tanaka et al. (1985) studied the photolysis of chlorpropham (50 mg/L) in aqueous solution using UV light (λ = 300 nm) or sunlight. After 10 hours of ir-radiation, 40% degraded yielding <1% of a hydroxylated biphenyl product (2'-hy-droxy-3,4'-biphenylcarbamic acid diisopropyl ester) and hydrogen chloride. The biphenyl product probably formed from the coupling of photoexcited chlorpropham with the intermediate compound isopropyl 3-hydroxycarbanilate (Tanaka et al., 1985).

Chemical/Physical. Emits toxic phosgene fumes when heated to decomposition (Sax and Lewis, 1987). In a 0.50 N sodium hydroxide solution at 20°C, chlorpropham was hydrolyzed to aniline derivatives. The half-life of this reaction was 3.5 days (El-Dib and Aly, 1976). Simple hydrolysis leads to the formation of 3-chlorophenylcarbamic acid and 2-propanol. The acid is very unstable and is spontaneously converted to 3-chloroaniline and carbon dioxide (Still and Herrett, 1976).

Exposure Limits: An experimental carcinogen and neoplastigen.

Formulation Types: Fogging concentrate; emulsifiable concentrate (3 and 4 lb/gal); granules (10 and 20%); dustable powder.

Toxicity: LC_{50} (48-hour) for bass 10 mg/L (Hartley and Kidd, 1987) and bluegill sunfish 8 mg/L (Verschueren, 1983); acute oral LD_{50} for rats 1,200 mg/kg (RTECS, 1985).

Uses: Preemergent and postemergent herbicide used to regulate plant growth and control of weeds in carrot, onion, garlic and other crops.

CHLORPYRIFOS

Synonyms: Brodan; Chlorpyrifos-ethyl; Detmol U.A.; *O,O*-Diethyl-*O*-3,5,6-tri-chloro-2-pyridyl phosphorothioate; Dowco 179; Dursban; Dursban F; ENT 27311; Eradex; Lorsban; NA 2783; OMS 971; **Phosphorothionic acid *O,O*-diethyl *O*-(3,5,6-trichloro-2-pyridyl)ester**; Pyrinex.

Designations: CAS Registry Number: 2921-88-2; DOT: 2783; mf: $C_9H_{11}Cl_3NO_3PS$; fw: 350.59; RTECS: TF6300000.

Properties: White to amber granular crystals with a mild mercaptan-like odor. Mp: 41.5–43.5°C; bp: begins to decompose at 160°C; ρ: 1.398 at 43.5/4°C; H-$t_{1/2}$: 22.8 days (pH 8.1), 35.3 days (pH 6.9), 62.7 days (pH 4.7); K_H: 4.16×10^{-6} atm · m^3/mol at 25°C; log K_{oc}: 3.77–4.13; log K_{ow}: 5.2 (Schimmel et al., 1983); P-$t_{1/2}$: 52.45 hours (absorbance $\lambda = 229.5$ nm, concentration on glass plates = 6.7 µg/cm^2), 11.0, 12.2 and 7.8 days in buffered, distilled water (9×10^{-7} M, 25°C) at pH 5.0, 6.9 and 8.0, respectively; So (kg/kg): acetone (6.5), benzene (7.9), chloroform (6.3), 2,2,4-trimethylpentane (isooctane) (79), methanol (43), ethanol, *n*-propanol, xylene and many other organic solvents; S_w: 450, 730 and 1,300 µg/L at 10, 20 and 30°C, respectively; vp: 1.9×10^{-5} mmHg at 25°C (Melnikov, 1971).

Soil properties and adsorption data

Soil	K_d (mL/g)	f_{oc} (%)	K_{oc} (mL/g)	pH (meq/100 g)	CEC
Catlin	99.70	2.01	5,000	6.20	—
Clarion soil	161.81	2.64	6,119	5.00	21.02
Commerce	49.50	0.68	7,300	6.70	—
Harps soil	397.19	3.80	10,450	7.30	37.83
Kanuma high clay	13.40	1.35	995	5.70	—
Sarpy fine sandy loam	28.38	0.51	5,560	7.30	5.71
Thurman loamy fine sand	46.88	1.07	4,395	6.83	6.10
Tracy	65.60	1.12	5,900	6.20	—
Tsukuba clay loam	116.20	4.24	2,740	6.50	—

Source: Felsot and Dahm, 1979; McCall et al., 1981; Kanazawa, 1989.

Environmental Fate

Biological. From the first-order biotic and abiotic rate constants of chlorpyrifos in estuarine water and sediment/water systems, the estimated biodegradation half-lives were 3.5–41 and 11.9–51.4 days, respectively (Walker et al., 1988).

Soil. Hydrolyzes in soil to 3,5,6-trichloro-2-pyridinol (Somasundaram et al., 1991). The half-lives in a silt loam and clay loam were 12 and 4 weeks, respectively (Getzin, 1981). In another study, Getzin (1981a) reported the hydrolysis half-lives in a Sultan silt loam at 5, 15, 25, 35 and 45°C were >20, >20, 8, 3 and 1 day, respectively. The only breakdown product identified was the hydrolysis product 3,5,6-trichloro-2-pyridinol. Degrades in soil forming oxychlorpyrifos, 3,5,6-trichloro-2-pyridinol (hydrolysis product), carbon dioxide, soil-bound residues and water-soluble products (Racke et al., 1988).

Leoni et al. (1981) reported that the major degradation product of chlorpyrifos in soil is 3,5,6-trichloro-2-pyridinol. The major factors affecting the rate of degradation include chemical hydrolysis in moist soils, clay-catalyzed hydrolysis on dry soil surfaces, microbial degradation and volatility (Davis and Kuhr, 1976; Felsot and Dahm, 1979; Miles et al., 1979; Chapman and Harris, 1980; Getzin, 1981; Chapman et al., 1984; Miles et al., 1983, 1984; Getzin, 1985; Chapman and Chapman, 1986). Getzin (1981) reported that catalyzed hydrolysis and microbial degradation were the major factors of chlorpyrifos disappearance in soil. The reported half-lives in sandy and muck soils were 2 and 8 weeks, respectively (Chapman and Harris, 1980).

Plant. The half-life of chlorpyrifos in Bermuda grasses was 2.9 days (Leuck et al., 1975). The concentration and the formulation of application of chlorpyrifos will determine the rate of evaporation from leaf surfaces. Reported foliar half-lives on tomato, orange and cotton leaves were 15–139, 1.4–96 and 5.5–57 hours, respectively (Veierov et al., 1988).

Dislodgable residues of chlorpyrifos on cotton leaf 0, 24, 48, 72 and 96 hours after application (1.1 kg/ha) were 3.64, 0.13, 0.071, 0.055 and 0.034 µg/m^2, respectively (Buck et al., 1980).

Surface Water. In an estuary, the half-life of chlorpyrifos was 24 days (Schimmel et al., 1983).

Photolytic. Photolysis of chlorpyrifos in water yielded 3,5,6-trichloro-2-pyridinol. Continued photolysis yielded chloride ions, carbon dioxide, ammonia and possibly poly-hydroxychloropyridines. The following photolytic half-lives in water at north 40° latitude were reported: 31 days during midsummer at a depth of 10^{-3} cm; 345 days during midwinter at a depth of 10^{-3} cm; 43 days at a depth of one meter; 2.7 years during midsummer at a depth of one meter in river water (Dilling et al., 1984). The combined photolysis-hydrolysis products identified in buffered, distilled water were *O*-ethyl *O*-(3,5,6-trichloro-2-pyridyl)phosphorothioate, 3,5,6-trichloro-2-pyridinol, and five radioactive unknowns (Meikle et al., 1983).

Chemical/Physical. Hydrolyzes in water forming 3,5,6-trichloro-2-pyridinol, *O*-ethyl *O*-hydrogen-*O*-(3,5,6-trichloro-2-pyridyl)phosphorothioate and *O,O*-dihydrogen-*O*-(3,5,6-trichloro-2-pyridyl)phosphorothioate. Reported half-lives in buffered distilled water at 25°C at pH values of 8.1, 6.9 and 4.7 are 22.8, 35.3 and 62.7 days, respectively (Meikle and Youngson, 1978). The hydrolysis half-life in three different natural waters was approximately 48 days at 25°C (Macalady and Wolfe, 1985). Freed et al. (1979) reported hydrolysis half-lives of 120 and 53 days at pH 6.1 and pH 7.4, respectively, at 25°C. At 25°C and a pH range of 1–7, the hydrolysis half-life was about 78 days (Macalady and Wolfe, 1983). However, the alkaline hydrolysis rate of chlorpyrifos in the sediment-sorbed phase was found to be considerably slower (Macalady and Wolfe, 1985). Over the pH range of 9–13, 3,5,6-trichloro-2-pyridinol and *O,O*-diethyl phosphorothioic acid formed as major hydrolysis products (Macalady and Wolfe, 1983).

The hydrolysis half-lives of chlorpyrifos in a sterile 1% ethanol/water solution at 25°C and pH values of 4.5, 5.0, 6.0, 7.0 and 8.0, were 11, 11, 7.0, 4.2 and 2.7 weeks, respectively (Chapman and Cole, 1982).

Chlorpyrifos is stable to hydrolysis over the pH range of 5–6 (Mortland and Raman, 1967). However, in the presence of a Cu(II) salt (as cupric chloride) or when present as the exchangeable Cu(II) cation in montmorillonite clays, chlorpyrifos is completely hydrolyzed via first-order kinetics in <24 hours at 20°C. It was suggested that chlorpyrifos decomposition in the presence of Cu(II) was a result of coordination of molecules to the copper atom with subsequent cleavage of the side chain containing the phosphorus atom forming 3,5,6-trichloro-2-pyridinol and *O,O*-ethyl-*O*-phosphorothioate (Mortland and Raman, 1967).

Emits very toxic fumes of chlorides and oxides of nitrogen, phosphorus and sulfur when heated to decomposition (Lewis, 1990).

Exposure Limits: OSHA PEL: TWA 0.2 mg/m^3, STEL 0.6 mg/m^3; ACGIH TLV: TWA 0.2 mg/m^3, STEL 0.6 mg/m^3.

Formulation Types: Emulsifiable concentrate; wettable powder; granules; suspension concentrate; dustable powder; pellets.

Toxicity: LC$_{50}$ (96-hour) for rainbow trout 3 μg/L (Hartley and Kidd, 1987), estuarine mysid 0.035 μg/L, sheapshead minnow 136 μg/L, longnose killifish 4.1 μg/L, Atlantic silverside 1.7 μg/L, striped mullet 5.4 μg/L (Schimmel et al., 1983) and bluegill sunfish 2.6 μg/L (Verschueren, 1983); acute oral LD$_{50}$ for male and female rats is 163 and 135 mg/kg, respectively (Verschueren, 1983), 82 mg/kg (RTECS, 1985).

Uses: Chlorpyrifos is an organophosphorus insecticide used to control insects on a wide variety of crops including fruits, vegetables, ornamentals and forestry.

CHLORSULFURON

Synonyms: **2-Chloro-*N*-(((4-methoxy-6-methyl-1,3,5-triazin-2-yl)amino)carbo-nyl)benzenesulfonamide**; 1-((*o*-Chlorophenyl)sulfonyl)-3-(4-methoxy-6-methyl-*s*-tri-azin-2-yl) urea; DPX 4189; Finesse; Glean; Glean 20DF; Telar.

Designations: CAS Registry Number: 64902-72-3; mf: $C_{12}H_{12}ClN_5O_4S$; fw: 357.80; RTECS: YS6640000.

Properties: Odorless, colorless to white crystals. Mp: 174–178°C; bp: decomposes at 192°C; fl p: nonflammable; pK_a: 3.6 at 25°C and pH 7; H-$t_{1/2}$: 4–8 weeks at 20°C and pH 5.7–7.0; K_H: 3.55 × 10^{-16} atm · m^3/mol at 25°C (approximate — calculated from water solubility and vapor pressure); log K_{oc}: 1.02 (Flanagan silt loam); log K_{ow}: 0.74 at pH 5, −1.34 at pH 7 (Beyer et al., 1988); pK_a: 3.6 (Beyer et al., 1988); So (g/L at 22°C): acetone (57), hexane (0.01), methanol (14), methylene chloride (102), toluene (3); Sw at 25°C: 60 mg/L at pH 5, 7,000 mg/L at pH 7; vp: 2.33 × 10^{-11} mmHg at 25°C.

Soil properties and adsorption data

Soil	K_d (mL/g)	f_{oc} (%)	K_{oc} (mL/g)	pH (meq/100 g)	CEC
Acredale silt loam	2.4	1.42	169	4.6	234.0
Cullen clay loam	1.1	0.24	458	5.6	13.9
Kenansville loamy sand	1.0	0.16	625	6.9	3.2
Roanoke sandy loam	2.0	0.35	571	6.4	10.6

Source: Mersie and Foy, 1986.

Environmental Fate

Soil. Degrades in soil via hydrolysis followed by microbial degradation forming low molecular weight, inactive compounds. The estimated half-life was reported to range from 4 to 6 weeks (Hartley and Kidd, 1987; Cremlyn, 1991). Microorganisms capable of degrading chlorsulfuron are *Aspergillis niger, Streptomyces griseolus* and *Penicillium* sp. (Humburg et al., 1989). One transformation product reported in field soils is 2-chloroben-zenesulfonamide (Smith, 1988).

The reported dissipation rate of chlorsulfuron in surface soil is 0.024/day (Walker and Brown, 1983). The persistence of chlorsulfuron decreased when soil temperature and moisture were increased (Walker and Brown, 1983; Thirunarayanan et al., 1985).

Plant. Chlorsulfuron is metabolized by plants to hydroxylated, nonphytotoxic com-pounds including 2-chloro-*N*-(((4-methoxy-6-methyl-1,3,5-triazin-2-yl)-amino)carbo-nyl)benzenesulfonamide (Duke et al., 1991). Devine and Born (1985) and Peterson and

111

Swisher (1985) reported that the uptake of chlorsulfuron in Canada thistle leaves ranged from 23 to 43% after 2 days. The uptake in roots is higher under slightly acidic conditions. Fredrickson and Shea (1986) reported 12% of ^{14}C-chlor-sulfuron was taken up in the roots at soil pH 5.9.

Photolytic. The reported photolysis half-lives of chlorsulfuron in distilled water, methanol and natural creek water at λ >290 nm were 18, 92 and 18 hours, respectively. In all cases, 2-chlorobenzene sulfonamide, 2-methoxy-4-methyl-6-amino-1,3,5-triazine and trace amounts of the tentatively identified compound nitroso-2-chlorophenylsulfone formed as photoproducts (Herrmann et al., 1985).

Symptoms of Exposure: May cause eye, nose, throat and skin irritation.

Formulation Types: Water dispersible granules; dry flowable formulation (60%).

Toxicity: LC_{50} (96-hour) for rainbow trout, carp and bluegill sunfish >250 mg/L (Hartley and Kidd, 1987); acute oral LD_{50} for male and female rats 5,545 and 6,293 mg/kg, respectively (Hartley and Kidd, 1987).

Uses: Triazine urea herbicide used to control broad-leaved weeds and some annual grass weeds.

CHLORTHAL-DIMETHYL

Synonyms: Chlorothal; DAC 893; Dacthal; Dacthalor; DCPA; Dimethyl 2,3,5,6-tetra-chloro-1,4-benzenedicarboxylate; Dimethyl tetrachloroterephthalate; Dimethyl 2,3,5,6-tet-rachloroterephthalate; Fatal; **2,3,5,6-Tetrachloro-1,4-benzenedicar-boxylic acid dimethyl ester**; Tetrachloroterephthalic acid dimethyl ester; 2,3,5,6-Tetrachloroterephthalic acid dimethyl ester.

$$
\begin{array}{c}
\text{COOCH}_3 \\
\text{Cl} \quad \quad \text{Cl} \\
\\
\text{Cl} \quad \quad \text{Cl} \\
\text{COOCH}_3
\end{array}
$$

Designations: CAS Registry Number: 1861-32-1; mf: $C_{10}H_6Cl_4O_4$; fw: 331.99; RTECS: WZ1500000.

Properties: Colorless to beige crystals with a slight aromatic odor. Mp: 156°C; bp: decomposes at approximately 360–370°C; ρ: 1.70 at 20/4°C; fl p: nonflammable; K_H: 2.2 × 10^{-6} atm · m³/mol at 25°C (approximate — calculated from water solu-bility and vapor pressure); log K_{oc}: 3.81 (calculated); log K_{ow}: 4.87 (calculated); So (wt %): acetone (10), benzene (25), carbon tetrachloride (7), 1,4-dioxane (12), toluene (17), xylene (14); S_w: ≈ 500 µg/L at 25°C; vp: 2.5 × 10^{-6} mmHg at 25°C.

Environmental Fate

Soil. Bartha and Pramer (1967) reported that DCPA was degraded by soil microorgan-isms via cleavage of the herbicide molecule into propionic acid and 3,4-dichloroaniline. The acid was mineralized to carbon dioxide and water and two molecules of 3,4-dichlo-roaniline were condensed to form 3,3'4,4'-tetrachloro-azobenzene. Metabolites identified in soil and turfgrass thatch are monomethyl tetrachloroterephthalate and 2,3,5,6-tetrachlo-roterephthalic acid (Hartley and Kidd, 1987; Krause and Niemczyk, 1990). Residual activity in soil and the half-life in soil were reported to be approximately 3 months (Hartley and Kidd, 1987; Worthing and Hance, 1991).

Formulation Types: Granules; wettable powder (75%); suspension concentrate.

Toxicity: Nontoxic to fish (Hartley and Kidd, 1987); acute oral LD$_{50}$ for rats >3,000 mg/kg (Hartley and Kidd, 1987).

Uses: Selective, nonsystemic, preemergent herbicide to control most annual grasses and many broad-leaved weeds.

CROTOXYPHOS

Synonyms: Ciodrin; Ciovap; Cyodrin; Cypona E.C.; **3-((Dimethoxyphosphinyl)oxy)-2-butenoic acid 1-phenylethyl ester**; *O,O*-Dimethyl *O*-(1-methyl-2-carboxy-α-phenylethyl)vinyl phosphate; Decrotox; Dimethyl-*cis*-1-methyl-2-(1-phenylethoxycarbonyl)vinyl phosphate; Dimethyl phosphate of α-methylbenzyl 3-hydroxy-*cis*-crotonate; Duo-kill; Duravos; ENT 24717; 1-Methylbenzyl-3-(dimethoxyphosphinyloxo)isocrotonate; α-Methylbenzyl-3-(dimethoxyphosphinyloxy)-*cis*-crotonate; α-Methylbenzyl 3-hydroxycrotonate dimethyl phosphate; Pantozol 1; *cis*-2-(1-Phenylethoxy)carbonyl-1-methylvinyl dimethyl phosphate; SD 4294; Shell SD 4294; Volfazol.

Designations: CAS Registry Number: 7700-17-6; mf: $C_{14}H_{19}O_6P$; fw: 314.28; RTECS: GQ5075000.

Properties: Pale straw-colored liquid. Mp: <25°C; bp: 135°C at 0.03 mmHg; ρ: 1.19 at 25/4°C; K_H: 6.2 × 10^{-9} atm · m³/mol at 20–25°C (approximate — calculated from water solubility and vapor pressure); log K_{oc}: 2.23; log K_{ow}: 1.28 (calculated); S_o: soluble in acetone, chloroform, ethanol, chlorinated hydrocarbons; S_w: 1 g/L at 25°C; vp: 1.4 × 10^{-5} mmHg at 20°C.

Environmental Fate

Chemical/Physical. Emits toxic fumes of phosphorus oxides when heated to decomposition.

Toxicity: LC_{50} (96-hour) for bluegill sunfish 250 μg/L, largemouth bass 1.10 mg/L, rainbow trout 55 μg/L and channel catfish 2.50 mg/L (Verschueren, 1983); acute oral LD_{50} for rats approximately 125 mg/kg (Verschueren, 1983), 74 mg/kg (RTECS, 1985).

Uses: Insecticide.

CYANAZINE

Synonyms: Bladex; Bladex 90DF; Bladex 4L; Bladex 80WP; 2-Chloro-4-(1-cyano-1-methylethylamino)-6-ethylamino-1,3,5-triazine; 2-(4-Chloro-6-ethylamino-s-triazine-2-ylamino)-2-methylpropionitrile; 2-((4-Chloro-6-(ethylamino)-s-triazin-2-yl)amino)-2-methylpropionitrile; **2-((4-Chloro-6-(ethylamino)-1,3,5-triazin-2-yl)amino)-2-methyl-propanenitrile**; DW 3418; Fortrol; Payze; SD 15418; WL 19805.

Designations: CAS Registry Number: 21725-46-2; mf: $C_9H_{13}ClN_6$; fw: 240.70; RTECS: UG1490000.

Properties: Colorless to white crystals. Mp: 167.5–169°C; ρ: 0.35 g/mL (fluffed technical material), 0.45 g/mL (packed technical material); fl p: nonflammable; pK_a: 0.63, 1.1; K_H: 2.78×10^{-2} atm · m³/mol at 20–25°C (approximate — calculated from water solubility and vapor pressure); log K_{oc}: 1.58–2.63; log K_{ow}: 2.04 (Liu and Qian, 1995); So (g/L at 25°C): acetone (195), benzene (15), carbon tetra-chloride (<10), chlorobenzene (<100), chloroform (210), ethanol (45), hexane (15), methylcyclohexanone (210), methylene chloride (145), xylene (<100); S_w: 171 mg/L at 25°C; vp: 1.6×10^{-9} mmHg at 20°C.

Soil properties and adsorption data

Soil	K_d (mL/g)	f_{oc} (%)	K_{oc} (mL/g)	pH (meq/100 g)	CEC
Ginseng field soil I	0.65	5.81	78	8.2	16.0
Ginseng field soil II	0.92	5.15	161	8.3	15.6
Ginseng field soil III	0.63	3.23	133	7.4	19.9
Ginseng field soil IV	0.72	3.30	70	7.8	10.6
Hickory Hill silt	5.98	3.27	183	—	—
Monona clay loam	4.30	1.67	257	6.5	21.2
Rhinebeck silty clay loam	1.20	3.13	38	6.7	—
Valentine loamy fine sand	3.40	0.80	425	6.6	10.1
Valois silt loam	1.20	1.64	73	5.9	—

Source: Majka and Lavy, 1977; Brown and Flagg, 1981; Gamerdinger et al., 1991; Liu and Qian, 1995.

Environmental Fate

Soil. In sandy loam soils ($f_{oc} = 0.01$), the half-life is 12–15 days. However, in silt loam ($f_{oc} = 0.028$) and clay loam ($f_{oc} = 0.03$) soils, the reported half-life is 20–25 days (Humburg et al., 1989). The half-life of cyanazine is longer in alkaline soils (pH >7.5) than in acidic soils (pH <5.5) (Humburg et al., 1989).

Groundwater. According to the U.S. EPA (1986) cyanazine has a high potential to leach to groundwater. Cyanazine amide was identified as a metabolite in groundwater in corn fields (Muir and Baker, 1976).

Plant. In plants, cyanazine is degraded by elimination of the ethyl group, hydration of the cyano group and the removal and replacement of the chlorine atom by a hydroxyl group (Humburg et al., 1989).

Chemical/Physical. In laboratory tests, the nitrile group was hydrolyzed to the corresponding carboxylic acid. The rate of hydrolysis is faster under higher temperatures and low pHs (Grayson, 1980). The chlorine atom may be replaced by a hydroxyl group forming 2((4-hydroxy-6-(ethylamino)-1,3,5-triazin-2-yl)amino)-2-methylpropanenitrile (Hartley and Kidd, 1987).

Formulation Types: Wettable powder; granules; suspension concentrate.

Toxicity: LC_{50} (48-hour) for sheepshead minnow 18 mg/L and harlequin fish 10 mg/L (Worthing and Hance, 1991); acute oral LD_{50} for rats 182–334 mg/kg (Hartley and Kidd, 1987), 149 mg/kg (RTECS, 1985).

Uses: Herbicide used for control of annual grasses and broad-leaved weeds in cereals, cotton, maize, onions, peanuts, peas, potatoes, soybeans, sugarcane and wheat fallow.

CYCLOATE

Synonyms: *S*-**Ethyl cyclohexylethylcarbamothioate**; *S*-Ethyl *N*-ethyl *N*-cyclohexylthi-olcarbamate; Eurex; Hexylthiocarbam; R 2063; Ro-neet; Ronit.

$$NCOSCH_2CH_3$$
$$CH_2CH_3$$

Designations: CAS Registry Number: 1134-23-2; mf: $C_{11}H_{21}NOS$; fw: 215.37; RTECS: GU7200000.

Properties: Clear, colorless liquid with an aromatic odor. Mp: 11.5°C; bp: 145–146°C at 10 mmHg; ρ: 1.0156 at 30/4°C; fl p: 139°C (open cup); K_H: 2.4 × 10^{-5} atm · m³/mol at 20°C (approximate — calculated from water solubility and vapor pressure); log K_{oc}: 2.54; log K_{ow}: 4.11; So at 20°C: miscible with kerosene, methyl isobutyl ketone and xylene; S_w: 75 mg/L at 20°C; vap d: 8.80 g/L at 25°C, 7.46 (air = 1); vp: 6.22 × 10^{-3} mmHg at 20°C.

Environmental Fate

Soil. The reported half-life in soil is approximately 4–8 weeks (Hartley and Kidd, 1987).

Groundwater. According to the U.S. EPA (1986) cycloate has a high potential to leach to groundwater.

Plant. Cycloate is rapidly metabolized in sugarbeets to carbon dioxide, ethylcyclohex-ylamine, sugars, amino acids and other natural constituents (Humburg et al., 1989).

Chemical/Physical. In the gas phase, cycloate reacts with hydroxyl and NO_3 radicals but not with ozone. With hydroxy radicals, cleavage of the cyclohexyl ring was suggested leading to the formation of a compound tentatively identified as $C_2H_5(CHO)NC(O)SC_2H_5$. The calculated photolysis lifetimes of cycloate in the troposphere with hydroxyl and NO_3 radicals are 5.2 hours and 1.4 days, respectively (Kwok et al., 1992).

Formulation Types: Granules (10%), emulsifiable concentrate (6 lb/gal).

Toxicity: LC_{50} (96-hour) for rainbow trout 4.5 mg/L (Hartley and Kidd, 1987), for mosquito fish 10 ppm (Humburg et al., 1989); acute oral LD_{50} of technical cycloate for male and female rats 2,000–3,190 and 3,160–4,100 mg/kg, respectively (Hartley and Kidd, 1987), 1,678 mg/kg (RTECS, 1985).

Uses: Herbicide used to control several broad-leaved weeds and many annual grasses in sugar beets, table beets and spinach.

CYFLUTHRIN

Synonyms: Bay FCR 1272; Baythroid; Baythroid H; **Cyano(4-fluoro-3-phenoxyphenyl)methyl 3-(2,2-dichloroethenyl)-2,2-dimethylcyclopropanecarboxylate.**

Designations: CAS Registry Number: 68359-37-5; mf: $C_{22}H_{18}Cl_2FNO_3$; fw: 434.30; RTECS: GZ1253000.

Properties: Yellow oil or paste. Mp: 60°C; ρ: 1.27–1.28 (supercooled at 20°C); K_H: 9.41 × 10^{-3} atm · m^3/mol at 20–25°C (approximate — calculated from water solubility and vapor pressure); log K_{oc}: 5.13 (calculated); log K_{ow}: 5.91, 6.00; S_o: soluble in 2-propanol, methylene chloride and toluene; S_w: 2.0 μg/L at 25°C; vp: 3.3 × 10^{-5} mmHg at 20°C.

Formulation Types: Emulsifiable concentrate; wettable powder; granules; water-oil emulsion.

Toxicity: LC_{50} (96-hour) for rainbow trout 0.6 μg/L, golden orfe 3.2 μg/L, bluegill sunfish 1.5 μg/L and carp 22 μg/L (Hartley and Kidd, 1987); acute oral LD_{50} for male rats 590 mg/kg (Hartley and Kidd, 1987), 251 mg/kg (RTECS, 1985).

Uses: Insecticide.

CYPERMETHRIN

Synonyms: Ammo; Ardap; Avicade; Barricade; CCN52; (+)-α-Cyano-3-phenoxy-benzyl 2,2-dimethyl-3-(2,2-dichlorovinyl)cyclopropane carboxylate; **Cyano(3-phenoxyphenyl)methyl 3-(2,2-dichloroethenyl)-2,2-dimethylcyclopropane-carboxylate**; Cymbush; Cyperkill; FMC 30980; FMC 45497; FMC 45806; Imperator; JF 5705F; Kafil super; NRDC 149; NRDC 160; NRDC 166; PP 383; Ripcord; Siperin; Stockade; WL 43467.

Designations: CAS Registry Number: 52315-07-8; mf: $C_{22}H_{19}Cl_2NO_3$; fw: 416.30; RTECS: GZ1250000.

Properties: Viscous, yellowish-brown liquid. Mp: 60–80°C (technical grade); K_H: 1.96×10^{-7} atm · m³/mol at 20–25°C (approximate — calculated from water solubility and vapor pressure); log K_{oc}: 4.00–4.53; log K_{ow}: 6.60; P-$t_{1/2}$: 179.82 hours (*cis*) (absorbance λ = 227.5 nm, concentration on glass plates = 6.7 µg/cm²); So (g/L): acetone (>450), ethanol (337), hexane (103); S_w: 4.0 µg/L at 25°C; vp: 1.43×10^{-9} mmHg at 20°C (extrapolated).

Soil properties and adsorption data

Soil	K_d (mL/g)	f_{oc} (%)	K_{oc} (mL/g)	pH
Loam	2	3.02	66	7.1
Loamy sand	20	1.22	1,639	5.4
Sandy loam	70	1.97	3,553	6.5
Sandy loam	260	1.97	13,198	6.5
Silt loam	4	1.16	345	5.6

Source: U.S. Department of Agriculture, 1990.

Environmental Fate

Soil. The major soil metabolite was reported to be 3-phenoxybenzoic acid (Hartley and Kidd, 1987).

Formulation Types: Emulsifiable concentrate; wettable powder; granules.

Toxicity: LC_{50} (96-hour) for brown trout 2.0–2.8 µg/L (Hartley and Kidd, 1987); acute oral LD_{50} for rats 200–800 mg/kg (Hartley and Kidd, 1987).

Uses: Insecticide.

CYROMAZINE

Synonyms: *N*-**Cyclopropyl-1,3,5-triazine-2,4,6-triamine**; 2-Cyclopropylamino-4,6-diamino-*s*-triazine; CGA 72662; Vetrazine.

$$H_2N \diagup \overset{N}{\underset{N \diagdown \underset{NH_2}{N}}{\bigcirc}} NH - \triangleleft$$

Designations: CAS Registry Number: 66215-27-8; DOT: 2763; mf: $C_6H_{10}N_6$; fw: 166.19.

Properties: Colorless crystals. Mp: 219–222°C; ρ: 1.35 at 20/4°C; K_H: 6.16 × 10^{-14} atm · m^3/mol at 25°C (approximate — calculated from water solubility and vapor pressure); log K_{oc}: 3.07 (calculated); log K_{ow}: -0.155; pK_a: >7; S_o: methanol (17 g/L); S_w: 11 g/L at 25°C (pH 7.5); vp: 3.3 × 10^{-9} mmHg at 25°C.

Environmental Fate

Chemical/Physical. Cyromazine will react with mineral acids (e.g., hydrochloric acid, sulfuric acid) forming water-soluble salts.

Formulation Types: Water soluble granules or powder; wettable powder.

Toxicity: LC_{50} (96-hour) for rainbow trout and carp >100 mg/L and bluegill sunfish >90 mg/L (Hartley and Kidd, 1987); acute oral LD_{50} for rats 3,387 mg/kg (Hartley and Kidd, 1987).

Uses: Insect growth regulator.

2,4-D

Synonyms: Agrotect; Agroxone; Amidox; Amoxone; Aqua-kleen; BH 2,4-D; Brush-rhap; B-Selektonon; Chipco turf herbicide 'D'; Chloroxone; Crop rider; Crotilin; D 50; 2,4-D acid; Dacamine; Debroussaillant 600; Decamine; Dedweed; Dedweed LV-69; Desormone; Dichlorophenoxyacetic acid; **(2,4-Dichlorophen-oxy)acetic acid**; Dicopur; Dicotox; Dinoxol; DMA-4; Dormone; Emulsamine BK; Emulsamine E-3; ENT 8538; Envert 171; Envert DT; Esteron; Esteron 76 BE; Esteron 44 weed killer; Esteron 99; Esteron 99 concentrate; Esteron brush killer; Esterone 4; Estone; Farmco; Fernesta; Fernimine; Fernoxone; Ferxone; Foredex 75; Formula 40; Hedonal; Herbidal; Ipaner; Krotiline; Lawn-keep; Macrondray; Miracle; Monosan; Moxone; NA 2765; Netagrone; Netagrone 600; NSC 423; Pennamine; Pennamine D; Phenox; Pielik; Planotox; Plant-gard; RCRA waste number U240; Rhodia; Salvo; Spritz-hormin/2,4-D; Spritz-hor-mit/2,4-D; Super D weedone; Transamine; Tributon; Trinoxol; U 46; U 46 D; U 5043; U 46DP; Vergemaster; Verton; Verton D; Verton 2D; Vertron 2D; Vidon 638; Visko-rhap; Visko-rhap drift herbicides; Visko-rhap low volatile 4L; Weedar; Weddar-64; Weddatul; Weed-b-gon; Weedez wonder bar; Weedone; Weedone LV4; Weed-rhap; Weed tox; Weedtrol.

Designations: CAS Registry Number: 94-75-7; DOT: 2765; mf: $C_8H_6Cl_2O_3$; fw: 221.04; RTECS: AG6825000.

Properties: White to pale yellow prismatic crystals with a faint phenolic-like odor. Mp: 140–141°C (free acid), 179–180°C (ammonium salt), 157–159°C (methylammonium salt), 85–87°C (dimethylammonium salt), 145–147°C (ethanolamine salt), 142–144°C (trietha-nolamine salt). Sodium salt decomposes at 215°C. Bp: 160°C at 0.4 mmHg, decomposes at 760 mmHg; ρ: 1.416 at 25/4°C, 1.565 at 30/4°C; fl p: nonflammable (free acid and salts); pK_a: 2.64–3.31; K_H: 1.95×10^{-2} atm · m³/mol at 20–25°C (approximate — calculated from water solubility and vapor pressure); log K_{oc}: 1.68–2.73; log K_{ow}: 1.47–4.88; P-$t_{1/2}$: 2–4 days (aqueous solution irradiated at λ = 356 nm); So (g/kg at 25°C unless noted otherwise): acetone (850), benzene (10.7 at 28°C), carbon disulfide (5.0 at 29°C), carbon tetrachloride (1), *o*-dichlorobenzene (4), diesel oil (1), diesel oil + kerosene (3.5), 1,4-dioxane (785 at 31°C), ethyl ether (270), ethyl alcohol (1,300), *n*-heptane (1.15), methyl isobutyl ketone (312.7), 2-propanol (316 at 31°C), toluene (0.67), xylene (5.8), insoluble in petroleum oils; S_w: 890 ppm at 25°C (free acid), 45 g/L at 20°C (sodium salt); vp: 4.7 × 10⁻³ mmHg at 20°C.

Soil properties and adsorption data

Soil	K_d (mL/g)	f_{oc} (%)	K_{oc} (mL/g)	pH (meq/100 g)	CEC
A-horizon	2.21	1.41	160	3.23	4.8
AB-horizon	2.38	5.11	50	3.88	13.0
Agricultural soil	2.48	1.64	150	5.40	14.0
Alumina	0.83	—	—	—	6.0
Asquith sandy loam[a]	0.14	1.02	14	7.50	—
B-horizon	5.45	2.58	210	3.59	9.6
Bernard-Fagne silt loam	13.18	2.42	545	3.60	8.8
Bullingen silt loam	1.83	3.16	58	3.55	8.2
C-horizon	2.70	1.82	150	4.07	7.0
C-horizon	0.16	0.09	180	4.95	1.6
C-horizon	0.14	0.15	90	4.21	1.3
Catlin	0.33	2.01	48	6.20	—
Clayey till	0.13	0.13	100	7.64	40.5
Commerce	0.33	0.68	48	6.70	—
Ferrod	4.30	3.56	112	3.88	—
Fleron silty clay loam	2.04	3.23	63	3.75	12.3
Heverlee III sandy loam	0.85	1.45	59	5.84	10.7
Humic acid	79.43	—	—	—	6.0
Illite	10.47	—	—	—	6.0
Indian Head loam[a]	0.53	2.34	23	7.80	—
Indian Head loam	0.44	2.35	19	7.80	—
Lubbeck II sand	0.09	0.07	129	6.46	2.4
Lubbeck II sand	0.05	0.07	71	6.43	2.3
Lubbeck II sandy loam	0.31	0.53	58	6.71	4.5
Lubbeck I silt loam	0.77	1.15	67	6.62	9.5
Lubbeck III silt loam	0.43	0.65	66	6.91	7.0
Meerdael silt loam	7.05	3.59	196	4.00	11.7
Melfort loam	3.38	6.05	56	5.90	—
Melfort loam[a]	0.99	6.05	16	5.90	—
Meltwater sand	0.27	0.05	540	6.14	1.4
Montmorillonite	0.65	—	—	—	6.0
Nodebais silt loam	0.39	0.73	53	6.20	8.4
Regina clay	0.31	2.40	13	7.70	—
Regina heavy clay[a]	0.19	2.39	8	7.70	—
Sand	28.44	—	—	—	6.0
Sandy till	0.23	0.06	380	4.71	9.1
Silica gel	1.29	—	—	—	6.0
Soignes silt loam	16.18	4.94	327	3.40	16.9
Spa silty clay loam	23.89	3.89	614	3.25	12.1
Stavelot silt loam	7.60	2.53	300	3.90	5.6
Stookrooie II loamy sand	1.81	1.07	169	5.64	2.9
Tracy	0.85	1.12	76	6.20	—
Udalf	1.10	0.76	145	7.45	—
Weyburn Oxbow loam	0.61	3.72	16	6.50	—
Weyburn Oxbow loam[a]	0.45	3.72	12	6.50	—

Soil properties and adsorption data (continued)

Soil	K_d (mL/g)	f_{oc} (%)	K_{oc} (mL/g)	pH (meq/100 g)	CEC
Zolder sand	0.42	0.07	600	4.73	0.7
Zolder sand I	10.36	1.86	557	3.84	1.7
Zolder sand II	0.95	0.19	500	4.23	0.5

Source: Haque and Sexton, 1968; Grover and Smith, 1974; Grover, 1977; Moreale and Van Bladel, 1980; McCall et al., 1981; Rippen et al., 1982; Løkke, 1984. a) Dimethylamine salt.

Environmental Fate

Biological. 2,4-D degraded in anaerobic sewage sludge to 4-chlorophenol (Mikesell and Boyd, 1985). In moist nonsterile soils, degradation of 2,4-D occurs via cleavage of the carbon-oxygen bond at the 2-position on the aromatic ring (Foster and McKercher, 1973). In filtered sewage water, 2,4-D underwent complete mineralization but degradation was much slower in oligotrophic water, especially when 2,4-D was present in high concentrations (Rubin et al., 1982). In a primary digester sludge under methanogenic conditions, 2,4-D did not display any anaerobic biodegradation after 60 days (Battersby and Wilson, 1988, 1989).

In clear water and muddy water, hydrolysis half-lives of 18 to 50 days and 10–25 days, respectively, were reported (Nesbitt and Watson, 1980).

Soil. In moist soils, 2,4-D degraded to 2,4-dichlorophenol and 2,4-dichloro-anisole as intermediates followed by complete mineralization to carbon dioxide (Wilson and Cheng, 1978; Stott, 1983; Smith, 1985). 2,4-Dichlorophenol also was reported as a hydrolysis metabolite (Somasundaram et al., 1989, 1991; Somasundaram and Coats, 1991). In a soil pretreated with its hydrolysis metabolite, 80% of the applied [^{14}C]2,4-D mineralized to $^{14}CO_2$ within 4 days. In soils not treated with the hydrolysis product (2,4-dichlorophenol), only 6% of the applied [^{14}C]2,4-D degraded to $^{14}CO_2$ after 4 days (Somasundaram et al., 1989). Steenson and Walker (1957) reported that the soil microorganisms *Flavobacterium peregrinum* and *Achromobacter* both degraded 2,4-D yielding 2,4-dichlorophenol and 4-chlorocatechol as metabolites. The microorganisms *Gloeosporium olivarium, Gloeosporium kaki* and *Schisophyllum communs* also degraded 2,4-D in soil forming 2-(2,4-dichlorophenoxy)ethanol as the major metabolite (Nakajima et al., 1973). Microbial degradation of 2,4-D was more rapid under aerobic conditions (half-life 1.8 to 3.1 days) than under anaerobic conditions (half-life 69 to 135 days) (Liu et al., 1981). In a 5-day experiment, ^{14}C-labeled 2,4-D applied to soil-water suspensions under aerobic and anaerobic conditions gave $^{14}CO_2$ yields of 0.5 and 0.7%, respectively (Scheunert et al., 1987). Degradation was observed to be lowest at low redox potentials (Gambrell et al., 1984) but is enhanced with the addition of inorganic phosphorus and organic amendments (Duah-Yentumi and Kuwatsuka, 1982).

The reported half-lives for 2,4-D in soil ranged from 4 days in a laboratory experiment (McCall et al., 1981a) to 15 days (Jury et al., 1987). Residual activity in soil is limited to approximately 6 weeks (Hartley and Kidd, 1987).

After one application of 2,4-D to soil, the half-life was reported to be approximately 80 days but can be as low as 2 weeks after repeated applications (Cullimore, 1971). The half-lives for 2,4-D in soil incubated in the laboratory under aerobic conditions ranged from 4 to 34 days with an average of 16 days (Altom and Stritzke, 1973; Foster and McKercher, 1973; Yoshida and Castro, 1975). In field soils, the disappearance half-lives were lower and ranged from approximately 1 to 15 days with an average of 5 days (Radosevich and Winterlin, 1977; Wilson and Cheung, 1976; Stewart and Gaul, 1977). Under aerobic conditions, the mineralization half-lives of 2,4-D in soil ranged from 11 to 25 days (Ou et al., 1978; Wilson and Cheung, 1978). The half-lives of 2,4-D in a sandy loam, clay loam and an organic amended soil under nonsterile conditions were 722–2,936, 488–3,609 and 120–1,325 days, respectively (Schoen and Winterlin, 1987).

Groundwater. According to the U.S. EPA (1986) 2,4-D has a high potential to leach to groundwater.

Plant. Reported metabolic products in bean and soybean plants include 4-*O*-β-gluco-sides of 4-hydroxy-2,5-dichlorophenoxyacetic acid, 4-hydroxy-2,3-dichlorophenoxyacetic acid, *N*-(2,4-dichlorophenoxyacetyl)-L-aspartic acid and *N*-(2,4-dichlorophenocyacetyl)-L-glutamic acid. Metabolites identified in cereals and strawberries include 1-*O*-(2,4-dichlo-rophenoxyacetyl)-β-D-glucose and 2,4-dichlorophenol, respectively (Verschueren, 1983). In alfalfa, the side chain in the 2,4-D molecule was found to be lengthened by two and four methylene groups resulting in the formation of (2,4-dichlorophenoxy)butyric acid and (2,4-dichlorophenoxy)-hexanoic acid, respectively (Linscott and Hagin, 1970). During side chain degradation, 2,4-dichloroanisole also was reported as a possible intermediate (Luckwill and Lloyd-Jones, 1960). In several resistant grasses, however, the side chain increased by one methylene group forming (2,4-dichlorophenoxy)propionic acid (Hagin and Linscott, 1970).

2,4-D was metabolized by soybean cultures forming 2,4-dichlorophenoxyacetyl deriv-atives of alanine, leucine, phenylalanine, tryptophan, valine, aspartic and glutamic acids (Feung et al., 1971, 1972, 1973). When 2,4-D was applied to resistant grasses, 3-(2,4-dichlorophenoxy)propionic acid formed (Hagin et al., 1970). On bean plants, 2,4-D degraded via β-oxidation and ring hydroxylation to form 2,4-dichloro-4-hydroxyphenoxy-acetic acid, 2,3-dichloro-4-hydroxyphen-oxyacetic acid (Hamilton et al., 1971) and 2-chloro-4-hydroxyphenoxyacetic acid (Fleeker and Steen, 1971).

2,5-Dichloro-4-hydroxyphenoxyacetic acid was the predominant product identified in several weed species as well as smaller quantities of 2-chloro-4-hydroxyphenoxyacetic acid in wild buckwheat, yellow foxtail and wild oats (Fleeker and Steen, 1971).

Esterification of 2,4-D with plant constituents via conjugation formed the β-D-glucose ester of 2,4-D (Thomas et al., 1964).

Photolytic. Photolysis of 2,4-D in distilled water using mercury arc lamps ($\lambda = 254$ nm) or by natural sunlight yielded 2,4-dichlorophenol, 4-chlorocatechol, 2-hydroxy-4-chlorophenoxyacetic acid, 1,2,4-benzenetriol and brown polymeric humic acids. The half-life for this reaction was 50 minutes (Crosby and Tutass, 1966). Half-lives of 2 to 4 days were reported when an aqueous solution was irradiated with UV light ($\lambda = 356$ nm) (Baur and Bovey, 1974).

Bell (1956) reported that the composition of photodegradation products formed were dependent upon the initial 2,4-D concentration and pH of the solutions. 2,4-D undergoes reductive dechlorination when various polar solvents (methanol, *n*-butanol, isobutyl alco-hol, *t*-butyl alcohol, *n*-octanol, ethylene glycol) are irradiated at wavelengths between 254 and 420 nm. Photoproducts formed included 2,4-di-chlorophenol, 2,4-dichloroanisole, 4-chlorophenol, 2- and 4-chlorophenoxyacetic acid (Que Hee and Sutherland, 1981).

Irradiation of a 2,4-D sodium salt solution by a 660 W mercury discharge lamp produced 2,4-dichlorophenol within 20 minutes. Further irradiation resulted in further decomposition. The irradiation times required for 50% decomposition of the 2,4-D sodium salt at pH values of 4.0, 7.0 and 9.0 are 71, 50 and 23 minutes, respectively (Aly and Faust, 1964).

Surface Water. In filtered lake water at 29°C, 90% of 2,4-D (1 mg/L) mineralized to carbon dioxide. The half-life was <5 days. At low concentrations (0.2 mg/L), no mineralization was observed (Wang et al., 1984). Subba-Rao et al. (1982) reported that 2,4-D in very low concentrations mineralized in one of three lakes tested. Mineralization did not occur when concentrations were at the picogram level.

Chemical/Physical. In a helium pressurized reactor containing ammonium nitrate and polyphosphoric acid at temperatures of 121 and 232°C, 2,4-D was oxidized to carbon dioxide, water and hydrochloric acid (Leavitt and Abraham, 1990). Carbon dioxide, chloride, aldehydes, oxalic and glycolic acids, were reported as ozonation products of 2,4-D in water at pH 8 (Struif et al., 1978). Reacts with alkali metals and amines forming water soluble salts (Hartley and Kidd, 1987).

When 2,4-D was heated to 900°C, carbon monoxide, carbon dioxide, chlorine, hydrochloric acid and oxygen were produced (Kennedy et al., 1972, 1972a). Total mineralization of 2,4-D was observed when a solution containing the herbicide and Fenton's reagent (Fe^{3+} and hydrogen peroxide) was subjected to UV light (λ = 300–400 nm). One intermediate compound identified was oxalic acid (Sun and Pignatello, 1993). Emits very toxic chloride fumes when heated to decomposition (Lewis, 1990).

The solubilities of the calcium and magnesium salts of 2,4-D acid at 25°C are 9.05 and 25.1 mM, respectively (Aly and Faust, 1964).

2,4-D will not hydrolyze to any reasonable extent (Kollig, 1993).

Exposure Limits: NIOSH REL: TWA 10 mg/m^3, IDLH 100 mg/m^3; OSHA PEL: TWA 10 mg/m^3; ACGIH TLV: TWA 10 mg/m^3.

Symptoms of Exposure: Weak stupor, hypoflexia, muscle twitch, convulsions, dermatitis.

Formulation Types: Emulsifiable concentrate; water soluble powder or granules; soluble concentrate.

Toxicity: EC50 (24-hour) for *Daphnia magna* 249 mg/L, *Daphnia pulex* 324 mg/L (Lilius et al., 1995); LC_{50} (48-hour) for bluegill sunfish 0.9 ppm, rainbow trout 1.1 ppm (Verschueren, 1983); acute oral LD_{50} for rats 375 mg/kg (Hartley and Kidd, 1987).

Uses: 2,4-D is a chlorine-substituted phenoxyacetic acid herbicide used for postemergence control of annual and perennial broad-leaved weeds in fruits, vegetables, turfs and ornamentals.

DALAPON-SODIUM

Synonyms: Basfapon F; Dalapon; Dalapon sodium salt; **2,2-Dichloropropionic acid sodium salt**; α,α-Dichloropropionic acid sodium salt; 2,2-DPA; Dowpon; Gramevin; Radapon; Sodium dalapon; Sodium 2,2-dichloropropanoate; Sodium α,α-dichloropropionate; Unipon.

$$\begin{array}{c} Cl \\ | \\ CH_3CCOONa \\ | \\ Cl \end{array}$$

Designations: CAS Registry Number: 127-20-8; mf: $C_3H_3Cl_2NaO_2$; fw: 164.95; RTECS: UF1225000.

Properties: Light-colored, very hygroscopic powder. The free acid is a colorless, odorless liquid. Mp: decomposes at 174–176°C; bp: 185–190°C (free acid), 98–99°C at 20 mmHg; ρ: 1.389 at 22.8/4°C (free acid); fl p: nonflammable; H-$t_{1/2}$: unknown but hydrolyzes readily; K_H: 6×10^{-6} atm · m^3/mol (calculated value for the free acid reported by Reinert and Rodgers (1987); pK_a: 1.84 (free acid); log K_{oc}: 0.37–2.18 (calculated); log K_{ow}: 0.76; So (g/kg at 25°C): acetone (1.4), benzene (0.02), ethanol (185), ethyl ether (0.16), methanol (179); S_w: 450–900 g/L at 25°C; vp: not applicable.

Environmental Fate

Soil. Undergoes dechlorination and the liberation of carbon dioxide in soil. The residual activity is limited to approximately 3–4 months (Hartley and Kidd, 1987). The average half-life for dalapon-sodium in soil incubated in the laboratory under aerobic conditions was 15 days (Namdeo, 1972).

Photolytic. Dalapon (free acid) is subject to photodegradation. When an aqueous solution (0.25 M) was irradiated with UV light at 253.7 nm at 49°C, 70% degraded in 7 hours. Pyruvic acid is formed which is subsequently decarboxylated to acetaldehyde, carbon dioxide and small quantities of 1,1-dichloroethane (2–4%) and a water-insoluble polymer (Kenaga, 1974). The photolysis of an aqueous solution of dalapon (free acid) by UV light ($\lambda = 2537$ Å) yielded chloride ions, carbon dioxide, carbon monoxide and methyl chloride at quantum yields of 0.29, 0.10, 0.02 and 0.02, respectively (Baxter and Johnston, 1968).

Chemical/Physical. Slowly reacts with moisture at room temperature forming pyruvic acid (Frank and Demint, 1969; Kenaga, 1974), hydrochloric acid and sodium chloride (Kenaga, 1974; Wolfe et al., 1990). The reported hydrolysis half-life of dalapon sodium salt at low concentrations (<1%) and temperatures less than 25°C is several months (Kenaga, 1974).

Products reported from the combustion of the free acid (dalapon) at 900°C include carbon monoxide, carbon dioxide, chlorine and hydrochloric acid (Kennedy et al., 1972a).

Symptoms of Exposure: Irritating to eyes and skin.

Formulation Types: Granules; wettable powder; water-soluble powder.

Toxicity: LC_{50} (96-hour) for rainbow trout, goldfish and channel catfish >100 mg/L, carp >500 mg/L, guppies >1,000 mg/L (Hartley and Kidd, 1987), fathead minnow 290 mg/L and bluegill sunfish 290 mg/L (Verschueren, 1983); LC_{50} (48-hour) for coho salmon 340 mg/L (Verschueren, 1983); acute oral LD_{50} for male and female rats 9,330 and 7,570 mg/kg, respectively (Hartley and Kidd, 1987).

Uses: Selective, systemic herbicide used to control perennial and annual grasses on noncrop land, fruits, vegetables and some aquatic weeds.

DAZOMET

Synonyms: Basamid; Basamid G; Basamid-granular; Basamid P; Basamid-puder; Carbothialdin; Carbothialdine; Crag 974; Crag fungicide 974; Crag nematocide; Crag 85W; Dazomet-powder BASF; Dimethylformocarbothialdine; 3,5-Dimethyltetrahydro-1,3,5-thiadiazine-2-thione; 3,5-Dimethyl-1,2,3,5-tetrahydro-1,3,5-thiadiazinethione-2; 3,5-Dimethyltetrahydro-1,3,5-2*H*-thiadiazine-2-thione; 3,5-Dimethyl-1,3,5,2*H*-tetrahydrothiadiazine-2-thione; 3,5-Dimethyltetrahydro-2*H*-1,3,5-thiadiazine-2-thione; Dimethyl-2*H*-1,3,5-tetrahydrothiadiazine-2-thione; 3,5-Dimethyl-2-thionotetrahydro-1,3,5-thiadiazine; DMTT; Fennosan B 100; Micofume; Mylon; Mylone; Mylone 85; NA 521; Nalcon 243; Nefusan; Prezervit; Stauffer *N* 521; Tetrahydro-2*H*-3,5-dimethyl-1,3,5-thiadiazine-2-thione; **Tetrahydro-3,5-dimethyl-2*H*-1,3,5-thiadiazine-2-thione**; Thiazon; Thiazone; 2-Thio-3,5-dimethyltetrahydro-1,3,5-thiadiazine; Tiazon; Troysan 142; UC 974.

Designations: CAS Registry Number: 533-74-4; mf: $C_5H_{10}N_2S_2$; fw: 162.28; RTECS: XI2800000.

Properties: Colorless to white, almost odorless crystalline solid. Mp: decomposes at 104–105°C; ρ: 1.37 at room temperature; fl p: nonflammable; K_H: 2×10^{-10} atm · m³/mol at 20°C (approximate — calculated from water solubility and vapor pressure); log K_{oc}: 0.48 (calculated); log K_{ow}: 0.15; So (g/kg at 25°C): acetone (173), benzene (51), chloroform (391), cyclohexane (400), ethanol (15), ethyl ether (6); S_w: 3 g/kg at 20°C, 1.2 g/kg at 30°C; vp: 2.78×10^{-6} mmHg at 20°C.

Environmental Fate

Soil. Soil metabolites include formaldehyde, hydrogen sulfide, methylamine and methyl(methylaminomethyl)dithiocarbamic acid (Hartley and Kidd, 1987) which further decomposes to methyl isothiocyanate (Harley and Kidd, 1987; Ashton and Monaco, 1991; Cremlyn, 1991). The rate of decomposition is dependent upon the soil type, temperature and humidity (Cremlyn, 1991).

Chemical/Physical. Hydrolyzes in acidic solutions forming carbon disulfide, methylamine and formaldehyde (Hartley and Kidd, 1987; Humburg et al., 1989). These compounds are probably formed following the decomposition of dazomet with alcohol and water (Windholz et al., 1983). Emits toxic fumes of nitrogen and sulfur oxides when heated to decomposition (Sax and Lewis, 1987).

Symptoms of Exposure: Skin and eye irritant. Symptoms of ingestion include nausea, irritation of the gastrointestinal tract, vomiting, cramps and diarrhea.

Formulation Types: Dustable powder; granules; wettable powder.

Toxicity: Toxic to fish (Hartley and Kidd, 1987); acute oral LD_{50} for male albino mice 650 mg/kg (Ashton and Monaco, 1991); acute oral LD_{50} for male and female rats 500 and 400 mg/kg, respectively (Verschueren, 1983), 320 mg/kg (RTECS, 1985).

Uses: Soil fungicide; nematocide; herbicide; insecticide; soil sterilant.

p,p'-DDD

Synonyms: 1,1-Bis(4-chlorophenyl)-2,2-dichloroethane; 1,1-Bis(*p*-chlorophenyl)-2,2-dichloroethane; 2,2-Bis(4-chlorophenyl)-1,1-dichloroethane; 2,2-Bis(*p*-chlorophenyl)-1,1-dichloroethane; DDD; 4,4'-DDD; 1,1-Dichloro-2,2-bis(*p*-chlorophenyl)ethane; 1,1-Dichloro-2,2-di-(4-chlorophenyl)ethane; 1,1-Dichloro-2,2-di-(*p*-chlorophenyl)ethane; Dichlorodiphenyldichloroethane; 4,4'-Dichlorodiphenyldichloroethane; *p,p'*-Dichlorodiphenyldichloroethane; **1,1'-(2,2-Dichloroethylidene)bis(4-chlorobenzene)**; Dilene; ENT 4225; ME 1700; NA 2761; NCI-C00475; RCRA waste number U060; Rhothane; Rhothane D-3; Rothane; TDE; 4,4'-TDE; *p,p'*-TDE; Tetrachlorodiphenylethane.

$$Cl\text{—}\underset{}{\bigcirc}\text{—}\overset{\displaystyle CHCHCl_2}{\underset{}{C}}\text{—}\underset{}{\bigcirc}\text{—}Cl$$

Designations: CAS Registry Number: 72-54-8; DOT: 2761; mf: $C_{14}H_{10}Cl_4$; fw: 320.05; RTECS: KI0700000.

Properties: White crystalline solid. Mp: 88–90°C, 109–112°C; bp: 193°C; ρ: 1.476 at 20/4°C; fl p: not pertinent; H-$t_{1/2}$: 28 years at 25°C and pH 7; K_H: 2.16 × 10^{-5} atm · m³/mol at 25°C (approximate — calculated from water solubility and vapor pressure); log K_{oc}: 5.38; log K_{ow}: 5.061–6.217; So (g/L at 22°C): acetone (50–100), dimethylsulfoxide (50–100), 95% ethanol (10–50); S_w: 20–90 μg/L at 25°C; vap d: 17.2 ng/L at 30°C; vp: 4.68 × 10^{-6} mmHg at 25°C.

Environmental Fate

Biological. It was reported that *p,p'*-DDD, a major biodegradation product of *p,p'*-DDT, was degraded by *Aerobacter aerogenes* under aerobic conditions to yield 1-chloro-2,2-bis(*p*-chlorophenyl)ethylene, 1-chloro-2,2-bis-(*p*-chlorophenyl)ethane and 1,1-bis(*p*-chlorophenyl)ethylene. Under anaerobic conditions, however, four additional compounds were identified: bis(*p*-chlorophenyl)acetic acid, *p,p'*-dichlorodiphenylmethane, *p,p'*-dichlorobenzhydrol and *p,p'*-dichlorobenzophenone (Fries, 1972). Under reducing conditions, indigenous microbes in Lake Michigan sediments degraded DDD to 2,2-bis(*p*-chlorophenyl)ethane and 2,2-bis(*p*-chlorophenyl)ethanol (Leland et al., 1973). Incubation of *p,p'*-DDD with hematin and ammonia gave 4,4'-dichlorobenzophenone, 1-chloro-2,2-bis-(*p*-chlorophenyl)ethylene and bis(*p*-chlorophenyl)acetic acid methyl ester (Quirke et al., 1979). Using settled domestic wastewater inoculum, *p,p'*-DDD (5 and 10 mg/L) did not degrade after 28 days of incubation at 25°C (Tabak et al., 1981).

Chemical/Physical. The hydrolysis rate constant for *p,p'*-DDD at pH 7 and 25°C was determined to be 2.8 × 10^{-6}/hour, resulting in a half-life of 28.2 years (Ellington et al., 1987). 2,2-Bis(4-chlorophenyl)-1-chloroethene and hydrochloric acid were reported as hydrolysis products (Kollig, 1993).

Symptoms of Exposure: Contact with eyes causes irritation. Ingestion causes vomiting.

Formulation Types: Wettable powders; dusts; emulsions.

Toxicity: EC50 (96-hour) for catfish <2.6 mg/L; LC_{50} (24-hour) for brown shrimp 6.8 μg/L (salt water); acute oral LD_{50} for rats 3,400 mg/kg (Verschueren, 1983), 113 mg/kg (RTECS, 1985).

Uses: Contact control of leaf rollers and other insects on vegetables and tobacco.

p,p'-DDE

Synonyms: 2,2-Bis(4-chlorophenyl)-1,1-dichloroethene; 2,2-Bis(p-chlorophenyl)-1,1-dichloroethene; 1,1-Bis(4-chlorophenyl)-2,2-dichloroethylene; 1,1-Bis(p-chlorophenyl)-2,2-dichloroethylene; DDE; 4,4'-DDE; DDT dehydrochloride; 1,1-Dichloro-2,2-bis(p-chlorophenyl)ethylene; Dichlorodiphenyldichloroethylene; p,p'-Dichlorodiphenyldichloroethylene; **1,1'-(Dichloroethenylidene)bis(4-chlorobenzene)**; NCI-C00555.

Designations: CAS Registry Number: 72-55-9; DOT: 2761; mf: $C_{14}H_8Cl_4$; fw: 319.03; RTECS: KV9450000.

Properties: Solid. Mp: 88–90°C; K_H: 1.22×10^{-3} atm · m^3/mol at 23°C; log K_{oc}: 5.386; log K_{ow}: 5.69–6.956; P-$t_{1/2}$: 206.66 hours (absorbance $\lambda = 248.5$ nm, concentration on glass plates = 6.7 µg/cm²); S_o: cod liver oil (245.5, 269.2 and 398.1 mM at 4, 12 and 20°C, respectively), n-octanol (44.0 and 63.7 g/L at 4 and 20°C, respectively), triolein (73.1, 96.3 and 105.6 g/L, at 4, 12 and 20°C, respectively); S_w: 40 at 20°C, 1.3 µg/L at 25°C (Metcalf et al., 1973); 40 and 65 µg/L at 20 and 24°C, respectively; vap d: 109 ng/L at 30°C; vp: 1.57×10^{-5} mmHg at 25°C.

Environmental Fate

Biological. In four successive 7-day incubation periods, p,p'-DDE (5 and 10 mg/L) was recalcitrant to degradation in a settled domestic wastewater inoculum (Tabak et al., 1981).

Photolytic. When an aqueous solution of p,p'-DDE (0.004 µM) in natural water samples collected from California and Hawaii was irradiated (maximum $\lambda = 240$ nm) for 120 hours, 62% was photooxidized to p,p'-dichlorobenzophenone (Ross and Crosby, 1985). In an air-saturated distilled water medium irradiated with monochromic light ($\lambda = 313$ nm), p,p'-DDE degraded to p,p'-dichlorobenzophenone, 1,1-bis(p-chlorophenyl)-2-chloroethylene (DDMU) and 1-(4-chlorophenyl)-1-(2,4-dichlorophenyl)-2-chloroethylene (o-chloro DDMU). Identical photoproducts also were observed using tap water containing Mississippi River sediments (Miller and Zepp, 1972). The photolysis half-life under sunlight irradiation was reported to be 1.5 days (Mansour et al., 1989).

When p,p'-DDE in water was irradiated at 313 nm, a quantum yield of 0.3 was achieved. A photolysis half-life of 0.9 days in summer and 6.1 days in winter by direct sunlight at 40° latitude was observed. Photolysis products included DDMU (yield 20%), o-chloro DDMU (yield 15%), and a dichlorobenzophenone (Zepp et al., 1976, 1977). Quantum yields of 0.26 and 0.24 were reported for the photolysis of p,p'-DDE in hexane at wavelengths of 254 and 313 nm, respectively (Mosier et al., 1969; Zepp et al., 1977).

When *p,p'*-DDE in a methanol solvent was photolyzed at 260 nm, a dichloro-benzophenone, a dichlorobiphenyl, DDMU and 3,6-dichlorofluorenone (yield 10%) formed as the major products (Plimmer et al., 1970).

Chemical/Physical. May degrade to bis(chlorophenyl)acetic acid (DDA) and hydrochloric acid in water (Verschueren, 1983) or oxidize to *p,p'*-dichlorobenzophenone using UV light as a catalyst (U.S. Department of Health and Human Services, 1989).

Toxicity: Acute oral LD_{50} for rats 880 mg/kg (RTECS, 1985).

Uses: Military product; chemical research. Degradation product of *p,p'*-DDT.

p,p'-DDT

Synonyms: Agritan; Anofex; Arkotine; Azotox; 2,2-Bis(4-chlorophenyl)-1,1,1-trichloro-ethane; 2,2-Bis(*p*-chlorophenyl)-1,1,1-trichloroethane; α,α-Bis(*p*-chlorophenyl)-β,β,β-trichloroethane; 1,1-Bis(*p*-chlorophenyl)-2,2,2-trichloroethane; Bosan Supra; Bovidermol; Chlorophenothan; Chlorophenothane; Chlorophenotoxum; Citox; Clofenotane; DDT; 4,4'-DDT; Dedelo; Deoval; Detox; Detoxan; Dibovan; Dichlorodiphenyltrichloroethane; *p,p'*-Dichlorodiphenyltrichloroethane; 4,4'-Dichlorodiphenyltrichloroethane; Dicophane; Didi-gam; Didimac; Diphenyltrichloroethane; Dodat; Dykol; ENT 1506; Estonate; Genitox; Gesafid; Gesapon; Gesarex; Gesarol; Guesapon; Gyron; Havero-extra; Ivoran; Ixodex; Kopsol; Mutoxin; NCI-C00464; Neocid; Parachlorocidum; PEB1; Pentachlorin; Pentech; PPzeidan; RCRA waste number U061; Rukseam; Santobane; Trichlorobis(4-chlorophe-nyl)ethane; Trichlorobis(*p*-chlorophenyl)ethane; 1,1,1-Trichloro-2,2-bis(*p*-chlorophe-nyl)ethane; 1,1,1-Trichloro-2,2-di(4-chlorophenyl)ethane; 1,1,1-Trichloro-2,2-di(*p*-chlo-rophenyl)ethane; **1,1'-(2,2,2-Trichloroethylidene)bis(4-chlorobenzene)**; Zeidane; Zerdane.

Designations: CAS Registry Number: 50-29-3; DOT: 2761; mf: $C_{14}H_9Cl_5$; fw: 354.49; RTECS: KJ3325000.

Properties: Odorless to slightly fragrant colorless crystals or white powder. Mp: 108.5°C; bp: 185°C at 0.05 mmHg (decomposes); ρ: 1.56 at 15/4°C; fl p: 72.2–77.2°C; K_H: 1.29 × 10^{-5} atm · m³/mol at 23°C; log K_{oc}: 5.146–6.26; log K_{ow}: 4.89–6.914; P-t$_{1/2}$: 192.31 hours (absorbance λ = 237.5 nm, concentration on glass plates = 6.7 μg/cm²); S_o: soluble in cyclohexane, morpholine, pyridine, 1,4-dioxane, acetone (580 g/L), benzene (780 g/L), benzyl benzoate (420 g/L), carbon tetrachloride (450 g/L), chlorobenzene (740 g/L), cod liver oil (151.4, 162.2 and 204.2 mM at 4, 12 and 20°C, respectively), cyclohexanone (1,160 g/L), ethyl ether (280 g/L), gasoline (100 g/L), isopropanol (30 g/L), kerosene (80–100 g/L), methylene chloride (850 g/L at 27°C), morpholine (750 g/L), *n*-octanol (26.9, 43.6 and 57.5 g/L at 4, 12 and 20°C, respectively), peanut oil (110 g/L), pine oil (100–160 g/L), tetralin (610 g/L), tributyl phosphate (500 g/L), trichloroethylene (720 g/L at 27°C), triolein (134.9, 182.0 and 323.6 mM at 4, 12 and 20°C, respectively), xylene (600 g/L at 27°C); S_w: 1.2–5.5 μg/L at 25°C; vap d: 13.6 ng/L at 30°C; vp (10^7): 2.2, 4.3, 9.3, 40, 150, 480, 1,500 and 4,500 at 20, 25, 30, 40, 50, 60, 70 and 80°C, respectively (Rothman, 1980).

Soil properties and adsorption data

Soil	K_d (mL/g)	f_{oc} (%)	K_{oc} (mL/g)	pH
Catlin	2,800	2.01	152,000	6.2
Commerce	1,070	0.68	157,000	6.7
Gilat	6,468	0.73	886,027	—
Humic acid	300,000	57.00	526,316	—
Kinneret-A	34,504	4.55	758,330	—
Kinneret-G	15,012	2.55	588,706	—
Malkiya	9,637	3.38	285,118	—
Marine sediment (sea water)	48,000	2.70	1,777,778	—
Mivtahim	4,546	0.26	1,748,462	—
Montcalm sandy loam	1,300	1.00	130,000	—
Montmorillonite clay	40,000	0.20	20,000,000	—
Neve Yaar	8,121	1.64	495,182	—
Sims clay	13,700	3.81	360,000	—
Tracy	1,830	1.12	147,000	6.2

Source: Shin et al., 1970; Pierce et al., 1974; McCall et al., 1981; Gerstl and Mingelgrin, 1984.

Environmental Fate

Biological. In four successive 7-day incubation periods, *p,p′*-DDT (5 and 10 mg/L) was recalcitrant to degradation in a settled domestic wastewater inoculum (Tabak et al., 1981).

The white rot fungus *Phanerochaete chrysosporium* degraded *p,p′*-DDT yielding the following metabolites: 1,1-dichloro-2,2-bis(4-chlorophenyl)ethane (*p,p′*-DDD), 2,2,2-trichloro-1,1-bis(4-chlorophenyl)ethanol (dicofol), 2,2-dichloro-1,1-bis(4-chlorophenyl)ethanol, 4,4′-dichlorobenzophenone and carbon dioxide (Bumpus et al., 1985; Bumpus and Aust, 1987). Mineralization began between the third and sixth day of incubation. The production of carbon dioxide was highest from 3 to 18 days of incubation, after which the rate of carbon dioxide produced decreased until the 30th day. It was suggested that the metabolism of *p,p′*-DDT was dependent on the extracellular lignin-degrading enzyme system of this fungus (Bumpus et al., 1985).

Mineralization of *p,p′*-DDT by the white rot fungi *Pleurotus ostreatus*, *Phellinus weirri* and *Polyporus versicolor* was also demonstrated (Bumpus and Aust, 1987). *Aerobacter aerogenes* degraded *p,p′*-DDT under aerobic conditions to *p,p′*-DDD, *p,p′*-DDE, 1-chloro-2,2-bis(*p*-chlorophenyl)ethylene, 1-chloro-2,2-bis-(*p*-chlorophenyl)ethane and 1,1-bis(*p*-chlorophenyl)ethylene (Fries, 1972). Under anaerobic conditions the same organism produced four additional compounds. These were bis(*p*-chlorophenyl)acetic acid, *p,p′*-dichlorodiphenylmethane, *p,p′*-dichlorobenzhydrol and *p,p′*-dichlorobenzophenone. Other degradation products of *p,p′*-DDT under aerobic and anaerobic conditions in soils by various cultures not previously mentioned include 1,1-bis(*p*-chlorophenyl)-2,2,2-trichloroethanol (Kelthane) and *p*-chlorobenzoic acid (Fries, 1972).

Under aerobic conditions, the amoeba *Acanthamoeba castellanii* (Neff strain ATCC 30.010) degraded *p,p′*-DDT to *p,p′*-DDE, *p,p′*-DDD and dibenzophenone (Pollero and dePollero, 1978).

Incubation of *p,p'*-DDT with hematin and ammonia gave *p,p'*-DDD, *p,p'*-DDE, bis(*p*-chlorophenyl)acetonitrile, 1-chloro-2,2-bis(*p*-chlorophenyl)ethylene, 4,4'-dichloroben-zophenone and the methyl ester of bis(*p*-chlorophenyl)acetic acid (Quirke et al., 1979).

Chacko et al. (1966) reported DDT dechlorinated to DDD by six actino-mycetes (*Norcardia* sp., *Streptomyces albus, Streptomyces antibioticus, Streptomyces auerofaciens, Streptomyces cinnamoneus, Streptomyces viridochromogenes*) but not by 8 fungi. The maximum degradation observed was 25% in 6 days.

Thirty-five microorganisms isolated from marine sediment and marine water samples taken from Hawaii and Houston, TX were capable of degrading *p,p'*-DDT. *p,p'*-DDD was identified as the major metabolite. Minor transformation products included 2,2-bis(*p*-chlorophenyl)ethanol (DDOH), 2,2-bis(*p*-chlorophenyl)ethane (DDNS) and *p,p'*-DDE (Patil et al., 1972). Similarly, Matsumura et al. (1971) found that *p,p'*-DDT was degraded by numerous aquatic microorganisms isolated from water and silt samples collected from Lake Michigan and its tributaries in Wisconsin. The major metabolites identified were TDE, DDNS and DDE. *p,p'*-DDT was metabolized by the following microorganisms under laboratory conditions to *p,p'*-DDD: *Escherichia coli* (Langlois, 1967), *Aerobacter aerogenes* (Wedemeyer, 1966; Plimmer et al., 1968) and *Proteus vulgaris* (Wedemeyer, 1966). In addition, *p,p'*-DDT was degraded to *p,p'*-DDD, *p,p'*-DDE and dicofol by *Trichoderma viride* (Matsumura and Boush, 1968) and *p,p'*-DDD and *p,p'*-DDE by *Ankistrodemus amalloides* (Neudorf and Khan, 1975).

Jensen et al. (1972) studied the anaerobic degradation of *p,p'*-DDT (100 mg) in 1 L of sewage sludge containing *p,p'*-DDD (4.0%) and *p,p'*-DDE (3.1%) as contaminants. The sludge was incubated at 20°C for 8 days under a nitrogen atmosphere. The parent compound degraded rapidly (half-life 7 hours) forming *p,p'*-DDD, *p,p'*-dichlorodiphenylben-zophenone (DBP), 1,1-bis(*p*-chlorophenyl)-2-chloroethylene (DDMU) and bis(*p*-chlorophenyl)acetonitrile. After 48 hours, the original amount of *p,p'*-DDD added to the sewage sludge had completely reacted. In a similar study, Pfaender and Alexander (1973) observed the cometabolic conversion of DDT (0.005%) in unamended sewage sludge to give DDD, DDE and DBP. When the sewage sludge was amended with glucose (0.10%), the rate of DDD formation was enhanced. However, with the addition of diphenylmethane, the rate of formation of both DDD and DBP was reduced. The diphenylmethane-amended sewage sludge showed the greatest abundance of bacteria capable of cometabolizing DDT, whereas the unamended sewage showed the fewest number of bacteria. Zoro et al. (1974) also reported that *p,p'*-DDT in untreated sewage sludge was converted to *p,p'*-DDD especially in the presence of sodium dithionate, a widely used reducing agent.

In an *in vitro* fermentation study, rumen microorganisms metabolized both isomers of ^{14}C-labeled DDT (*o,p'*- and *p,p'*-) to the corresponding DDD isomers at a rate of 12%/hour. With *p,p'*-DDT, 11% of the ^{14}C detected was an unidentified polar product associated with microbial and substrate residues (Fries et al., 1969). In another *in vitro* study, extracts of *Hydrogenomonas* sp. cultures degraded DDT to DDD, 1-chloro-2,2-bis(*p*-chlorophenyl)ethane (DDMS), DBP and several other products under anaerobic conditions. Under aerobic conditions containing whole cells, one of the rings is cleaved and *p*-chlorophenyl-lacetic acid is formed (Pfaender and Alexander, 1972).

Soil. p,p'-DDD and *p,p'*-DDE are the major metabolites of *p,p'*-DDT in the environment (Metcalf, 1973). In soils under anaerobic conditions, *p,p'*-DDT is rapidly converted to *p,p'*-DDD via reductive dechlorination (Johnsen, 1976) and very slowly to *p,p'*-DDE under aerobic conditions (Guenzi and Beard, 1967; Kearney and Kaufman, 1976). The aerobic degradation of *p,p'*-DDT under flooded conditions is very slow with *p,p'*-DDE forming as the major metabolite. Dicofol was also detected in minor amounts (Lichtenstein et al.,

1971). In addition to *p,p'*-DDD and *p,p'*-DDE, 2,2-bis(*p*-chlorophenyl)acetic acid (DDA), bis(*p*-chlorophenyl)methane (DDM), *p,p'*-dichlorobenzhydrol (DBH), DBP and *p*-chlorophenylacetic acid (PCPA) were also reported as metabolites of *p,p'*-DDT in soil under aerobic conditions (Subba-Rao and Alexander, 1980).

The anaerobic conversion of *p,p'*-DDT to *p,p'*-DDD in soil was catalyzed by the presence of ground alfalfa or glucose (Burge, 1971). Under flooded conditions, *p,p*-DDT was rapidly converted to TDE via reductive dehalogenation and other metabolites (Guenzi and Beard, 1967; Castro and Yoshida, 1971). Degradation was faster in flooded soil than in upland soil and was faster in soils containing high organic matter (Castro and Yoshida, 1971). Other reported degradation products under aerobic and anaerobic conditions by various soil microbes include 1,1'-bis(*p*-chlorophenyl)-2-chloroethane, 1,1'-bis(*p*-chlorophenyl)-2-hydroxyethane and *p*-chlorophenyl acetic acid (Kobayashi and Rittman, 1982). It was also reported that *p,p'*-DDE formed by hydrolyzing *p,p'*-DDT (Wolfe et al., 1977). The clay-catalyzed reaction of DDT to form DDE was reported by Lopez-Gonzales and Valenzuela-Calahorro (1970). They observed that DDT adsorbed of sodium bentonite clay surfaces was transformed more rapidly than on the corresponding hydrogen-bentonite clay. In one day, *p,p'*-DDT reacted rapidly with reduced hematin forming *p,p'*-DDD and unidentified products (Baxter, 1990). In an Everglades muck, *p,p'*-DDT was slowly converted to *p,p'*-DDD and *p,p'*-DDE (Parr and Smith, 1974). The reported half-life in soil is 3,800 days (Jury et al., 1987).

Oat plants were grown in two soils treated with [^{14}C]*p,p'*-DDT. Most of the residues remained bound to the soil. Metabolites identified were *p,p'*-DDE, *o,p'*-DDT, TDE, DBP, dicofol and DDA (Fuhremann and Lichtenstein, 1980).

In a 42-day experiment, ^{14}C-labeled *p,p'*-DDT applied to soil-water suspensions under aerobic and anaerobic conditions gave $^{14}CO_2$ yields of 0.8 and 0.7%, respectively (Scheunert et al., 1987).

The half-lives for *p,p'*-DDT in field soils ranged from 10^6 days to 15.5 years with an average half-life of 4.5 years (Lichtenstein and Schulz, 1959; Lichtenstein et al., 1960; Nash and Woolson, 1967; Lichtenstein et al., 1971; Stewart and Chisolm, 1971; Suzuki et al., 1977). The average half-life of *p,p'*-DDT in a variety of anaerobic soils was 692 days (Burge, 1971; Glass, 1972; Guenzi and Beard, 1976).

p,p'-DDT is very persistent in soil. The percentage of the initial dosage (1 ppm) remaining after 8 weeks of incubation in an organic and mineral soil were 76 and 79%, respectively, while in sterilized controls 100 and 92% remained, respectively (Chapman et al., 1981).

Photolytic. Photolysis of *p,p'*-DDT in nitrogen-sparged methanol solvent by UV light (λ = 260 nm) produced DDD and DDMU. But photolysis of *p,p'*-DDT at 280 nm in an oxygenated methanol solution yielded a complex mixture containing the methyl ester of 2,2-bis(*p*-chlorophenyl)acetic acid (Plimmer et al., 1970). *p,p'*-DDT in an aqueous solution containing suspended titanium dioxide as a catalyst and irradiated with UV light (λ >340 nm) formed chloride ions. Based on the amount of chloride ions generated, carbon dioxide and hydrochloric acid were reported as the end products (Borello et al., 1989).

When *p,p'*-DDT on quartz was subjected to UV radiation (2537 Å) for 2 days, 80% of *p,p'*-DDT degraded to 4,4'-dichlorobenzophenone, 1,1-dichloro-2,2-bis(*p*-chlorophenyl)ethane and 1,1-dichloro-2,2-bis(*p*-chlorophenyl)ethene. Irradiation of *p,p'*-DDT in a hexane solution yielded 1,1-dichloro-2,2-bis(*p*-chlorophenyl)ethane and hydrochloric (Mosier et al., 1969). When an aqueous solution containing *p,p'*-DDT was photooxidized by UV light at 90–95°C, 25, 50 and 75% degraded to carbon dioxide after 25.9, 66.5 and 120.0 hours, respectively (Knoevenagel and Himmelreich, 1976).

Chemical/Physical. In alkaline solutions and temperatures >108.5°C, *p,p'*-DDT undergoes dehydrochlorination via hydrolysis releasing hydrochloric acid to give the noninsecticidal *p,p'*-DDE (Hartley and Kidd, 1987; Kollig, 1993; Worthing and Hance, 1991). This reaction is also catalyzed by ferric and aluminum chlorides and UV light (Worthing and Hance, 1991).

Castro (1964) reported that iron(II) porphyrins in dilute aqueous solution was rapidly oxidized by DDT to form the corresponding iron(III) chloride complex (hematin) and DDE, respectively. Incubation of *p,p'*-DDT with hematin and ammonia gave *p,p'*-DDD, *p,p'*-DDE, bis(*p*-chlorophenyl)acetonitrile, 1-chloro-2,2-bis(*p*-chlorophenyl)ethylene, 4,4'-dichlorobenzophenone and the methyl ester of bis(*p*-chlorophenyl)acetic acid (Quirke et al., 1979).

When *p,p'*-DDT was heated at 900°C, carbon monoxide, carbon dioxide, chlorine, hydrochloric acid and other unidentified substances were produced (Kennedy et al., 1972, 1972a). Emits hydrochloric acid and chlorine when incinerated (Sittig, 1985).

At 33°C, 35% relative humidity and a 2 mile/hour wind speed, the volatility losses of *p,p'*-DDT as a thick film, droplet on grass, droplets on leaves and formulation film on glass after 48 hours were 97.1, 46.0, 40.6 and 5.5%, respectively (Que Hee et al., 1975).

Exposure Limits: NIOSH REL: 0.5 mg/m^3, IDLH 500 mg/m^3; OSHA PEL: TWA 1 mg/m^3; ACGIH TLV: TWA 1 mg/m^3.

Symptoms of Exposure: Paresthesia tongue, lips, face; tremor; apprehension, confusion, malaise; headache; convulsions; paresis of the hands; vomiting; irritates eyes, skin.

Formulation Types: Emulsifiable concentrate; wettable powder; granules; dustable powder; aerosol.

Toxicity: LC_{50} (96-hour) for fathead minnow 19 µg/L, bluegill sunfish 8 µg/L, largemouth bass 2 µg/L, rainbow trout 7 µg/L, brown trout 2 µg/L, coho salmon 4 µg/L, perch 9 µg/L, channel catfish 16 µg/L, black bullhead 5 µg/L, goldfish 21 µg/L, minnow 19 µg/L (Verschueren, 1983); LC_{50} (48-hour) for killifish 2.8 µg/L; EC50 for freshwater bass 3.9 and 1.8 µg/L at 24 and 96 hours, respectively; acute oral LD_{50} for rats 113 mg/kg (Hartley and Kidd, 1987), 87 mg/kg (RTECS, 1985).

Uses: Use as an insecticide is now prohibited.

DESMEDIPHAM

Synonyms: Bentanex; Betanal AM; Betanex; EP 475; 3-Ethoxycarbonylaminophenyl-*N*-phenylcarbamate; **Ethyl 3-phenylcarbamoyloxyphenylcarbamate**; Schering 38107; SN 38107.

$$\text{(structure)} \quad \text{NHCOO} \quad \text{NHCOOCH}_2\text{CH}_3$$

Designations: CAS Registry Number: 13684-56-5; mf: $C_{16}H_{16}N_2O_4$; fw: 300.32; RTECS: FD0425000.

Properties: Colorless to pale yellow crystals. Mp: 120°C; H-$t_{1/2}$: 31 days (pH 5), 14 hours (pH 7), 20 minutes (pH 9); fl p: 68°C (open cup); K_H: 1.7×10^{-10} atm · m^3/mol at 20°C (approximate — calculated from water solubility and vapor pressure); log K_{oc}: 3.18 (calculated); log K_{ow}: 3.39 (pH 3.9); So (g/L at 20°C): acetone (400), benzene (1.6), chloroform (80), ethyl acetate (149), hexane (0.5), methanol (180), toluene (1.2); S_w: 7 mg/L at 20°C; vp: 3×10^{-9} mmHg at 20°C.

Environmental Fate

Soil. Degrades in soil forming the intermediate 3-hydroxycarbanilate (Worthing and Hance, 1991). The reported half-lives in soil are 70 days, 20 hours and 10 minutes at pH values of 5, 7 and 9, respectively (Worthing and Hance, 1991).

Plant. In sugar beets, *m*-aminophenol and ethyl-*N*-(3-hydroxyphenyl)-carbamate were identified as metabolites (Hartley and Kidd, 1987).

Symptoms of Exposure: May cause skin irritation. Symptoms of poisoning include hypoactivity and muscular weakness.

Formulation Types: Emulsifiable concentrate (1.3 and 4 lb/gal); granules (10%).

Toxicity: LC$_{50}$ (96-hour) for rainbow trout 3.8 mg/L and bluegill sunfish 13.4 mg/L (Hartley and Kidd, 1987); acute oral LD$_{50}$ of pure desmedipham and the formulated product for rats 10,300 and 3,700 mg/kg, respectively (Ashton and Monaco, 1991).

Uses: Selective systemic herbicide used to control broad-leaved weeds such as buckwheat, chickweed, fiddleneck, kochnia, mustard, pigweed and ragweed.

DIALIFOS

Synonyms: *S*-(2-Chloro-1-(1,3-dihydro-1,3-dioxo-2*H*-isoindol-2-yl)ethyl) *O,O*-diethyl phosphorodithioate; *S*-(2-Chloro-1-phthalimidoethyl) *O,O*-diethyl phosphorodithioate; Dialifor; *O,O*-Diethyl *S*-(2-chloro-1-phthalimidoethyl) phosphorodithioate; ENT 27320; Hercules 14503; **Phosphorodithioic acid *S*-(2-chloro-1-(1,3-dihydro-1,3-dioxo-2*H*-isoindol-2-yl)ethyl) *O,O*-diethyl ester**; Phosphorodithioic acid *S*-(2-chloro-1-phthalimi-doethyl) *O,O*-diethyl ester; Torax.

Designations: CAS Registry Number: 10311-84-9; mf: $C_{14}H_{17}ClNO_4PS_2$; fw: 393.84; RTECS: TD5165000.

Properties: Colorless to white crystalline solid. Mp: 67–69°C; H-$t_{1/2}$: 14 hours at 20°C and pH 7.4, 1.8 hours at 37.5°C and pH 7.4; K_H: 1.4×10^{-6} atm · m³/mol at 20°C; log K_{oc}: 4.05 (calculated); log K_{ow}: 4.69 (Freed et al., 1979a); So (g/kg at 20°C): acetone (760), chloroform (620), ethanol (<10), ethyl ether (500), isophorone (400), xylene (570); S_w: 0.18 mg/L at room temperature; vp: 6.2×10^{-8} mmHg at 20–25°C (Freed et al., 1979a).

Environmental Fate

Chemical/Physical. Though no products were identified, the hydrolysis half-lives at 20°C were 15 days and 14 hours at pH 6.1 and pH 7.4, respectively (Freed et al., 1979, 1979a).

Formulation Types: Emulsifiable concentrate.

Toxicity: LC_{50} (96-hour) for rainbow trout 0.55–1.08 mg/L and goldfish 1.80–8.30 mg/L (Hartley and Kidd, 1987); acute oral LD_{50} for male and female rats 45–53 and 5 mg/kg, respectively (Hartley and Kidd, 1987).

Uses: Nonsystemic insecticide and acaricide to control chewing and sucking insects and spider mites on a wide variety of fruits and vegetables.

DIALLATE

Synonyms: Avadex; **Bis(1-methylethyl)carbamothioic acid *S*-(2,3-dichloro-2-prope-nyl) ester**; CP 15336; DATC; 2,3-DCDT; Dichloroallyl diisopropylthiocarbamate; *S*-Dichloroallyl diisopropylthiocarbamate; Dichloroallyl *N,N*-diisopropylthiolcarbamate; *S*-2,3-Dichlorodiisopropylthiocarbamate; Pyradex; RCRA waste number U062.

$$(CH_3)_2 \diagdown$$
$$\diagup NCOSCH_2CCl{=}CHCl$$
$$(CH_3)_2 \diagup$$

Designations: CAS Registry Number: 2303-16-4; mf: $C_{10}H_{17}Cl_2NOS$; fw: 270.21; RTECS: EZ8225000.

Properties: Amber to brown, oily liquid. Mp: 25–30°C; bp: 108 and 150°C at 0.25 and 9.0 mmHg, respectively. Decomposes >200°C; ρ: 1.188 at 25/15.6°C; fl p: 144°C (technical), 38°C (emulsifiable concentrate); H-$t_{1/2}$: 6.6 years at 25°C and pH 7; K_H: 2.5×10^{-6} atm · m³/mol at 20–25°C (approximate — calculated from water solubility and vapor pressure); log K_{oc}: 2.28; log K_{ow}: 3.29 (calculated); S_o: soluble or miscible with acetone, benzene, ethanol, ethyl acetate, ethyl ether, hexane, heptane, kerosene, methanol, 2-propanol, xylene and many other organic solvents; S_w: 14 mg/L at 25°C; vap d: 11.04 g/L at 25°C, 9.36 (air = 1); vp: 1.5×10^{-4} mmHg at 20°C.

Environmental Fate

Soil. Soil metabolites include 2,3-dichloroallyl alcohol, 2,3-dichloroallyl mercaptan (Hartley and Kidd, 1987) and carbon dioxide (Smith, 1988). The formation of the alcohol occurs via hydrolysis of the ester linkage and transthiolation of the allylic group (Kaufman, 1967). In an agricultural soil, $^{14}CO_2$ was the only biodegradation identified; however, bound residue and traces of benzene and water-soluble radioactivity were also detected in large amounts (Anderson and Domsch, 1980). The reported half-lives in soil range from 2 to 4 weeks to (Smith and Fitzpatrick, 1970) to approximately 30 days (Hartley and Kidd, 1987). Diallate did not migrate deeper than 5 cm on test field plots (Smith, 1970). In four microbially-active agricultural soils, the half-life was 4 weeks but in sterilized soil the half-life was 20 weeks (Anderson and Domsch, 1976).

Plant. In plants, diallate is metabolized and carbon dioxide is released (Hartley and Kidd, 1987). Diallate undergoes *cis/trans* isomerization and oxidative cleavage when irradiated at 300 nm forming 2,3-dichloroacrolein and 2-chloroacrolein as products (Ruzo and Casida, 1985).

Photolytic. Irradiation of diallate was also conducted in oxygenated chloroform or water until a 10% conversion was obtained. Products formed included acetaldehyde, 2,3-dichloroacrolein and trace amounts of 2-chloroacrolein (Ruzo and Casida, 1985).

Chemical/Physical. Emits toxic fumes of chlorine, nitrogen and sulfur oxides when heated to decomposition (Sax and Lewis, 1987). Though no products were reported, the calculated hydrolysis half-life at 25°C and pH 7 is 6.6 years (Ellington et al., 1988).

141

Symptoms of Exposure: Eye, skin and mucous membrane irritant.

Formulation Types: Emulsifiable concentrate (4 lb/gal); granules (10%).

Toxicity: LC_{50} (96-hour) for rainbow trout 7.9 mg/L and bluegill sunfish 5.9 mg/L (Hartley and Kidd, 1987); LC_{50} (96-hour for technical grade) for rainbow trout 3.2 mg/L, for bluegill sunfish 2.5 mg/L (Humburg et al., 1989); LC_{50} (48-hour for technical grade) for harlequin fish 8.2 mg/L and for *Daphnia magna* 7.5 mg/L (Humburg et al., 1989); acute oral LD_{50} for rats 395 mg/kg (RTECS, 1985).

Uses: Preemergent, selective herbicide used to control wild oats and blackgrass in barley, corn, flax, lentils, peas, potatoes, soybeans and sugar beets.

DIAZINON

Synonyms: Alfatox; Basudin; Basudin 10 G; Bazinon; Bazuden; Dazzel; Desapon; Dianon; Diaterr-fos; Diazatol; Diazide; Diethyl 4-(2-isopropyl-6-methylpyrimidinyl)phosphorothioate; *O,O*-Diethyl *O*-(2-isopropyl-4-methyl-6-pyrimidinyl)phosphorothioate; *O,O*-Diethyl *O*-(2-isopropyl-6-methyl-4-pyrimidinyl)phosphorothioate; *O,O*-Diethyl *O*-(2-isopropyl-4-methyl-6-pyrimimidinyl)thionophosphate; *O,O*-Diethyl 2-isopropyl-4-methylpyrimidinyl-6-thiophosphate; *O,O*-Diethyl *O*-6-methyl-2-isopropyl-4-pyrimidinyl phosphorothioate; ***O,O*-Diethyl *O*-(6-methyl-2-(1-methylethyl)-4-pyrimidinyl) phosphorothioate**; Dimpylate; Dipofene; Dizinon; Dyzol; ENT 19507; G 301; G 24480; Gardentox; Geigy 24480; *O*-2-Isopropyl-4-methylpyrimidinyl-*O,O*-diethyl phosphorothioate; Isopropylmethylpyrimidinyl diethyl thiophosphate; Kayazinon; Kayazol; NA 2763; NCI-C08673; Nedcisol; Neocidol; Nipsan; Nucidol; Sarolex; Spectracide.

Designations: CAS Registry Number: 333-41-5; DOT: 2783; mf: $C_{12}H_{21}N_2O_3PS$; fw: 304.35; RTECS: TF3325000.

Properties: Clear, colorless liquid with a faint ester-like odor. Technical grades are yellow. The reported odor threshold concentration in air and the taste threshold concentration in water are 40 and >55 ppb, respectively (Young et al., 1996). Mp: decomposes >120°C; bp: 83–84°C at 0.0002 mmHg; ρ: 1.116–1.118 at 20/4°C; fl p: difficult to burn; pK_a: <2.5 (estimated); H-$t_{1/2}$ (20°C): 11.77 hours (pH 3.1), 185 days (pH 7.4), 136 days (pH 9.0), 6.0 days (pH 10.4); K_H: 1.12×10^{-7} atm · m³/mol at 23°C (Fendinger and Glotfelty, 1988); log BCF: topmouth gudgeon (*Pseudorasbora parva*) 2.18, silver crucian carp (*Cyprinus auratus*) 1.56, carp (*Cyprinus carpio*) 1.81, guppy (*Labistes reticulatus*) 1.24, crayfish (*Procambarus clarkii*) 0.69, red snail (*Indoplanorbis exustus*) 1.23, pond snail (*Cipangopoludina malleata*) 0.77 (Kanazawa, 1978); log K_{oc}: 3.00–3.27; log K_{ow}: 3.02–3.81; S_o: miscible with acetone, benzene, cyclohexane, dimethyl sulfoxide, ethyl ether, ethanol, ketones, methylene chloride, *n*-octanol, petroleum ether, toluene; S_w: 71.1, 53.5 and 43.7 mg/L at 10, 20 and 30°C, respectively; vp: 8.47×10^{-5} mmHg at 20°C, 1.5×10^{-4} mmHg at 25°C.

Soil properties and adsorption data

Soil	K_d (mL/g)	f_{oc} (%)	K_{oc} (mL/g)	pH (meq/100 g)	CEC
Batcombe silt loam	4.6	2.05	224	6.1	—
Kanuma high clay	2.4	1.35	175	5.7	—
Kaolinite	3.6	0.00	—	4.7	5
Muck	0.9	23.60	38	7.5	66

Soil properties and adsorption data *(continued)*

Soil	K_d (mL/g)	f_{oc} (%)	K_{oc} (mL/g)	pH (meq/100 g)	CEC
Rothamsted Farm	3.4	1.51	225	5.1	—
Sand	15.0	0.81	1,847	7.0	—
Sandy loam	17.9	1.16	1,543	5.4	—
Silt loam	8.2	0.81	1,010	7.0	—
Soil 1	3.6	0.35	1,028	6.3	—
Soil 2	42.4	5.05	840	5.1	—
Soil 3	8.8	2.73	322	5.1	—
Soil 4	1.89	0.46	411	7.5	—
Tsukuba clay loam	13.9	4.24	327	6.5	—

Source: Briggs, 1981; Burnside and Lavy, 1966; Lord et al., 1980; Kanazawa, 1989; U.S. Department of Agriculture, 1990; Arienzo et al., 1993.

Environmental Fate

Biological. Sethunathan and Yoshida (1973a) isolated a *Flavobacterium* sp. (ATCC 27551) from rice paddy water that metabolized diazinon as the sole carbon source. Diazinon was readily hydrolyzed to 2-isopropyl-4-methyl-6-hydroxypyrimidine under aerobic conditions but less rapidly under anaerobic conditions. This bacterium as well as enrichment cultures isolated from a diazinon-treated rice field mineralized the hydrolysis product to carbon dioxide (Sethunathan and Pathak, 1971; Sethunathan and Yoshida, 1973). Rosenberg and Alexander (1979) demonstrated that two strains of *Pseudomonas* grew on diazinon and produced diethyl phosphorothioate as the major end product. The rate of microbial degradation increased in the presence of an enzyme (parathion hydrolase), produced by a mixed culture of *Pseudomonas* sp. (Honeycutt et al., 1984).

Soil. Hydrolyzes in soil to 2-isopropyl-4-methyl-2-hydroxypyrimidine, diethylphosphorothioic acid, carbon dioxide (Getzin, 1967; Lichtenstein et al., 1968; Sethunathan and Yoshida, 1969; Sethunathan and Pathak, 1972; Bartsch, 1974; Wolfe et al., 1976; Somasundaram and Coats, 1991) and tetraethylpyrophosphate (Paris and Lewis, 1973). The half-life of diazinon in soil was observed to be inversely proportional to temperature and soil moisture content (Getzin, 1968). Seven months after diazinon was applied on a sandy loam (2 kg/ha), only 1% of the total applied amount remained and 10% was detected in a peat loam (Suett, 1971).

The reported half-life in soil is 32 days (Jury et al., 1987). Reported half-lives in soil following incubation of 10 ppm diazinon in sterile sand loam, sterile organic soil, nonsterile sandy loam and nonsterile organic soil are 12.5, 6.5, <1 and 2 weeks, respectively (Miles et al., 1979). The reported half-life of diazinon in sterile soil at pH 4.7 was 43.8 days (Sethunathan and MacRae, 1969). Major metabolites identified were diethyl thiophosphoric acid, 2-isopropyl-4-methyl-6-hydroxypyrimidine and carbon dioxide (Konrad et al., 1967). When soil is sterilized, the persistence of diazinon increased more so than changes in soil moisture, soil type or rate of application (Bro-Rasmussen et al., 1968). The half-lives for diazinon in flooded soil incubated in the laboratory ranged from 4 to 17 days with an average half-life of 10 days (Sethunathan and MacRae, 1969; Sethunathan and Yoshida, 1969; Laanio et al., 1972). The mineralization half-life for diazinon in soil was 5.1 years (Sethunathan and MacRae, 1969; Sethunathan and Yoshida, 1969).

The half-lives of diazinon in a sandy loam, clay loam and an organic amended soil under nonsterile conditions were 66–1,496, 49–1,121 and 14–194 days, respectively, while under sterile conditions the half-lives were 57–1,634, 46–1,550 and 14–226 days, respectively (Schoen and Winterlin, 1987).

In a silt loam and sandy loam, reported R_f values were 0.86 and 0.88, respectively (Sharma et al., 1986).

Surface Water. In estuarine water, the half-life of diazinon ranged from 8.2 to 10.2 days (Lacorte et al., 1995).

Groundwater. According to the U.S. EPA (1986) diazinon has a high potential to leach to groundwater.

Plant. Diazinon was rapidly absorbed by and translocated in rice plants. Metabolites identified in both rice plants and a paddy soil were 2-isopropyl-4-methyl-6-hydroxypyrimidine (hydrolysis product), 2-(1'-hydroxy-1'-methyl)ethyl-4-methyl-6-hydroxypyrimidine and other polar compounds (Laanio et al., 1972). Oxidizes in plants to diazoxon (Ralls et al., 1966; Laanio et al., 1972; Wolfe et al., 1976) although 2-isopropyl-4-methyl-6-pyrimidin-6-ol was identified in bean plants (Kansouh and Hopkins, 1968) and as a hydrolysis product in soil (Somasundaram et al., 1991) and water (Suffet et al., 1967). Five days after spraying, pyrimidine ring-labeled ^{14}C-diazinon was oxidized to oxodiazinon which was then hydrolyzed to 2-isopropyl-4-methylpyrimidin-6-ol which in turn, was further metabolized to carbon dioxide (Ralls et al., 1966). Diazinon was transformed in field-sprayed kale plants to form hydroxydiazinon {*O,O*-diethyl-*O*-[2-(2'-hydroxy-2'-propyl)-4-methyl-6-pyrimidinyl] phosphorothioate} which was not previously reported (Pardue et al., 1970).

Diazinon rapidly disappeared from cotton leaves within 24 hours. The concentrations of diazinon residues on cotton leaf following application after 0, 24, 48, 72 and 96 hours were 32.9, 7.56, 4.51, 2.24 and 3.32 mg/m², respectively (Ware et al., 1978).

Photolytic. Direct photolysis of diazinon should occur since it absorbs UV light at wavelengths greater then 290 nm (Gore et al., 1971). *O,O*-Diethyl-*O*-[2-(2'-propyl)-4-methyl-6-pyrimidinyl]phosphorothioate (hydroxydiazinon) was reported as a product of UV irradiation of diazinon (Paris and Lewis, 1973). When diazinon in an aqueous buffer solution (25°C and pH 7.0) was exposed to filtered UV light (λ >290 nm) for 24 hours, 36% decomposed to 2-isopropyl-6-methylpyrimidin-4-ol. The photolysis half-life for this reaction was calculated to be 15 days (Burkhard and Guth, 1979).

Chemical/Physical. In water, diazinon is hydrolyzed following first-order kinetics forming 2-isopropyl-4-methyl-6-hydroxypyrimidine and diethyl thiophosphoric acid or diethyl phosphoric acid in the pH range 3.1 to 10.4. At pH values of 7.4 and 10.4, the persistence of diazinon is 185 and <6 days, respectively (Gomaa et al., 1969). Cowart et al. (1971) reported a half-life of approximately 2–3 weeks in a neutral solution at room temperature. Chapman and Cole (1982) reported the following hydrolysis half-lives of diazinon in a sterile 1% ethanol/water solution at 25°C: 0.45, 2.0, 7.8, 10 and 7.7 weeks at pH values of 4.5, 5.0, 6.0, 7.0 and 8.0, respectively. Diazoxon was also found in fogwater collected near Parlier, CA (Glotfelty et al., 1990).

It was suggested that diazinon was oxidized in the atmosphere during daylight hours prior to its partitioning from the vapor phase into the fog. On 12 January 1986, the distribution of diazinon (1.6 ng/m³) in the vapor phase, dissolved phase, air particles and water particles were 76.1, 19.8, 3.7 and 0.4%, respectively. For diazoxon (0.82 ng/m³), the distribution in the vapor phase, dissolved phase, air particles and water particles were 13.4, 81.7, 4.9 and 0.02%, respectively (Glotfelty et al., 1990).

Diazinon begins to decompose at a temperature of 100°C. During distillation procedures at this temperature, <0.5% is decomposed to 2-isopropyl-4-methyl-6-hydroxypyrimidine. When technical diazinon is dissolved in 20% hydrochloric acid, triethyl thiophosphate is formed (Gysin and Margot, 1958). Above 120°C (Windholz et al., 1983), diazinon decomposes and emits toxic fumes of phosphorus, nitrogen and sulfur oxides (Sax and Lewis, 1987; Lewis, 1990).

In excess water under acidic conditions, diazinon is hydrolyzed to 2-isopropyl-4-methyl-6-hydroxypyrimidine and diethylthiophosphoric acid. With insufficient water, tetraethyl monothiopyrophosphate is formed (Sittig, 1985).

Diazinon is stable to hydrolysis over the pH range of 5–6 (Mortland and Raman, 1967). However, in the presence of a Cu(II) salt (as cupric chloride) or when present as the exchangeable Cu(II) cation in montmorillonite clays, diazinon is completely hydrolyzed via second-order kinetics in <24 hours at 20°C. The calculated half-life of diazinon (0.008 mmole) at 20°C for this reaction is 4.0 hours. It was suggested that decomposition in the presence of Cu(II) was a result of coordination of molecules to the copper atom with subsequent cleavage of the side chain containing the phosphorus atom forming O,O-ethyl-O-phosphorothioate and a compound with the empirical formula $C_7H_{12}N_2$ (Mortland and Raman, 1967).

Diazinon in a buffered, aqueous solution is rapidly degraded by sodium hypochlorite (Clorox, 5% aqueous solution). Initially, diazinon degraded to diazoxon (via oxidation of P=S bond to P=O), 2-isopropyl-4-methyl-6-pyrimidinol and sulfate ion. The pyrimidinol compound, probably formed during the hydrolysis of diazoxon, underwent further degradation forming three compounds including acetic acid. The final reaction products included trichloroacetate ion and chloroform. The rate of hydrolysis was pH dependent. In a carbonate-phosphate buffer solution (0.5–1.0 M), the half-lives were 3,000, 110 and 78 seconds at pH 11.4, 10.3 and 10.0, respectively. At pH 13.1, diazinon did not react after two hours using sodium hydroxide as the buffering agent. At pH 6.8, 5.8 and 4.0, the reaction half-lives were 5,000, 430 and 600 seconds, respectively (Dennis et al., 1979).

Exposure Limits: OSHA PEL: TWA 0.1 mg/m³; ACGIH TLV: TWA 0.1 mg/m³.

Symptoms of Exposure: Irritates eyes and skin.

Formulation Types: Emulsifiable concentrate; wettable powder; granules; aerosol; dry seed treatment; capsule suspension, smoke tablet; coating agent; microcapsule suspension.

Toxicity: LC_{50} (96-hour) for rainbow trout 16 mg/L, bluegill sunfish 2.6–3.2 mg/L, carp 7.6–23.4 mg/L (Worthing and Hance, 1991), fathead minnow 2.7–10.0 mg/L (Verschueren, 1983); LC_{50} (24-hour) for bluegill sunfish 52 ppb and rainbow trout 380 ppb (Verschueren, 1983); acute oral LD_{50} for rats 240–480 mg/kg (Hartley and Kidd, 1987), 66 mg/kg (RTECS, 1985).

Uses: Nonsystemic contact insecticide used against flies, aphids and spider mites in soil, fruit, vegetables and ornamentals.

1,2-DIBROMO-3-CHLOROPROPANE

Synonyms: BBC 12; 1-Chloro-2,3-dibromopropane; 3-Chloro-1,2-dibromopropane; DBCP; Dibromochloropropane; Fumagon; Fumazone; Fumazone 86; Fumazone 86E; NCI-C00500; Nemabrom; Nemafume; Nemagon; Nemagon 20; Nemagon 20G; Nemagon 90; Nemagon 206; Nemagon soil fumigant; Nemanax; Nemapaz; Nemaset; Nematocide; Nematox; Nemazon; OS 1987; Oxy DBCP; RCRA waste number U066; SD-1897; UN 2872.

Designations: CAS Registry Number: 96-12-8; DOT: 2872; mf: $C_3H_5Br_2Cl$; fw: 236.36; RTECS: TX8750000.

Properties: Colorless to brown liquid with a mildly pungent odor. Mp: 6.1°C; bp: 196°C; ρ: 2.05 at 20/4°C; fl p: 76.7°C (open cup); H-$t_{1/2}$: 38 ± 4 years at 25°C and pH 7; K_H: 2.49 × 10^{-4} atm · m³/mol at 20°C (approximate — calculated from water solubility and vapor pressure); log K_{oc}: 1.49–2.16; log K_{ow}: 2.63 (calculated); S_o: miscible with oils, acetone, dichloropropane, dimethylsulfoxide, 95% ethanol, halogenated hydrocarbons, isopropanol, liquid hydrocarbons and methanol; S_w: 1,230 mg/L at 20°C; vap d: 9.66 g/L at 25°C, 8.16 (air = 1); vp: 0.8 mmHg at 20°C.

Soil properties and adsorption data

Soil	K_d (mL/g)	f_{oc} (%)	K_{oc} (mL/g)	pH (meq/100 g)	CEC
Fresno aquifer solid (53.6 m)	0.06	0.02	305	7.7	—
Fresno aquifer solid (109.7 m)	0.07	0.02	355	7.3	—
Panoche clay loam	0.20	0.65	31	—	22.2
Panoche clay loam	0.39	0.27	144	—	17.2
Panoche clay loam	0.30	0.29	103	—	17.4

Source: Biggar et al., 1984; Deeley et al., 1991.

Environmental Fate

Biological. Biodegradation is not expected to be significant in removing 1,2-dibromo-3-chloropropane. In aerobic soil columns, no degradation was observed after 25 days (Wilson et al., 1981a).

Soil. Soil water cultures converted 1,2-dibromo-3-chloropropane to *n*-propanol, bromide and chloride ions. Precursors to the alcohol formation include allyl chloride and allyl alcohol (Castro and Belser, 1968). The reported half-life in soil is 6 months (Jury et al., 1987).

Groundwater. According to the U.S. EPA (1986) 1,2-dibromo-3-chloropropane has a high potential to leach to groundwater.

Chemical/Physical. 1,2-Dibromo-3-chloropropane is subject to both neutral and base-mediate hydrolysis (Kollig, 1993). Under neutral conditions, the chlorine or bromine atoms may be displaced by hydroxyl ions. If nucleophilic attack occurs at the carbon-chlorine bond, 2,3-dibromopropanol is formed which reacts further to give 2,3-dihydroxybromopropane via the intermediate epibromohydrin. 2,3-Dihydroxybromopropane will undergo hydrolysis via the intermediate 1-hydroxy-2,3-propylene oxide which further reacts with water to give glycerol. If the nucleophilic attack occurs at the carbon-bromine bond, 2-bromo-3-chloropropanol is formed which further reacts forming the end product glycerol (Kollig, 1993). If hydrolysis of 1,2-dibromo-2-chloropropane occurs under basic conditions, the compound will undergo dehydrohalogenation to form 2-bromo-3-chloropropene and 2,3-dibromo-1-propene as intermediates. Both compounds are subject to further attack forming 2-bromo-3-hydroxypropene as the end product (Burlinson et al., 1982; Kollig, 1993). The hydrolysis half-life at pH 7 and 25°C was calculated to be 38 years (Burlinson et al., 1982; Ellington et al., 1986).

Emits toxic chloride and bromide fumes when heated to decomposition (Lewis, 1990).

Exposure Limits: OSHA PEL: TWA 1 ppb.

Formulation Types: Fumigant.

Toxicity: Acute oral LD_{50} for male rats 173 mg/kg (Verschueren, 1983), 170 mg/kg (RTECS, 1985).

Uses: Soil fumigant, nematocide and pesticide.

DI-*n*-BUTYL PHTHALATE

Synonyms: 1,2-Benzenedicarboxylate; **1,2-Benzenedicarboxylic acid dibutyl ester**; *o*-Benzenedicarboxylic acid dibutyl ester; Benzene-*o*-dicarboxylic acid di-*n*-butyl ester; Butyl phthalate; *n*-Butyl phthalate; Celluflex DPB; DBP; Dibutyl-1,2-benzenedicarboxylate; Dibutyl phthalate; Elaol; Hexaplas M/B; Palatinol C; Phthalic acid dibutyl ester; Polycizer DBP; PX 104; RCRA waste number U069; Staflex DBP; Witicizer 300.

Designations: CAS Registry Number: 84-74-2; DOT: 9095; mf: $C_{16}H_{22}O_4$; fw: 278.35; RTECS: TI0875000.

Properties: Colorless, oily, viscous liquid with a mild aromatic or ammoniacal odor. Mp: −35°C; bp: 340°C; ρ: 1.0459 at 20/4°C; fl p: 157°C, 171°C (open cup); lel: 0.5% at 235°C; uel: 2.5% (calculated); K_H: 6.3 × 10⁻⁵ atm · m³/mol at 20–25°C (approximate — calculated from water solubility and vapor pressure); log BCF: 3.61 (*Chlorella pyrenoidosa*, Yan et al., 1995); log K_{oc}: 3.14; log K_{ow}: 4.31–4.79; S_o: soluble in ethanol, benzene, ethyl ether; very soluble in acetone; miscible with acetone, dimethylsulfoxide, 95% ethanol and most organic solvents; S_w: 10.1 mg/L at 20°C; vap d: 11.38 g/L at 25°C, 9.61 (air = 1); vp: 4.2 × 10⁻⁵ mmHg at 25°C.

Environmental Fate

Biological. Under aerobic conditions using a freshwater hydrosol, mono-*n*-butyl phthalate and phthalic acid were produced. Under anaerobic conditions, phthalic acid was not present (Verschueren, 1983). In anaerobic sludge, di-*n*-butyl phthalate degraded as follows: monobutyl phthalate to phthalic acid to protocatechuic acid followed by ring cleavage and mineralization (Shelton et al., 1984). Engelhardt et al. (1975) reported that a variety of microorganisms were capable of degrading of di-*n*-butyl phthalate and suggested the following degradation scheme: di-*n*-butyl phthalate to mono-*n*-butyl phthalate to phthalic acid to 3,4-dihydroxybenzoic acid and other unidentified products. Di-*n*-butyl phthalate was degraded to benzoic acid by tomato cell suspension cultures (*Lycopericon lycopersicum*) (Pogány et al., 1990).

In a static-culture-flask screening test, di-*n*-butyl phthalate showed significant biodegradation with rapid adaptation. The ester (5 and 10 mg/L) was statically incubated in the dark at 25°C with yeast extract and settled domestic wastewater inoculum. After 7 days, 100% biodegradation was achieved (Tabak et al., 1981).

Soil. Under aerobic conditions using a fresh-water hydrosol, mono-*n*-butyl phthalate and phthalic acid were produced. Under anaerobic conditions, however, phthalic acid was not formed (Verschueren, 1983).

Photolytic. An aqueous solution containing titanium dioxide and subjected to UV radiation (λ >290 nm) produced hydroxyphthalates and dihydroxyphthalates as intermediates (Hustert and Moza, 1988).

Chemical/Physical. Pyrolysis of di-*n*-butyl phthalate in the presence of polyvinyl chloride at 600°C gave the following compounds: indene, methylindene, naphthalene, 1-methylnaphthalene, 2-methylnaphthalene, biphenyl, dimethylnaphthalene, acenaphthene, fluorene, methylacenaphthene, methylfluorene and six unidentified compounds (Bove and Dalven, 1984).

Under alkaline conditions, di-*n*-butyl phthalate will initially hydrolyze to *n*-butyl hydrogen phthalate and *n*-butanol. The monoester will undergo further hydrolysis forming *o*-phthalic acid and *n*-butanol (Kollig, 1993).

Exposure Limits: NIOSH REL: TWA 5 mg/m^3, IDLH 4,000 mg/m^3; OSHA PEL: TWA 5 mg/m^3; ACGIH TLV: TWA 5 mg/m^3.

Symptoms of Exposure: Irritates nasal passages, stomach, upper respiratory system; light sensitivity.

Formulation Types: Aerosol.

Toxicity: Acute oral LD$_{50}$ for rats 8,000 mg/kg (RTECS, 1985).

Use: Insect repellent.

DICAMBA

Synonyms: Banex; Banvel; Banvel CST; Banvel D; Banvel herbicide; Banvel II herbicide; Banvel 4S; Banvel 4WS; Brush buster; Compound B dicamba; Dianate; Dicambe; 3,6-Dichloro-*o*-anisic acid; **2,5-Dichloro-6-methoxybenzoic acid**; 3,6-Dichloro-2-methoxy-benzoic acid; MDBA; Mediben; 2-Methoxy-3,6-dichlorobenzoic acid; NA 2769; Velsicol compound 'R'; Velsicol 58-CS-11.

Designations: CAS Registry Number: 1918-00-9; DOT: 2769; mf: $C_8H_6Cl_2O_3$; fw: 221.04; RTECS: DG7525000.

Properties: Odorless, colorless to pale buff crystals. Mp: 114–116°C; bp: decomposes >200°C; ρ: 1.57 at 25/4°C; fl p: nonflammable; pK_a: 1.90, 1.95; K_H: 1.2 × 10^{-9} atm · m^3/mol at 20–25°C (approximate — calculated from water solubility and vapor pressure); log K_{oc}: -0.40, 0.34; log K_{ow}: 0.48; So (g/L at 25°C): acetone (810), 1,4-diacetone alcohol (910), cyclohexanone (916), diacetone alcohol (910), 1,4-dioxane (1,180), ethanol (922), heavy aromatic naphthalene solvent (52), methylene chloride (260), toluene (130), xylene (78); S_w: 6.5 g/L at 25°C; vp: 3.38 × 10^{-5} mmHg at 20°C.

Soil properties and adsorption data

Soil	K_d (mL/g)	f_{oc} (%)	K_{oc} (mL/g)	pH (meq/100 g)	CEC
Indian Head loam	0.03	2.35	1.28	7.8	—
Melfort loama	0.08	6.05	1.32	5.9	—
Melfort loam	0.11	6.05	1.82	5.9	—
Weyburn Oxbow loama	0.07	3.72	1.88	6.5	—

Source: Grover and Smith, 1974; Grover, 1977. a) Dimethylamine salt.

Environmental Fate

Biological. In a model ecosystem containing sand, water, plants and biota, dicamba was slowly transformed to 5-hydroxydicamba (10% after 32 days) which slowly underwent decarboxylation (Yu et al., 1975).

Soil. Smith (1974) studied the degradation of ^{14}C-ring- and ^{14}C-carboxyl-labeled dicamba in moist prairie soils at 25°C. After 4 weeks, >50% of the herbicide degraded to the principal products 3,6-dichlorosalicylic acid and carbon dioxide (Smith, 1974).

The half-lives for dicamba in soil incubated in the laboratory under aerobic conditions ranged from 0 to 32 days (Altom and Stritzke, 1973; Smith, 1973, 1974; Smith and Cullimore, 1975). In field soils, the half-lives for dicamba ranged from 6 to 10 days with an average half-life of 7 days (Scifres and Allen, 1973; Stewart and Gaul, 1977). The mineralization half-lives for dicamba in soil ranged from 147 to 309 days (Smith, 1974;

151

Smith and Cullimore, 1975). In a Regina heavy clay, the loss of dicamba was rapid. Approximately 10% of the applied dosage was recovered after 5 weeks. At the end of 5 weeks, approximately 28% was transformed to 3,6-dichlorosalicylic acid and carbon dioxide (Smith, 1973a).

Groundwater. According to the U.S. EPA (1986) dicamba has a high potential to leach to groundwater.

Plant. Dicamba is hydrolyzed in wheat and Kentucky bluegrass plants to 5-hydroxy-2-methoxy-3,6-dichlorobenzoic acid and 3,6-dichlorosalicylic acid at yields of 90 and 5%, respectively. The remaining 5% was unreacted dicamba (Broadhurst et al., 1966). Dicamba was absorbed from treated soils, translocated in corn plants and then converted to 3,6-dichlorosalicylic acid, *p*-aminobenzoic acid and benzoic acid (Krumzdorf, 1974).

Photolytic. When dicamba on silica gel plates was exposed to UV radiation ($\lambda = 254$ nm), it slowly degraded to the 5-hydroxy analog and water solubles (Humburg et al., 1989).

Chemical/Physical. Reacts with alkalies (Hartley and Kidd, 1987), amines and alkali metals (Worthing and Hance, 1991) forming very water-soluble salts.

When dicamba was heated at 900°C, carbon monoxide, carbon dioxide, chlorine, hydrochloric acid, oxygen and ammonia were produced (Kennedy et al., 1972, 1972a).

Formulation Types: Granules (10%); water-soluble liquid (1 and 4 lb/gal).

Toxicity: LC_{50} (96-hour) for rainbow trout and bluegill sunfish 135 mg/L (Hartley and Kidd, 1987); LC_{50} (48-hour) for bluegill sunfish 20.0 mg/L (Verschueren, 1983), for *Daphnia magna* 110.7 mg/L, for grass shrimp >100.0 mg/L, for both sheaps-head minnow and fiddler crab >180.0 mg/L (Humburg et al., 1989); acute oral LD_{50} of technical dicamba for rats 1,700 mg/kg (Ashton and Monaco, 1991), pure dicamba 1,039 mg/kg (RTECS, 1985).

Uses: Selective, systemic preemergence and postemergence herbicide used to control both annual and perennial broad-leaved weeds, chickweed, mayweed and bindweed in cereals and other related crops.

DICHLOBENIL

Synonyms: Barrier 2G; Barrier 50W; Carsoron; Casaron; Casoron; Casoron 133; Casoron G; Casoron G-4; Casoron G-10; Casoron W-50; Code H 133; 2,6-DBN; DCB; Decabane; **2,6-Dichlorobenzonitrile**; Dusprex; Dyclomec; Dyclomec G2; Dyclomec 4G; H 133; H 1313; NA 2769; NIA 5996; Niagara 5006; Niagara 5996; Norosac; Norosac 4G; Norosac 10G.

Designations: CAS Registry Number: 1194-65-6; DOT: 2769; mf: $C_7H_3Cl_2N$; fw: 172.02; RTECS: DI3500000.

Properties: Colorless, white to off-white crystals with an aromatic odor. The reported odor threshold concentration in air is 40 ppb (Young et al., 1996). Mp: 144–145°C; bp: 270–270.1°C; fl p: nonflammable; K_H: 6.6×10^{-6} atm · m^3/mol at 20–25°C (approximate — calculated from water solubility and vapor pressure); log K_{oc}: 2.03–2.32; log K_{ow}: 2.90; So (g/L at 8°C unless noted otherwise): acetone (50), benzene (50), 2-butanone (7 g/L at 15–50°C), cyclohexanone (70 g/L at 15–20°C), ethanol (50), furfural (70), methylene chloride (10 g/L at 20°C), 2-propanol, tetrahydrofuran (90), toluene (40), xylene (50); S_w: 18 mg/L at 20°C, 25 mg/L at 25°C; vp: 5.5×10^{-4} mmHg at 25°C, 1.5×10^{-2} mmHg at 50°C, 1.1 mmHg at 100°C.

Soil properties and adsorption data

Soil	K_d (mL/g)	f_{oc} (%)	K_{oc} (mL/g)	pH
Amenia silt loam	5.0	3.4	147	6.2
Camroden silt loam	6.6	4.8	138	4.6
Colonie loamy fine sand	1.9	0.9	211	5.0
Covington silty clay	6.7	5.3	126	5.7
Croghan loamy fine sand	3.7	2.7	137	5.4
Empeyville stony loam	7.8	5.6	139	4.9
Granby fine sandy loam	5.1	3.0	170	6.8
Howard gravelly loam	4.4	3.1	142	4.6
Lackawanna stony loam	6.0	5.6	107	4.6
Lima gravelly silt	4.2	2.6	162	6.3
Mardin silt loam	4.5	2.9	155	5.9
Sodus gravelly loam	5.3	3.1	171	4.8
Troy gravelly silt loam	3.9	2.4	163	3.8
Vergennes clay	4.1	2.6	158	4.8

Soil properties and adsorption data *(continued)*

Soil	K_d (mL/g)	f_{oc} (%)	K_{oc} (mL/g)	pH
Vergennes silt loam	4.5	2.7	167	4.6
Williamson silt loam	4.3	2.6	165	3.9

Source: Briggs and Dawson, 1970.

Environmental Fate

Biological. A cell suspension of *Arthrobacter* sp., isolated from a hydrosol, degraded dichlobenil to 2,6-dichlorobenzamide (71% yield) and several unidentified water soluble metabolites (Miyazaki et al., 1975). This microorganism was capable of rapidly degrading dichlobenil in aerobic sediment-water suspensions and in enrichment cultures (Miyazaki et al., 1975).

Soil. The major soil metabolite is 2,6-dichlorobenzamide which undergoes further degradation to form 2,6-dichlorobenzoic acid. The estimated half-lives ranged from 1 to 12 months (Hartley and Kidd, 1987). Under field conditions, dichlobenil persists from 2 to 12 months (Ashton and Monaco, 1991). The disappearance of dichlobenil from a hydrosol and pond water was primarily due to volatilization and biodegradation. The time required for 50 and 90% dissipation of the herbicide from a hydrosol were approximately 20 and 50 days, respectively (Rice et al., 1974). Dichlobenil has a high vapor pressure and volatilization should be an important process. Williams and Eagle (1979) found that the half-life of dichlobenil was 4 weeks in soil 4–8 weeks after application. After 1 year following application, the half-life increased to 1 year.

Plant. In plants, dichlobenil is transformed into glucose conjugates, insoluble residues and hydroxy products that are phytotoxic (Ashton and Monaco, 1991). These include three phytotoxic compounds, namely 2,6-dichlorobenzonitrile, 3-hydroxy-2,6-dichlorobenzonitrile and 4-hydroxy-2,6-dichlorobenzonitrile (Duke et al., 1991). Massini (1961) provided some evidence that dichlobenil is metabolized by plants. French dwarf beans, tomatoes, gherkin and oat plants were all exposed to a saturated atmosphere of dichlobenil at room temperature for 4 days. Most of the herbicide was absorbed and translocated by the plants in 3 days. After 6 days of exposure, bean seedlings were analyzed for residues using thin-layer plate chromatography. In addition to dichlobenil, another compound was found but it was not 2,6-dichlorobenzoic acid (Massini, 1963).

Surface Water. The time required for 50 and 90% dissipation of the herbicide from New York pond water was approximately 21 and 60 days, respectively (Rice et al., 1974).

Photolytic. When dichlobenil was irradiated in methanol with a 450-W mercury lamp and a Corex filter for 8 hours, *o*-chlorobenzonitrile and benzonitrile formed as the major and minor products, respectively (Plimmer, 1970).

Chemical/Physical. Dichlobenil is hydrolyzed, especially in the presence of alkali, to 2,6-dichlorobenzamide (Briggs and Dawson, 1970; Worthing and Hance, 1991).

Emits toxic fumes of nitrogen oxides and chlorine when heated to decomposition (Sax and Lewis, 1987).

Formulation Types: Granules (4 and 10%).

Toxicity: LC$_{50}$ (48-hour) for guppies >18 mg/L (Worthing and Hance, 1991), *Daphnia magna* 10 mg/L (Sanders, 1970), rainbow trout 22.0 mg/L and bluegill sunfish 20 mg/L

(Verschueren, 1983); acute oral LD_{50} for rats 3,100 mg/kg (Ashton and Monaco, 1991), 2,710 mg/kg (RTECS, 1985).

Uses: Soil-applied herbicide used to control many annual and perennial broad-leaved weeds.

DICHLONE

Synonyms: Algistat; Compound 604; **2,3-Dichloro-1,4-naphthalenedione**; 2,3-Dichloro-1,4-naphthaquinone; Dichloronaphthoquinone; 2,3-Dichloronaphthoquinone; 2,3-Dichloro-1,4-naphthoquinone; 2,3-Dichloro-α-naphthoquinone; 2,3-Dichloronaphthoquinone-1,4; ENT 3776; NA 2761; Phygon; Phygon paste; Phygon seed protectant; Phygon XL; Quintar; Quintar 540F; Sanquinon; U.S. Rubber 604; Uniroyal; USR 604.

Designations: CAS Registry Number: 117-80-6; DOT: 2761; mf: $C_{10}H_4Cl_2O_2$; fw: 227.06; RTECS: QL7525000.

Properties: Golden yellow leaflets or crystals. Mp: 193°C (sublimes >32°C); bp: 275°C at 2 mmHg; log K_{oc}: 4.19 (calculated); log K_{ow}: 5.62 (calculated); S_o: moderately soluble in acetone, acetic acid, benzene, *o*-dichlorobenzene (≈ 4 wt %), *N,N*-dimethylformamide, 1,4-dioxane, ethanol, ethylbenzene, ethyl ether, ethyl acetate, toluene, xylene (≈ 4 wt %) and many other organic solvents; S_w: 1.0 mg/L at 25°C.

Environmental Fate

Plant. In plants, dichlone loses both chlorine atoms and are replaced by sulphydryl groups to give a substituted dimercapto compound (Hartley and Kidd, 1987).

Photolytic. The UV absorption band for dichlone is 330 nm (Gore et al., 1971). Irradiation of dichlone in a variety of organic solvents (benzene, isopropanol, ethanol) using UV light produced a number of dehalogenated compounds. In the absence or presence of oxygen, 2-chloro-1,4-naphthoquinone, 1,4-naphthoquinone and 1,4-naphthalenediol were produced. Further irradiation in the presence of oxygen yielded phthalic acid and phthalic anhydride as the major products. In a mixture of benzene and isopropanol, dichlone degraded to the minor products: 2-chloro-3-hydroxy-1,4-naphthoquinone, 2-chloro-3-phenoxy-1,4-naphthoquinone, 2,3-dichloro-4-hydroxy-1-keto-2-phenyl-1,2-dihydronaphthalene and isopropyl-1-chloro-2,3-dioxo-1-indanecarboxylate (Ide et al., 1979).

Chemical/Physical. Emits toxic fumes of chlorine when heated to decomposition (Sax and Lewis, 1987).

Symptoms of Exposure: Irritates skin, eyes and mucous membranes; central nervous system depressant.

Formulation Types: Wettable powder.

Toxicity: LC_{50} (96-hour) for brown trout 310 µg/L (Hartley and Kidd, 1987); LC_{50} (48-hour) for bluegill sunfish 70 µg/L, largemouth bass 120 µg/L, *Daphnia magna* 25 µg/L

(Verschueren, 1983); acute oral LD_{50} for rats 1,300 mg/kg (Hartley and Kidd, 1987), 160 mg/kg (RTECS, 1985).

Uses: Fungicide used on fruits, field crops and vegetables.

1,2-DICHLOROPROPANE

Synonyms: α,β-Dichloropropane; ENT 15406; NCI-C55141; Propylene chloride; Propylene dichloride; α,β-Propylene dichloride; RCRA waste number U083.

Designations: CAS Registry Number: 78-87-5; DOT: 1279; mf: $C_3H_6Cl_2$; fw: 112.99; RTECS: TX9625000.

Properties: Colorless, mobile liquid with a sweet, chloroform-like odor. Mp: −100.4°C, −70°C; bp: 96.4°C; ρ: 1.560 at 20/4°C; fl p: 15.6°C; lel: 3.4%; uel: 14.5%; H-$t_{1/2}$: 15.8 years at 25°C and pH 7; K_H: 2.94 × 10^{-3} atm · m³/mol at 25°C; IP: 10.87 eV; log K_{oc}: 1.431, 1.71; log K_{ow}: 2.28; S_o: miscible with acetone, dimethylsulfoxide, 95% ethanol and most other organic solvents; S_w: 2,700 mg/L at 20°C; vap d: 4.62 g/L at 25°C, 3.90 (air = 1); vp: 42 mmHg at 20°C.

Environmental Fate

Biological. 1,2-Dichloropropane showed significant degradation with gradual adaptation in a static-culture flask-screening test (settled domestic wastewater inoculum) conducted at 25°C. At concentrations of 5 and 10 mg/L, percent losses after 4 weeks of incubation were 89 and 81, respectively. The amount lost due to volatilization was only 0–3% (Tabak et al., 1981).

Soil. Boesten et al. (1992) investigated the transformation of ^{14}C-labeled 1,2-dichloropropane under laboratory conditions of three sub-soils collected from the Netherlands (Wassenaar low-humic sand, Kibbelveen peat, Noord-Sleen humic sand podsol). The groundwater saturated soils were incubated in the dark at 9.5–10.5°C. In the Wassenaar soil, no transformation of 1,2-dichloropropane was observed after 156 days of incubation. After 608 and 712 days, however, >90% degraded to nonhalogenated volatile compounds which were detected in the headspace above the soil. These investigators postulated these compounds could be propylene and *n*-propane in a ratio of 8:1. Degradation of 1,2-dichloropropane in the Kibbelveen peat and Noord-Sleen humic sand podsoil was not observed possibly because the soil redox potentials in both soils (50–180 and 650–670 mV, respectively) were higher than the redox potential in the Wassenaar soil (10–20 mV).

Groundwater. According to the U.S. EPA (1986) 1,2-dichloropropane has a high potential to leach to groundwater.

Photolytic. Distilled water irradiated with UV light (λ = 290 nm) yielded the following photolysis products: 2-chloro-1-propanol, allyl chloride, allyl alcohol and acetone. The half-lives in distilled water and distilled water containing hydrogen peroxide were 50 and 30 minutes, respectively (Milano et al., 1988).

Chemical/Physical. Hydrolysis of 1,2-dichloropropane in distilled water at 25°C produced 1-chloro-2-propanol and hydrochloric acid (Milano et al., 1988). The calculated hydrolysis half-life at 25°C and pH 7 is 15.8 years (Ellington et al., 1988). Ozonolysis yielded carbon dioxide at low ozone concentrations (Medley and Stover, 1983).

Emits toxic chloride fumes when heated to decomposition (Lewis, 1990).

Exposure Limits: NIOSH REL: IDLH 400 ppm; OSHA PEL: TWA 75 ppm; ACGIH TLV: TWA 75 ppm, STEL 110 ppm.

Symptoms of Exposure: Eye and skin irritation, drowsiness, light-headedness.

Toxicity: LC_{50} (7-day) for guppies 116 ppm (Verschueren, 1983); acute oral LD_{50} for rats 2,196 mg/kg (RTECS, 1985).

Uses: Soil fumigant for nematodes.

cis-1,3-DICHLOROPROPYLENE

Synonyms: *cis*-1,3-Dichloropropene; *cis*-1,3-Dichloro-1-propene; (Z)-1,3-Dichloropropene; **(Z)-1,3-Dichloro-1-propene**; 1,3-Dichloroprop-1-ene; *cis*-1,3-Dichloro-1-propylene.

Designations: CAS Registry Number: 10061-01-5; DOT: 2047 (isomeric mixture); mf: $C_3H_4Cl_2$; fw: 110.97; RTECS: UC8325000.

Properties: Colorless to amber-colored liquid with a sweet, penetrating, chloroform-like odor. Mp: –84°C (isomeric mixture), bp: 104.3°C; ρ: 1.224 at 20/4°C; fl p: 35°C (isomeric mixture); lel: 5.3% (isomeric mixture); uel: 14.5% (isomeric mixture); K_H: 3.55 × 10^{-3} atm · m^3/mol; log K_{oc}: 1.36, 1.68; log K_{ow}: 1.41; S_o: soluble in benzene, chloroform, 95% ethanol, ethyl ether, *n*-octane, toluene; S_w: 2,700 mg/L at 25°C; vap d: 4.54 g/L at 25°C, 3.83 (air = 1); vp: 25 mmHg at 20°C.

	$K_d{}^a$	f_{oc}	$K_{oc}{}^a$	pH	
Soil	**(mL/g)**	**(%)**	**(mL/g)**	**(meq/100 g)**	**CEC**
Humus sand	14.0	3.17	442	—	—
Humus sand	22.0	3.17	694	—	—
Humus sand	38.0	3.17	1,199	—	—
Peaty sand	47.0	10.39	452	—	—
Peaty sand	78.0	10.39	751	—	—
Peaty sand	130.0	10.39	1,251	—	—

Soil properties and adsorption data

Source: Leistra, 1970. a) The reported K_d and calculated K_{oc} values are expressed on a soil/vapor relationship, i.e., (μg absorbed per g of dry soil)/(μg vapor per mL vapor phase). The author reported an average K_{om} (μg absorbed per g of organic matter)/(μg dissolved per mL water phase) of 14 which is approximately equal to a K_{oc} value of 26.

Environmental Fate

Biological. cis-1,3-Dichloropropylene was reported to hydrolyze to 3-chloro-2-propen-1-ol and can be biologically oxidized to 3-chloropropenoic acid, which is further oxidized to formylacetic acid. Decarboxylation of this compound yields carbon dioxide (Connors et al., 1990). The isomeric mixture showed significant degradation with gradual adaptation in a static-culture flask-screening test (settled domestic wastewater inoculum) conducted at 25°C. At concentrations of 5 and 10 mg/L, percent losses after 4 weeks of incubation were 85 and 84, respectively. Ten days into the incubation study, 7–19% was lost due to volatilization (Tabak et al., 1981).

Soil. Hydrolyzes in wet soil forming *cis*-3-chloroallyl alcohol (Castro and Belser, 1966).

Chemical/Physical. Hydrolyzes in distilled water at 25°C forming 2-chloro-3-propenol and hydrochloric acid. The half-life was 1 day (Milano et al., 1988; Kollig, 1993). Chloroacetaldehyde, formyl chloride and chloroacetic acid were formed from the ozonation of dichloropropylene at about 23°C and 730 mmHg. Chloroacetaldehyde and formyl chloride also formed from the reaction of dichloropropylene with hydroxyl radicals (Tuazon et al., 1984).

The evaporation half-life of *cis*-1,3-dichloropropylene (1 mg/L) from water at 25°C using a shallow-pitch propeller stirrer at 200 rpm at an average depth of 6.5 cm is 29.6 minutes (Dilling, 1977).

Emits chlorinated acids when incinerated. Incomplete combustion may release toxic phosgene (Sittig, 1985).

Formulation Types: Liquid.

Toxicity: LC_{50} (96-hour) for bluegill sunfish 7.1 mg/L and rainbow trout 7.1 mg/L (Worthing and Hance, 1991); acute oral LD_{50} of the isomeric mixture for male and female rats is 713 and 470 mg/kg, respectively (Verschueren, 1983).

Uses: The isomeric mixture is used as a soil fumigant and a nematocide.

trans-1,3-DICHLOROPROPYLENE

Synonyms: (*E*)-1,3-Dichloropropene; *trans*-1,3-Dichloropropene; **(*E*)-1,3-Dichloro-1-propene**; *trans*-1,3-Dichloro-1-propene; 1,3-Dichloroprop-1-ene; *trans*-1,3-Dichloro-1-propylene.

Designations: CAS Registry Number: 10061-02-6; DOT: 2047 (isomeric mixture); mf: $C_3H_4Cl_2$; fw: 110.97; RTECS: UC8320000.

Properties: Clear, colorless liquid with a chloroform-like odor. Mp: −84°C (isomeric mixture), bp: 77°C at 757 mmHg, 112.1°C; ρ: 1.1818 at 20/4°C; fl p: 5.3°C (isomeric mixture); lel: 5.3% (isomeric mixture); uel: 14.5% (isomeric mixture); K_H: 3.55×10^{-3} atm · m³/mol; log K_{oc}: 1.415, 1.68; log K_{ow}: 1.41; S_o: soluble in acetone, benzene, chloroform, dimethylsulfoxide, 95% ethanol, ethyl ether, octane, toluene; S_w: 2,800 mg/L at 25°C; vap d: 4.54 g/L at 25°C, 3.83 (air = 1); vp: 25 mmHg at 20°C.

Soil properties and adsorption data

Soil	K_d^a (mL/g)	f_{oc} (%)	$K_{oc}^{\ a}$ (mL/g)	pH (meq/100 g)	CEC
Humus sand	14.0	3.17	442	—	—
Humus sand	22.0	3.17	694	—	—
Humus sand	38.0	3.17	1,199	—	—
Peaty sand	47.0	10.39	452	—	—
Peaty sand	78.0	10.39	751	—	—
Peaty sand	130.0	10.39	1,251	—	—

Source: Leistra, 1970. a) The reported K_d and calculated K_{oc} values are expressed on a soil/vapor relationship, i.e., (μg absorbed per g of dry soil)/(μg vapor per mL vapor phase). The author reported an average K_{om} (μg absorbed per g of organic matter)/(μg dissolved per mL water phase) of 14 which is approximately equal to a K_{oc} value of 24.

Environmental Fate

Biological. The isomeric mixture showed significant degradation with gradual adaptation in a static-culture flask-screening test (settled domestic wastewater inoculum) conducted at 25°C. At concentrations of 5 and 10 mg/L, percent losses after 4 weeks of incubation were 85 and 84, respectively. Ten days into the incubation study, 7–19% was lost due to volatilization (Tabak et al., 1981).

Chemical/Physical. Hydrolysis in distilled water at 25°C produced *trans*-3-chloro-2-propen-1-ol and hydrochloric acid. The reported half-life for this reaction is only 2 days

(Kollig, 1993; Milano et al., 1988). *trans*-1,3-Dichloropropylene was reported to hydrolyze to 3-chloro-2-propen-1-ol and can be biologically oxidized to 3-chloropropenoic acid which is further oxidized to formylacetic acid. Decarboxylation of this compound yields carbon dioxide (Connors et al., 1990). Chloroacetaldehyde, formyl chloride and chloroacetic acid were formed from the ozonation of dichloropropylene at approximately 23°C and 730 mmHg. Chloroacetaldehyde and formyl chloride also formed from the reaction of dichloropropylene and hydroxyl radicals (Tuazon et al., 1984).

The evaporation half-life of *trans*-1,3-dichloropropylene (1 mg/L) from water at 25°C using a shallow-pitch propeller stirrer at 200 rpm at an average depth of 6.5 cm is 24.6 minutes (Dilling, 1977).

Emits chlorinated acids when incinerated. Incomplete combustion may release toxic phosgene (Sittig, 1985).

Formulation Types: Liquid.

Toxicity: LC_{50} (96-hour) for bluegill sunfish 7.1 mg/L and rainbow trout 7.1 mg/L (Worthing and Hance, 1991); acute oral LD_{50} of the isomeric mixture for male and female rats 713 and 470 mg/kg, respectively (Verschueren, 1983).

Uses: The isomer mixture is used as a soil fumigant and a nematocide.

DICHLORVOS

Synonyms: Apavap; Astrobot; Atgard; Atgard C; Atgard V; Bay 19149; Benfos; Bibesol; Brevinyl; Brevinyl E50; Canogard; Cekusan; Chlorvinphos; Cyanophos; Cypona; DDVF; DDVP; Dedevap; Deriban; Derribante; Devikol; Dichlorman; 2,2-Dichloroethenyl dimethyl phosphate; 2,2-Dichloroethenyl phosphoric acid dimethyl ester; Dichlorophos; 2,2-Dichlorovinyl dimethyl phosphate; 2,2-Dichlorovinyl dimethyl phosphoric acid ester; Dichlorovos; Dimethyl 2,2-dichloroethenyl phosphate; Dimethyl dichlorovinyl phosphate; Dimethyl 2,2-dichlorovinyl phosphate; O,O-Dimethyl O-(2,2-dichlorovinyl)phosphate; Divipan; Duo-kill; Duravos; ENT 20738; Equigard; Equigel; Estrosel; Estrosol; Fecama; Fly-die; Fly fighter; Herkal; Herkol; Krecalvin; Lindan; Mafu; Mafu strip; Marvex; Mopari; NA 2783; NCI-C00113; Nerkol; Nogos; Nogos 50; Nogos G; No-pest; No-pest strip; NSC 6738; Nuva; Nuvan; Nuvan 100EC; Oko; OMS 14; **Phosphoric acid 2,2-dichloroethenyl dimethyl ester**; Phosphoric acid 2,2-dichlorovinyl dimethyl ester; Phosvit; SD-1750; Szklarniak; Tap 9VP; Task; Task tabs; Tenac; Tetravos; UDVF; Unifos; Unifos 50 EC; Vapona; Vaponite; Vapora II; Verdican; Verdipor; Vinyl alcohol 2,2-dichlorodimethyl phosphate; Vinylofos; Vinylophos.

$$CH_3O\diagdown \overset{\displaystyle O}{\underset{\displaystyle \diagup}{P}} - O - CH = CCl_2$$
$$CH_3O$$

Designations: CAS Registry Number: 62-73-7; DOT: 2783; mf: $C_4H_7Cl_2O_4P$; fw: 220.98; RTECS: TC0350000.

Properties: Colorless to yellow liquid with an aromatic odor. Mp: <25°C; bp: 35, 120 and 140°C at 0.05, 14 and 20 mmHg, respectively; ρ: 1.44 and 1.415 at 20/4 and 25/4°C, respectively; fl p: practically nonflammable; H-$t_{1/2}$: 462 minutes (pH 7), 30 minutes (pH 8); K_H: 5 × 10^{-3} atm · m³/mol; log K_{oc}: 1.70 (calculated); log K_{ow}: 1.40–2.29; P-$t_{1/2}$: 7.25 hours (absorbance λ = 226.0 nm, concentration on glass plates = 6.7 µg/cm²); S_o: soluble in glycerol (≈ 5 g/L) but miscible with alcohol and most non-polar solvents; S_w: 16.0 g/L at 20°C; vap d: 9.03 g/L at 25°C, 7.63 (air = 1); vp: 0.0527 mmHg at 25°C (Kim et al., 1984).

Environmental Fate

Biological. When dichlorvos was incubated with sewage sludge for one week at 29°C, it was converted to dichloroethanol, dichloroacetic acid, ethyl dichloroacetate and an inorganic phosphate. In addition, dimethyl phosphate formed in the presence or absence of microorganisms (Lieberman and Alexander, 1983). Dichlorvos degraded fastest in nonsterile soils and decomposed faster in soils that were sterilized by gamma radiation than in soils that were sterilized by autoclaving. After one day of incubation, the percent of dichlorvos degradation that occurred in autoclaved, irradiated and nonsterile soils were 17, 88 and 99, respectively (Getzin and Rosefield, 1968).

Soil. In a silt loam and sandy loam, reported R_f values were 0.79 and 0.80, respectively (Sharma et al., 1986).

Plant. Metabolites identified in cotton leaves include dimethyl phosphate, phosphoric acid, methyl phosphate and *O*-demethyl dichlorvos (Bull and Ridgway, 1969).

Photolytic. Dichlorvos should not undergo direct photolysis since it does not absorb UV light at wavelengths >240 nm (Gore et al., 1971).

Chemical/Physical. Releases very toxic fumes of phosphorus oxides and chlorine when heated to decomposition (Sax and Lewis, 1987).

Slowly hydrolyzes in water and in acidic media but is more rapidly hydrolyzed under alkaline conditions to dimethyl hydrogen phosphate and dichloroacetaldehyde (Capel et al., 1988; Hartley and Kidd, 1987; Worthing and Hance, 1991). In the Rhine River (pH 7.4), the hydrolysis half-life of dichlorvos was 6 hours (Capel et al., 1988).

Atkinson and Carter (1984) estimated a half-life of 320 days for the reaction of dichlorvos with ozone in the atmosphere.

Exposure Limits: NIOSH REL: TWA 1 mg/m^3, IDLH 100 mg/m^3; OSHA PEL: TWA 1 ppm; ACGIH TLV: TWA 0.1 ppm.

Symptoms of Exposure: Miosis, ache eyes; rhinorrhea; headache; chest, wheezing, laryngeal spasm, salivation; cyanosis; anorexia, nausea, vomiting, diarrhea, sweating; muscle fasiculation, paralysis, ataxia; convulsions; low blood pressure.

Formulation Types: Emulsifiable concentrate; granules; hot and cold fogging concentrates; oil-miscible liquid; aerosol; impregnated strip.

Toxicity: LC_{50} (96-hour) for bluegill sunfish 869 μg/L (Verschueren, 1983), blood clam 1,790 μg/L (Bharathi, 1994); LC_{50} (72-hour) for blood clam 3,500 μg/L (Bharathi, 1994); LC_{50} (48-hour) for blood clam 4,720 μg/L (Bharathi, 1994); LC_{50} (24-hour) for bluegill sunfish 1.0 mg/L (Hartley and Kidd, 1987), blood clam 6,200 μg/L (Bharathi, 1994); acute oral LD_{50} for white male rats 56 mg/kg, white female rats 80 mg/kg (Mattson et al., 1955).

Uses: Organophosphorus insecticide and fumigant used against flies, mosquitoes and moths.

DICLOFOP-METHYL

Synonyms: 2-(4-(2,4-Dichlorophenoxy)phenoxy)methyl propionoate; **2-(4-(2,4-Dichlorophenoxy)phenoxy)propanoic acid methyl ester**; Hoe 23408; Hoe-Grass; Hoelon; Hoelon 3EC; Illoxan; Iloxan; Methyl 2-(4-(2,4-dichlorophenoxy)phenoxy)propanoate.

$$Cl\text{-}\langle\bigcirc\rangle\text{-}O\text{-}\langle\bigcirc\rangle\text{-}O\text{-}\underset{\displaystyle CH_3}{\overset{|}{CHCOOCH_3}}$$

Designations: CAS Registry Number: 51338-27–3; mf: $C_{16}H_{14}Cl_2O_4$; fw: 341.20; RTECS: UF1180000.

Properties: Colorless and odorless crystals. Mp: 39–41°C; bp: 175–176°C at 0.1 mmHg, decomposes at 288.4°C; ρ: 1.30 at 40/4°C; H-$t_{1/2}$: 1 day (pH 3), 265 days (pH 5), 29 days (pH 7); log K_{oc}: 4.89; log K_{ow}: 4.58; So (kg/L at 20°C): acetone (2.49), ethanol (0.11), ethyl ether (2.28), xylene (2.53); S_w: 3 g/L at 22°C; vp: 2.6×10^{-6} mmHg at 20°C.

Environmental Fate

Biological. From the first-order biotic and abiotic rate constants of diclofop-methyl in estuarine water and sediment/water systems, the estimated biodegradation half-lives were 0.24–12.4 and 0.11–2.2 days, respectively (Walker et al., 1988).

Soil. Under aerobic conditions, diclofop-methyl decomposes in soil forming diclofop-acid (Smith, 1977, 1979; Hartley and Kidd, 1987; Humburg et al., 1989) which undergoes further degradation to 4-(2,4-dichlorophenoxy)phenol (Smith, 1979), 4-(2,4-dichlorophenoxy)ethoxybenzene (Smith, 1977, 1979) and hydroxylated free acids (Hartley and Kidd, 1987; Humburg et al., 1989). The half-lives in sandy soils and sandy clay soils were reported to be 10 and 30 days, respectively (Ashton and Monaco, 1991).

Formulation Types: Emulsifiable concentrate (3 lb/gal).

Toxicity: LC$_{50}$ (96-hour) for rainbow trout 350 µg/L (Hartley and Kidd, 1987); acute oral LD$_{50}$ for rats 557–580 (sesame oil) mg/kg (Hartley and Kidd, 1987), 563 mg/kg (RTECS, 1985).

Uses: Postemergence herbicide used to control annual grasses, including wild oats, in flax, lentils, peas, soybeans, wheat and barley crops.

DICOFOL

Synonyms: Acarin; 1,1-Bis(chlorophenyl)-2,2,2-trichloroethanol; 1,1-Bis(4-chlorophenyl)-2,2,2-trichloroethanol; 1,1-Bis(*p*-chlorophenyl)-2,2,2-trichloroethanol; Carbax; Cekudifol; **4-Chloro-α-(4-chlorophenyl)-α-(trichloromethyl)benzene-methanol**; CPCA; Decofol; Dichlorokelthane; Di(*p*-chlorophenyl)trichloromethylcarbinol; 4,4'-Dichloro-α-(trichloromethyl)benzhydrol; DTMC; ENT 23648; FW 293; Hifol; Hifol 18.5 EC; Keltane; Kelthane; *p,p'*-Kelthane; Kelthane A; Kelthane dust base; Kelthanehanol; Milbol; Mitigan; NA 2761; NCI-C00486; 2,2,2-Trichloro-1,1-bis(*p*-chlorophenyl)ethanol; 2,2,2-Trichloro-1,1-di(4-chlorophenyl)ethanol.

Designations: CAS Registry Number: 115-32-2; DOT: 2761; mf: $C_{14}H_9Cl_5O$; fw: 370.51; RTECS: DC8400000.

Properties: Crystalline solid. Technical grades (80%) are brown, viscous oils. Mp: 77–78°C; bp: 180°C at 0.1 mmHg (pure), 193°C at 360 mmHg (technical grade); ρ: 1.45 at 25/4°C (technical); log K_{oc}: 3.93 (sand), 3.91 (sandy loam), 3.77 (silty loam), 3.77 (clay loam); log K_{ow}: 5.28; So (g/L): acetone (400), ethyl acetate (400), hexane (30), methanol (36), 2-propanol (30), toluene (400); S_w: 800 μg/L at 20°C, 1,200 μg/L at 24°C; vp: 3.98 × 10^{-7} mmHg at 20°C.

Environmental Fate

Plant. The major metabolite reported on apples is 4,4'-dichlorobenzophenone. On apples, dicofol concentrations decreased from 702 ppm to 436 ppm after 15 days. 4,4'-Dichlorobenzophenone increased from 25.5 ppm at time of spraying to 25.5 ppm 15 days after spraying (Archer, 1974). Four days after spraying cucumbers with dicofol, residues decreased from 0.95–1.6 ppm to 0.4–1.5 ppm. No residues were detected 8 days after spraying (Nazer and Masoud, 1986). A half-life of 6 days was reported for dicofol in alfalfa (Akesson and Yates, 1964).

Chemical/Physical. When dicofol was exposed to sunlight for 20 days, a 10% yield of 4,4'-dichlorobenzophenone was obtained. Solvents containing dicofol and exposed to UV light resulted in the formation of chlorobenzilic acid esters (Vaidyanathaswamy et al., 1981). Hydrolyzes in water forming 4,4'-dichlorobenzophenone (Walsh and Hites, 1979; Humburg et al., 1989). The hydrolysis half-lives of dicofol (400 μg/L initial concentration) at pH 8.2 and 10.2 were 1 hour and 3 minutes, respectively (Walsh and Hites, 1979).

Toxicity: LC_{50} (96-hour) for channel catfish 360 μg/L, bluegill sunfish 520 μg/L, largemouth bass 395 μg/L; LC_{50} (24-hour) for rainbow trout 110 μg/L; acute oral LD_{50} for male rats 595 mg/kg (Worthing and Hance, 1991), guinea pigs 1,810 mg/kg, rabbits 1,879 mg/kg (Humburg et al., 1989).

Formulation Types: Emulsifiable concentrate; wettable and dustable powders.

Uses: Nonsystemic acaricide to control mites in citrus fruits, grapes, cotton, pome and stone fruit.

DICROTOPHOS

Synonyms: Bidirl; Bidrin; C 709; Carbicron; CIBA 709; **(*E*)-3-(Diethylamino)-1-methyl-3-oxo-1-propenyl dimethyl phosphate**; 3-(Dimethoxyphosphinyloxy)-*N*,*N*-dimethyl-*cis*-crotonamide; 3-(Dimethoxyphosphinyloxy)-*N*,*N*-dimethylisocrotonamide; 3-(Dimethyl-amino)-1-methyl-3-oxo-1-propenyl dimethyl phosphate; *cis*-2-Dimethylcarbamoyl-1-methylvinyl dimethyl phosphate; *O*,*O*-Dimethyl *O*-(*N*,*N*-dimethylcarbamoyl-1-methylvi-nyl)phosphate; *O*,*O*-Dimethyl-*O*-(1,4-dimethyl-3-oxo-4-azapent-1-enyl)phosphate; Dimethyl phosphate of 3-hydroxy-*N*,*N*-dimethyl-*cis*-crotonamide; Dimethyl phosphate ester with 3-hydroxy-*N*,*N*-dimethyl-*cis*-crotonamide; Dimethyl phosphate ester with (*E*)-3-hydroxy-*N*,*N*-dimethylcrotonamide; ENT 24482; 3-Hydroxy-*N*,*N*-dimethyl-*cis*-crotona-mide dimethyl phosphate; SD 3562; Shell SD-3562.

$$CH_3O\!\!-\!\!\underset{CH_3O}{\overset{\overset{O}{\|}}{P}}\!\!-\!\!O\!\!-\!\!\underset{H_3C}{\overset{}{C}}\!\!=\!\!\underset{CON(CH_3)_2}{\overset{H}{C}}$$

Designations: CAS Registry Number: 141-66-2; DOT: 2783; mf: $C_8H_{16}NO_5P$; fw: 237.20; RTECS: TC3850000.

Properties: Yellow to brown liquid with a mild ester-like odor. Mp: <25°C; bp: 130°C at 0.1 mmHg (pure), 400°C (technical); ρ: 1.216 at 20/4°C; fl p: >93.3°C; H-$t_{1/2}$ (20°C): 88 days (pH 5), 23 days (pH 9); log K_{oc}: 1.04–2.27; log K_{ow}: -0.50; S_o: slightly soluble in kerosene but miscible with acetone, ethanol, hexylene glycol, isobutyl alcohol, xylene; S_w: miscible; vp: 6.98 × 10^{-5} mmHg at 20°C.

Soil properties and adsorption data

Soil	K_d (mL/g)	f_{oc} (%)	K_{oc} (mL/g)	pH (meq/100 g)	CEC
Catlin silt loam	0.92	2.32	40	5.7	13.0
Georgia sand	0.07	0.64	11	6.7	2.0
Handford sandy loam	0.40	0.75	53	6.4	7.1
Sharkey clay loam	3.58	1.91	187	5.9	23.5

Source: Lee et al., 1989.

Environmental Fate

Soil. The dimethylamino group is converted to an *N*-oxide then to -CH$_2$OH and aldehyde groups which further degrade via demethylation and hydrolysis (Hartley and Kidd, 1987). Dicrotophos is rapidly degraded under aerobic and anaerobic conditions forming *N*,*N*-dimethylacetoacetamide and 3-hydroxy-*N*,*N*-dimethylbutyramide as the major metabolites. Other metabolites included carbon dioxide and unextractable residues. The half-life of dicrotophos in a Hanford sandy loam soil was 3 days (Lee et al., 1989).

Chemical/Physical. Dicrotophos emits toxic fumes of phosphorus and nitrogen oxides when heated to decomposition (Sax and Lewis, 1987; Lewis, 1990).

Dicrotophos is hydrolyzed in sodium hydroxide solutions forming dimethylamine. The hydrolysis half-lives at 38°C and pH values of 1.1 and 9.1 are 100 and 50 days, respectively (Sittig, 1985). Lee et al. (1989) reported that the hydrolysis half-lives of dicrotophos in pH 5, 7 and 9 buffer solutions at 25°C were 117, 72 and 28 days, respectively. *N,N*-Dimethylacetoacetamide and *O*-desmethyldicrotophos were the major products identified.

Exposure Limits: OSHA PEL: TWA 0.25 mg/m^3; ACGIH TLV: TWA 0.25 mg/m^3.

Symptoms of Exposure: Headache, anorexia, nausea, vertigo, weakness, abdominal cramps, diarrhea, salivation, lacrimation, ataxia, cyanosis, pulmonary edema, convulsions, coma, shock.

Formulation Types: Emulsifiable concentrate; water soluble concentrate.

Toxicity: LC$_{50}$ (24-hour) for harlequin fish >1,000 mg/L and mosquito fish 200 mg/L (Hartley and Kidd, 1987); acute oral LD$_{50}$ for rats 17–22 mg/kg (Hartley and Kidd, 1987), 13 mg/kg (RTECS, 1985), for male and female rats, 21 and 16 mg/kg, respectively (Windholz et al., 1983).

Uses: Contact and systemic insecticide and acaricide used to control pests on rice, cotton, maize, soybeans, coffee, citrus and potatoes.

DIELDRIN

Synonyms: Alvit; Compound 497; Dieldrite; Dieldrix; ENT 16225; HEOD; Hexachloroepoxyoctahydro-*endo,exo*-dimethanonaphthalene; 1,2,3,4,10,10-Hexachloro-6,7-epoxy-1,4,4a,5,6,7,8,8a-octahydro-1,4-*endo,exo*-5,8-dimethanonaphthalene; **3,4,5,6,9,9-Hexachloro-1a,2,2a,3,6,6a,7,7a-octahydro-2,7:3,6-dimethanonaphth[2,3-*b*]oxirene**; Illoxol; Insecticide 497; NA 2761; NCI-C00124; Octalox; Panoram D-31; Quintox; RCRA waste number P037.

Designations: CAS Registry Number: 60-57-1; DOT: 2761; mf: $C_{12}H_8Cl_6O$; fw: 380.91; RTECS: IO1750000.

Properties: A stereoisomer of endrin, dieldrin may be white crystals or pale tan flakes with an odorless to mild chemical odor. Odor threshold in water is 41 µg/kg. Mp: 175–176°C; bp: decomposes; ρ: 1.75 at 20/4°C; fl p: nonflammable; H-$t_{1/2}$: 10.5 years at 25°C and pH 7 (Ellington et al., 1987); K_H: 5.8 × 10⁻⁵ atm · m³/mol at 25°C; log BCF: freshwater clam (*Corbicula manilensis*) 3.55 (Hartley and Johnston, 1983); log K_{oc}: 4.08–4.55; log K_{ow}: 3.692-6.2; P-$t_{1/2}$: 153.84 hours (absorbance λ = 227.0 nm, concentration on glass plates = 6.7 µg/cm²); S_o (g/L at 20°C): benzene (400), carbon tetrachloride (380), cod liver oil (57.7, 61.8, 77.8 g/L at 4, 12 and 20°C, respectively), methylene chloride (480), octanol (91.2, 39.9 and 40.8 g/L at 4, 12 and 20°C, respectively), triolein (49.1, 63.2 and 72.6 at 4, 12 and 20°C, respectively); S_w: 80, 140 and 200 µg/L at 10, 20 and 30°C, respectively; vap d: 54 ng/L at 20°C; vp: 1.78 × 10⁻⁷ mmHg at 20°C, 2.4 × 10⁻⁵ mmHg at 25°C.

Soil properties and adsorption data

Soil	K_d (mL/g)	f_{oc} (%)	K_{oc} (mL/g)	pH (meq/100 g)	CEC
Batcombe silt loam	268	2.10	12,762	6.1	—
Bryce silty clay loam	297	7.50	3,960	—	55.5
Bryce-Swygert silty clay	265	2.70	9,815	—	34.4
Clay loam	195	1.42	14,000	5.9	12.4
Drummer silty clay loam	198	3.10	6,387	—	24.8
Flanagan silt loam	260	3.50	7,429	—	27.7
Gilford-Hoopeston-Ade sandy loam	147	1.20	12,250	—	7.5
Houghton muck	1,507	16.80	8,970	—	72.4

Soil properties and adsorption data (continued)

Soil	K_d (mL/g)	f_{oc} (%)	K_{oc} (mL/g)	pH (meq/100 g)	CEC
Plainfield-Bloomfield sand	39	0.40	9,750	—	1.7
Rothamsted Farm	194	1.51	12,847	5.1	—

Source: Felsot and Wilson, 1980; Lord et al., 1980; Briggs, 1981; Kishi et al., 1990.

Environmental Fate

Biological. Identified metabolites of dieldrin from solution cultures containing *Pseudomonas* sp. in soils include aldrin and dihydroxydihydroaldrin. Other unidentified byproducts included a ketone, an aldehyde and an acid (Matsumura et al., 1968; Kearney and Kaufman, 1976). A pure culture of the marine alga, namely *Dunaliella* sp., degraded dieldrin to photodieldrin and an unknown metabolite at yields of 8.5 and 3.2%, respectively.

Photodieldrin and the diol were also identified as metabolites in field-collected samples of marine water, sediments and associated biological materials (Patil et al., 1972). At least 10 different types of bacteria comprising a mixed anaerobic population degraded dieldrin, via monodechlorination at the methylene bridge carbon, to give *syn-* and *anti*-monodechlorodieldrin. Three isolates, *Clostridium bifermentans, Clostridium glycolium* and *Clostridium* spp., were capable of dieldrin dechlorination but the rate was much lower than that of the mixed population (Maule et al., 1987). Using settled domestic wastewater inoculum, dieldrin (5 and 10 mg/L) did not degrade after 28 days of incubation at 25°C in four successive 7-day incubation periods (Tabak et al., 1981).

Chacko et al. (1966) reported that cultures of six actinomycetes (*Norcardia* sp., *Streptomyces albus, Streptomyces antibioticus, Streptomyces auerofaciens, Streptomyces cinnamoneus, Streptomyces viridochromogenes*) and 8 fungi had no effect on the degradation of dieldrin. Matsumura et al. (1970) reported microorganisms isolated from soil and Lake Michigan water converted dieldrin to photodieldrin.

The percentage of dieldrin remaining in a Congaree sandy loam soil after 7 years was 50% (Nash and Woolson, 1967).

Soil. Dieldrin is very persistent in soil under both aerobic and anaerobic conditions (Castro and Yoshida, 1971; Sanborn and Yu, 1973). Reported half-lives in soil ranged from 175 days to 3 years (Howard et al., 1991).

Groundwater. According to the U.S. EPA (1986) dieldrin has a high potential to leach to groundwater.

Photolytic. Photolysis of an aqueous solution by sunlight for 3 months resulted in a 70% yield of photodieldrin (Henderson and Crosby, 1968). A solid film of dieldrin exposed to sunlight for 2 months resulted in a 25% yield of photodieldrin (Benson, 1971). In addition to sunlight, UV light converts dieldrin to photodieldrin (Georgacakis and Khan, 1971). Solid dieldrin exposed to UV light (λ <300 nm) under a stream of oxygen yielded small amounts of photodieldrin (Gäb et al., 1974). Many other investigators reported photodieldrin as a photolysis product of dieldrin under various conditions (Rosen et al., 1966; Robinson et al., 1966; Rosen and Carey, 1968; Ivie and Casida, 1970, 1971, 1971a; Crosby and Moilanen, 1974). One of the photoproducts identified besides photodieldrin was photoaldrin chlorohydrin [1,1,2,3,3a,5(or 6),7a-heptachloro-6(or 5)-hydroxydecahydro-2,4,7-metheno-1*H*-cyclopenta[*a*]pentalene] (Lombardo et al., 1972). After a 1 hour exposure to sunlight, dieldrin was converted to photodieldrin. Photodecomposition was accelerated by a number of photosensitizing agents (Ivie and Casida, 1971a). When an aqueous solution containing dieldrin was photooxidized by UV light at 90–95°C, 25, 50 and 75% degraded to carbon dioxide after 2.9, 4.8 and 12.5 hours, respectively (Knoevenagel and Himmelreich, 1976).

Chemical/Physical. The hydrolysis rate constant for dieldrin at pH 7 and 25°C was determined to be 7.5×10^{-6}/hour, resulting in a half-life of 10.5 years (Ellington et al., 1987). The epoxide moiety undergoes nucleophilic substitution with water forming dieldrin diol (Kollig, 1993). At higher temperatures, the hydrolysis half-lives decreased significantly. At 69°C and pH values of 3.13, 7.22 and 10.45, the calculated hydrolysis half-lives were 19.5, 39.5 and 29.2 days, respectively (Ellington et al., 1986).

Products reported from the combustion of dieldrin at 900°C include carbon monoxide, carbon dioxide, hydrochloric acid, chlorine and unidentified compounds (Kennedy et al., 1972a). When dieldrin is heated to decomposition (>175°C), very toxic chloride fumes are emitted (Lewis, 1990).

At 33°C, 35% relative humidity and a 2-mile/hour wind speed, the volatility losses of dieldrin as a thick film, droplets on glass, droplets on leaves and formulation film on glass after 48 hours were 78.6, 70.3, 53.5 and 12.5%, respectively (Que Hee et al., 1975).

Exposure Limits: NIOSH REL: TWA 0.25 mg/m³, IDLH 50 mg/m³; OSHA PEL: TWA 0.25 mg/m³; ACGIH TLV: TWA 0.25 mg/m³.

Symptoms of Exposure: Headache, dizziness, nausea, vomiting, malaise, sweating, myoclonic limb jerks, clonic and tonic convulsions, coma, respiratory failure.

Formulation Types: Emulsifiable concentrate; wettable powder; granules; dustable powder.

Toxicity: LC_{50} (96-hour) for goldfish 37 µg/L (Hartley and Kidd, 1987), bluegill sunfish 8 µg/L, fathead minnow 16 µg/L (Henderson et al., 1959), rainbow trout 10 µg/L, coho salmon 11 µg/L, chinook 6 µg/L (Katz, 1961), pumpkinseed 6.7 µg/L, channel catfish 4.5 µg/L (Verschueren, 1983); LC_{50} (48-hour) for mosquito fish 8 ppb (Verschueren, 1983); LC_{50} (24-hour) for bluegill sunfish 170 ppb and fathead minnow 24 ppb (Verschueren, 1983); acute oral LD_{50} for rats 37–87 g/kg (Hartley and Kidd, 1987), 38,300 mg/kg (RTECS, 1985).

Use: Insecticide.

DIENOCHLOR

Synonyms: Bis(pentachlor-2,4-cyclopentadien-1-yl); Bis(pentachloroclopentadienyl); Bis(pentachloro-2,4-cyclopentadien-1-yl); Decachlor; Decachlorobi-2,4-cyclopentadien-1-yl; **1,1′,2,2′,3,3′,4,4′,5,5′-Decachlorobi-2,4-cyclopentadien-1-yl**; ENT 25718; Hooker HRS 16; Hooker HRS 1654; HRS 16; HRS 16A; HRS 1654; Pentac; Pentac WP.

Designations: CAS Registry Number: 2227-17-0; mf: $C_{10}Cl_{10}$; fw: 474.60; RTECS: DT8225000.

Properties: Pale yellow powder. Mp: 111–128°C; bp: decomposes >130°C; ρ: 1.923 at 20/4°C; log K_{oc}: 3.59 (silty clay loam, pH 6.8), 4.26 (silty loam, pH 6.1), 4.62 (sandy loam, pH 6.9), 4.83 (loamy sand, pH 7.5); log K_{ow}: 3.15–3.30; S_o: slightly soluble in acetone, aliphatic hydrocarbons, ethanol; S_w: <10 ng/L; vp: 2.86 × 10⁻⁹ mmHg at 25°C (extrapolated).

Environmental Fate

Plant. On plants, dienochlor was converted by sunlight to form perchloro ketones (Quistad and Mulholland, 1983).

Chemical/Physical. Dienochlor is unstable when exposed to sunlight. When dienochlor applied as a thin film on glass plates was exposed to sunlight, nonpolar products, a tricyclic chlorocarbon and 3 isomeric perchloro ketones were formed at yields of 25, 10 and 14%, respectively (Quistad and Mulholland, 1983).

Dienochlor begins to decompose at 130°C (Worthing and Hance, 1991).

Toxicity: Acute oral LD_{50} for male albino rats >3.16 g/kg, bobwhite quail 705 mg/kg (Worthing and Hance, 1991).

Use: Acaricide used for control of mites on ornamentals.

DIFENZOQUAT METHYL SULFATE

Synonyms: AC 84777; Avenge; **1,2-Dimethyl-3,5-diphenyl-1*H*-pyrazolium methyl sulfate**; Finaven; Mataven; Superaven; Yeh-Yan-Ku.

Designations: CAS Registry Number: 43222-48-6; mf: $C_{18}H_{20}N_2O_4S$; fw: 360.40; RTECS: UQ9820000.

Properties: Colorless to off-white, odorless, hygroscopic solid. Mp: 156.5–158°C with decomposition; ρ: 1.13 at 25°C; fl p: >82°C (technical grade and formulation); K_H: 5.66 × 10^{-14} atm · m^3/mol at 20–25°C (approximate — calculated from water solubility and vapor pressure); log K_{oc}: 4.49–5.80; log K_{ow}: 0.65 (pH 5), –0.62 (pH 7), –0.32 (pH 9); S_o (g/L at 25°C unless noted otherwise): acetone (130 g/L at 20°C), chlorobenzene (0.4), chloroform (5,000 g/L at 20°C), dichloroethane (80), methanol (6,200), 2-propanol (0.7), xylene (<0.1 g/L at 20°C); S_w: 817 g/L at 25°C; vp: 9.75 × 10^{-8} mmHg at 20°C.

Soil properties and adsorption data

Soil	K_d (mL/g)	f_{oc} (%)	K_{oc} (mL/g)	pH
Clayey loam	2,680	2.90	92,414	7.7
Sandy loam	181	0.58	31,207	6.9
Sandy clay loam	636	1.80	35,333	6.4
Silt loam	1,093	1.74	62,816	5.2

Source: U.S. Department of Agriculture, 1990.

Environmental Fate

Soil. Though no products were reported, the half-life in soil is approximately 3 months (Hartley and Kidd, 1987).

Photolytic. Degrades photolytically to the volatile monomethyl pyrazole (Hartley and Kidd, 1987).

Chemical/Physical. May react with aluminum releasing hydrogen (Hartley and Kidd, 1987).

Symptoms of Exposure: Do not get in eyes — causes irreversible eye damage.

Formulation Types: Soluble concentrate (aqueous solution for cation = 2 lb/gal); water-soluble powder.

Toxicity: LC_{50} (96-hour) for rainbow trout 694 mg/L and bluegill sunfish 696 mg/L (Worthing and Hance, 1991); acute oral LD_{50} for male rats 270 mg/kg (Ashton and Monaco, 1991).

Uses: Selective postemergence herbicide used to control wild oats in wheat, barley, flax, maize, rye grass, vetches and rye grass crops.

DIFLUBENZURON

Synonyms: *N*-(((4-Chlorophenyl)amino)carbonyl)-2,6-difluorobenzamide; 1-(4-Chlorophenyl)-3-(2,6-difluorobenzoyl)urea; Dfluron; Dimilin; DU 112307; ENT 29054; OMS 1804; PDD 60401; PH 60-40; Philips-duphar PH 60-40; TH 6040; Thompson-Hayward TH6040.

Designations: CAS Registry Number: 35367-38-5; mf: $C_{14}H_9ClF_2N_2O_2$; fw: 310.69; RTECS: YS6200000.

Properties: Colorless crystals when pure. Technical grade is off-white to yellow crystalline solid. Mp: 230–232°C (pure), 210–230°C (technical); bp: decomposes; K_H: 7.3×10^{-9} atm · m³/mol at 20–25°C (approximate — calculated from water solubility and vapor pressure); log K_{oc}: 3.01 (calculated); log K_{ow}: 3.29 (calculated); S_o (g/L at 25°C): *N,N*-dimethylformamide (104), 1,4-dioxane (20); S_w: 14 mg/L at 25°C; vp: 2.5×10^{-7} mmHg at 20°C.

Environmental Fate

Soil. The half-life in soil is <1 week (Hartley and Kidd, 1987). Diflubenzuron degrades more rapidly in neutral or basic conditions but more slowly under acidic conditions (pH <6) (Ivie et al., 1980).

Chemical/Physical. Hydrolyzes in water to 4-chlorophenylurea (Verschueren, 1983).

Formulation Types: Granules; wettable powder.

Toxicity: LC_{50} (96-hour) for rainbow trout 140 mg/L, bluegill sunfish 135 mg/L (Hartley and Kidd, 1987), coho salmon and juvenile rainbow trout >150 mg/L (Verschueren, 1983); acute oral LD_{50} for mice 4,640 mg/kg (RTECS, 1985).

Uses: Nonsystemic insecticide used to control leaf-eating larvae and leaf miners in forestry, woody ornamentals and fruit trees.

DIMETHOATE

Synonyms: AC 12880; AC 18682; American Cyanamid 12880; BI 58; BI 58 EC; 8014 Bis HC; Cekuthoate; Chemathoate; CL 12880; Cygon; Cygon 4E; Cygon insecticide; Daphene; De-Fend; Demos-L40; Devigon; Dimate 267; Dimetate; Dimethoate-267; Dimethoat tecvhnisch 95%; Dimethogen; *O,O*-Dimethyl dithiophosphorylacetic acid *N*-monomethylamide salt; *O,O*-Dimethyl *S*-(2-(methylamino)-2-oxoethyl) phosphorodithioate; *O,O*-Dimethyl *S*-(*N*-methylcarbamoylethyl)dithiophosphate; *O,O*-Dimethyl *S*-methylcarbamoylmethyl phosphorodithioate; *O,O*-Dimethyl *S*-(*N*-methylcarbamoylmethyl) phosphorodithioate; *O,O*-Di-methyl S-(*N*-methylcarbamylmethyl) thiothionophosphate; *O,O*-Dimethyl S-(*N*-monomethyl)carbamyl methyldithiophosphate; Dimeton; Dimevur; EI-12880; ENT 24650; Experimental insecticide 12880; Ferkethion; Fip; Fortion NM; Fosfamid; Fosfotox; Fosfotox R; Fosfotox R 35; Fostion MM; L 395; Lurgo; *S*-Methyl-carbamoylmethyl *O,O*-dimethyl phosphorodithioate; *N*-Monomethylamide of *O,O*-dimethyldithiophosphoryllacetic acid; NC 262; NCI-C00135; PEI 35; Perfecthion; Perfekthion; Perfektion; Phosphamid; Phosphamide; Phosphorodithoic acid *O,O*-dimethyl ester, ester with 2-mercapto-*N*-methylacetamide; **Phosphorodithoic acid *O,O*-dimethyl-*S*-(2-(methylamino)-2-oxoethyl) ester**; Racusan; RCRA waste number P044; Rebelate; Rogodial; Rogor; Rogor 40; Rogor L; Rogor 20L; Rogor P; Roxion; Roxion U.A.; Sinoratox; Trimetion.

$$CH_3O \underset{CH_3O}{\overset{}{\diagdown}} \overset{\overset{S}{\parallel}}{P} - SCH_2CONHCH_3$$

Designations: CAS Registry Number: 60-51-5; DOT: 2783; mf: $C_5H_{12}NO_3PS_2$; fw: 229.30; RTECS: TE1750000.

Properties: Colorless crystals with a mercaptan-like odor. Technical grades (96%) are white to grayish crystals. Mp: 52–52.5°C; bp: 117°C at 0.1 mmHg (technical grade); ρ: 1.281 and 1.277 at 50/4 and 65/4°C, respectively; fl p: burns readily on contact with flame; H-$t_{1/2}$: 118 hours at 25°C and pH 7; K_H: 2.63 × 10^{-11} atm · m^3/mol at 20–21°C (approximate — calculated from water solubility and vapor pressure); log K_{oc}: 0.96; log K_{ow}: -0.294 (Freed et al., 1979a), 0.508-0.78; P-$t_{1/2}$: 64.10 hours (absorbance λ = 226.5 nm, concentration on glass plates = 6.7 µg/cm^2); S_o: very soluble in lower alcohols, benzene, carbon tetrachloride, chloroform, ethylbenzene, methyl chloride, methylene chloride, methyl ethyl ketone, toluene, xylene and many other common organic solvents except saturated hydrocarbons; S_w: 25 g/L at 21°C; vp: 5.06 × 10^{-6} mmHg at 20°C.

Soil properties and adsorption data

Soil	K_d (mL/g)	f_{oc} (%)	K_{oc} (mL/g)	pH
Alluvial-Rio Nacimiento	1.51	0.91	166	8.0
Batcombe silt loam	0.20	2.05	10	6.1
Brown Clay-Almanzora Alto	4.21	1.17	359	8.5
Brown Lime-Almanzora Bajo	3.15	1.48	213	8.9
Brown Lime-Los Velez	8.94	2.06	434	8.1
Coarse sand (Jutland, Denmark)	0.08	0.21	38	5.3
Desert-Campo de Tabernas	2.21	0.33	670	7.9
Kanuma high clay	0.50	1.35	36	5.7
Rendzine-Andarax	2.35	0.65	361	8.1
Saline-Campo de Dalías	1.81	1.65	110	8.2
Sandy loam (Jutland, Denmark)	0.05	0.15	33	6.4
Tsukuba clay loam	0.80	4.24	18	6.5
Volcanic-Campo de Nijar	1.06	0.37	286	8.7

Source: Briggs, 1981; Valverde-García et al., 1988; Kanazawa, 1989; Kjeldsen et al., 1990.

Environmental Fate

Soil. Duff and Menzer (1973) reported that in moist soils, dimethoate is converted to the oxygen analog, dimethoate carboxylic acid (dimethoxon) and two unidentified metabolites. The degradation rate of dimethoate in three different soils increased almost twofold with a 10°C increase in temperature (Kolbe et al., 1991). The reported half-lives of dimethoate in a humus-rich sandy soil, clay loam and heavy clay soil at 10 and 20°C are 15.3, 10.3, 15.5 days and 9.7, 4.8 and 8.5 days, respectively. Degradates included dimethoxon (*O,O*-dimethyl-*S*-(*N*-methylcarbamoylmethyl)phosphorothiolate) and unidentified polar compounds (Kolbe et al., 1991).

In a silt loam and sandy loam, reported R_f values were 0.42 and 0.45, respectively (Sharma et al., 1986).

Plant. In plants, oxidation/hydrolysis leads to the formation of the phosphorothioate. Other hydrolysis products in plants include *O,O*-dimethylphosphorodithioate and *O,O*-dimethylphosphorophosphate which occurs via demethylation and hydrolytic cleavage of the methylamino group (Hartley and Kidd, 1987). In bean plants, dimethoate degraded to *N*-hydroxymethyl dimethoate which further degraded to des-*N*-methyl dimethoate (*N,N*-dimethyl *S*-(carbamoylmethyl)phosphorodithioate). Other metabolites identified in various plants including corn, cotton, pea and potato, include des-*O*-methyl carboxylic acid, dimethoate carboxylic acid, dimethyl phosphorothioic acid, dimethyl phosphorodithioic acid, *N*-hydroxymethyl dimethoate (*O,O*-dimethyl *S*-(*N*-hydroxymethylcarbamoylmethyl) phosphorothioate), an oxygen analog (*O,O*-dimethyl *S*-(*N*-methylcarbamoylmethyl) phosphorothiolate), a des-*N*-methyl oxygen analog (*O,O*-dimethyl *S*-(carbamoylmethyl) phosphorothiolate), *N*-hydroxymethyl oxygen analog (*O,O*-dimethyl *S*-(*N*-hydroxymethylcarbamoylmethyl) phosphorothiolate) and three unknown substances (Lucier and Menzer, 1968, 1970; Garner and Menzer, 1986). These compounds were not detected on grapes treated with dimethoate 28 days after application (Steller and Brand, 1974).

The half-life in Bermuda grass was reported to be 3.1 days (Beck et al., 1966). The disappearance half-lives of dimethoate on bean, tomato, cucumber and cotton plants were 4.3, 6.0, 3.8 and 3.3 days, respectively (Belal and Gomaa, 1979).

Surface Water. Though no products were identified, the half-life in raw river water was 8 weeks (Eichelberger and Lichtenberg, 1971).

Photolytic. Dichlorvos should not undergo direct photolysis since it does not absorb UV light at wavelengths greater than 290 nm (Gore et al., 1971).

Chemical/Physical. On heating, dimethoate is converted to the *O,S*-dimethyl analog (Worthing and Hance, 1991). Burns readily in contact with flame releasing toxic fumes of phosphorus, nitrogen and sulfur oxides (Sax and Lewis, 1987).

The calculated hydrolysis half-life at 25°C and pH 7 is 118 hours (Ellington et al., 1988).

Formulation Types: Emulsifiable concentrate; wettable powder; granules; aerosol; dustable powder.

Toxicity: LC_{50} (96-hour) for mosquito fish 40–60 mg/L (Hartley and Kidd, 1987); LC_{50} (24-hour) for bluegill sunfish 28.0 mg/L and rainbow trout 20.0 mg/L (Verschueren, 1983); acute oral LD_{50} for rats 500–680 mg/kg (Hartley and Kidd, 1987), 250 mg/kg (Windholz et al., 1983).

Uses: Systemic and contact organophosphorus insecticide and acaricide used to control thrips and red spider mites on many agricultural crops, sawflies on apples and plums, wheat bulb and olive flies.

DIMETHYL PHTHALATE

Synonyms: Avolin; **1,2-Benzenedicarboxylic acid dimethyl ester**; Dimethyl-1,2-ben-zenedicarboxylate; Dimethylbenzeneorthodicarboxylate; DMP; ENT 262; Fermine; Methyl phthalate; Mipax; NTM; Palatinol M; Phthalic acid dimethyl ester; Phthalic acid methyl ester; RCRA waste number U102; Solvanom; Solvarone.

$$\text{(benzene ring)}\quad \text{COOCH}_3 \quad \text{COOCH}_3$$

Designations: CAS Registry Number: 131-11-3; mf: $C_{10}H_{10}O_4$; fw: 194.19; RTECS: TI1575000.

Properties: Clear, colorless, odorless, moderately viscous liquid. Technical grades have a slight aromatic odor. Mp: 5.5°C; bp: 283.8°C; ρ: 1.1905 at 20/4°C; fl p: 146°C; lel: 1.2% at 146°C; K_H: 4.2×10^{-7} atm · m^3/mol; IP: 9.75 eV; log BCF: 2.21 (*Chlorella pyrenoidosa*, Yan et al., 1995); log K_{oc}: 0.88–2.28; log K_{ow}: 1.53–2.00; S$_o$: soluble in acetone, dimethylsulfoxide, benzene, mineral oil (0.34 wt % at 20°C), miscible with ethanol, ethyl ether, chloroform; S$_w$: 4,320 mg/L at 25°C; vap d: 7.94 g/L at 25°C, 6.70 (air = 1); vp: 8.93×10^{-3} mmHg at 25°C.

Environmental Fate

Biological. In anaerobic sludge, degradation occurred as follows: monomethyl phthalate to phthalic acid to protocatechuic acid followed by ring cleavage and mineralization (Shelton et al., 1984). In a static-culture-flask screening test, dimethyl phthalate showed significant biodegradation with rapid adaptation. The ester (5 and 10 mg/L) was statically incubated in the dark at 25°C with yeast extract and settled domestic wastewater inoculum. After 7 days, 100% biodegrada-tion was achieved (Tabak et al., 1981).

Photolytic. An aqueous solution containing titanium dioxide and subjected to UV light (λ >290 nm) yielded mono- and dihydroxyphthalates as intermediates (Hustert and Moza, 1988).

Chemical/Physical. Hydrolyzes in water forming phthalic acid and methyl alcohol (Wolfe et al., 1980).

Exposure Limits: NIOSH REL: TWA 5 mg/m^3, IDLH 2,000 mg/m^3; OSHA PEL: TWA 5 mg/m^3; ACGIH TLV: TWA 5 mg/m^3.

Symptoms of Exposure: Irritates nasal passages, upper respiratory system, stomach; eye pain. Ingestion may cause central nervous system depression.

Formulation Types: Liquid; aerosol.

Toxicity: LC$_{50}$ (8-day) for grass shrimp larvae 100 ppm (Verschueren, 1983); acute oral LD$_{50}$ for rats 6,900 mg/kg (Verschueren, 1983), 6,800 mg/kg (RTECS, 1985).

Use: Insect repellant.

4,6-DINITRO-o-CRESOL

Synonyms: Antinonin; Antinonnon; Arborol; Capsine; Chemsect DNOC; Degrassan; Dekrysil; Detal; Dinitrocresol; Dinitro-o-cresol; 2,4-Dinitro-o-cresol; 3,5-Dinitro-o-cresol; Dinitrodendtroxal; 3,5-Dinitro-2-hydroxytoluene; Dinitrol; Dinitromethyl cyclohexyltrienol; 2,4-Dinitro-2-methylphenol; 2,4-Dinitro-6-methylphenol; 4,6-Dinitro-2-methylphenol; Dinitrosol; Dinoc; Dinurania; DN; DNC; DN-dry mix no. 2; DNOC; Effusan; Effusan 3436; Elgetol; Elgetol 30; Elipol; ENT 154; Extrar; Hedolit; Hedolite; K III; K IV; Kresamone; Krezotol 50; Lipan; **2-Methyl-4,6-dinitrophenol**; 6-Methyl-2,4-dinitrophenol; Nitrador; Nitrofan; Prokarbol; Rafex; Rafex 35; Raphatox; RCRA waste number P047; Sandolin; Sandolin A; Selinon; Sinox; Trifina; Trifocide; Winterwash.

Designations: CAS Registry Number: 534-52-1; DOT: 1598; mf: $C_7H_6N_2O_5$; fw: 198.14; RTECS: GO9625000.

Properties: Yellow, odorless crystals. Mp: 86.5°C; bp: 312°C; pK_a: 4.35–4.46; K_H: 1.4 × 10^{-6} atm · m³/mol at 25°C; log K_{oc}: 2.64 (calculated); log K_{ow}: 2.12–2.85; So (mg/L at 15°C): methanol (7.33), ethanol (9.12), chloroform (37.2), acetone (100.6); slightly soluble in petroleum ether; S_w: 198 mg/L at 20°C; vp: 5 × 10^{-5} mmHg at 20°C.

Environmental Fate

Soil/Plant. In plants and soils, the nitro groups reduced to amino groups (Hartley and Kidd, 1987). When 4,6-dinitro-o-cresol was statically incubated in the dark at 25°C with yeast extract and settled domestic wastewater inoculum, no significant biodegradation and necessary acclimation for optimum biooxidation within the 4-week incubation period was observed (Tabak et al., 1981).

Chemical/Physical. 4,6-Dichloro-o-cresol will react with amines and alkali metals forming water-soluble salts which are indicative of phenols (Morrison and Boyd, 1971).

Exposure Limits: NIOSH REL: TWA 0.2 mg/m³, IDLH 5 mg/m³; OSHA PEL: TWA 0.2 mg/m³.

Symptoms of Exposure: Sense of well-being, headache, fever, lassitude, profuse sweating, excess thirst, tachycardia, coughing, shortness of breath, coma; eye and skin irritant.

Formulation Types: Emulsifiable concentrate; soluble concentrate; suspension concentrate; wettable powder; liquid cream.

Toxicity: LC_{50} for carp 6–13 mg/L (Hartley and Kidd, 1987); acute oral LD_{50} for rats 25–40 mg/kg (Hartley and Kidd, 1987), 10 mg/kg (RTECS, 1985).

Uses: Dormant ovicidal spray for fruit trees (highly phytotoxic and cannot be used successfully on actively growing plants); herbicide; insecticide.

DINOSEB

Synonyms: Aretit; Basanite; Butaphene; BNP 30; 2-*sec*-Butyl-2,4-dinitrophenol; Caldon; Chemox general; Chemox P.E.; Dinitro; Dinitro-3; Dinitrobutylphenol; 2-Dinitro-6-*sec*-butylphenol; 4,6-Dinitro-2-*sec*-butylphenol; 4,6-Dinitro-*o*-*sec*-butylphenol; 4,6-Dinitro-2-(1-methyl-*n*-propyl)phenol; DN 289; DNBP; Dnosbp; DNSBP; Dow general; Dow general weed killer; Dow selective weed killer; Elgetol; Elgetol 318; ENT 1122; Gebutox; Helfire; Kiloseb; **2-(1-Methylpropyl)-4,6-dinitrophenol**; Nitropone P; Phenotan; Premerge; Premerge 3; RCRA waste number P020; Sinox general; Sparic; Spurge; Subitex; Unicrop DNBP; Vertac dinitro weed killer; Vertac general weed killer; Vertac selective weed killer.

$$O_2N \quad \overset{OH}{\underset{NO_2}{\bigcirc}} \quad \overset{CH_3}{\underset{CHCH_2CH_3}{|}}$$

Designations: CAS Registry Number: 88-85-7; mf: $C_{10}H_{12}N_2O_5$; fw: 240.22; RTECS: SJ9800000.

Properties: Dark brown solid or orange liquid with a pungent odor. Mp: 38–42°C; ρ: 1.2647 at 45/4°C; fl p: 177°C; pK_a: 4.62; K_H: 5×10^{-4} atm · m³/mol at 20°C (calculated); log K_{oc}: 2.09, 2.70; K_d = 54 mL/g on a Cs⁺-kaolinite (Haderlein and Schwarzenbach, 1993); log K_{ow}: 2.29; S_o (g/kg): ethanol (480), *n*-heptane (270); S_w: 52 mg/L at 20°C; vp: 1 mmHg at 151.1°C.

Environmental Fate

Biological. When ¹⁴C-labeled dinoseb (5 ppm) was incubated in soil at 25°C for 60 days, 36.0% of the applied amount degraded to $^{14}CO_2$ (Doyle et al., 1978). Thom and Agg (1975) reported that dinoseb is unlikely to be degraded in conventional sewage treatment processes.

Groundwater. According to the U.S. EPA (1986) dinoseb has a high potential to leach to groundwater.

Plant. When dinoseb on bean leaves was exposed to sunlight, photodegradation resulted in the formation of persistent, polar compounds. The compounds could not be identified by TLC (Matsuo and Casida, 1970).

Chemical/Physical. Reacts with organic and inorganic bases forming water-soluble salts (Worthing and Hance, 1991).

Emits toxic fumes of chlorine when heated to decomposition (Sax and Lewis, 1987).

Symptoms of Exposure: Eye and skin irritant. Symptoms of poisoning include sweating, increased body temperature, excessive fatigue, excessive thirst and nausea.

Formulation Types: Emulsifiable concentrate; aqueous solution; soluble concentrate; water-in-oil emulsion.

Toxicity: Toxic to fish (Hartley and Kidd, 1987); LC_{100} (24-hour) for goldfish 0.4 ppm (Humburg et al., 1989); acute oral LD_{50} for rats 58 mg/kg (Hartley and Kidd, 1987), 25 mg/kg (RTECS, 1985).

Uses: The amine, ammonium salt or acetate ester is used as a contact herbicide for postemergence weed control in cereals, cotton, peas, beans, potatoes, pumpkins, soybeans and strawberries.

DIPHENAMID

Synonyms: Dif 4; Diamide; *N,N*-Dimethyldiphenylacetamide; *N,N*-Dimethyl-2,2-diphenylacetamide; *N,N*-Dimethyl-α,α-diphenylacetamide; 2,2-Dimethyl-*N,N*-dimethylacetamide; **N,N-Dimethyl-α-phenylbenzeneacetamide**; Dimid; Diphenamide; Diphenylamide; Dymid; Enide; Enide 50; Enide 90; FDN; Fenam; L-34314; Lilly 34314; Nor-Am; U 4513; 80W.

Designations: CAS Registry Number: 957-51-7; mf: $C_{16}H_{17}NO$; fw: 239.30; RTECS: AB8050000.

Properties: Colorless, nearly odorless crystals. Mp: 134.5–135.5°C; bp: decomposes at 210°C; ρ: 1.17 at 23.3/4°C; fl p: nonflammable; log K_{oc}: 2.31 (calculated); log K_{ow}: 1.92 (calculated); S_o (g/L at 27°C): acetone (189); *N,N*-dimethylformamide (165), xylene (65); S_w: 260 mg/L at 27°C; vp: negligible at 20°C.

Environmental Fate

Soil. Degradation of diphenamid in soils was reported to form desmethyldiphenamid via monodemethylation and a bidemethylated product of diphenamid (Somasundaram and Coats, 1991). The persistence of diphenamid under warm-moist soil conditions ranged from 3 to 6 months (Ashton and Monaco, 1991).

Groundwater. According to the U.S. EPA (1986), diphenamid has a high potential to leach to groundwater.

Plant. In strawberries, diphenamide was transformed via *N*-demethylation yielding *N*-methyl-2,2-diphenylacetamide (Golab et al., 1966).

Photolytic. *N*-Methyl-2,2-diphenylacetamide and benzoic acid were reported as major photoproducts following the UV irradiation of diphenamid in distilled water (Cessna and Muir, 1991).

Chemical/Physical. Emits toxic fumes of nitrogen oxides when heated to decomposition (Sax and Lewis, 1987).

Formulation Types: Wettable powder (50 and 90%); granules (5%); liquid dispersion (4 lb/gal).

Toxicity: Slightly toxic to fish (Hartley and Kidd, 1987); LC_{50} (48-hour) for *Daphnia magna* 56 mg/L (Sanders, 1970); acute oral LD_{50} for rats 1,050 mg/kg (Hartley and Kidd, 1987), 685 mg/kg (RTECS, 1985).

Uses: Selective preemergence herbicide used to control many broad-leaved weeds and most grass weeds in okra, cotton, peanuts tomatoes, sweet potatoes, potatoes, tobacco, fruits, turf and ornamentals.

DIPROPETRYN

Synonyms: Cotofor; Dipropetryne; 2-Ethylthio-4,6-bis(isopropylamino)-*s*-triazine; **6-(Ethylthio)-*N*,*N*′-bis(1-methylethyl)-1,3,5-triazine-2,4-diamine**; GS 16068; Sancap 80W.

$$CH_3CH_2S \quad N \quad NHCH(CH_3)_2$$

$$N \quad N$$

$$NHCH(CH_3)_2$$

Designations: CAS Registry Number: 4147-51-7; mf: $C_{11}H_{21}N_5S$; fw: 255.40; RTECS: XY4100000.

Properties: Colorless to white crystalline powder. Mp: 104–106°C; ρ: 1.120 at 20/4°C; fl p: nonflammable; H-$t_{1/2}$ (25°C): 24–28 days (pH 1), >2.5 years (7 <pH <13); K_H: 1.53 × 10^{-8} atm · m³/mol at 20°C (approximate — calculated from water solubility and vapor pressure); log K_{oc}: 2.58–2.95; log K_{ow}: 3.45, 3.81; So (g/L at 20°C): acetone (270), benzene (540), ethanol (180), hexane (9), kerosene (10), methanol (190), methylene chloride (300), octanol (130), toluene (220), xylene (220); S_w: 16 mg/L at 20°C; vp: 7.28 × 10^{-7} mmHg at 20°C.

Soil properties and adsorption data

Soil	K_d (mL/g)	f_{oc} (%)	K_{oc} (mL/g)	pH (meq/100 g)	CEC
Brewer clay loam	18.50	1.61	1,149	5.8	13.5
Cobb sand	1.32	0.34	388	7.3	3.8
Cobb sand + 2% muck	32.50	1.21	2,686	5.3	9.0
Loam	24.60	2.78	885	5.9	—
Port silty clay	8.91	1.04	857	6.3	17.9
Sand	1.50	0.35	431	7.7	—
Sandy loam	5.00	0.70	718	7.6	—
Sandy loam	8.90	2.32	384	7.8	—
Teller fine sandy loam	6.18	0.75	824	5.7	8.6

Source: Murray et al., 1975; U.S. Department of Agriculture, 1990.

Environmental Fate

Soil. Degradation of dipropetryn includes dealkylation of the side chain(s), ring opening and the evolution of carbon dioxide (Hartley and Kidd, 1987). The reported half-life in soil is approximately 100 days (Worthing and Hance, 1991).

Formulation Types: Wettable powder; suspension concentrate.

Toxicity: LC_{50} (96-hour) for rainbow trout 2.7 mg/L and bluegill sunfish 1.6 mg/L (Hartley and Kidd, 1987); acute oral LD_{50} for rats 3,900–5,000 mg/kg (Hartley and Kidd, 1987), 7,144 mg/kg (RTECS, 1985).

Uses: Preemergence herbicide used to control weeds in cotton and melon crops.

DIQUAT

Synonyms: Aquacide; Deiquat; Dextrone; 9,10-Dihydro-8a,10-diazoniaphenanthrene dibromide; Dihydro-8a,10a-diazoniaphenanthrene-(1,1'-ethylene-2,2'-bipyridy-lium)dibromide; 5,6-Dihydrodipyrido-(1,2a;2,1c)pyrazinium dibromide; **6,7-Dihydropy-rido[1,2-*a*:2′,1′-c]pyrazine-dium dibromide**; 1,1-Ethylene-2,2-bipyridylium dibromide; 1,1'-Ethylene-2,2'-bipyridylium dibromide; FB/2; NA 2781; Ortho; Pathclear; Preeglone; Reglon; Reglone; Weedol; Weedtrine-D.

Designations: CAS Registry Number: 85-00-7; mf: $C_{12}H_{12}Br_2N_2$; fw: 344.06; RTECS: JM5690000.

Properties: Colorless to pale yellow crystals. The reported odor threshold concentration in air and the taste threshold concentration in water are >8.9 ppm and >56 ppb, respectively (Young et al., 1996). Mp: decomposes at 320°C; ρ: 1.22–1.27 at 20/4°C; fl p: aqueous salt solutions are nonflammable; H-$t_{1/2}$: 74 days under simulated sunlight at pH 7; K_H: <6.3 × 10^{-14} atm · m³/mol at 20–25°C (approximate — calculated from water solubility and vapor pressure); log K_{oc}: 0.42 (calculated); log K_{ow}: –4.60; S_o: slightly soluble in alcohols and hydroxylic solvents; S_w: 700 g/L at 25°C; vp: <9.75 × 10^{-8} mmHg at 20°C.

Environmental Fate

Biological. Under aerobic and anaerobic conditions, the rate of diquat mineralization in eutrophic water and sediments was very low. After 65 days, only 0.88 and 0.21% of the applied amount (5 μg/mL) evolved as carbon dioxide (Simsiman and Chesters, 1976). Diquat is readily mineralized to carbon dioxide in nutrient solutions containing microorganisms. The addition of montmorillonite clay in an amount equal to adsorb one-half of the diquat decreased the amount of carbon dioxide by 50%. Additions of kaolinite clay had no effect on the amount of diquat degraded by microorganisms (Weber and Coble, 1968).

Photolytic. Diquat has an absorption maximum of 310 nm (Slade and Smith, 1967). The sunlight irradiation of a diquat solution (0.4 mg/100 mL) yielded 1,2,3,4-tetrahydro-1-oxopyrido[1,2-*a*]-5-pyrazinium chloride (TOPPS) as the principal metabolite. This compound also formed when diquat absorbed on filter paper, silica gel and on leaf surfaces is subjected to sunlight irradiation (Slade and Smith, 1974). When an aqueous diquat solution (5 μg/L) contained in a borosilicate glass beaker was exposed to sunlight, TOPPS, *o*-picolinic acid and picolinamide formed as major products (Smith and Grove, 1969). When the diquat solution was exposed to sunlight in May and June for 3 weeks, 70% was degraded giving rise only to TOPPS and *o*-picolinic acid. The sunlight photolysis half-life of diquat in aqueous solution was estimated to be about 14 days (Smith and Grove, 1969). Extensive photodegradation was also observed as thin films on leaf surfaces. *o*-

Picolinic acid and other products formed in small amounts (Smith and Grove, 1969). Funderburk and Bozarth (1967) reported that dry diquat was decomposed by UV light (half-life 48 hours). However in an aqueous solution, degradation was complete after 8 days.

Chemical/Physical. Decomposes at 320°C (Windholz et al., 1983) emitting toxic fumes of bromides and nitrogen oxides (Lewis, 1990). Diquat absorbs water forming well-defined, pale yellow crystalline hydrate (Calderbank and Slade, 1976).

In aqueous alkaline solutions, diquat decomposes forming complex colored products including small amounts of dipyridone (Calderbank and Slade, 1976).

Exposure Limits: ACGIH TLV: TWA 0.5 mg/m^3.

Symptoms of Exposure: Eye and skin irritant; nose bleeding if inhaled. Symptoms of poisoning include diarrhea, vomiting and general malaise.

Formulation Types: Soluble concentrate; gel; aqueous solution (2 lb/gal).

Toxicity: LC$_{50}$ (96-hour) for rainbow trout 21 mg/L, mirror carp 67 mg/L (Worthing and Hance, 1991), fathead minnow 14 mg/L, largemouth bass 7.8 mg/L (Surber and Pickering, 1962), bluegill sunfish 35 mg/L, walleye 2.1 mg/L (Gilderhus, 1967) and striped bass 0.25 ppm (Wellborn, 1969); LC$_{50}$ (96-hour) for northern pike 16 mg/L, rainbow trout 11.2 mg/L (Gilderhus, 1967); acute oral LD$_{50}$ for rats is 230 mg/kg (Ashton and Monaco, 1991), 120 mg/kg (RTECS, 1985).

Uses: Nonselective contact herbicide used to control broad-leaved weeds in fruit and vegetable crops.

DISULFOTON

Synonyms: Bay 19639; Bayer 19639; *O,O*-Diethyl *S*-(2-eththioethyl) phosphorodithioate; *O,O*-Diethyl *S*-(2-ethylthioethyl) thiothionophosphate; *O,O*-Diethyl *S*-(2-ethylmercaptoethyl) dithiophosphate; **O,O-Diethyl S-(2-(ethylthio)ethyl) phosphorodithioate**; Dimaz; Disulfaton; Disyston; Disystox; Dithiodemeton; Dithiosystox; ENT 23437; *O,O*-Ethyl *S*-2-((ethylthio)ethyl) phosphorodithioate; S-2-(Ethylthio)ethyl *O,O*-diethyl ester of phosphorodithioic acid; Frumin AL; Frumin G; M 74; NA 2783; RCRA waste number P039; S 276; Solvirex; Thiodemeton; Thiodemetron.

$$CH_3CH_2O \begin{array}{c} S \\ \backslash || \\ P-SCH_2CH_2SCH_2CH_3 \\ / \end{array} CH_3CH_2O$$

Designations: CAS Registry Number: 298-04-4; DOT: 2783; mf: $C_8H_{19}O_2PS_3$; fw: 274.40; RTECS: TD9275000.

Properties: Colorless to yellowish oil with a characteristic odor. Mp: $<-25°C$; bp: $128°C$ at 0.1 mmHg; ρ: 1.144 at 20/4°C; H-$t_{1/2}$: 3.04 years at 20°C and pH 1–5, 1.2–103 days at 25°C and pH 7, 7.2 hours at 20°C and pH 9; K_H: 5.42×10^{-6} atm · m³/mol at 20–22°C (approximate — calculated from water solubility and vapor pressure); log K_{oc}: 2.76–2.89; log K_{ow}: 3.95, 4.02; S_o: miscible with most solvents; S_w: 12 mg/L at 22°C; vp: 1.80×10^{-4} mmHg at 20°C.

Soil properties and adsorption data

Soil	K_d (mL/g)	f_{oc} (%)	K_{oc} (mL/g)	pH
Adventurers	134.60	31.00	434	6.9
Bottisham	58.60	8.80	660	7.7
Broadbalk loam	5.80	0.90	644	8.1
Broadbalk loam	21.50	2.70	796	7.8
Elkhorn sandy loam	5.01	0.86	583	6.0
Hugo gravelly sand loam	3.03	0.12	2,525	5.5
Isleham	49.10	7.60	646	7.5
Isleham	55.50	2.80	1,982	6.3
Loam	7.06	1.22	579	6.7
Moulton	21.10	1.70	1,241	8.1
Oakington	16.00	1.80	889	7.2
Peacock	70.30	11.00	639	7.6
Prickwillow	100.50	15.00	670	5.1
Sand	14.29	2.15	665	6.9
Sandy loam	13.43	1.74	772	5.5

Soil properties and adsorption data *(continued)*

Soil	K_d (mL/g)	f_{oc} (%)	K_{oc} (mL/g)	pH
Spinney	95.70	12.00	798	7.2
Stretham	14.70	1.40	1,050	7.5
Sweeney sandy clay loam	11.49	0.65	1,768	6.3
Tierra heavy clay loam	36.86	0.33	11,170	6.2
Wicken	20.30	1.70	1,194	8.0
Woburn	20.00	1.80	1,111	6.5
Woburn	14.90	1.10	1,354	6.8
Woburn	20.50	1.30	1,577	6.8
Worlington	5.30	0.70	757	8.1

Source: Graham-Bryce, 1967; King and McCarty, 1968; U.S. Department of Agriculture, 1990.

Environmental Fate

Soil/Plant. Disulfoton was metabolized in soil and plants to the corresponding sulfoxide and sulfone via oxidation of the thioether sulfur atoms (Metcalf et al., 1957; Getzin and Shanks, 1970; Takase et al., 1972; Clapp et al., 1976; Worthing and Hance, 1991), the corresponding phosphorothioate analogs and then to derivatives of *O,O*-diethyl hydrogen phosphate and 2-ethylthioethyl mercaptan (Worthing and Hance, 1991). Disulfoton is rapidly oxidized in soil to its sulfoxide and sulfone with disulfoton oxon sulfoxide and disulfoton oxon sulfone appearing in small amounts (Szeto et al., 1983). In a Portneuf silt loam soil, the persistence of the sulfoxide and sulfone was 32 and >64 days, respectively (Clapp et al., 1976).

Disulfoton was translocated from a sandy loam soil into asparagus tips. Disulfoton sulfoxide, disulfoton sulfone, disulfoton oxon sulfoxide and disulfoton oxon sulfone were recovered as metabolites (Szeto and Brown, 1982; Szeto et al., 1983). Disulfoton sulfoxide and disulfoton sulfone were also identified in spinach plants 5.5 months after application (Menzer and Dittman, 1968). In a later study, Menzer et al. (1970) reported that degradation of disulfoton in soil degraded at a higher rate in the winter months than in the summer months. They postulated that soil type, rather than temperature, had a greater influence on the rate of decomposition of disulfoton. Soils used in the winter and summer months were an Evesboro loamy sand and Chillum silt loam, respectively (Menzer et al., 1970). The half-life in soil is approximately 5 days (Jury et al., 1987).

Groundwater. According to the U.S. EPA (1986) disulfoton has a high potential to leach to groundwater.

Photolytic. Disulfoton was rapidly oxidized to disulfoton sulfoxide and trace amounts (<5% yield) of disulfoton sulfone when sorbed on soil and exposed to sunlight (half-life 1–4 days) (Gohre and Miller, 1986). The photosensitized oxidation was probably due to the presence of singlet oxygen (Gohre and Miller, 1986; Zepp et al., 1981). The degradation rate was higher in soils containing the lowest organic carbon (Gohre and Miller, 1986).

Chemical/Physical. Emits toxic fumes of phosphorus and sulfur oxides when heated to decomposition (Sax and Lewis, 1987; Lewis, 1990).

When fertilizers containing superphosphate and ammonium nitrate were impregnated with disulfoton, the latter chemically degraded to form disulfoton sulfone and disulfoton sulfoxide (Ibrahim et al., 1969).

The reported half-lives for abiotic hydrolysis, photochemical transformation and primary degradation in Rhine River water samples were 170, 1,000 and 7–41 days, respectively (Wanner et al., 1989). Hydrolysis half-lives at 69.0°C and pH values of 3.03 and 6.99 were 0.60 and 0.61 days, respectively. At a temperature of 48.0°C and pH of 11, however, the half-life was reduced to 0.12 hours (Ellington et al., 1986).

Exposure Limits: OSHA PEL: TWA 0.1 mg/m^3; ACGIH TLV: TWA 0.1 mg/m^3.

Symptoms of Exposure: Headache, anorexia, nausea, diarrhea, salivation, lacrimation, sweating, tremor, shortness of breath, cyanosis, fever, pulmonary edema, convulsions, coma, shock.

Formulation Types: Emulsifiable concentrate; granules; dry seed treatment.

Toxicity: LC$_{50}$ (96-hour) for rainbow trout 1.85 mg/L, bluegill sunfish 39 μg/L, goldfish 6.5 mg/L, guppies 0.25 mg/L (Hartley and Kidd, 1987) and fathead minnow 3.70 mg/L and bluegill sunfish 63 μg/L (Pickering et al., 1962); acute oral LD$_{50}$ for rats 2–12 mg/kg (Hartley and Kidd, 1987), male and female rats, 2.3 and 6.8 mg/kg, respectively (Windholz et al., 1983).

Uses: Systemic insecticide and acaricide for control of sucking insects and mites in fruits, vegetables, cotton and forestry nurseries.

DIURON

Synonyms: AF 101; Cekiuron; Crisuron; Dailon; DCMU; Diater; Dichlorfenidim; 3-(3,4-Dichlorophenol)-1,1-dimethylurea; 3-(3,4-Dichlorophenyl)-1,1-dimethyl-urea; **N'-(3,4-Dichlorophenyl)-N,N-dimethylurea**; 1,1-Dimethyl-3-(3,4-dichlorophenyl)urea; Dion; Direx; Direx 4L; Diurex; Diurol; Diuron 4L; Diuron 80W; DMU; Dynex; Farmco diuron; Herbatox; HW 920; Karmex; Karmex diuron herbicide; Karmex DW; Marmer; NA 2767; Sup'r flo; Telvar; Telvar Diuron Weed Killer; Unidron; Urox D; USAF P-7; USAF XR-42; Vonduron.

Designations: CAS Registry Number: 330-54-1; DOT: 2767; mf: $C_9H_{10}Cl_2N_2O$; fw: 233.11; RTECS: YS8925000.

Properties: White, odorless, crystalline solid. Mp: 150–155°C; bp: 180°C (decomposes); ρ: 1.385 (calculated); fl p: nonflammable; pK_a: –1 to –2; K_H: $1.46 × 10^{-9}$ atm · m³/mol at 25–30°C (approximate — calculated from water solubility and vapor pressure); log K_{oc}: 2.21–2.87; log K_{ow}: 2.58 (Liu and Qian, 1995); So (ppm at 27°C): acetone (53,000), benzene (1,200), butyl stearate (1,400), refined cottonseed oil (900); S_w: 40 ppm at 20°C; vp: $2 × 10^{-7}$ mmHg at 30°C.

Soil properties and adsorption data

Soil	K_d (mL/g)	f_{oc} (%)	K_{oc} (mL/g)	pH (meq/100 g)	CEC
Aguadilla loamy sand	4.73	1.44	328	7.4	10.0
Aguirre clay loam	4.63	1.84	252	9.0	13.8
Alonso clay	4.63	1.84	252	5.1	13.8
Altura loam	5.36	2.13	252	8.0	27.6
Ascalon sandy clay loam	1.33	0.85	156	7.3	12.7
Asquith sandy loam	6.90	1.02	676	7.5	—
Adkins loamy sand	1.91	0.40	657	7.3	—
Barnes clay loam	8.51	3.98	214	7.4	33.8
Baymón sandy clay loam	2.29	0.98	234	4.7	5.0
Begbroke sandy loam	2.70	1.90	142	7.1	—
Beltsville silt loam	1.67	1.40	119	4.3	4.2
Benevola clay (subsoil)	1.49	1.30	115	7.6	20.1
Benevola silty clay (topsoil)	3.62	2.70	134	7.7	19.5
Berkley clay (subsoil)	1.25	0.99	126	7.3	34.4
Berkley silty clay (topsoil)	5.87	4.63	127	7.1	33.7
Bosket silt loam	1.25	0.57	219	5.8	8.4

Soil properties and adsorption data *(continued)*

Soil	K_d (mL/g)	f_{oc} (%)	K_{oc} (mL/g)	pH (meq/100 g)	CEC
Boxworth clay	15.00	2.08	721	7.9	22.0
Bridgets silt loam	12.00	3.09	388	8.0	24.0
Catalina clay	1.86	1.09	171	4.7	11.8
Catlin	4.70	2.01	419	6.2	—
Cataño sand	3.90	1.21	322	7.9	6.9
Cayaguá sandy loam	2.96	1.15	257	5.2	7.3
Cecil sandy loam	2.40	0.40	600	5.8	—
Cecil sandy clay loam	1.58	1.09	145	5.3	3.6
Chester loam	2.81	1.67	168	4.9	5.2
Chillum silt loam	4.62	2.54	182	4.6	7.6
Christiana loam	0.88	0.57	155	4.4	5.6
Cialitos clay	12.40	2.82	440	5.4	18.6
Coloso clay loam	12.16	2.13	571	5.7	23.0
Commerce	2.20	0.68	325	6.7	—
Coto clay	9.36	1.84	509	7.7	14.0
Crosby silt loam	4.26	1.90	224	4.8	11.5
Dundee silty clay loam	2.25	0.96	234	5.0	18.1
Eustis fine sand	6.60	0.74	891	—	—
Fe clay loam	5.86	1.96	299	7.5	27.6
Fortuna silty clay loam	13.40	1.90	705	5.4	23.3
Fraternidad clay	15.89	2.92	544	5.9	58.0
Fraternidad clay	69.60	1.21	575	6.3	36.0
Garland clay	1.09	0.65	169	7.7	23.2
Great Horse sandy loam	75.00	12.00	625	6.3	18.0
Guánica clay	9.39	2.77	339	8.1	52.1
Hagerstown silty clay loam	2.14	1.30	165	7.5	8.8
Hagerstown silty clay loam	4.26	2.48	172	5.5	12.5
Hawaiian gray hydromorphic	17.00	—	—	—	—
Hawaiian alluvial	9.60	—	—	—	—
Hawaiian alluvial	13.50	—	—	—	—
Hawaiian regosoil	1.70	—	—	—	—
Hawaiian low humic latosol	0.94	—	—	—	—
Hawaiian low humic latosol	1.35	—	—	—	—
Hawaiian low humic latosol	1.65	—	—	—	—
Humata silty clay loam	2.13	0.98	217	4.5	10.1
Indian Head clay loam	13.30	2.34	568	7.8	—
Iredell clay (subsoil)	0.95	6.17	15	5.6	20.9
Irdell silt loam (topsoil)	6.90	3.04	227	5.4	17.0
Josefa silt loam	12.06	1.90	635	6.0	16.8
Juncos silty clay	10.10	1.55	652	6.2	13.4
Keyport silt loam	4.00	1.21	328	5.4	—
Kirton sandy loam	10.00	1.50	667	7.6	13.0
Lakeland sandy loam	1.10	1.88	58	6.2	2.9
Liscombe sandy loam	25.00	3.45	725	6.2	13.0

Soil properties and adsorption data *(continued)*

Soil	K_d (mL/g)	f_{oc} (%)	K_{oc} (mL/g)	pH (meq/100 g)	CEC
Mabí clay	14.46	2.82	513	7.0	55.2
Mabí clay loam	10.33	1.38	749	5.7	31.0
Melfort loam	83.30	6.05	1,377	5.9	—
Mercedita silty clay	5.46	2.19	249	8.1	19.9
Moca clay	12.03	1.90	633	5.8	31.0
Monona clay loam	14.30	1.67	856	6.5	21.2
Montalto clay (subsoil)	0.32	0.86	37	5.9	8.4
Múcara loam	6.53	3.06	213	5.8	19.6
Nipe clay loam	15.10	1.15	1,313	5.7	11.9
Norfolk sandy loam	0.38	0.08	475	5.1	0.2
Ooster silt loam	2.14	1.31	164	4.7	6.8
Pantura sandy loam	4.73	2.02	234	5.7	7.7
Regina heavy clay	13.40	2.39	561	7.7	—
Río Piedras silty clay	3.63	2.02	180	4.9	11.5
Rosemaunde silty clay loam	10.20	1.76	580	6.7	14.0
Ruston sandy loam	1.49	1.05	142	5.1	3.4
San Antón loam	15.80	1.55	1,019	6.7	26.1
Semiahmoo mucky peat	244.34	27.8	879	5.4	—
Sharkey clay	6.36	2.25	283	6.2	40.2
Sterling clay loam	1.41	0.94	150	7.7	22.5
Talante sandy loam	2.36	0.80	295	5.1	4.0
Terrington silt loam	14.00	1.54	909	8.0	15.0
Thurlow clay loam	2.14	1.25	171	7.7	21.6
Tifton sandy loam	0.62	0.56	110	4.9	2.4
Toa loam	3.56	1.15	310	5.3	13.0
Toa sandy loam	1.26	0.34	371	6.0	8.0
Toledo silty clay	5.64	2.80	201	5.5	29.8
Toll Farm HP	53.00	11.70	453	7.4	41.0
Tracy	4.70	1.12	419	6.2	—
Trawscoed silty clay loam	16.00	3.69	434	6.2	12.0
Tripp loam	1.17	0.86	136	7.6	14.7
Truckton sandy loam	0.56	0.25	222	7.0	4.4
Valentine loamy fine sand	6.50	0.80	813	6.6	10.1
Vega Alta sandy loam	6.30	2.02	312	5.0	5.6
Via loam	5.13	1.32	389	5.1	39.9
Webster silty clay loam	24.40	3.34	733	7.3	22.0
Weed Res. sandy loam	13.60	1.93	705	7.1	11.0
Wehadkee silt loam	1.41	1.11	127	5.6	10.2
Weyburn Oxbow loam	26.90	3.72	723	6.5	—
Wooster silt loam	3.33	1.32	252	4.7	6.8

Source: Yuen and Hilton, 1962; Hance, 1965; Harris and Sheets, 1965; Grover and Hance, 1969; Liu et al., 1970; Rhodes et al., 1970; Grover, 1975; Majka and Lavy, 1977; McCall et al., 1981; Nkedi-Kizza et al., 1983; Madhun et al., 1986; Wood et al., 1990.

Environmental Fate

Biological. Degradation of radiolabeled diuron (20 ppm) was not observed after 2 weeks of culturing with *Fusarium* and two unidentified microorganisms. After 80 days, only 3.5% of the applied amount evolved as $^{14}CO_2$ (Lopez and Kirkwood, 1974). In 8 weeks, <20% of diuron in soil (60 ppm) was detoxified (Corbin and Upchurch, 1967). 3,4-Dichloroaniline was reported as a minor degradation product of diuron in water (Drinking Water Health Advisory, 1989) and soils (Duke et al., 1991).

Under aerobic conditions, mixed cultures isolated from pond water and sediment degraded diuron (10 µg/mL) to CPDU, 3,4-dichloroaniline, 3-(3,4-dichlorophenyl)-1-methylurea, carbon dioxide and a monodemethylated product. The extent of biodegradation varied with time, glycerol concentration and microbial population. The degradation half-life was <70 days at 30°C (Ellis and Camper, 1982).

Thom and Agg (1975) reported that diuron is amenable to biological treatment with acclimation.

Soil. Several degradation pathways were reported. The major products and reaction pathways include formation of 1-methyl-3-(3,4-dichlorophenol) urea and 3-(3,4-dichlorophenyl) urea via *N*-dealkylation, a 6-hydroxy derivative via ring hydroxylation, and formation of 3,4-dichloroaniline, 3,4-dichloronitroaniline and 3,4-dichloronitrobenzene via hydrolysis and oxidation (Geissbühler et al., 1975).

Incubation of diuron in soils releases carbon dioxide (Madhun and Freed, 1987). The rate of carbon dioxide formation nearly tripled when the soil temperature was increased from 25 to 35°C. Reported half-lives in an Adkins loamy sand are 705, 414 and 225 days at 25, 30 and 35°C, respectively. However, in a Semiahoo mucky peat, the half-lives were considerably higher: 3,991, 2,164 and 1,165 days at 25, 30 and 35°C, respectively (Madhun and Freed, 1987). Under aerobic conditions, biologically active, organic-rich, diuron-treated pond sediment (40 µg/mL) converted diuron exclusively to 3-(3-chlorophenyl)-1,1-dimethylurea (CPDU) (Attaway et al., 1982, 1982a; Stepp et al., 1985). At 25 and 30°C, 90% degradation was observed after 55 and 17 days, respectively (Attaway, 1982a).

The half-lives for diuron in field soils ranged from 133 to 212 days with an average half-life of 328 days (Hill et al., 1955). Hill et al. (1955) studied the degradation of diuron using a Cecil loamy sand (1 ppm) and Brookstone silty clay loam (5 ppm) in the laboratory maintained at 27°C and 60% relative humidity. In both soils, diuron was applied on four separate occasions over 22 weeks. In both instances, the investigators observed 40% of the applied amount degraded in both soils.

In a field application study, diuron did not leach below 5 cm in depth despite repeated applications or water addition (Majka and Lavy, 1977).

Groundwater. According to the U.S. EPA (1986) diuron has a high potential to leach to groundwater.

Photolytic. Tanaka et al. (1985) studied the photolysis of diuron (40 mg/L) in aqueous solution using UV light (λ = 300 nm) or sunlight. After 25 days of exposure to sunlight, diuron degraded to 2,2,3'-trichloro-2,4'-di-*N,N*-dimethylurea biphenyl (yield = 1.3%) and hydrogen chloride (Tanaka et al., 1985). Diuron should undergo direct photolysis since it absorbs UV light at wavelengths greater than 290 nm (Gore et al., 1971).

Chemical/Physical. Diuron decomposes at 180 to 190°C releasing dimethylamine and 3,4-dichlorophenylisocyanate. Dimethylamine and 3,4-dichloroaniline are produced when hydrolyzed or when acids or bases are added at elevated temperatures (Sittig, 1985). The hydrolysis half-life of diuron in a 0.5 N sodium hydroxide solution at 20°C is 150 days (El-Dib and Aly, 1976).

When diuron was pyrolyzed in a helium atmosphere between 400 and 1,000°C, the following products were identified: dimethylamine, chlorobenzene, 1,2-dichlorobenzene, benzonitrile, a trichlorobenzene, aniline, 4-chloroaniline, 3,4-dichlorophenyl isocyanate, bis(1,3-(3,4-dichlorophenyl)urea), 3,4-dichloroaniline and monuron (3-(4-chlorophenyl)-1,1-dimethylurea) (Gomez et al., 1982). Products reported from the combustion of diuron at 900°C include carbon monoxide, carbon dioxide, chlorine, nitrogen oxides and hydrochloric acid (Kennedy et al., 1972a).

Diuron is stable in aqueous solutions and does not hydrolyze (Hance, 1967a).

Symptoms of Exposure: May irritate eyes, skin, nose and throat.

Exposure Limits: NIOSH REL: TWA 10 mg/m^3.

Formulation Types: Liquid concentrate (4 lb/gal); wettable powder (80%).

Toxicity: LC_{50} (96-hour) for rainbow trout 5.6 mg/L, bluegill sunfish 5.9 mg/L and guppies 25 mg/L (Hartley and Kidd, 1987); LC_{50} (48-hour) for bluegill sunfish 7.4 ppm, rainbow trout 4.3 ppm and coho salmon 16.0 mg/L (Verschueren, 1983); acute oral LD_{50} for rats 3,400 mg/kg (Hartley and Kidd, 1987), 1,017 mg/kg (RTECS, 1985), 437 mg/kg (Windholz et al., 1983).

Uses: Diuron is a urea compound used as a preemergence herbicide in soils to control germinating broad-leaved grasses and weeds in crops such as apples, cotton, grapes, pears, pineapple and alfalfa; sugar cane flowering depressant.

α-ENDOSULFAN

Synonyms: Benzoepin; Beosit; Bio 5462; Chlorthiepin; Crisulfan; Cyclodan; Endocel; Endosol; Endosulfan; Endosulfan I; Endosulphan; ENT 23979; FMC 5462; 1,2,3,7,7-Hexachlorobicyclo[2.2.1]-2-heptene-5,6-bisoxymethylene sulfite; α,β-1,2,3,7,7-Hexachlorobicyclo[2.2.1]-2-heptene-5,6-bisoxymethylene sulfite; Hexachlorohexahydromethano-2,4,3-benzodioxathiepin-3-oxide; **(3α,5αβ,6α,9α,9αβ)-6,7,8,9,10,10-Hexachloro-1,5,5a,6,9,9a-hexahydro-6,9-methano-2,4,3-benzodioxathiepin-3-oxide**; 1,4,5,6,7,7-Hexachloro-5-norborene-2,3-dimethanol cyclic sulfite; Hildan; Hoe 2671; Insectophene; Kopthiodan; Malix; NCI-C00566; NIA 5462; Niagara 5462; OMS-570; RCRA waste number P050; Thifor; Thimul; Thiodan; Thiofor; Thiomul; Thionex; Thiosulfan; Tionel; Tiovel.

Designations: CAS Registry Number: 959-98-8; DOT: 2761; mf: $C_9H_6Cl_6O_3S$; fw: 406.92; RTECS: RB9275000.

Properties: Colorless to brown crystals with a sulfur dioxide odor. Mp: 106°C, 70–100°C (technical grade containing both α and β isomers); ρ: 1.745 at 20/4°C; H-$t_{1/2}$: 218 hours at 25°C and pH 7; K_H: 1.01×10^{-4} atm · m^3/mol at 25°C; log K_{oc}: 3.31 (calculated); log K_{ow}: 3.55; S_o: soluble in acetone, benzene, ethyl ether, 95% ethanol, toluene, xylene and most other organic solvents; S_w: 530 ppb at 25°C; vp: 4.58×10^{-5} mmHg at 25°C.

Environmental Fate

Soil. Metabolites of endosulfan identified in soils were endosulfandiol (1,4,5,6,7,7-hexachlorobicyclo[2.2.1]hept-5-ene-2,3-dimethanol), endosulfan ether, endosulfan lactone (4,5,6,7,8,8-hexachloro-1,3,3a,4,7,7a-hexahydro-4,7-methane-isobenzofuran-1-one) and endosulfan sulfate (6,7,8,9,10,10-hexachloro-1,5,5a,6,9,9a-hexahydro-6,9-methano-2,3,4-benzodioxathiepin-3,3-dioxide) (Martens, 1977; Dreher and Podratzki, 1988). These compounds, including endosulfan ether, were also reported as metabolites identified in aquatic systems (Day, 1991). Endosulfan sulfate was the major biodegradation product in soils under aerobic, anaerobic and flooded conditions (Martens, 1977). In flooded soils, endolactone was detected only once whereas endodiol and endohydroxy ether were identified in all soils under these conditions. Under anaerobic conditions, endodiol formed in low amounts in two soils (Martens, 1977).

Indigenous microorganisms obtained from a sandy loam degraded α-endosulfan to endosulfandiol. This diol was converted to endosulfan α-hydroxy ether and trace amounts of endosulfan ether and both were degraded to endosulfan lactone (Miles and Moy, 1979). Using settled domestic wastewater inoculum, α-endosulfan (5 and 10 mg/L) did not degrade after 28 days of incubation at 25°C (Tabak et al., 1981).

Plant. Endosulfan sulfate was formed when endosulfan was translocated from the leaves to roots in both bean and sugar beet plants (Beard and Ware, 1969). In tobacco leaves, α-endosulfan is hydrolyzed to endosulfandiol (Chopra and Mahfouz, 1977). Stewart and Cairns (1974) reported the metabolite endosulfan sulfate was identified in potato peels and pulp at concentrations of 0.3 and 0.03 ppm, respectively. They also reported that the half-life for the conversion of α-endosulfan to β-endosulfan was 60 days. On apple leaves, direct photolysis of endosulfan by sunlight yielded endosulfan sulfate (Harrison et al., 1967).

In carnation plants, the half-lives of α-endosulfan stored under four different conditions, non-washed and exposed to open air, washed and exposed to open air, non-washed and placed in an enclosed container and under greenhouse conditions were 6.79, 6.38, 10.45 and 4.22 days, respectively (Céron et al., 1995).

Surface Water. Endosulfan sulfate was also identified as a metabolite in a survey of 11 agricultural watersheds located in southern Ontario, Canada (Frank et al., 1982). When endosulfan (α- and β- isomers, 10 µg/L) was added to Little Miami River water, sealed and exposed to sunlight and UV light for 1 week, a degradation yield of 70% was observed. After 2 and 4 weeks, 95% and 100% of the applied amount degraded. The major degradation product was identified as endosulfan alcohol by IR spectrometry (Eichelberger and Lichtenberg, 1971).

Photolytic. Thin films of endosulfan on glass and irradiated by UV light (λ >300 nm) produced endosulfandiol with minor amounts of endosulfan ether, a lactone, an α-hydroxyether and other unidentified compounds (Archer et al., 1972). When an aqueous solution containing endosulfan was photooxidized by UV light at 90–95°C, 25, 50 and 75% degraded to carbon dioxide after 5.0, 9.5 and 31.0 hours, respectively (Knoevenagel and Himmelreich, 1976).

Chemical/Physical. Endosulfan slowly hydrolyzes forming endosulfandiol and endosulfan sulfate (Kollig, 1993; Martens, 1976; Worthing and Hance, 1991). The hydrolysis rate constant for α-endosulfan at pH 7 and 25°C was determined to be 3.2×10^{-3}/hour, resulting in a half-life of 9.0 days (Ellington et al., 1988). The hydrolysis half-lives are reduced significantly at varying pHs and temperature. At temperatures (pH) of 87.0 (3.12), 68.0 (6.89) and 38.0°C (8.69), the half-lives were 4.3, 0.10 and 0.08 days, respectively (Ellington et al., 1986). Greve and Wit (1971) reported the hydrolysis half-lives of α-endosulfan at 20°C and pH values of 7 and 5.5 were 36 and 151 days, respectively.

Emits toxic fumes of chlorides and sulfur oxides when heated to decomposition (Lewis, 1990).

Exposure Limits: ACGIH TLV: TWA 0.1 mg/m³.

Formulation Types: Emulsifiable concentrate; wettable powder; granules; dustable powder; smoke tablet.

Toxicity: LC$_{50}$ (96-hour) for golden orfe 2 µg/L (Hartley and Kidd, 1987), rainbow trout 0.3 µg/L and white sucker 3.0 µg/L (Verschueren, 1983).

Use: Insecticide for vegetable crops.

β-ENDOSULFAN

Synonyms: Benzoepin; Beosit; Bio 5462; Chlorthiepin; Crisulfan; Cyclodan; Endocel; Endosol; Endosulfan; Endosulfan II; Endosulphan; ENT 23979; FMC 5462; 1,2,3,7,7-Hexachlorobicyclo[2.2.1]-2-heptene-5,6-bisoxymethylene sulfite; α,β-1,2,3,7,7-Hexachlorobicyclo[2.2.1]-2-heptene-5,6-bisoxymethylene sulfite; Hexachlorohexahy-dromethano-2,4,3-benzodioxathiepin-3-oxide; **(3α,5aα,6β,9β,9aα)-6,7,8,9,10,10-Hexachloro-1,5,5a,6,9,9a-hexahydro-6,9-methano-2,4,3-benzodioxathiepin-3-oxide;** 1,4,5,6,7,7-Hexachloro-5-norborene-2,3-dimethanol cyclic sulfite; Hildan; Hoe 2671; Insectophene; Kopthiodan; Malix; NCI-C00566; NIA 5462; Niagara 5462; OMS-570; RCRA waste number P050; Thifor; Thimul; Thiodan; Thiofor; Thiomul; Thionex; Thio-sulfan; Tionel; Tiovel.

Designations: CAS Registry Number: 33213-65-9; DOT: 2761; mf: $C_9H_6Cl_6O_3S$; fw: 406.92; RTECS: RB9275000.

Properties: Colorless to brown crystals with a sulfur dioxide odor. Mp: 207–209°C; ρ: 1.745 at 20/20°C; H-$t_{1/2}$: 187 hours at 25°C and pH 7; K_H: 1.91×10^{-5} atm · m³/mol at 25°C (approximate — calculated from water solubility and vapor pressure); log K_{oc}: 3.37 (calculated); log K_{ow}: 3.62; S_o: soluble in acetone, benzene, ethyl ether, 95% ethanol, toluene, xylene and most other organic solvents; S_w: 280 ppb at 25°C; vp: 2.40×10^{-5} mmHg at 25°C.

Environmental Fate

Soil. Metabolites of endosulfan identified in soils included endosulfandiol, endosul-fanhydroxy ether, endosulfan lactone and endosulfan sulfate (Martens, 1977; Dreher and Podratzki, 1988). These compounds, including endosulfan ether, were also reported as metabolites identified in aquatic systems (Day, 1991). In aerobic soils, β-endosulfan is converted to the corresponding alcohol and ether (Perscheid et al., 1973). Endosulfan sulfate was the major biodegradation product in soils under aerobic, anaerobic and flooded conditions (Martens, 1977). In flooded soils, endolactone was detected only once whereas endodiol and endohydroxy ether were identified in all soils under these conditions. Under anaerobic conditions, endodiol formed in low amounts in two soils (Martens, 1977). Indigenous microorganisms obtained from a sandy loam degraded β-endosulfan to endosulfan diol. This diol was converted to endosulfan α-hydroxy ether and trace amounts of endosulfan ether and both were degraded to endosulfan lactone (Miles and Moy, 1979).

Plant. In addition, endosulfan sulfate was formed when endosulfan was translocated from the leaves to roots in both bean and sugar beet plants (Beard and Ware, 1969). In tobacco leaves, β-endosulfan hydrolyzed into endosulfandiol (Chopra and Mahfouz, 1977).

Stewart and Cairns (1974) reported the metabolite endosulfan sulfate was identified in potato peels and pulp at concentrations of 0.3 and 0.03 ppm, respectively. They also reported that the half-life for the oxidative conversion of β-endosulfan to endosulfan sulfate was 800 days.

In carnation plants, the half-lives of β-endosulfan stored under four different conditions, non-washed and exposed to open air, washed and exposed to open air, non-washed and placed in an enclosed container and under greenhouse conditions were 23.40, 12.64, 37.42 and 7.62 days, respectively (Ceron et al., 1995).

Surface Water. Endosulfan sulfate was also identified as a metabolite in a survey of 11 agricultural watersheds located in southern Ontario, Canada (Frank et al., 1982). When endosulfan (α- and β- isomers, 10 µg/L) was added to Little Miami River water, sealed and exposed to sunlight and UV light for 1 week, a degradation yield of 70% was observed. After two and four weeks, 95% and 100% of the applied amount degraded. The major degradation product was identified as endosulfan alcohol by IR spectrometry (Eichelberger and Lichtenberg, 1971).

Photolytic. Thin films of endosulfan on glass and irradiated by UV light (λ >300 nm) produced endosulfan diol with minor amounts of endosulfan ether, lactone, α-hydroxyether and other unidentified compounds (Archer et al., 1972). Gaseous β-endosulfan subjected to UV light (λ >300 nm) produced endosulfan ether, endosulfan diol, endosulfan sulfate, endosulfan lactone, α-endosulfan and a dechlorinated ether (Schumacher et al., 1974). Irradiation of β-endosulfan in *n*-hexane by UV light produced the photoisomer α-endosulfan (Putnam et al., 1975). When an aqueous solution containing endosulfan was photooxidized by UV light at 90–95°C, 25, 50 and 75% degraded to carbon dioxide after 5.0, 9.5 and 31.0 hours, respectively (Knoevenagel and Himmelreich, 1976).

Chemical/Physical. Endosulfan detected in Little Miami River, OH was readily hydrolyzed to a compound tentatively identified as endosulfan diol (Eichelberger and Lichtenberg, 1971). Sulfuric acid is also an end product of hydrolysis (Kollig, 1993). The hydrolysis half-lives at pH values (temperature) of 3.32 (87.0°C), 6.89 (68.0°C) and 8.69 (38.0°C) were calculated to be 2.7, 0.07 and 0.04 days, respectively (Ellington et al., 1988). Greve and Wit (1971) reported the hydrolysis half-lives of β-endosulfan at 20°C and pH values of 7 and 5.5 were 37 and 187 days, respectively.

Exposure Limits: ACGIH TLV: TWA 0.1 mg/m³.

Formulation Types: Emulsifiable concentrate; wettable powder; granules; dustable powder; smoke tablet.

Toxicity: LC_{50} (96-hour) for golden orfe 2 µg/L (Hartley and Kidd, 1987), rainbow trout 0.3 µg/L and white sucker 3.0 µg/L (Verschueren, 1983).

Use: Insecticide for vegetable crops.

ENDOSULFAN SULFATE

Synonyms: **6,7,8,9,10,10-Hexachloro-1,5,5a,6,9,9a-hexahydro-3,3-dioxide**; 6,9-Methano-2,4,3-benzodioxathiepin.

Designations: CAS Registry Number: 1031-07-8; DOT: 2761; mf: $C_9H_6Cl_6O_4S$; fw: 422.92.

Properties: Solid. Mp: 198–201°C; log K_{oc}: 3.37 (calculated); log K_{ow}: 3.66; S_w: 117 ppb; vp: 9.75×10^{-6} mmHg at 25°C.

Environmental Fate

Soil. A mixed culture of soil microorganisms biodegraded endosulfan sulfate to endosulfan ether, endosulfan-α-hydroxy ether and endosulfan lactone (Verschueren, 1983). Indigenous microorganisms obtained from a sandy loam degraded endosulfan sulfate (a metabolite of α- and β-endosulfan) to endosulfan diol. This diol was converted to endosulfan α-hydroxy ether and trace amounts of endosulfan ether and both were degraded to endosulfan lactone (Miles and Moy, 1979). Using settled domestic wastewater inoculum, endosulfan sulfate (5 and 10 mg/L) did not degrade after 28 days of incubation at 25°C (Tabak et al., 1981).

Plant. In tobacco leaves, endosulfan sulfate was converted to α-endosulfan which subsequently hydrolyzed into endosulfandiol (Chopra and Mahfouz, 1977).

Uses: Not known. Compound is described here because it is a degradate of endosulfan, a widely used insecticide.

ENDOTHALL

Synonyms: Accelerate; Aquathol; Des-i-cate; 3,6-*endo*-Epoxy-1,2-cyclohexane-dicarboxylic acid; Endothal; 3,6-Endooxohexahydrophthalic acid; 3,6-Epoxycyclohexane-1,2-dicarboxylic acid; Hydout; Hydrothal; Hydrothal-47; Hydrothal-191; **7-Oxabicyclo[2.2.1]heptane-2,3-dicarboxylic acid**; Pennout; RCRA waste number P088; Ripenthal; Triendothal.

Designations: CAS Registry Number: 145-73-3; mf: $C_8H_{10}O_5$; fw: 186.16; RTECS: RN7875000.

Properties: Odorless, colorless crystals (monohydrate). Mp: 144°C (monohydrate); ρ: 1.431 at 20/4°C; fl p: nonflammable; pK_1 = 3.4, pK_2 = 6.7; log K_{oc}: 2.04, 2.14; log K_{ow}: 1.91 (Reinert and Rodgers, 1984); S_o (g/kg at 20°C): acetone (70), benzene (0.1), 1,4-dioxane (76), ethyl ether (1), methanol (280), 2-propanol (17); S_w: 100 g/kg at 25°C; vp: negligible at 20°C.

Soil properties and adsorption data

Soil	K_d (mL/g)	f_{oc} (%)	K_{oc} (mL/g)	pH
Pat Mayse Lake sediments	0.94	0.68	138	6.5–7.5
Roselawn Cemetery Pond sediments	1.42	1.29	110	8.3–8.5

Source: Reinert and Rogers, 1984.

Environmental Fate

Biological. Incubation of ^{14}C-ring labeled endothall (10 μg/mL) by *Arthrobacter* sp., which was isolated from pond water and a hydrosol, in aerobic sediment-water suspensions revealed that after 30 days, 40% evolved as $^{14}CO_2$. Glutamic acid was the major transformation product. Minor products were alanine, citric and aspartic acids and unidentified products, some of which were tentatively identified as phosphate esters (Sikka and Saxena, 1973). In pond water treated with endothall (2 and 4 ppm), detectable levels were found after 7 days (Sikka and Rice, 1973). Biodegradation was rapid in an Ontario soil sample. After 1 week, 70% of endothall added was converted to carbon dioxide (Simsiman et al., 1976).

Chemical/Physical. Reacts with bases forming water-soluble salts. Above 90°C, endothall is slowly converted to the anhydride (Windholz et al., 1983; Hartley and Kidd, 1987) and water (Humburg et al., 1989). Endothall is stable to light (Hartley and Kidd, 1987).

Symptoms of Exposure: Strong irritant to eyes, nose, throat and skin. Ingestion may cause vomiting and diarrhea.

Formulation Types: Granules; soluble concentrate.

Toxicity: LC_{50} (96-hour) using disodium salt: bluegill sunfish 125 mg/L, chinook 136 mg/L, fathead minnow 110 mg/L, largemouth bass 120 mg/L (Verschueren, 1983); acute oral LD_{50} for rats 38–51 mg/kg (free acid) (Hartley and Kidd, 1987), 182–197 mg/kg (sodium salt) (Verschueren, 1983).

Uses: Preemergence and postemergence herbicide for control of broad-leaved weeds and annual grass in vegetable crops. The disodium and dipotassium salts are used as defoliants and herbicides.

ENDRIN

Synonyms: Compound 269; Endrex; ENT 17251; Experimental insecticide 269; Hexachloroepoxyoctahydro-*endo,endo*-dimethanonaphthalene; 1,2,3,4,10,10-Hexachloro-6,7-epoxy-1,4,4a,5,6,7,8,8a-octahydro-*endo,endo*-1,4:5,8-dimethanonaphthalene; **3,4,5,6,9,9-Hexachloro-1a,2,2a,3,6,6a,7,7a-octahydro-2,7:3,6-dimethanonaphth[2,3-*b*]oxirene**; Hexadrin; Isodrin epoxide; Mendrin; NA 2761; NCI-C00157; Nendrin; RCRA waste number P051.

Designations: CAS Registry Number: 72-20-8; DOT: 2761; mf: $C_{12}H_8Cl_6O$; fw: 380.92; RTECS: IO1575000.

Properties: White, odorless, crystalline solid when pure; light tan color with faint chemical odor for technical grade. Mp: 200°C; bp: 245°C (decomposes); ρ: 1.70 at 25/4°C (pure), 1.65 at 25/4°C (technical); fl p: >26.6°C (xylene solution); lel: 1.1% in xylene; uel: 7.0% in xylene; K_H: 5×10^{-7} atm · m³/mol; log BCF: 3.28 (mussels, Ernst, 1977); log K_{ow}: 3.209–5.339; S_o (g/L at 25°C): acetone (170), benzene (138), carbon tetrachloride (33), cod liver oil (76.0, 83.3 and 91.4 g/L at 4, 12 and 20°C, respectively), hexane (71), octanol (36.4, 38.1 and 43.7 g/L at 4, 12 and 20°C, respectively), triolein (61.8, 72.6 and 87.3 g/L at 4, 12 and 20°C, respectively), xylene (183); S_w: 220–260 ppb at 25°C; vp (mmHg): 3.0 $\times 10^{-6}$ at 20°C (Nash, 1983); 7×10^{-7} mmHg at 25°C.

Environmental Fate

Biological. In four successive 7-day incubation periods, endrin (5 and 10 mg/L) was recalcitrant to degradation in a settled domestic wastewater inoculum (Tabak et al., 1981).

Soil. Microbial degradation of endrin in soil formed several ketones and aldehydes of which *keto*-endrin was the only metabolite identified (Kearney and Kaufman, 1976). In eight Indian rice soils, endrin degraded rapidly to low concentrations after 55 days. Degradation was highest in a pokkali soil and lowest in a sandy soil (Gowda and Sethunathan, 1976).

Under laboratory conditions, endrin degraded to other compounds in a variety of soils maintained at 45°C. Except for Rutledge sand, endrin disappeared or was transformed in the following soils after 24 hours: Lynchburg loamy sand, Magnolia sandy loam, Magnolia sandy clay loam, Greenville sandy clay and Susquehanna sandy clay. No products were identified (Bowman et al., 1965).

The disappearance half-lives for endrin in field soils under flooded and nonflooded conditions were 130 and 468 days, respectively (Guenzi et al., 1971). The average disappearance half-life in flooded soils under laboratory conditions was 31 days (Gowda and Sethunathan, 1976, 1977).

The percentage of endrin remaining in a Congaree sandy loam soil after 14 years was 41% (Nash and Woolson, 1967).

Groundwater. According to the U.S. EPA (1986) endrin has a high potential to leach to groundwater.

Plant. In plants, endrin is converted to the corresponding sulfate (Hartley and Kidd, 1987).

Surface Water. Algae isolated from a stagnant fish pond degraded 24.4% of the applied endrin to ketoendrin (Patil et al., 1972).

Photolytic. Photolysis of thin films of solid endrin using UV light (λ = 254 nm) produced δ-ketoendrin, endrin aldehyde and other compounds (Rosen et al., 1966). Endrin exposed to a hot California sun for 17 days completely isomerized to δ-ketoendrin or 1,8-*exo*-9,10,11,11-hexachlorocyclo[6.2.1.1^{3,6}.0^{2,7}.0^{4,10}]dodecan-5-one (Burton and Pollard, 1974). Irradiation of endrin by UV light (λ = 253.7 nm and 300 nm) or by natural sunlight in cyclohexane and *n*-hexane solution resulted in an 80% yield of 1,8-*exo*-9,11,11-pen-tachloropentacyclo[6.2.1.1^{3,6}.0^{2,7}.0^{4,10}]dodecan-5-one (Zabik et al., 1971). When an aqueous solution containing endrin was photooxidized by UV light at 90–95°C, 25, 50 and 75% degraded to carbon dioxide after 15.0, 41.0 and 172.0 hours, respectively (Knoevenagel and Himmelreich, 1976).

Chemical/Physical. At 230°C, endrin isomerizes to an aldehyde and a ketone. When heated to decomposition, hydrogen chloride and phosgene may be released (NIOSH, 1994) but residues containing an aldehyde (15–20%), a ketone (55–60%), a caged alcohol (5%), and other volatile products (15–20%) were reported (Phillips et al., 1962).

In water, endrin will undergo nucleophilic attack at the epoxide moiety forming endrin diol (Kollig, 1993).

At 50°C, endrin was not unaffected by the oxidants chlorine, permanganate and persulfate (Leigh, 1969).

Exposure Limits: NIOSH REL: TWA 0.1 mg/m^3, IDLH 2 mg/m^3; OSHA PEL: TWA 0.1 mg/m^3; ACGIH TLV: TWA 0.1 mg/m^3.

Symptoms of Exposure: Epileptiform convulsions, stupor, headache, dizziness, abdominal discomfort, nausea, vomiting, insomnia, aggressive confusion, lethargy, weakness, anorexia.

Formulation Types: Emulsifiable concentrate; wettable powder; granules; dustable powder.

Toxicity: LC$_{50}$ (28-hour) for bullhead fish 0.010 μg/L (Anderson and DeFoe, 1980); LC$_{50}$ (96-hour) for bluegill sunfish 0.6 μg/L, fathead minnow 1.0 μg/L (Henderson et al., 1959), rainbow trout 0.6 μg/L, coho salmon 0.5 μg/L and chinook 1.2 μg/L (Katz, 1961); acute oral LD$_{50}$ for rats 7–15 mg/kg (Hartley and Kidd, 1987), 3 mg/kg (RTECS, 1985), male and female rats 18 and 7.5 mg/kg, respectively (Windholz et al., 1983).

Use: Insecticide.

EPN

Synonyms: ENT 17798; EPN 300; Ethoxy-4-nitrophenoxy phenylphosphine sulfide; Ethyl *p*-nitrophenyl benzenethionophosphate; Ethyl *p*-nitrophenyl benzenethiophosphonate; Ethyl *p*-nitrophenyl ester; *O*-Ethyl *O*-4-nitrophenyl phenylphosphonothioate; Ethyl *p*-nitrophenyl phenylphosphonothioate; *O*-Ethyl *O*-*p*-nitrophenyl phenylphosphonothioate; Ethyl *p*-nitrophenyl thionobenzenephosphate; Ethyl *p*-nitrophenyl thionobenzenephosphonate; *O*-Ethyl phenyl *p*-nitro-phenyl phenylphosphorothioate; *O*-Ethyl phenyl *p*-nitrophenyl thiophosphonate; Phenylphosphonothioic acid *O*-ethyl *O*-*p*-nitrophenyl ester; **Phosphonothioic acid *O,O*-diethyl *O*-(3,5,6-trichloro-2-pyridinyl) ester**; Pin; Santox.

Designations: CAS Registry Number: 2104-64-5; DOT: 2783; mf: $C_{14}H_{14}NO_4PS$; fw: 323.31; RTECS: TB1925000.

Properties: Yellow solid or crystals to brown liquid with an aromatic odor. Mp: 36°C; bp: 215°C at 5 mmHg; ρ: 1.268 and 1.5978 at 25/4 and 30/4°C, respectively; log K_{oc}: 3.12; log K_{ow}: 3.85, 5.07; S_o: miscible with acetone, benzene, methanol, isopropanol, toluene, xylene and many other aromatic solvents; vp: 0.126 mPa at 25°C.

Soil properties and adsorption data

Soil	K_d (mL/g)	f_{oc} (%)	K_{oc} (mL/g)	pH
Kanuma high clay	7.8	1.35	706	5.7
Tsukuba clay loam	60.7	4.24	1,997	6.5

Source: Kanazawa, 1989.

Environmental Fate

Biological. From the first-order biotic and abiotic rate constants of EPN in estuarine water and sediment/water systems, the estimated biodegradation half-lives were 6.2 and 9.2 days, respectively (Walker et al., 1988).

Soil. Though no products were reported, the half-life in soil is 15–30 days (Hartley and Kidd, 1987).

Photolytic. EPN may undergo direct photolysis since the insecticide showed some absorption when a 1,4-dioxane was irradiated with UV light (λ >290 nm) (Gore et al., 1971).

Chemical/Physical. On heating, EPN is converted to the *S*-ethyl isomer (Worthing and Hance, 1991). Releases toxic fumes of phosphorus, nitrogen and sulfur oxides when heated to decomposition (Sax and Lewis, 1987; Lewis, 1990). Rapidly hydrolyzed in alkaline solutions to *p*-nitrophenol, alcohol and benzene thiophosphoric acid (Sittig, 1985).

Exposure Limits: NIOSH REL: TWA 0.5 mg/m³, IDLH 5 mg/m³; OSHA PEL: TWA 0.5 mg/m³; ACGIH TLV: TWA 0.5 mg/m³.

Symptoms of Exposure: Miosis, irritates eyes; rhinorrhea; headache; tight chest, wheezing, laryngeal spasm; salivation; cyanosis; anorexia, nausea, abdominal cramps, diarrhea; paralysis convulsions; low blood pressure.

Formulation Types: Emulsifiable concentrate; wettable powder; granules.

Toxicity: LC_{50} for rainbow trout 0.21 mg/L (Worthing and Hance, 1991), bluegill sunfish 100 µg/L (Sanders and Cope, 1968), fathead minnow 110 mg/L (Solon and Nair, 1970); acute oral LD_{50} for male and female rats, 36 and 7.7 mg/kg, respectively (Windholz et al., 1983), 7 mg/kg (RTECS, 1985).

Uses: Insecticide; acaricide.

EPTC

Synonyms: Dipropylcarbamothioic acid *S*-ethyl ester; *N,N*-Dipropylthiocarbamic acid *S*-ethyl ester; Eptam; *S*-Ethyl dipropylcarbamothioate; *S*-Ethyl dipropylthiocarbamate; *S*-Ethyl di-*n*-propylthiocarbamate; *S*-Ethyl-*N,N*-di-*n*-propylthiolcarbamate; FDA 1541; R 1608.

$$CH_3CH_2S - \overset{\overset{\textstyle O}{\|}}{C} - N \overset{\diagup CH_2CH_2CH_3}{\diagdown CH_2CH_2CH_3}$$

Designations: CAS Registry Number: 759-94-4; mf: $C_9H_{19}NOS$; fw: 189.32; RTECS: FA4550000.

Properties: Colorless to pale yellow liquid with an amine-like odor. Mp: <25°C; bp: 127°C at 20 mmHg, 235°C (extrapolated); ρ: 0.960 at 25/25°C, 0.9546 at 30/4°C; fl p: 116°C (open cup); K_H: 1×10^{-5} atm · m^3/mol at 20–25°C (approximate — calculated from water solubility and vapor pressure); log K_{oc}: 2.38; log K_{ow}: 3.20; S$_o$: miscible with most organic solvents, e.g., acetone, benzene, ethanol, ethylbenzene, isopropanol, kerosene, methanol, methyl isobutyl ketone, toluene, xylene; S$_w$: 375 mg/L at 25°C; vap d: 7.74 g/L at 25°C, 6.56 (air = 1); vp: 3.4×10^{-2} mmHg at 20°C.

Soil properties and adsorption data

Soil	K_d (mL/g)	f_{oc} (%)	K_{oc} (mL/g)	pH (cmol/kg)	CEC
Hanford sandy loam	0.47	0.43	109	6.0	6.0
Hanford sandy loam	0.72	0.43	167	6.0	6.0
Hanford sandy loam	1.81	0.43	421	6.0	6.0
Tujunga loamy sand	0.36	0.33	109	6.3	0.4
Tujunga loamy sand	0.80	0.33	242	6.3	0.4
Tujunga loamy sand	1.90	0.33	576	6.3	0.4

Source: Singh et al., 1990.

Environmental Fate

Soil. EPTC is rapidly degraded by soil microbes and fungi yielding carbon dioxide, ethyl mercaptan and amino residues (Kaufman, 1967; Lee, 1984; Hartley and Kidd, 1987). Degradation occurs via hydrolysis of the ester linkage forming the corresponding mercaptan, alkylamine (*n*-dipropylamine) and carbon dioxide. Transthiolation and oxidation of the mercaptan forms the alcohol which is further oxidized to afford a metabolic pool (Kaufman, 1967). EPTC partially degraded in both sterile and nonsterile clay soils. Mineralization was not observed since carbon dioxide was not detected (MacRae and Alexander, 1965). EPTC sulfoxide was also reported as a metabolite identified in soil (Soma-sundaram and Coats, 1991) and in corn plants (Casida et al., 1974). The rapid formation of carbon dioxide was also observed from the microbial degradation of EPTC by a microbial metabolite isolated from Jimtown loam soil and designated JE1 (Dick et al., 1990). These researchers proposed that EPTC hydroxylated at the α-propyl carbon forming the unstable α-hydroxypropyl EPTC which degrades to N-depropyl EPTC and propional-dehyde. Metabolization of N-depropyl EPTC yields s-ethyl formic acid and propylamine. Demethylation of s-ethyl formic acid gives s-methyl formic acid. Propylamine and s-methyl formic acid probably degrades to ammonia and methyl mercaptan, respectively and carbon dioxide (Dick et al., 1990). The reported half-lives in soil ranged from 2 to 4 weeks in two agricultural soils (Regina heavy clay and Weyburn loam) (Smith and Fitz-patrick, 1970) to 30 days (Jury et al., 1987). Rajagopal et al. (1989) reported that the persistence of EPTC in soil ranged from <4 to 6 weeks when applied at recommended rates.

Plant. EPTC is rapidly metabolized by plants to carbon dioxide and naturally occurring plants constituents (Humburg et al., 1989).

Chemical/Physical. Emits toxic fumes of phosphorus and sulfur oxides when heated to decomposition (Sax and Lewis, 1987).

In the gas phase, EPTC reacts with hydroxyl and/or NO_3 radicals but not with ozone. With hydroxy radicals in the presence of NO_3, S-ethyl N-formyl-N-propylthiocarbamate formed as the major product. With NO_3 radicals only, the major product was $C_3H_7(CHO)NC(O)SC_2H_5$. In addition, a minor product formed tentatively identified as $CH_3CH_2CH_2(CH_3CH_2C(O))NC(O)SCH_2CH_3$. The calculated photolysis lifetimes of EPTC in the troposphere with hydroxyl and NO_3 radicals are 5.8 hours and 5.0 days, respectively (Kwok et al., 1992).

Formulation Types: Granules (10%); emulsifiable concentrate (7 lb/gal).

Toxicity: LC_{50} (48-hour) for bluegill sunfish 27 mg/L and rainbow trout 19 mg/L (Hartley and Kidd, 1987); LC_{50} (24-hour) for blue crab 10 mg/L (Humburg et al., 1989); acute oral LD_{50} for male rats and male mice 1,700 and 3,200 mg/kg, respectively (Ashton and Monaco, 1991), 1,325 mg/kg (RTECS, 1985), 1,631 mg/kg (Windholz et al., 1983).

Uses: Selective systemic herbicide used for preemergence control of perennial and annual grasses such as johnsongrass, nutgrass and quackgrass. Also for control of some broad-leaved weeds such as chickweed, henbit, lambsquarters, pigweed and purslane. EPTC is also used in vegetable crops, alfalfa, cotton, flax, pineapple, almonds and walnuts.

ESFENVALERATE

Synonyms: (*S*-(*R***,*R***))-Cyano(3-phenoxyphenyl)methyl 4-chloro-α-(1-methyl-ethyl)benzeneacetate.

Designations: CAS Registry Number: 66230-04-4; mf: $C_{25}H_{22}ClNO_3$; fw: 419.90.

Properties: Brown solid. Mp: 59.0–60.2°C; bp: 151–157°C; ρ: 1.26 at 26/4°C; K_H: 9.26 × 10^{-7} atm · m³/mol at 25°C (approximate — calculated from water solubility and vapor pressure); log K_{oc}: 3.93 (calculated); log K_{ow}: 6.22; S_o: very soluble in acetone, acetonitrile, chloroform, ethyl acetate, *N,N*-dimethylformamide, dimethyl sulfoxide, 4-methyl-2-pentanone, xylene; S_w: 0.3 mg/L at 25°C; vp: 5.03 × 10^{-7} mmHg at 25°C.

Environmental Fate

Chemical/Physical. May hydrolyze in aqueous solutions forming acetic acid and other compounds.

Formulation Types: Emulsifiable concentrate; suspension concentrate.

Toxicity: LC_{50} (96-hour) for fathead minnows 0.69 µg/L (Worthing and Hance, 1991); acute oral LD_{50} for rats 75 mg/kg (Hartley and Kidd, 1987).

Use: Insecticide.

ETHEPHON

Synonyms: Amchem 68-250; Bromoflor; Camposan; CEP; Cepha; 2-CEPA; Cepha 10LS; Cerone; Chlorethephon; 2-Chloroethanephosphonic acid; **(2-Chloroethyl)-phosphonic acid**; Ethel; Etheverse; Ethrel; Flordimex; Florel; G 996; Kamposan; Prep; Rollfruct; Tomathrel.

$$ClCH_2CH_2PO(OH)_2$$

Designations: CAS Registry Number: 16672-87-0; mf: $C_2H_6ClO_3P$; fw: 144.50; RTECS: SZ7100000.

Properties: Colorless to grayish-white waxy solid. Needles from benzene are very hygroscopic. Mp: 74–75°C; ρ: 1.2–1.3; fl p: nonflammable; $H\text{-}t_{1/2}$: 24 hours at 33°C and pH 7; K_H: 1.11×10^{-11} atm · m³/mol at 20–23°C (approximate — calculated from water solubility and vapor pressure); log K_{oc}: 0.29 (calculated); log K_{ow}: -0.22; pK_a: <7.0; S_o: very soluble in acetone, chloroform, ethanol, ethylene glycol, methanol, hexane, methylene chloride and propanol but only slightly soluble in benzene, toluene and xylene; S_w: 1.24 kg/L at 23°C; vp: 7.50×10^{-5} mmHg at 20°C.

Environmental Fate

Soil. Degrades rapidly in soil to phosphoric acid, ethylene and chloride ions (Hartley and Kidd, 1987) and naturally occurring substances (Humburg et al., 1989).

Chemical/Physical. In an aqueous solution at pH 3.5, ethepon begins to hydrolyze, releasing ethylene (Windholz et al., 1983).

Symptoms of Exposure: Irritates eyes and skin.

Formulation Types: Emulsifiable concentrate; soluble concentrate.

Toxicity: LC_{50} (96-hour) for rainbow trout 254 mg/L and bluegill sunfish 222 mg/L (Worthing and Hance, 1991); acute oral LD_{50} for rats 4,229 mg/kg (24% solution in propylene glycol) (Hartley and Kidd, 1987), 3,400 mg/kg (RTECS, 1985).

Uses: Accelerates the preharvest ripening of fruits and vegetables.

ETHIOFENCARB

Synonyms: Bay-Hox-1901; Croneton; Ethiophencarp; 2-Ethylmercapotomethylphenyl-*N*-methylcarbamate; 2-((Ethylthio)methyl)phenol methylcarbamate; **2-Ethylthiomethylphenyl methylcarbamate**; 2-Ethylthiomethylphenyl-*N*-methylcarbamate; α-Ethylthio-*o*-tolyl methylcarbamate; HOX 1901.

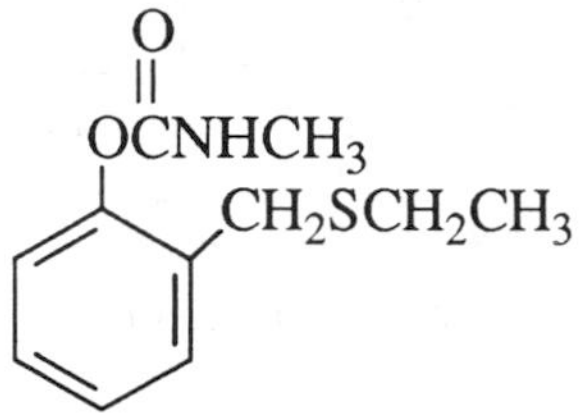

$$
\begin{array}{c}
\text{O} \\
\parallel \\
\text{OCNHCH}_3 \\
\text{CH}_2\text{SCH}_2\text{CH}_3
\end{array}
$$

Designations: CAS Registry Number: 29973-13-5; mf: $C_{11}H_{15}NO_2S$; fw: 225.31; RTECS: FC2826000.

Properties: Colorless crystals. Mp: 33.4°C; bp: decomposes; ρ: 1.1473 at 20/4°C; H-$t_{1/2}$ (isopropanol/water = 1:1 at 37–40°C): 330 days (pH 2), 450 hours (pH 7), 5 minutes (pH 11.4); K_H: 5.3 × 10^{-10} atm · m³/mol at 20°C (approximate — calculated from water solubility and vapor pressure); log K_{oc}: 1.84 (calculated); log K_{ow}: 0.98 (calculated); S_o: >600 g/kg in methylene chloride, 2-propanol, toluene; S_w: 1.9 g/L at 20°C; vp: 3.38 × 10^{-6} mmHg at 20°C.

Environmental Fate
Plant. Degrades in plants to the sulfone and sulfoxide (Hartley and Kidd, 1987).

Formulation Types: Granules, emulsifiable concentrate.

Toxicity: LC$_{50}$ (96-hour) for carp 10–20 mg/L, golden orfe 8–10 mg/L, goldfish 20–40 mg/L and rudd 10–20 mg/L (Hartley and Kidd, 1987); acute oral LD$_{50}$ for rats 411–499 mg/kg (Hartley and Kidd, 1987), 200 mg/kg (RTECS, 1985).

Uses: Systemic insecticide used to control aphids on fruit crops.

ETHION

Synonyms: AC 3422; Bis(*S*-(dimethoxyphosphinothioyl)mercapto)methane; Bladan; Diethion; Embathion; ENT 24105; Ethanox; Ethiol; Ethodan; Ethyl methylene phosphorodithioate; FMC 1240; Fosfono 50; Hylemox; Itopaz; Kwit; Methanedithiol-*S,S*-diester with *O,O*-diethyl phosphorodithioate; **S,S′-Methylene bis(*O,O*-diethyl phosphorodithioate);** *S,S′*-Methylene *O,O,O′,O′*-tetraethyl phosphorodithioate; NA 2783; NIA 1240; Niagara 1240; Nialate; Phosphorodithioic acid *O,O*-diethyl ester, *S,S*-diester with methanedithiol; *O,O,O,O*-Tetraethyl *S,S′*-methylenebisdithiophosphate; Phosphotox E; Rhodiacide; Rhodocide; Rodocid; RP 8167; Soprathion; *O,O,O′,O′*-Tetraethyl *S,S′*-methylenebisphosphordithioate; *O,O,O′,O′*-Tetraethyl *S,S′*-methylenebisphosphorodithioate; *O,O,O′,O′*-Tetraethyl *S,S′*-methylenebisphosphorothiolothionate; *O,O,O′,O′*-Tetraethyl *S,S′*-methylene di(phosphorodithioate); Vegfru fosmite.

$$CH_3CH_2O \overset{\overset{\displaystyle S}{||}}{\underset{CH_3CH_2O}{\diagup}} P - SCH_2S - \overset{\overset{\displaystyle S}{||}}{\underset{OCH_2CH_3}{\diagdown}} P \overset{OCH_2CH_3}{\diagdown}$$

Designations: CAS Registry Number: 563-12-2; DOT: 2783; mf: $C_9H_{22}O_4P_2S_4$; fw: 384.48; RTECS: TE4550000.

Properties: Colorless to amber-colored liquid with a very disagreeable odor. Mp: −15 to −12°C; ρ: 1.220 at 20/4°C; H-$t_{1/2}$: 5 to 63 days at pH 6 and 25°C; K_H: 3.79 × 10⁻⁷ atm · m³/mol at 25°C; log K_{oc}: 3.54–4.34; log K_{ow}: 4.28, 5.07; P-$t_{1/2}$: 85.47 hours (absorbance λ = 232.5 nm, concentration on glass plates = 6.7 µg/cm²); S_o: soluble in most organic solvents including acetone, chloroform, methylene chloride, xylene and kerosene + 1% benzene; S_w: 570, 680 and 760 µg/L at 10, 20 and 30°C, respectively; vap d: 15.71 g/L at 25°C, 13.27 (air = 1); vp: 1.50 × 10⁻⁶ mmHg at 25°C.

Soil properties and adsorption data

Soil	K_d (mL/g)	f_{oc} (%)	K_{oc} (mL/g)	pH
Sand	167	0.75	22,149	6.2
Sandy loam	215	1.74	12,356	7.0
Silt loam	105	1.80	5,833	7.1
Silt loam	47	1.33	3,534	7.5

Source: U.S. Department of Agriculture, 1990.

Environmental Fate

Biological. Ethion degraded in lagoonal sediments obtained at various sites in the Indian River between Cape Kennedy and Vero Beach, FL. In 14 sediment samples enriched with ethion, 8 exhibited iron sulfide (precursor hydrogen sulfide) production following 20 days of incubation at room temperature. The bacteria responsible for the degradation of ethion, a reducing agent, was tentatively identified as *Clostridium* (Sherman et al., 1974).

Soil. The half-lives of ethion in an organic soil varied from 16 to 49 weeks; however, repeated applications each spring resulted in increased residues of unreacted ethion (Chapman et al., 1984).

Photolytic. Ethion in hexane did not exhibit absorption at UV wavelengths >260 nm (Gore et al., 1971).

Chemical/Physical. Emits toxic fumes of phosphorus and sulfur oxides when heated to decomposition (Sax and Lewis, 1987; Lewis, 1990).

The hydrolysis half-lives of ethion in a sterile 1% ethanol/water solution at 25°C and pH values of 4.5, 5.0, 6.0, 7.0 and 8.0 were 99, 63, 58, 24 and 8.4 weeks, respectively (Chapman and Cole, 1982).

Exposure Limits: NIOSH REL: TWA 0.4 mg/m^3; ACGIH TLV: TWA 0.4 mg/m^3.

Symptoms of Exposure: Headache, anorexia, nausea, weakness, dizziness, blurred vision, salivation, lacrimation, sweating, shortness of breath, ataxia, fever, cyanosis, pulmonary edema, convulsions, shock, heart block, respiratory failure.

Formulation Types: Emulsifiable concentrate; wettable powder; granules; dustable powder; seed treatment.

Toxicity: LC$_{50}$ (96-hour) for bluegill sunfish 220 µg/L, largemouth bass 150 µg/L, rainbow trout 560 µg/L, cutthroat trout 720 µg/L, channel catfish 7.50 mg/L (Verschueren, 1983); acute oral LD$_{50}$ for rats 208 mg/kg (pure) and 96 mg/kg (technical) (Hartley and Kidd, 1987), 13 mg/kg (RTECS, 1985), male and female rats, 65 and 27 mg/kg, respectively (Windholz et al., 1983).

Uses: Nonsystemic insecticide and acaricide used on apples.

ETHOPROP

Synonyms: ENT 27318; Ethoprophos; *O*-Ethyl *S,S*-dipropyl phosphorodithioate; Jolt; Mobil V-C 9-104; Mocap; **Phosphorodithioic acid *O*-ethyl *S,S*-dipropyl ester**; Prophos; V-C 9-104; V-C chemical V-C 9-104; Virginia-Carolina VC 9-104.

$$CH_3CH_2CH_2S \diagdown \overset{\overset{O}{\|}}{P} - OCH_2CH_3$$
$$CH_3CH_2CH_2S \diagup$$

Designations: CAS Registry Number: 13194-48-4; DOT: 2784; mf: $C_8H_{19}O_2PS_2$; fw: 242.33; RTECS: TE4025000.

Properties: Clear, pale yellow liquid. Mp: 20°C; bp: 86°C at 0.2 mmHg; ρ: 1.094 at 20/4°C; K_H: 1.59 × 10^{-7} atm · m³/mol at 20–25°C (approximate — calculated from water solubility and vapor pressure); log K_{oc}: 1.82–2.27; log K_{ow}: 3.59 (21°C); S_o: miscible with acetone, hexane, xylene; S_w: 700 mg/L at 20°C; vap d: 9.90 g/L at 25°C, 8.39 (air = 1); vp: 3.49 × 10^{-4} mmHg at 20°C.

Soil properties and adsorption data

Soil	K_d (mL/g)	f_{oc} (%)	K_{oc} (mL/g)	pH
Loamy sand	2.44	3.71	66	7.0
Loamy sand	1.08	0.58	186	7.2
Loamy sand	1.24	1.10	113	5.3
Sandy loam	1.61	1.86	87	5.7
Silt loam	2.10	1.33	158	5.6

Source: U.S. Department of Agriculture, 1990.

Environmental Fate

Soil. The reported half-life in humus-containing soil (pH 4.5) and a sandy loam (pH 7.2–7.3) are 87 and 14–28 days, respectively (Hartley and Kidd, 1987). The rate of degradation increased in soils that had been treated annually four times (Smelt et al., 1987).

Chemical/Physical. Emits toxic fumes of phosphorus and sulfur oxides when heated to decomposition (Sax and Lewis, 1987).

Symptoms of Exposure: Tightness across the chest, nausea, salivation, vomiting, abdominal cramps, diarrhea, abnormal heart rates, arm and leg weakness, constriction of pupils, involuntary urination.

Formulation Types: Emulsifiable concentrate; granules.

Toxicity: LC_{50} (96-hour) for rainbow trout 13.8 mg/L, bluegill sunfish 2.1 mg/L and goldfish 13.6 mg/L (Hartley and Kidd, 1987); acute oral LD_{50} for rats 262 mg/kg (Hartley and Kidd, 1987), 34 mg/kg (RTECS, 1985).

Uses: Nonsystemic, nonfumigant nematocide and soil insecticide for control of insects in ornamentals, potatoes, sweet potatoes, tomatoes, strawberries, bananas, pineapples, sugar cane, turf and many other crops.

ETHYLENE DIBROMIDE

Synonyms: Acetylene dibromide; Bromofume; Celmide; DBE; Dibromoethane; **1,2-Dibromoethane**; *sym*-Dibromoethane; α,β-Dibromoethane; Dowfume 40; Dowfume EDB; Dowfume W-8; Dowfume W-85; Dowfume W-90; Dowfume W-100; EDB; EDB-85; E-D-BEE; ENT 15349; Ethylene bromide; Ethylene bromide glycol dibromide; 1,2-Ethylene dibromide; Fumogas; Glycol bromide; Glycol dibromide; Iscobrome D; Kopfume; NCI-C00522; Nephis; Pestmaster; Pestmaster EDB-85; RCRA waste number U067; Soilbrom-40; Soilbrom-85; Soilbrom-90; Soilbrom-90EC; Soilbrom-100; Soilbrome-85; Soilfume; UN 1605; Unifume.

$$Br\, CH_2\, CH_2\, Br$$

Designations: CAS Registry Number: 106-93-4; DOT: 1605; mf: $C_2H_4Br_2$; fw: 187.86; RTECS: KH9275000.

Properties: Clear, colorless liquid with a sweet, chloroform-like odor. Mp: 9.8°C; bp: 131.3°C; ρ: 2.1792 at 20/4°C; fl p: nonflammable; H-$t_{1/2}$: 8 years at 25°C and pH 7; K_H: 7.06×10^{-4} atm · m^3/mol at 25°C; IP: 9.45 eV; log K_{oc}: 1.56–2.21; log K_{ow}: 1.76; S_o: miscible with most organic solvents and thinners; S_w: 4,321 mg/L at 20°C; vap d: 7.68 g/L at 25°C, 6.49 (air = 1); vp: 11 mmHg at 20°C, 17.4 mmHg at 30°C.

Environmental Fate

Biological. Complete biodegradation of ethylene dibromide by soil cultures yielded ethylene and bromide ions (Castro and Belser, 1968). A mutant of strain *Acinetobacter* sp. GJ70 isolated from activated sludge degraded ethylene dibromide to ethylene glycol and bromide ions (Janssen et al., 1987). When *Methanococcus thermolithotrophicus, Methanococcus deltae* and *Methanobacterium thermoautotrophicum* were grown with H_2-CO_2 in the presence of ethylene dibromide, methane and ethylene were produced (Belay and Daniels, 1987).

In a shallow aquifer material, ethylene dibromide aerobically degraded to carbon dioxide, microbial biomass and nonvolatile water-soluble compound(s) (Pignatello, 1986, 1987).

Soil. In soil and water, chemical- and biological-mediated reactions transform ethylene dibromide in the presence of hydrogen sulfides to ethyl mercaptan and other sulfur-containing compounds (Alexander, 1981).

Groundwater. According to the U.S. EPA (1986) ethylene dibromide has a high potential to leach to groundwater.

Chemical/Physical. In an aqueous phosphate buffer solution (0.05 M) containing hydrogen sulfide ions, ethylene dibromide was transformed into 1,2-dithioethane and vinyl bromide. The hydrolysis half-lives for solutions with and without sulfides present ranged from 37 to 70 days and 0.8 to 4.6 years, respectively (Barbash and Reinhard, 1989).

Dehydrobromination of ethylene dibromide to vinyl bromide was observed in various aqueous buffer solutions (pH 7–11) over the temperature range of 45 to 90°C. The estimated half-life for this reaction at 25°C and pH 7 is 2.5 years (Vogel and Reinhard, 1986).

Ethylene dibromide may hydrolyze via two pathways. In the first pathway, ethylene dibromide undergoes nucleophilic attack at the carbon-bromine bond by water forming hydrogen bromide and 2-bromoethanol. The alcohol may react further through the formation of ethylene oxide forming ethylene glycol (Kollig, 1993; Leinster et al., 1978). In the second pathway, dehydrobromination of ethylene dibromide to vinyl bromide was observed in various aqueous buffer solutions (pH 7–11) over the temperature range of 45 to 90°C. The estimated hydrolysis half-life for this reaction at 25°C and pH 7 was 2.5 years (Vogel and Reinhard, 1986).

The hydrolysis rate constant for ethylene dibromide at pH 7 and 25°C was determined to be 9.9×10^{-6}/hour, resulting in a half-life of 8.0 years (Ellington et al., 1988). At pH 5 and temperatures of 30, 45 and 60°C, the hydrolysis half-lives were 180, 29 and 9 days, respectively. When the pH was raised to pH 7, the half-lives increased slightly to 410, 57 and 11 days at temperatures of 30, 45 and 60°C, respectively. At pH 9, the hydrolysis half-lives were nearly identical to those determined under acidic conditions (Ellington et al., 1986).

Anticipated products from the reaction of ethylene dibromide with ozone or hydroxyl radicals in the atmosphere include bromoacetaldehyde, formaldehyde, bromoformaldehyde and bromide radicals (Cupitt, 1980). In the atmosphere, ethylene dibromide is slowly oxidized by peroxides and ozone. The half-life for these reactions is generally >100 days (Leinster et al., 1978).

Exposure Limits: NIOSH REL: TWA 0.045 ppm, 15-min C 0.13 ppm, IDLH 100 ppm; OSHA PEL: TWA 20 ppm, C 30 ppm, 5-min peak 50 ppm; ACGIH TLV: suspected human carcinogen.

Symptoms of Exposure: Irritation of the respiratory system, eyes and skin; dermatitis with vesiculation.

Formulation Types: Fumigant.

Toxicity: LC_{50} (48-hour) for bluegill sunfish 18 mg/L (Davis and Hardcastle, 1959); acute oral LD_{50} for male and female rats 148 and 117 mg/kg, respectively (Verschueren, 1983), 108 mg/kg (RTECS, 1985).

Uses: Grain and fruit fumigant; insecticide.

FENAMIPHOS

Synonyms: Bay 68138; ENT 27572; **Ethyl 3-methyl-4-(methylthio)phenyl (1-methyl-ethyl)phosphoramidate**; Ethyl-4-methylthio-*m*-tolyl isopropyl phosphoramidate; Isopropylamino-*o*-ethyl-(4-methylmercapto)-3-methylphenyl)phosphate; 1-(Methylethyl)ethyl 3-methyl-4-(methylthio)phenyl phosphoramidate; Phenamiphos.

Designations: CAS Registry Number: 22224-92-6; mf: $C_{13}H_{22}NO_3PS$; fw: 303.40; RTECS: TB3675000.

Properties: Colorless solid. Mp: 49.2°C; ρ: 1.15 at 20/4°C; pK_a: 10.5 at 25°C; K_H: 9.5 × 10^{-10} atm · m³/mol at 30°C (approximate — calculated from water solubility and vapor pressure); log K_{oc}: 2.27–3.20; log K_{ow}: 3.23, 3.25; S_o: miscible with acetone, dimethylsulfoxide, 95% ethanol and many other common organic solvents; S_w: 306, 329 and 419 mg/L at 10, 20 and 30°C, respectively; vp: 9.98 × 10^{-7} mmHg at 30°C.

Soil properties and adsorption data

Soil	K_d (mL/g)	f_{oc} (%)	K_{oc} (mL/g)	pH
Arredondo sand	1.18	0.80	148	6.8
Batcombe silt loam	6.74	2.05	329	6.1
Cecil sandy loam	1.77	0.90	197	5.6
Clayey loam	5.78	2.90	199	7.9
Clayey loam	4.59	0.29	1,583	6.0
Loam	9.62	2.32	415	7.3
Loamy sand	3.05	1.62	188	6.6
Rothamsted Farm	4.99	1.51	330	5.1
Sand	1.18	0.46	254	6.8
Sandy loam	1.77	0.52	339	5.6
Sand	1.18	0.46	254	6.8
Sandy loam	1.77	0.52	339	5.6
Webster silty clay loam	9.62	3.97	249	7.3

Source: Lord et al., 1980; Briggs, 1981; Bilkert and Rao, 1985; U.S. Department of Agriculture, 1990.

Environmental Fate

Soil. Oxidizes in soil to the corresponding sulfone and sulfoxide (Lee et al., 1986). Fenamiphos rapidly degraded in Arredondo soil to fenamiphos sulfone and at the same time to the corresponding phenol. The half-life in this soil is 38–67 days (Ou and Rao, 1986).

Surface Water. In estuarine water, the half-life of fenamiphos was 1.80 days (Lacorte et al., 1995).

Chemical/Physical. Emits toxic fumes of phosphorus, nitrogen and sulfur oxides when heated to decomposition (Sax and Lewis, 1987; Lewis, 1990).

Exposure Limits: OSHA PEL: TWA 0.1 mg/m^3; ACGIH TLV: TWA 0.1 mg/m^3.

Formulation Types: Emulsifiable concentrate; granules.

Toxicity: LC$_{50}$ (96-hour) for rainbow trout 72.1 µg/L, bluegill sunfish 9.6 µg/L and goldfish 3,200 µg/L (Hartley and Kidd, 1987); acute oral LD$_{50}$ for male and female rats 15.3 and 19.4 mg/kg, respectively (Hartley and Kidd, 1987), 8 mg/kg (RTECS, 1985).

Use: Nematocide.

FENBUTATIN OXIDE

Synonyms: Bendex; Bis(tris(β,β-dimethylphenethyl)tin)oxide; Bis(tris(2-methyl-2-phenylpropyl)tin)oxide; Di(tri(2,2-dimethyl-2-phenylpropyl)tin)oxide; ENT 27738; Hexakis(β,β-dimethylphenethyl)distannoxane; **Hexakis(2-methyl-2-phenylpropyl)distannoxane**; SD 14114; Shell SD-14114; Torque; Vendex.

Designations: CAS Registry Number: 13356-08-6; mf: $C_{60}H_{78}OSn_2$; fw: 1052.70; RTECS: JN8770000.

Properties: Colorless to white crystals. Mp: 138'39°C; log K_{oc}: 4.91 (calculated); log K_{ow}: 5.10; S_o (g/L): acetone (6), benzene (140), methylene chloride (380); S_w: 5 µg/L at 23°C.

Environmental Fate
Chemical/Physical. Reacts with moisture forming tris(2-methyl-2-phenylpropyl)tin hydroxide (Worthing and Hance, 1991).

Formulation Types: Suspension concentrate; wettable powder.

Toxicity: $LC_{50}LC_{50}$ (48-hour) for rainbow trout 0.27 mg/L (Hartley and Kidd, 1987); acute oral LD_{50} for rats 2,631 mg/kg (Hartley and Kidd, 1987).

Use: Acaricide.

FENOXAPROP-ETHYL

Synonyms: (±)-2-(4-(6-Chlorobenzoxazolyl)oxy)phenoxy)propionic acid; (±)-2-(4-(6-Chlorobenzoxazol-2-yloxy)phenoxy)propionic acid; (±)-2-(4-(6-Chloro-1,3-benzoxazol-2-yloxy)phenoxy)propionic acid; (±)-Ethyl 2-(4-((6-chloro-2-benzoxazolyl)oxy)phenoxy) propanoate; Hoe 33171.

Designations: CAS Registry Number: 66441-23-4; mf: $C_{16}H_{12}ClNO_5$; fw: 333.70.

Properties: Colorless solid to light beige to brown powder with a faint aromatic odor. Mp: 84–85°C; bp: 300°C; ρ: 1.3 at 20/4°C; log K_{ow}: 4.12; S_o (g/L at 20°C): acetone (510), ethanol (20), hexane (5), ethyl acetate (240); S_w: 900 mg/L at 20; vp: 1.9×10^{-11} mmHg at 20°C.

Environmental Fate

Soil. Hydrolyzes rapidly in soil forming fenoxaprop acid, ethyl alcohol, 6-chloro-2,3-dihydrobenzoxazole-2-one and 4-(6-chloro-2-benzoxazolyloxy) phenol (Wink and Luley, 1988; Humburg et al., 1989). Under aerobic and anaerobic conditions, the half-life was less than 24 hours (Humburg et al., 1989).

Plant. The major degradation product identified in crabgrass, oats and wheat was 6-chloro-2,3-dihydrobenzoxazol (Lefsrud and Hall, 1989).

Photolytic. Susceptible to degradation by UV light (Humburg et al., 1989).

Toxicity: LC_{50} (96-hour) for bluegill sunfish 3.34 mg/L and *Daphnia magna* 11.15 mg/L; acute oral LD_{50} for male and female rats 3.31 and 3.40 g/kg, respectively (Humburg et al., 1989).

Use: Selective, postemergence herbicide for control of annual and perennial grass weeds in crops such as beans, beets, cotton, groundnuts, potatoes and other vegetables.

FENSULFOTHION

Synonyms: Bay 25141; Bayer 25141; Bayer S 767; Chemagro 25,141; Dasanit; *O,O*-Diethyl *O*-4-methylsulphinylphenyl phosphorothioate; *O,O*-Diethyl *O-p*-methylsulphinylphenyl phosphorothioate; *O,O*-Diethyl *O-p*-methylsulphinylphenyl thiophosphate; DMSP; ENT 24945; OMS 37; **Phosphorothioic acid *O,O*-diethyl *O*-(*p*-(methylsulfinyl)phenyl) ester**; Terracur P.

Designations: CAS Registry Number: 115-90-2; DOT: 2765; mf: $C_{11}H_{17}O_4PS_2$; fw: 308.35; RTECS: TF3850000.

Properties: Yellowish-brown oil. Mp: <25°C; bp: 133–141°C at 0.01 mmHg; ρ: 1.202 at 20/4°C; log K_{oc}: 1.89 (calculated); log K_{ow}: 2.23; S_o: miscible with most solvents except aliphatics; S_w: 1.54 g/L at 25°C; vap d: 12.60 g/L at 25°C, 9.02 (air = 1).

Environmental Fate

Soil. In soils, the bacterium *Klebsiella pneumoniae* degraded fensulfothion to fensulfothion sulfide (Timms and MacRae, 1982, 1983). The following microorganisms were also capable of degrading the parent compound to the corresponding sulfide: *Escherichia coli, Pseudomonas fluorescens, Nocardia opaca, Lactobacillus plantarum* and *Leuconostoc mesenteroides* (Timms and MacRae, 1983).

Plant. Readily oxidized in plants to the corresponding sulfone (Hartley and Kidd, 1987).

Chemical/Physical. Emits toxic fumes of phosphorus and sulfur oxides when heated to decomposition (Sax and Lewis, 1987; Lewis, 1990).

Isomerizes readily to the *O,S*-diethyl isomer (Worthing and Hance, 1991). The hydrolysis half-lives of fensulfothion in a sterile 1% ethanol/water solution at 25°C and pH values of 4.5, 6.0, 7.0 and 8.0, were 69, 77, 87 and 58 weeks, respectively (Chapman and Cole, 1982).

Exposure Limits: ACGIH TLV: TWA 0.1 mg/m³.

Formulation Types: Emulsifiable concentrate; dustable powder; granules; wettable powder.

Toxicity: LC_{50} (96-hour) for bluegill sunfish 0.12 mg/L, golden orfe 6.8 mg/L and rainbow trout 8.8 mg/L (Hartley and Kidd, 1987); acute oral LD_{50} for male and female rats 10.5 and 2.2 mg/kg, respectively (Hartley and Kidd, 1987), 2 mg/kg (RTECS, 1985).

Uses: Nematocide and pesticide used to control free-living, cyst-forming and root-knot nematotodes and soil insects in vegetable and fruit crops.

FENTHION

Synonyms: B 29493; Bay 29493; Baycid; Bayer 9007; Bayer 24493; Bayer S 1752; Baytex; *O,O*-Dimethyl *O*-4-(methylmercapto)-3-methylphenyl phosphorothioate; *O,O*-Dimethyl *O*-4-(methylmercapto)-3-methylphenyl thiophosphate; *O,O*-Dimethyl *O*-(3-methyl-4-methylmercaptophenyl) phosphorothioate; *O,O*-Dimethyl *O*-(3-methyl-4-methylthiophenyl) phosphorothioate; *O,O*-Dimethyl *O*-(4-methylthio-3-methylphenyl) phosphorothioate; *O,O*-Dimethyl *O*-(4-methylthio-*m*-tolyl) phosphorothioate; DMTP; ENT 25540; Entex; Lebaycid; Mercaptophos; 4-Methylmercapto-3-methylphenyl dimethyl thiophosphate; MPP; NCI-C08651; OMS 2; **Phosphorothioic acid *O,O*-dimethyl *O*-(3-methyl-4-(methylthio)phenyl) ester**; Queletox; S 1752; Spottan; Talodex; Tiguvon.

Designations: CAS Registry Number: 55-38-9; DOT: 2784; mf: $C_{10}H_{15}O_3PS_2$; fw: 278.33; RTECS: TF9625000.

Properties: Colorless to amber liquid or brown oil with a faint garlic-like odor. Mp: 7.0°C; bp: 87°C at 0.01 mmHg; ρ: 1.250 at 20/4°C; K_H: 5.49 × 10^{-6} atm · m^3/mol at 25°C; log K_{oc}: 0.89–1.58; log K_{ow}: 4.09, 4.84; P-$t_{1/2}$: 55.83 hours (absorbance λ = 268 nm, concentration on glass plates = 6.7 μg/cm^2); S_o: miscible with methylene chloride, 2-propanol, and many other solvents; S_w: 6.4, 9.3 and 11.3 mg/L at 10, 20 and 30°C, respectively; vap d: 11.37 g/L at 25°C, 9.64 (air = 1); vp: 3 × 10^{-5} mmHg at 20°C.

Soil properties and adsorption data

Soil	K_d (mL/g)	f_{oc} (%)	K_{oc} (mL/g)	pH
Sand	36.2	2.15	1,684	6.9
Sandy loam	7.7	0.81	948	7.7
Sandy loam	38.0	1.74	2,184	5.5
Silt loam	19.8	1.22	1,623	6.7
Silt loam	12.4	1.16	1,069	6.3

Source: U.S. Department of Agriculture, 1990.

Environmental Fate

Biological. From the first-order biotic and abiotic rate constants of fenthion in estuarine water and sediment/water systems, the estimated biodegradation half-lives were 22.4 and 3.9–14.5 days, respectively (Walker et al., 1988).

Surface Water. In estuarine water, the half-life of fenthion was 4.6 days (Lacorte et al., 1995).

Plant. In plants, fenthion oxidizes to the mesulfenfos and sulfone which further degrades to the sulfone phosphate before undergoing hydrolysis (Hartley and Kidd, 1987).

Photolytic. Fenthion was oxidized to the corresponding sulfoxide and trace amounts (<5% yield) of sulfone when sorbed on soil and exposed to sunlight. The photosensitized oxidation was probably due to the presence of singlet oxygen. The degradation rate was higher in soils containing the lowest organic carbon (Gohre and Miller, 1986).

Chemical/Physical. Stable at temperatures below 210°C (Worthing and Hance, 1991). Emits very toxic fumes of phosphorus and sulfur oxides when heated to decomposition (Sax and Lewis, 1987; Lewis, 1990).

Fenthion hydrolyzes in water forming *O,O*-dimethyl-*O*-(4-(methylthio)-*m*-tolyl) phosphate (bayoxon) and 3-methyl-4-methylthiophenol (Suffet et al., 1967).

Exposure Limits: OSHA PEL: TWA 0.2 mg/m^3; ACGIH TLV: TWA 0.2 mg/m^3.

Formulation Types: Emulsifiable concentrate; wettable powder; granules; dustable powder; fogging concentrate.

Toxicity: LC_{50} (96-hour) for fathead minnow 2.44 mg/L, largemouth bass 1.54 mg/L, brown trout 1.33 mg/L, coho salmon 1.32 mg/L, perch 1.65 mg/L, channel catfish 1.68 mg/L, black bullhead 1.62 mg/L, rainbow trout 0.93 mg/L, bluegill sunfish 1.4 mg/L, goldfish 3.4 mg/L, perch 1.7 mg/L and carp 1.2 mg/L (Macek and McAllister, 1970); LC_{50} (48-hour) for goldfish 1.9 mg/L (Hartley and Kidd, 1987); acute oral LD_{50} for male and female rats 375 and 290 mg/kg, respectively (Hartley and Kidd, 1987), 180 mg/kg (RTECS, 1985).

Uses: Insecticide for control of fruit flies, leaf hoppers and cereal bugs. Fenthion is also used as an acaracide.

FENVALERATE

Synonyms: Belmark; α-Cyano-3-phenoxybenzyl-2-(4-chlorophenyl)-3-methylbutyrate; **Cyano(3-phenoxyphenyl)methyl 4-chloro-α-(1-methylethyl)benzene-acetate**; Phenoxybenzyl-2-(4-chlorophenyl)isovalerate; S 5602; Sanmarton; SD 43775; Sumicidin; Sumifly; Sumipower; WL 43775.

Designations: CAS Registry Number: 51630-58-1; mf: $C_{25}H_{22}ClNO_3$; fw: 419.92; RTECS: CY1576350.

Properties: Clear yellow or brown, viscous liquid. Mp: <23°C; ρ: 1.175 at 25/25°C; K_H: 1.5×10^{-7} atm · m³/mol at 20–25°C (approximate — calculated from water solubility and vapor pressure); log K_{oc}: 3.64 (calculated); log K_{ow}: 6.2 (Schimmel et al., 1983); P-$t_{1/2}$: 168.85 hours (absorbance λ = 228.0 nm, concentration on glass plates = 6.7 µg/cm²); S_o: miscible with acetone, chloroform, cyclohexanone, ethanol, xylene; S_w in seawater: 24 µg/L at 22°C; vap d: 17.16 g/L at 25°C, 14.54 (air = 1); vp: 2.78×10^{-7} mmHg at 25°C.

Environmental Fate

Soil. Fenvalerate is moderately persistent in soil. The percentage of the initial dosage (1 ppm) remaining after 8 weeks of incubation in an organic and mineral soil were 58 and 12%, respectively, while in sterilized controls 100 and 91% remained, respectively (Chapman et al., 1981).

In a sugarcane runoff plot, fenvalerate was applied at a rate of 0.22 kg/ha 4 times each year in 1980 and 1981. Runoff losses in 1980 and 1981 were 0.08 and 0.56 of the applied amount, respectively (Smith et al., 1983).

Plant. Dislodgable residues of fenvalerate on cotton leaf 0, 24, 48, 72 and 96 hours after application (0.22 kg/ha) were 0.85, 0.36, 0.38, 0.28 and 0.28 µg/m², respectively (Buck et al., 1980).

Surface Water. In an estuary, the half-life of fenvalerate was 27–42 days (Schimmel et al., 1983).

Chemical/Physical. Undergoes hydrolysis at the ester bond (Hartley and Kidd, 1987). Decomposes gradually at 150–300°C (Windholz et al., 1983) probably releasing toxic fumes of nitrogen and chlorine.

Formulation Types: Emulsifiable concentrate; suspension concentrate.

Toxicity: LC_{50} (96-hour) for rainbow trout 3.6 µg/L (Worthing and Hance, 1991), estuarine mysid 0.008 µg/L, pink shrimp 0.84 µg/L, sheepshead minnow 5.0 µg/L, Atlantic silverside 0.31 µg/L, striped mullet 0.58 µg/L, Gulf toadfish 5.4 µg/L (Schimmel et al., 1983); LC_{50} (48-hour) for carp <100 µg/L (Hartley and Kidd, 1987); LC_{50} (24-hour) for rainbow trout 76.0 ppb (technical) and 21.0 ppb (formulated product) (Coats and O'Don-nell-Jeffrey, 1979); acute oral LD_{50} for rats 451 mg/kg (dimethylsulfoxide) (Hartley and Kidd, 1987).

Use: Insecticide used against a wide variety of pests.

FERBAM

Synonyms: Aafertis; Bercema Fertam 50; Carbamate; Dimethylcarbamodithioc acid iron complex; Dimethylcarbamodithioc acid iron(3+) salt; Dimethyldithiocarbamic acid iron salt; Dimethyldithiocarbamic acid iron(3+) salt; ENT 14689; Ferbam 50; Ferbame; Ferbam, iron salt; Ferbeck; Ferberk; Fermate; Fermate ferbam fungicide; Fermocide; Ferradow; Ferric dimethyldithiocarbamate; Fuklasin; Fuklasin ultra; Hexaferb; Hokmate; Iron flowable; Iron tris(dimethyldithiocarbamate); Karbam black; Trifungol; **Tris(dimethylcarbamodithioato-*S,S'*)iron**; Tris(dimethyldithiocarbamato)iron; Vancide FE95.

$$[(CH_3)_2NCS_2]_3Fe$$

Designations: CAS Registry Number: 14484-64-1; mf: $C_9H_{18}FeN_3S_6$; fw: 416.50; RTECS: NO8750000.

Properties: Black, fluffy powder. Mp: >180°C with decomposition; pK_a: unknown but pH of saturated solution is 5.0; log K_{oc}: 0.83 (calculated); log K_{ow}: −1.00 (calculated); S_o: soluble in acetone, acetonitrile, benzene, carbon tetrachloride, chloroform, ethanol, methanol, *n*-propanol, pyridine, toluene, xylene and many other organic solvents; S_w: 120–130 g/L at 20°C (pH of solution = 5.0); vp: negligible at 20–25°C.

Environmental Fate

Plant. Decomposes in plants to ethylene thiourea, ethylene thiuram monosulfide, ethylene thiuram disulfide and sulfur (Hartley and Kidd, 1987).

Chemical/Physical. Hydrolyzes in acidic media releasing carbon disulfide. Decomposes in water forming ethylene thiourea (Hartley and Kidd, 1987).

Decomposes >180°C (Windholz et al., 1983) emitting toxic fumes of nitrogen and sulfur oxides (Lewis, 1990; Sax and Lewis, 1987).

Exposure Limits: NIOSH 10 mg/m³, IDLH 800 mg/m³; OSHA PEL: TWA 15 mg/m³; ACGIH TLV: TWA 10 mg/m³.

Symptoms of Exposure: May cause irritation of skin and mucous membranes and renal damage.

Formulation Types: Wettable powders.

Toxicity: Moderately toxic to fish (Hartley and Kidd, 1987); acute oral LD_{50} for rats >17,000 mg/kg (Hartley and Kidd, 1987), 4,000 mg/kg (RTECS, 1985).

Uses: Nonphytotoxic fungicide used to control scab on fruits and other crops.

FLUCYTHRINATE

Synonyms: AC 222705; (*RS*)-Cyano-(3-phenoxyphenyl)methyl (*S*)-4-(difluo-romethoxy)phenyl)-α-(1-methylethyl)benzeneacetate; **4-(Difluoromethoxy)-α-(1-meth-ylethyl)benzeneacetic acid cyano(3-phenoxyphenyl)methyl ester**; Cybolt; Pay-off.

Designations: CAS Registry Number: 70124-77-5; mf: $C_{26}H_{23}F_2NO_4$; fw: 451.48; RTECS: CY1578620.

Properties: Dark amber, viscous liquid with an ester-like odor. Mp: <25°C; bp: 108°C at 0.35 mmHg; ρ: 1.189 at 22/4°C; H-$t_{1/2}$ (27°C): 40 days (pH 3), 52 days (pH 5), 6.3 days (pH 9); K_H: 8.08×10^{-2} atm · m³/mol at 21–25°C (approximate — calculated from water solubility and vapor pressure); log K_{oc}: 3.81 (calculated); log K_{ow}: 4.70; S_o (g/L at 21°C): acetone (820), hexane (90), 2-propanol (780), xylene (1,810); S_w: 0.5 mg/L at 21°C; vp: 6.80×10^{-2} mmHg at 25°C.

Environmental Fate
Surface Water. The half-life of flucythrinate in an estuarine environment is 34 days (Schimmel et al., 1983).

Chemical/Physical. Hydrolyzes in aqueous solutions forming acetic acid and other compounds.

Formulation Types: Emulsifiable concentrate; water-dispersible granules.

Toxicity: LC$_{50}$ (96-hour) for rainbow trout 0.32 µg/L, bluegill sunfish 0.71 µg/L, channel catfish 0.51 µg/L, sheepshead minnow 1.6 µg/L (Hartley and Kidd, 1987), estuarine mysid 0.008 µg/L, pink shrimp 0.22 µg/L and sheepshead minnow 1.1 µg/L (Schimmel et al., 1983); acute oral LD$_{50}$ for male and female rats is 81 and 67 mg/kg, respectively (Hartley and Kidd, 1987).

Use: Nonsystemic insecticide.

FLUOMETURON

Synonyms: C 2059; CIBA 2059; Cotoran; Cotoran multi 50WP; Cottonex; 1,1-Dimethyl-3-(3-trifluoromethylphenyl)urea; **N,N-Dimethyl-N'-(3-(trifluoromethyl)phenyl)urea**; 1,1-Dimethyl-3-(α,α,α-trifluoro-m-tolyl)urea; Herbicide C-2059; Lanex; Meturon; Meturon 4L; NCI-C08695; Pakhtaran; N-(m-Trifluoromethylphenyl)-N',N'-dimethylurea; N-(3-Trifluoromethylphenyl)-N',N'-dimethylurea; 3-(m-Trifluoromethylphenyl)-1,1-dimethylurea.

$$\text{(structure: benzene ring with } CF_3 \text{ and } -NHCON(CH_3)_2 \text{ substituents)}$$

Designations: CAS Registry Number: 2164-17-2; mf: $C_{10}H_{11}F_3N_2O$; fw: 232.21; RTECS: YT1575000.

Properties: Odorless, white, crystalline powder. Mp: 163–164.5°C, 155°C (technical — 95%); ρ: 1.39 at 20/4°C; fl p: nonflammable; H-$t_{1/2}$ (20°C): 1.6 years (pH 1), 2.4 years (pH 5), 2.8 years (pH 9); K_H: <2.79 × 10^{-6} atm · m^3/mol at 20–25°C (approximate — calculated from water solubility and vapor pressure); log K_{oc}: 1.46–2.08; log K_{ow}: 2.23, 2.38; P-$t_{1/2}$: 1.2 days at 23°C; S_o (g/L at 20°C): acetone (105), hexane (170), methanol (110), methylene chloride (23) and octanol (220); S_w: 80 mg/L at 25°C; vp: 5 × 10^{-7} mmHg at 20°C.

Soil properties and adsorption data

Soil	K_d (mL/g)	f_{oc} (%)	K_{oc} (mL/g)	pH
Loam	1.64	2.78	59	5.9
Sand	0.15	0.52	29	6.5
Sandy loam	0.74	0.70	106	7.6
Sandy loam	2.81	2.32	121	7.8

Source: U.S. Department of Agriculture, 1990.

Environmental Fate

Soil. In soils, fluometuron rapidly degrades (half-life approximately 30 days) to carbon dioxide, polar and nonextractable compounds (Hartley and Kidd, 1987; Humburg et al., 1989).

Groundwater. According to the U.S. EPA (1986) fluometuron has a high potential to leach to *groundwater.*

Plant. In plants, fluometuron degrades to a demethylated intermediate which subsequently is degraded to the aniline moiety (possibly m-trifluoromethylaniline) (Hartley and Kidd, 1987; Humburg et al., 1989). Duke et al. (1991) reported that fluometuron degrades in plants via the following degradative pathway: fluometuron to N-methyl-N'-(3-(trifluoromethyl)phenyl)urea which undergoes demethylation to 3-(trifluoromethyl)phenylurea followed by deamination and elimination of the ketone group to form 3-trifluoromethyla-niline.

Chemical/Physical. Emits toxic fumes of nitrogen oxides and fluorine when heated to decomposition (Sax and Lewis, 1987).

Symptoms of Exposure: Dust may cause eye irritation.

Formulation Types: Suspension concentrate (4 lb/gal); wettable powder (80%).

Toxicity: LC_{50} (96-hour) for rainbow trout 47 mg/L, bluegill sunfish 96 mg/L, catfish 55 mg/L and crucian carp 170 mg/L (Hartley and Kidd, 1987); acute oral LD_{50} of the 80% formulation for rats is 1,800 mg/kg (Ashton and Monaco, 1991), 6,416 mg/kg (RTECS, 1985).

Uses: Herbicide used to control many annual broad-leaved weeds in sugarcane and cotton.

FONOFOS

Synonyms: Difonate; Dyfonate; Dyphonate; ENT 25796; *O*-Ethyl *S*-phenyl ethyldithiophosphonate; *O*-Ethyl *S*-phenyl ethylphosphonodithioate; **Ethylphosphonodithioic acid *O*-ethyl *S*-phenyl ester**; Fonophos; *N*-2790; Stauffer NA 2790.

$$CH_3CH_2-P \overset{\overset{S}{\parallel}}{\underset{S}{\Big<}} \begin{matrix} O-CH_2CH_3 \\ \\ \end{matrix}$$

Designations: CAS Registry Number: 944-22-9; mf: $C_{10}H_{15}OPS_2$; fw: 246.32; RTECS: TA5950000.

Properties: Colorless to pale yellow liquid with an aromatic odor. Mp: <25°C; bp: 130°C at 0.1 mmHg; ρ: 1.154 at 20/20°C; H-$t_{1/2}$ (40°C): 101 days (pH 4), 74–127 (pH 7, buffer dependent), 1.8 days (pH 10); K_d: 15.3 (loam soil); K_H: 5.2 × 10⁻⁶ atm · m³/mol at 25°C (approximate — calculated from water solubility and vapor pressure); log K_{oc}: 3.03 (calculated); log K_{ow}: 3.89, 3.90; S_o: miscible with many organic solvents; S_w: 13 mg/L at room temperature; vp: 2.1 × 10⁻⁴ mmHg at 25°C.

Environmental Fate

Soil. The half-life for fonofos in soil incubated in the laboratory under aerobic conditions ranged was 25 days (Lichtenstein et al., 1977). In field soils, the half-lives for fonofos ranged from 24 to 102 days (Kiigemgi and Terriere, 1971; Schulz and Lichtenstein, 1971; Mathur et al., 1976; Talekar et al., 1977).

Plant. In plants, fonofos is oxidized to the phosphonothioate (Hartley and Kidd, 1987). Oat plants were grown in two soils treated with [¹⁴C]fonofos. Most of the residues remained bound to the *soil.* Less than 2% of the applied [¹⁴C]fonofos was recovered from the oat leaves. Metabolites identified in both soils and leaves were methyl phenyl sulfone, 2-, 3- and 4-hydroxymethyl phenyl sulfone, thiophenol, diphenyl disulfide and fonofos oxon (Fuhremann and Lichtenstein, 1980; Lichtenstein et al., 1982).

Chemical/Physical. The hydrolysis half-lives of fonofos in a sterile 1% ethanol/water solution at 25°C and pH values of 4.5, 5.0, 6.0, 7.0 and 8.0, were 87, 50, 41, 22 and 6.9 weeks, respectively (Chapman and Cole, 1982). Emits toxic nitrogen and phosphorus oxide fumes when heated to decomposition (Sax and Lewis, 1987; Lewis, 1990).

Exposure Limits: NIOSH REL: 0.1 mg/m³; ACGIH TLV: 0.1 mg/m³.

Formulation Types: Granules; emulsifiable concentrate.

Toxicity: LC_{50} (24-hour) for bluegill sunfish 45 µg/L and rainbow trout 110 µg/L (Verschueren, 1983); acute oral LD_{50} for rats 8–17.5 mg/kg (Hartley and Kidd, 1987), 3 mg/kg (RTECS, 1985).

Uses: Soil insecticide used to control rootworms, wireworms, crickets and similar crop pests in vegetables, sorghum, ornamentals, cereals, maize, vines, olives, sugar beet, sugar-cane, potatoes, groundnuts, tobacco, turf and fruit crops.

FORMALDEHYDE

Synonyms: BFV; FA; Fannoform; Formalin; Formalin 40; Formalith; Formic aldehyde; Formol; Fyde; HOCH; Ivalon; Karsan; Lysoform; Methanal; Methyl aldehyde; Methylene glycol; Methylene oxide; Morbicid; NCI-C02799; Oxomethane; Oxymethylene; Paraform; Polyoxymethylene glycols; RCRA waste number U122; Superlysoform; UN 1198; UN 2209.

HCHO

Designations: CAS Registry Number: 50-00-0; DOT: 1198; mf: CH_2O; fw: 30.03; RTECS: LP8925000.

Properties: Clear, colorless liquid with a pungent, suffocating odor and burning taste. Mp: $-92°C$; bp: $-21°C$, $98-99°C$ (40% aqueous solution); ρ: 0.815 at 20/4°C, 1.081–1.085 at 25/25°C (37% aqueous solution); fl p: 50°C (15% methanol-free); lel: 7.0%; uel: 73.0%; K_H: 1.67×10^{-7} atm · m^3/mol; IP: 10.88 eV; log K_{oc}: 0.56 (calculated); log K_{ow}: 0.00; S_o: acetone (>100 mg/mL), benzene, dimethylsulfoxide (>100 mg/mL), 95% ethanol (>100 mg/mL), ethyl ether; S_w: miscible at 25°C; vap d: 1.23 g/L at 25°C, 1.067 (air = 1); vp: 760 mmHg at $-19.5°C$.

Environmental Fate

Biological. Biodegradation products reported include formic acid and ethanol, each of which can further degrade to carbon dioxide (Verschueren, 1983).

Photolytic. Major products reported from the photooxidation of formaldehyde with nitrogen oxides are carbon monoxide, carbon dioxide and hydrogen peroxide (Altshuller, 1983). In synthetic air, photolysis of formaldehyde gave hydrochloric acid and carbon monoxide (Su et al., 1979). Calvert et al. (1972) reported, however, that formaldehyde photodecomposed by direct sunlight in the atmosphere yielding hydrogen, formyl radicals and carbon monoxide. Photooxidation of formaldehyde in the absence of nitrogen oxides in air ($\lambda = 2900-3500$ Å) gave hydrogen peroxide, alkylhydroperoxides, carbon monoxide and lower molecular weight aldehydes. In the presence of nitrogen oxides, photooxidation products reported include ozone, hydrogen peroxide and peroxyacyl nitrates (Kopczynski et al., 1974).

Irradiation of gaseous formaldehyde containing an excess of nitrogen dioxide over chlorine yielded ozone, carbon monoxide, nitrogen pentoxide, nitryl chloride, nitric acid and hydrochloric acid. Peroxynitric acid was the major photolysis product when chlorine exceeded nitrogen dioxide concentrations (Hanst and Gay, 1977).

Chemical/Physical. Oxidizes in air to formic acid (Hartley and Kidd, 1987). Trioxymethylene may precipitate under cold temperatures (Sax, 1984). Polymerizes easily (Windholz et al., 1983). Anticipated products from the reaction of formaldehyde with ozone or hydroxyl radicals in air are carbon monoxide and carbon dioxide (Cupitt, 1980). Major products reported from the photooxidation of formaldehyde with nitrogen oxides are carbon monoxide, carbon dioxide and hydrogen peroxide (Altshuller, 1983).

Reacts with hydrochloric acid in moist air forming bis(chloromethyl)ether. This compound may also form from an acidic solution containing chloride ion and formaldehyde (Frankel et al., 1974). In an aqueous solution at 25°C, nearly all the formaldehyde added is hydrated forming a gemdiol (Bell and McDougall, 1960). May polymerize in an aqueous solution to trioxymethylene (Hartley and Kidd, 1987).

Exposure Limits: NIOSH REL: TWA 0.016 ppm, 15-min C 0.1 ppm, IDLH 20 ppm; OSHA PEL: TWA 0.75 ppm, STEL 2 ppm.

Symptoms of Exposure: Irritates eyes, nose, throat; lacrimation; cough, bronchospasm; pulmonary irritation; dermatitis, nausea, vomiting; loss of consciousness.

Formulation Types: Aqueous solutions.

Toxicity: Toxic to fish (Hartley and Kidd, 1987); acute oral LD_{50} for rats 550–800 mg/kg (Hartley and Kidd, 1987).

Uses: Fungicide; bactericide.

FOSAMINE-AMMONIUM

Synonyms: Ammonium salt, ammonium ethyl(aminocarbonyl)phosphonate; Ammonium ethyl carbamoylphosphonate solution; DPX 1108; Krenite; Krenite brush control agent.

$$C_2H_5 - O - \overset{\overset{\displaystyle O}{\|}}{\underset{\underset{\displaystyle O^- \ \overset{+}{N}H_4}{|}}{P}} - CONH_2$$

Designations: CAS Registry Number: 25954-13-6; mf: $C_3H_{11}N_2O_4P$; fw: 170.10; RTECS: BQ4112000.

Properties: Colorless to white, crystalline solid. Mp: 175°C; ρ: 1.33; fl p: nonflammable; K_H: 4.98×10^{-13} atm · m³/mol at 20–25°C (approximate — calculated from water solubility and vapor pressure); log K_{oc}: 0.20 (calculated); log K_{ow}: –2.90; S_o (g/kg at 25°C): acetone (0.3), benzene (0.4), chloroform (0.04), *N,N*-dimethylformamide (1.4), ethanol (12.0), hexane (0.2), methanol (158.0); S_w: 1,790 g/L at 25°C; vp: 4×10^{-6} mmHg at 20°C.

Environmental Fate

Soil. Fosamine is rapidly degraded to carbon dioxide by microorganisms in soil (Humburg et al., 1989). The reported half-life in soil is approximately 7 to 10 days (Hartley and Kidd, 1987; Worthing and Hance, 1991).

Plant. After a 2.5% application to multiflora rose plant leaves, 19, 21, 22, and 43% of the applied amount was absorbed after 1, 4, 8, 16 and 32 days, respectively (Mann et al., 1986). Degrades rapidly in plants with a half-life of 2 to 3 weeks (Humburg et al., 1989; Ashton and Monaco, 1991).

Chemical/Physical. Fosamine-ammonium is stable as a dilute solution but decomposes in weak acidic media (Worthing and Hance, 1991), probably forming water-soluble salts.

Symptoms of Exposure: May irritate eyes, nose, throat and skin.

Formulation Types: Soluble concentrate (4 lb/gal).

Toxicity: LC_{50} (96-hour) for bluegill sunfish 0.67 g/L, rainbow trout and fathead minnows >1 g/L (Hartley and Kidd, 1987); acute oral LD_{50} for rats 24.4 g/kg (Ashton and Monaco, 1991).

Uses: Nonselective contact herbicide used to control many woody and brush species on noncrop land.

FOSETYL-ALUMINUM

Synonyms: Aliette; **Aluminum tris(O-ethyl phosphonate)**; Efosite-AL; Epal; Fosetyl AL; LS 74783; Phosethyl; Phosethyl AL; RP 32545.

$$\left[CH_3CH_2O - \overset{\overset{\displaystyle O}{\|}}{\underset{|}{P}} - H \right]_3^{-} \quad Al^{3+}$$

Designations: CAS Registry Number: 39148-24-8; mf: $C_6H_{18}AlO_9P_3$; fw: 354.10; RTECS: SZ9640000.

Properties: Colorless powder. Mp: decomposes >200°C; log K_{oc}: 2.49; log K_{ow}: –2.70 (pH 4); S_o (mg/L): acetone (13), acetonitrile (5), methanol (920); S_w: 120 g/L at room temperature; vp: 9.75×10^{-8} mmHg at 25°C.

Environmental Fate

Plant. Felsot and Pedersen (1991) reported that fosetyl-aluminum degrades in plants forming phosphonic acid which ionizes to the dianion phosphonate, HPO_3^{-2}.

Formulation Types: Wettable powder.

Toxicity: LC_{50} (96-hour) for rainbow trout 428 mg/L (Hartley and Kidd, 1987); acute oral LD_{50} for rats 5,800 mg/kg (Hartley and Kidd, 1987), 5,400 mg/kg (RTECS, 1985).

Use: Fungicide.

GLYPHOSATE

Synonyms: Mon 0573; *N*-(Phosphonomethyl)glycine.

$$\text{HOOCCH}_2\text{NHCH}_2 - \overset{\displaystyle \overset{O}{\|}}{\underset{\displaystyle \underset{OH}{|}}{P}} - OH$$

Designations: CAS Registry Number: 1071-83-6; mf: $C_3H_8NO_5P$; fw: 169.08; RTECS: MC1075000.

Properties: Colorless to white, odorless crystals or powdery solid. Mp: 230°C (decomposes); ρ: 1.74 g/mL; fl p: nonflammable (water-based formulations only); pK_1: 2.32, pK_2 = 5.86, pK_3 = 10.86; K_H: 1.39×10^{-10} atm · m³/mol at 25°C (approximate — calculated from water solubility and vapor pressure); log K_{oc}: 3.43–4.53; log K_{ow}: –1.60; S_o: insoluble in most organic solvents; S_w: 12 g/L at 25°C; vp: 7.50×10^{-6} mmHg at 25°C.

Soil properties and adsorption data

Soil	K_d (mL/g)	f_{oc} (%)	K_{oc} (mL/g)	pH (meq/100 g)	CEC
Acid brown silty sand	41	3.73	1,099	5.8	18.3
Dupo silt loam	33	—	—	—	—
Drummer silty clay	324	—	—	—	—
Houston clay loam	76	1.56	4,871	7.5	29
Illite	115	—	—	—	115
Kaolinite	8	—	—	—	6
Kaolinite	100	—	—	2.0	16
Kaolinite	519	—	—	4.5	16
Kaolinite	66	—	—	7.0	16
Kaolinite	69	—	—	11.0	16
Montmorillonite	138	—	—	—	117
Montmorillonite (Al^{3+})	516	—	—	2.0	98
Montmorillonite (Al^{3+})	1,156	—	—	4.5	98
Montmorillonite (Al^{3+})	290	—	—	7.0	98
Montmorillonite (Al^{3+})	198	—	—	11.5	98
Montmorillonite (Ca^{2+})	64	—	—	2.0	98
Montmorillonite (Ca^{2+})	113	—	—	4.5	98
Montmorillonite (Ca^{2+})	361	—	—	7.0	98
Montmorillonite (Ca^{2+})	198	—	—	11.5	98
Montmorillonite (Na^+)	152	—	—	2.0	98
Montmorillonite (Na^+)	101	—	—	4.5	98
Montmorillonite (Na^+)	197	—	—	7.0	98
Montmorillonite (Na^+)	234	—	—	11.5	98

Soil properties and adsorption data *(continued)*

Soil	K_d (mL/g)	f_{oc} (%)	K_{oc} (mL/g)	pH (meq/100 g)	CEC
Nontronite (Al^{3+})	738	—	—	2.0	142
Nontronite (Al^{3+})	1,621	—	—	4.5	142
Nontronite (Al^{3+})	421	—	—	7.0	142
Nontronite (Al^{3+})	644	—	—	11.5	142
Nontronite (Ca^{2+})	287	—	—	2.0	142
Nontronite (Ca^{2+})	237	—	—	4.5	142
Nontronite (Ca^{2+})	778	—	—	7.0	142
Nontronite (Ca^{2+})	244	—	—	11.5	142
Nontronite (Na^+)	236	—	—	2.0	142
Nontronite (Na^+)	96	—	—	4.5	142
Nontronite (Na^+)	831	—	—	7.0	142
Nontronite (Na^+)	344	—	—	11.5	142
Podzol sand	51	9.23	554	4.6	31
Podol silt	153	0.45	33,967	8.3	11
Muskingum silt loam	56	1.64	3,414	5.8	11
Sassafras sandy loam	33	1.24	2,661	5.6	7
Spinks loamy sand	660	—	—	—	—

Source: McConnell and Hossner, 1985; Glass, 1987; Piccolo et al., 1994; U.S. Department of Agriculture, 1990.

Environmental Fate

Soil. Degrades microbially in soil releasing phosphoric acid, *N*-nitrosoglyphosate (Newton et al., 1984), ammonia (Cremlyn, 1991), *N,N*-dimethylphosphinic acid, *N*-methylphosphinic acid, aminoacetic acid (glycine), *N*-methylaminoacetic acid (sarcosine), hydroxymethylphosphonic acid (Duke et al., 1991), aminomethylphosphonic acid (Normura and Hilton, 1977; Rueppel et al., 1977; Hoagland, 1980; Duke et al., 1991; Muir, 1991) and carbon dioxide (Sprankle et al., 1975; Cremlyn, 1991). *N*-Nitrosoglyphosate also formed from the nitrosation of glyphosate in soil solutions containing nitrite ions (Young and Kahn, 1978).

The reported half-life of glyphosate in soil is <60 days (Hartley and Kidd, 1987). In the laboratory, experimentally determined dissipation rates of glyphosate in a Lintonia sandy loam, Drummer silty clay, Norfolk sandy loam and Raye silty loam were 0.028, 0.02, 0.0006 and 0.022/day, respectively. The half-lives ranged from 3 to 130 days with an average of 38 days (Rueppel et al., 1977). The mineralization half-lives for glyphosate in soil ranged from 30 days to 14.2 years with an average half-life of 4.75 years (Sprankle et al., 1975a; Normura and Hilton, 1977; Moshier and Penner, 1978). In agricultural loam and fine silt soils from Finland, the calculated half-lives of glyphosate were 69 and 127 days, respectively (Muller et al., 1981).

Plant. In a forest brush field ecosystem, the half-life of glyphosate in foliage and litter ranged from 10.4 to 26.6 days, respectively (Newton et al., 1984).

Photolytic. When an aqueous solution of glyphosate (1 ppm) was exposed to outdoor sunlight for 9 weeks (from August 12 through October 15, 1983), aminomethylphosphonic acid and ammonia formed as major and minor photoproducts, respectively (Lund-Høie and Friestad, 1986). More than 90% degradation was observed after only 4 weeks of exposure. Photodegradation was also observed when an aqueous solution was exposed indoors to UV light (λ = 254 nm). The reported half-lives of this reaction at starting concentrations of 1.0 and 2,000 ppm were 4 days and 3–4 weeks, respectively. When aqueous solutions were exposed indoors to sodium light (λ = 550–650 nm) and mercury light (λ = 400–600 nm), no photo-degradation occurred (Lund-Høie and Friestad, 1986).

Chemical/Physical. Under laboratory conditions, the half-life of glyphosate in natural waters was 7–10 weeks (Muir, 1991). A 1% aqueous solution has a pH of 2.5 (Keith and Walters, 1992). This suggests glyphosate will react with alkalies and amines forming water-soluble salts.

At 200–230°C, glyphosate undergoes dehydration to form N,N'-diphosphonomethyl-2,5-diketopiperazine (Duke, 1988).

Formulation Types: Soluble concentrate (free acid 3 lb/gal, isopropylamine salt 4 lb/gal); water-soluble powder.

Toxicity: LC_{50} (96-hour) for rainbow trout 86 mg/L, bluegill sunfish 120 mg/L (Hartley and Kidd, 1987), for harlequin fish 168 ppm, for Atlantic oyster >10 mg/L, for shrimp 281 ppm and fiddler crab 934 mg/L (Humburg et al., 1989); LC_{50} (48-hour) for *Daphnia magna* 780 mg/L (Humburg et al., 1989); acute oral LD_{50} of pure glyphosate and Roundup formulation for rats 5,600 and 5,400 mg/kg, respectively (Ashton and Monaco, 1991), 470 mg/kg (RTECS, 1985).

Uses: Nonselective, postemergence, broad spectrum herbicide used to control annual and perennial grasses, sedges, broad-leaved and emerged aquatic weeds. This herbicide is also used to control insects on fruit trees.

HEPTACHLOR

Synonyms: Aahepta; Agroceres; Basaklor; 3-Chlorochlordene; Drinox; Drinox H-34; E 3314; ENT 15152; GPKh; H 34; Heptachlorane; 3,4,5,6,7,8,8-Heptachlorodicyclopentadiene; 3,4,5,6,7,8,8a-Heptachlorodicyclopentadiene; 1(3a),4,5,6,7,8,8-Heptachloro-3a(1),4,7,7a-tetrahydro-4,7-methanoindene; 1,4,5,6,7,8,8-Heptachloro-3a,4,7,7a-tetrahydro-4,7-methanoindene; **1,4,5,6,7,8,8-Heptachloro-3a,4,7,7a-tetrahydro-4,7-methanol-1*H*-indene**; 1,4,5,6,7,8,8-Heptachloro-3a,4,7,7a-tetrahydro-4,7-*endo*-methanoindene; 1,4,5,6,7,8,8a-Heptachloro-3a,4,7,7a-tetrahydro-4,7-methanoindene; 1,4,5,6,7,8,8-Heptachloro-3a,4,7,7a-tetrahydro-4,7-methyleneindene; 1,4,5,6,7,10,10-Heptachloro-4,7,8,9-tetrahydro-4,7-methyleneindene; 1,4,5,6,7,10,10-Heptachloro-4,7,8,9-tetrahydro-4,7-*endo*-methylenein-dene; 3,4,5,6,7,8,8a-Heptachloro-α-dicyclopentadiene; Heptadichloro-cyclopenta-diene; Heptagran; Heptagranox; Heptamak; Heptamul; Heptasol; Heptox; NA 2761; NCI-C00180; Soleptax; RCRA waste number P059; Rhodiachlor; Velsicol 104; Velsicol heptachlor.

Designations: CAS Registry Number: 76-44-8; DOT: 2761; mf: $C_{10}H_5Cl_7$; fw: 373.32; RTECS: PC0700000.

Properties: White to light tan, waxy solid or crystals with a camphor-like odor. Mp: 95–96°C (pure), 46–74°C (technical); bp: 135–145°C at 1–1.5 mmHg, decomposes at 760 mmHg; ρ: 1.66 at 20/4°C; K_H: 2.3×10^{-3} atm · m³/mol; log BCF: 3.76–3.92 (*Leiostomus xanthurus*, Schimmel et al., 1976), freshwater clam (*Corbicula manilensis*) 4.03 (Hartley and Johnston, 1983); log K_{oc}: 4.38; log K_{ow}: 4.40–5.5; S_o (g/L at 27°C unless noted otherwise): acetone (750), benzene (1,060), carbon tetrachloride (1,120), cod liver oil (93.8, 115.4, 178.7 g/L at 4, 12 and 20°C, respectively), cyclohexane (1,190), cyclohexanone (1,190), deodorized kerosene (263), ethanol (45), kerosene (1,890), octanol (67.9, 81.7 and 87.5 g/L at 4, 12 and 20°C, respectively), triolein (83.6, 129.4 and 132.4 g/L at 4, 12 and 20°C, respectively), xylene (1,020); S_w: 180 ppb at 25°C; vp: 3×10^{-4} mmHg at 20°C, 2.33×10^{-4} mmHg at 25°C.

Environmental Fate

Biological. Many soil microorganisms were found to oxidize heptachlor to heptachlor epoxide (Miles et al., 1969). In addition, hydrolysis produced hydroxychlordene with subsequent epoxidation yielding 1-hydroxy-2,3-epoxychlordene (Kearney and Kaufman, 1976). Heptachlor reacted with reduced hematin forming chlordene, which decomposed to hexachlorocyclopentadiene and cyclopentadiene (Baxter, 1990). In a model ecosystem containing plankton, *Daphnia magna*, mosquito larva (*Culex pipiens quinquefasciatus*), fish (*Cambusia affinis*), alga (*Oedogonium cardiacum*) and snail (*Physa* sp.), heptachlor degraded to 1-hydroxychlordene, 1-hydroxy-2,3-epoxychlordene, hydroxychlordene epoxide, heptachlor epoxide and five unidentified compounds (Lu et al., 1975). In four successive 7-day incubation periods, heptachlor (5 and 10 mg/L) was recalcitrant to degradation in a settled domestic wastewater inoculum (Tabak et al., 1981). When heptachlor (10 ppm) in sewage sludge was incubated under anaerobic conditions at 53°C for 24 hours, complete degradation was achieved (Hill and McCarty, 1967).

In a mixed bacterial culture under aerobic conditions, heptachlor was transformed to chlordene, 1-hydroxychlordene, heptachlor epoxide and chlordene epoxide in low yields (Miles et al., 1971). Heptachlor rapidly degraded when incubated with acclimated, mixed microbial cultures under aerobic conditions. After 4 weeks, 95.3% of the applied dosage was removed (Leigh, 1969).

Soil. Heptachlor reacted with reduced hematin forming chlordene which decomposed to hexachlorocyclopentadiene and cyclopentadiene (Baxter, 1990). The reported half-life in soil is 9–10 months (Hartley and Kidd, 1987).

Heptachlor was rapidly converted to 1-hydroxychlordene in eight dry soils having low moisture contents. Under these conditions, heptachlor epoxide was not identified (Bowman et al., 1965).

Following application to an Ohio soil, only 7% of the applied amount had volatilized after 170 days (Glotfelty et al., 1984). Harris (1969) concluded that heptachlor has a very low tendency to leach in soils.

Groundwater. According to the U.S. EPA (1986) heptachlor has a high potential to leach to *groundwater.*

Plant. Heptachlor is converted to its epoxide on plant surfaces (Gannon and Decker, 1958).

Photolytic. Sunlight and UV light converts heptachlor to photoheptachlor (Georgacakis and Khan, 1971). This is in agreement with Gore et al. (1971) who reported that heptachlor exhibited weakly absorption of UV light at wavelengths above 290 nm. Eichelberger and Lichtenberg (1971) reported heptachlor (10 µg/L) in river water, kept in a sealed jar under sunlight and fluorescent light, was completely converted to 1-hydroxychlordene. Under the same conditions, but in distilled water, 1-hydroxychlordene and heptachlor epoxide formed in yields of 60 and 40%, respectively (Eichelberger and Lichtenberg, 1971).

The photolysis of heptachlor in various organic solvents afforded different photoproducts (McGuire et al., 1972). Photolysis at 253.7 nm in hydrocarbon solvents yields two olefinic monodechlorination isomers: 1,4,5,7,8,8-hexachloro-3a,4,7,7a-tetrahydro-4,7-methanoindene and 1,4,6,7,8,8-hexachloro-3a,4,7,7a-tetrahydro-4,7-methanoindene. Irradiation at 300 nm in acetone, 1,2,3,6,9,10,10-heptachloropentacyclo[5.3.0.0^{2,5}.0^{3,9}.0^{4,8}]decane is the only product formed. This compound and a C-1 cyclohexyl adduct are formed when heptachlor in a cyclohexane/acetone solvent system is irradiated at 300 nm (McGuire et al., 1972).

Heptachlor reacts with photochemically produced hydroxyl radicals in the atmosphere. At a concentration of 5×10^5 hydroxyl radicals/cm^3, the atmospheric half-life was estimated to be about 6 hours (Atkinson, 1987).

Chemical/Physical. Slowly releases hydrogen chloride in aqueous media (Hartley and Kidd, 1987; Kollig, 1993). The hydrolysis half-lives of heptachlor in a sterile 1% ethanol/water solution at 25°C and pH values of 4.5, 5.0, 6.0, 7.0 and 8.0 were 0.77, 0.62, 0.64, 0.64 and 0.43 weeks, respectively (Chapman and Cole, 1982). Chemical degradation of heptachlor give heptachlor epoxide (Newland et al., 1969). Heptachlor degraded in aqueous saturated calcium hypochlorite solution to 1-hydroxychlordene. Although further degradation occurred, no other metabolites were identified (Kaneda et al., 1974).

Heptachlor (1 mM) in ethyl alcohol (30 mL) underwent dechlorination in the presence of nickel boride (generated by the reacted nickel chloride and sodium borohydride). The catalytic dechlorination of heptachlor by this method yielded a pentachloro derivative as the major product having the empirical formula $C_{10}H_9Cl_5$ (Dennis and Cooper, 1976).

Exposure Limits: NIOSH REL: TWA 0.5 mg/m³, IDLH 35 mg/m³; OSHA PEL: TWA 0.5 mg/m³; ACGIH TLV: TWA 0.5 mg/m³.

Symptoms of Exposure: In animals: tremors, convulsions, liver damage.

Formulation Types: Emulsifiable concentrate; wettable powder; granules; dustable powder; seed treatment.

Toxicity: LC_{50} (96-hour) for rainbow trout 7 μg/L, bluegill sunfish 26 μg/L, fathead minnow 78–130 μg/L (Hartley and Kidd, 1987), sheepshead minnow 2.7–8.8 μg/L and marine pin perch 0.20–4.4 μg/L (Verschueren, 1983); LC_{50} (24-hour) for sheepshead minnow 1.22–4.3 μg/L (Verschueren, 1983); acute oral LD_{50} for rats 147–220 mg/kg (Hartley and Kidd, 1987), 40 mg/kg (RTECS, 1985).

Use: Insecticide for termite control.

HEPTACHLOR EPOXIDE

Synonyms: ENT 25584; Epoxy heptachlor; HCE; 1,4,5,6,7,8,8-Heptachloro-2,3-epoxy-2,3,3a,4,7,7a-hexahydro-4,7-methanoindene; 1,2,3,4,5,6,7,8,8-Heptachloro-2,3-epoxy-3a,4,7,7a-tetrahydro-4,7-methanoindene; **5a,6,6a-Hexahydro-2,5-methano-2*H*-indeno[1,2-*b*]oxirene**; 2,3,4,5,6,7,7-Heptachloro-1a,1b,5,5a,6,6a-hexahydro-2,5-methano-2*H*-oxireno[*a*]indene; Velsicol 53-CS-17.

Designations: CAS Registry Number: 1024-57-3; DOT: 2761; mf: $C_{10}H_5Cl_7O$; fw: 389.32; RTECS: PB9450000.

Properties: Liquid. Mp: 157–160°C; K_H: 3.2×10^{-5} atm · m³/mol at 25°C; log BCF: freshwater clam (*Corbicula manilensis*) 3.37 (Hartley and Johnston, 1983); log K_{oc}: 4.32 (calculated); log K_{ow}: 3.65, 5.40; S_o: cod liver oil (60.3 g/L at 4°C), *n*-octanol (33.9, 30.3 and 46.8 g/L at 4, 12 and 20°C, respectively), triolein (75.9 and 77.7 g/L at 12 and 20°C, respectively); S_w: 275 µg/L at 25°C; vp: 2.6×10^{-6} mmHg at 20°C.

Environmental Fate

Biological. In a model ecosystem containing plankton, *Daphnia magna*, mosquito larva (*Culex pipiens quinquefasciatus*), fish (*Cambusia affinis*), alga (*Oedogonium cardiacum*) and snail (*Physa* sp.), heptachlor epoxide degraded to hydroxychlordene epoxide (Lu et al., 1975). Using settled domestic wastewater inoculum, heptachlor epoxide (5 and 10 mg/L) did not degrade after 28 days of incubation at 25°C (Tabak et al., 1981). This is consistent with the findings of Bowman et al. (1965). They observed that under laboratory conditions, heptachlor epoxide did not show any evidence of degradation when incubated in a variety of soils maintained at 45°C for 8 days. The soils used in this experiment included Lakeland sand, Lynchburg loamy sand, Magnolia sandy loam, Magnolia sandy clay loam, Greenville sandy clay and Susquehanna sandy clay (Bowman et al., 1965). When heptachlor epoxide was incubated in a sandy loam soil at 28°C, however, 1-hydroxychlordene formed at yields of 2.8, 5.8 and 12.0% after 4, 8 and 12 weeks, respectively (Miles et al., 1971).

Photolytic. Irradiation of heptachlor epoxide by a 450-W high-pressure mercury lamp gave two half-cage isomers, each containing a ketone functional group (Ivie et al., 1972). Benson et al. (1971) reported a degradation yield of 99% when an acetone solution containing heptachlor epoxide was photolyzed at >300 nm for 11 hours. An identical degradation yield was achieved in only 60 minutes when the UV wavelength was reduced to >290 nm.

Graham et al. (1973) reported that when solid heptachlor epoxide was exposed to July sunshine for 23.2 days, 59.3% degradation was achieved. In powdered form, however, only 5 days were required for complete degradation to occur.

Chemical/Physical. Heptachlor epoxide will hydrolyze via nucleophilic attack at the epoxide moiety forming heptachlor diol which may undergo further hydrolysis forming heptachlor triol and hydrogen chloride (Kollig, 1993).

Toxicity: Acute oral LD_{50} for rats 47 mg/kg (RTECS, 1985).

Uses: Not known. A degradation product of heptachlor.

HEXACHLOROBENZENE

Synonyms: Amatin; Anticarie; Bunt-cure; Bunt-no-more; Coop hexa; Granox NM; HCB; Hexa C.B.; Julin's carbon chloride; No bunt; No bunt 40; No bunt 80; No bunt liquid; Pentachlorophenyl chloride; Perchlorobenzene; Phenyl perchloryl; RCRA waste number U127; Sanocide; Smut-go; Snieciotox; UN 2729.

Designations: CAS Registry Number: 118-74-1; DOT: 2729; mf: C_6Cl_6; fw: 284.78; RTECS: DA2975000.

Properties: White monoclinic crystals or crystalline solid. Mp: 227–230°C; bp: 323–326°C (sublimes); ρ: 2.049 and 2.044 at 20/4 and 23/4°C, respectively; fl p: 242°C; H-$t_{1/2}$: none observed after 13 days at 85°C and pH values of 3, 7 and 11; K_H: 7.1×10^{-3} atm · m³/mol at 20°C; log BCF: 6.42 (Atlantic croakers), 6.71 (blue crabs), 5.96 (spotted sea trout), 5.98 (blue catfish) (Pereira et al., 1988); log K_{oc}: 2.56–4.54; log K_{ow}: 3.93–6.42; S_o: soluble in acetone (1–5 mg/mL), benzene, carbon disulfide, chloroform, dimethylsulfoxide (<1 mg/mL), 95% ethanol (<1 mg/mL), ethyl ether, chloroform, cod liver oil (3.1, 4.2 and 5.8 g/L at 4, 12 and 20°C, respectively), octanol (2.4, 3.1 and 3.9 g/L at 4, 12 and 20°C, respectively), triolein (3.4, 4.7 and 7.3 g/L at 4, 12 and 20°C, respectively) and many other organic solvents; S_w: 110 at 24°C, 6 at 25°C (Metcalf et al., 1973); 40 µg/L at 20°C; vp: 1.089×10^{-5} mmHg at 20°C, 1.19×10^{-3} mmHg at 25°C.

Soil properties and adsorption data

Soil	K_d (mL/g)	f_{oc} (%)	K_{oc} (mL/g)	pH
Rothamsted Farm	462.8	1.51	30,649	5.1

Source: Lord et al., 1980.

Environmental Fate

Biological. Reductive monodechlorination occurred in an anaerobic sewage sludge yielding principally 1,3,5-trichlorobenzene. Other compounds identified included pentachlorobenzene, 1,2,3,5-tetrachlorobenzene and dichlorobenzenes (Fathepure et al., 1988). In activated sludge, only 1.5% of the applied hexachlorobenzene mineralized to carbon dioxide after 5 days (Freitag et al., 1985). In a 5-day experiment, ^{14}C-labeled hexachlorobenzene applied to soil-water suspensions under aerobic and anaerobic conditions gave [14]CO_2 yields of 0.4 and 0.2%, respectively (Scheunert et al., 1987).

When hexachlorobenzene was statically incubated in the dark at 25°C with yeast extract and settled domestic wastewater inoculum, no significant biodegradation was observed. At a concentration of 5 mg/L, percent losses after 7, 14, 21 and 28-day incubation periods were 56, 30, 8 and 5, respectively. At a concentration of 10 mg/L, only 21 and 3% losses were observed after the 7 and 14-day incubation periods, respectively. The decrease in concentration over time suggests biodegradation followed a deadaptive process (Tabak et al., 1981).

Groundwater. According to the U.S. EPA (1986) hexachlorobenzene has a high potential to leach to *groundwater.*

Photolytic. Solid hexachlorobenzene exposed to artificial sunlight for 5 months photolyzed at a very slow rate with no decomposition products identified (Plimmer and Klingebiel, 1976). The sunlight irradiation of hexachlorobenzene (20 g) in a 100 mL borosilicate glass-stoppered Erlenmeyer flask for 56 days yielded 64 ppm pentachlorobiphenyl (Uyeta et al., 1976). A carbon dioxide yield <0.1% was observed when hexachlorobenzene adsorbed on silica gel was irradiated with light (λ >290 nm) for 17 hours (Freitag et al., 1985).

Irradiation (λ ≥285 nm) of hexachlorobenzene (1.1–1.2 mM/L) in an acetonitrile-water mixture containing acetone (concentration = 0.553 mM/L) as a sensitizer gave the following products (% yield): pentachlorobenzene (71.0), 1,2,3,4-tetrachlorobenzene (0.6), 1,2,3,5-tetrachlorobenzene (2.2) and 1,2,4,5-tetrachlorobenzene (3.7) (Choudhry and Hutzinger, 1984). Without acetone, the identified photolysis products (% yield) included 1,2,3,4,5-pentachlorobenzene (76.8), 1,2,3,5-tetrachlorobenzene (1.2), 1,2,4,5-tetrachlorobenzene (1.7) and 1,2,4-trichlorobenzene (0.2) (Choudhry and Hutzinger, 1984).

In another study, the irradiation (λ = 290–310 nm) of hexachlorobenzene in aqueous solution gave only pentachlorobenzene and possibly pentachlorophenol as the transformation products. The photolysis rate increased with the addition of naturally occurring substances (tryptophan and pond proteins) and abiotic sensitizers (diphenylamine and skatole) (Hirsch and Hutzinger, 1989).

When an aqueous solution containing hexachlorobenzene (150 nM) and a nonionic surfactant micelle (0.50 M Brij 58, a polyoxyethylene cetyl ether) was illuminated by a photoreactor equipped with 253.7 nm monochromatic UV lamps, significant concentrations of pentachlorobenzene, all tetra-, tri- and dichlorobenzenes, chlorobenzene, benzene, phenol, hydrogen and chloride ions were formed. Two compounds, namely 1,2-dichlorobenzene and 1,2,3,4-tetrachlorobenzene, formed in minor amounts (<40 ppb). The half-life for this reaction, based on the first-order photodecomposition rate of 1.44×10^{-2}/sec, is 48 seconds (Chu and Jafvert, 1994).

Chemical/Physical. Hydrolysis is not considered an environmentally relevant process. No hydrolysis was observed after 13 days at 85°C and pH values of 3, 7 and 11 (Ellington et al., 1987).

Emits toxic chloride fumes when heated to decomposition (Lewis, 1990).

Formulation Types: Dry seed treatment.

Toxicity: LC_{50} for five freshwater species 0.05-0.2 mg/L (Hartley and Kidd, 1987); acute oral LD_{50} for rats 10,000 mg/kg (RTECS, 1985).

Use: Seed fungicide.

HEXAZINONE

Synonyms: **3-Cyclohexyl-6-(dimethylamino)-1-methyl-1,3,5-triazine-2,4(1H,3H)-dione**; 3-Cyclohexyl-6-(dimethylamino)-1-methyl-s-triazine-2,4(1H,3H)-dione; DPX 3674; Velpar; Velpar K; Velpar L; Velpar RP; Velpar ULW; Velpar weed killer.

Designations: CAS Registry Number: 51235-04-2; mf: $C_{12}H_{20}N_4O_2$; fw: 252.30; RTECS: XY7850000.

Properties: Colorless to white, odorless crystals. Mp: 115–117°C; bp: decomposes; ρ: 1.25; fl p: nonflammable; log K_{oc}: 1.30–1.43; log K_{ow}: 1.36 (Liu and Qian, 1995); S_o (g/kg at 25°C): acetone (790), benzene (940), chloroform (3,880), N,N-dimethylformamide (836), hexane (3), methanol (2,650), toluene (386); S_w: 33 g/kg at 25°C; vp: 2×10^{-7} mmHg at 20°C (extrapolated), 6.4×10^{-5} mmHg at 86°C.

Soil properties and adsorption data

Soil	K_d (mL/g)	f_{oc} (%)	K_{oc} (mL/g)	pH (meq/100 g)	CEC
Fallsington sandy loam	0.2	0.81	25	5.6	4.8
Flanagan silt loam	1.0	2.33	173	5.0	23.4

Source: Rhodes, 1980a.

Environmental Fate

Soil/Plant. Biodegrades in soil and natural waters releasing carbon dioxide. The reported half-life in soil is 1 to 6 months (Hartley and Kidd, 1987). Rhodes (1980a) found that the persistence of hexazinone varied from 4 weeks in a Delaware sandy loam to 24 weeks in a Mississippi silt loam.

Hexazinone is subject to microbial degradation (Rhodes, 1980a; Feng, 1987). Metabolites identified in soils, alfalfa and/or sugarcane include 3-(4-hydroxycyclohexyl)-6-(dimethylamino)-1-methyl-1,3,5-triazine-2,4(1H,3H)-dione, 3-cyclohexyl-6-(methylamino)-1-methyl-1,3,5-triazine-2,4(1H,3H)-dione (Rhodes, 1980a; Holt, 1981; Roy et al., 1989), 3-(4-hydrocyclohexyl)-6-(methylamino)-1-methyl-1,3,5-triazine-2,4(1H,3H)dione, 3-cyclohexyl-1-methyl-1,3,5-triazine-2,4,6-(1H,3H,5H)trione, 3-(4-hydrocyclohexyl)-1-methyl-1,3,5-triazine-2,4,6(1H,3H,5H)trione, 3-cyclohexyl-6-amino-1-methyl-1,3,5-triazine-2,4(1H,3H)dione and 3-cyclohexyl-6-(methylamino)-1,3,5-triazine-2,4(1H,3H)-dione (Rhodes, 1980a; Holt, 1981). The half-lives in a sandy soil and clay were 1 and 6 months, respectively (Rhodes, 1980a). In a sandy loam soil, the half-life ranged from 10 to 30 days (Neary et al., 1983).

Feng (1987) monitored the persistence and degradation of hexazinone in a silt loam soil 104 days after treatment of the herbicide. After 104 days, 66% of the hexazinone degraded via hydroxylation to form the major metabolite 3-(4-hydroxycyclohexyl)-6-(dimethylamino)-1-methyl-1,3,5-triazine-2,4(1*H*,3*H*)dione (30–50% yield). A minor metabolite, 3-cyclohexyl-6-(methylamino)-1-methyl-1,3,5-triazine-2,4(1*H*,3*H*)-dione which formed via demethylation, accounted for only 0–12% of the applied herbicide. Hexazinone and its metabolites were not detected in soils at depths of 15–30 cm (Feng, 1987).

No traces of hexazinone or its metabolites were detected on treated blueberries (Jenson and Kimball, 1985).

Photolytic. Photodegradation products identified in aqueous hexazinone solutions following exposure to UV light (λ = 300–400 nm) were 3-(4-hydroxycyclohexyl)-6-(dimethylamino)-1-methyl-1,3,5-triazine-2,4(1*H*,3*H*)-dione, 3-cyclohexyl-6-(methylamino)-1-methyl-1,3,5-triazine-2,4(1*H*,3*H*)dione and 3-cyclohexyl-6-(dimethylamino)-1,3,5-triazine-2,4(1*H*,3*H*)dione (Rhodes, 1980).

Symptoms of Exposure: May irritate eyes, nose, skin and throat.

Formulation Types: Soluble concentrate; water-soluble powder (90%); granules; miscible liquid (2 lb/gal).

Toxicity: LC_{50} (96-hour) for rainbow trout 320–420 mg/L, fathead minnow 274 mg/L and bluegill sunfish 370–420 mg/L (Hartley and Kidd, 1987); acute oral LD_{50} for rats 1,700 mg/kg (Ashton and Monaco, 1991).

Uses: Hexazinone is a triazine compound used as a preemergence or postemergence herbicide to control many annual grasses and broad-leaved weeds in noncropped land and certain crops such as alfalfa, blueberries, coffee, pecans and sugarcane.

HEXYTHIAZOX

Synonyms: *trans*-5-(4-Chlorophenyl)-*N*-cyclohexyl-4-methyl-2-oxo-3-thiazolidine-carboxamide.

Designations: CAS Registry Number: 78587-05-0; mf: $C_{17}H_{21}ClN_2O_2S$; fw: 352.90.

Properties: Colorless crystals. Mp: 108–108.5°C; H-$t_{1/2}$ (aqueous solution in sunlight): 16.7 days; K_H: 2.4×10^{-8} atm · m³/mol at 20°C (approximate — calculated from water solubility and vapor pressure); log K_{oc}: 3.81 (calculated); log K_{ow}: 2.53; S_o (g/L at 20°C): acetone (160), acetonitrile (28.6), chloroform (1,380), hexane (3.9), methanol (206), xylene (362); S_w: 0.5 mg/L at 20°C; vp: 2.55×10^{-8} mmHg at 20°C.

Environmental Fate

Soil. Though no products were reported, the half-life in a clay loam at 15°C is 8 days (Worthing and Hance, 1991).

Formulation Types: Emulsifiable concentrate; wettable powder.

Toxicity: LC_{50} (96-hour) for rainbow trout >300 mg/L and bluegill sunfish 11.6 mg/L (Hartley and Kidd, 1987); LC_{50} (48-hour) for carp 3.7 mg/L (Worthing and Hance, 1991); acute oral LD_{50} for mice and rats >5,000 mg/kg (Hartley and Kidd, 1987).

Use: Nonsystemic acaricide.

IPRODIONE

Synonyms: Chipco 26019; **3-(3,5-Dichlorophenyl)-*N*-(1-methylethyl)-2,4-dioxo-1-imi-dazolidinecarboxamide**; Glycophen; Glycophene; 1-Isopropyl carbamoyl-3-(3,5-dichlo-rophenyl)hydantoin; LFA 2043; MRC 910; Promidione; Rop 500 F; Rovral; RP 26019.

Designations: CAS Registry Number: 36734-19-7; mf: $C_{13}H_{13}Cl_2N_3O_3$; fw: 330.17; RTECS; NI8870000.

Properties: Colorless, odorless crystals. Mp: 136°C; K_H: 3.3 × 10^{-8} atm · m³/mol at 20°C (approximate — calculated from water solubility and vapor pressure); log K_{oc}: 1.48; log K_{ow}: 3.10; S_o (g/L at 20°C): acetone (300), ethanol (25), methanol (25), methylene chloride (500), *N,N*-dimethylformamide (500); S_w: 13 mg/L at 20°C; vp: <10^{-6} mmHg at 20°C.

Soil properties and adsorption data

Soil	K_d (mL/g)	f_{oc} (%)	K_{oc} (mL/g)	pH
Loam	30.4	2.67	1,139	6.0

Source: U.S. Department of Agriculture, 1990.

Environmental Fate

Soil. Readily degrades in soil (half-life 20 to 160 days) releasing carbon dioxide 3,5-dichloroaniline (Walker, 1987) and (Hartley and Kidd, 1987; Worthing and Hance, 1991). The rate of degradation increases with repeated applications of this fungicide. In a clay loam, the half-life was 1 week. After the second and third applications, the half-lives were 5 and 2 days, respectively (Walker et al., 1986).

Plant. Translocation and uptake by potato plants were reported (Cayley and Hide, 1980). Iprodione is rapidly metabolized in plants to 3,5-dichloroaniline (Cayley and Hide, 1980; Hartley and Kidd, 1987).

Chemical/Physical. In an aqueous solution at pH 8.7, iprodione hydrolyzed to *N*-(3,5-dichloroanilinocarbonyl)-*N*-(isopropylaminocarbonyl)glycine (Belafdal et al., 1986). At pH 8.7, complete hydrolysis occurred after 14 hours (Cayley and Hide, 1980).

Gomez et al. (1982) studied the pyrolysis of iprodione in an helium atmosphere at 400–1,000°C. Decomposition began at 300°C producing isopropyl isocyanate and 3-(3,5-dichlorophenyl)hydantoin. Above 600°C, the hydantoin ring began to decompose forming the following products: 3-chloroaniline, 3,5-dichloroaniline, chlorinated benzenes and benzonitrile. From 800 to 1,000°C, the hydantoin ring was completed destroyed which led to the formation of aryl isocyanates, anilines and the corresponding diarylureas, namely 3-(3,5-dichlorophenyl)urea and 1-(3-chlorophenyl)-3-(3,5-dichlorophenyl)urea (Gomez et al., 1982).

Formulation Types: Suspension concentrate; hot fogging concentrate; wettable powder; dustable powder.

Toxicity: LC_{50} (96-hour) for rainbow trout 6.7 mg/L and bluegill sunfish 2.25 mg/L (Hartley and Kidd, 1987); acute oral LD_{50} for rats 3,500 mg/kg (Hartley and Kidd, 1987), 4,400 mg/kg (RTECS, 1985).

Use: Fungicide.

ISOFENPHOS

Synonyms: Amaze; Bay 92114; Baysra-12869; **2-((Ethoxy((1-methylethyl)amino)-phosphinothioyl)oxy)benzoic acid 1-methylethyl ester**; *O*-Ethyl *O*-(2-isopropoxycarbonyl)phenyl isopropylphosphoramidothioate; Isophenphos; Isopropyl salicylate *O*-ester with *O*-ethylisopropylphosphoramidothioate; 1-Methylethyl-2-((ethoxy((1-methylethyl)amino)phosphinothioyl)oxy)benzoate; Oftanol; SD 40; SRA 12869.

$$(CH_3)_2CHNH \diagdown \underset{\underset{CH_3CH_2O}{\diagup}}{\overset{\overset{S}{\overset{\|}{}}}{P}} - O - \underset{COOCH(CH_3)_2}{\bigcirc}$$

Designations: CAS Registry Number: 25311-71-1; mf: $C_{15}H_{24}NO_4PS$; fw: 345.40; RTECS: VO4395500.

Properties: Colorless oil. Mp: −12°C; bp: 120°C at 0.01 mmHg; ρ: 1.134 at 20/4°C; H-t$_{1/2}$ (37°C): 263 days (pH 2), 525 days (pH 7), 52 hours (pH 11.5); K_H: 7.6 × 10^{-6} atm · m³/mol at 20°C; log K_{oc}: 2.73, 2.79; log K_{ow}: 3.30, 4.12; S_o: soluble in acetone, benzene, cyclohexanone, ethanol, ethyl ether, methylene chloride; S_w: 23.8 mg/L at 20°C; vap d: 14.12 g/L at 25°C, 11.97 (air = 1); vp: 4 × 10^{-6} mmHg at 20°C.

Soil properties and adsorption data

Soil	K_d (mL/g)	f_{oc} (%)	K_{oc} (mL/g)	pH
Sandy loam	10.3	1.68	613	2.5
Silt loam	5.6	1.04	538	2.6

Source: U.S. Department of Agriculture, 1990.

Environmental Fate

Soil. Rapidly degraded by microbes via oxidative desulfuration in soils forming isofenphos oxon (Abou-Assaf et al., 1986; Abou-Assaf and Coats, 1987; Somasundarum et al., 1989), isopropyl salicylate and carbon dioxide (Somasundaram et al., 1989). The formation of isofenphos oxon is largely dependent upon the pH, moisture and temperature of the soil. The degradation rate of isofenphos decreased with a decrease in temperature (35°C >25°C >15°C), moisture content (22.5% >30% >15%) and in acidic and alkaline soils (pH 6 and 8 >pH 7). After isofenphos was applied to soil at a rate of 1,12 kg/ai/ha, concentrations of 8.3, 7.2, 5.1 and 1.0 ppm were found after 5, 21, 43 and 69 days, respectively. Following a second application, 4.9, 1.55, 0.25 and 0.10 ppm of isofenphos were found after 5, 21, 43 and 69 days, respectively (Abou-Assaf and Coats, 1987).

A pure culture of *Arthrobacter* sp. was capable of degrading isofenphos at different soil concentrations (10, 50 and 100 ppm) in less than 6 hours. In previously treated soils, isofenphos could be mineralized to carbon dioxide by indigenous microorganisms (Racke and Coats, 1987). Hydrolyzes in soil to salicylic acid (Somasundaram et al., 1991).

Plant. Two weeks following application to thatch, 65% of the applied amount was present (Sears et al., 1987).

Surface Water. In estuarine water, the half-life of isofenphos ranged from 9.8 to 11.9 days (Lacorte et al., 1995).

Photolytic. Irradiation of an isofenphos (500 mg) in hexane and methanol (100 mL) using a high pressure mercury lamp (λ = 254–360 nm) for 24 hours yielded the following products: isofenphos oxon and *O*-ethyl hydrogen-*N*-isopropylphosphoroamidothioate and its methylated product and *O*-ethyl hydrogen-*N*-isopropylphosphoramidate and its methylated product. Photolysis products identified following the irradiation (λ = 253 nm) of a sandy loam soil for 2 days were isofenphos oxon, isopropyl salicylic acid and *O*-ethyl hydrogen-*N*-isopropylphosphoramidate. When isofenphos was uniformly coated on borosilicate glass and irradiated using a low pressure mercury lamp for 5 days, the following products formed: isofenphos oxon, isopropyl salicylic acid and *O*-ethyl hydrogen-*N*-isopropylphosphoramidothioate (Dureja et al., 1989).

Formulation Types: Emulsifiable concentrate; wettable powder; granules; seed treatment.

Toxicity: LC_{50} (96-hour) for carp 2–4 mg/L, orfe 1–2 mg/L, goldfish 2 mg/L and rudd 1 mg/L (Hartley and Kidd, 1987); acute oral LD_{50} for rats 28.0–38.7 mg/kg (Hartley and Kidd, 1987).

Uses: Isofenphos is an organophosphorus insecticide used in soil to control leaf-eating and soil-dwelling pests in vegetables, fruits, turf and field crops.

KEPONE

Synonyms: Chlordecone; CIBA 8514; Compound 1189; 1,2,3,5,6,7,8,9,10,10-Decachloro[5.2.1.0^{2,6}.0^{3,9}.0^{5,8}]decano-4-one; Decachloroketone; Decachloro-1,3,4-metheno-2H-cyclobuta[cd]pentalen-2-one; Decachlorooctahydrokepone-2-one; Decachlorooctahydro-1,3,4-methene-2H-cyclobuta[cd]pentalen-2-one; **1,1a,3,3a,4,5,5a,5b,6-Decachlorooctahydro-1,3,4-metheno-2H-cyclobuta-[cd]pentalen-2-one**; Decachloropentacyclo-[5.2.1.0^{2,6}.0^{3,9}.0^{5,8}]decan-3-one; Decachloropentacyclo-[5.2.1.0^{2,6}.0^{4,10}.0^{5,9}]decan-3-one; Decachlorotetracyclodecanone; Decachlorotetrahydro-4,7-methanoindeneone; ENT 16391; GC 1189; General chemicals 1189; Merex; NA 2761; NCI-C00191; RCRA waste number U142.

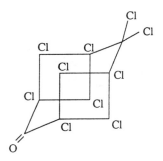

Designations: CAS Registry Number: 143-50-0; DOT: 2761; mf: $C_{10}Cl_{10}O$; fw: 490.68; RTECS: PC8575000.

Properties: Odorless, white to tan crystals. Mp: sublimes; bp: 350°C (decomposes); K_H: 3.11×10^{-2} atm · m^3/mol at 24–25°C (approximate — calculated from water solubility and vapor pressure); log K_{oc}: 4.74 (calculated); log K_{ow}: 4.07 (calculated); S$_o$: acetic acid (10–50 mg/mL at 20°C), benzene, corn oil, dimethylsulfoxide (>100 mg/mL at 20°C), 95% ethanol (>100 mg/mL at 20°C), ketones, light petroleum and fats; S$_w$: 7,600 µg/L at 24°C (note: solubility increases with pH); vp: 2.25×10^{-7} mmHg at 25°C.

Environmental Fate

Photolytic. Kepone-contaminated soils from a site in Hopewell, VA were analyzed by GC/MS. 8-Chloro and 9-chloro homologs identified suggested these were photodegradation products of kepone (Borsetti and Roach, 1978). Products identified from the photolysis of kepone in cyclohexane were 1,2,3,4,6,7,9,10,10-nonachloro-5,5-dihydroxypentacyclo[5.3.0.0^{2,6}.0^{3,9}.0^{4,8}]decane for the hydrate and 1,2,3,4,6,7,9,10,10-nonachloro-5,5-dimethoxypentacyclo[5.3.0.0^{2,6}.0^{3,9}.0^{4,8}]decane (Alley et al., 1974).

Chemical/Physical. Readily reacts with moisture forming hydrates (Hollifield, 1979). Decomposes at 350°C (Windholz et al., 1983) probably emitting toxic chlorine fumes.

Exposure Limits: NIOSH REL: TWA 1 µg/m^3.

Symptoms of Exposure: May cause tremors.

Toxicity: LC_{50} (96-hour) for sunfish 0.14 ppm, trout 0.02 ppm, bluegill sunfish 0.051, rainbow trout 0.036 ppm (Verschueren, 1983); LC_{50} (48-hour) for sunfish 0.27 ppm, trout 38 ppb (Verschueren, 1983); LC_{50} (24-hour) for bluegill sunfish 257 ppb, trout 66 ppb, bluegill sunfish 257 ppb and rainbow trout 156 ppb (Verschueren, 1983); acute oral LD_{50} for male and female rats 132 and 126 mg/kg, respectively (Verschueren, 1983), 95 mg/kg (RTECS, 1985).

Uses: Insecticide; fungicide.

LINDANE

Synonyms: Aalindan; Aficide; Agrisol G-20; Agrocide; Agrocide 2; Agrocide 6G; Agrocide 7; Agrocide III; Agrocide WP; Agronexit; Ameisenatod; Ameisenmittel merck; Aparasin; Aphtiria; Aplidal; Arbitex; BBX; Benhex; Bentox 10; Benzenehexachloride; Benzene-γ-hexachloride; γ-Benzenehexachloride; Bexol; BHC; γ-BHC; Celanex; Chloran; Chloresene; Codechine; DBH; Detmol-extrakt; Detox 25; Devoran; Dol granule; ENT 7796; Entomoxan; Exagama; Forlin; Gallogama; Gamacid; Gamaphex; Gamene; Gamiso; Gammahexa; Gammalin; Gammexene; Gammopaz; Gexane; HCCH; HCH; γ-HCH; Heclotox; Hexa; γ-Hexachlor; Hexachloran; γ-Hexachloran; Hexachlorane; γ-Hexachlorane; γ-Hexachlorobenzene; 1,2,3,4,5,6-Hexachlorocyclohexane; **1α,2α,3β,4α,5α,6β-Hexachlorocyclohexane**; 1,2,3,4,5,6-Hexachloro-γ-cyclohexane; γ-Hexachlorocyclohexane; γ-1,2,3,4,5,6-Hexachlorocyclohexane; Hexatox; Hexaverm; Hexicide; Hexyclan; HGI; Hortex; Inexit; γ-Isomer; Isotox; Jacutin; Kokotine; Kwell; Lendine; Lentox; Lidenal; Lindafor; Lindagam; Lindagrain; Lindagranox; γ-Lindane; Lindapoudre; Lindatox; Lindosep; Lintox; Lorexane; Milbol 49; Mszycol; NA 2761; NCI-C00204; Neoscabicidol; Nexen FB; Nexit; Nexit-stark; Nexol-E; Nicochloran; Novigam; Omnitox; Ovadziak; Owadziak; Pedraczak; Pflanzol; Quellada; RCRA waste number U129; Silvanol; Spritz-rapidin; Spruehpflanzol; Streunex; Tap 85; TBH; Tri-6; Viton.

Designations: CAS Registry Number: 58-89-9; DOT: 2761; mf: $C_6H_6Cl_6$; fw: 290.83; RTECS: GV4900000.

Properties: Colorless, odorless to slightly musty solid with a bitter taste. Mp: 112.5°C; bp: 323.4°C; ρ: 1.5691 and 1.87 at 23.6/4 and 20/4°C, respectively; fl p: nonflammable; H-$t_{1/2}$: 206 days at 25°C and pH 7; K_H: 1.99×10^{-6} atm · m³/mol at 23°C (Fendinger and Glotfelty, 1988); log BCF: 1.92 (pink shrimp), 2.34 (pin fish), 1.80 (grass shrimp), 2.69 (sheepshead minnow) (Schimmel et al., 1977), freshwater clam (*Corbicula manilensis*) 3.42 (Hartley and Johnston, 1983), 2.08 mussel (*Mytilus edulis*) (Renberg et al., 1985); log K_{oc}: 2.38–3.52; log K_{ow}: 3.20–3.89; P-$t_{1/2}$: 103.95 hours (absorbance λ ≤300 nm, concentration on glass plates = 6.7 µg/cm²); S_o (wt % at 20°C unless noted otherwise): acetone (30.31), benzene (22.42), chloroform (19.35), cod liver oil (58.0, 69.8 and 82.0 g/L at 4, 12 and 20°C, respectively), ethyl ether (17.22), ethanol (6.02), hexane (12.6 g/L at 20°C) (Chiou et al., 1985), octanol (29.1, 41.1 and 52.9 g/L at 4, 12 and 20°C, respectively), triolein (52.9, 68.2 and 78.3 g/L at 4, 12 and 20°C, respectively); S_w: 7.52 ppm at 25°C, 17.0 mg/L at 24°C; vap d: 518 ng/L at 20°C; vp: 9.4×10^{-6} mmHg at 20°C.

Soil properties and adsorption data

Soil	K$_d$ (mL/g)	f$_{oc}$ (%)	K$_{oc}$ (mL/g)	pH (meq/100 g)	CEC
Brookston sandy loam	22.7	2.10	1,081	—	—
Ca-Bentonite (20°C)	2.9	—	—	—	—
Ca-Bentonite (30°C)	2.7	—	—	—	—
Ca-Staten peaty muck (10°C)	377.0	12.76	2,954	—	—
Ca-Staten peaty muck (20°C)	331.0	12.76	2,594	—	—
Ca-Staten peaty muck (30°C)	269.0	12.76	2,108	—	—
Ca-Staten peaty muck (40°C)	197.0	12.76	1,544	—	—
Catlin	10.9	2.01	44	6.20	—
Ca-Venado clay (10°C)	53.8	—	—	—	—
Ca-Venado clay (20°C)	45.7	—	—	—	—
Ca-Venado clay (30°C)	41.3	—	—	—	—
Ca-Venado clay (40°C)	36.1	—	—	—	—
Clay loam	9.8	1.42	690	5.91	12.4
Clay loam	79.4	10.40	760	4.89	35.0
Commerce	49.5	0.68	7,300	6.70	—
Fox sandy loam	17.3	1.84	940	—	—
Honeywood loam	20.4	1.67	1,222	—	—
Kanuma high clay	1.1	1.35	85	5.70	—
Light clay	20.0	1.51	1,300	5.18	13.2
Light clay	37.4	3.23	1,200	5.26	28.3
Sandy loam	75.8	7.91	960	5.51	26.3
Silica gel (20°C)	6.9	—	—	—	—
Silica gel (30°C)	4.6	—	—	—	—
Tracy	10.1	1.12	901	6.20	—
Tsukuba clay loam	20.0	4.24	398	6.50	—
Udalf	7.7	0.76	1,010	7.45	—

Source: Kay and Elrick, 1967; Mills and Biggar, 1969; McCall et al., 1981; Rippen et al., 1982; Kanazawa, 1989; Kishi et al., 1990.

Environmental Fate

Biological. In a laboratory experiment, a strain of *Pseudomonas putida* culture transformed lindane to γ-3,4,5,6-tetrachlorocyclohexane (γ-TCCH), γ-pentachlorocyclohexane (γ-PCCH) and α-BHC (Benezet and Matsumura, 1973). γ-TCCH was also reported as a product of lindane degradation by *Clostridium sphenoides* (MacRae et al., 1969; Heritage and MacRae, 1977, 1977a), an anaerobic bacterium isolated from flooded soils (MacRae et al., 1969; Sethunathan and Yoshida, 1973a). Lindane degradation by *Escherichia coli* also yielded γ-PCCH (Francis et al., 1975). Evidence suggests that degradation of lindane in anaerobic cultures or flooded soils amended with lindane occurs via reductive dehalogenation producing chlorine-free volatile metabolites (Sethunathan and Yoshida, 1973a).

After a 30-day incubation period, the white rot fungus *Phanerochaete chrysosporium* converted lindane to carbon dioxide. Mineralization began between the third and sixth day of incubation. The production of carbon dioxide was highest between 3 to 18 days of incubation, after which the rate of carbon dioxide produced decreased until the 30th day. It was suggested that the metabolism of lindane and other compounds, including *p,p'-*

DDT, TCDD and benzo[a]pyrene, was dependent on the extracellular lignin-degrading enzyme system of this fungus (Bumpus et al., 1985).

Beland et al. (1976) studied the degradation of lindane in sewage sludge under anaerobic conditions. Lindane underwent reductive hydrodechlorination forming 3,4,5,6-tetrachlorocyclohex-1-ene (γ-BTC). The amount of γ-BTC that formed reached a maximum concentration of 5% after 2 weeks. Further incubation with sewage sludge resulted in decreased concentrations. The evidence suggested that γ-BTC underwent further reduction affording benzene (Beland et al., 1976). Hill and McCarty (1967) reported that lindane (1 ppm) in sewage sludge completely degraded in 24 hours. Under anaerobic conditions, the rate of degradation was higher.

When lindane was incubated in aerobic and anaerobic soil suspensions for 3 weeks, 0 and 63.8% was lost, respectively (MacRae et al., 1984). Using settled domestic wastewater inoculum, lindane (5 and 10 mg/L) did not degrade after 28 days of incubation at 25°C (Tabak et al., 1981).

Soil. In moist soils, lindane biodegraded to γ-pentachlorocyclohexene (Elsner et al., 1972; Kearney and Kaufman, 1976; Fuhremann and Lichtenstein, 1980). Under anaerobic conditions, degradation by soil bacteria yielded γ-BTC and α-BHC (Kobayashi and Rittman, 1982). Other reported biodegradation products include penta- and tetrachloro-1-cyclohexanes and penta- and tetrachlorobenzenes (Moore and Ramamoorthy, 1984).

Lindane degraded rapidly in flooded rice soils (Raghu and MacRae, 1966). Incubation of lindane for 6 weeks in a sandy loam soil under flooded conditions yielded γ-TCCH, γ-2,3,4,5,6-pentachlorocyclohex-1-ene and small amounts of 1,2,4-trichlorobenzene, 1,2,3,4-tetrachlorobenzene, 1,2,3,5- and/or 1,2,4,5-tetrachlorobenzene (Mathur and Saha, 1975). Incubation of lindane in moist soil for 8 weeks yielded the following metabolites: γ-BTC, γ-1,2,3,4,5-pentachlorocyclohex-1-ene, pentachlorobenzene, 1,2,3,4-, 1,2,3,5- and/or 1,2,4,5-tetrachlorobenzene, 1,2,4-trichlorobenzene, 1,3,5-trichlorobenzene, *m*-and/or *p*-dichlorobenzene (Mathur and Saha, 1977). Microorganisms isolated from a loamy sand soil degraded lindane and some of the metabolites identified were pentachlorobenzene, 1,2,4,5-tetrachlorobenzene, 1,2,3,5-tetrachlorobenzene, γ-PCCH, γ-TCCH and β-3,4,5,6-tetrachloro-1-cyclohexane (β-TCCH) (Tu, 1976). γ-PCCH was also reported as a metabolite of lindane in an Ontario soil that was pretreated with *p,p'*-DDT, dieldrin, lindane and heptachlor (Yule et al., 1967). The reported half-life in soil is 266 days (Jury et al., 1987).

Indigenous microbes in soil partially degraded lindane to carbon dioxide (MacRae et al., 1967). In a 42-day experiment, ^{14}C-labeled lindane applied to soil-water suspensions under aerobic and anaerobic conditions gave $^{14}CO_2$ yields of 1.9 and 3.0%, respectively (Scheunert et al., 1987).

In a moist Hatboro silt loam, volatilization yields of 50 and 90% were found after 6 hours and 6 days, respectively. In a dry Norfolk sand loam, 12% volatilization was reported after 50 hours (Glotfelty et al., 1984).

The average half-lives for lindane in aerobic and flooded soils under laboratory conditions were 276 and 114 days, respectively (Mathur and Saha, 1977). In field soils, the half-lives for lindane ranged from 88 days to 3.2 years with an average half-life of 426 days (Lichtenstein and Schulz, 1959, 1959a; Lichtenstein et al., 1971; Voerman and Besemer, 1975; Mathur and Saha, 1977).

Groundwater. According to the U.S. EPA (1986), lindane has a high potential to leach to groundwater.

Plant. Lindane appeared to be metabolized by several grasses to hexa-chlorobenzene and α-BHC, the latter isomerizing to β-BHC (Steinwandter, 1978; Steinwandter and

Schluter, 1978). Oat plants were grown in two soils treated with [^{14}C]lindane. 2,4,5-Trichlorophenol and possibly γ-PCCH were identified in soils but no other compounds other than lindane were identified in the oat roots or tops (Fuhremann and Lichtenstein, 1980). The half-life of lindane in alfalfa was 3.3 days (Treece and Ware, 1965).

Surface Water. Lindane degraded in simulated lake impounds under aerobic (15%) and anaerobic (90%) conditions forming α-BHC with trace amounts of δ-BHC (Newland et al., 1969).

Photolytic. Photolysis of lindane in aqueous solutions gives β-BHC (U.S. Department of Health and Human Services, 1989). When an aqueous solution containing lindane was photooxidized by UV light at 90–95°C, 25, 50 and 75% degraded to carbon dioxide after 3.0, 17.4 and 45.8 hours, respectively (Knoevenagel and Himmelreich, 1976).

Chemical/Physical. In basic aqueous solutions, lindane dehydrochlorinates to form pentachlorocyclohexene, then to trichlorobenzenes. In a buffered aqueous solution at pH 8 and 5°C, the calculated hydrolysis half-life was determined to be 42 years (Ngabe et al., 1993). The hydrolysis rate constant for lindane at pH 7 and 25°C was determined to be 1.2×10^{-4}/hour, resulting in a half-life of 241 days (Ellington et al., 1987). In weakly basic media, lindane undergoes *trans*-dehydrochlorination of the axial chlorines to give the intermediate 1,3,4,5,6-pentachlorocyclohexane. This compound further reacts with water to give 1,2,4-trichlorobenzene, 1,2,3 trichlorobenzene and hydrochloric acid. Three molecules of the acid are produced for every molecule of lindane that reacts (Cremlyn, 1991; Kollig, 1993).

Exposure Limits: NIOSH REL: TWA 0.5 mg/m^3, IDLH 50 mg/m^3; OSHA PEL: TWA 0.5 mg/m^3; ACGIH TLV: TWA 0.5 mg/m^3.

Symptoms of Exposure: Irritates eyes, nose, throat and skin; headache, nausea, clonic convulsions, respiratory problems, cyanosis, aplastic anemia and muscle spasms.

Formulation Types: Emulsifiable concentrate; wettable powder; fumigant; suspension concentrate; granules; dustable powder.

Toxicity: EC$_{50}$ (24-hour) for *Daphnia magna* 14.5 mg/L (Lilius et al., 1995); LC$_{50}$ (48-hour) for guppies 0.16-0.3 mg/L (Hartley and Kidd, 1987); LC$_{50}$ (96-hour) for bullhead 64 μg/L, carp 90 μg/L, catfish 44 μg/L, coho salmon 41 μg/L, goldfish 131 μg/L, perch 68 μg/L, bass 32 μg/L, minnow 87 μg/L, bluegill sunfish 68 μg/L and rainbow trout 27 μg/L (Macek and McAllister, 1970); acute oral LD$_{50}$ for rats 88–91 mg/kg (Reuber, 1979), 76 mg/kg (RTECS, 1985).

Uses: Pesticide and insecticide.

LINURON

Synonyms: Afalon; Afalon inuron; Aphalon; Cephalon; 3-(3,4-Dichlorophenyl)-1-methoxymethylurea; 3-(3,4-Dichlorophenyl)-1-methoxy-1-methylurea; **N'-(3,4-Dichlorophenyl)-N-methoxy-N-methylurea**; N-(3,4-Dichlorophenyl)-N'-methyl-N'-methoxyurea; Du Pont 326; Du Pont herbicide 326; Garnitan; Herbicide 326; Hoe 2810; Linex 4L; Linorox; Linurex; Linuron 4L; Lorex; Lorox; Lorox DF; Lorox L; Lorox linuron weed killer; Methoxydiuron; 1-Methoxy-1-methyl-3-(3,4-dichlorophenyl)urea; Premalin; Sarclex; Scarclex; Sinuron.

Designations: CAS Registry Number: 330-55-2; mf: $C_9H_{10}Cl_2N_2O_2$; fw: 249.10; RTECS: YS9100000.

Properties: Colorless to white, odorless crystals or crystalline solid. The odor threshold concentration in air is >9.7 ppm (Young et al., 1996). Mp: 93–94°C; fl p: nonflammable; K_H: 6.1×10^{-8} atm · m³/mol at 20–25°C (approximate — calculated from water solubility and vapor pressure); log K_{oc}: 2.70–2.78; log K_{ow}: 2.19, 3.00; S_o (g/kg at 25°C): acetone (500), benzene (150), ethanol (150), heptane (150), xylene (130); S_w: 75–81 mg/L at 25°C; vp: 1.5×10^{-5} mmHg at 20°C.

Soil properties and adsorption data

Soil	K_d (mL/g)	f_{oc} (%)	K_{oc} (mL/g)	pH	Salinity	TOC (mg/L)	CEC[a]
Annandale	11.39	1.70	670	5.8	—	—	11.3
Annandale	12.00	1.70	706	5.9	—	—	11.3
Asquith sandy loam	6.90	1.02	676	7.5	—	—	—
Begbroke sandy loam	10.80	1.90	568	7.1	—	—	—
Bermudian	13.70	1.60	856	6.0	—	—	13.2
Bermudian	14.01	1.60	876	6.0	—	—	13.2
Collington	14.18	2.60	545	5.0	—	—	12.8
Collington	15.05	2.60	579	4.7	—	—	12.8
Colts Neck	1.49	1.20	124	4.2	—	—	7.7
Colts Neck	12.00	1.20	1,000	4.2	—	—	7.7
Dark sandy loam	17.00	12.00	142	6.3	—	—	18.0
Dutchess	11.74	2.90	405	5.5	—	—	12.7
Dutchess	15.42	2.90	532	5.3	—	—	12.7
Great House sandy loam	73.00	12.00	608	6.3	—	—	18.0

Soil properties and adsorption data *(continued)*

Soil	K_d (mL/g)	f_{oc} (%)	K_{oc} (mL/g)	pH	Salinity	TOC (mg/L)	CEC[a]
Indian Head clay loam	17.80	2.34	761	7.8	—	—	—
Lakewood	10.82	0.50	2,163	3.5	—	—	1.8
Lakewood	13.39	0.50	2,678	3.6	—	—	1.8
Melfort loam	97.70	6.05	1,615	5.9	—	—	—
Patuxent River, MD	—	—	6,760	—	13.5	49.0	—
Patuxent River, MD	—	—	6,210	—	14.5	52.5	—
Regina heavy clay	18.20	2.39	762	7.7	—	—	—
Rosemaunde sandy clay loam	35.00	1.76	1,989	6.7	—	—	14.0
Sandy loam	11.40	1.93	591	7.1	—	—	11.0
Sassafras	12.00	2.00	600	5.2	—	—	7.7
Sassafras	10.29	2.00	514	5.2	—	—	7.7
Squires	9.79	1.70	576	6.5	—	—	7.0
Squires	11.39	1.70	670	6.6	—	—	7.0
Toll Farm HP	63.00	11.70	538	7.4	—	—	41.0
Trawscoed silty clay loam	50.00	3.69	1,355	6.2	—	—	12.0
Washington	13.39	2.40	558	6.0	—	—	11.2
Washington	13.39	2.40	558	6.2	—	—	11.2
Weed RES. sandy loam	47.00	1.93	2,435	7.1	—	—	11.0
Weyburn Oxbow loam	19.10	3.72	513	6.5	—	—	—
Whippany	8.31	1.90	437	5.6	—	—	9.4
Whippany	8.90	1.90	479	5.6	—	—	9.4

Source: Hance, 1965; Grover and Hance, 1969; MacNamara and Toth, 1970; Grover, 1975.
a) meq/100 g.

Environmental Fate

Soil. Linuron degraded in soil forming the common metabolite 3,4-dichloroaniline (Duke et al., 1991). In an aerobic, biologically active, organic-rich, pond sediment, linuron was converted to the intermediate 3-(3-chlorophenyl)-1-methoxymethylurea. This compound further degraded to unidentified compounds (Stepp et al., 1985).

Linuron was degraded by *Bacillus sphaericus* in soil forming *N,O*-dimethylhydroxylamine and carbon dioxide (Engelhardt et al., 1972). Only 1 ppm 3,4-dichloroaniline was identified in soils after incubation of soils containing 500 ppm linuron (Belasco and Pease, 1969). Boerner (1967) reported that the soil microorganism *Aspergillus niger* degraded linuron to phenylmethylurea, phenylmethoxyurea, chloroaniline, ammonia and carbon dioxide. In an earlier study, Boerner (1965) found 3,4-chloroaniline as a soil metabolite but the microorganism(s) were not identified.

The half-lives for linuron in soil incubated in the laboratory under aerobic conditions ranged from 56 to 88 days with an average of 75 days (Moyer et al., 1972; Hance, 1974; Usorol and Hance, 1974). In field soils, the average half-life for linuron was 88 days (Smith and Edmond, 1975).

Plant. Undergoes demethylation and demethoxylation in plants (Hartley and Kidd, 1987). Metabolites identified in carrots 117 days after treatment were 3,4-dichlorophenylurea, 3-(3,4-dichlorophenyl)-1-methylurea and 3,4-dichloroaniline. About 87% of the linuron remained unreacted (Loekke, 1974).

Photolytic. When an aqueous solution of linuron was exposed to summer sunlight for 2 months, 3-(3-chloro-4-hydroxyphenyl)-1-methoxy-1-methylurea, 3,4-dichlorophenylurea and 3-(3,4-dichlorophenyl)-1-methylurea formed at yields of 13, 10 and 2%, respectively. The photolysis half-life of this reaction was approximately 97 days (Rosen et al., 1969). In a more recent study, Tanaka et al. (1985) studied the photolysis of linuron (75 mg/L) in aqueous solution using UV light (λ = 300 nm) or sunlight. After 24 days of exposure to sunlight, linuron degraded to a trichlorinated biphenyl (1% yield) with the concomitant loss of hydrogen chloride (Tanaka et al., 1985).

Chemical/Physical. Linuron can be hydrolyzed to an aromatic amine by refluxing in an alkaline medium (Humburg et al., 1989). The hydrolysis half-life of linuron in 0.5 N sodium hydroxide solution at 20°C is one day (El-Dib and Aly, 1976).

Symptoms of Exposure: May irritate eyes, nose, skin and throat.

Formulation Types: Suspension concentrate; emulsifiable concentrate; flowable powder (50%); liquid concentrate (4 lb/gal); wettable powder.

Toxicity: LC_{50} (96-hour) for rainbow trout and bluegill sunfish 16 mg/L (Worthing and Hance, 1991); acute oral LD_{50} for rats approximately 1,500 mg/kg (Ashton and Monaco, 1991), 1,146 mg/kg (RTECS, 1985).

Uses: Selective preemergence and postemergence herbicide used on a wide variety of food crops to control many annual broad-leaved and grass weeds.

MALATHION

Synonyms: American Cyanamid 4049; *S*-1,2-Bis(carbethoxy)ethyl-*O,O*-dimethyl dithiophosphate; *S*-1,2-Bis(ethoxycarbonyl)ethyl-*O,O*-dimethyl phosphorodithioate; *S*-1,2-Bis(ethoxycarbonyl)ethyl-*O,O*-dimethyl thiophosphate; Calmathion; Carbethoxy malathion; Carbetovur; Carbetox; Carbofos; Carbophos; Celthion; Chemathion; Cimexan; Compound 4049; Cython; Detmol MA; Detmol MA 96%; *S*-1,2-Dicarbethoxyethyl-*O,O*-dimethyl dithiophosphate; Dicarboethoxyethyl-*O,O*-dimethyl phosphorodithioate; 1,2-Di(ethoxycarbonyl)ethyl-*O,O*-dimethyl phosphorodithioate; *S*-1,2-Di(ethoxycarbonyl)ethyl dimethyl phosphorothiolothionate; Diethyl (dimethoxyphosphinothioylthio) butanedioate; Diethyl (dimethoxyphosphinothioylthio) succinate; Diethyl mercaptosuccinate, *O,O*-dimethyl phosphorodithioate; Diethyl mercaptosuccinate, *O,O*-dimethyl thiophosphate; Diethyl mercaptosuccinic acid *O,O*-dimethyl phosphorodithioate; **((Dimethoxyphosphinothioyl)thio)butanedioic acid diethyl ester**; *O,O*-Dimethyl-*S*-1,2-bis(ethoxycarbonyl)ethyldithiophosphate; *O,O*-Dimethyl-*S*-(1,2-dicarbethoxyethyl)dithiophosphate; *O,O*-Dimethyl-*S*-(1,2-dicarbethoxyethyl)phosphorodithioate; *O,O*-Dimethyl-*S*-(1,2-dicarbethoxyethyl)thiothionophosphate; *O,O*-Di-methyl-*S*-1,2-di(ethoxycarbamyl)ethyl phosphorodithioate; *O,O*-Dimethyldithiophosphate dimethylmercaptosuccinate; EL 4049; Emmatos; Emmatos extra; ENT 17034; Ethiolacar; Etiol; Experimental insecticide 4049; Extermathion; Formal; Forthion; Fosfothion; Fosfotion; Fyfanon; Hilthion; Hilthion 25WDP; Insecticide 4049; Karbofos; Kopthion; Kypfos; Malacide; Malafor; Malakill; Malagran; Malamar; Malamar 50; Malaphele; Malaphos; Malasol; Malaspray; Malathion E50; Malathion LV concentrate; Malathion ULV concentrate; Malathiozoo; Malathon; Malathyl LV concentrate & ULV concentrate; Malatol; Malatox; Maldison; Malmed; Malphos; Maltox; Maltox MLT; Mercaptosuccinic acid diethyl ester; Mercaptothion; MLT; Moscardia; NA 2783; NCI-C00215; Oleophosphothion; Orthomalathion; Phosphothion; Prioderm; Sadofos; Sadophos; SF 60; Siptox I; Sumitox; Tak; TM-4049; Vegfru malatox; Vetiol; Zithiol.

$$CH_3O \diagdown \overset{\displaystyle S}{\underset{\displaystyle \diagup}{\underset{\displaystyle CH_3O}{P}}} - SCHCOOCH_2CH_3$$
$$\underset{\displaystyle CH_2COOCCH_2CH_3}{|}$$

NOTE: Brown et al. (1993) reported the following impurities in technical grade malathion concentrate (%): isomalathion (0.20, malaoxon (0.10), diethyl fumarate (0.90), *O,S,S*-trimethyl phosphorodithioate (0.003), *O,O,S*-trimethyl phosphorothioate (0.04), *O,O,S*-trimethyl phosphorodithioate (1.2), *O,O,O*-trimethyl phosphorothioate (0.45), diethyl hydroxysuccinate (0.05), ethyl nitrite (0.03), diethyl mercaptosuccinate (0.15) and diethyl methylthiosuccinate (1.0).

Designations: CAS Registry Number: 121-75-5; DOT: 2783; mf: $C_{10}H_{19}O_6PS_2$; fw: 330.36; RTECS: WM8400000.

Properties: Yellow to dark brown liquid with a garlic (technical) or mercaptan-like odor. Mp: 2.85°C; bp: 156–157°C at 0.7 mmHg (decomposes); ρ: 1.23 at 25/4°C; fl p: >162.8°C (open cup); H-t$_{1/2}$: ≈ 9 days at pH 6; K$_H$: 4.89 × 10^{-9} atm · m^3/mol at 25°C; log K$_{oc}$: 2.61; log K$_{ow}$: 2.36–2.89; P-t$_{1/2}$: 51.28 hours (absorbance λ = 210.0 nm, concentration on glass plates = 6.7 μg/cm^2), 15 hours in Suwannee River, FL at pH 4.7 under September sunlight (latitude 34° N); S$_o$: miscible with alcohols, chloroform, dimethylsulfoxide, esters, ethers, hexane, ketones, vegetable oils, aromatic and alkylated aromatic hydrocarbons; S$_w$: 141, 145 and 164 mg/L at 10, 20 and 30°C, respectively; vap d: 13.50 g/L at 25°C, 11.40 (air = 1); vp: 1.25 × 10^{-6} mmHg at 20°C.

Soil properties and adsorption data

Soil	K$_d$ (mL/g)	foc (%)	K$_{oc}$ (mL/g)	pH	CEC
Annandale	33.48	1.70	1,969	6.2	11.3
Annandale	76.96	1.70	4,527	6.1	11.3
Bermudian	40.63	1.60	2,540	6.4	13.2
Bermudian	66.92	1.60	4,183	6.4	13.2
Coarse sand (Jutland, Denmark)	0.45	0.21	214	5.3	2.2
Collington	36.51	2.60	1,404	5.7	12.8
Collington	56.67	2.60	2,180	5.6	12.8
Colts Neck	6.94	1.20	579	4.5	7.7
Colts Neck	8.86	1.20	739	4.8	7.7
Dutchess	47.14	2.90	1,626	5.7	12.7
Dutchess	66.92	2.90	2,308	5.8	12.7
Lakewood	4.23	0.50	847	4.7	1.8
Lakewood	4.29	0.50	857	4.5	1.8
Sandy loam (Jutland, Denmark)	0.75	0.15	500	6.4	7.3
Sassafras	24.48	2.00	1,224	5.3	7.7
Sassafras	26.70	2.00	1,335	5.1	7.7
Squires	36.51	1.70	2,148	6.4	7.0
Squires	64.07	1.70	3,769	6.6	7.0
Washington	16.67	2.40	694	6.0	11.2
Whippany	13.26	1.90	698	5.7	9.4
Whippany	16.67	1.90	877	5.7	9.4
Washington	34.44	2.40	1,435	6.1	11.2

Source: MacNamara and Toth, 1970; Kjeldsen et al., 1990.

Environmental Fate

Biological. Walker (1976) reported that 97% of malathion added to both sterile and nonsterile estuarine water was degraded after incubation in the dark for 18 days. Complete degradation was obtained after 25 days. Malathion degraded fastest in nonsterile soils and decomposed faster in soils that were sterilized by gamma radiation than in soils that were sterilized by autoclaving. After 1 day of incubation, the amounts of malathion degradation that occurred in autoclaved, irradiated and nonsterile soils were 7, 90 and 97%, respectively (Getzin and Rosefield, 1968). Degradation of malathion in organic-rich soils was 3 to 6 times higher than in soils not containing organic matter. The half-life in an organic-rich soil was about 1 day (Gibson and Burns, 1977). Malathion was degraded by soil micro-cosms isolated from an agricultural area on Kauai, HI. Degradation half-lives in the laboratory and field experiments were 8.2 and 2 hours, respectively. Dimethyl phosphorodithioic acid and diethyl fumarate were identified as degradation products (Miles and Takashima, 1991). Mostafa et al. (1972) found the soil fungi *Penicillium notatum, Aspergillus niger, Rhizoctonia solani, Rhizobium trifolii* and *Rhizobium leguminosarum* converted malathion to the following metabolites: malathion diacid, dimethyl phosphorothioate, dimethyl phosphorodithioate, dimethyl phosphate, monomethyl phosphate and thiophosphates. Malathion also degraded in groundwater and seawater but at a slower rate (half-life 4.7 days). Microorganisms isolated from paper-mill effluents were responsible for the formation of malathion monocarboxylic acid (Singh and Seth, 1989).

Paris et al. (1975) isolated a heterogenous bacterial population that was capable of degrading low concentrations of malathion to β-malathion monoacid. About 1% of the original malathion concentration degraded to malathion dicarboxylic acid, *O,O*-dimethyl phosphorodithioic acid and diethyl maleate. The major metabolite in soil is β-malathion monoacid (Paris et al., 1981). Rosenberg and Alexander (1979) demonstrated that two strains of *Pseudomonas* used malathion as the sole source of phosphorus. It was suggested that degradation of malathion resulted from an induced enzyme or enzyme system that catalytically hydrolyzed the aryl P-O bond, forming dimethyl phosphorothioate as the major product.

Matsumura and Bousch (1966) isolated carboxylesterase(s) enzymes from the soil fungus *Trichoderma viride* and a bacterium *Pseudomonas* sp. obtained from Ohio soil samples that were capable of degrading malathion. Compounds identified included diethyl maleate, desmethyl malathion, carboxyelsterase products, other hydrolysis products and unidentified metabolites. The authors found that these microbial populations did not have the capability to oxidize malathion due to the absence of malaoxon. However, the major degradative pathway appeared to be desmethylation and the formation of carboxylic acid derivatives.

Soil. In soil, malathion was degraded by *Arthrobacter* sp. to malathion monoacid, malathion dicarboxylic acid, potassium dimethyl phosphorothioate and potassium dimethyl phosphorodithioate. After 10 days, degradation yields in sterile and nonsterile soils were 8, 5, 19% and 92, 94, 81%, respectively (Walker and Stojanovic, 1974). Chen et al. (1969) reported that the microbial conversion of malathion to malathion monoacid was a result of demethylation of the *O*-methyl group. Malathion was converted by unidentified microorganisms in soil to thiomalic acid, dimethyl thiophosphoric acid and diethylthiomaleate (Konrad et al., 1969).

The half-lives for malathion in soil incubated in the laboratory under aerobic conditions ranged from 0.2 to 2.1 days with an average of 0.8 days (Konrad et al., 1969; Walker and Stojanovic, 1973; Gibson and Burns, 1977).

In a silt loam and sandy loam, reported R_f values were 0.88 and 0.90, respectively (Sharma et al., 1986).

Plant. When malathion on ladino clover seeds (10.9 ppm) was exposed to UV light (2537 Å) for 168 hours, malathion was the only residue detected. It was reported that 66.1% of the applied amount was lost due to volatilization (Archer, 1971).

Residues identified on field-treated kale besides malathion included the oxygen analog and the impurity identified by nuclear magnetic resonance and mass spectrometry as ethyl butyl mercaptosuccinate, *S*-ester with *O,O*-dimethyl phosphorodithioate. This compound did not form as an alteration product of malathion but was present in the 50% emulsifiable concentrate (Gardner et al., 1969).

Dogger and Bowery (1958) reported a half-life of malathion in alfalfa of 4.1 days. Foliar half-life of 0.3–4.9 days was reported by many investigators (Brett and Bowery,

1958; Dogger and Bowery, 1958; Waites and Van Middelem, 1958; Smith et al., 1960; Wheeler et al., 1967; Gardner et al., 1969; Polles and Vinson, 1969; Saini and Dorough, 1970; Nigg et al., 1981).

Surface Water. In raw river water (pH 7.3–8.0), 90% degraded within 2 weeks, presumably by biological activity (Eichelberger and Lichtenberg, 1971). The half-life of malathion in estuarine water ranged from 4.4 to 4.9 days (Lacorte et al., 1995).

Photolytic. Malathion absorbs UV light at wavelengths greater than 290 nm indicating direct photolysis should occur (Gore et al., 1971). When malathion was exposed to UV light, malathion monoacid, malathion diacid O,O-diethyl phosphorothioic acid, dimethyl phosphate and phosphoric acid were formed (Mosher and Kadoum, 1972).

In sterile water and river water, photolytic half-lives of 41.25 days and 16 hours, respectively, were reported (Archer, 1971).

Chemical/Physical. Hydrolyzes in water forming *cis*-diethyl fumarate, *trans*-diethyl fumarate (Suffet et al., 1967), thiomalic acid and dimethyl thiophosphate (Mulla et al., 1981). The reported hydrolysis half-lives at pH 7.4 at 20 and 37.5°C were 10.5 and 1.3 days, respectively (Freed et al., 1977).

Day (1991) reported that the hydrolysis products are dependent upon pH. In basic solutions, malathion hydrolyzes to diethyl fumarate and dimethyl phosphorodithioic acid (Bender, 1969; Day, 1991). Dimethyl phosphorothionic acid and 2-mercaptodiethyl succinate formed in acidic solutions (Day, 1991). The hydrolysis half-lives of malathion in a sterile 1% ethanol/water solution at 25°C and pH values of 4.5, 6.0, 7.0 and 8.0, were 18, 5.8, 1.7 and 0.53 weeks, respectively (Chapman and Cole, 1982). The reported hydrolysis half-lives at pH 7.4 at 20 and 37.5°C were 10.5 and 1.3 days, respectively. At 20°C and pH 6.1, the hydrolysis half-life is 120 days (Freed et al., 1979). Konrad et al. (1969) reported that after 7 days at pH values of 9.0 and 11.0, 25 and 100% of the malathion was hydrolyzed. Hydrolysis of malathion in acidic and alkaline (0.5 M sodium hydroxide) conditions gives $(CH_3O)_2P(S)Na$ and $(CH_3O)_2P(S)OH$ (Sittig, 1985).

Malaoxon and phosphoric acid were reported as ozonation products of malathion in drinking water (Richard and Bréner, 1984).

At 87°C and pH 2.5, malathion degraded in water to malathion α-monoacid and malathion β-monoacid. From the extrapolated acid degradation constant at 27°C, the half-life was calculated to be >4 years (Wolfe et al., 1977a). Under alkaline conditions (pH 8 and 27°C), malathion degraded in water to malathion monoacid, diethyl fumarate, ethyl hydrogen fumarate and O,O-dimethyl phosphorodithioic acid. At pH 8, the reported half-lives at 0, 27 and 40°C are 40 days, 36 and 1 hour, respectively (Wolfe et al., 1977a). However, under acidic conditions, it was reported that malathion degraded into diethyl thiomalate and O,O-dimethyl phosphorothionic acid (Wolfe et al., 1977a).

When applied as an aerial spray, malathion was converted to malaoxon and diethyl fumarate via oxidation and hydrolysis, respectively (Brown et al., 1993).

Emits toxic fumes of nitrogen and phosphorus oxides when heated to decomposition (Lewis, 1990). Products reported from the combustion of malathion at 900°C include carbon monoxide, carbon dioxide, chlorine, sulfur oxides, nitrogen oxides, hydrogen sulfide and oxygen (Kennedy et al., 1972a).

Exposure Limits: NIOSH REL: TWA 10 mg/m³, IDLH 250 mg/m³; OSHA PEL: TWA 15 mg/m³; ACGIH TLV: TWA 10 mg/m³.

Symptoms of Exposure: Miosis; eye, skin irritation; rhinorrhea; headache; tight chest, wheezing, laryngeal spasm; salivation; anorexia, nausea, vomiting, abdominal cramps; diarrhea, ataxia.

Formulation Types: Emulsifiable concentrate; wettable powder; aerosol; dustable powder; oil solutions.

Toxicity: EC_{50} (24-hour) for *Daphnia magna* 353 µg/L, *Daphnia pulex* 6.6 µg/L (Lilius et al., 1995); LC_{50} (96-hour) for bluegill sunfish 100 µg/L (Hartley and Kidd, 1987), largemouth bass 285 µg/L (Worthing and Hance, 1991), coho salmon 100 µg/L, brown trout 200 µg/L, channel catfish 9.0 mg/L, channel black bullhead 12.9 mg/L, fathead minnow 8.7 mg/L, rainbow trout 170 µg/L, perch 260 µg/L (Macek and McAllister, 1970), green sunfish 120 µg/L (Verschueren, 1983); LC_{50} (24-hour) for bluegill sunfish 120 ppb and rainbow trout 100 ppb (Verschueren, 1983); acute oral LD_{50} for rats 2,800 mg/kg (Hartley and Kidd, 1987), 370 mg/kg (RTECS, 1985).

Uses: Insecticide for control of sucking and chewing insects and spider mites on vegetables, fruits, ornamentals, field crops, greenhouses, gardens and forestry.

MALEIC HYDRAZIDE

Synonyms: Burtolin; Chemform; De-cut; 1,2-Dihydroxypyridazine-3,6-dione; **1,2-Dihydro-3,6-pyradizinedione**; 1,2-Dihydro-3,6-pyridizinedione; Drexel-super P; ENT 18870; Fair-2; Fair-30; Fairplus; Fair PS; 6-Hydroxy-2*H*-pyridazin-3-one; 6-Hydroxy-3(2*H*)-pyridazinone; KMH; MAH; Maintain 3; Maleic acid hydrazide; Maleic hydrazide 30%; Malein 30; *N,N*-Maleohydrazine; Malzid; Mazide; MH; MH 30; MH 40; MH 36 Bayer; RCRA waste number U148; Regulox; Regulox W; Regulox 50 W; Retard; Royal MH-30; Royal Slo-Gro; Slo-Grow; Sprout/Off; Sprout-Stop; Stunt-Man; Sucker-Stuff; Super-De-Sprout; Super Sprout Stop; Super Sucker-Stuff; Super Sucker-Stuff HC; 1,2,3,6-Tetrahydro-3,6-dioxopyridazine; Vondalhyde; Vondrax.

Designations: CAS Registry Number: 123-33-1; mf: $C_4H_4N_2O_2$; fw: 112.10; RTECS: UR5950000.

Properties: Colorless to white, odorless crystals. Mp: 292–298°C; bp: decomposes at 260°C; ρ: 1.60 at 25/4°C; pK_a: 5.62 at 20°C (monobasic acid); log K_{oc}: 1.56 (calculated); log K_{ow}: –1.96; S_o (g/kg at 25°C): *N,N*-dimethylformamide (24), acetone, ethanol and xylene (<1); S_w: 6 g/kg at 25°C.

Environmental Fate

Soil. The half-life in soil was reported to be 2–8 weeks (Hartley and Kidd, 1987). When maleic hydrazide was applied to muck, sand and clay at concentrations of 0.7 and 2.7, 1.0 and 3.75 and 0.85 and 3.4 ppm, 86 and 100, 87 and 100 and 47 and 67% degradation yields were obtained, respectively (Hoffman et al., 1962).

Plant. Major plant metabolites include fumaric, succinic and maleic acids (Hartley and Kidd, 1987).

Photolytic. In water, maleic hydrazide showed an absorption maximum at 300 nm indicating direct photolysis should occur (Gore et al., 1971).

Chemical/Physical. Reacts with alkalies and amines forming water-soluble salts (Hartley and Kidd, 1987). Decomposed by oxidizing acids releasing nitrogen (Worthing and Hance, 1991). Decomposes at 260°C (Windholz et al., 1983) releasing toxic fumes of nitrogen oxides (Sax and Lewis, 1987).

Formulation Types: Soluble concentrate; water-soluble granules.

Toxicity: LC_{50} (96-hour) for bluegill sunfish 1,608 mg/L, rainbow trout 1,435 mg/L (Hartley and Kidd, 1987), harlequin fish 125 mg/L (Verschueren, 1983); acute oral LD_{50} for rats >5,000 mg/kg (free acid) (Hartley and Kidd, 1987), 3,800 mg/kg (RTECS, 1985).

Uses: Growth regulator used to control the sprouting of potatoes and onions and to prevent sucker development on tobacco, fruits, ornamentals, vines, field crops and in forestry. Also used to control insects in warehouses, storerooms, empty sacks and in animal and poultry houses.

MANCOZEB

Synonyms: Carbamic acid ethylenebis(dithio-, manganese zinc complex); Carmazine; Dithane M 45; Dithane S 60; Dithane SPC; Dithane ultra; **((1,2-Ethanediylbis(carbamodithioato))(2-))manganese mixture with ((1,2-ethanediyl-bis(carbamodithioato))(2-))zinc**; F 2966; Fore; Green-daisen M; Karamate; Mancofol; Maneb-zinc; Manoseb; Manzate 200; Manzeb; Manzin; Manzin 80; Nemispor; Policar MZ; Policar S; Triziman; Triziman D; Zimanat; Zimaneb; Zimmandithane; Vondozeb.

$$[-\underset{\underset{S}{\|}}{S}CNHCH_2CH_2NH\underset{\underset{S}{\|}}{C}SMn-]_x (Zn^{2+})_y$$

Designations: CAS Registry Number: 8018-01-7; mf: $[C_4H_6N_2S_2Mn]_xZn_y$; fw: variable; RTECS: ZB3200000.

Properties: Grayish-yellow powder. Mp: 192–194°C (decomposes); fl p: 137.8°C (open cup); H-$t_{1/2}$ (20°C): 20 days (pH 5), 17 hours (pH 7), 34 hours (pH 9); log K_{oc}: 2.93–3.21 (calculated); log K_{ow}: 3.12–3.70 (calculated); S_o: slightly soluble in polar solvents; S_w: 6–20 mg/L.

Environmental Fate

Plant. Undergoes metabolism in plants to ethyl thiourea, ethylene thiuram disulfide, thiuram monosulfide and sulfur (Hartley and Kidd, 1987).

Chemical/Physical. Decomposes in acids releasing carbon disulfide. In oxygenated waters, mancozeb degraded to ethylene thiuram monosulfide, ethylene diisocyanate, ethylene thiourea, ethylenediamine and sulfur (Worthing and Hance, 1991).

Formulation Types: Suspension concentrate; wettable powder; dry seed treatment; dustable powder.

Toxicity: LC_{50} (48-hour) for carp 4.0 mg/L (Hartley and Kidd, 1987), catfish 5.2 mg/L, goldfish 9.0 mg/L (21°C) and rainbow trout 2.2 mg/L (17°C) (Worthing and Hance, 1991); acute oral LD_{50} for rats >8,000 mg/kg (Hartley and Kidd, 1987), mouse 600 mg/kg (RTECS, 1985).

Use: Fungicide used as a foliar or seed treatment to control a wide variety of pathogens in many field crops, fruits, ornamentals and vegetables.

MCPA

Synonyms: Agritox; Agroxon; Agroxone; Anicon Kombi; Anicon M; BH MCPA; Bordermaster; Brominal M & Plus; B-Selektonon M; Chiptox; 4-Chloro-*o*-cresolxyacetic acid; **4-Chloro-2-methylphenoxyacetic acid**; 4-Chloro-*o*-toloxyacetic acid; Chwastox; Cornox; Cornox-M; Dedweed; Dicopur-M; Dicotex; Dikotes; Dikotex; Dow MCP amine weed killer; Emcepan; Empal; Hedapur M 52; Hederax M; Hedonal M; Herbicide M; Hormotuho; Hornotuho; Kilsem; 4K-2M; Krezone; Legumex DB; Leuna M; Leyspray; Linormone; M 40; 2M-4C; 2M-4CH; MCP; 2,4-MCPA; Mephanac; Metaxon; Methoxone; 2-Methyl-4-chlorophenoxyacetic acid; 2M-4KH; Netazol; Okultin M; Phenoxylene 50; Phenoxylene Plus; Phenoxylene Super; Raphone; Razol dock killer; Rhomenc; Rhomene; Rhonox; Shamrox; Seppic MMD; Soviet technical herbicide 2M-4C; Trasan; U 46; U 46 M-fluid; Ustinex; Vacate; Vesakontuho MCPA; Verdone; Weedar; Weedar MCPA concentrate; Weedone; Weedone MCPA ester; Weedrhap; Zelan.

OCH$_2$COOH

Cl

Cl

Designations: CAS Registry Number: 94-74-6; mf: C$_9$H$_9$ClO$_3$; fw: 200.63; RTECS: AG1575000.

Properties: Colorless to light-brown solid. The reported odor threshold concentration in air and the taste threshold concentration in water are 460 and 4.1 ppb, respectively (Young et al., 1996). Mp: 120°C (pure), 99–107°C (technical); ρ: 1.56 at 25/15.5°C; pK$_a$: 3.05–3.13; log K$_{oc}$: 2.03–2.07 (calculated); log K$_{ow}$: 1.37–1.43 (calculated); S$_o$ (g/L at 25°C): ethanol (1,530), ethyl ether (770), *n*-heptane (5), toluene (62), xylene (49); S$_w$: 730–825 mg/L at 25°C, 270 g/L (sodium salt); vp: 1.5 × 10^{-6} mmHg at 20°C.

Environmental Fate

Biological. Cell-free extracts isolated from *Pseudomonas* sp. in a basal salt medium degraded MCPA to 4-chloro-*o*-cresol and glyoxylic acid (Gamar and Gaunt, 1971).

Soil. Residual activity in soil is limited to approximately 3–4 months (Hartley and Kidd, 1987).

Plant. The penetration, translocation and metabolism of radiolabeled MCPA in a cornland weed (*Galium aparine*) was studied by Leafe (1962). Carbon dioxide was identified as a metabolite but this could only account 7% of the applied MCPA. Though no additional compounds were identified, it was postulated that MCPA was detoxified in the weed via loss of both carbon atoms of the side chain (Leafe, 1962).

Photolytic. When MCPA in dilute aqueous solution was exposed to summer sunlight or an indoor photoreactor (λ >290 nm), 2-methyl-4-chlorophenol formed as the major product as well as *o*-cresol and 4-chloro-2-formylphenol (Soderquist and Crosby, 1975). Clapés et al. (1986) studied the photodecomposition of aqueous solution of MCPA (120 ppm, pH 5.4, 25°C) in a photoreactor equipped with a high pressure mercury lamp. After three minutes of irradiation, 4-chloro-2-methylphenol formed as an intermediate which degraded to 2-methylphenol. Both compounds were not detected after 6 minutes of irradiation; however, 1,4-dihydroxy-2-methylbenzene and 2-methyl-2,5-cyclohexadiene-1,4-dione formed as major and minor photodecomposition products, respectively. The same experiment was conducted using simulated sunlight (λ <300 nm) in the presence of riboflavin, a known photosensitizer. 4-Chloro-2-methylphenol and 4-chloro-2-methylbenzyl formate formed as major and minor photoproducts, respectively (Clapés et al., 1986). Ozone degraded MCPA in dilute aqueous solution with and without UV light (λ >300 nm) (Benoit-Guyod et al., 1986).

Chemical/Physical. Reacts with alkalies forming water soluble salts (Hartley and Kidd, 1987). Ozonolysis of MCPA in the dark yielded the following benzenoid intermediates: 4-chloro-2-methylphenol, its formate ester, 5-chlorosalicyaldehyde, 5-chlorosalicyclic acid and 5-chloro-3-methylbenzene-1,2-diol. Ozonolysis occurred much more rapidly in the presence of UV light. Based upon the intermediate compounds formed, the suggested degradative pathway in the dark included ring hydroxylation followed by cleavage of the ozone molecule and under irradiation, oxidation of the side chains by hydroxyl radicals (Benoit-Guyod et al., 1986).

Formulation Types: Aqueous solutions; emulsifiable concentrate; soluble concentrate; water-soluble powder.

Toxicity: LC_{50} (48-hour) for bluegill sunfish 100 ppm (Verschueren, 1983); LC_{50} (24-hour) for bluegill sunfish 1.5 mg/L and rainbow trout 117 mg/L (Worthing and Hance, 1991); acute oral LD_{50} for rats 700 mg/kg (RTECS, 1985).

Uses: Systemic postemergence herbicide used to control annual and perennial weeds in cereals, rice, flax, vines, peas, potatoes, asparagus, grassland and turf.

MEFLUIDIDE

Synonyms: *N*-(2,4-Dimethyl-5-((((trifluoromethyl)sulfonyl)amino)phenyl)acetamide; Embark; MBR 12325; Mowchem; 5′-(1,1,1-Trifluoromethanesulphonamido)acet-2′,4′-xylidide; VEL 3973; Vistar; Vistar herbicide.

Designations: CAS Registry Number: 53780-34-0; mf: $C_{11}H_{13}F_3N_2O_3S$; fw: 310.29; RTECS: AE2460000.

Properties: Colorless, odorless crystals. Mp: 183–185°C; fl p: nonflammable (water formulation); pK_a: 4.6; K_H: 1.7×10^{-7} atm · m³/mol at 23–25°C (approximate — calculated from water solubility and vapor pressure); log K_{oc}: 2.40 (calculated); log K_{ow}: 2.08 (calculated); S_o (g/L at 23°C): acetone (350), acetonitrile (64), benzene (0.31), ethyl acetate (50), ethyl ether (3.9), methanol (310), methylene chloride (2.1), *n*-octanol (17), petroleum benzene (0.002), xylene (0.12); S_w: 180 mg/L at 23°C; vp: $<7.5 \times 10^{-5}$ mmHg at 25°C.

Environmental Fate

Soil. Rapidly degrades in soil forming 5-amino-2,4-dimethyltrifluoromethane sulfone anilide as the major metabolite (half-life <1 week) (Hartley and Kidd, 1987).

Symptoms of Exposure: Mild eye irritant.

Formulation Types: Aqueous solution.

Toxicity: LC_{50} (96-hour) for bluegill sunfish and rainbow trout >100 mg/L (Hartley and Kidd, 1987); LC_{50} (4-day) for bluegill sunfish 1,600 ppm and trout >1,200 ppm (Humburg et al., 1989); acute oral LD_{50} for rats >4,000 mg/kg (Hartley and Kidd, 1987).

Uses: Plant growth regulator and herbicide used in lawns, turf, grassland, industrial areas and areas where grass cutting is difficult such as embankments. Also used to enhance sucrose content in sugarcane.

METALAXYL

Synonyms: Apron 2E; CGA 117; CGA 48988; N-(2,6-Dimethylphenyl)-N-methoxy-acetyl)alanine methyl ester; N-(2,6-Dimethylphenyl)-N-methoxyacetyl)-DL-alanine methyl ester; Metalaxil; **Methyl N-(2,6-dimethylphenyl)-N-(methoxyacetyl)-DL-alaninate**; Ridomil; Ridomil E; Subdue; Subdue 2E; Subdue 5SP.

Designations: CAS Registry Number: 57837-19-1; mf: $C_{15}H_{21}NO_4$; fw: 279.30; RTECS: AY6910000.

Properties: Colorless crystals. Mp: 71.8–72.3°C; ρ: 1.21 at 20/4°C; H-$t_{1/2}$ (20°C): >200 days (pH 1), 115 days (pH 9), 12 days (pH 10); K_H: 1.1×10^{-10} atm · m³/mol at 20°C (approximate — calculated from water solubility and vapor pressure); log K_{oc}: 1.53–1.84; log K_{ow}: 1.52; S_o (g/L at 20°C): benzene (550), methanol (650), methylene chloride (750), n-octanol (130), 2-propanol (270); S_w: 7.1 g/L at 20°C; vp: 2.20×10^{-6} mmHg at 20°C.

Soil properties and adsorption data

Soil	K_d (mL/g)	f_{oc} (%)	K_{oc} (mL/g)	pH
Loamy sand	0.43	1.28	34	7.8
Sand	0.48	0.70	69	6.3
Sandy loam	1.40	3.25	43	6.7
Silt loam	0.87	2.09	42	6.1
Soil #1 (0–30 cm)	3.47	1.46	238	6.0
Soil #1 (30–50 cm)	17.84	0.93	1,918	6.1
Soil #1 (50–91 cm)	11.93	1.24	962	6.1
Soil #1 (91–135 cm)	2.17	0.70	310	6.4
Soil #1 (135–165 cm)	1.51	0.17	889	6.5
Soil #1 (>165 cm)	1.64	0.07	2,343	6.7
Soil #2 (0–27 cm)	3.06	1.50	204	5.9
Soil #2 (27–53 cm)	3.15	0.16	1,971	6.3
Soil #2 (53–61 cm)	4.89	0.14	3,490	6.4
Soil #2 (61–81 cm)	8.53	0.28	3,046	6.4
Soil #2 (81–157 cm)	13.30	0.50	2,660	6.0
Soil #2 (>157 cm)	13.95	1.30	1,073	5.9

Source: U.S. Department of Agriculture, 1990; Sukop and Cogger, 1992.

Environmental Fate

Soil. Little information is available on the degradation of metalaxyl in soil; however, Sharom and Edgington (1986) reported metalaxyl acid as a possible metabolite. Repeated applications of metalaxyl decreases its persistence. Following an initial application, the average half-life was 28 days. After repeated applications, the half-life decreased to 14 days (Bailey and Coffey, 1985).

Carsel et al. (1986) studied the persistence of metalaxyl in various soil types. The application rate was 2.2 kg/ha. In a fine sand, metalaxyl concentrations at soil depths of 15, 20, 45 and 60 cm were 100, 150, 100 and 75 ppb, respectively, 55 days after application and 35, 90, 60 and 0 ppb, 85 days after application. In a fine sandy loam (upper 15 cm), metalaxyl concentrations detected after 15, 106 and 287 days after application were 1.65, 1.25 and 0.2 ppb, respectively.

Plant. In plants, metalaxyl undergoes ring oxidation, methyl ester hydrolysis, ether cleavage, ring methyl hydroxylation and N-dealkylation (Owen and Donzel, 1986). Metalaxyl acid was identified as a hydrolysis product in both sunflower leaves and lettuce treated with the fungicide. Metalaxyl acid was further metabolized to form a conjugated, benzyl alcohol derivative of metalaxyl (Businelli et al., 1984; Owen and Donzel, 1986).

In pigeon peas, metalaxyl may persist up to 12 days (Indira et al., 1981; Chaube et al., 1984).

Formulation Types: Emulsifiable concentrate; wettable powder; granules; seed treatment.

Toxicity: LC_{50} (96-hour) for rainbow trout, carp and bluegill sunfish >100 mg/L (Hartley and Kidd, 1987); acute oral LD_{50} for rats 669 mg/kg (RTECS, 1985).

Use: Systemic fungicide used to control a variety of diseases on a wide range of temperate, subtropical and tropical crops.

METALDEHYDE

Synonyms: Acetaldehyde homopolymer; Acetaldehyde polymer; Metacetaldehyde; **2,4,6,8-Tetramethyl-1,3,5,7-tetraoxacyclooctane.**

Designations: CAS Registry Number: 9002-91-9; mf: $C_8H_{16}O_4$; fw: 176.20.

Properties: Colorless, odorless crystals. Mp: 246°C (sealed tube), sublimes at 110–120°C; S_o: soluble in benzene and chloroform but only sparingly soluble in ethyl ether and ethanol; S_w: 200 mg/L at 17°C; vp: $<7.5 \times 10^{-5}$ mmHg at 25°C.

Environmental Fate

Plant. When applied to citrus rinds, 50% was lost after 4.6 days for the first 33 days and an additional 25% was lost 14 days for the subsequent 26 days (Iwata et al., 1982).

Chemical/Physical. Metaldehyde can be converted to acetaldehyde by heating to 150°C for 4–5 hours or by the reaction of concentrated hydrochloric acid (6 M) for a couple of minutes (Booze and Oehme, 1985).

Toxicity: Acute oral LD_{50} for dogs 0.6–1.0 g/kg (Worthing and Hance, 1991).

Use: Molluscicide.

281

METHAMIDOPHOS

Synonyms: Acephate-met; Bay 71628; Bayer 71268; Chevron 9006; Chevron ortho 9006; *O,S*-Dimethyl ester amide of aminothioate; *O,S*-Dimethyl phosphoramidothioate; ENT 27396; Hamidop; Metamidofos estrella; Monitor; MTD; NSC 190987; Ortho 9006; **Phosphoramidothioic acid *O,S*-dimethyl ester**; Pillaron; SRA 5172; Tahmabon; Tamaron.

$$CH_3O \quad \overset{O}{\underset{}{\underset{CH_3S}{\diagup}} \overset{\|}{P} - NH_2}$$

Designations: CAS Registry Number: 10265-92-6; mf: $C_2H_8NO_2PS$; fw: 141.13; RTECS: TB4970000.

Properties: Colorless crystals. Mp: 44.5°C; ρ: 1.31 at 44.5/4°C; H-$t_{1/2}$: 120 hours at 37°C and pH 9, 140 hours at 40°C and pH 2; log K_{oc}: 0.70; log K_{ow}: -0.79; S_o: soluble in ethanol, 2-propanol, methylene chloride; S_w: miscible; vp: 3.2×10^{-5} mmHg at 20°C.

Soil properties and adsorption data

Soil	K_d (mL/g)	f_{oc} (%)	K_{oc} (mL/g)	pH
Silt loam	0.12	2.32	5	6.1

Source: *U.S.* Department of Agriculture, 1990.

Environmental Fate

Chemical/Physical. Emits toxic fumes of phosphorus, nitrogen and sulfur oxides when heated to decomposition (Sax and Lewis, 1987).

Formulation Types: Emulsifiable concentrate; soluble concentrate.

Toxicity: LC_{50} (96-hour) for rainbow trout 51 mg/L, guppies 46 mg/L, carp and goldfish approximately 100 mg/L (Hartley and Kidd, 1987); acute oral LD_{50} for rats approximately 30 mg/kg (Hartley and Kidd, 1987), 7,500 µg/kg (RTECS, 1985).

Uses: Insecticide and acaricide.

METHIDATHION

Synonyms: Ciba-Geigy CS 13005; *S*-(2-Dihydro-5-methoxy-2-oxo-1,3,4-thiadiazol-3-methyl)dimethyl phosphorothiolothionate; *S*-2-Dihydro-5-methoxy-2-oxo-1,3,4-thiadiazol-3-ylmethyl *O*,*O*-dimethyl phosphorodithionate; *O*,*O*-Dimethyl *S*-(5-methoxy-1,3,4-thiadiazolinyl-3-methyl) dithiophosphate; *O*,*O*-Dimethyl *S*-(2-methoxy-1,3,4-thiadiazol-5(4*H*)-onyl-4-methyl) phosphorodithioate; DMTP; ENT 27193; Fisons NC 2964; Geigy 13005; Geigy GS-13005; GS 13005; Methidathion 50S; *S*-((5-Methoxy-2-oxo-1,3,4-thiadiazol-3(2*H*)-yl)methyl) *O*,*O*-dimethyl phosphorodithioate; **Phosphorodithioic acid *O*,*O*-dimethyl ester, *S*-ester with 4-(mercaptomethyl)-2-methoxy-Δ2-1,3,4-thiadiazolin-5-one**; Somonil; Surpracide; Ultracide.

Designations: CAS Registry Number: 950-37-8; mf: $C_6H_{11}N_2O_4PS_3$; fw: 302.31; RTECS: TE2100000.

Properties: Colorless crystals. Mp: 39–40°C; ρ: 1.495 at 20/4°C; H-$t_{1/2}$: 30 minutes at 25°C and pH 13; K_H: 2.2×10^{-9} atm · m³/mol at 20°C; log K_{oc}: 2.29–2.76; log K_{ow}: 2.22, 2.42; S_o (g/kg at 20°C): acetone (690), cyclohexanone (850), ethanol (260), xylene (600); S_w: 250 mg/L at 20°C, 240 mg/L at 25°C; vp: 1.40×10^{-6} mmHg at 20°C.

Soil properties and adsorption data

Soil	K_d (mL/g)	f_{oc} (%)	K_{oc} (mL/g)	pH
Sand	2.48	1.28	194	7.8
Sandy loam	14.83	3.25	456	6.7
Silt loam	4.53	2.09	217	6.1

Source: U.S. Department of Agriculture, 1990.

Environmental Fate

Photolytic. When methidathion in an aqueous buffer solution (25°C and pH 7.0) was exposed to filtered UV light (λ >290 nm) for 24 hours, 17% decomposed to 5-methoxy-3*H*-1,3,4-thiadiazol-2-one. At 50°C, 56% was degraded after 24 hours. Degradation occurred via hydrolysis of the thiol bond of the phosphorodithioic ester. Under acidic and alkaline conditions, hydrolytic cleavage occurred at the C-S and P-S bonds, respectively (Burkhard and Guth, 1979). Smith et al. (1978) demonstrated that methidathion degraded to methidathion oxon at a faster rate in six air-dried soils than in moist soils. Half-lives for methidathion in the air-dried soils ranged from 19 to 110 days. In addition, methidathion degraded faster in an air-dried soil exposed to ozone (half-lives 2.5 to 7.0 days).

Chemical/Physical. Emits toxic fumes of phosphorus, nitrogen and sulfur oxides when heated to decomposition (Sax and Lewis, 1987). Methidathion oxon was also found in fogwater collected near Parlier, CA (Glotfelty et al., 1990). It was suggested that methidathion was oxidized in the atmosphere during daylight hours prior to its partitioning from the vapor phase into the fog. On 12 January 1986, the distribution of parathion (0.45 ng/m³) in the vapor phase, dissolved phase, air particles and water particles were 57.5, 25.4, 16.8 and 0.3%, respectively. For methidathion oxon (0.84 ng/m³), the distribution in the vapor phase, dissolved phase, air particles and water particles were <7.1, 20.8, 78.6 and 0.1%, respectively (Glotfelty et al., 1990).

Symptoms of Exposure: Eye irritant.

Formulation Types: Emulsifiable concentrate; wettable powder.

Toxicity: LC_{50} (96-hour) for rainbow trout 10 μg/L and bluegill sunfish 2 μg/L (Hartley and Kidd, 1987); acute oral LD_{50} for rats 25–54 mg/kg (Hartley and Kidd, 1987), 20 mg/kg (RTECS, 1985).

Uses: Insecticide and acaricide.

METHIOCARB

Synonyms: B 37344; Bay 9026; Bay 37344; Bayer 37344; **3,5-Dimethyl-4-(methyl-thio)phenyl methylcarbamate**; 3,5-Dimethyl-4-methylthiophenyl *N*-methylcarbamate; Draza; ENT 25726; H 321; Mercaptodimethur; Mesurol; Methyl carbamic acid 4-(methylthio)-3,5-xylyl ester; 4-Methylmercapto-3,5-dimethylphenyl *N*-methyl carbamate; 4-Methylthio-3,5-dimethylphenyl methyl carbamate; 4-Methylthio-3,5-xylyl isomethylcarbamate; Metmercapturon; NA 2757; OMS 93.

Designations: CAS Registry Number: 2032-65-7; DOT: 2757; mf: $C_{11}H_{15}NO_2S$; fw: 225.31; RTECS: FC5775000.

Properties: White crystalline powder. Mp: 121.5°C; H-$t_{1/2}$ (20°C): >1 year (pH 4), <35 days (pH 7), 6 hours (pH 9); log K_{oc}: 2.71; log K_{ow}: 2.97, 3.04; S_o (g/L at 20°C): methylene chloride (500), 2-propanol (80), toluene (50–100); S_w: 10–30 mg/L at 20°C.

Soil properties and adsorption data

Soil	K_d (mL/g)	f_{oc} (%)	K_{oc} (mL/g)	pH
Batcombe silt loam	—	0.63–1.46	208	6.7–7.5
Sandy loam	12.6	2.44	516	5.9

Source: Briggs, 1981; U.S. Department of Agriculture, 1990.

Environmental Fate

Soil. Methiocarb was oxidized, probably by singlet oxygen, to the corresponding sulfoxide and trace amounts (<5% yield) of sulfone when sorbed on soil and exposed to sunlight. The photosensitized oxidation was faster in soils containing the lowest organic carbon content (Gohre and Miller, 1986).

Plant. On and/or in bean plants, the methylthio group is rapidly oxidized to the sulfoxide and sulfone (Abdel-Wahab et al., 1966) followed by hydrolysis yielding the corresponding thiophenol, methylsulfoxide phenol and methylsulphonyl phenol (Hartley and Kidd, 1987).

Photolytic. When methiocarb in ethanol was irradiated by UV light, only a few unidentified cholinesterase inhibitors were formed (Crosby et al., 1965).

Chemical/Physical. Emits toxic fumes of nitrogen and sulfur oxides when heated to decomposition (Sax and Lewis, 1987).

Formulation Types: Granular bait; seed treatment; wettable powder; dustable powder.

Toxicity: LC_{50} (96-hour) for rainbow trout 0.64 mg/L, bluegill sunfish 0.21-0.75 mg/L and golden orfe 3.8 mg/L; acute oral LD_{50} for rats 100–130 mg/kg (Hartley and Kidd, 1987), 15 mg/kg (RTECS, 1985).

Uses: Insecticide, acaricide and bird repellent.

METHOMYL

Synonyms: Du Pont 1179; Du Pont insecticide 1179; ENT 27341; IN 1179; Insecticide 1179; Lannate; Lannate L; Mesomile; *N*-(((Methylamino)carbonyl)oxy)ethanimidothioate; **N-(((Methylamino)carbonyl)oxy)ethanimidothioic acid methyl ester**; *N*-((Methylcarbamoyl)oxy)thioacetimidic acid methyl ester; Methyl-*N*-(((methylamino)carbonyl)oxy)ethanimidothioate; Methyl-*N*-((methylcarbamoyl)oxy)thioacetimidate; *S*-Methyl *N*-(methylcarbamoyloxy)thioacetimidate; Methyl *O*-(methylcarbamoyl)thioacetohydroxamate; Nubait II; Nudrin; RCRA waste number P066; SD 14999; WL 18236.

$$CH_3\underset{\underset{SCH_3}{|}}{C} = NO\overset{\overset{O}{||}}{C}NHCH_3$$

Designations: CAS Registry Number: 16752-77-5; mf: $C_5H_{10}N_2O_2S$; fw: 162.20; RTECS: AK2975000.

Properties: Colorless crystals with a faint sulfur-like odor. Mp: 78–79°C; ρ: 1.2946 at 24/4°C; H-$t_{1/2}$: 262 days at 25°C and pH 7; K_H: 6.4 × 10^{-10} atm · m^3/mol at 25°C (approximate — calculated from water solubility and vapor pressure); log K_{oc}: 1.86, 2.20; log K_{ow}: 0.13, 1.08; P-$t_{1/2}$: 48.41 hours (absorbance λ = 257.5 nm, concentration on glass plates = 6.7 µg/cm^2); S_o (wt %): acetone (73), ethanol (42), methanol (100), 2-propanol (22) and many other organic solvents; S_w: 57.9 g/kg at 25°C; vp: 4.99 × 10^{-5} mmHg at 25°C.

Environmental Fate

Biological. From the first-order biotic and abiotic rate constants of methomyl in estuarine water and sediment/water systems, the estimated biodegradation half-lives were 75–165 and 39–134 days, respectively (Walker et al., 1988).

Soil. Harvey and Pease (1973) reported that methomyl dissipated rapidly in fine sand and loamy sand soils. One month following application to a Delaware soil, 1.8% of the applied dosage was recovered and after 1 year, methomyl was not detected. The hydrolysis product (*S*-methyl *N*-hydroxythioacetimidate) and a trace mixture of very polar compounds were the intermediate degradation compounds before forming the principal end product, carbon dioxide.

Groundwater. According to the U.S. EPA (1986) methomyl has a high potential to leach to *groundwater.*

Plant. The reported half-lives of methomyl on cotton plants, mint plants and Bermuda grass were 0.4–8.5, 0.8–1.2 and 2.5 days, respectively (Willis and McDowell, 1987).

Chemical/Physical. Emits toxic fumes of nitrogen and sulfur oxides when heated to decomposition (Sax and Lewis, 1987; Lewis, 1990).

Though no products were reported, the calculated hydrolysis half-life at 25°C and pH 7 is 262 days (Ellington et al., 1988). The hydrolysis half-lives of methomyl in a sterile 1% ethanol/water solution at 25°C and pH values of 4.5, 6.0, 7.0 and 8.0 were 56, 54, 38 and 20 weeks, respectively (Chapman and Cole, 1982). In both soils and water, chemical-

and biological-mediated reactions can transform methomyl into two compounds — a nitrile and a mercaptan (Alexander, 1981).

Methomyl degraded rapidly in distilled water containing free chlorine. The degradation rate increased with a decrease in pH, increase in temperature and increase in chlorine concentrations. The reaction rate with free chlorine was 1,000 times faster than with chloramine. In chlorinated water, the half-lives for methomyl ranged from 24 seconds at pH 7.6 to 12 minutes at pH 8.9. The half-life for the reaction of methomyl with chloramine was 19 hours between the pH range of 7 to 9. Methomyl degraded to methomyl sulfoxide and N-chloromethomyl before degrading to acetic acid, methane sulfonic acid and dichloromethylamine (Miles and Oshiro, 1990).

Exposure Limits: NIOSH REL: TWA 2.5 mg/m^3; ACGIH TLV: TWA 2.5 mg/m^3.

Formulation Types: Emulsifiable concentrate; wettable powder; soluble concentrate.

Toxicity: LC$_{50}$ (96-hour) for rainbow trout 3.4 mg/L and bluegill sunfish 0.9 mg/L (Hartley and Kidd, 1987); acute oral LD$_{50}$ for male and female rats 17 (RTECS, 1985) and 24 mg/kg (Hartley and Kidd, 1987), respectively.

Use: Insecticide.

METHOXYCHLOR

Synonyms: 2,2-Bis(*p*-anisyl)-1,1,1-trichloroethane; 1,1-Bis(*p*-ethoxyphenyl)-2,2,2-trichloroethane; 2,2-Bis(*p*-methoxyphenyl)-1,1,1-trichloroethane; Chemform; 2,2-Di-*p*-anisyl-1,1,1-trichloroethane; Dimethoxy-DDT; *p,p'*-Dimethoxydiphenyltrichloroethane; Dimethoxy-DT; 2,2-Di(*p*-methoxyphenyl)-1,1,1-trichloroethane; Di(*p*-methoxyphenyl)trichloromethyl methane; DMDT; 4,4'-DMDT; *p,p'*-DMDT; DMTD; ENT 1716; Maralate; Marlate; Marlate 50; Methoxcide; Methoxo; 4,4'-Methoxychlor; *p,p'*-Methoxychlor; Methoxy-DDT; Metox; Moxie; NCI-C00497; RCRA waste number U247; 1,1,1-Trichloro-2,2-bis(*p*-anisyl)ethane; 1,1,1-Trichloro-2,2-bis(*p*-methoxyphenol)ethanol; 1,1,1-Trichloro-2,2-bis(*p*-methoxyphenyl)ethane; 1,1,1-Trichloro-2,2-di(4-methoxyphenyl)ethane; **1,1'-(2,2,2-Trichloroethylidene)bis(4-methoxybenzene).**

Designations: CAS Registry Number: 72-43-5; DOT: 2761; mf: $C_{16}H_{15}Cl_3O_2$; fw: 345.66; RTECS: KJ3675000.

Properties: White, gray or pale yellow crystals or powder. May be dissolved in an organic solvent or petroleum distillate for application. Pungent to mild fruity odor. May turn color on exposure to light. Mp: 86–88°C, 89–98°C; bp: decomposes; ρ: 1.41 at 25/4°C; log BCF: 4.08 mussel (*Mytilus edulis*) (Renberg et al., 1985); H-$t_{1/2}$: 7–18 days in natural surface water samples; log K_{oc}: 4.90, 4.95; log K_{ow}: 3.31–5.08; S_o: soluble in ethanol, chloroform, xylene (440 g/kg at 20°C); S_w: 40 μg/L at 24°C.

Soil properties and adsorption data

Soil	K_d (mL/g)	f_{oc} (%)	K_{oc} (mL/g)	pH
Doe Run clay	2,400	3.29	72,948	—
Doe Run coarse silt	2,200	2.78	79,137	—
Doe Run fine silt	2,300	2.89	79,585	—
Doe Run medium silt	1,700	2.34	72,650	—
Hickory Hill clay	1,100	1.20	91,667	—
Hickory Hill coarse silt	2,600	3.27	79,510	—
Hickory Hill fine silt	1,400	1.34	104,407	—
Hickory Hill medium silt	1,800	1.98	90,909	—
Hickory Hill sand	53	0.13	40,769	—
Oconee River coarse silt	2,500	2.92	85,616	—
Oconee River fine sand	2,100	2.26	92,920	—
Oconee River medium silt	2,000	1.99	100,502	—
Oconee River sand	95	0.57	16,667	—

Source: Karickhoff et al., 1979.

Environmental Fate

Biological. Degradation by the microorganism *Aerobacter aerogenes* under aerobic or anaerobic conditions yielded 1,1-dichloro-2,2-bis(*p*-methoxyphenyl)ethylene and 1,1-dichloro-2,2-bis(*p*-methoxyphenyl)ethane (Mendel and Walton, 1966; Kobayashi and Rittman, 1982). Methoxychlor degrades at a faster rate in flooded/anaerobic soils than in nonflooded and aerobic soils (Fogel et al., 1982; Golovleva et al., 1984). In anaerobic soil, 90% of the applied dosage was lost after 3 months. In aerobic soil, only 0.3% was lost as carbon dioxide after 410 days (Fogel et al., 1982). In a modified river die-away test, methoxychlor underwent aerobic biodegradation at a rate of 0.0024/hour which corresponds to a half-life of 12 days (Cripe et al., 1987).

In a model aquatic ecosystem, methoxychlor degraded to ethanol, dihydroxyethane, dihydroxyethylene and unidentified polar metabolites (Metcalf et al., 1971). Kapoor et al. (1970) also studied the biodegradation of methoxychlor in a model ecosystem containing snails, plankton, mosquito larvae, *Daphnia magna* and mosquito fish (*Gambusia affinis*). The following metabolites were identified: 2-(*p*-methoxyphenyl)-2-(*p*-hydroxyphenyl)-1,1,1-trichloroethane, 2,2-bis(*p*-hydroxyphenyl)-1,1,1-trichloroethane, 2,2-bis(*p*-hydroxyphenyl)-1,1,1-trichloroethylene and polar metabolites (Kapoor et al., 1970).

From the first-order biotic and abiotic rate constants of methoxychlor in estuarine water and sediment/water systems, the estimated biodegradation half-lives were 208–8,837 and 12–45 days, respectively (Walker et al., 1988).

Paris and Lewis (1976) reported that the microorganism *Aspergillis* sp. accumulated methoxychlor slowly, requiring 16 hours to reach equilibrium. Some microorganisms, such as *Flavobacterium harrisonii, Bacillus subtilis* and *Chlorella pyrenoidoda* accumulated methoxychlor and reached equilibrium in only 30 minutes.

Groundwater. According to the U.S. EPA (1986) methoxychor has a high potential to leach to *groundwater.*

Plant. Brett and Bowery (1958) and Johansen (1954) reported foliar half-lives of 1.8 and 6.3 days in collards and cherries, respectively.

Photolytic. In air-saturated distilled water, direct photolysis of methoxychlor (λ >280 nm) produced 1,1-bis(*p*-methoxyphenyl)-2,2-dichloroethylene (DMDE) which photolyzed to *p*-methoxybenzaldehyde (estimated half-life 4.5 months) (Zepp et al., 1976). Methoxychlor-DDE and *p,p*-dimethoxybenzophenone were formed when methoxychlor in water was irradiated by UV light (Paris and Lewis, 1973). Compounds reported from the photolysis of methoxychlor in aqueous, alcoholic solutions were *p,p*-dimethoxybenzophenone, *p*-methoxybenzoic acid and *p*-methoxyphenol (Wolfe et al., 1976). However, when methoxychlor in milk was irradiated by UV light (λ = 220 and 330 nm), methoxyphenol, methoxychlor-DDE, *p,p*-dimethoxybenzophenone and 1,1,4,4-tetrakis(*p*-methoxyphenyl)-1,2,3-butatriene were formed (Li and Bradley, 1969).

Chemical/Physical. Hydrolysis at common aquatic pHs produced anisoin, anisil and 2,2-bis(*p*-methoxyphenyl)-1,1-dichloroethylene (estimated half-life 270 days at 25°C and pH 7.1) (Wolfe et al., 1977). Above pH 10, 2,2-bis(*p*-methoxyphenyl)-1,1-dichloroethylene is the only reported product. Below pH 10, anisoin is formed (Kollig, 1993). The estimated hydrolysis half-life in water at 25°C and pH 7.1 is 270 days (Mabey and Mill, 1978).

Emits toxic chloride fumes when heated to decomposition (Lewis, 1990).

Exposure Limits: NIOSH REL: IDLH 5,000 mg/m³; OSHA PEL: TWA 15 mg/m³; ACGIH TLV: TWA 10 mg/m³.

Symptoms of Exposure: Slightly irritating to skin.

Formulation Types: Aerosol; emulsifiable concentrate; wettable powder; granules; dustable powder.

Toxicity: LC_{50} (96-hour) for fathead minnow 7.5 µg/L, bluegill sunfish 62.0 µg/L, rainbow trout 62.6 µg/L, coho salmon 66.2 µg/L, chinook 27.9 µg/L, perch 20.0 µg/L (Verschueren, 1983); LC_{50} (24-hour) for rainbow trout 52 µg/L, bluegill sunfish 67 µg/L (Worthing and Hance, 1991), isopod (*Asellus communis*) 0.42 µg/L (Anderson and DeFoe, 1980); acute oral LD_{50} for rats 6,000 mg/kg (Hartley and Kidd, 1987), 5,000 mg/kg (RTECS, 1985).

Uses: Insecticide used to control mosquito larvae, house flies and other insect pests in field crops, fruits and vegetables; to control ecto-parasites on cattle, sheep and goats; recommended for use in dairy barns.

METHYL BROMIDE

Synonyms: Brom-o-gas; Brom-o-gaz; **Bromomethane**; Celfume; Dawson 100; Dow-fume; Dowfume MC-2; Dowfume MC-2 soil fumigant; Dowfume MC-33; Edco; Embafume; Fumigant-1; Halon 1001; Iscobrome; Kayafume; MB; M-B-C Fumigant; MBX; MEBR; Metafume; Methogas; Monobromomethane; Pestmaster; Profume; R 40B1; RCRA waste number U029; Rotox; Terabol; Terr-o-gas 100; UN 1062; Zytox.

$$CH_3Br$$

Designations: CAS Registry Number: 74-83-9; DOT: 1062; mf: CH_3Br; fw: 94.94; RTECS: PA4900000.

Properties: Colorless liquid or gas with a sweetish odor similar to chloroform at high concentrations. Odorless at low concentrations. Burning taste. Mp: –93.6°C; bp: 3.55°C; ρ: 1.732 at 0/4°C, 1.6755 at 20/4°C; lel: 10%; uel: 16%; H-$t_{1/2}$: 20 days at 25°C and pH 7: K_H: 3.18×10^{-2} atm · m^3/mol; IP: 10.54 eV; log K_{oc}: 1.92 (calculated); log K_{ow}: 1.00–1.19; S_o: soluble in 95% ethanol, ethyl ether, chloroform, carbon disulfide, carbon tetrachloride, benzene, esters and ketones; S_w: 13.200 g/L at 25°C; vap d: 3.88 g/L at 25°C, 3.28 (air = 1); vp: 1,420 at 20°C.

Environmental Fate
Photolytic. When methyl bromide and bromine gas (concentration = 3%) was irradiated at 1850 Å, methane was produced (Kobrinsky and Martin, 1968).

Chemical/Physical. Methyl bromide hydrolyzes in water forming methanol and hydro-bromic acid. The estimated hydrolysis half-life in water at 25°C and pH 7 is 20 days (Mabey and Mill, 1978). Forms a voluminous crystalline hydrate at 0–5°C (Keith and Walters, 1992).

When methyl bromide was heated to 550°C in the absence of oxygen, methane, hydrogen, bromine, ethyl bromide, anthracene, pyrene and free radicals were produced (Chaigneau et al., 1966).

Emits toxic bromide fumes when heated to decomposition (Lewis, 1990).

Exposure Limits: NIOSH REL: IDLH 250 ppm; OSHA PEL: C 20 ppm; ACGIH TLV: TWA 5 ppm.

Symptoms of Exposure: Inhalation causes headache, visual disturbance, vertigo, nausea, vomiting, malaise, hand tremor, convulsions, dyspnea. Eye and skin irritation.

Formulation Types: Vapor.

Toxicity: Acute oral LD_{50} for rats is 100 mg/kg and inhalation LD_{50} for rats is 3,150 mg/L (Ashton and Monaco, 1991).

Uses: Soil, space and food fumigant; disinfestation of potatoes, tomatoes and other crops.

METHYL ISOTHIOCYANATE

Synonyms: EP 161E; **Isothiocyanatomethane**; Isothiocyanic acid methyl ester; Methyl mustard oil; MIC; MIT; MITC; Morton EP 161E; Mustard oil; Trapex; Trapexide; UN 2477; Vorlex; Vortex; WN 12.

$$CH_3NCS$$

Designations: CAS Registry Number: 556-61-6; DOT: 2477; mf: C_2H_3NS; fw: 73.12; RTECS: PA7625000.

Properties: Colorless crystals with a pungent, horseradish-like odor. Mp: 35–36°C; bp: 119°C; ρ: 1.048 at 24/4°C; fl p: 32°C; H-$t_{1/2}$ (25°C): 85 hours (pH 5), 490 hours (pH 7), 110 hours (pH 9); K_H: 2.4 × 10^{-4} atm · m³/mol at 20°C (approximate — calculated from water solubility and vapor pressure); log K_{oc}: 1.51 (calculated); log K_{ow}: 1.37; S_o: very soluble in most organic solvents; S_w: 7.6 g/L at 20°C; vp: 19 mmHg at 20°C.

Environmental Fate

Soil. Though no products were reported, the reported half-life in soil is <14 days (Worthing and Hance, 1991).

Chemical/Physical. Emits toxic fumes of nitrogen and sulfur oxides when heated to decomposition (Sax and Lewis, 1987).

Symptoms of Exposure: Highly irritating to eyes, skin and mucous membranes.

Formulation Types: Emulsifiable concentrate.

Toxicity: LC$_{50}$ (96-hour) for bluegill sunfish 130 µg/L, carp 370 µg/L and rainbow trout 370 µg/L (Hartley and Kidd, 1987); acute oral LD$_{50}$ for rats 175 mg/kg (Hartley and Kidd, 1987), 97 mg/kg (RTECS, 1985).

Uses: Pesticide and soil fumigant used to control insects, soil fungi, nematodes.

METOLACHLOR

Synonyms: Bicep; CGA 24705; **2-Chloro-*N*-(2-ethyl-6-methylphenyl)-*N*-(2-methoxy-1-methylethyl)acetamide**; 2-Chloro-6′-ethyl-*N*-(2-methoxy-1-methylethyl)acet-*o*-toluidide; α-Chloro-2′-ethyl-6′-methyl-*N*-(1-methyl-2-methoxyethyl)acetanilide; 2-Chloro-*N*-(2-ethyl-6-methylphenyl)-*N*-(2-methoxy-1-methylethyl)acetamide; Codal; Cotoran multi; Dual; Dual 8E; Dual 25G; 2-Ethyl-6-methyl-1-*N*-(2-methoxy-1-methylethyl)chloroacetanilide; Metelilachlor; Milocep; Ontrack 8E; Pennant; Pennant 5G; Primagram; Primextra.

Designations: CAS Registry Number: 51218-45-2; mf: $C_{15}H_{22}ClNO_2$; fw: 283.81; RTECS: AN3430000.

Properties: Clear, colorless to tan, odorless liquid. Mp: <25°C; bp: 100°C at 0.001 mmHg; ρ: 1.12 at 20/4°C; fl p: 93.3°C (Dual 8E formulation); H-$t_{1/2}$ (20°C): >200 days at 1 ≤pH ≤9; K_H: 9.2×10^{-9} atm · m³/mol at 20°C (approximate — calculated from water solubility and vapor pressure); log K_{oc}: 2.08–2.49; log K_{ow}: 2.93–3.45; S_o: miscible with acetone, benzene, butyl cellosolve, cyclohexanone, 1,2-dichloroethane, *N,N*-dimethylformamide, ethanol, methanol, methyl cellosolve, methylene chloride, toluene, xylene; S_w: 530 mg/L at 20°C; vp: 1.3×10^{-5} mmHg at 20°C.

Soil properties and adsorption data

Soil	K_d (mL/g)	f_{oc} (%)	K_{oc} (mL/g)	pH	CEC (meq/100 g)
Appling	1.10	0.81	136	6.8	6.9
Augusta	0.50	0.29	172	5.7	3.2
Cape Fear sandy loam	8.52	5.34	160	—	10.3
Cape Fear sandy loam	10.90	5.05	216	5.1	10.3
Cecil sandy loam	1.00	0.99	101	5.4	3.1
Goldsboro	1.40	0.70	200	5.3	3.3
Lynchburg	2.50	1.45	172	5.5	6.6
Norfolk fine loamy sand	2.20	0.99	222	—	2.3
Norfolk fine loamy sand	0.50	0.29	172	5.4	2.3
Portsmouth	3.30	2.55	129	5.4	10.6
Rains fine loamy sand	2.40	0.99	242	6.0	7.1
Rains fine loamy sand	3.20	1.45	229	—	7.0
Sand	1.54	1.28	120	7.8	—
Sand	1.69	0.70	243	6.3	—
Sandy loam	10.00	3.25	308	6.7	—
Sandy silt loam	3.18	2.09	152	6.1	—

Soil properties and adsorption data *(continued)*

Soil	K_d (mL/g)	f_{oc} (%)	K_{oc} (mL/g)	pH	CEC (meq/100 g)
Taloka silt loam (0–8 cm)	0.66	0.64	103	6.2	—
Taloka silt loam (8–15 cm)	0.35	0.58	60	6.4	—
Taloka silt loam (15–22 cm)	0.28	0.58	48	6.1	—
Taloka silt loam (22–30 cm)	0.22	0.46	48	5.4	—
Taloka silt loam (30–38 cm)	0.13	0.46	28	4.9	—

Source: Kozak et al., 1983; LeBaron et al., 1988; U.S. Department of Agriculture, 1990; Barnes et al., 1992.

Environmental Fate

Soil. Metolachlor and its degradation products combine with humic acids in soils and small quantities are degraded to carbon dioxide (Ashton and Monaco, 1991). In soil, the fungus *Chaetomium globosum* degraded metolachlor to 2-chloro-*N*-(2-ethyl-6-methylphenyl)acetamide, 2-hydroxy-*N*-(2-methylvinylphenyl)-*N*-(methoxyprop-2-yl)acetamide, 3-hydroxy-8-methyl-*N*-(methoxyprop-2-yl)-2-oxo-1,2,3,4-tetrahydraquinoline, 2-chloro-*N*-(2-ethyl-6-methylphenyl)-*N*-(hydroxyprop-2-yl)acetamide and the tentatively identified compounds 3-hydroxy-1-isopropyl-8-methyl-2-oxo-1,2,3,4-tetrahydroquinoline, *N*-(methoxyprop-2-yl)-8-methyl-2-oxo-1,2,3,4-tetrahydroquinoline, *N*-(methoxyprop-2-yl)-*N*-(2-methyl-6-vinyl)aniline and 1-(methoxyprop-2-yl)-7-methyl-2,3-dihydroindole (McGahen 8). Metolachlor was transformed by a strain of soil actinomycetes to the following products: 2-chloro-*N*-(2-ethyl-6-methylphenyl)-*N*-(hydroxyprop-2-yl)acetamide, 2-chloro-*N*-2-(1-hydroxyethyl)-6-(methylphenyl)-*N*-(hydroxyprop-2-yl)acetamide, 2-chloro-*N*-(2-ethyl-6-hydroxymethylphenyl)-*N*-(hydroxy-prop-2-yl)-acetamide, diastereoisomers of 2-chloro-*N*-(2-ethyl-6-hydroxymethylphenyl)-*N*-(methoxyprop-2-yl)acetamide and 2-chloro-*N*-2-(hydroxyethyl)-6-hydroxymethylphenyl)-*N*-(methoxyprop-2-yl)acetamide. These products were formed via hydroxylation of both the *N*-alkyl and alkyl side chains (Krause et al., 1985). In sterilized soil, metolachlor did not degrade after 4 months (Bouchard et al., 1982).

Zimdahl and Clark (1982) reported half-lives of 15–38 and 33–100 days for the herbicide in clay loam soil and sandy loam soil, respectively. They also reported that soil moisture increased the dissipation rate. At 20°C, the dissipation rates of metolachlor in the clay loam and sandy loam soils at 20, 50 and 80% soil moisture contents were 0.028, 0.053, 0.062 and 0.016, 0.028 and 0.037/day, respectively. The half-lives of metolachlor in soil at maintained at temperatures of 30 and 40°C were approximately 3.85 and 2.75 weeks, respectively (Bravermann et al., 1986). The reported half-lives of metolachlor in soil is approximately 6 days (Worthing and Hance, 1991) and 3–4 weeks (Bowman, 1988).

Groundwater. According to the U.S. EPA (1986) metolachlor has a high potential to leach to *groundwater.*

Plant. Metabolizes in plants forming water soluble, polar, nonvolatile products (Hartley and Kidd, 1987) and glutathione conjugates (Breaux et al., 1987).

Chemical/Physical. The volatilization half-life of metolachlor in an unstirred solution was 20 days at 40°C. Volatilization is not significant at temperatures < 25°C (Lau et al., 1995).

Symptoms of Exposure: Eye and skin irritant.

Formulation Types: Emulsifiable concentrate (8 lb/gal); granules (5 and 25%).

Toxicity: LC_{50} (96-hour) for rainbow trout 2 mg/L, bluegill sunfish 15 mg/L, carp 4.9 mg/L (Hartley and Kidd, 1987) and channel catfish 4.9 ppm (Humburg et al., 1989); acute oral LD_{50} of technical metolachlor for rats 2,800 mg/kg (Ashton and Monaco, 1991), pure metolachlor 2,534 mg/kg (RTECS, 1985).

Uses: Preemergence herbicide used to control most annual grasses and many annual weeds in beans, chickpeas, corn, cotton, milo, okra, peanuts, peas, potatoes, safflower, soybeans and some ornamentals.

METRIBUZIN

Synonyms: 4-Amino-6-(1,1-dimethylethyl)-3-(methylthio)-1,2,4-triazin-5-(4*H*)-one; Bay 61597; Bay dic 1468; Bayer 94337; Bayer 6159H; Bayer 6443H; DIC 1468; Lexone; Lexone DF; Lexone 4L; Preview; Sencor; Sencor 4; Sencor DF; Sencoral; Sencorer; Sencorex.

Designations: CAS Registry Number: 21087-64-9; mf: $C_8H_{14}N_4OS$; fw: 214.28; RTECS: XZ2990000.

Properties: White crystalline solid with a slight sulfur-like odor. Mp: 125–126.5°C; ρ: 1.31 at 20/4°C; fl p: nonflammable; pK_a: 1.00; H-$t_{1/2}$: ≈ one week in pond water; K_H: 1.2 × 10⁻¹⁰ atm · m³/mol at 20°C (approximate — calculated from water solu-bility and vapor pressure); log K_{oc}: 1.81–2.72; log K_{ow}: 1.60, 1.70; S_o (g/kg at 20°C): acetone (820), benzene (220), *n*-butanol (150), chloroform (850), cyclohexane (1,000), cyclohexanone (1,000), *N,N*-dimethylformamide (1,780), ethanol (190), hexane (2), kerosene (<10), methanol (450), methylene chloride (333), toluene (120), xylene (90); S_w: 1,050 mg/L at 20°C; vp: 4.35 × 10⁻⁷ mmHg at 20°C.

Soil properties and adsorption data

Soil	K_d (mL/g)	f_{oc} (%)	K_{oc} (mL/g)	pH	CEC (cmol/kg)
Bashaw clay loam	2.13	0.58	367	6.2	35.1
Chehalis sandy loam	2.42	1.39	174	6.0	19.5
Clayey loam	1.90	2.90	66	7.9	—
Clayey loam	1.53	0.29	528	6.0	—
Crooked sandy loam	1.11	0.64	173	8.2	13.7
Loamy sand	1.32	1.62	81	6.6	—
Ontko loam	3.37	3.19	106	6.2	44.2
Woodburn silty clay loam	7.00	1.39	504	4.6	13.7

Source: Peek and Appleby, 1989; U.S. Department of Agriculture, 1990.

Environmental Fate

Soil. In soils, metribuzin undergoes deamination and further degradation forming water soluble conjugates (Hartley and Kidd, 1987). Metribuzin degrades rapidly in soil (Kempson-Jones and Hance, 1979; LaFleur, 1980). The half-lives in soil ranged from 4 to 9 weeks at 22°C and 60% moisture content to 9 to 11 weeks at 10°C and 10% moisture content. The highest half-life reported ranged from 15 to 43 weeks in a soil having 60% moisture content maintained at 10°C (Kempson-Jones and Hance, 1979).

The dissipation rates (day^{-1}) for metribuzin on Hillsdale silty loam, Cecil silty loam, Regina heavy clay, Melfort clay loam and White City silt loam were 0.013-0.017, 0.055-0.131, >0.017, >0.0016 and >0.023, respectively.

Groundwater. According to the U.S. EPA (1986) metribuzin has a high potential to leach to *groundwater.*

Plant. Metribuzin is metabolized in soybean plants to a deaminated diketo derivative which is nonphytoxic (Duke et al., 1991).

Photolytic. The simulated sunlight (λ >230 nm) photolysis as a thin film on silica gel or sand yielded 6-(1,1-dimethylethyl)-3-(methylthio)-1,2,4-triazin-5-(4*H*)-one and two additional photoproducts. In both of the unnamed photoproducts, the methylthio group in the parent compound is replaced by oxygen and one of the compounds also underwent *N*-deamination (Bartl and Korte, 1975).

Chemical/Physical. Emits toxic fumes of nitrogen and sulfur oxides when heated to decomposition (Lewis, 1990).

Exposure Limits: NIOSH REL: 5 mg/m^3; ACGIH TLV: TWA 5 mg/m^3.

Formulation Types: Water-dispersible granules; emulsifiable concentrate; suspension concentrate (4 lb/gal); flowable powder (75%); wettable powder for application to soil surface (50%).

Toxicity: LC$_{50}$ (96-hour) for rainbow trout 64 mg/L and bluegill sunfish 80 mg/L (Worthing and Hance, 1991); acute oral LD$_{50}$ of technical metribuzin for male and female rats 1,100 and 1,200 mg/kg, respectively (Ashton and Monaco, 1991).

Uses: Selective herbicide used for preemergence broad-leaved weed control in potatoes, tomatoes, lucerne, raspberry and sugarcane, and postemergence weed control in beets. Also used for selective control of annual grasses.

METSULFURON-METHYL

Synonyms: Ally; Ally 20DF; Escort; **2-(((((4-Methoxy-6-methyl-1,3,5-triazin-2-yl)amino)carbonyl)amino)sulfonyl)benzoic acid.**

Designations: CAS Registry Number: 74223-64-6; mf: $C_{14}H_{15}N_5O_6S$; fw: 381.40.

Properties: Colorless crystals with a slight ester-like odor. Mp: 158°C, decomposes at 172°C; fl p: nonflammable; pK_a: 3.3 at 25°C (imide group); H-$t_{1/2}$ (25°C): 15 hours (pH 2), 33 days (pH 5), >41 days (pH 7–9) and at 45°C: 2 hours (pH 2), 2 days (pH 5), 33 days (pH 9); K_H: 3.1×10^{-9} atm · m³/mol at pH 7 and 25°C (approximate — calculated from water solubility and vapor pressure); log K_{oc}: 1.54, 2.31 (Flanagan silt loam); log K_{ow}: 0.00 (pH 5), –1.85 (pH 7); S_o (g/L at 20°C): acetone (36), ethanol (2.3), hexane (0.00079), methanol (7.3), methylene chloride (121), xylene (0.58); S_w (g/L at 25°C): 0.27 at pH 4.6, 1.1 at pH 5, 1.75 at pH 5.4, 9.5 g/L at pH 7.0; vp: 5.8×10^{-5} mmHg at 25°C.

Environmental Fate

Soil/Plant. Hydrolyzes in soil and plants to nontoxic products (Hartley and Kidd, 1987). The half-life in soil varies from 7 days to 1 month (Hartley and Kidd, 1987).

Ismail and Lee (1995) studied the persistence of metsulfuron-methyl in a sandy loam (pH 5.1) and clay soil (3.1) under laboratory conditions. Degradation was more rapid in non-sterilized than in sterilized *soil*. In non-sterilized soil, the rate of degradation increased with increasing soil moisture content. When the moisture level in the sandy loam and clay soil was increased from 20 to 80% of field capacity at 35°C, the half-lives were reduced from 9.0 to 5.7 and 11.2 to 4.6, days, respectively. The investigators concluded that the disappearance of metsulfuron-methyl in soil resulted from microbial degradation and chemical hydrolysis.

Symptoms of Exposure: May irritate eyes, nose, skin and throat.

Formulation Types: Water-dispersible granules (20 and 60%).

Toxicity: LC_{50} (96-hour) for rainbow trout and bluegill sunfish >125 mg/L (Worthing and Hance, 1991); acute oral LD_{50} for rats >5,000 mg/kg (Ashton and Monaco, 1991).

Uses: Metsulfuron-methyl is a triazine urea herbicide used to control broad-leaved weeds in barley and wheat.

MEVINPHOS

Synonyms: Apavinphos; 2-Butenoic acid 3-((dimethoxyphosphinyl)oxy)methyl ester; 2-Carbomethoxy-1-methylvinyl dimethyl phosphate; α-Carbomethoxy-1-methylvinyl dimethyl phosphate; 2-Carbomethoxy-1-propen-2-yl dimethyl phosphate; CMDP; Compound 2046; **3-((Dimethoxyphosphinyl)oxy)-2-butenoic acid methyl ester**; O,O-Dimethyl-O-(2-carbomethoxy-1-methylvinyl)phosphate; Dimethyl-1-carbomethoxy-1-propen-2-yl phosphate; O,O-Dimethyl 1-carbomethoxy-1-propen-2-yl phosphate; Dimethyl 2-methoxycarbonyl-1-methylvinyl phosphate; Dimethyl methoxycarbonylpropenyl phosphate; Dimethyl (1-methoxycarboxypropen-2-yl)phosphate; O,O-Dimethyl O-(1-methyl-2-carboxyvinyl)phosphate; Dimethyl phosphate of methyl-3-hydroxy-cis-crotonate; Duraphos; ENT 22324; Fosdrin; Gesfid; Gestid; 3-Hydroxycrotonic acid methyl ester dimethyl phosphate; Meniphos; Menite; 2-Methoxycarbonyl-1-methylvinyl dimethyl phosphate; cis-2-Methoxycarbonyl-1-methylvinyl dimethyl phosphate; 1-Methoxycarbonyl-1-propen-2-yl dimethyl phosphate; Methyl 3-(dimethoxyphosphinyloxy)crotonate; NA 2783; OS 2046; PD 5; Phosdrin; cis-Phosdrin; Phosfene; Phosphoric acid (1-methoxycarboxypropen-2-yl) dimethyl ester.

$$\begin{array}{c} CH_3O \\ \searrow \overset{\displaystyle O}{\underset{}{\|}} \\ CH_3O \nearrow \end{array} P - O - \overset{\displaystyle CH_3}{\underset{}{|}} C = CHCOOCH_3$$

Designations: CAS Registry Number: 7786-34-7; DOT: 2783; mf: $C_7H_{13}O_6P$; fw: 224.16; RTECS: GQ5250000.

Properties: Colorless to pale yellow liquid. Mp: –56.1°C; bp: 106–107.5°C at 1 mmHg; ρ: 1.25 at 20/4°C; fl p: 79.4°C (open cup); H-$t_{1/2}$: 120 days (pH 6), 35 days pH 7), 3 days (pH 9), 1.4 hours (pH 11); P-$t_{1/2}$: 23.79 hours (absorbance λ = 230.5 nm, concentration on glass plates = 6.7 μg/cm^2); S$_o$: miscible with acetone, benzene, carbon tetrachloride, chloroform, ethanol, isopropanol, methylene chloride, toluene and xylene; soluble in carbon disulfide and kerosene (50 g/L); S$_w$: miscible; vap d: 9.16 g/L at 25°C, 7.74 (air = 1); vp: 2.2 × 10^{-3} mmHg at 20°C.

Environmental Fate

Plant. In plants, mevinphos is hydrolyzed to phosphoric acid dimethyl ester, phosphoric acid and other less toxic compounds (Hartley and Kidd, 1987). In one day, the compound is almost completely degraded in plants (Cremlyn, 1991). Casida et al. (1956) proposed two degradative pathways of mevinphos in bean plants and cabbage. In the first degradative pathway, cleavage of the vinyl phosphate bond affords methylacetoacetate and acetoacetic acid which may be precursors to the formation of the end products dimethyl phosphoric acid, methanol, acetone and carbon dioxide. In the other degradative pathway, direct hydrolysis of the carboxylic ester would yield vinyl phosphates as intermediates. The half-life of mevinphos in bean plants was 12 hours (Casida et al., 1956). In alfalfa, the half-life was 17 hours (Huddelston and Gyrisco, 1961).

Chemical/Physical. The reported hydrolysis half-lives of *cis*-mevinphos and *trans*-mevinphos at pH 11.6 are 1.8 and 3.0 hours, respectively (Casida et al., 1956). The volatility half-lives for the *cis* and *trans* forms at 28°C were 21 and 24 hours, respectively (Casida et al., 1956). Worthing and Hance (1991) reported that at pH values of 6, 7, 9 and 11, the hydrolysis half-lives were 120 days, 35 days, 3.0 days and 1.4 hours, respectively (Worthing and Hance, 1991).

Emits toxic phosphorus oxide fumes when heated to decomposition (Lewis, 1990).

Exposure Limits: NIOSH REL: TWA 0.01 ppm, STEL 0.03 ppm, IDLH 4 ppm; OSHA PEL: 0.1 mg/m³; ACGIH TWA: TLV 0.01 ppm, STEL 0.03 ppm.

Symptoms of Exposure: Miosis, rhinorrhea, headache, wheezing, laryngeal spasm; salivation, cyanosis; anorexia, nausea, abdominal cramps, diarrhea, paralysis, ataxia, convulsions, low blood pressure; irritates eyes and skin.

Formulation Types: Emulsifiable concentrate; soluble concentrate.

Toxicity: LC_{50} (96-hour) for bluegill sunfish 70 µg/L, largemouth bass 110 µg/L, mummichog 65–300 µg/L, striped killifish 75 µg/L, striped mullet 300 µg/L, Atlantic silverside 320 µg/L, Atlantic eel 65 µg/L, bullhead 74 µg/L, northern puffer 800 µg/L (Verschueren, 1983); LC_{50} (24-hour) for bluegill sunfish 41 ppb and rainbow trout 34 ppb (Verschueren, 1983); acute oral LD_{50} for rats 3–12 mg/kg (Hartley and Kidd, 1987), 1,350 µg/kg (RTECS, 1985).

Uses: Contact insecticide and acaricide for control of chewing insects and spider mites in fruits, vegetables and ornamentals.

MOLINATE

Synonyms: *S*-Ethyl azepane-1-carbothioate; **S-Ethyl hexahydro-1*H*-azepine-1-carboth-ioate**; *S*-Ethyl *N,N*-hexamethylenethiocarbamate; *S*-Ethyl perhydroazepine-1-thiocarbox-ylate; Felan; Higalnate; Hydram; Jalan; Molmate; Ordram; Ordram 8E; Ordram 10G; Ordram 15G; R 4572; Stauffer R 4572; Sakkimol; Yalan; Yulan.

Designations: CAS Registry Number: 2212-67-1; mf: $C_9H_{17}NOS$; fw: 187.32; RTECS: CM2625000.

Properties: Clear liquid with an aromatic odor. Mp: <25°C; bp: 117°C at 10 mmHg; ρ: 1.0643 at 20/20°C; fl p: 139°C (open cup); K_H: 1.6×10^{-6} atm · m³/mol at 25°C (approx-imate — calculated from water solubility and vapor pressure); log K_{oc}: 1.93–1.97; log K_{ow}: 2.88; S_o: miscible with most organic solvents; S_w: 880 mg/L at 20°C; vap d: 7.66 g/L at 25°C, 6.49 (air = 1); vp: 5.6×10^{-3} mmHg at 25°C.

Soil properties and adsorption data

Soil	K_d (mL/g)	f_{oc} (%)	K_{oc} (mL/g)	pH
Kanuma high clay	1.1	1.35	80	5.7
Tsukuba clay loam	3.7	4.24	89	6.5

Source: Kanazawa, 1989.

Environmental Fate

Soil. Hydrolyzes in soil forming ethyl mercaptan, carbon dioxide and dialkylamine (half-life approximately 2–5 weeks) (Hartley and Kidd, 1987). At recommended rates of application, the half-life of molinate in moist loam soils at 21–27°C is approximately 3 weeks (Humburg et al., 1989). Rajagopal et al. (1989) reported that under flooded condi-tions, molinate was hydroxylated at the 3- and 4-position with subsequent oxidation forming many compounds including molinate sulfoxide, carboxymethyl molinate, hexahy-droazepine-1-carbothioate, 4-hydroxymolinate, 4-hydroxymolinate sulfoxide, hexahy-droazepine, *S*-methyl hexahydroazepine-1-carbothioate, 4-ketomolinate, 4-hydroxy-hexahydroazepine, 4-hydroxy-*N*-acetyl-hexahydroazepine, carbon dioxide and bound residues.

Plant. Molinate is rapidly metabolized by plants releasing carbon dioxide and naturally occurring plant constituents (Humburg et al., 1989).

Photolytic. Molinate in a hydrogen peroxide solution (120 μM) was irradiated by UV light (λ = 290 nm) at 23°C. The major photooxidation products were the two isomers of 2-oxomolinate (20% yield) and *s*-molinate oxide (5% yield) (Draper and Crosby, 1984).

Half-lives of 180 and 120 hours were observed using one and two equivalents of hydrogen peroxide, respectively (Draper and Crosby, 1984). Molinate has a UV absorption maximum at 225 nm and no absorption at wavelengths >290 nm. Therefore, molinate is not expected to undergo aqueous photolysis under natural sunlight ($\lambda = 290$ nm). In the presence of tryptophan, a naturally occurring photosensitizer, molinate in aqueous solution photodegraded to form 1-((ethylsulfinyl)carbonyl)hexahydro-1H-azepine, S-ethyl hexahydro-2-oxo-1H-azepine-1-carbothioate and hexamethyleneimine (Soderquist et al., 1977).

   Chemical/Physical. Metabolites identified in tap water were molinate sulfoxide, 3- and 4-hydroxymolinate, ketohexamethyleneimine and 4-ketomolinate (Verschueren, 1983).

Formulation Types: Emulsifiable concentrate (8 lb/gal); granules (10 and 15%).

Toxicity: LC_{50} (96-hour) for bluegill sunfish 29–30 mg/L, goldfish 30 mg/L, mosquito fish 16.4 mg/L, rainbow trout 0.2–1.3 mg/L (Hartley and Kidd, 1987); acute oral LD_{50} of technical molinate for rats and mice 720 and 795 mg/kg, respectively (Ashton and Monaco, 1991), 501 mg/kg (RTECS, 1985).

Uses: Selective herbicide used to control the germination of annual grasses and broadleaved weeds in rice crops.

MONALIDE

Synonyms: 4'-Chloro-2,2-dimethylvaleranilide; **N-(4-Chlorophenyl)-2,2-dimethylpen-tanamide**; *N*-(4-Chlorophenyl)-2,2-dimethylvaleramide; 4'-Chloro-α,α-dimethylvalera-nilide; D-90-A; Schering 35830; Potablan; SN 35830.

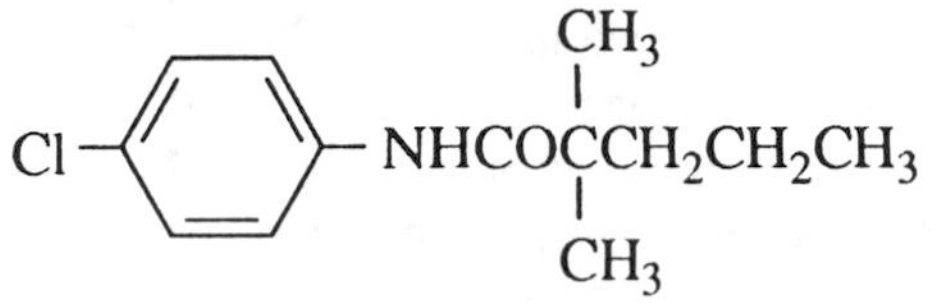

$$Cl-\left\langle\bigcirc\right\rangle-NHCOCCH_2CH_2CH_3$$

with CH_3 above and CH_3 below the central carbon.

Designations: CAS Registry Number: 7287-36-7; mf: $C_{13}H_{18}ClNO$; fw: 239.75; RTECS: YV6010000.

Properties: Colorless crystals. Mp: 87–88°C; H-$t_{1/2}$ at room temperature: 154 days (pH 5), 116 days (pH 8.95); K_H: 2.5×10^{-8} atm · m³/mol at 23–25°C (approximate — calculated from water solubility and vapor pressure); log K_{oc}: 2.89 (calculated); log K_{ow}: 3.06 (cal-culated); S_o (g/kg): cyclohexane (≈500), petroleum ether (<10), xylene (100); S_w: 22.8 mg/L at 23°C; vp: 1.8×10^{-6} mmHg at 20°C.

Environmental Fate

Biological. In the presence of suspended natural populations from unpolluted aquatic systems, the second-order microbial transformation rate constant determined in the labo-ratory was reported to be 6×10^{-13} L/organisms-hour (Steen, 1991).

Soil. Probably degrades via ring hydroxylation and subsequent ring cleavage. Persis-tence in soil is limited to approximate 6–8 weeks (Hartley and Kidd, 1987). Under laboratory conditions, the half-lives in soil were 30, 48 and 59 days at pH 4.85, 5.2 and 10.8, respectively (Worthing and Hance, 1991).

Formulation Types: Emulsifiable concentrate.

Toxicity: LC_{50} for guppy >100 mg/L (Worthing and Hance, 1991); acute oral LD_{50} for rats 2,600 mg/kg (RTECS, 1985).

Uses: Preemergence and postemergence control of broad-leaved weeds in parsley, carrots, dill and celery.

MONOCROTOPHOS

Synonyms: Apadrin; Azodrin; Azodrin insecticide; Biloborb; Bilobran; C 1414; CIBA 1414; Crisodrin; 3-(Dimethoxyphosphinyloxy)-*N*-methylisocrotonamide; Dimethyl-1-methyl-2-(methylcarbamoyl)vinyl phosphate; *O,O*-Dimethyl-*O*-(2-*N*-methylcarbamoyl-1-methylvinyl) phosphate; (*E*)-Dimethyl 1-methyl-3-(methylamino)-3-oxo-1-propenyl phosphate; Dimethyl (*E*)-1-methyl-2-(methylcarbamoyl)vinyl phosphate; Dimethyl phosphate ester of 3-hydroxy-*N*-methyl-*cis*-crotonamide; ENT 27129; Hazodrin; 3-Hydroxy-*N*-methyl-*cis*-crotonamide dimethyl phosphate; *cis*-1-Methyl-2-(methylcarbamoyl) vinyl phosphate; Monocil 40; Monocron; Nuvacron; Pandar; **Phosphoric acid dimethyl (1-methyl-3-(methylamino)-3-oxo-1-propenyl) ester**; Pillardrin; Plantdrin; SD 9129; Shell SD 9129; Susvin.

Designations: CAS Registry Number: 6923-22-4; mf: $C_7H_{14}NO_5P$; fw: 223.16; RTECS: TC4375000.

Properties: Colorless to reddish-brown, hygroscopic crystals with a mild ester-like odor. Technical grade is a reddish-brown to dark brown semisolid. Mp: 54–55°C (pure), 25–35°C (technical); bp: 125°C at 5×10^{-4} mmHg; ρ: 1.33 at 20/4°C; H-$t_{1/2}$: 96 days (pH 5), 66 days (pH 7), 17 days (pH 9); K_d: 0.077-0.615; log K_{ow}: –1.97 (calculated); S_o (g/kg): acetone (700), methanol (1,000), methylene chloride (800), *n*-octanol (60), toluene (60); S_w: miscible; vp: 6.75×10^{-5} mmHg at 20°C.

Environmental Fate

Plant. Decomposes in plants forming the *N*-hydroxy compound in small amounts (Hartley and Kidd, 1987).

Chemical/Physical. Hua et al. (1995) studied the titanium dioxide-mediated photooxidation of monocrotophos using a recirculating photoreactor. The initial concentration and flow rate used were 50 nM and 30 mL/minute, respectively. The irradiation source was a black light that emits radiation between 300 and 420 nm with a maximum radiation occurring at 350 nm. Degradation yields of 51 and 95% were achieved after 1 and 3 hours, respectively. The rate of degradation was enhanced with the addition of oxygen, hydrogen peroxide and sulfate ions but decreased when chloride, perchlorate, nitrate and phosphate ions were present. In the presence of Cu^{2+} ions, the rate of reaction was enhanced when the concentration was $<10^{-5}$ M but was hindered at higher concentrations.

Emits toxic fumes of phosphorus and nitrogen oxides when heated to decomposition (Sax and Lewis, 1987; Lewis, 1990).

Exposure Limits: OSHA PEL: TWA 0.25 mg/m³; ACGIH TLV: TWA 0.25 mg/m³.

Formulation Types: Soluble concentrate; granules.

Toxicity: LC_{50} (24-hour) for bluegill sunfish 23 mg/L, rainbow trout 12 mg/L (Worthing and Hance, 1991), blood clam 9.31 mg/L (Bharathi, 1994); LC_{50} (48-hour) for rainbow trout 7 mg/L (Hartley and Kidd, 1987), blood clam 6.19 mg/L (Bharathi, 1994); LC_{50} (72-hour) for blood clam 5.15 mg/L (Bharathi, 1994); LC_{50} (96-hour) for blood clam 3.53 mg/L (Bharathi, 1994); acute oral LD_{50} for male and female rats 18 and 20 mg/kg, respectively (Hartley and Kidd, 1987), 8 mg/kg (RTECS, 1985).

Uses: Systemic insecticide and acaricide used to control pests in cotton, sugarcane, coffee, tobacco, olives, rice, hops, sorghum, maize, deciduous fruits, citrus fruits, potatoes, sugarbeet, tomatoes, soya beans and ornamentals.

MONURON

Synonyms: Chlorfenidim; 1-(4-Chlorophenyl)-3,3-dimethylurea; 1-(p-Chlorophenyl)-3,3-dimethylurea; 3-(4-Chlorophenyl)-1,1-dimethylurea; 3-(p-Chlorophenyl)-1,1-dimethylurea; *N'*-**(4-Chlorophenyl)**-*N,N*-**dimethylurea**; *N*-(4-Chlorophenyl)-*N',N'*-dimethylurea; *N'*-(p-Chlorophenyl)-*N,N*-dimethylurea; CMU; *N,N*-Dimethyl-*N'*-(4-chlorophenyl)urea; *N,N*-Dimethyl-*N'*-(p-chlorophenyl)urea; Karmex; Lirobetarex; Monurex; Monurox; Monuuron; NCI-C02846; Rosuran; Telvar; Telvar Monuron Weed Killer; Urox; USAF P-8; USAF XR-41.

$$NHCON(CH_3)_2$$

Designations: CAS Registry Number: 150-68-5; mf: $C_9H_{11}ClN_2O$; fw: 198.66; RTECS: YS6300000.

Properties: Colorless to white, odorless crystals. Mp: 174–175°C; bp: 185–200°C (decomposes); ρ: 1.27 at 20/4°C; pK_a: unknown but saturated aqueous solution has a pH of 6.26; K_H: 3×10^{-8} atm · m³/mol at 20–25°C (approximate — calculated from water solubility and vapor pressure); log K_{oc}: 1.99, 2.33; log K_{ow}: 1.54 (Liu and Qian, 1995); S_o (g/kg at 27°C): acetone (52), benzene (3); S_w: 230 mg/L at 25°C (pH = 6.26); vp: 4.5×10^{-7} mmHg at 20°C.

Soil properties and adsorption data

Soil	K_d (mL/g)	f_{oc} (%)	K_{oc} (mL/g)	pH	CEC (meq/100 g)
Cecil loamy sand	0.40	0.41	98	5.8	—
Chillum silt loam	3.30	2.55	129	4.6	7.6
Keyport silt loam	2.60	1.22	213	5.4	—
Hagerstown silty clay loam	4.00	2.49	160	5.5	12.5
Hawaiian alluvial	4.80	—	—	—	—
Hawaiian alluvial	6.00	—	—	—	—
Hawaiian gray hydromorphic	8.00	—	—	—	—
Hawaiian low humic latosol	0.70	—	—	—	—
Hawaiian low humic latosol	0.90	—	—	—	—
Hawaiian low humic latosol	1.00	—	—	—	—
Hawaiian regosoil	1.20	—	—	—	—
Lakeland sandy loam	1.10	1.91	58	6.2	2.9
Wehadkee silt loam	1.90	1.10	173	5.6	7.6

Source: Yuen and Hilton, 1962; Harris, 1966; Rhodes et al., 1970.

Environmental Fate

Biological. Monuron was mineralized in sewage samples obtained from a water treatment plant in Ithica, NY. (4-Chlorophenyl)urea and 4-chloroaniline were tentatively identified as metabolites (Wang et al., 1985).

Soil/Plant. In soils and plants, monuron is demethylated at the terminal nitrogen atom coupled with ring hydroxylation forming 3-(2-hydroxy-4-chlorophenyl)urea and 3-(3-hydroxy-4-chlorophenyl)urea (Hartley and Kidd, 1987). Wallnöefer et al. (1973) reported that the soil microorganism *Rhizopus japonicus* degraded monuron 3-(*p*-chlorophenyl)-1-methylurea. However, in the presence of *Pseudomonas* or *Arthrobacter* sp., monuron degraded to 2,4-dichloroaniline, *sym*-bis(3,4-dichlorophenyl)urea and unidentified metabolites (Janko et al., 1970). The reported half-life in soil is 166 days (Jury et al., 1987).

Photolytic. When an aqueous solution of monuron was exposed to sunlight or simulated sunlight, the major degradative pathways observed were the photooxidation and demethylation of the *N*-methyl groups (Crosby and Tang, 1969; Tanaka et al., 1982a), hydroxylation of the aromatic ring and polymerization. Products identified by TLC and confirmed using IR, mass spectra and/or chromatographic methods were 3-(*p*-chlorophenyl)-1-formyl-1-methylurea, 1-(*p*-chlorophenyl)-3-methylurea, 3-(4-chloro-2-hydroxyphenyl)-1,1-dimethylurea, 4,4'-dichlorocarbanilide and the tentatively identified compounds *p*'-chloroformanilide, 1-(*p*-chlorophenyl)-3-formylurea, *p*-chloroaniline (Crosby and Tang, 1969), formaldehyde, formic acid and carbon dioxide (Tanaka et al., 1982a). A similar study was performed by Rosen et al. (1969). An aqueous solution of monuron was exposed to a late summer sun for 17 days. The solution contained 3-(4-hydroxyphenyl)-1,1-dimethylurea and other unidentified compounds (Rosen et al., 1969). Tanaka et al. (1977) also studied the simulated sunlight photolysis of a saturated aqueous solution (200 ppm) of monuron. The main degradative processes were ring hydroxylation, methyl oxidation, *N*-demethylation, dechlorination and dimerization. Photoproducts identified by TLC were 3-(4-chlorophenyl)-1-methylurea, 3-(4-chlorophenyl)-1-formyl-1-methylurea, 3-(4-chloro-2-hydroxyphenyl)-1,1-dimethylurea, 3-(4-hydroxyphenyl)-1,1-dimethylurea, 3-(4-hydroxyphenyl)-1-formyl-1-methylurea, 4,4'-dichlorocarbanilide, 3-{4-[*N*-(*N'*,*N'*-dimethylaminocarbonyl)-4'-chloroanilino]phenyl}-1,1'-dimethylurea (dimer), a monodemethylated dimer, a hydroxylated dimer, a dihydroxylated dimer and a trimer (Tanaka et al., 1977). The rate of photolysis of monuron in aqueous solutions increased with the addition of nonionic surfactants (Tergitol TMN-6, Tergitol TMN-10, Triton X-100, Triton X-405) at concentrations in excess of the critical micelle concentration (Tanaka et al., 1979). The surfactants eliminated ring hydroxylation reactions but enhanced reductive dechlorination reactions with simultaneous formation of biphenyls. Photoproducts included 3-(4-chlorophenyl)-1,1-dimethylurea, 3-(4-chlorophenyl)-1-methylurea (monomethyl monuron), 3-phenyl-1,1-dimethylurea, 3-phenyl-1-methylurea, 3-{4-[*N*-(*N'*,*N'*-dimethylaminocarbonyl)-4'-chloroanilino]phenyl}-1,1-dimethylurea (monuron dimer), 3-{4-[*N*-(*N'*,*N'*-dimethylaminocarbonyl)anilino]phenyl}-1,1'-dimethylurea (fenuron dimer), formaldehyde and a polymeric material (Tanaka et al., 1979).

Tanaka et al. (1981) studied the photolysis of monuron in dilute aqueous solutions in order to fully characterize a substituted diphenylamine that was observed in an earlier investigation (Tanaka et al., 1977). They identified this compound as an isomeric mixture containing 92% 2-chloro-4',5-bis(*N'*,*N'*-dimethylureido)biphenyl and 8% 5-chloro-2,4'-bis(*N'*,*N'*-dimethylureido)biphenyl (Tanaka et al., 1981).

Tanaka et al. (1982) undertook a study to identify the several biphenyls formed in earlier photolysis studies (Tanaka et al., 1979, 1981). They identified these compounds as 2,4'-, 3,4'- and 4,4'-bis-(*N'*,*N'*-dimethylureido)biphenyls (fenuron biphenyls) (Tanaka et al., 1982, 1984). When an aqueous solution containing a surfactant was irradiated by UV light, photodegradation was the major degradative pathway. The compounds isolated were fenuron (3-phenyl-1,1-dimethylurea) (Tanaka et al., 1984) and the three fenuron biphenyls (Tanaka et al., 1982, 1984). In a later study, Tanaka et al. (1985) studied the photolysis of monuron (175 mg/L) in aqueous solution using UV light (λ = 300 nm) or sunlight. After 15 days of exposure to sunlight, monuron degraded forming a monochlorinated biphenyl product (1.5% yield) with the concomitant loss of hydrogen chloride (Tanaka et al., 1985). When a methanolic solution containing monuron was irradiated by UV light (λ = 253.7 nm) under aerobic conditions, 3-phenyl-1,1-dimethylurea formed as the major product. Methyl-*p*-chlorophenylcarbamate also formed but in small quantities (Mazzochi and Rao, 1972).

Chemical/Physical. Hydrolyzes in acidic media forming 4-chloroaniline (Hartley and Kidd, 1987). Under alkaline conditions (0.50 *N* sodium hydroxide) at 20°C, the half-life of monuron was 177 days (El-Dib and Aly, 1976).

Emits toxic fumes of nitrogen oxides and chlorine when heated to decomposition (Sax and Lewis, 1987).

Symptoms of Exposure: Causes anemia and methemoglobinemia in experimental animals.

Formulation Types: Granules; wettable powder; oil-miscible liquid.

Toxicity: LC_{50} (96-hour) for rainbow trout 76 mg/L; LC_{50} (48-hour) for coho salmon 110 mg/L; acute oral LD_{50} for rats 3,600 mg/kg (Hartley and Kidd, 1987), 1,053 mg/kg (RTECS, 1985).

Uses: Herbicide; sugarcane flowering suppressant.

NALED

Synonyms: Arthodibrom; Bromchlophos; Bromex; Dibrom; 1,2-Dibromo-2,2-dichloro-ethyldimethyl phosphate; Dimethyl 1,2-dibromo-2,2-dichloroethyl phosphate; *O,O*-Dimethyl-*O*-(1,2-dibromo-2,2-dichloroethyl)phosphate; *O,O*-Dimethyl *O*-(2,2-dichloro-1,2-dibromoethyl)phosphate; ENT 24988; Hibrom; NA 2783; Ortho 4355; Orthodibrom; Orthodibromo; **Phosphoric acid 1,2-dibromo-2,2-dichloroethyl dimethyl ester**; RE 4355.

$$CH_3O \quad \underset{\underset{CH_3O}{\diagup}}{\overset{\overset{O}{\|}}{\underset{}{P}}} - O - \underset{\underset{H}{|}}{\overset{\overset{Br}{|}}{C}} - \underset{\underset{Cl}{|}}{\overset{\overset{Br}{|}}{C}} - Cl$$

Designations: CAS Registry Number: 300-76-5; DOT: 2783; mf: $C_4H_7Br_2Cl_2O_4P$; fw: 380.79; RTECS: TB9450000.

Properties: Colorless to pale yellow liquid or solid with a slight pungent odor. Mp: 26.5–27.5°C; bp: 110°C at 0.5 mmHg; ρ: 1.96 at 25/4°C; S_o: freely soluble in ketone, alcohols, aromatic and chlorinated hydrocarbons but sparingly soluble in petroleum solvents and mineral oils; S_w: 10 mg/L; vp: 2×10^{-3} at 20°C.

Environmental Fate

Chemical/Physical. Completely hydrolyzed in water within 2 days (Windholz et al., 1983). In the presence of metals or reducing agents, naled loses bromine, forming dichlorvos (Hartley and Kidd, 1987).

Naled emits toxic fumes of bromines, chlorides and phosphorus oxides when heated to decomposition (Lewis, 1990).

Exposure Limits: NIOSH REL: TWA 3 mg/m³, IDLH 200 mg/m³; OSHA PEL: TWA 3 mg/m³; ACGIH TLV: TWA 3 mg/m³.

Formulation Types: Dustable powder; emulsifiable concentrate.

Toxicity: LC_{50} for bluegill sunfish 180 µg/L, rainbow trout 132 µg/L (Verschueren, 1983); LC_{50} (24-hour) for goldfish 2–4 mg/L (Hartley and Kidd, 1987) and crabs 0.33 mg/L (Worthing and Hance, 1991); acute oral LD_{50} for rats 430 mg/kg (Melnikov, 1971), 250 mg/kg (RTECS, 1985).

Uses: Insecticide used for control of spider mites, sucking and chewing insects in fruits, vegetables and ornamentals. Its use may be restricted.

NAPROPAMIDE

Synonyms: Devrinol; Devrinol 2EC; Devrinol 10G; Devrinol 50WP; *N,N*-**Diethyl-2-(1-naphthalenyloxy)propionamide**; *(RS)-N,N*-Diethyl-2-(1-naphthyloxy)propionamide; 2-(α-Napthoxy)-*N,N*-diethylpropionamide; R 7465; R 7475.

Designations: CAS Registry Number: 15299-99-7; mf: $C_{17}H_{21}NO_2$; fw: 271.36; RTECS: UE3600000.

Properties: Colorless crystals (pure); brown solid (technical). Mp: 74.8–77.5°C (pure), 68–70°C (technical); fl p: 191°C (open cup); H-$t_{1/2}$: 25.7 minutes (aqueous solution exposed to artificial sunlight); K_H: 2.9 × 10^{-8} atm · m^3/mol at 20°C (approximate — calculated from water solubility and vapor pressure); log K_{oc}: 2.83; log K_{ow}: 3.36; S_o (g/L at 20°C): acetone (>1,000), ethanol (>1,000), hexane (15), kerosene (62), xylene (505); S_w: 73 mg/L at 20°C; vp: 4 × 10^{-6} mmHg at 20°C.

Soil properties and adsorption data

Soil	K_d (mL/g)	f_{oc} (%)	K_{oc} (mL/g)	pH
Bermeo soil	53.60	—	—	5.3
Bermeo soil	46.00	—	—	6.0
Bermeo soil	39.80	—	—	6.4
Bet Dagan	2.23	0.84	265	—
Bet Dagan I	1.40	0.40	350	7.9
Bet Dagan II	2.96	1.01	293	7.8
Dead Sea sediment	—	2.13	9,769	—
Dead Sea sediment	—	2.13	5,231	2.0
Dead Sea sediment	—	2.13	2,242	6.0
Gilat	1.92	0.55	349	7.8
Gilat	1.85	0.73	253	—
Golan	3.16	0.68	465	—
Kinneret	7.60	1.79	426	—
Kinneret-A	30.60	4.55	673	—
Kinneret-F	—	5.05	414	—
Kinneret-F	—	5.05	3,437	2.0
Kinneret-F	—	5.05	2,493	6.0
Kinneret-F	27.70	4.31	643	—
Kinneret-G	—	2.84	517	—
Kinneret-G	—	2.84	4,121	2.0

Soil properties and adsorption data *(continued)*

Soil	K_d (mL/g)	f_{oc} (%)	K_{oc} (mL/g)	pH
Kinneret-G	—	2.84	1,883	6.0
Kinneret-G	21.20	2.55	831	—
Malkiya	—	2.21	353	—
Malkiya	—	2.21	1,411	3.0
Malkiya	—	2.21	760	6.0
Malkiya	7.70	3.38	228	—
Mivtahim	0.27	0.06	450	8.5
Mivtahim	0.47	0.26	181	—
Netanya	0.90	0.10	900	—
Neve Yaar	2.94	1.18	249	7.7
Neve Yaar	—	1.68	196	—
Neve Yaar	—	1.68	1,819	3.0
Neve Yaar	—	1.68	760	6.0
Neve Yaar	4.38	1.64	267	—
Oxford soil	—	9.61	246	—
Shefer	2.86	0.82	349	—
Tujunga loamy sand	2.01	0.58	346	—
Tujunga loamy sand	1.14	—	—	6.7
Tujunga loamy sand	1.08	—	—	7.4
Tujunga loamy sand	0.96	—	—	8.2

Source: Gerstl and Yaron, 1983; Gerstl and Mingelgrin, 1984; Jury et al., 1986; Gerstl and Kliger, 1990; Lee et al., 1990.

Environmental Fate

Soil. Degrades slowly in soil to 1-naphthol, 1,4-naphthoquinone, 2-(α-naphthoxy)-*N*-ethylpropionamide and 2-(α-naphthoxy)propionamide (Hartley and Kidd, 1987). In moist loam or sandy-loam soils at 70–90°C, the half-life was 8–12 weeks. However, the persistence may be as long as 9 months under conditions where microbial growth is limited (Ashton and Monaco, 1991). Gerstl and Yaron (1983a) reported that the half-life of napropamide in soil can range from 34 to 201 days depending upon the organic matter content, soil texture, soil moisture content and analytical method.

Plant. Rapidly metabolized in tomatoes and several fruit trees forming the water-soluble hexose conjugates of 4-hydroxynapropamide (Humburg et al., 1989; Ashton and Monaco, 1991).

Photolytic. Napropamide decomposes under UV light. Under laboratory conditions, irradiation of an aerated aqueous solution with UV light gave 2-hydroxypropananilide, 2-propenanilide and pyruvinanilide as the major products. Minor photoproducts include 2-(2-naphthoxy)propanoic acid and aniline (Tsao and Eto, 1990). In a deaerated aqueous solution, the photoproducts identified were 2-hydroxypropananilide, 2′-amino-2-hydroxypropiophenone, 2′-amino-2-naphthoxypropiophenone and 4′-amino-2-napthoxypropiophenone (Tsao and Eto, 1990).

Formulation Types: Emulsifiable concentrate (2 lb/gal); granules (10%); wettable powder (50%); suspension concentrate.

Toxicity: LC_{50} (96-hour) for bluegill sunfish 30 mg/L, goldfish >10 mg/L and rainbow trout 16.6 mg/L (Hartley and Kidd, 1987); acute oral LD_{50} for rats >5,000 mg/kg (RTECS, 1985).

Uses: Preemergence control of annual grasses and certain broad-leaved weeds in mint, orchards, small fruits, tobacco, turf, vegetable crops and ornamentals.

NAPTALAM

Synonyms: ACP 322; Alanap; Alanap L; Alanape; Alanap 10G AT; Dyanap; Morcran; **2-((1-Naphthalenylamino)carbonyl)benzoic acid**; α-Naphthylphthalamic acid; *N*-1-Naphthylphthalamic acid; Naptalame; Nip-A-Thin; NPA; PA; Peach-Thin; 6Q8; Solo.

Designations: CAS Registry Number: 132-66-1; mf: $C_{18}H_{13}NO_3$; fw: 291.29; RTECS: TH7350000.

Properties: Purple crystals with an unpleasant odor. Mp: 203°C (pure), 175–185°C (technical); ρ: 1.362 at 20/4°C (acid), 1.386 (sodium salt); fl p: nonflammable (water formulation); K_H: <1.9 × 10^{-3} atm · m^3/mol at 20°C (approximate — calculated from water solubility and vapor pressure); log K_{oc}: 2.37 (calculated); log K_{ow}: 2.04 (calculated); S_o (g/kg at 25°C): acetone (5), 2-butanone (4), carbon tetrachloride (0.1), *N,N*-dimethylformamide (39), dimethyl sulfoxide (32), methyl ethyl ketone (3.7), 2-propanol (2); S_w: 200 mg/L at 20°C; vp: <1 mmHg at 20°C.

Environmental Fate

Soil/Plant. Degrades in soils and plants forming 1-naphthylamine and phthalic acid (Hartley and Kidd, 1987; Humburg et al., 1989). Residual activity in soil is limited to approximately 3 to 4 months (Hartley and Kidd, 1987).

Chemical/Physical. Forms *N*-(1-naphthyl)phthalimide at elevated temperatures (Worthing and Hance, 1991). Naptalam will precipitate as the free acid in very acidic waters or in extremely hard waters (Humburg et al., 1989).

Formulation Types: Soluble concentrate (sodium salt = 2 lb/gal); granules (sodium salt = 10%); wettable powder.

Toxicity: LC$_{50}$ (96-hour) for bluegill sunfish 354 mg/L and rainbow trout 76 mg/L (Hartley and Kidd, 1987); acute oral LD$_{50}$ of the free acid and sodium salt for rats >8,200 and 1,800 mg/kg, respectively (Ashton and Monaco, 1991).

Uses: Selective preemergence herbicide used to control some grasses and many broadleaved weeds in soybeans, cucurbits, asparagus, groundnuts, potatoes and established woody ornamentals.

NEBURON

Synonyms: 1-Butyl-3-(3,4-dichlorophenyl)-1-methylurea; **N-Butyl-N'-(3,4-dichlorophenyl)-N-methylurea**; Granurex; Kloben; Neburea; Neburex.

Designations: CAS Registry Number: 555-37-3; mf: $C_{12}H_{16}Cl_2N_2O$; fw: 275.18; RTECS: YS3810000.

Properties: Colorless crystals. Mp: 101.5–103°C; log K_{oc}: 3.49; log K_{ow}: 4.10 (Liu and Qian, 1995); S_o: sparingly soluble in hydrocarbons; S_w: 4.8 mg/L at 24°C; vp: negligible at room temperature.

Soil properties and adsorption data

Soil	K_d (mL/g)	f_{oc} (%)	K_{oc} (mL/g)	pH	CEC (meq/100 g)
Great House sandy loam	325	12.00	2,708	6.3	18.0
Rosemaunde sandy clay loam	58	1.76	3,296	6.7	14.0
Toll Farm HP	240	11.70	2,051	7.4	41.0
Transcoed silty clay loam	139	3.69	3,767	6.2	12.0
Weed Res. sandy loam	72	1.93	3,731	7.1	11.0

Source: Hance, 1965.

Environmental Fate

Soil. Neburon undergoes dealkylation of the terminal nitrogen atom, ring hydroxylation and degradation to dichlorodihydroxyaniline (Hartley and Kidd, 1987). Residual activity in soil is limited to approximately 3 to 4 months (Hartley and Kidd, 1987).

Chemical/Physical. Neburon hydrolyzes under basic conditions forming aniline derivatives. The half-life of neburon in 0.50 N sodium hydroxide at 20°C is 167 days (El-Dib and Aly, 1976).

Formulation Types: Wettable powder.

Toxicity: LC90 (96-hour) for four fish species is 600–900 μg/L (Hartley and Kidd, 1987); acute oral LD$_{50}$ for rats >11,000 mg/kg (Hartley and Kidd, 1987), 1,100 mg/kg (RTECS, 1985).

Uses: Preemergence control of grasses and broad-leaved weeds in peas, beans, lucerne, garlic, beets, cereals, strawberries, ornamentals and forestry.

NITRAPYRIN

Synonyms: 2-Chloro-6-(trichloromethyl)pyridine; Dowco 163; N-serve; N-serve nitrogen stabilizer.

Designations: CAS Registry Number: 1929-82-4; mf: $C_6H_3Cl_4N$; fw: 230.90; RTECS: US7525000.

Properties: Colorless, crystalline solid. Mp: 62–63°C; ρ: 1.744 at 20/4°C (estimated); K_H: 2.13×10^{-3} atm · m³/mol at 22–25°C (approximate — calculated from water solubility and vapor pressure); log K_{oc}: 2.62–2.68; log K_{ow}: 3.02–3.41; S_o (kg/kg at 20–26°C): acetone (1.98), ethanol (0.29), methylene chloride (1.85), toluene (1.39), 1,1,1-trichloroethane (0.80), xylene (1.04); S_w: 40 mg/kg at 22°C; vp: 2.8×10^{-3} mmHg at 20°C.

Soil properties and adsorption data

Soil	K_d (mL/g)	f_{oc} (%)	K_{oc} (mL/g)	pH
Batcombe silt loam	—	0.63–1.46	172	6.7–7.5
British Columbia silty clay loam	4.59	1.97	233	7.6
California clay	2.01	1.04	193	7.7
California loam	8.11	1.80	451	7.8
California loam	44.45	18.68	238	5.9
California sandy loam	1.46	0.46	317	7.3
California sandy loam	0.32	0.17	188	7.5
California silt loam	20.21	6.21	325	5.3
Catlin	9.21	2.01	460	6.2
Commerce	3.01	0.68	440	6.7
Minnesota loam	10.55	7.19	147	8.1
Minnesota loam	9.11	3.25	311	7.8
Texas clay	4.89	1.57	311	6.8
Tracy	5.31	1.12	474	6.2

Source: Goring, 1962; Briggs, 1981; McCall et al., 1981.

Environmental Fate

Biological. 6-Chloropicolinic acid and carbon dioxide were reported as biodegradation products (Verschueren, 1983).

Soil. Hydrolyzes in soil to 6-chloropyridine-2-carboxylic acid (Worthing and Hance, 1991).

Photolytic. Photolysis of nitrapyrin in water yielded 6-chloropicolinic acid, 6-hydroxypicolinic acid and an unidentified polar material (Verschueren, 1983).

Chemical/Physical. Emits toxic fumes of nitrogen oxides and chlorides when heated to decomposition (Sax and Lewis, 1987; Lewis, 1990).

Exposure Limits: NIOSH REL: 10 mg/m^3, STEL 20 mg/m^3 (total), 5 mg/m^3 (respirable fraction); OSHA PEL: 15 mg/m^3 (total), 5 mg/m^3 (respirable fraction); ACGIH TLV: TWA 10 mg/m^3.

Toxicity: LC$_{50}$ for channel catfish 5.8 mg/L (Worthing and Hance, 1991); acute oral LD$_{50}$ for rats 1,230 mg/kg (Verschueren, 1983), 940 mg/kg (RTECS, 1985).

Uses: Bactericide used to inhibit *Nitrosomonas* spp. from oxidizing ammonium ions in *soil*.

OXADIAZON

Synonyms: 2-*tert*-Butyl-4-(2-2,4-dichloro-5-isopropyloxyphenyl)-1,3,4-oxadiazolin-5-one; Chipco Ronstar G; Chipco Ronstar 50WP; **3-(2,4-Di-chloro-5-(1-methylethoxy)phenyl)-5-(1,1-dimethylethyl)-1,3,4-oxadiazol-2(3***H***)-one**; Ronstar; Ronstar 25EC; Ronstar 2G; Ronstar 12L; RP 17623.

Designations: CAS Registry Number: 19666-30-9; mf: $C_{15}H_{18}Cl_2N_2O_3$; fw: 345.22; RTECS: RO0874000.

Properties: Odorless, white crystals. Mp: 88–90°C; fl p: nonflammable; K_H: $<6.5 \times 10^{-10}$ atm · m³/mol at 20°C (approximate — calculated from water solubility and vapor pressure); log K_{oc}: 0.70–2.99; log K_{ow}: 4.70; S_o (g/L at 20°C): acetone (600), benzene (1,000), cyclohexanone (200), ethanol (100); S_w: 700 mg/L at 20°C; vp: $<10^{-6}$ mmHg at 20°C.

Soil properties and adsorption data

Soil	K_d (mL/g)	f_{oc} (%)	K_{oc} (mL/g)	pH
Batcombe silt loam	—	0.63–1.46	5	6.7–7.5
Kanuma high clay	5.2	1.35	387	5.7
Tsukuba clay loam	41.3	4.24	973	6.5

Source: Briggs, 1981; Kanazawa, 1989.

Environmental Fate

Soil. The reported half-life in soil is approximately 3–6 months (Hartley and Kidd, 1987). Oxadiazon degraded slowly in both moist and flooded soils. After 25 weeks, only 0.1–3.5% degraded to carbon dioxide and 0.5–1.1% as volatile products. Metabolites identified included oxadiazonphenol, oxadiazon acid and methoxyoxadiazon (Ambrosi et al., 1977).

Plant. Reported half-lives of oxadiazon in rice and orchard fruit are 1–2 and.3–6 months, respectively (Hartley and Kidd, 1987).

Symptoms of Exposure: May be irritating to skin.

Formulation Types: Wettable powder (50%); granules (5%).

Toxicity: LC_{50} (96-hour) for carp 1.76 mg/L and channel catfish ≥15.4 mg/L (Worthing and Hance, 1991); acute oral LD_{50} of technical oxadiazon for rats >8,000 mg/kg (Ashton and Monaco, 1991), 3,500 mg/kg (RTECS, 1985).

Uses: Preemergence herbicide used for controlling certain annual grasses (e.g., bluegrass, barnyardgrass, crabgrass, goosegrass, sprangletop) and broad-leaved weeds (e.g., cudweed, dayflower, filaree, groundsel, jimsonweed, morningglory, mustards, pigweed, redmaids, smartweed, sowthistle, velvetleaf) in turf, lawns, orchards and ornamentals.

OXAMYL

Synonyms: D-1410; 2-(Dimethylamino)-*N*-(((methylamino)carbonyl)oxy)-2-oxoetha-nimidothioic acid methyl ester; 2-Dimethylamino-1-(methylthio)glyoxal *O*-methylcar-bamoylmonoxime; *N*,*N*-Dimethyl-α-methylcarbamoyloxyimino-α-(methylthio)aceta-mide; *N'*,*N'*-Dimethyl-*N*-((methylcarbamoyl)oxy)-1-thiooxamimidic acid methyl ester; DPX 1410; Insecticide-nematocide 1410; **Methyl 2-(dimethylamino)-*N*-(((methyl-amino)carbonyl)oxy)-2-oxoethanimidothioate**; Methyl-1-(dimethylcarbamoyl)-*N*-((methylcarbamoyl)oxy)thioformimidate; *S*-Methyl 1-(dimethylcarbamoyl)-*N*-((methyl-carbamoyl)oxy)thioformimidate; Methyl-*N'*,*N'*-dimethyl-*N*-((methylcarbamoyl)oxy)-1-thiooxamimidate; Thiozamyl; Vydate; Vydate L insecticide/nematocide; Vydate L oxamyl insecticide/nematocide.

$$(CH_3)_2N\overset{\overset{\displaystyle O}{\|}}{C} - \underset{\underset{\displaystyle SCH_3}{|}}{C}=N-O\overset{\overset{\displaystyle O}{\|}}{C}NHCH_3$$

Designations: CAS Registry Number: 23135-22-0; mf: $C_7H_{13}N_3O_3S$; fw: 219.25; RTECS: RP2300000.

Properties: Crystalline solid with a sulfur-like odor. Mp: 100–102°C which changes to a dimorphic form melting at 108–110°C; bp: decomposes; ρ: 0.97 at 25/4°C; K_H: 2.6×10^{-6} atm · m³/mol at 20–25°C (approximate — calculated from water solubility and vapor pressure); log K_{oc}: –0.70 to 1.40; log K_{ow}: –0.4; P-$t_{1/2}$: 55.38 hours (absorbance λ = 223.0 nm, concentration on glass plates = 6.7 μg/cm²); S_o (g/L at 25°C): acetone (670), ethanol (330), methanol (1,440), 2-propanol (110), toluene (10); S_w: 280 g/L at 25°C; vp: 2.33×10^{-4} mmHg at 20°C.

Soil properties and adsorption data

Soil	K_d (mL/g)	f_{oc} (%)	K_{oc} (mL/g)	pH
Arredondo sand	0.06	0.80	7.5	6.8
Cecil sandy loam	0.05	0.90	5.6	5.6
Cu-Montmorillonite	11.50	—	—	—
Devizes	0.12	4.47	2.7	7.8
Golan	0.23	0.68	33.8	—
Kinneret-A	0.41	4.55	9.0	—
Kinneret-F	0.41	4.31	9.5	—
Kinneret-G	0.51	2.55	20.0	—
Malkiya	0.24	3.38	7.1	—
Mepal	0.62	11.60	5.3	7.2
Pitstone	0.17	2.18	7.8	8.0
Sutton Veany	0.35	9.05	3.9	7.6

Soil properties and adsorption data *(continued)*

Soil	K_d (mL/g)	f_{oc} (%)	K_{oc} (mL/g)	pH
Webster silty clay loam	0.40	3.97	10.1	7.3
Woburn sandy loam	0.02	0.78	2.6	7.0
Woburn sandy loam	0.30	3.43	87.5	6.3
Zn-montmorillonite	6.70	—	—	—

Source: Bromilow et al., 1980; Khan and Bansal, 1980; Gerstl and Mingelgrin, 1984; Bilkert and Rao, 1985.

Environmental Fate

Soil. Oxamyl rapidly degraded in a loamy sand and fine sand soil at 25°C to carbon dioxide and the intermediate methyl *N*-hydroxy-*N*,*N*-dimethyl-1-thiooxaminidate (Rajagopol et al., 1984). The reported half-life in soil is approximately one week (Worthing and Hance, 1991). Ou and Rao (1986) reported a half-life in soil of 8–50 days. The reported half-lives of oxamyl in Pitstone, Devizes, Sutton Veany and Mepal soils at 15°C were reported to be 10.2–13.1, 6.2, 7.1 and 17.8 days, respectively (Bromilow et al., 1980). Smelt et al. (1987) reported that oxamyl degraded at a higher rate in field plots after repeated applications of this nematocide than in soils that received no treatment. The repeated applications of oxamyl to soils probably induced microbial activity, which resulted in the accelerated disappearance of this compound.

Harvey and Han (1978) reported a half-life of 8 days for oxamyl in soil.

Groundwater. According to the U.S. EPA (1986) oxamyl has a high potential to leach to *groundwater.*

Plant. Dislodgable residues of oxamyl on cotton leaf 0, 24, 48, 72 and 96 hours after application (0.41 kg/ha) were 1.5, 1.1, 1.2, 0.85 and 0.76 µg/m², respectively (Buck et al., 1980).

Chemical/Physical. The hydrolysis half-lives of oxamyl in a sterile 1% ethanol/water solution at 25°C and pH values of 4.5, 6.0, 7.0 and 8.0, were 300, 17, 1.6 and 0.20 weeks, respectively (Chapman and Cole, 1982). Under alkaline conditions, oxamyl hydrolyzed following first-order kinetics (Bromilow et al., 1980). Emits toxic fumes of nitrogen and sulfur oxides when heated to decomposition (Sax and Lewis, 1987).

Formulation Types: Soluble concentrate; granules.

Toxicity: LC_{50} (96-hour) for rainbow trout 4.2 mg/L, bluegill sunfish 5.6 mg/L and goldfish 27.5 mg/L (Hartley and Kidd, 1987); acute oral LD_{50} for rats 5.4 mg/kg (Hartley and Kidd, 1987), 2,500 µg/kg (RTECS, 1985).

Uses: Insecticide, nematocide, acaricide.

OXYDEMETON-METHYL

Synonyms: Bay 21097; Bayer 21097; Demeton-methyl sulfoxide; Demeton-*O*-methyl sulfoxide; Demeton-*S*-methyl sulfoxide; *O,O*-Dimethyl *S*-(2-eththionylethyl) phosphorothioate; Dimethyl *S*-(2-eththionylethyl) thiophosphate; *O,O*-Dimethyl *S*-2-(ethylsulfinyl)ethyl phosphorothioate; *O,O*-Dimethyl *S*-2-(ethylsulfinyl)ethyl thiophosphate; *O,O*-Dimethyl *S*-ethylsulphinylethyl phosphorothioate; ENT 24964; **S-2-(Ethylsulfinyl)ethyl O,O-dimethyl phosphorothioate**; Isomethylsystox sulfoxide; Metaisosystox sulfoxide; Metasystemox; Metasystox-R; Methyl demeton-*O*-sulfoxide; Metilmercaptofosoksid; Oxydemetonmethyl; Phosphothioic acid *O,O*-dimethyl *S*-2-(ethylsulfinyl)ethyl ester; R 2170.

$$CH_3CH_2SCH_2CH_2S\overset{\displaystyle O}{\overset{\displaystyle \|}{P}}(OCH_3)_2$$

Designations: CAS Registry Number: 301-12-2; mf: $C_6H_{15}O_4PS_2$; fw: 246.29; RTECS: TG1420000.

Properties: Clear, amber-colored liquid. Mp: $<-10°C$; bp: $>80°C$ (decomposes); ρ: 1.289 at 20/4°C; log K_{oc}: 0.70–1.49; log K_{ow}: -1.97 (calculated); S_o: miscible with many aromatic and halogenated solvents; S_w: miscible; S_w: miscible; vp: 2.85×10^{-5} mmHg at 20°C.

Soil properties and adsorption data

Soil	K_d (mL/g)	f_{oc} (%)	K_{oc} (mL/g)	pH
Sand	0.10	2.15	5	6.9
Sandy loam	0.54	1.74	31	5.5
Silt loam	0.19	1.22	16	6.7

Source: U.S. Department of Agriculture, 1990.

Environmental Fate

Soil. The sulfoxide group is oxidized to the sulfone and oxidative and hydrolytic cleavage of the side chain gives dimethylphosphoric and phosphoric acids (Hartley and Kidd, 1987). Oxamyl was degraded by the microorganism *Pseudomonas putida* in a laboratory study using cultured bacteria (Zeigler, 1980).

Plant. In asparagus, oxydemeton-methyl was converted to the corresponding sulfone (Szeto and Brown, 1982).

Chemical/Physical. Oxydemeton-methyl can be converted to the corresponding sulfone by hydrogen peroxide (Cremlyn, 1991). Emits toxic fumes of phosphorus and sulfur oxides when heated to decomposition (Sax and Lewis, 1987).

Formulation Types: Soluble concentrate; emulsifiable concentrate.

Toxicity: LC_{50} (96-hour) for rainbow trout 4.0 mg/L (Walker, 1964); LC_{50} (24-hour) for rainbow trout and bluegill sunfish 10 mg/L (Hartley and Kidd, 1987); acute oral LD_{50} for rats 65–75 mg/kg (Hartley and Kidd, 1987), 30 mg/kg (RTECS, 1985).

Uses: Systemic and contact insecticide and acaricide used to control spider mites and other insects on vegetables and some ornamentals.

PARATHION

Synonyms: AAT; AATP; AC 3422; ACC 3422; Alkron; Alleron; American Cyanamid 3,422; Aphamite; B 404; Bay E-605; Bladan; Bladan F; Compound 3422; Corothion; Corthion; Corthione; Danthion; DDP; *O,O*-Diethyl-*O*-4-nitrophenyl phosphorothioate; *O,O*-Diethyl *O-p*-nitrophenyl phosphorothioate; Diethyl-4-nitrophenyl phosphorothionate; Diethyl-*p*-nitrophenyl thionophosphate; *O,O*-Diethyl-*O*-4-nitrophenyl thionophosphate; *O,O*-Diethyl-*O-p*-nitrophenyl thionophosphate; Diethyl-*p*-nitrophenyl thiophosphate; *O,O*-Diethyl-*O*-p-nitrophenyl thiophosphate; Diethyl-parathion; DNTP; DPP; Drexel parathion 8E; E 605; Ecatox; Ekatin WF & WF ULV; Ekatox; ENT 15108; Ethlon; Ethyl parathion; Etilon; Folidol; Folidol E605; Folidol E & E 605; Fosfermo; Fosferno; Fosfex; Fosfive; Fosova; Fostern; Fostox; Gearphos; Genithion; Kolphos; Kypthion; Lethalaire G 54; Lirothion; Murfos; NA 2783; NCI-C00226; Niran; Niran E-4; Nitrostigmine; Nitrostygmine; Niuif-100; Nourithion; Oleofos 20; Oleoparaphene; Oleoparathion; Orthophos; Pac; Panthion; Paradust; Paraflow; Paramar; Paramar 50; Paraphos; Paraspray; Parathene; Parathionethyl; Parawet; Penphos; Pestox plus; Pethion; Phoskil; Phosphemol; Phosphenol; **Phosphorothioic acid *O,O*-diethyl *O*-(4-nitrophenyl) ester**; Phosphostigmine; RB; RCRA waste number P089; Rhodiasol; Rhodiatox; Rhodiatrox; Selephos; Sixty-three special E.C. insecticide; SNP; Soprathion; Stabilized ethyl parathion; Stathion; Strathion; Sulphos; Super rodiatox; T-47; Thiofos; Tiophos; Thiophos 3422; Tox 47; Vapophos; Vitrex.

Designations: CAS Registry Number: 56-38-2; DOT: 2783; mf: $C_{10}H_{14}NO_5PS$; fw: 291.27; RTECS: TF4550000.

Properties: Light yellow to dark brown liquid with a garlic-like odor. Mp: 6.1°C; bp: 375°C, 157–162°C at 0.6 mmHg; ρ: 1.26 at 25/4°C; fl p: 174°C; H-$t_{1/2}$: ≈ 3.5 weeks at pH 6; K_H: 8.56 × 10^{-8} atm · m³/mol at 25°C; log K_{oc}: 2.50–4.20 (Mingelgrin and Gerstl, 1983); log K_{ow}: 2.15–3.93; P-$t_{1/2}$: 57.6 hours at 90–95°C (irradiation of aqueous solution by UV light); S_o: freely soluble in lower alcohols, aromatic and saturated hydrocarbons, esters, ethers, ketones and many common other organic solvents but only slightly soluble in petroleum ether and kerosene, hexane (57.4 and 85.6 g/L at 20 and 30°C, respectively (Chiou et al., 1985); S_w: 10.3, 12.9 and 15.2 mg/L at 10, 20 and 30°C, respectively; vap d: 11.91 g/L at 25°C, 10.06 (air = 1); vp: 4 × 10^{-4} mmHg at 20°C.

Soil properties and adsorption data

Soil	K_d (mL/g)	f_{oc} (%)	K_{oc} (mL/g)	pH	CEC (meq/100 g)
Amarillo silt loam	5.13	0.29	1,769	7.7	8.00
Bet Dagan	5.19	0.84	618	—	—

Soil properties and adsorption data *(continued)*

Soil	K_d (mL/g)	f_{oc} (%)	K_{oc} (mL/g)	pH	CEC (meq/100 g)
Ca-montmorillonite	120.80	—	—	—	—
Clarion soil	33.81	2.64	1,278	5.0	21.02
Elkhorn sandy loam	18.19	0.09	21,151	6.0	—
Fe-montmorillonite	605.80	—	—	—	—
Gilat	5.35	0.73	733	—	—
Golan	3.63	0.68	534	—	—
Harps soil	41.12	3.80	1,083	7.3	37.84
Hugo gravelly sand loam	6.90	0.12	5,750	5.5	—
Katy silt loam	6.05	0.58	1,043	5.1	—
Kinneret	30.00	1.79	1,676	—	—
Kinneret-A	63.70	4.55	1,400	—	—
Kinneret-F	64.90	4.31	1,506	—	—
Kinneret-G	40.10	2.55	1,573	—	—
Malkiya	28.30	3.38	837	—	—
Mivtahim	1.37	0.26	527	—	—
Nacogdoches clay	2.26	0.23	983	5.0	14.00
Na-montmorillonite	119.80	—	—	—	—
Netanya	0.78	0.10	780	—	—
Neve Yaar	12.30	1.64	750	—	—
Peat soil	254.68	18.36	1,388	7.0	77.34
Rothamsted Farm	15.68	1.51	1,038	5.1	—
Sarpy fine sandy loam	5.68	0.51	1,114	7.3	5.71
Shefer	4.99	0.82	609	—	—
Soil #8	12.30	0.94	1,308	6.3	26.60
Soil #10	7.67	0.44	1,743	6.2	18.60
Soil #11	38.02	1.67	2,276	6.3	42.80
Soil #13	125.90	3.20	3,934	5.2	19.20
Soil #14	457.10	14.28	3,201	3.3	28.90
Soil #15	213.80	4.76	4,492	3.5	21.20
Sweeney sandy clay loam	22.16	0.65	3,409	6.3	—
Thurman loamy fine sand	11.07	1.07	1,034	6.8	6.10
Tierra heavy clay loam	40.56	0.33	12,291	6.2	—

Source: King and McCarty, 1968; Bowman and Sans, 1977; Wahid and Sethunathan, 1978; Felsot and Dahm, 1979; Lord et al., 1980; Gerstl and Mingelgrin, 1984.

Environmental Fate

Biological. Initial hydrolysis products include diethyl-*O*-thiophosphoric acid, *p*-nitrophenol (Sethunathan, 1973, 1973a; Munnecke and Hsieh, 1976; Sethunathan et al., 1977; Verschueren, 1983) and the biodegradation products *p*-aminoparathion and *p*-aminophenol (Sethunathan, 1973; Laplanche et al., 1981; Nelson, 1982). Mixed bacterial cultures were capable of growing on technical parathion as the sole carbon and energy source (Munnecke and Hsieh, 1976). Three oxidative pathways were reported. The primary degradative pathway is initial hydrolysis to yield *p*-nitrophenol and diethylthiophosphoric acid. The secondary pathway involves the formation of paraoxon (diethyl *p*-nitrophenyl phosphate) which subsequently undergoes hydrolysis to yield *p*-nitrophenol and diethylphosphoric acid. The third degradative pathway involved reduction of parathion under low oxygen conditions to yield *p*-amino-parathion followed by hydrolysis to *p*-aminophenol and diethylphosphoric acid. Other potential degradation products include hydroquinone, 2-hydroxyhydroquinone, ammonia and polymeric substances (Munnecke and Hsieh, 1976). The reported half-life in soil is 18 days (Jury et al., 1987).

A *Flavobacterium* sp. (ATCC 27551), isolated from rice paddy water, degraded parathion to *p*-nitrophenol. The microbial hydrolysis half-life of this reaction was <1 hour (Sethunathan and Yoshida, 1973; Forrest, 1981). Sharmila et al. (1989) isolated a *Bacillus* sp. from a laterite soil that degraded parathion in the presence of yeast extracts. At yeast concentrations of (1) 0.05%, (2) 0.1 and 0.25% and (3) 0.5%, parathion degraded via (1) hydrolysis, (2) hydrolysis and nitro group reduction and (3) exclusively by nitro group reduction, respectively. Aminoparathion and *p*-nitrophenol were the identified metabolites under these conditions (Sharmila et al., 1989). Rosenberg and Alexander (1979) demonstrated that two strains of *Pseudomonas* used parathion as the sole source of phosphorus. It was suggested that degradation of parathion resulted from an induced enzyme or enzyme system that catalytically hydrolyzed the aryl P-O bond, forming dimethyl phosphorothioate as the major product.

In both soils and water, chemical- and biological-mediated reactions transform parathion to paraoxon (Alexander, 1981). Parathion was reported to biologically hydrolyze to *p*-nitrophenol in different soils under flooded conditions (Sudhakar-Barik and Sethunathan, 1978; Ferris and Lichtenstein, 1980).

p-Nitrophenol, paraoxon and three unidentified metabolites were identified in a model ecosystem containing algae, *Daphnia magna*, fish, mosquito and snails (Yu and Sanborn, 1975).

Soil. A *Pseudomonas* sp. (ATCC 29354), isolated from parathion-amended treated soil, degraded *p*-nitrophenol to *p*-nitrocatechol which was recalcitrant to further degradation. In an unsterilized soil, however, *p*-nitrocatechol was further degraded to nitrites and other unidentified compounds (Sudhakar-Barik et al., 1978a). *Pseudomonas* sp. and *Bacillus* sp., isolated from a parathion-amended flooded soil, degraded *p*-nitrophenol (parathion hydrolysis product) to nitrite ions (Siddaramappa et al., 1973; Sudhakar-Barik et al., 1976) and carbon dioxide (Sudhakar-Barik et al., 1976).

When parathion was equilibrated with aerobic soils, virtually no degradation was observed (Adhya et al., 1981a). However, in flooded (anaerobic) acid sulfate soils or low sulfate soils, aminoparathion and desethyl aminoparathion formed as the major metabolites (Adhya et al., 1981). However, under flooded acid conditions, parathion hydrolyzed to form *p*-nitrophenol (Sethunathan, 1973a). The rate of hydrolysis was found to be higher in soils containing higher organic matter content. The bacterium *Bacillus* sp., isolated from parathion-amended flooded alluvial soil, decomposed *p*-nitrophenol as a sole carbon source (Sethunathan, 1973a). In a similar study, Rajaram and Sethunathan (1975) were able to increase the rate of hydrolysis of parathion in soil by a variety of organic materials. They found that the rate of hydrolysis was as follows: glucose >rice straw >algal crust >farmyard manure >unamended *soil*. However, these organic materials inhibited the rate of hydrolysis when the soils were inoculated with a parathion-hydrolyzing bacterial culture.

p-Nitrophenol was identified as a hydrolysis product in soil (Suffet et al., 1967; Miles et al., 1979; Camper, 1991; Somasundaram et al., 1991). The rate of hydrolysis in soil is accelerated following repeated applications of parathion (Ferris and Lichtenstein, 1980) or the surface catalysis of clays (Mingelgrin et al., 1977). The reported hydrolysis half-lives at pH 7.4 (at 20 and 37.5°C) were 130 and 26.8 days, respectively. At pH 6.1 and

20°C, the hydrolysis half-life is 170 days (Freed et al., 1979). When equilibrated with a prereduced pokkali soil (acid sulfate), parathion instantaneously degraded to aminoparathion. The quick rate of reaction was reportedly due to soil enzymes and/or other heat labile substances. Desethyl aminoparathion was also identified as a metabolite in two separate studies (Wahid and Sethunathan, 1979; Wahid et al., 1980).

Aminoparathion also was formed when parathion (500 ppm) was incubated in a flooded alluvial soil. The amounts of parathion remaining after 6 and 12 days were 43.0 and 0.09%, respectively (Freed et al., 1979). It was observed that the degradation of parathion in flooded soil was more rapid in the presence of ferrous sulfate. Parathion degraded to aminoparathion and four unknown products (Rao and Sethunathan, 1979). In flooded alluvial soils, parathion degraded to aminoparathion via nitro group reduction. The rate of degradation remained constant despite variations in the redox potential of the soils (Adhya et al., 1981a). In soil, parathion may degrade via two oxidative pathways. The primary pathway is hydrolysis to p-nitrophenol (Sudhakar-Barik et al., 1979) and diethylthiophosphoric acid (Miles et al., 1979). The other pathway involves oxidation to paraoxon but aminoparathion is formed under anaerobic (Miles et al., 1979) and flooded conditions (Sudhakar-Barik et al., 1979). The degradation pathway as well as the rate of degradation of parathion in a flooded soil changed following each successive application of parathion (Sudhakar-Barik et al., 1979). After the first application, nitro group reduction gave aminoparathion as the major product. After the second application, both the hydrolysis product (p-nitrophenol) and aminoparathion were found. Following the third addition of parathion, p-nitrophenol was the only product detected. It was reported that the change from nitro group reduction to hydrolysis occurred as a result of rapid proliferation of parathion-hydrolyzing microorganisms that utilized p-nitrophenol as the carbon source (Sudhakar-Barik et al., 1979).

In a cranberry soil pretreated with p-nitrophenol, parathion was rapidly mineralized to carbon dioxide by indigenous microorganisms (Ferris and Lichtenstein, 1980). The half-lives of parathion (10 ppm) in a nonsterile sandy loam and a nonsterile organic soil were <1 and 1.5 weeks, respectively (Miles et al., 1979). Walker (1976) reported that 16–23% of parathion added to both sterile and nonsterile estuarine water was degraded after incubation in the dark for 40 days.

The half-lives for parathion in soil incubated in the laboratory under aerobic conditions ranged from 18 to 78 days with an average of 35 days (Lichtenstein and Schulz, 1964; Sacher et al., 1971; Katan et al., 1976). In flooded soils incubated in the laboratory, the half-lives for parathion ranged from 2 to 22 days with an average half-life of 11 days (Sethunathan, 1973, 1973b; Rajaram and Sethunathan, 1975). In field soils, the disappearance half-lives ranged from 3 to 29 days with an average of 18 days (Lichtenstein and Schulz, 1964; Sacher et al., 1971; Spencer et al., 1975).

Parathion degrades more readily in nonsaline soil than in saline soil under flooded conditions. In nonsaline soil, no parathion was detected after 20 days. Degradation yields observed in various soils with varying amounts of salinity ranged from 10 to 50% after the 20-day period (Reddy and Sethunathan, 1985). Six years after applying parathion to soil at concentrations of 30,000–95,000 ppm, very small concentrations were detected at the 9 cm below the surface (Wolfe et al., 1973).

The percentages of the initial dosage (1 ppm) of parathion remaining after 8 weeks of incubation in an organic and mineral soil were 6 and <2%, respectively, while in sterilized controls 95 and 80% remained, respectively (Chapman et al., 1981).

Six years after applying parathion to soil at concentrations of 30,000–95,000 ppm, very small concentrations were detected at 9 cm below the surface (Wolfe et al., 1973).

In a silt loam and sandy loam, reported R_f values were 0.87 and 0.88, respectively (Sharma et al., 1986).

Plant. Oat plants were grown in two soils treated with [^{14}C]parathion. Less than 2% of the applied [^{14}C]parathion was translocated to the oat *plant*. Metabolites identified in both soils and leaves were paraoxon, aminoparaoxon, aminoparathion, *p*-nitrophenol and an aminophenol (Fuhremann and Lichtenstein, 1980).

The following metabolites were identified in a soil-oat system: paraoxon, aminoparathion, *p*-nitrophenol and *p*-aminophenol (Lichtenstein, 1980; Lichtenstein et al., 1982). Mick and Dahm (1970) reported that *Rhizobium* sp. converted 85% ^{14}C-labeled parathion to aminoparathion in one day and 10% diethylphosphorothioic acid.

One month after application of [^{14}C]parathion to cotton plants, 6.5–10.5% of the total reactivity was found to be unreacted [^{14}C]parathion. Photoalteration products identified included *S*-ethyl parathion, *S*-phenyl parathion, paraoxon and *p*-nitrophenol (Joiner and Baetcke, 1973). Reddy and Sethunathan (1983) studied the mineralization of ring-labeled [2,6-^{14}C]parathion in the rhizosphere of rice seedlings under flooded and nonflooded soil conditions. In unplanted soil, only 5.5% of the ^{14}C in the parathion was evolved as ^{14}CO$_2$ in 15 days under flooded and nonflooded conditions. However, in soils planted with rice, 9.2 and 22.6% of the ^{14}C in the parathion evolved as ^{14}CO$_2$ under nonflooded and flooded conditions, respectively (Reddy and Sethunathan, 1983). In an earlier study, the presence of rice straw in a flooded alluvial soil inoculated with an enrichment culture greatly inhibited the hydrolysis of parathion to *p*-nitrophenol and *O,O*-diethylphosphorothioic acid. In uninoculated soils, however, rice straw enhanced the degradation of parathion via nitro group reduction to *p*-aminoparathion and a compound possessing a P=S bond (Sethunathan, 1973).

In cabbage and wheat, parathion was converted to paraoxon (David and Aldridge, 1957).

Photolytic. *p*-Nitrophenol and paraoxon were formed from the irradiation of parathion in water, aqueous methanol and aqueous *n*-propyl alcohol solutions by a low-pressure mercury lamp. Degradation was more rapid in water than in organic solvent/water mixtures with *p*-nitrophenol forming as the major product (Mansour et al., 1983). When parathion in aqueous tetrahydrofuran or ethanol solutions (80%) was irradiated at 2537 Å, *O,O,S*-triethylthiophosphate formed as the major product (Grunwell and Erickson, 1973). Minor photoproducts included *O,O,O*-triethylthiophosphate, triethylphosphate, paraoxon and traces of ethanethiol and *p*-nitrophenol.

Parathion degraded on both glass surfaces and on bean plant leaves. Metabolites reported were paraoxon, *p*-nitrophenol and a compound tentatively identified as *s*-ethyl parathion (El-Refai and Hopkins, 1966). Upon exposure to high intensity UV light, parathion was altered to the following photoproducts: paraoxon, *O,S*-diethyl *O*-4-nitrophenyl phosphorothioate, *O,O*-diethyl *S*-4-nitrophenyl phosphorothioate, *O,O*-bis(4-nitrophenyl) *O*-ethyl phosphorothioate, *O,O*-bis(4-nitro-phenyl) *O*-ethyl phosphate, *O,O*-diethyl *O*-phenyl phosphorothioate and *O,O*-diethyl *O*-phenyl phosphate (Joiner et al., 1971).

When parathion was released in the atmosphere on a sunny day, it was rapidly converted to the photochemical paraoxon (estimated half-life 2 minutes) (Woodrow et al., 1978). The reaction involving the oxidation of parathion to paraoxon is catalyzed in the presence of UV light, ozone, soil dust, or clay minerals (Spencer et al., 1980, 1980a). Ozone and UV light alone could not oxide parathion. However, in the presence of ozone (300 ppb) and UV light, dry kaolinite clays were more effective than the montmorillonite clays in the oxidation of parathion. The Cu-saturated clays were most effective oxidizing parathion to paraoxon and the Ca-saturated clays were the least effective (Spencer et al., 1980).

When applied as thin films on leaf surfaces, parathion was converted to paraoxon and p-nitrophenol. The photodegradation half-life was reported to be 88 hours (Hazardous Substances Data Bank, 1989).

When an aqueous solution containing parathion was photooxidized by UV light at 90–95°C, 25, 50 and 75% degraded to carbon dioxide after 15.1, 57.6 and 148.8 hours, respectively (Knoevenagel and Himmelreich, 1976). p-Nitrophenol and paraoxon were the major products identified following the sunlight irradiation of parathion in distilled water and river water. The photolytic half-life of parathion in river water was 15 hours (Mansour et al., 1989). Kotronarou et al. (1992) studied the degradation of parathion-saturated deionized water solution at 30°C by ultrasonic irradiation (sonolysis). After 2 hours of sonolysis, all the parathion degraded to the following final end products: sulfate ions, phosphate ions, nitrate ions, hydrogen ions and carbon dioxide. Precursors or intermediate compounds to the final end products included p-nitrophenol, diethylmonothiophosphoric acid, hydroquinone, benzoquinone, formic acid, oxalic acid, 4-nitrocatechol, nitrite ions and ethanol (Kotronarou et al., 1992). In a photoreactor, the photolysis half-life of parathion was 41 minutes. In the presence of ozone, the rate of photolysis increases. When ozone is present at a concentration >1 ppm, the photolysis half-life is reduced to 23 minutes (Woodrow et al., 1983).

Chemical/Physical. The reported hydrolysis half-lives at pH 7.4 at 20 and 37.5°C were 130 and 26.8 days, respectively (Freed et al., 1977). The hydrolysis half-lives of parathion in a sterile 1% ethanol/water solution at 25°C and pH values of 4.5, 5.0, 6.0, 7.0 and 8.0, were 39, 43, 33, 24 and 15 weeks, respectively (Chapman and Cole, 1982). Kollig (1993) reported the following products of hydrolysis: *O*-ethyl-*O*-(p-nitrophenyl)phosphorothioic acid, *O*-(p-nitrophenyl)phosphorothioic acid, phosphorothioic acid, phosphoric acid, p-nitrophenol, *O,O*-diethylphosphorothioic acid, *O*-ethylphosphorothioic acid and ethanol.

Paraoxon was also found in fogwater collected near Parlier, CA (Glotfelty et al., 1990). It was suggested that parathion was oxidized in the atmosphere during daylight hours prior to its partitioning in the fog. On January 12, 1986, the distribution of parathion (9.4 ng/m^3) in the vapor phase, dissolved phase, air particles and water particles were 78, 10, 11 and 0.6%, respectively. For paraoxon (2.3 ng/m^3), the distribution in the vapor phase, dissolved phase, air particles and water particles were 7.8, 35.5, 56.7 and 0.09%, respectively (Glotfelty et al., 1990).

Reported ozonation products of parathion in drinking water include sulfuric acid (Richard and Bréner, 1984), paraoxon, 2,4-dinitrophenol, picric and phosphoric acids (Laplanche et al., 1984).

At 130°C, parathion isomerizes to *O,S*-diethyl *O-p*-nitrophenyl phosphorothioate (Hartley and Kidd, 1987; Worthing and Hance, 1991). An 85% yield of this compound was reported when parathion was heated at 150°C for 24 hours (Wolfe et al., 1976). Emits toxic oxides of nitrogen, sulfur and phosphorus when heated to decomposition (Lewis, 1990).

Exposure Limits: NIOSH REL: TWA 0.05 mg/m^3, IDLH 10 mg/m^3; OSHA PEL: TWA 0.1 mg/m^3.

Symptoms of Exposure: Acute effects include miosis, rhinorrhea, headache, bronchoconstriction, chest wheezing, laryngeal spasm, excessive salivation, cyanosis, anorexia, abdominal cramps, diarrhea, sweating, muscle fasiculation and twitching, nausea, weakness, paralysis, ataxia, convulsions, low blood pressure, dermatitis, pupillary constriction.

Formulation Types: Emulsifiable concentrate; wettable powder; granules; dustable powder; aerosol.

Toxicity: LC_{50} (96-hour) for rainbow trout 1.5 mg/L, golden orfe 0.57 mg/L (Hartley and Kidd, 1987), fathead minnow 1.4–2.7 mg/L (Worthing and Hance, 1991), bluegill sunfish 65 µg/L, green sunfish 425 µg/L and largemouth bass 190 µg/L (Verschueren, 1983); acute oral LD_{50} for rats 3.6–13 mg/kg (Hartley and Kidd, 1987).

Uses: Insecticide and acaricide for control of sucking and chewing insects and mites in fruits, vegetables, ornamentals and field crops.

PEBULATE

Synonyms: Butylethylthiocarbamic acid *S*-propyl ester; PEBC; *S*-Propyl butylethyl-thiocarbamate; Propylethylbutylthiocarbamate; Propyl-*N*-ethyl-*n*-butylthiocarbamate; *S*-(*n*-Propyl)-*N*-ethyl-*N*,*N*-butylthiocarbamate; *n*-Propyl-*N*-ethyl-*N*-(*n*-butyl)thiocarbamate; Propyl ethyl-*n*-butylthiolcarbamate; Propylethylbutylthiolcarbamate; Propyl ethyl-*n*-butylthiolcarbamate; PEBC; R 2061; Stauffer 2061; Tillam; Timmam-6-E.

$$CH_3CH_2 \diagdown \diagup_{CH_3CH_2} NCOSCH_2CH_2CH_3$$

Designations: CAS Registry Number: 1114-71-2; mf: $C_{10}H_{21}NOS$; fw: 203.36; RTECS: EZ0400000.

Properties: Colorless to yellow liquid with an amine-like odor. Mp: <25°C; bp: 142°C at 20 mmHg; ρ: 0.9555 at 20/20°C, 0.9458 at 30/4°C; fl p: 124°C; K_H: 1.15 × 10^{-4} atm · m³/mol at 20°C (approximate — calculated from water solubility and vapor pressure); log K_{oc}: 2.80; log K_{ow}: 3.84; S_o: miscible with but not limited to acetone, benzene, ethanol, kerosene, methanol, 2-propanol, toluene, xylene; S_w: 60 mg/L at 20°C; vp: 6.8 × 10^{-2} mmHg at 20°C.

Environmental Fate

Soil. Pebulate rapidly degrades in soil, forming carbon dioxide, ethylbutylamine and mercaptan (Hartley and Kidd, 1987). The half-life in a moist loam soil at 21–27°C was reported to be about 2 weeks (Humburg et al., 1989).

Plant. Rapidly transformed in plants to carbon dioxide (Worthing and Hance, 1991) and naturally occurring plant constituents (Humburg et al., 1989).

Chemical/Physical. Emits toxic fumes of nitrogen and sulfur oxides when heated to decomposition (Sax and Lewis, 1987).

Symptoms of Exposure: May cause violent vomiting when accompanied by alcohol ingestion.

Formulation Types: Emulsifiable concentrate (6 lb/gal); granules.

Toxicity: LC_{50} (96-hour) for bluegill sunfish and rainbow trout about 7.4 mg/L (Worthing and Hance, 1991); LC_{50} (48-hour) for killifish 7.78 mg/L and silver mullet 6.25 mg/L (Hartley and Kidd, 1987); acute oral LD_{50} for male rats and male mice 0.9–1.1 and 1.5–1.8 g/kg, respectively (Ashton and Monaco, 1991).

Uses: Selective preemergence herbicide used to control some broad-leaved weeds and annual grasses in tomatoes, sugar beet and tobacco.

PENDIMETHALIN

Synonyms: AC 92553; *N*-(1-Ethylpropyl)-3,4-dimethyl-2,6-dinitrobenzenamine; **N-(1-Ethylpropyl)-2,6-dinitro-3,4-xylidine**; Herbadox; Horbadox; Pay-off; Penoxalin; Penoxaline; Penoxyn; Phenoxalin; Prowl; Stomp; Stomp 300D; Stomp 300E; Tendimethalin.

Designations: CAS Registry Number: 40487-42-1; mf: $C_{13}H_{19}N_3O_4$; fw: 281.31; RTECS: BX5470000.

Properties: Orange-yellow crystalline solid with a faint nutty odor. Mp: 54–58°C; bp: 330°C; ρ: 1.19 at 25/4°C; H-$t_{1/2}$: <21 days under continuous irradiation; K_H: 8.56×10^{-7} atm · m³/mol at 25°C; log K_{oc}: 1.48–2.93; log K_{ow}: 5.18; S_o (g/L at 26°C): acetone (700), dimethyl sulfoxide (214), *n*-heptane (138), methanol (55), 2-propanol (77), xylene (628); S_w: 0.3 mg/L at 20°C; vp: 3×10^{-5} mmHg at 25°C.

Soil properties and adsorption data

Soil	K_d (mL/g)	f_{oc} (%)	K_{oc} (mL/g)	pH
Loam	301	2.20	13,682	7.0
Loam	854	2.90	29,448	6.5
Sand	30	0.46	6,466	7.6
Sandy loam	110	0.93	11,853	6.4
Silt loam	380	2.73	13,919	7.0

Source: U.S. Department of Agriculture, 1990.

Environmental Fate

Biological. In an *in vitro* study, the soil fungi *Fusarium oxysporum* and *Paecilomyces varioti* degraded pendimethalin to *N*-(1-ethylpropyl)-3,3-dimethyl-2-nitrobenzene-1,6-diamine and 3,4-dimethyl-2,6-dinitroaniline. The latter compound was the only metabolite identified by another soil fungus; namely, *Rhizoctonia bataticola* (Singh and Kulshrestha, 1991).

Kole et al. (1994) studied the degradation of pendimethalin by an unadapted strain of *Azotobacter chroococcum* Beijerinck in pure culture condition. Degradation was rapid — 45 and 55% of the parent compound was degraded after 10 and 20 days, respectively. Six metabolites were identified. Oxidative complete *N*-dealkylation of the pendimethalin gave 3,4-dimethyl-1,5-dinitroaniline. This metabolite underwent acylation of the anilino nitrogen and direct elimination of the nitro group at the *ortho* position forming *N*-(2,6-dinitro-3,4-dimethyl) phenyl acetamide and 3,4-dimethyl-6-nitroaniline, respectively. Reduction of the nitro group at the C-6 position yielded 2-nitro-6-amino-*N*-(1-ethylpropyl)-3,4-

xylidine. At the C-3 position on the parent compound, the methyl group oxidized to -CH_2OH which underwent partial reduction of the 2-nitro group and *N*-dealkylation followed by oxidative cyclization to give 2-methyl-4-nitro-5-*N*-(1-cyclopropyl)-6-nitrosobenzyl alcohol (Kole et al., 1994).

Soil/Plant. In soil, the 4-methyl group on the benzene ring is oxidized to a hydroxy group which undergoes further oxidation to the carboxylic acid. The major plant metabolite is 4-(1-ethylpropyl)amino-2-methyl-3,5-dinitrobenzyl alcohol (Hartley and Kidd, 1987). Kulshrestha and Singh (1992) investigated the influence of soil moisture and microbial activity on the degradation of pendimethalin in a sandy loam soil. In both nonsterile, nonflooded and flooded soils, degradation followed first-order kinetics. The observed half-lives in flooded and nonflooded conditions in nonsterile and sterile soils were 33 and 45; 52 and 67 days, respectively. In the nonsterile, nonflooded soil, pendimethalin underwent *N*-dealkylation and reduction of the less hindered nitro group affording *N*-(1-ethylpropyl)-3,4-dimethyl-2-nitrobenzene-1,6-diamine and 3,4-dimethyl-2,6-dinitroaniline as major products. Under flooded conditions, only one degradation product was formed: *N*-(1-ethylpropyl)-5,6-dimethyl-7-nitrobenzimidazole. Under flooded conditions, pendimethalin was completely degraded after 4 days (Kulshrestha and Singh, 1992).

The reported half-lives of pendimethalin in a Sharkey clay soil and Bosket silt loam are 6 and 4 days, respectively (Savage and Jordan, 1980).

Photolytic. Irradiation of a sandy loam soil with a high pressure mercury lamp for 48 hours gave the following products (% yield): 2-amino-6-nitro-*N*-(1-ethylpropyl)-3,4-xylidine (35), 2,6-dinitro-3,4-xylidine (21), 2,2-diethyl-7-nitro-4,5-dimethyl-3-hydroxy-2,3-dihydrobenzimidazole (13) and 2-ethyl-7-nitro-4,5-dimethyl-3-hydroxy-2,3-dihydro-benzimidazole (8) (Dureja and Walia, 1989).

Parochetti and Dec (1978) reported that when pendimethalin on clay loam thin-layer plates was exposed to direct sunlight for 7 days, 9.9% of the applied amount (1 kg/ha) had decomposed.

Symptoms of Exposure: Eye irritant.

Formulation Types: Emulsifiable concentrate (4 lb/gal); wettable powder (50%); granules (2 and 5%).

Toxicity: LC_{50} (96-hour) for rainbow trout 138 µg/L and bluegill sunfish 199 µg/L; acute oral LD_{50} of the technical product for male and female rats 1,250 and 1,050 mg/kg, respectively (Ashton and Monaco, 1991).

Uses: Herbicide used to control many annual grasses and broad-leaved weeds in crops such as corn, cotton, peanuts, potatoes, rice, sorghum, sunflower and tobacco.

PENTACHLOROBENZENE

Synonyms: QCB; RCRA waste number U183.

Designations: CAS Registry Number: 608-93-5; mf: C_6HCl_5; fw: 250.34; RTECS: DA6640000.

Properties: White crystals or needles from alcohol. Mp: 86°C; bp: 277°C; ρ: 1.8342 at 16.5/4°C; H-$t_{1/2}$: >900 years at 25°C and pH 7; K_H: 0.0071 atm · m^3/mol at 20°C; log BCF: 3.63 mussel (*Mytilus edulis*) (Renberg et al., 1985), 6.42 (Atlantic croakers), 6.71 (blue crabs), 5.96 (spotted sea trout), 5.98 (blue catfish) (Pereira et al., 1988); log K_{oc}: 6.3; log K_{ow}: 4.88–5.75; S_o: slightly soluble in benzene, carbon disulfide, chloroform, dimethylsulfoxide (1–10 mg/mL at 22°C), 95% ethanol (1–10 mg/mL at 22°C), ethyl ether; S_w: 5.32 µM/L at 25°C; vp: 6×10^{-3} mmHg at 20–30°C.

Environmental Fate

Biological. In activated sludge, <0.1% mineralized to carbon dioxide after 5 days (Freitag et al., 1985). From the first-order biotic and abiotic rate constants of pentachlorobenzene in estuarine water and sediment/water systems, the estimated biodegradation half-lives were 4.6–6.5 and 6.0–7.6 days, respectively (Walker et al., 1988).

Photolytic. UV irradiation ($\lambda = 2537$ Å) of pentachlorobenzene in a hexane solution for 3 hours produced a 50% yield of 1,2,4,5-tetrachlorobenzene and a 13% yield of 1,2,3,5-tetrachlorobenzene (Crosby and Hamadmad, 1971). Irradiation ($\lambda \geq 285$ nm) of pentachlorobenzene (1.1–1.2 mM/L) in an acetonitrile-water mixture containing acetone (0.553 mM/L) as a sensitizer gave the following products (% yield): 1,2,3,4-tetrachlorobenzene (6.6), 1,2,3,5-tetrachlorobenzene (52.8), 1,2,4,5-tetrachlorobenzene (15.1), 1,2,4-trichlorobenzene (1.9), 1,3,5-trichlorobenzene (5.3), 1,3-dichlorobenzene (0.9), 2,2',3,3',4,4',5,6,6'-nonachlorobiphenyl (2.08), 2,2',3,3',4,4',5,5',6-nonachlorobiphenyl (0.34), 2,2',3,3',4,5,5',6,6'-nonachlorobiphenyl (trace), one octachlorobiphenyl (0.53) and one heptachlorobiphenyl (0.49) (Choudhry and Hutzinger, 1984). Without acetone, the identified photolysis products (% yield) included 1,2,3,4-tetrachlorobenzene (3.7), 1,2,3,5-tetrachlorobenzene (13.5), 1,2,4,5-tetrachlorobenzene (2.8), 1,2,4-trichlorobenzene (12.7), 1,3,5-trichlorobenzene (1.0) and 1,4-dichlorobenzene (6.7) (Choudhry and Hutzinger, 1984).

A carbon dioxide yield of 2.0% was achieved when pentachlorobenzene adsorbed on silica gel was irradiated with light (λ >290 nm) for 17 hours (Freitag et al., 1985).

The experimental first-order decay rate for pentachlorobenzene in an aqueous solution containing a nonionic surfactant micelle (Brij 58, a polyoxyethylene cetyl ether) and illuminated by a photoreactor equipped with 253.7-nm monochromatic UV lamps, is 1.47 $\times$ 10^{-2}/sec. The corresponding half-life is 47 seconds. Photoproducts reported include, all tetra-, tri- and dichlorobenzenes, chlorobenzene, benzene, phenol, hydrogen and chloride ions (Chu and Jafvert, 1994).

Chemical/Physical. Emits chlorinated acids when incinerated. Incomplete combustion may release toxic phosgene (Sittig, 1985).

Based on an assumed base-mediated 1% disappearance after 16 days at 85°C and pH 9.70 (pH 11.26 at 25°C), the hydrolysis half-life was estimated to be >900 years (Ellington et al., 1988).

Toxicity: LC_{50} (14-day) for guppy 178 ppb (Verschueren, 1983); acute oral LD_{50} for rats 1,080 mg/kg (RTECS, 1985).

Use: Agrochemical research.

PENTACHLORONITROBENZENE

Synonyms: Avicol; Batrilex; Botrilex; Brassicol; Earthcide; Fartox; Folosan; Fomac 2; Fungiclor; GC 3944-3-4; Kobu; Kobutol; KP 2; NCI-C00419; Olpisan; PCNB; Pentachloronitrobenzol; Pentagen; PKhNB; Quintocene; Quintozen; Quintozene; RCRA waste number U185; Saniclor; Saniclor 30; Terrachlor; Terraclor; Terrafun; Tilcarex; Tri-PCNB; Tritisan.

$$\text{Cl}_5\text{C}_6\text{NO}_2 \text{ structure}$$

Designations: CAS Registry Number: 82-68-8; mf: $C_6Cl_5NO_2$; fw: 295.36; RTECS: DA665000.

Properties: Colorless to white crystals or needles from alcohol. Mp: 144–146°C; bp: 328°C (some decomposition); ρ: 1.718 at 25/4°C; K_H: 2.88×10^{-3} atm · m³/mol at 25°C; S_o: freely soluble in benzene, carbon disulfide, 95% ethanol (20 g/kg); S_w: 32 μg/L at 25°C; vp: 2.38×10^{-3} mmHg at 25°C.

Environmental Fate

Biological. Pentachloronitrobenzene is rapidly degraded in flooded soils forming pentachloroaniline. When pentachloronitrobenzene was incubated in a submerged soil and moist soil for 3 weeks, the percentages of the applied dosage remaining were <1 and 82%, respectively. Chacko et al. (1966) reported microorganisms can convert pentachloronitrobenzene to pentachloroaniline and pentachlorothioanisole.

Soil. The half-lives of pentachloronitrobenzene in a Columbia fine sandy loam, Sacramento clay and Staten peaty muck from California were 4.7, 7.6 and 9.7 months, respectively. Degradation products were pentachloroaniline and pentachlorothioanisole (Wang and Broadbent, 1972). In the laboratory, half-lives ranged from 213 to 699 and 435 to 535 days at application rates of 60 and 10 kg/ha, respectively (Beck and Hansen, 1974). In anaerobic soils, the half-life of pentachloronitrobenzene was 21 days. After 3 weeks of incubation, <1% of the applied amount (10 μg/g soil) remained in submerged soil but 82% was found in moist *soil.* (Ko and Farley, 1969). In a Hagerstown silty clay loam, pentachloroaniline, pentachlorothioanisole and pentachlorophenol were reported as metabolites (Murthy and Kaufman, 1978).

When heated to decomposition (330°C), emits toxic fumes of nitrogen oxides and chlorine (Sax and Lewis, 1987).

Symptoms of Exposure: Moderately toxic by ingestion.

Toxicity: Acute oral LD_{50} for rats (aqueous solution) >12,000 mg/kg (Verschueren, 1983).

Use: Fungicide for seed and soil treatment.

PENTACHLOROPHENOL

Synonyms: Acutox; Chempenta; Chemtol; Chlorophen; Cryptogil OL; Dowcide 7; Dowicide 7; Dowicide EC-7; Dowicide G; Dow pentachlorophenol DP-2 antimicrobial; Durotox; EP 30; Fungifen; Fungol; Glazd penta; Grundier arbezol; Lauxtol; Lauxtol A; Liroprem; Monsanto penta; Moosuran; NCI-C54933; NCI-C55378; NCI-C56655; PCP; Penchlorol; Penta; Pentachlorofenol; Pentachlorophenate; Pentachlorphenol; 2,3,4,5,6-Pentachlorophenol; Pentacon; Pentakil; Pentasol; Penwar; Peratox; Permacide; Permaguard; Permasan; Permatox DP-2; Permatox Penta; Permite; Priltox; RCRA waste number U242; Santobrite; Santo-phen; Santophen 20; Sinituho; Term-i-trol; Thompson's wood fix; Weedone; Witophen P.

Designations: CAS Registry Number: 87-86-5; DOT: 2020; mf: C_6HCl_5O; fw: 266.34; RTECS: SM6300000.

Properties: White to dark-colored flakes or beads with a characteristic pungent odor and taste. Darkens on exposure to air. Mp: 190–191°C (anhydrous), 174°C (hydrous); bp: 310°C with decomposition; ρ: 1.978 at 22/4°C; fl p: nonflammable; pK_a: 4.74; K_H: 2.1–3.4 $\times 10^{-7}$ atm · m³/mol; log K_{oc}: 2.47–4.40; log K_{ow}: 3.32–5.86; S_o: acetone, carbitol, cellosolve, cod liver oil (184.2, 188.5 and 98.9 g/L at 4, 12 and 20°C, respectively), ethanol, ethyl ether, octanol (266.3, 260.2 and 206.7 g/L at 4, 12 and 20°C, respectively) and triolein (168.0, 168.0 and 106.0 g/L at 4, 12 and 20°C, respectively) but slightly soluble in alkanes, carbon tetrachloride, cold petroleum ether, methylene chloride and ligroin; S_w: 14 mg/L at 20°C at pH 5; vp: 1.7×10^{-4} mmHg at 20°C.

Soil properties and adsorption data

Soil	K_d (mL/g)	f_{oc} (%)	K_{oc} (mL/g)	pH
Aquifer material	200.00	0.84	23,800	—
Coarse sand	1.20	0.21	571	5.3
Humaquept sand	37.50	0.50	7,500	5.2
Lake sediment	3,670.00	9.40	39,000	—
Loamy sand	0.45	0.15	300	6.4
River sediment	930.00	2.60	35,800	—

Source: Schellenberg, 1984; Kjeldsen et al., 1990; LaFrance et al., 1994.

Environmental Fate

Biological. Under aerobic conditions, microbes in estuarine water partially dechlorinated pentachlorophenol to trichlorophenol (Hwang et al., 1986). The disappearance of

338

pentachlorophenol was studied in four aquaria with and without mud under aerobic and anaerobic conditions. Potential biological and/or chemical products identified include pentachloroanisole, 2,3,4,5-, 2,3,4,6- and 2,3,5,6-tetrachlorophenol (Boyle et al., 1980). Pentachlorophenol was also subject to methylation by a culture medium containing *Trichoderma viride* affording pentachloroanisole (Cserjesi and Johnson, 1972).

Pentachlorophenol degraded in anaerobic sludge to 3,4,5-trichlorophenol which was further reduced to 3,5-dichlorophenol (Mikesell and Boyd, 1985). In activated sludge, only 0.2% of the applied amount was mineralized to carbon dioxide after 5 days (Freitag et al., 1985).

Pentachlorophenol was statically incubated in the dark at 25°C with yeast extract and settled domestic wastewater inoculum. Significant biooxidation was observed but with a gradual adaptation over a 14-day period to achieve complete degradation at 5 mg/L substrate cultures. At a concentration of 10 mg/L, it took 28 days for pentachlorophenol to degrade completely (Tabak et al., 1981).

Melcer and Bedford (1988) studied the fate of pentachlorophenol in municipal activated sludge reactor systems that were operated at solids retention times of 10 to 20 days and hydraulic retention times of 120 days. Under these conditions, pentachlorophenol concentrations decreased from 0.1 and 12 mg/L to <10 μg/L. At solid retentions times of 5 days or less, pentrachlorophenol degradation was incomplete.

Soil. Under anaerobic conditions, pentachlorophenol may undergo sequential dehalogenation to produce tetra-, tri-, di- and *m*-chlorophenol (Kobayashi and Rittman, 1982). In aerobic and anaerobic soils, pentachloroanisole was the major metabolite along with 2,3,6-trichlorophenol, 2,3,4,5- and 2,3,5,6-tetrachlorophenol (Murthy et al., 1979). Degradation was rapid in estuarine sediments (pHs 6.5, 8.0). Following a 17-day lag period, 70% of pentachlorophenol degraded (DeLaune, 1983). After 160 days in aerobic soil, only 6% biodegradation was observed (Baker and Mayfield, 1980). Knowlton and Huggins (1983) reported 2,3,7,8-tetrachlorophenol and carbon dioxide as the major degradation products in *soil.*

Weiss et al. (1982) studied the fate of ^{14}C-pentachlorophenol added to flooded rice soil in a plant growth chamber. After one growing season, the following residues were observed (% of applied radioactivity): unidentified unextractable/bound compounds (28.61%), pentachlorophenol (0.51%), conjugated pentachlorophenol (0.61%), 2,3,6-, 2,4,5-, 2,4,6-, 2,3,4-, 2,3,5- and 3,4,5-tri-chlorophenols (1.27%), 2,3,4,5-tetrachlorophenol (0.38%), 2,3,4-, 2,3,6-, 2,4,6- and 3,4,5-trichloroanisoles (0.08%), 2,3,4,5-tetrachloroanisole (0.02%), pentachloroanisole (0.02%) and unidentified conversion products (including highly polar hydrolyzable and nonhydrolyzable compounds) (4.74%).

Metabolites identified in soil beneath a sawmill environment where pentachlorophenol was used as a wood preservative include pentachloroanisole, 2,3,4,6-tetrachloroanisole, tetrachlorocatechol, tetrachlorohydroquinone, 3,4,5-trichlorocatechol, 2,3,6-trichlorohydroquinone, 3,4,6-trichlorocatechol and 2,3,4,6-tetrachlorophenol (Knuutinen et al., 1990).

In a terrestrial ecosystem, the distribution of pentachlorophenol in soil, plant, air and leachate were 65, 9, 25 and 0.3%, respectively (Gile and Gillet, 1979).

The average disappearance half-lives for pentachlorophenol in flooded soils under laboratory and field conditions were 174 and 15 days, respectively (Kauwatsuka and Igarashi, 1975; Watanabe, 1978). In aerobic soils incubated under laboratory conditions, the disappearance half-lives ranged from 18 to 97 days with an average half-life of 48 days (Kauwatsuka and Igarashi, 1975).

Groundwater. According to the U.S. EPA (1986) pentachlorophenol has a high potential to leach to groundwater.

Photolytic. When pentachlorophenol in distilled water is exposed to UV irradiation, it is photolyzed to give tetrachlorophenols, trichlorophenols, chlorinated dihydroxybenzenes and dichloromaleic acid (Hwang et al., 1986). Wood treated with pure pentachlorophenol did not photolyze under natural sunlight or laboratory-induced UV radiation. However, in the presence of an antimicrobial (Dowcide EC-7), pure pentachlorophenol degraded to chlorinated dibenzo-*p*-dioxin. Wood containing composited technical grade pentachlorophenol yielded similar results (Lamparski et al., 1980).

An aqueous solution containing pentachlorophenol and exposed to sunlight or laboratory UV light yielded tetrachlorocatechol, tetrachlororesorcinol and tetrachlorohydroquinone. These compounds were air-oxidized to chloranil, hydroxyquinones and 2,3-dichloromaleic acid (DCM). Other compounds identified include a cyclic dichlorodiketone, 2,3,4,6- and 2,3,5,6-tetrachlorphenol and trichlorophenols (Munakata and Kuwahara, 1969; Wong and Crosby, 1981).

Photodecomposition of pentachlorophenol was observed when an aqueous solution was exposed to sunlight for 10 days. The violet-colored solution contained 3,4,5-trichloro-6-(2'-hydroxy-3',4',5',6'-tetrachlorophenoxy)-*o*-benzoquinone as the major product. Minor photodecomposition products (% yield) included tetrachlororesorcinol (0.10%), 2,5-dichloro-3-hydroxy-6-pentachlorophenoxy-*p*-benzoquinone (0.16%) and 3,5-dichloro-2-hydroxy-5-2',4',5',6'-tetrachloro-3-hydroxyphenoxy-*p*-benzoquinone (0.08%) (Plimmer, 1970).

UV irradiation ($\lambda = 2537$ Å) of pentachlorophenol in a *n*-hexane solution for 32 hours produced a 30% yield of 2,3,5,6-tetrachlorophenol and about a 10% yield of a compound tentatively identified as an isomeric tetrachlorophenol (Crosby and Hamadmad, 1971).

A carbon dioxide yield of 50.0% was achieved when pentachlorophenol adsorbed on silica gel was irradiated with light ($\lambda > 290$ nm) for 17 hours (Freitag et al., 1985). In soil, photodegration was not significant (Baker et al., 1980).

When an aqueous solution containing pentachlorophenol (4.5×10^{-5} M) and a suspension of titanium dioxide (2 g/L) was irradiated with UV light, carbon dioxide and hydrochloric acid formed in quantitative amounts (half-life 9 minutes at 45–50°C) (Barbeni et al., 1985). When an aqueous solution containing pentachlorophenol was photooxidized by UV light at 90–95°C, 25, 50 and 75% degraded to carbon dioxide after 31.7, 66.0 and 180.7 hours, respectively (Knoevenagel and Himmelreich, 1976). The photolysis half-lives of pentachlorophenol under sunlight irradiation in distilled water and river water were 27 and 53 hours, respectively (Mansour et al., 1989).

In a similar study, pentachlorophenol (47 µM) in an air-saturated solution containing titanium dioxide suspension was irradiated with UV light ($\lambda = 330$–370 nm). Major chemical intermediates included *p*-choranil, tetrachlorohydroquinone, hydrogen peroxide and *o*-chloranil. The intermediate compounds were attacked by hydroxyl radicals during the latter stages of irradiation forming HCO_2^-, acetate and formate ions, carbon dioxide and hydrochloric acid (Mills and Hoffman, 1993).

Petrier et al. (1992) studied the sonochemical degradation of pentachlorophenol in aqueous solutions saturated with different gases at 24°C. Ultrasonic irradiation of solutions saturated with air or oxygen resulted in the liberation of chloride ions and mineralization of the parent compound to carbon dioxide. When the solution is saturated with argon, pentachlorophenol completely degraded to carbon monoxide and chloride ions.

Chemical/Physical. Wet oxidation of pentachlorophenol at 320°C yielded formic and acetic acids (Randall and Knopp, 1980). In a dilute aqueous solution at pH 6.0, pentachlorophenol reacted with excess hypochlorous acid forming 2,3,5,6-tetrachlorobenzoquinone (chloranil), hexachloro-5-cyclohexadienone and two other chlorinated compounds at yields of 3, 20 and 27%, respectively (Smith et al., 1976).

Reacts with amines and alkali metals forming water-soluble salts (Sanborn et al., 1977).

Hexachlorobenzene, octachlorodiphenylene dioxide and other polymeric compounds were formed when pentachlorophenol was heated to 300°C for 24 hours (Sanderman et al., 1957).

Emits very toxic chloride fumes when heated to decomposition (Lewis, 1990).

Exposure Limits: NIOSH REL: IDLH 0.5 mg/m³, IDLH 2.5 mg/m³; OSHA PEL: TWA 0.5 mg/m³; ACGIH TLV: TWA 0.5 mg/m³.

Symptoms of Exposure: Ingestion causes fluctuation in blood pressure, respiration; fever; urinary output; motor weakness; convulsions and possibly death.

Formulation Types: Wettable powder; granules; oil-miscible liquid.

Toxicity: EC_{50} (24-hour) for *Daphnia magna* 858 µg/L (Lilius et al., 1995); LC_{50} (48-hour) for rainbow trout 0.17 mg/L (sodium salt) (Hartley and Kidd, 1987); LC_{50} (24-hour) for goldfish 270 ppb (Verschueren, 1983); acute oral LD_{50} for rats 210 mg/kg (Hartley and Kidd, 1987), 27 mg/kg (RTECS, 1985).

Uses: Insecticide; fungicide; herbicide.

PERMETHRIN

Synonyms: AI3-29158; Ambush; BW-21-Z; **3-(2,2-Dichloroethenyl)-2,2-dimethylcyclo-propanecarboxylic acid (3-phenoxyphenyl)methyl ester**; Ectiban; Exmin; FMC 33297; FMC 41665; ICI-PP 557; Kestrel; NDRC-143; NIA 33297; Niagara 33297; Outflank; Outflanf-stockade; (3-Phenoxyphenyl)methyl 3-(2,2-dichloroethenyl)-2,2-dimethylcyclo-propane carboxylate; Pounce; PP 557; S 3151; SBP-1513; Talcord; WL 43479.

Designations: CAS Registry Number: 52645-53-1; mf: $C_{21}H_{20}Cl_2O_3$; fw: 391.29; RTECS: GZ1255000.

Properties: Colorless crystals to a light yellow, viscous liquid. Technical material contains 60% *trans*- and 40% *cis*- isomers. Mp: 34–39°C; bp: 220°C at 0.05 mmHg; ρ: 1.190–1.272 at 20/4°C; K_H: 4.8 × 10^{-8} atm · m³/mol at 20°C (approximate — calculated from water solubility and vapor pressure); log K_{oc}: 1.32–2.79; log K_{ow}: 2.88–6.10; P-$t_{1/2}$: 177.51 hours (*cis*), 121.03 (*trans*) (absorbance λ = 227.5 nm, concentration on glass plates = 6.7 μg/cm²); S_o: soluble in methanol (258 g/kg at 25°C) and miscible with many organic solvents except ethylene glycol; S_w: 0.2 mg/L at 20°C; vp: 1.88 × 10^{-8} mmHg at 20°C.

Soil properties and adsorption data

Soil	K_d (mL/g)	f_{oc} (%)	K_{oc} (mL/g)	pH
Loam	63.3	—	—	—
Loamy sand	20.7	—	—	—
Sandy loam	439.4	—	—	—
Sandy loam	611.0	—	—	—
Silt loam	118.0	—	—	—

Source: U.S. Department of Agriculture, 1990.

Environmental Fate

Soil. Permethrin biodegraded rapidly via hydrolysis yielding 3-(2,2-dichloroethenyl)-2,2-dimethylcyclopropanecarboxylic acid and 3-phenoxybenzyl alcohol (Kaufman et al., 1981). The reported half-life in soil containing 1.3–51.3% organic matter and pH 4.2–7.7 is <38 days (Worthing and Hance, 1991).

In lake water, permethrin degraded more rapidly than in flooded sediment to *trans*- and *cis*-(dichlorovinyl)dimethylcyclopropanecarboxylic acid. The *cis* isomer was more stable toward biological and chemical degradation than the *trans* isomer (Sharom and Solomon, 1981).

Plant. Metabolites identified in cotton leaves included *trans*-hydroxypermethrin, 2'-hydroxypermethrin, 4'-hydroxypermethrin, dichlorovinyl acid, dichlorovinyl acid conjugates, hydroxydichlorovinyl acid, hydroxydichlorovinyl acid conjugates, phenoxybenzyl alcohol and phenoxybenzyl alcohol conjugates. Degradation of permethrin was mainly by ester cleavage and conjugation of the acid and alcohol intermediates (Gaughan and Casida, 1978).

Dislodgable residues of permethrin on cotton leaves 0, 24, 48, 72 and 96 hours after application (1.1 kg/ha) were (cis:trans): 0.26:0.38, 0.24:0.34, 0.22:0.32, 0.16:0.24 and 0.15:0.21 µg/m², respectively (Buck et al., 1980).

Photolytic. Photolysis of permethrin in aqueous solutions containing various solvents (acetone, hexane and methanol) under UV light (λ >290 nm) or on soil in sunlight initially resulted in the isomerization of the cyclopropane moiety and ester cleavage. Photolysis products identified included 3-phenoxybenzyldimethyl acrylate, 3-phenoxybenzaldehyde, 3-phenoxybenzoic acid, monochlorovinyl acids, *cis*- and *trans*-dichlorovinyl acids, benzoic acid, 3-hydroxybenzoic acid, 3-hydroxybenzyl alcohol, benzyl alcohol, benzaldehyde, 3-hydroxybenzaldehyde and 3-hydroxybenzoic acid (Holmstead et al., 1978).

Chemical/Physical. Stable to light and heat (Windholz et al., 1983; Worthing and Hance, 1991).

Symptoms of Exposure: Eye and skin irritant.

Formulation Types: Aerosol; emulsifiable concentrate; wettable powder; water-dispersible granules; dustable powder; aerosol; fumigant.

Toxicity: LC_{50} (48-hour) for rainbow trout 5.4 µg/L and bluegill sunfish 1.8 µg/L (Hartley and Kidd, 1987); acute oral LD_{50} for rats approximately 4,000 mg/kg (*cis:trans* = 40:60) (Hartley and Kidd, 1987); acute inhalation LC_{50} for rats 2.3 mg/L (Worhing and Hance, 1991).

Use: Insecticide.

PHENMEDIPHAM

Synonyms: Beetomax; Beetup; Betanal; Betanal E; Betosip; EP 452; Fender; Fenmedifam; Goliath; Gusto; *m*-Hydroxycarbanilic acid methyl ester *m*-methyl-carbanilate; Kemifam; 3-Methoxycarbonylaminophenyl 3-methylcarbanilate; 3-((Methoxycarbonyl)amino)phenyl (3-methylphenyl)carbamate; Methyl *m*-hydroxy-carbanilate *m*-methylcarbanilate; Methyl 3-(3-methylcarbaniloyloxy)carbanilate; **3-(Methylphenyl)carbamic acid 3-((methoxycarbonyl)amino)phenyl ester**; Methyl 3-(*m*-tolylcarbamoyloxy)phenylcarbamate; Pistol; Protrum K; Schering 38584; SN 4075; SN 38584; Spin-Aid; Vangard.

Designations: CAS Registry Number: 13684-63-4; mf: $C_{16}H_{16}N_2O_4$; fw: 300.32; RTECS: FD9050000.

Properties: Colorless to white crystals or crystalline solid. Mp: 139–142°C (pure), 143–144°C (technical); ρ: 0.25-0.30 at 20/4°C; fl p: 74°C; H-$t_{1/2}$ (22°C): 70 days (pH 5), one day (pH 7), 10 minutes (pH 9); K_d: 28–314; K_H: 8.4 × 10⁻¹¹ atm · m³/mol at 20°C (approximate — calculated from water solubility and vapor pressure); log K_{oc}: 3.21–3.27 (calculated); log K_{ow}: 3.59 at pH 4; S_o (g/L at 20°C): acetone (≈ 200), benzene (2.5), chloroform (20), cyclohexanone (≈ 200), hexane (≈ 0.5), ethanol (≈ 50), methanol (50), methylene chloride (16.7), toluene (0.97), 2,2,4-trimethylpentane (0.16); S_w: 4.7–6.0 mg/L at 20°C; vp: 10⁻⁹ mmHg at 20°C.

Environmental Fate

Soil. Phenmedipham degraded in soil forming methyl *N*-(3-hydroxyphenyl)-carbamate and *m*-aminophenol (Hartley and Kidd, 1987). Hydrolysis yields *m*-aminophenol (Rajagopal et al., 1989). The reported half-life in soil is approximately 20 days (Rajagopal et al., 1989) and 26 days (Worthing and Hance, 1991).

Plant. In plants, methyl *N*-(3-hydroxyphenyl)carbamate is the major metabolite (Hartley and Kidd, 1987).

Photolytic. Bussacchini et al. (1985) studied the photolysis (λ = 254 nm) of phenmedipham in ethanol, ethanol/water and hexane as solvents. In their proposed free radical mechanism, homolysis of the carbon-oxygen bond of the carbamate linkage gave the following photoproducts: 3-(hydroxyphenyl)carbamic acid methyl ester, *m*-toluidine, 2-hydroxy-4-aminomethyl benzoate, 3-hydroxy-5-aminomethyl benzoate, 2-amino-4-hydroxymethyl benzoate and 2-amino-6-hydroxymethyl benzoate.

Symptoms of Exposure: Ingestion may cause hyperactivity and muscle spasms.

Formulation Types: Emulsifiable concentrate (1.3 lb/gal).

Toxicity: LC_{50} (96-hour) for bluegill sunfish 3.98 mg/L, rainbow trout 1.4–3.0 mg/L, *Daphnia magna* 3.2 mg/L (Worthing and Hance, 1991), harlequin fish 16.5 mg/L (Hartley and Kidd, 1987); acute oral LD_{50} of pure phenmedipham and the formulated product for rats 3,700 and 10,300 mg/kg, respectively (Ashton and Monaco, 1991).

Uses: Postemergence herbicide used to control weeds such as chickweed, dogfennel, foxtail, kochia, nightshade and yellow mustard, in strawberries, beet crops and spinach.

PHORATE

Synonyms: AC 3911; American Cyanamid 3911; *O,O*-Diethyl *S*-ethylmercaptomethyl dithiophosphonate; *O,O*-Diethyl *S*-ethylthiomethyl dithiophosphonate; *O,O*-Diethyl *S*-ethylthiomethyl phosphorodithioate; EI-3911; ENT 24042; Experimental insecticide 3911; Granutox; L 11/6; Phorate-10G; **Phosphorodithioic acid *O,O*-diethyl *S*-((ethylthio)methyl) ester**; Rampart; RCRA waste number P094; Thimet; Timet; Vegfru; Vergfru foratox.

$$CH_3CH_2O \diagdown \overset{\overset{S}{\parallel}}{P} - SCH_2SCH_2CH_3$$
$$CH_3CH_2O \diagup$$

Designations: CAS Registry Number: 298-02-2; mf: $C_7H_{17}O_2PS_3$; fw: 260.40; RTECS: TD9450000.

Properties: Clear liquid. Mp: $<-15°C$; bp: 118–120°C at 0.8 mmHg, 125–127°C at 2.0 mmHg; ρ: 1.156 at 25/4°C; H-$t_{1/2}$: 96 hours at 25°C and pH 7; K_H: 6.4×10^{-6} atm · m³/mol at 20–24°C (approximate — calculated from water solubility and vapor pressure); log K_{oc}: 2.51–2.80; log K_{ow}: 2.91–3.92; S_o: miscible with carbon tetrachloride, di-*n*-butyl phthalate, methylene chloride, vegetable oils, 1,4-dioxane, xylene and many other organic solvents; S_w: 20 mg/L at 24°C; vp: 8.4×10^{-4} mmHg at 20°C, 8.25×10^{-4} mmHg at 25°C.

Soil properties and adsorption data

Soil	K_d (mL/g)	f_{oc} (%)	K_{oc} (mL/g)	pH	CEC (meq/100 g)
Batcombe silt loam	4.51	2.05	220	6.10	—
Clarion soil	9.62	2.64	364	5.00	21.02
Elkhorn sandy loam	3.94	0.09	4,581	6.00	—
Harps soil	16.14	3.80	424	7.30	37.84
Hugo gravelly sandy loam	1.96	0.12	1,633	5.50	—
Loam	12.00	2.20	545	7.00	—
Peat soil	73.79	18.36	402	6.98	77.34
Rothamsted Farm	9.83	1.51	651	5.10	—
Sand	1.80	0.58	310	6.00	—
Sandy loam	4.00	0.64	627	6.90	—
Silt loam	5.60	1.74	322	5.20	—
Sarpy fine sandy loam	2.32	0.51	455	7.30	5.71
Sweeney sandy clay loam	13.71	1.65	2,109	6.30	—
Thurman loamy fine sand	5.48	1.07	514	6.83	6.10
Tierra clay loam	15.50	0.33	4,697	6.20	—

Source: Felsot and Dahm, 1979; Lord et al., 1980; Briggs, 1981; U.S. Department of Agriculture, 1990.

Environmental Fate

Biological. [^{14}C]Phorate degraded in a model ecosystem consisting of soil, plants and water (Lichtenstein et al., 1974). Under both non-percolating and percolating water conditions, 12% of the applied amount migrated downward as the corresponding sulfone and sulfoxide. Phorate was absorbed in the roots of corn and was transformed primarily to the sulfone with trace amounts of the sulfoxide. Translocation of radioactive insecticide to the leaves was also observed but the major products were identified as phoratoxon sulfone and phoratoxon sulfoxide (Lichtenstein et al., 1974).

From the first-order biotic and abiotic rate constants of phorate in estuarine water and sediment/water systems, the estimated biodegradation half-lives were 1.1–1.6 and 0.7–1.6 days, respectively (Walker et al., 1988).

Soil. The corresponding sulfoxide and sulfone and their phosphorothioate analogs are major soil metabolites (Lichtenstein et al., 1974). The phosphorothioate analogs may hydrolyze forming dithio-, thio- and orthophosphoric acids (Hartley and Kidd, 1987). Phorate sulfoxide, a microbial metabolite of phorate and much more toxic, was degraded by microbes to phorate oxon (Ahmed and Casida, 1958; Dewey and Parker, 1965). Menzer et al. (1970) reported that degradation of disulfoton in soil degraded at a higher rate in the winter months than in the summer months. They postulated that soil type, rather than temperature, had greater influence on the rate of decomposition of disulfoton. Soils used in the winter and summer months were an Evesboro loamy sand and Chillum silt loam, respectively (Menzer et al., 1970). Suett (1971) reported that the sulfone and sulfoxide had a residual activity of 5 weeks in a peaty loam.

Phorate was moderately persistent in *soil*. Way and Scopes (1968) reported residues comprised 10% of applied dosage 540 days after application. The reported half-life in soil is 82 days (Jury et al., 1987) and 68 days in a sandy soil (Way and Scopes, 1968). The half-lives for phorate in soil incubated in the laboratory under aerobic conditions ranged from 7 to 82.5 days with an average half-life of 75 days (Getzin and Chapman, 1960; Getzin and Shark, 1970). Approximately three days after applying phorate to a sandy soil, 41% was lost to volatilization (Burns, 1971).

Plant. Oat plants were grown in two soils treated with [^{14}C]phorate. Most of the residues remained bound to the soil. Less than 2% of the applied [^{14}C]phorate was recovered from the oat leaves. The major residues in soil were phorate and the corresponding oxon, sulfone, oxon sulfone, sulfoxide and oxon sulfoxide (Fuhremann and Lichtenstein, 1980). These compounds were also found in asparagus tissue and soil treated with the insecticide (Szeto and Brown, 1982). In soil and plants, phorate is oxidized to the corresponding sulfone which is further oxidized to the sulfoxide (Metcalf et al., 1957; Fukuto and Metcalf, 1969; Getzin and Shanks, 1970). Both metabolites were identified in spinach 5.5 months after application (Menzer and Dittman, 1968). The half-lives in coastal Bermuda grass and alfalfa are 1.4 (Leuck and Bowman, 1970) and 3.6 days (Dobson et al., 1960), respectively.

Chemical/Physical. Emits toxic fumes of nitrogen and sulfur oxides when heated to decomposition (Lewis, 1990).

Tirey et al. (1993) evaluated the degradation of phorate at three different temperatures. When oxidized at temperatures of 200, 250 and 275°C, the following reaction products were identified by GC/MS: ethanol, ethanethiol, methanethiol, 1,2,4-trithiolane, 1,1-thiobis-ethane, 1,1'-(methylenebis(thio))bis-ethane, 1,3,5-trithiane, *O,O*-diethyl-*S*-pentenyl phosphorodithioic acid, ethylthioacetic acid, diethyl disulfide, 2,2'-dithiobis ethanol, ethyl-(1-methylpropyl) disulfide, sulfur dioxide, carbon monoxide, carbon dioxide, sulfuric acid and phosphine.

The hydrolysis half-lives of phorate in a sterile 1% ethanol/water solution at 25°C and pH values of 4.5, 5.0, 6.0, 7.0 and 8.0, were 0.55, 0.60, 0.55, 0.57 and 0.68 weeks, respectively (Chapman and Cole, 1982). The calculated hydrolysis half-life at 25°C and pH 7 is 96 hours (Ellington et al., 1988).

Exposure Limits: ACGIH TLV: TWA 0.05 mg/m³ ppm, STEL 0.2 mg/m³.

Formulation Types: Emulsifiable concentrate; granules.

Toxicity: LC_{50} (96-hour) for rainbow trout 13 µg/L and channel catfish 280 µg/L (Worthing and Hance, 1991); acute oral LD_{50} for male and female rats 3.7 and 1.6 mg/kg, respectively (Hartley and Kidd, 1987), 1 mg/kg (RTECS, 1985).

Uses: Systemic insecticide for control of mites, chewing and sucking insects in fruits and vegetables, cotton and some ornamentals.

PHOSALONE

Synonyms: Azofene; *S*-((3-Benzoxazolinyl-6-chloro-2-oxo)methyl) *O,O*-diethyl phosphorodithioate; Benzphos; Benzophosphate; Chipman 11974; *S*-(6-Chloro-3-(mercaptomethyl)-2-benzoxazolinone) *O,O*-diethyl phosphorothiolothionate; *S*-((6-Chloro-2-oxo-3(2*H*)-benzoxazolyl)methyl) *O,O*-diethyl phosphorodithioate; *O,O*-Diethyl *S*-(6-chlorobenzoxazolinyl-3-methyl) dithiophosphate; *O,O*-Diethyl *S*-((6-chloro-2-oxobenzoxazolin-3-yl)methyl) phosphorodithioate; *O,O*-Diethyl *S*-(6-chloro-2-oxobenzoxazolin-3-yl)methyl phosphorothiolothionate; 3-Diethyldithiophosphorymethyl-6-chlorobenzoxazolone-2; *O,O*-Diethyl phosphorodithioate, *S*-ester with 6-chloro-3-(mercaptomethyl)-2-benzoxazolinone; ENT 27163; Fozalon; NIA 9241; Niagara 9241; NPH 1091; P-974; Phasolon; Phosalon; Phozalon; **Phosphorodithioic acid *S*-((6-chloro-2-oxo-3(2*H*)-benzoxazolyl)methyl) *O,O*-diethyl ester**; Phosphorodithioic acid *S*-ester of 6-chloro-3-mercaptomethylbenzoxazyol-2-one; Rhodia RP 11974; RP 11974; Rubitox; Zolon; Zolon PM; Zolone.

Designations: CAS Registry Number: 2310-17-0; mf: $C_{12}H_{15}ClNO_4PS_2$; fw: 367.80; RTECS: TD5175000.

Properties: Colorless crystals with a garlic-like odor. Mp: 45–48°C; K_H: 7.6 × 10^{-8} atm · m^3/mol at 25°C (approximate — calculated from water solubility and vapor pressure); log K_{oc}: 3.41 (calculated); log K_{ow}: 3.77–4.38; P-t$_{1/2}$: 71.37 hours (absorbance λ = 241.5 nm, concentration on glass plates = 6.7 µg/cm^2); S_o (g/L): acetone (1,000), acetonitrile (1,000), ethanol (200), hexane (11); S_w: 1.2, 2.6 and 3.7 mg/L at 10, 20 and 30°C, respectively; vp: 5.03 × 10^{-7} mmHg at 25°C.

Environmental Fate

Plant. Degrades in plants to chlorbenzoxazolone, formaldehyde and diethyl phosphorodithioate (Hartley and Kidd, 1987).

Soil. Ambrosi et al. (1977a) studied the persistence and metabolism of phosalone in both moist and flooded Matapeake loam and Monmouth fine sandy loam. Phosalone rapidly degraded (half-life 3 to 7 days) but mineralization to carbon dioxide accounted for only 10% of the loss. The primary degradative pathway proceeded by oxidation of phosalone to give phosalone oxon. Subsequent cleavage of the *O,O*-diethyl methyl phos-phorodithioate linkage gave 6-chloro-2-benzoxazolinone. Although 2-amino-5-chlorophenol was not detected in this study, they postulated that the condensation of this compound yielded phenoxazinone.

Chemical/Physical. Emits toxic fumes of chlorine, phosphorus, nitrogen and sulfur oxides when heated to decomposition (Sax and Lewis, 1987).

Formulation Types: Emulsifiable concentrate; wettable powder; dustable powder.

Toxicity: LC_{50} for goldfish 2 mg/L, bluegill sunfish 0.11 mg/L and rainbow trout 0.3-0.63 mg/L (Worthing and Hance, 1991); acute oral LD_{50} for male and female rats is 120–175 and 135–170 mg/kg, respectively (Hartley and Kidd, 1987), 85 mg/kg (RTECS, 1985).

Uses: Insecticide; acaracide.

PHOSMET

Synonyms: Appa; *N*-Dimethoxyphosphinothioylthiomethyl)phthalimide; *O,O*-Dimethyl phosphorodithoate *S*-ester with *N*-(mercaptomethyl)phthalimide; *O,O*-Dimethyl *S*-phthal-imidomethyl phosphorodithioate; *S*-((1,3-Dioxo-2*H*-isoindol-2-yl)methyl) *O,O*-dimethyl phosphorodithioate; ENT 25705; Imidan; Kemolate; **Phosphorodithioic acid *S*-((1,3-dihydro-1,3-dioxo-2*H*-isoindol-2-yl)methyl) *O,O*-dimethyl ester**; Phosphorodithioic acid *O,O*-dimethyl ester with *N*-(mercaptomethyl)phthalimide; Phthalimidomethyl-*O,O*-dimethylphosphorodithioate; Phtalofos; PMP; Prolate; R 1504.

Designations: CAS Registry Number: 732-11-6; mf: $C_{11}H_{12}NO_4PS_2$; fw: 317.33; RTECS: TE2275000.

Properties: Colorless crystals. Technical grades are off-white to pale pink crystals with an offensive odor. Mp: 72.0–72.7°C (pure), 66.5–69.5°C (technical grade, 95–98% purity); bp: decomposes rapidly >100°C; H-$t_{1/2}$ (buffered solution at 20°C): 13 days (pH 4.5), <12 hours (pH 7), <4 hours (pH 8.3); K_H: 9.4×10^{-9} atm · m^3/mol at 25–30°C (approximate — calculated from water solubility and vapor pressure); log K_{oc}: 2.87–2.90 (calculated); log K_{ow}: 2.78–3.04; P-$t_{1/2}$: 53.25 hours (absorbance $\lambda = 243.0$ nm, concentration on glass plates = 6.7 μg/cm^2); S_o (g/L at 25°C): acetone (650), benzene (600), kerosene (5), methanol (50), 4-methyl-2-pentanone (300), toluene (300), xylene (250); S_w: 22–25 mg/L at 25°C; vp: 4.52×10^{-7} mmHg at 30°C, 10^{-3} mmHg at 50°C.

Environmental Fate

Soil. In soils, phosmet is rapidly hydrolyzed to phthalimide (Camazano and Martin, 1980; Sánchez Camazano and Sánchez Martin, 1983). The rate of hydrolysis is greater in the presence of various montmorillonite clays and chloride salts. The calculated hydrolysis half-lives of phosmet in the presence of calcium, barium, copper, magnesium and nickel montmorillonite clays were 0.084, 0.665, 10.025, 16.926 and 28.738 days, respectively. Similarly, the half-lives of phosmet in the presence of copper, calcium, magnesium and barium chlorides were <0.020, 5.731, 10.680 and 12.242 days, respectively. In comparison, the hydrolysis of phosmet in a neutral water solution was 46.210 days (Sánchez Camazano and Sánchez Martin, 1983).

Plant. In plants, phosmet is degraded to nontoxic compounds (Hartley and Kidd, 1987). Dorough et al. (1966) reported the half-life in Bermuda grass was 6.5 days. Half-life values ranged from 1.2 to 6.5 days (Dorough et al., 1966; Leuck and Bowman, 1968; Miller et al., 1969; Menzies et al., 1971; Pree et al., 1981).

Chemical/Physical. Emits toxic fumes of phosphorus, nitrogen and sulfur oxides when heated to decomposition (Sax and Lewis, 1987). Though no products were identified, the hydrolysis half-lives at 20°C were 7.0 days and 7.1 hours at pH 6.1 and pH 7.4, respectively. At 37.5°C and pH 7.4, the hydrolysis is 1.1 hours (Freed et al., 1979).

Formulation Types: Dustable powder; emulsifiable concentrate; wettable powder; soluble concentrate.

Toxicity: LC_{50} (96-hour) for bluegill sunfish 70 μg/L, channel catfish 11,000 μg/L, chinook salmon 150 μg/L, fathead minnow 7,300 μg/L, rainbow trout 560 μg/L and small mouth bass 150 μg/L (Verschueren, 1983); acute oral LD_{50} for male and female rats 113 and 160 mg/kg, respectively (Hartley and Kidd, 1987).

Uses: Nonsystemic acaricide and insecticide used on citrus, fruit, grape and potato crops.

PHOSPHAMIDON

Synonyms: Apamidon; C 570; 2-Chloro-3-(diethylamino)-1-methyl-3-oxo-1-propenyl dimethyl phosphate; 2-Chloro-2-diethylcarbamoyl-1-methylvinyl dimethylphosphate; CIBA 570; Dimecron; Dimecron 100; Dimethyl 2-chloro-2-diethylcarbamoyl-1-methylvinyl phosphate; *O,O*-Dimethyl *O*-(2-chloro-2-(*N,N*-diethylcarbamoyl)-1-methylvinyl) phosphate; Dimethyl phosphate of chloro-*N,N*-diethyl-3-hydroxycrotonamide; Dixon; ENT 25515; Famfos; ML 97; NCI-C00588; OR 1191; **Phosphoric acid 2-chloro-3-(dimethylamino)-1-methyl-3-oxo-1-propenyl dimethyl ester.**

$$CH_3O \diagdown \overset{\overset{O}{\|}}{P} - O\overset{\overset{CH_3}{|}}{C} = \underset{\underset{Cl}{|}}{C}CON \diagup \overset{CH_2CH_3}{\diagdown CH_2CH_3}$$

Designations: CAS Registry Number: 13171-21-6; mf: $C_{10}H_{19}ClNO_5P$; fw: 299.69; RTECS: TC2800000.

Properties: Pale yellow, oily liquid with a faint odor. Mp: –45°C; bp: 120°C at 0.001 mmHg, 162°C at 1.5 mmHg; ρ: 1.2132 at 25/4°C; H-$t_{1/2}$ at 20°C: 60 days (pH 5), 54 days (pH 7), 12 days (pH 9); K_H: 3.6×10^{-6} atm · m³/mol at 20–22°C (approximate — calculated from water solubility and vapor pressure); log K_{oc}: 0.63-0.98; log K_{ow}: 0.795; S_o: miscible with most organic solvents except saturated hydrocarbons such as hexane (33.3 g/kg); S_w: miscible at 22°C; vp: 2.5×10^{-5} mmHg at 20°C.

Soil properties and adsorption data

Soil	K_d (mL/g)	f_{oc} (%)	K_{oc} (mL/g)	pH
Loamy sand	0.034	0.80	4.25	7.4
Sand	0.057	0.60	9.55	6.5
Silt loam	0.078	1.50	5.20	6.2
Silt loam	0.390	5.40	7.22	7.3
Silt	1.360	25.00	5.44	6.9
Silt loam	5.600	1.74	321.84	5.2

Source: U.S. Department of Agriculture, 1990.

Environmental Fate

Chemical/Physical. Emits toxic fumes of chlorine, phosphorus and nitrogen oxides when heated to decomposition (Sax and Lewis, 1987).

Formulation Types: Emulsifiable concentrate; suspension concentrate; soluble concentrate.

Toxicity: LC_{50} (96-hour) for fathead minnow 100.0 mg/L, bluegill sunfish 4.50 mg/L, channel catfish 70.0 mg/L (Verschueren, 1983), blood clam 4.26 mg/L (Bharathi, 1994); LC_{50} (72-hour) for blood clam 6.33 mg/L (Bharathi, 1994); LC_{50} (48-hour) for blood clam 8.38 mg/L (Bharathi, 1994); LC_{50} (24-hour) for rainbow trout 5.0 ppm (Verschueren, 1983), blood clam 11.53 mg/L (Bharathi, 1994); acute oral LD_{50} for rats 17.9–30 mg/kg (Hartley and Kidd, 1987), 10,900 µg/kg (RTECS, 1985).

Uses: Insecticide used to control sap-feeding insects and other pests in a wide variety of crops.

PICLORAM

Synonyms: Amdon grazon; 4-Amino-3,5,6-trichloropicolinic acid; 4-Amino-3,5,6-trichloropyridine-2-carboxylic acid; **4-Amino-3,5,6-trichloro-2-pyridinecarboxylic acid**; ATCP; Borolin; Grazon; K-pin; NCI-C00237; Tordon; Tordon 10K; Tordon 101 mixture.

Designations: CAS Registry Number: 1918-02-1; mf: $C_6H_3Cl_3N_2O_2$; fw: 241.48; RTECS: TJ7525000.

Properties: Colorless to white crystals with a chlorine-like odor. Mp: decomposes at 215–219°C without melting; fl p (open cup): 35°C (Tordon 101 formulation), 46°C (Tordon 22K formulation); pK_a: 2.3 (20°C), 3.6; K_H: 3.4×10^{-10} atm · m³/mol at 25–35°C (approximate — calculated from water solubility and vapor pressure); log K_{oc}: 1.41; log K_{ow}: 0.30; S_o (g/L at 25°C): acetone (19.8), acetonitrile (1.6), benzene (0.2), carbon disulfide (<0.05), ethanol (10.5), ethyl ether (1.2), kerosene (0.01), methylene chloride (0.6), 2-propanol (5.5); S_w: 400–430 mg/L at 25°C; vp: 6.16×10^{-7} mmHg at 35°C, 1.07×10^{-7} mmHg at 45°C.

Soil properties and adsorption data

Soil	K_d (mL/g)	f_{oc} (%)	K_{oc} (mL/g)	pH	CEC (meq/100 g)
Asquith sandy loam	0.030	1.02	2.91	7.5	—
Asquith sandy loam	0.090	1.04	8.65	6.9	—
Ephrata sandy loam	0.070	0.54	12.96	7.1	—
Fiddletown silt loam	0.976	2.42	40.33	5.6	—
Indian Head clay loam	0.240	2.36	10.17	8.1	—
Indian Head loam	0.100	2.35	4.26	7.8	—
Indian Head clay	0.230	2.70	8.52	8.1	—
Kentwood sandy loam	0.118	0.92	12.83	6.4	—
Kinney clay loam	1.600	2.88	55.56	5.0	16.1
Kinney clay loam	4.600	4.27	107.73	5.2	6.5
Kinney clay loam	2.300	1.44	159.72	5.2	8.9
Lacombe loam	0.750	7.20	10.42	7.9	—
Linne clay loam	0.409	1.38	29.64	7.4	—
Melfort loam	0.490	6.05	8.10	5.9	—
Melfort loam	0.490	6.00	8.17	6.5	—
Minam loam	0.300	2.19	13.70	7.3	24.4
Minam loam	0.600	3.81	15.75	7.0	28.3
Palouse silt loam	0.310	1.38	22.46	5.7	—

Soil properties and adsorption data *(continued)*

Soil	K_d (mL/g)	f_{oc} (%)	K_{oc} (mL/g)	pH	CEC (meq/100 g)
Palouse silt loam	0.553	2.07	26.72	5.7	—
Regina heavy clay	0.100	2.40	4.17	8.0	—
Weyburn Oxbow loam	0.240	3.72	6.45	6.5	—
Weyburn Oxbow loam	0.310	3.75	8.27	7.9	—
Weyburn light loam	0.240	2.50	9.60	8.2	—
Woodcock loam	0.300	4.44	6.76	6.1	12.9
Woodcock loam	0.600	2.48	24.19	5.8	3.6
Woodcock sandy loam	0.409	0.92	44.56	5.6	3.2

Source: Grover, 1971; Farmer and Aochi, 1974; Gaynor and Volk, 1976; Grover, 1977.

Environmental Fate

Soil. Degrades in soil via cleavage of the chlorine atom at the *m*-position to form 4-amino-5,6-dichloro-2-picolinic acid. Replacement of the chlorine at the *m*-posi-tion by a hydroxyl group yields 4-amino-3-hydroxy-5,6-dichloropicolinic acid (Hartley and Kidd, 1987). Other soil metabolites reported include carbon dioxide, chloride ions, 4-amino-6-hydroxy-3,5-dichloropicolinic acid (Meikle et al., 1974), 4-amino-3,5-dichloro-6-hydrox-ypicolinic acid and 4-amino-3,5,6-trichloropyridine (Goring and Hamaker, 1971). Young-son et al. (1967) reported that degradation increased with an increase in temperature and organic matter.

The half-lives for picloram in soil incubated in the laboratory under aerobic conditions ranged from 29 days to 3 years with an average of 201 days (Meikle et al., 1973; Yoshida and Castro, 1975; Merkle et al., 1976). In field soils, the half-lives for picloram ranged from 12 to 84 days with an average of 31 days (Bovey et al., 1969; Lutz et al., 1973; Sirons et al., 1977). The mineralization half-lives for ^{14}C-labeled picloram in soil between temperatures of 15 to 30°C ranged from 175 days to 175 years with an average half-life of 43.4 years (Meikle et al., 1966, 1973; Guenzi and Beard, 1976a).

Groundwater. According to the U.S. EPA (1986) picloram has a high potential to leach to *groundwater.*

Plant. Picloram degraded very slowly in cotton plants releasing carbon dioxide (Meikle et al., 1966). Metabolites identified in spring wheat were 4-amino-2,3,5-trichloropyridine, oxalic acid and 4-amino-3,5-dichloro-6-hydroxypicolinic acid (Redemann et al., 1968; Plimmer, 1970). In soil, 4-amino-3,5-dichloro-6-hydroxypicolinic acid was the only com-pound positively identified (Redemann et al., 1968).

Photolytic. The sodium salt of picloram in aqueous solution was readily decom-posed by UV light (λ = 300–380 nm). Two chloride ions were formed for each molecule of picloram that reacted. It was postulated that degradation proceeded via a free radical and an ionic mechanism (Mosier and Guenzi, 1973). Hall et al. (1968) exposed picloram (0.02 M and pH >7) to UV light (λ = 253.7 nm) at an intensity of 200 μW/cm^2. After 2 days, 20% degraded releasing two chloride ions per molecule, five unidentified products and the destruction of the pyridine ring. Replication of this experiment in sunlight produced similar, but less effective and variable results (Hall et al., 1968).

Picloram rapidly degraded in irradiated aqueous solutions (Kearney et al., 1968). Irradiation of picloram in aqueous solution by UV light (2,537 Å) released two chloride ions for each molecule of picloram degraded. Photodecomposition products included acidic compounds and five methylated derivatives (Plimmer, 1970). When picloram in an aqueous solution (25°C) was exposed by a high intensity monochromatic UV lamp, dechlorination occurred yielding 4-amino-3,5-dichloro-6-hydroxypicolinic acid which underwent decarboxylation to give 4-amino-3,5-dichloropyridin-2-ol. In addition, decarboxylation of picloram yielded 2,3,5-trichloro-4-pyridylamine which may undergo dechlorination to give 4-amino-3,5-dichloro-6-hydroxypicolinic acid (Burkhard and Guth, 1979).

Chemical/Physical. Releases carbon monoxide, carbon dioxide, chlorine and ammonia when heated to 900°C (Kennedy, 1972, 1972a).

Reacts with alkalies and amines forming water-soluble salts (Hartley and Kidd, 1987).

Exposure Limits: OSHA PEL: 15 mg/m³ (total), 5 mg/m³ (respirable fraction); ACGIH TLV: TWA 10 mg/m³, STEL 20 mg/m³.

Symptoms of Exposure: Mild eye and skin irritant. Ingestion of large amounts may cause nausea.

Formulation Types: Aqueous solution; granules; pellets; soluble concentrate.

Toxicity: LC_{50} (96-hour) for fathead minnow 55.3 mg/L and rainbow trout 19.3 mg/L (Worthing and Hance, 1991); acute oral LD_{50} for rats 8,200 mg/kg (Ashton and Monaco, 1991), 2,898 mg/kg (RTECS, 1985).

Uses: Systemic herbicide used to control most broad-leaved weeds on grassland and noncrop areas. Use as a pesticide is restricted.

PINDONE

Synonyms: Chemrat; **2-(2,2-Dimethyl-1-oxopropyl)-1*H*-indene-1,3(2*H*)-dione**; Pivacin; Pival; 2-Pivaloylindane-1,3-dione; 2-Pivaloyl-1,3-indanedione; Pivalyl; Pivalyl indandione; 2-Pivalyl-1,3-indandione; Pivalyl valone; Tri-Ban; UN 2472.

Designations: CAS Registry Number: 83-26-1; DOT: 2472; mf: $C_{14}H_{14}O_3$; fw: 230.25; RTECS: NK6300000.

Properties: Bright yellow crystals. Mp: 108–110°C; log K_{oc}: 2.95 (calculated); log K_{ow}: 3.18 (calculated); S_w: 18 ppm at 25°C.

Exposure Limits: NIOSH REL: TWA 0.1 mg/m³, IDLH 100 mg/m³; OSHA PEL: TWA 0.1 mg/m³; ACGIH TLV: TWA 0.1 mg/m³.

Symptoms of Exposure: Nosebleed, excess bleeding of minor cuts, bruises; smokey urine, black tarry stools; abdominal and back pain.

Formulation Types: Grain bait; tracking powder.

Toxicity: LC_{50} (96-hour) for rainbow trout 0.21 mg/L and bluegill sunfish 1.6 mg/L (Hartley and Kidd, 1987); acute oral LD_{50} for rats 280 mg/kg (RTECS, 1985).

Uses: Insecticide; rodenticide.

PROFENOFOS

Synonyms: *O*-(**4-Bromo-2-chlorophenyl**) *O*-ethyl *S*-propyl **phosphorothioate**; CGA 15324; Curacron; Phosphorothioic acid *O*-(4-bromo-2-chlorophenyl)-*O*-ethyl-*S*-propyl ester; Polycron; Selecron.

$$Br-\underset{Cl}{\underset{|}{\overset{\bigcirc}{\bigcirc}}}-\overset{\overset{O}{\|}}{\underset{\underset{SCH_2CH_2CH_3}{|}}{OP}}-OCH_2CH_3$$

Designations: CAS Registry Number: 41198-08-7; mf: $C_{11}H_{15}BrClO_3PS$; fw: 373.64; RTECS: TE6975000.

Properties: Pale yellow liquid. Mp: <25°C; bp: 110°C at 0.001 mmHg; ρ: 1.455 at 20/4°C; H-$t_{1/2}$ (20°C): 93 days (pH 5), 14.6 days (pH 7), 5.7 hours (pH 9); K_H: 2.4 × 10^{-7} atm · m³/mol at 20°C (approximate — calculated from water solubility and vapor pressure); log K_{oc}: 2.55–3.46; log K_{ow}: 1.9; S_o: miscible with most organic solvents including aromatic and chlorinated hydrocarbons; S_w: 20 mg/L at 20°C; vp: 9.75 × 10^{-6} mmHg at 20°C.

Soil properties and adsorption data

Soil	K_d (mL/g)	f_{oc} (%)	K_{oc} (mL/g)	pH
Sand	4.6	1.28	356	7.8
Sand	20.2	0.70	2,902	6.3
Sandy loam	22.2	2.09	1,062	6.1
Sandy-silt loam	55.6	3.25	1,711	6.7
Silt loam	5.6	1.74	322	5.20

Source: U.S. Department of Agriculture, 1990.

Environmental Fate

Plant. Dislodgable residues of profenofos on cotton leaf 0, 24, 48, 72 and 96 hours after application (1.1 kg/ha) were 3.5, 1.1, 0.74, 0.51 and 0.35 µg/m², respectively (Buck et al., 1980).

Photolytic. When profenofos in an aqueous buffer solution (pH 7.0) was exposed to filtered UV light (λ >290 nm) for 24 hours at 25 and 50°C, 29 and 61% respectively decomposed to 4-bromo-2-chlorophenol and 4-bromo-2-chlorophenyl ethyl hydrogen phosphate (Burkhard and Guth, 1979).

Chemical/Physical. Emits toxic fumes of bromine, chlorine phosphorus and sulfur oxides when heated to decomposition (Sax and Lewis, 1987).

Formulation Types: Water-dispersible granules; granules; suspension concentrate; wettable powder.

Toxicity: LC_{50} (96-hour) for rainbow trout 80 µg/L, bluegill sunfish 300 µg/L and crucian carp 90 µg/L (Hartley and Kidd, 1987); acute oral LD_{50} for rats 358 mg/kg (Hartley and Kidd, 1987), 400 mg/kg (RTECS, 1985).

Uses: Insecticide; acaricide.

PROMETON

Synonyms: 2,4-Bis(isopropylamino)-6-methoxy-*s*-triazine; 2,6-Diisopropylamino-4-methoxytriazine; *N,N'*-Diisopropyl-6-methoxy-1,3,5-triazine-2,4-diyldiamine; G 31435; Gesafram; Gesafram 50; 2-Methoxy-4,6-bis(isopropylamino)-1,3,5-triazine; **6-Methoxy-*N,N'*-bis(1-methylethyl)-1,3,5-triazine-2,4-diamine**; Methoxypropazine; Ontracic 800; Ontrack; Ontrack-WE-2; Pramitol; Primatol; Primatol 25E; Prometone.

$$\text{CH}_3\text{O} \quad\quad \text{NHCH(CH}_3)_2$$

$$\text{NHCH(CH}_3)_2$$

Designations: CAS Registry Number: 1610-18-0; mf: $C_{10}H_{19}N_5O$; fw: 225.30; RTECS: XY4200000.

Properties: Colorless to white crystals or powder. Mp: 91–92°C; ρ: 1.088 at 20/4°C; fl p: nonflammable; pK_a: 4.3 at 21°C; K_H: 8.9×10^{-10} atm · m³/mol at 20°C (approximate — calculated from water solubility and vapor pressure); log K_{oc}: 1.92–2.24; log K_{ow}: 2.55 (Liu and Qian, 1995); S_o (g/L): acetone (300), benzene (>250), methanol (600), methylene chloride (350), toluene (250); S_w at 26°C: 185, 44.4, 26.5, 30.1 and 29.7 M at pH values of 2.0, 3.0, 5.0, 7.0 and 10.0, respectively (Ward and Weber, 1968); vp: 2.3×10^{-6} mmHg at 20°C, 7.9×10^{-6} mmHg at 30°C.

Soil properties and adsorption data

Soil	K_d (mL/g)	f_{oc} (%)	K_{oc} (mL/g)	pH	CEC (meq/100 g)
Bates silt loam	0.60	0.80	75	6.5	9.3
Baxter silty clay loam	1.70	1.21	141	6.0	11.2
Chillum silt loam	5.40	2.54	213	4.6	7.6
Clarksville silty clay loam	2.20	0.80	275	5.7	5.7
Cumberland silt loam	0.50	0.69	72	6.4	6.5
Eldon silt loam	1.30	1.73	75	5.9	12.9
Gerald silt loam	6.00	1.55	387	4.7	11.0
Grundy silt clay loam	6.30	2.07	304	5.6	13.5
Hagerstown silty clay loam	4.30	2.48	173	5.5	12.5
Hanford sandy loam	0.15	0.43	35	6.0	6.0[a]
Hanford sandy loam	0.36	0.43	84	6.0	6.0[a]
Hanford sandy loam	1.68	0.43	391	6.0	6.0[a]
Knox silt loam	6.40	1.67	383	5.4	18.8
Lakeland sandy loam	1.00	1.90	53	6.2	2.9
Lebanon silt loam	7.80	1.04	750	4.9	7.7
Lindley loam	4.70	0.86	546	4.7	6.9
Lintonia loamy sand	0.70	0.34	206	5.3	3.2
Loam	2.90	1.68	173	6.9	—

Soil properties and adsorption data *(continued)*

Soil	K_d (mL/g)	f_{oc} (%)	K_{oc} (mL/g)	pH	CEC (meq/100 g)
Marian silt loam	14.90	0.80	1,863	4.6	9.9
Marshall silty clay loam	8.80	2.42	364	5.4	21.3
Menfro silt loam	1.20	1.38	87	5.3	9.1
Newtonia silt loam	2.40	0.92	261	5.2	8.8
Oswego silty clay loam	3.40	1.67	204	6.4	21.0
Putnam silt loam	2.80	1.09	257	5.3	12.3
Salix loam	4.60	1.21	380	6.3	17.9
Sandy loam	2.61	1.74	150	6.1	—
Sandy loam	1.20	1.22	98	7.0	—
Sandy clay loam	2.40	2.90	83	7.0	—
Sarpy loam	1.50	0.75	200	7.1	14.3
Sharkey clay	55.20	1.44	3,833	5.0	28.2
Shelby loam	22.30	2.07	1,077	4.3	20.1
Summit silty clay	11.60	2.82	411	4.8	35.1
Tujunga loamy sand	0.17	0.33	52	6.3	0.4[a]
Tujunga loamy sand	0.22	0.33	67	6.3	0.4[a]
Tujunga loamy sand	1.13	0.33	342	6.3	0.4[a]
Union silt loam	6.10	1.04	586	5.4	6.8
Wabash clay	17.00	1.27	1,339	5.7	40.3
Waverley silt loam	3.80	1.15	330	5.6	12.8
Wehadkee silt loam	3.90	1.09	358	5.6	10.2

Source: Talbert and Fletchall, 1965; Harris, 1966; Singh et al., 1990; U.S. Department of Agriculture, 1990. a) cmol/kg.

Environmental Fate

Soil. Degrades in soil yielding hydroxy metabolites and dealkylation of the side chains (Hartley and Kidd, 1987).

Photolytic. Pelizzetti et al. (1990) studied the aqueous photocatalytic degradation of prometon and other *s*-triazines (ppb level) using simulated sunlight (λ >340 nm) and titanium dioxide as a photocatalyst. Prometon rapidly degraded forming cyanuric acid, nitrates, the intermediate tentatively identified as 2,4-diamino-6-hydroxy-*N,N'*-bis(1-methylethyl)-1,3,5-triazine and other intermediate compounds similar to those found for atrazine. Mineralization of cyanuric acid to carbon dioxide was not observed (Pelizzetti et al., 1990).

Formulation Types: Emulsifiable concentrate; wettable powder.

Toxicity: LC_{50} (96-hour) for rainbow trout 12 mg/L, bluegill sunfish 40 mg/L, crucian carp 70 mg/L (Worthing and Hance, 1991) and goldfish 8.6 mg/L (Hartley and Kidd, 1987); acute oral LD_{50} of the Prometon EC formulation for rats 2,300 mg/kg (Ashton and Monaco, 1991), 503 mg/kg (RTECS, 1985).

Uses: Nonselective preemergence and postemergence herbicide used to control most annual and perennial broad-leaved weeds, grasses and brush weeds on noncrop land.

PROMETRYN

Synonyms: 2,4-Bis(isopropylamino)-6-methylmercapto-*s*-triazine; 2,4-Bis(isopropy-lamino)-6-methylthio-*s*-triazine; 2,4-Bis(isopropylamino)-6-methylthio-1,3,5-triazine; *N,N′*-**Bis(1-methylethyl)-6-(methylthio)-1,3,5-triazine-2,4-diamine**; Caparol; Caparol 80W; Cotton-Pro; *N,N′*-Diisopropyl-6-methylthio-1,3,5-triazine-2,4-diyldiamine; G 34161; Gesagard; Merkazin; 2-Methylmercapto-4,6-bis(isopropylamino)-*s*-triazine; 2-Methylthio-4,6-bis(isopropylamino)-*s*-triazine; Polisin; Primatol; Primatol Q; Prometrex; Prometrin; Prometryne; Selektin; Sesagard.

Designations: CAS Registry Number: 7287-19-6; mf: $C_{10}H_{19}N_5S$; fw: 241.37; RTECS: XY4390000.

Properties: Colorless to white crystalline solid. Mp: 118–120°C; ρ: 1.157 at 20/4°C; fl p: nonflammable; pK_a: 4.05; H-$t_{1/2}$ at 25°C: 22 days (0.1 N hydrochloric acid solution), ≈ 500 years (pH 7), ≈ 30 years (0.01 N sodium hydroxide solution); pK_a: 4.05 at 21°C; K_H: 4.9×10^{-9} atm · m³/mol at 20°C (approximate — calculated from water solubility and vapor pressure); log K_{oc}: 2.28–2.79; log K_{ow}: 3.34, 3.46; S_o (g/L at 20°C): acetone (240), hexane (5.5), methanol (160), methylene chloride (300), toluene (170); S_w at 26°C: 135, 8.53, 1.73, 1.67 and 1.73 M at pH values of 2.0, 3.0, 5.0, 7.0 and 10.0, respectively (Ward and Weber, 1968); vp: 10^{-6} mmHg at 20°C, 4×10^{-6} mmHg at 30°C.

Soil properties and adsorption data

Soil	K_d (mL/g)	f_{oc} (%)	K_{oc} (mL/g)	pH	CEC (meq/100 g)
Bates silt loam	1.60	0.80	200	6.5	9.3
Baxter silty clay loam	4.30	1.21	355	6.0	11.2
Brewer clay loam	9.95	1.61	618	5.8	13.5
Cape Fear sandy loam	844.00	5.34	15,805	—	10.3
Chillum silt loam	12.90	2.54	508	4.6	7.6
Clarksville silty clay loam	5.10	0.80	638	5.7	5.7
Cobb sand	0.66	0.34	194	7.3	3.8
Cobb sand + 2% muck	28.90	1.21	2,388	5.3	9.0
Cumberland silt loam	1.40	0.69	203	6.4	6.5
Eldon silt loam	3.60	1.73	208	5.9	12.9
Gerald silt loam	9.40	1.55	606	4.7	11.0
Grundy sandy clay loam	9.20	2.07	444	5.6	13.5
Hagerstown silty clay loam	10.30	2.48	415	5.5	12.5
Kebabib silt loam	9.00	1.04	865	4.9	7.7

Soil properties and adsorption data *(continued)*

Soil	K_d (mL/g)	f_{oc} (%)	K_{oc} (mL/g)	pH	CEC (meq/100 g)
Knox silt loam	8.40	1.67	503	5.4	18.8
Lakeland sandy loam	1.90	1.90	100	6.2	2.9
Lebanon silt loam	9.00	1.04	865	4.9	7.7
Lindley loam	7.90	0.86	919	4.7	6.9
Lintonia loamy sand	0.90	0.34	265	5.3	3.2
Loam	9.95	1.62	614	5.8	—
Loam	4.90	1.04	471	6.3	—
Marian silt loam	14.20	0.80	1,775	4.6	9.9
Marshall silty clay loam	12.30	2.42	508	5.4	21.3
Menfro silt loam	3.30	1.38	239	5.3	9.1
Newtonia silt loam	3.50	0.92	380	5.2	8.8
Norfolk fine sandy loam	87.00	0.99	8,788	—	2.3
Oswego silty clay loam	5.00	1.67	299	6.4	21.0
Port silty clay	4.90	1.04	471	6.3	17.9
Putnam silt loam	3.80	1.09	349	5.3	12.3
Rains fine sandy loam	139.00	1.45	9,586	—	7.1
Salix loam	6.40	1.21	529	6.3	17.9
Sand	0.66	0.35	190	7.3	—
Sandy loam	3.49	0.75	463	5.7	—
Sarpy loam	2.90	0.75	387	7.1	14.3
Sharkey clay	43.40	1.44	3,014	5.0	28.2
Shelby loam	22.10	2.07	1,068	4.3	20.1
Soil 1	3.10	0.51	608	—	10.6
Soil 2	5.00	1.56	321	—	10.3
Soil 3	5.70	1.76	324	—	22.3
Soil 4	10.40	3.82	272	—	26.9
Soil 5	39.80	13.61	292	—	31.2
Soil 6	65.40	19.98	327	—	83.1
Summit silty clay	17.70	2.82	628	4.8	35.1
Teller fine sandy loam	3.49	0.75	465	8.6	5.7
Union silt loam	8.60	1.04	827	5.4	6.8
Wabash clay	17.30	1.27	1,362	5.7	40.3
Waverley silt loam	5.80	1.15	504	6.4	12.8
Wehadkee silt loam	0.62	1.09	57	5.6	10.2

Source: Talbert and Fletchall, 1965; Harris, 1966; Walker and Crawford, 1970; Murray et al., 1975; Kozak et al., 1983; U.S. Department of Agriculture, 1990.

Environmental Fate

Soil. In soil and plants, the methylthio group is oxidized. The proposed degradative pathway is the formation of the corresponding sulfoxide and sulfone derivatives of prometryn followed by oxidation of the latter to forming hydroxypropazine (Kearney and Kaufman, 1976). Cook and Hütter (1982) reported that bacterial cultures degraded prometryne to form the corresponding hydroxy derivative (hydroxyprometryne). ^{14}C-Prometryn was incubated in an organic soil for one year. It was observed that 57.4% of the bound residues were comprised of prometryn (>50%), hydroxypropazine, mono-*n*-dealkylated prometryn, mono-N-dealkylated hydroxypropazine, a didealkylated compound [2-(methylthio)-4-amino-6-(isopropylamino)-s-triazine] and unidentified methanol soluble products (Khan and Hamilton, 1980; Khan, 1982). Hydroxyprometryn and ammeline were reported as hydrolysis metabolites in soil (Somasundaram et al., 1991). The reported half-life in soil is 60 days (Jury et al., 1987).

Photolytic. When prometryn in aqueous solution was exposed to UV light for 3 hours, the herbicide was completely converted to hydroxypropazine. Irradiation of soil suspensions containing prometryn was found to be more resistant to photodecomposition. About 75% of the applied amount was converted to hydroxypropazine after 72 hours of exposure (Khan, 1982). The UV ($\lambda = 253.7$ nm) photolysis of prometryn in water, methanol, ethanol, n-butanol and benzene, yielded 2-methylthio-4,6-bis(isoprop-ylamino)-s-triazine. At wavelengths >300 nm, photodegradation was not observed (Pape and Zabik, 1970). Khan and Gamble (1983) also studied the UV irradiation ($\lambda = 253.7$ nm) of prometryn in distilled water and dissolved humic substances. In distilled water, 2-hydroxy-4,6-bis(isopropyl-amino)-s-triazine and 4,6-bis(isopropylamino)-s-triazine formed as major products.

In the presence of humic/fulvic acids, 4-amino-6-(isopropylamino)-s-triazine was also formed (Khan and Gamble, 1983). Pelizzetti et al. (1990) studied the aqueous photocatalytic degradation of prometryn and other s-triazines (ppb level) using simulated sunlight (λ >340 nm) and titanium dioxide as a photocatalyst. Prometryn rapidly degraded forming cyanuric acid, nitrates, sulfates, the intermediate tentatively identified as 2,4-diamino-6-hydroxy-N,N'-bis(1-methyl-ethyl)-1,3,5-triazine and other intermediate compounds similar to those found for atrazine. It was suggested that the appearance of sulfate ions was due to the attack of the methylthio group at the number two position. Mineralization of cyanuric acid to carbon dioxide was not observed (Pelizzetti et al., 1990).

Chemical/Physical. Mascolo et al. (1995) studied the reaction of prometryn (100 µM) with sodium hypochlorite (10 mM) at 25°C and pH 7. Degradation followed the following pathway: prometryn → 2,4-(N,N'-diisopropyl)diamino-6-methylsulfinyl-1,3,5-triazine → 2,4-(N,N'-diisopropyl)diamino-6-methylsulfonyl-1,3,5-triazine → 2,4-(N,N'-diisopropyl)diamino-6-methanesulfonate ester 1,3,5-triazine → 2,4-(N,N'-diisopropyl)diamino-6-hydroxy-1,3,5-triazine.

Symptoms of Exposure: Eye and skin irritant.

Formulation Types: Suspension concentrate (4 lb/gal); wettable powder.

Toxicity: LC_{50} (96-hour) for rainbow trout 2.5 mg/L, bluegill sunfish 10.0 mg/L, goldfish 3.5 mg/L and carp 8 mg/L (Hartley and Kidd, 1987); acute oral LD_{50} of the 80% formulation for rats 3,800 mg/kg (Ashton and Monaco, 1991), 2,100 mg/kg (RTECS, 1985).

Uses: Selective herbicide used to control many annual grasses and broad-leaved weeds in cotton, celery and peas.

PROPACHLOR

Synonyms: Albrass; Bexton; Bexton 4L; 2-Chloro-*N*-isopropylacetanilide; α-Chloro-*N*-isopropylacetanilide; **2-Chloro-*N*-(1-methylethyl)-*N*-phenylacetamide**; CP 31393; *N*-Isopropyl-2-chloroacetanilide; *N*-Isopropyl-α-chloroacetanilide; Niticid; Propachlore; Prolex; Ramrod; Ramrod 65; Satecid.

Designations: CAS Registry Number: 1918-16-7; mf: $C_{11}H_{14}ClNO$; fw: 211.67; RTECS: AE1575000.

Properties: Pale tan solid. Mp: 77°C (pure), 67–76°C (technical); bp: 110°C at 0.03 mmHg, decomposes at 170°C; ρ: 1.242 at 25/4°C; fl p: nonflammable; K_H: 1.1×10^{-7} atm · m^3/mol at 25°C (approximate — calculated from water solubility and vapor pressure); log K_{oc}: 2.07–2.11 (calculated); log K_{ow}: 1.61; S_o (g/kg at 25°C): acetone (448), benzene (737), carbon tetrachloride (174), chloroform (602), ethanol (408), ethyl ether (219), toluene (342), xylene (239); S_w: 613–700 mg/L at 25°C; vp: 2.25×10^{-4} mmHg at 25°C.

Environmental Fate

Biological. In the presence of suspended natural populations from unpolluted aquatic systems, the second-order microbial transformation rate constant determined in the laboratory was reported to be 1.1×10^{-9} L/organisms-hour (Steen, 1991).

Groundwater. According to the U.S. EPA (1986) propachlor has a high potential to leach to *groundwater.*

Plant. In corn seedlings and excised leaves of corn, sorghum, sugarcane and barley, propachlor was metabolized to at least three water-soluble products. Two of these metabolites were identified as a γ-glutamylcysteine conjugate of propachlor and a glutathione conjugate of propachlor. It was postulated that both compounds were intermediate compounds in corn seedlings since they were not detected 3 days following treatment (Lamoureux et al., 1971).

Photolytic. When propachlor in an aqueous ethanolic solution was irradiated with UV light (λ = 290 nm) for 5 hours, 80% decomposed to the following cyclic photo-products: *N*-isopropyloxindole, *N*-isopropyl-3-hydroxyoxindole and a spiro compound. Irradiation of propachlor in an aqueous solution containing riboflavin as a sensitizer resulted in completed degradation of the parent compound. *m*-Hydroxypropachlor was the only compound identified in trace amounts which formed via ring hydroxylation (Rejtö et al., 1984). Hydrolyzes under alkaline conditions forming *N*-isopropylaniline (Sittig, 1985) which is also a product of microbial metabolism (Novick et al., 1986).

Chemical/Physical. Emits toxic fumes of nitrogen oxides and chlorine when heated to decomposition (Sax and Lewis, 1987). Hydrolyzes under alkaline conditions forming *N*-isopropylaniline (Sittig, 1985) which is also a product of microbial metabolism (Novick et al., 1986). Propachlor is rapidly hydrolyzed in Water (Yu et al., 1975a). The hydrolysis half-lives at 68.0°C and pH values of 3.10 and 10.20 were calculated to be 36.6 and 1.2 days, respectively (Ellington et al., 1986).

Symptoms of Exposure: May cause allergic skin reactions upon contact.

Formulation Types: Granules (20%); suspension concentrate or flowable liquid (4 lb/gal); wettable powder.

Toxicity: LC_{50} (96-hour) for bluegill sunfish >1.4 mg/L and rainbow trout 0.17 mg/L (Hartley and Kidd, 1987). For the Ramrod flowable formulation the LC_{50} (96-hour) for bluegill sunfish and rainbow trout is 1.6 and 0.42 mg/L, respectively (Humburg et al., 1989); acute oral LD_{50} for rats is 710 mg/kg (RTECS, 1985).

Uses: Selective preemergence herbicide used to control most annual grasses and some broad-leaved weeds in brassicas, corn, cotton, flax, leeks, maize, milo, onions, peas, roses, ornamental trees and shrubs, soybeans, sugarcane.

PROPANIL

Synonyms: Bay 30130; Chem-Rice; Crystal Propanil-4; DCPA; **N-(3,4-Dichlorophe-nyl)propanamide**; N-(3,4-Dichlorophenyl)propionamide; Dichloropropionanilide; 3',4'-Dichloropropionanilide; DPA; Dipram; Erban; Erbanil; Farmco propanil; FW 734; Grascide; Herbax; Herbax technical; Montrose Propanil; Propanex; Propanid; Propanide; Prop-Job; Propionic acid 3,4-dichloroanilide; Riselect; Rogue; Rosanil; S 10165; Stam; Stam F-34; Stam LV 10; Stam M-4; Stampede; Stampede 360; Stampede 3E; Stam Supernox; Strel; Supernox; Surcopur; Surpur; Synpran N; Vertac; Wham EZ.

$$NHOCCH_2CH_3$$

Designations: CAS Registry Number: 709-98-8; mf: $C_9H_9Cl_2NO$; fw: 218.09; RTECS: UE4900000.

Properties: Colorless to light brown to gray-black crystals. Technical grades are light-brown to grayish-black. Mp: 92–93°C (pure), 88–91°C (technical); ρ: 1.25 at 25/4°C; K_H: 3.6×10^{-8} atm · m^3/mol at 20°C (approximate — calculated from water solubility and vapor pressure); log K_{oc}: 2.19; log K_{ow}: 2.03, 2.29; So g/L at 20°C: 2-butanone (250), cyclohexanone (350), isophorone (5520), g/L at 25°C: acetone (1,700), benzene (70), ethanol (1,100), hexane (<1), methylene chloride (>200), 2-propanol (>200), toluene (50–100); S_w: 130 mg/L at 20°C, 225 mg/L at 25°C; vp: 2×10^{-7} mmHg at 20°C, 9×10^{-5} mmHg at 60°C.

Environmental Fate

Biological. In the presence of suspended natural populations from unpolluted aquatic systems, the second-order microbial transformation rate constant determined in the laboratory was reported to be 5×10^{-10} L/organisms-hour (Steen, 1991).

Soil. Propanil degrades in soil forming 3,4-dichloroaniline (Bartha, 1968; Bartha and Pramer, 1970; Chisaka and Kearney, 1970; Bartha, 1971; Duke et al., 1991; Pothuluri et al., 1991) which is further degraded by microbial peroxidases to 3,3',4,4'-tetrachloroa-zobenzene (Bartha and Pramer, 1967; Bartha et al., 1968; Chisaka and Kearney, 1970), 3,3',4,4'-tetrachloroazooxybenzene (Bartha and Pramer, 1970), 4-(3,4-dichloroanilo)-3,3',4,4'-tetrachloroazobenzene (Linke and Bartha, 1970) and 1,3-bis(3,4-dichlorophe-nyl)triazine (Plimmer et al., 1970a), propanoic acid, carbon dioxide and unidentified products (Chisaka and Kearney, 1970). Evidence suggests that 3,3',4,4'-tetrachloroazoben-zene reacted with 3,4-dichloroaniline forming a new reaction product, namely 4-(3,4-dichloroanilo)-3,3',4'-trichloroazobenzene (Chisaka and Kearney, 1970). Under aerobic conditions, propanil in a biologically active, organic-rich pond sediment underwent dechlo-rination at the *para-* position forming N-(3-chlorophenyl)propanamide (Stepp et al., 1985). Residual activity in soil is limited to approximately 3–4 months (Hartley and Kidd, 1987).

Plant. In rice plants, propanil is rapidly hydrolyzed via an aryl acylamidase enzyme isolated by Frear and Still (1968) forming the nonphototoxic compounds (Ashton and Monaco, 1991) 3,4-dichloroaniline, propionic acid (Matsunaka, 1969; Menn and Still, 1977; Hatzios, 1991) and a 3',4'-dichloroaniline-lignin complex. This complex was identified as a metabolite of *N*-(3,4-dichlorophenyl)glucosylamine, a 3,4-dichloroaniline saccharide conjugate and a 3,4-dichloroaniline sugar derivative (Yi et al., 1968). In a rice field soil under anaerobic conditions, however, propanil underwent amide hydrolysis and dechlorination at the *para* position forming 3,4-dichloroaniline and *m*-chloroaniline (Pettigrew et al., 1985). In addition, propanil may degrade indirectly via an initial oxidation step resulting in the formation of 3,4-dichlorolacetanilide which is further hydrolyzed to 3,4-dichloroaniline and lactic acid (Hatzios, 1991). In an earlier study, four metabolites were identified in rice plants, two of which were positively identified as 3,4-dichloroaniline and *N*-(3,4-dichlorophenyl)glucosylamine (Still, 1968).

Photolytic. Photoproducts reported from the sunlight irradiation of propanil (200 mg/L) in distilled water were 3'-hydroxy-4'-chloropropionanilide, 3'-chloro-4'-hydroxypropionanilide, 3',4'-dihydroxypropionanilide, 3'-chloropropionanilide, 4'-chloropropionanilide, propionanilide, 3,4-dichloroaniline, 3-chloroaniline, propionic acid, propionamide, 3,3',4,4'-tetrachloroazobenzene and a dark polymeric humic substance. The photolysis products resulted from the reductive dechlorination, replacement of chlorine substituents by hydroxyl groups, formation of propionamide, hydrolysis of the amide group and azobenzene formation (Moilanen and Crosby, 1972). Tanaka et al. (1985) studied the photolysis of propanil (100 mg/L) in aqueous solution using UV light ($\lambda = 300$ nm) or sunlight. After 26 days of exposure to sunlight, propanil degraded forming a trichlorinated biphenyl product (<1% yield) and hydrogen chloride (Tanaka et al., 1985).

Chemical/Physical. Hydrolyzes in acidic and alkaline media to propionic acid (Worthing and Hance, 1991) and 3,4-dichloroaniline (Sittig, 1985; Worthing and Hance, 1991). The half-life of propanil in a 0.50 N sodium hydroxide solution at 20°C was determined to be 6.6 days (El-Dib and Aly, 1976).

Symptoms of Exposure: Inhalation of vapors may irritate the nose and throat and cause drowsiness, stupor and unconsciousness. Irritates the eyes and skin. Ingestion causes dizziness, stupor, drowsiness, fever and blue lips and fingernails.

Formulation Types: Emulsifiable concentrate.

Toxicity: LC_{50} (48-hour) for carp 13 mg/L, goldfish 14 mg/L, Japanese killifish 14 mg/L and mosquitofish 11 mg/L (Hartley and Kidd, 1987); acute oral LD_{50} for rats 1,400 mg/kg (Ashton and Monaco, 1991), 367 mg/kg (RTECS, 1985).

Uses: Selective preemergence and postemergence herbicide used to control many grasses and broad-leaved weeds in potatoes, rice and wheat.

PROPAZINE

Synonyms: 2,4-Bis(isopropylamino)-6-chloro-*s*-triazine; 2-Chloro-4,6-bis(isopropy-lamino)-*s*-triazine; **6-Chloro-*N*,*N*′-bis(1-methylethyl)-1,3,5-triazine-2,4-diamine**; 6-Chloro-*N*,*N*′-diisopropyl-1,3,5-triazine-2,4-diyldiamine; G 30028; Geigy 30028; Gesamil; Maax; Milogard; Milogard 4L; Milogard 80W; Plantulin; Primatol P; Propasin; Propazin; Prozinex.

$$(CH_3)_2CHNH \quad\text{—}\quad N\text{—}Cl$$
$$N\text{—}N$$
$$NHCH(CH_3)_2$$

Designations: CAS Registry Number: 139-40-2; mf: $C_9H_{16}ClN_5$; fw: 230.09; RTECS: XY5300000.

Properties: Colorless, crystalline solid. Mp: 212°C; ρ: 1.162 at 20/4°C; fl p: nonflamma-ble; pK_a: 1.85 at 22°C; K_H: 9.9×10^{-9} atm · m^3/mol at 20°C (approximate — calculated from water solubility and vapor pressure); log K_{oc}: 1.69–2.56; log K_{ow}: 2.89 (Liu and Qian, 1995); P-$t_{1/2}$: 108.17 hours (absorbance λ = 222.5 nm, concentration on glass plates = 6.7 μg/cm^2); S_o (g/kg at 20°C): benzene (6.2), carbon tetrachloride (2.5), ethyl ether (5.0), toluene (6.2); S_w at 26°C: 620, 83, 21, 27, 20 and 22 μM at pH values of 1.0, 2.0, 3.0, 5.0, 7.0 and 10.0, respectively (Ward and Weber, 1968); vp: 2.9×10^{-8} mmHg at 20°C, 1.6×10^{-7} mmHg at 30°C.

Soil properties and adsorption data

Soil	K_d (mL/g)	f_{oc} (%)	K_{oc} (mL/g)	pH	CEC (meq/100 g)
Bates silt loam	0.70	0.80	88	6.5	9.3
Baxter silty clay loam	1.90	1.21	157	6.0	11.2
Chillum silt loam	4.60	2.54	181	4.6	7.6
Clarksville silty clay loam	2.10	0.80	263	5.7	5.7
Cumberland silt loam	0.70	0.69	101	6.4	6.5
Eldon silt loam	1.80	1.73	104	5.9	12.9
Evouettes	1.15	2.09	55	6.1	—
Gerald silt loam	1.80	1.55	116	4.7	11.0
Grundy silty clay loam	2.80	2.07	135	5.6	13.5
Hagerstown silty clay loam	3.70	2.48	149	5.5	12.5
Hickory Hill silt	11.90	3.27	363	—	—
Knox silt loam	2.70	1.67	162	5.4	18.8
Lakeland sandy loam	0.80	1.90	49	6.2	2.9
Lebanon silt loam	2.00	1.04	192	4.9	7.7
Lindley loam	2.20	0.86	256	4.7	6.9
Lintonia loamy sand	0.10	0.34	29	5.3	3.2

Soil properties and adsorption data *(continued)*

Soil	K_d (mL/g)	f_{oc} (%)	K_{oc} (mL/g)	pH	CEC (meq/100 g)
Marian silt loam	2.10	0.80	263	4.6	9.9
Marshall silty clay loam	3.00	2.42	124	5.4	21.3
Menfro silt loam	1.80	1.38	130	5.3	9.1
Newtonia silt loam	1.40	0.92	152	5.2	8.8
Oswego silty clay loam	1.90	1.67	114	6.4	21.0
Putnam silt loam	1.10	1.09	101	5.9	12.3
Salix loam	1.90	1.21	157	6.3	17.9
Sarpy loam	1.20	0.75	160	7.1	14.3
Sharkey clay	3.00	1.44	208	5.0	28.2
Shelby loam	2.80	2.07	135	4.3	20.1
Soil 1	0.90	0.51	176	—	10.6
Soil 2	3.00	1.56	192	—	10.3
Soil 3	3.00	1.76	170	—	22.3
Soil 4	3.50	3.82	92	—	26.9
Soil 5	19.70	13.61	145	—	31.2
Soil 6	20.50	19.98	103	—	83.1
Summit silty clay	3.40	2.82	121	4.8	35.1
Union silt loam	2.40	1.04	231	5.4	6.8
Vertroz	4.69	3.25	144	6.7	—
Wabash clay	3.10	1.27	244	5.7	40.3
Waverley silt loam	2.00	1.15	174	6.4	12.8
Wehadkee silt loam	1.60	1.09	147	5.6	10.2

Source: Talbert and Fletchall, 1965; Harris, 1966; Walker and Crawford, 1970; Burkhard and Guth, 1981.

Environmental Fate

Soil. Undergoes microbial degradation in soil forming hydroxypropazine (Harris, 1967). Dealkylation of both substituted amino groups is presumably followed by ring opening and decomposition (Hartley and Kidd, 1987). Under laboratory conditions, the half-lives of propazine in a Hatzenbühl soil (pH 4.8) and Neuhofen soil (pH 6.5) at 22°C were 62 and 127 days, respectively (Burkhard and Guth, 1981).

Groundwater. According to the U.S. EPA (1986) propazine has a high potential to leach to *groundwater.*

Photolytic. Irradiation of propazine in methanol afforded prometone (2-methoxy-4,6-bis(isopropylamino-s-triazine). Photodegradation of propazine in methanol did not occur when irradiated at wavelengths >300 nm (Pape and Zabik, 1970).

Chemical/Physical. In aqueous solutions, propazine is converted exclusively to hydroxypropazine (2-hydroxy-4,6-bisisopropylamino)-s-triazine by UV light ($\lambda = 253.7$ nm) (Pape and Zabik, 1970). In acidic aqueous soil-free systems, propazine hydrolysis is pH dependent and follows first-order kinetics. At 23.5°C, the estimated hydrolysis half-lives at pH 2.0, 3.0, 4.0 and 5.0 are 9, 36, 141 and 550 days, respectively. The rate of hydrolysis was also found to increase and decrease in the presence of organic matter and calcium salts, respectively (Nearpass, 1972).

Symptoms of Exposure: Eye and skin irritant.

Formulation Types: Water-dispersible granules; wettable powder; liquid (4 lb/gal).

Toxicity: LC_{50} (96-hour) for rainbow trout 17.5 mg/L, bluegill sunfish >100 mg/L and goldfish >32.0 mg/L (Hartley and Kidd, 1987); acute oral LD_{50} of the 80% formulation for rats >5,000 mg/kg (Ashton and Monaco, 1991), 3,840 mg/kg (RTECS, 1985).

Uses: Selective preemergence herbicide used to control annual grasses and broad-leaved weeds in milo and sweet sorghum.

PROPHAM

Synonyms: Ban-Hoe; Beet-Kleen; Carbanilic acid isopropyl ester; Chem-Hoe; IFC; IFK; INPC; IPC; IsoPPC; Isopropyl carbanilate; Isopropyl carbanilic acid ester; Isopropyl phenylcarbamate; Isopropyl *N*-phenylurethane; *O*-Isopropyl *N*-phenyl carbamate; 1-Methylethyl phenylcarbamate; Ortho grass killer; **Phenylcarbamic acid 1-methylethyl ester**; *N*-Phenyl isopropylcarbamate; Premalox; Profam; Prophame; Triherbide; Triherbide-IPC; Tuberit; Tuberite; USAF D-9; Y 2.

$$NHCOOCH(CH_3)_2$$

Designations: CAS Registry Number: 122-42-9; mf: $C_{10}H_{13}NO_2$; fw: 179.22; RTECS: FD9100000.

Properties: Colorless to light tan crystals. Technical grade is a light tan solid. Mp: 87–88°C; bp: sublimes but decomposes >150°C; ρ: ≈ 1.09 at 30/4°C; fl p: nonflammable; log K_{oc}: 1.71; log K_{ow}: 1.93 (calculated); S_o: soluble in most organic solvents; S_w: 32–250 mg/L at 20–25°C; vp: sublimes at room temperature.

Environmental Fate

Biological. Rajagopal et al. (1989) reported that *Achromobacter* sp. and an *Arthrobacter* sp. utilized propham as a sole carbon source. Metabolites identified were *N*-phenylcarbamic acid, aniline, catechol, monoisopropyl carbonate, 2-propanol and carbon dioxide (Rajagopal et al., 1989).

Soil. Readily degraded by soil microorganisms forming aniline and carbon dioxide (Humburg et al., 1989). The reported half-life in soil is approximately 15 and 5 days at 16 and 29°C, respectively (Hartley and Kidd, 1987).

Groundwater. According to the U.S. EPA (1986) propham has a high potential to leach to *groundwater.*

Plant. The major plant metabolite which was identified from soybean plants is isopropyl *N*-2-hydroxycarbanilate (Hartley and Kidd, 1987; Humburg et al., 1989).

Chemical/Physical. Hydrolysis of propham yields *N*-phenylcarbamic acid and 2-propanol. The acid is very unstable and is spontaneously decomposed to form aniline and carbon dioxide (Still and Herrett, 1976). Emits toxic fumes of nitrogen oxides when heated to decomposition (Sax and Lewis, 1987).

Formulation Types: Dustable powder; emulsifiable concentrate; granules (15%); suspension concentrate (3 and 4 lb/gal); wettable powder.

Toxicity: LC_{50} (48-hour) for bluegill sunfish 32 mg/L and guppies 35 mg/L (Hartley and Kidd, 1987); acute oral LD_{50} for rats 9,000 mg/kg (Ashton and Monaco, 1991).

Uses: Preemergence and postemergence herbicide used to control annual grass weeds in peas, beet crops, lucerne, clover, sugar beet, beans, lettuce, flax, safflowers and lentils.

PROPOXUR

Synonyms: Aprocarb; Arprocarb; Bay 9010; Bay 39007; Bayer 39007; Baygon; Bifex; Blattanex; Boygon; Brygou; Chemagro 90; ENT 25671; *o*-IMPC; Invisigard; Isocarb; *o*-Isopropoxyphenyl methylcarbamate; *o*-Isopropoxyphenyl *N*-methylcarbamate; **2-(1-Methylethoxy)phenol methylcarbamate**; *N*-Methyl-2-isopropoxyphenylcarbamate; OMS 33; PHC; Propotox M; Propoxure; Propyon; Sendran; Suncide; Tugon fliegenkugel; Unden.

$$OCONHCH_3$$
$$OCH(CH_3)_2$$

Designations: CAS Registry Number: 114-26-1; mf: $C_{11}H_{15}NO_3$; fw: 209.25; RTECS: FC3150000.

Properties: White to tan crystalline solid. Mp: 84–87°C; bp: decomposes; H-$t_{1/2}$: 290 days (pH 7), 17.9 days (pH 8), 40 minutes at 20°C and pH 10; K_H: 1.3×10^{-9} atm · m³/mol at 20°C (approximate — calculated from water solubility and vapor pressure); log K_{oc}: 0.48–1.97; log K_{ow}: 1.45–1.56; S_o: soluble in most solvents; S_w: 1.74, 1.93 and 2.44 g/L at 10, 20 and 30°C, respectively; vp: 3.0×10^{-6} mmHg at 20°C (Wright et al., 1981).

Soil properties and adsorption data

Soil	K_d (mL/g)	f_{oc} (%)	K_{oc} (mL/g)	pH
Clayey loam	0.27	0.29	93	6.0
Kanuma high clay	0.80	1.35	53	5.7
Loamy sand	0.05	1.62	3	6.6
Sandy loam	0.62	0.81	76	7.7
Silt loam	0.49	1.16	42	6.3
Silt loam	1.12	2.55	44	6.1
Silt loam	0.30	2.96	10	5.1
Tsukuba clay loam	1.70	4.24	41	6.5

Source: Kanazawa, 1989; U.S. Department of Agriculture, 1990.

Environmental Fate

Groundwater. According to the U.S. EPA (1986) propoxur has a high potential to leach to groundwater.

Photolytic. Though no products were identified, the half-life in UV irradiated water (λ >290 nm) was 87.9 hours (Jensen-Korte et al., 1987). When propoxur in ethanol was irradiated by UV light, only one unidentified cholinesterase inhibitor formed. Exposure to sunlight for 3 hours yielded no photodecomposition products (Crosby et al., 1965).

Chemical/Physical. Decomposes at elevated temperatures forming methyl isocyanate (Windholz et al., 1983) and nitrogen oxides (Lewis, 1990). Hydrolyzes in water to 1-naphthol and 2-isopropoxyphenol (Miles et al., 1988). At pH 6.9, half-lives of 78 and 124 days were reported under aerobic and anaerobic conditions, respectively (Kanazawa, 1987). Miles et al. (1988) studied the rate of hydrolysis of propoxur in phosphate-buffered water (0.01 M) at 26°C with and without a chlorinating agent (10 mg/L hypochlorite solution). The hydrolysis half-lives at pH 7 and 8 with and without chlorine were 3.5 and 10.3 days and 0.05 and 1.2 days, respectively (Miles et al., 1988). The reported hydrolysis half-lives of propoxur in water at pH 8, 9 and 10 were 16.0 days, 1.6 days and 4.2 hours, respectively (Aly and El-Dib, 1971). In a 0.50 *N* sodium hydroxide solution at 20°C, the hydrolysis half-life was reported to be 3.0 days (El-Dib and Aly, 1976).

Exposure Limits: OSHA TWA: 0.5 mg/m^3; ACGIH TLV: TWA 0.5 mg/m^3.

Formulation Types: Emulsifiable concentrate; wettable powder; fumigant; dustable powder; aerosol.

Toxicity: LC_{50} (96-hour) for rainbow trout 3.7 mg/L and bluegill sunfish 6.6 mg/L (Hartley and Kidd, 1987); acute oral LD_{50} for male and female rats 95 and 104 mg/kg, respectively (Hartley and Kidd, 1987).

Uses: Insecticide used for control of chewing and sucking insects in fruits, vegetables, ornamentals and forestry.

PROPYZAMIDE

Synonyms: Clanex; **3,5-Dichloro-*N*-(1,1-dimethylpropynyl)benzamide**; 3,5-Dichloro-*N*-(1,1-dimethyl-2-propynyl)benzamide; Kerb; Kerb 50W; Promamide; Pronamide; RCRA waste number U192; RH-315.

Designations: CAS Registry Number: 23950-58-5; mf: $C_{12}H_{11}Cl_2NO$; fw: 256.13; RTECS: CV3460000.

Properties: Colorless to white powder or crystalline solid with a mild odor. The reported odor threshold concentration in air is 700 ppb (Young et al., 1996). Mp: 155–156°C; bp: decomposes above mp; fl p: nonflammable; $H\text{-}t_{1/2}$: >700 days at 25°C and pH 7; K_H: 1.9 × 10^{-7} atm · m³/mol at 20–25°C (approximate — calculated from water solubility and vapor pressure); log K_{oc}: 2.30–2.34; log K_{ow}: 3.09–3.28; S_o (g/L): 2-butanone (200), cyclohexanone (200), dimethyl sulfoxide (300), methanol (150), methyl ethyl ketone (200), 2-propanol (150); S_w: 15 mg/L at 25°C; vp: 8.5 × 10^{-5} mmHg at 20°C.

Environmental Fate

Biological. In the presence of suspended natural populations from unpolluted aquatic systems, the second-order microbial transformation rate constant determined in the laboratory was reported to be 5 × 10^{-14} L/organisms-hour (Steen, 1991).

Soil. The major soil metabolite is 2-(3,5-dichlorophenyl)-4,4-dimethyl-5-methyleneoxazoline. The half-life in soil is approximately 30 days at 25°C (Hartley and Kidd, 1987). Residual activity in soil is limited to approximately 2–6 months (Hartley and Kidd, 1987).

Chemical/Physical. Emits toxic fumes of nitrogen oxides and chlorine when heated to decomposition (Sax and Lewis, 1987).

Propyzamide is hydrolyzed to 3,5-dichlorobenzoate by refluxing under strongly acidic conditions (Humburg et al., 1989).

Formulation Types: Granules; suspension concentrate; wettable powder.

Toxicity: LC_{50} (96-hour) for catfish 500 mg/L, goldfish 350 mg/L, guppies 150 mg/L and rainbow trout 72 mg/L (Hartley and Kidd, 1987); acute oral LD_{50} for male and female rats 8,350 and 5,620 mg/kg, respectively (Hartley and Kidd, 1987).

Uses: Preemergence or postemergence herbicide used to control many perennial and annual grasses and broad-leaved weeds in ornamental trees and shrubs, forestry, fruits and vegetables.

QUINOMETHIONATE

Synonyms: Bay 36205; Bayer 4964; Bayer 36205; Chinomethionat; Chinomethionate; Cyclic *S,S*-(6-methyl-2,3-quinoxalinediyl) dithiocarbonate; ENT 25606; Erade; Erazidon; Forstan; **6-Methyl-1,3-dithiolo[4,5-*b*]quinoxalin-2-one**; 6-Methyl-2-oxo-1,3-dithiolo[4,5-*b*]quinoxaline; 6-Methyl-2,3-quinoxaline dithiocarbonate; 6-Methyl-2,3-quinoxalinedithiol cyclic carbonate; 6-Methyl-2,3-quinoxalinedithiol cyclic dithiocarbonate; 6-Methyl-2,3-quinoxalinedithiol cyclic *S,S*-dithiocarbonate; 6-Methylquinoxaline-2,3-dithiolcyclocarbonate; *S,S*-(6-Methylquinoxaline-2,3-diyl) dithiocarbonate; Morestan; Morestane; Oxythioquinox; Quinomethoate; SS 2074.

Designations: CAS Registry Number: 2439-01-2; mf: $C_{10}H_6N_2OS_2$; fw: 234.29; RTECS: FG1400000.

Properties: Yellow crystals. Mp: 169.8–170°C; H-$t_{1/2}$ (22°C): 10 days (pH 4), 80 hours (pH 7), 225 minutes (pH 9); K_H: 1.2×10^{-9} atm · m³/mol at 20°C (approximate — calculated from water solubility and vapor pressure); log K_{oc}: 3.36–3.37; log K_{ow}: 3.78; S_o (g/L at 20°C): cyclohexanone (18), *N,N*-dimethylformamide (10), petroleum oils (4); S_w: 50 mg/L at 20°C; vp: 2.03×10^{-7} mmHg at 20°C.

Soil properties and adsorption data

Soil	K_d (mL/g)	f_{oc} (%)	K_{oc} (mL/g)	pH
Sandy loam	37	1.62	2,284	5.1
Silt loam	41	1.74	2,356	5.9

Source: U.S. Department of Agriculture, 1990.

Environmental Fate
Chemical/Physical. Reacts with ammonia forming 6-methyl-2,3-quinoxalinedithiol.

Formulation Types: Wettable powder; fumigant; dustable powder.

Toxicity: LC$_{50}$ (96-hour) for rainbow trout 220 µg/L and bluegill sunfish 120 µg/L (Hartley and Kidd, 1987); acute oral LD$_{50}$ for rats 2,500–3,000 mg/kg (Hartley and Kidd, 1987), 1,100 mg/kg (RTECS, 1985).

Uses: Acaricide; fungicide.

QUIZALOFOP-ETHYL

Synonyms: Assure; (±)-2-(4-((6-Chloro-2-quinoxalinyl)oxy)phenoxy)propanoic acid ethyl ester.

Designations: CAS Registry Number: 76578-14-8; mf: $C_{19}H_{17}ClN_2O_4$; fw: 372.80.

Properties: Colorless to white crystalline solid with a petroleum-like odor. Mp: 91.7–92.1°C; bp: 220°C at 0.2 mmHg (decomposes at 320°C); ρ: 1.35 at 20/4°C; fl p: 61°C; K_H: 4.8 × 10⁻⁷ atm · m³/mol at 20–25°C (approximate — calculated from water solubility and vapor pressure); log K_{oc}: ≈ 2.76; log K_{ow}: 4.20; P-$t_{1/2}$: 40 days (sandy loam soil); S_o (g/L at 20°C): acetone (110), benzene (290), ethanol (9), hexane (2.6), xylene (120); S_w: 310 µg/L at 25°C; vp: 3 × 10⁻⁷ mmHg at 20°C.

Environmental Fate
Plant. Rapidly hydrolyzed in plants to quizalofop and ethanol (Humburg et al., 1989).

Symptoms of Exposure: May irritate eyes, nose, skin and throat.

Formulation Types: Suspension concentrate; emulsifiable concentrate (0.79 lb/gal).

Toxicity: LC_{50} (96-hour) for rainbow trout 10.7 mg/L and bluegill sunfish 0.46–2.8 mg/L (Worthing and Hance, 1991); acute oral LD_{50} for male and female rats 1,700 and 1,500 mg/kg, respectively (Ashton and Monaco, 1991).

Uses: Selective postemergence herbicide used to control both annual and perennial grasses in soybeans and cotton.

RONNEL

Synonyms: Dermaphos; Dimethyl trichlorophenyl thiophosphate; *O,O*-Dimethyl-*O*-2,4,5-trichlorophenyl phosphorothioate; *O,O*-Dimethyl *O*-(2,4,5-trichlorophenyl)thiophosphate; Dow ET 14; Dow ET 57; Ectoral; ENT 23284; ET 14; ET 57; Etrolene; Fenchlorfos; Fenchlorophos; Fenchchlorphos; Karlan; Korlan; Korlane; Nanchor; Nanker; Nankor; **Phosphorothioic acid *O,O*-dimethyl *O*-(2,4,5-trichlorophenyl)ester**; Trichlorometafos; Trolen; Trolene; Viozene.

Designations: CAS Registry Number: 299-84-3; DOT: 2922; mf: $C_8H_8Cl_3O_3PS$; fw: 321.57; RTECS: TG0525000.

Properties: White to light brown, waxy solid. Mp: 41°C; bp: 97°C at 0.01 mmHg; ρ: 1.48 at 25/4°C; H-$t_{1/2}$: approximately 3 days at pH 6; K_H: 8.46 × 10^{-6} atm · m^3/mol at 20–25°C (approximate — calculated from water solubility and vapor pressure); log K_{oc}: 2.76 (calculated); log K_{ow}: 4.67–5.068; S_o: soluble in acetone, benzene, carbon tetrachloride, chloroform, ethyl ether, kerosene, methylene chloride, toluene, xylene and many other organic solvents; S_w: 40 mg/L at 25°C; vp: 3.37 × 10^{-5} mmHg at 20°C.

Environmental Fate

Chemical/Physical. Though no products were identified, the reported hydrolysis half-life at pH 7.4 and 70°C using a 1:4 ethanol/water mixture is 10.2–10.4 hours (Freed et al., 1977). Ronnel decomposed at elevated temperatures on five clay surfaces, each treated with hydrogen, calcium, magnesium, aluminum and iron ions. At temperatures <950°C (125, 300 and 750°C), bentonite clays impregnated with technical ronnel (18.6 wt %) decomposed to 2,4,5-trichlorophenol and a rearrangement product tentatively identified as *O*-methyl *S*-methyl-*O*-(2,4,5-trichlorophenyl) phosphorothioate (Rosenfield and Van Valkenburg, 1965). At 950°C, only the latter product formed. It was postulated that this compound resulted from an acid-catalyzed molecular rearrangement reaction. Ronnel also undergoes base-catalyzed hydrolysis at elevated temperatures. Products include methanol and a new compound that is formed via cleavage of a methyl group from one of the methoxy groups, which is then bonded to the sulfur atom (Rosenfield and Van Valkenburg, 1965).

Ronnel is stable to hydrolysis over the pH range of 5–6 (Mortland and Raman, 1967). However, in the presence of a Cu(II) salt (as cupric chloride) or when present as the exchangeable Cu(II) cation in montmorillonite clays, ronnel is completely hydrolyzed via first-order kinetics in <24 hours at 20°C. The calculated half-life of ronnel at 20°C for this reaction is 6.0 hours. It was suggested that decomposition in the presence of Cu(II) was a result of coordination of the copper atom through the oxygen or sulfur on the phosphorus atom resulting in the cleavage of the side chain containing the phosphorus atom forming O,O-ethyl-O-phosphorothioate and 1,2,4-trichlorobenzene (Mortland and Raman, 1967).

Emits very toxic fumes of chlorides, sulfur and phosphorus oxides when heated to decomposition (Lewis, 1990).

Exposure Limits: NIOSH REL: TWA 10 mg/m^3, IDLH 300 mg/m^3; OSHA PEL: TWA 15 mg/m^3; ACGIH TLV: TWA 10 mg/m^3.

Symptoms of Exposure: In animals: cholinesterase inhibition; irritates eyes; liver and kidney damage.

Toxicity: LC$_{50}$ (96-hour) for fathead minnow 305 µg/L (Verschueren, 1983); acute oral LD$_{50}$ for rats 1,740 mg/kg (Verschueren, 1983), 625 mg/kg (RTECS, 1985).

Use: Insecticide.

SIDURON

Synonyms: Du Pont herbicide 1318; H 1318; 1-(2-Methylcychohexyl)-3-phenyl-urea; *N*-(2-Methylcyclohexyl)-*N'*-phenylurea; Tupersan.

$$\text{cyclohexyl} - \text{NHCONH} - \text{phenyl}$$

$$CH_3$$

Designations: CAS Registry Number: 1982-49-6; mf: $C_{14}H_{20}N_2O$; fw: 232.30; RTECS: YT7350000.

Properties: Colorless to white, odorless, crystalline solid. Mp: 133–138°C; ρ: 1.08 at 25/4°C; fl p: nonflammable; K_H: 6.8 × 10⁻¹⁰ atm · m³/mol at 25°C (approximate — calculated from water solubility and vapor pressure); log K_{oc}: 2.88 (estimated); log K_{ow}: 3.09 (Liu and Qian, 1995); S_o (g/kg at 25°C): cellosolve (175), dimethylacetamide (367), *N,N*-dimethylformamide (260), ethanol (160), isophorone (118), methylene chloride (118); S_w: 18 mg/L at 25°C; vp: 4 × 10⁻⁸ mmHg at 25°C.

Environmental Fate

Soil. A fungus and two *Pseudomonas* spp. isolated from soil degraded siduron to form the major metabolites: 1-(4-hydroxy-2-methylcyclohexyl)-3-(*p*-hydroxyphenyl)urea, 1-(4-hydroxy-2-methylcyclohexyl)-3-phenylurea and 1-(4-hydroxyphenyl)-3-(2-methycyclohexyl)urea (Belasco and Langsdorf, 1969).

Plant. After 8 days following absorption in barley plants, no metabolites were detected (Splittstoesser and Hopen, 1968).

Symptoms of Exposure: May irritate eyes, nose, throat and skin.

Formulation Types: Wettable powder (50%).

Toxicity: LC_{50} (48-hour) for carp 18 mg/L and Japanese goldfish 10–40 mg/L (Worthing and Hance, 1991); acute oral LD_{50} for rats >7,500 mg/kg (Ashton and Monaco, 1991), 5,000 mg/kg (RTECS, 1985).

Uses: Selective preemergence herbicide used to control annual weed grasses and prevent growth of crabgrass, foxtail and barnyard grass in turf, cereals and other food crops.

SIMAZINE

Synonyms: A 2079; Aktinit S; Aquazine; Aquazine 80W; Batazina; 2,4-Bis(ethylamino)-6-chloro-*s*-triazine; Bitemol; Bitemol S 50; Caliper; Caliper 90; CAT; CDT; Cekusan; Cekuzina-S; 2-Chloro-4,6-bis(ethylamino)-1,3,5-triazine; 2-Chloro-4,6-bis(ethylamino)-*s*-triazine; **6-Chloro-*N'*,*N'*-diethyl-1,3,5-triazine-2,4-diamine**; 6-Chloro-N^2,N^4-diethyl-1,3,5-triazine-2,4-diyldiamine; Framed; G 27,692; Geigy 27692; Gesaran; Gesatop; Gesatop 50; H 1803; Herbazin; Herbazin 50; Herbex; Herboxy; Hungazin DT; Premazine; Primatol S; Princep; Princep 4G; Princep 4L; Princep 80W; Printop; Radocon; Radokor; Simadex; Simanex; Simazin; Simazine 80W; Sim-Trol; Symazine; Tafazine; Tafazine 50-W; Taphazine; Triazine A 384; W 6658; Weedex; Zeapur.

Designations: CAS Registry Number: 122-34-9; mf: $C_7H_{12}ClN_5$; fw: 201.66; RTECS: XY5250000.

Properties: Colorless to white crystalline solid. Mp: 225–227°C; ρ: 1.203 at 20/4°C; fl p: nonflammable; pK_a: 1.7 at 21°C; K_H: 3.4×10^{-9} atm · m³/mol at 20°C (approximate — calculated from water solubility and vapor pressure); log K_{oc}: 2.14; log K_{ow}: 1.94–2.26; P-$t_{1/2}$: 108.17 hours (absorbance λ = 53.25 nm, concentration on glass plates = 6.7 μg/cm²); S_o (mg/L at 20–25°C): chloroform (900), ethyl ether (300), light petroleum (2), methanol (400), pentane (3), petroleum ether (2); S_w at 26°C: 581, 78, 29, and 25 μM at pH values of 1.0, 2.0, 3.0, and 5.0–10.0, respectively (Ward and Weber, 1968); vp: 6.1×10^{-9} mmHg at 20°C, 3.6×10^{-8} mmHg at 20°C.

Soil properties and adsorption data

Soil	K_d (mL/g)	f_{oc} (%)	K_{oc} (mL/g)	pH	CEC (meq/100 g)
Aiken loam	2.59	2.94	88	5.2	20.6
Ascalon sandy clay loam	0.81	0.85	96	7.3	12.7
Barnes clay loam	3.62	3.98	91	7.4	33.8
Basinger fine sand	0.73	0.61	119	5.8	—
Batcombe silt loam	0.97	2.05	47	6.1	—
Bates sandy loam	1.00	0.80	125	6.5	9.3
Bautista silt loam	0.49	1.03	48	7.2	14.2
Bautista loamy sand	0.30	0.28	107	6.4	5.5
Baxter silty clay loam	2.30	1.21	190	6.0	11.2
Beltsville silt loam	1.94	1.40	139	4.3	4.2
Benevola clay (subsoil)	0.88	1.30	68	7.6	20.1
Benevola silty clay (topsoil)	1.58	2.70	58	7.7	19.5

Soil properties and adsorption data *(continued)*

Soil	K_d (mL/g)	f_{oc} (%)	K_{oc} (mL/g)	pH	CEC (meq/100 g)
Berkley clay (subsoil)	0.81	0.99	82	7.3	34.4
Berkley silty clay (topsoil)	2.94	4.63	63	7.1	33.7
Boca fine sand	0.85	1.67	71	7.1	—
Bosket silt loam	1.76	0.57	308	5.8	8.4
Brentwood clay	3.03	1.67	181	6.1	30.8
Broadbalk 2B	1.50	2.90	52	7.5	—
Broadbalk 3	0.50	1.00	50	7.9	—
Broadbalk 8	0.60	1.20	50	6.6	—
Cajon loam	1.65	1.27	130	6.2	14.6
Cecil sandy clay loam	0.75	1.09	68	5.3	3.6
Chester loam	1.58	1.67	94	4.9	5.2
Chillum silt loam	3.30	2.54	130	4.6	7.6
Chillum silt loam	2.58	2.54	101	4.6	7.6
Chino silt loam	3.60	2.36	152	7.4	14.2
Chobee fine sandy loam	1.06	1.39	76	7.2	—
Christiana loam	1.25	0.57	219	4.4	5.6
Clarksville silty clay loam	1.40	0.80	175	5.7	5.7
Coachella sandy loam	0.54	1.03	52	7.1	11.8
Collombey	0.64	1.28	50	7.8	—
Columbia clay loam	3.23	1.38	234	6.8	26.5
Columbia very fine silt loam	1.99	0.86	231	7.3	15.0
Crosby loam	2.57	1.88	137	4.9	10.2
Crosby silt loam	2.81	1.90	148	4.8	11.5
Cumberland silt loam	1.20	0.69	174	6.4	6.5
Davidson clay	2.57	1.93	133	5.4	11.4
Diablo sandy clay loam	0.91	0.69	132	6.1	35.8
Dinuba fine sandy loam	0.42	0.46	91	6.2	6.6
Ducor adobe clay loam	1.22	2.13	57	7.2	34.6
Dundee silty clay loam	6.62	0.96	690	5.0	18.1
Ecru sandy clay loam	0.92	0.44	209	5.9	9.3
Eldon silt loam	2.90	1.73	167	5.9	12.9
Elkhorn loamy sand	1.10	1.67	66	6.7	13.2
Elkhorn loamy sand	0.88	1.27	69	7.0	11.6
Evesboro sand	0.70	0.84	83	4.3	5.6
Evouettes	1.78	2.09	51	6.1	—
Exeter loamy sand	1.07	1.50	71	5.8	14.7
Exeter sandy loam	0.56	1.32	42	6.7	12.8
Fallbrook fine sandy loam	1.41	2.77	51	6.4	19.7
Garland clay	0.56	0.65	86	7.7	23.2
Gerald silt loam	4.20	1.55	271	4.7	11.0
Greenfield sandy loam	0.93	2.77	34	7.3	15.8
Greenfield sandy loam	0.69	1.38	50	6.5	12.6
Greenfield sandy loam	1.18	2.07	57	6.4	16.4
Groseclose loam	0.67	0.84	80	5.9	5.9

Soil properties and adsorption data *(continued)*

Soil	K_d (mL/g)	f_{oc} (%)	K_{oc} (mL/g)	pH	CEC (meq/100 g)
Grundy silty clay loam	6.50	2.07	314	5.6	13.5
Hagerstown silty clay loam	1.67	2.48	67	5.5	12.5
Hagerstown silty clay loam	1.10	1.30	84	7.5	8.8
Hagerstown silty clay loam	3.30	2.48	133	5.5	12.5
Hanford gravelly sandy loam	1.48	2.82	52	6.3	18.1
Hanford sandy loam	0.69	1.90	36	6.2	12.4
Hanford sandy loam	0.78	1.38	56	6.0	11.0
Hanford sandy loam	0.90	1.55	58	6.4	12.4
Hanford loamy sand	0.59	1.27	46	7.2	9.5
Hanford loamy sand	0.67	1.15	58	5.8	9.6
Hanford loamy sand	0.60	0.86	70	6.5	7.5
Hickory Hill silt	7.04	3.27	215	—	—
Holopaw fine sand	0.29	0.50	58	6.1	—
Hovey adobe loam	2.33	2.25	104	6.6	49.8
Indio loam	0.71	1.32	54	7.2	15.1
Indio loam	0.81	1.21	67	7.2	13.4
Iredell clay (subsoil)	1.68	6.17	26	5.6	20.9
Iredell silt loam (topsoil)	3.48	3.04	114	5.4	17.0
Knox silt loam	5.10	1.67	305	5.4	18.8
Lakeland sandy loam	0.68	1.88	36	6.2	2.9
Lakeland sandy loam	0.90	1.90	47	6.2	2.9
Las Posas sandy loam	1.20	1.09	110	7.3	11.8
Leadvale clay loam	1.04	0.60	173	5.5	10.0
Lebanon silt loam	2.80	1.04	269	4.9	7.7
Leon sand	9.00	1.96	459	3.9	9.8
Lindley loam	2.60	0.86	302	4.7	6.9
Lintonia loamy sand	1.00	0.34	294	5.3	3.2
Manzanita clay loam	1.42	1.21	117	5.9	13.8
Marian silt loam	3.50	0.80	438	4.6	9.9
Marshall silty clay loam	7.20	2.42	298	5.4	21.3
Menfro silt loam	2.50	1.38	181	5.3	9.1
Montalto clay (subsoil)	0.32	0.86	37	5.9	8.4
Montview loam	0.75	0.79	95	6.6	8.5
Montview loam	0.79	0.90	88	6.6	8.5
Newtonia silt loam	3.00	0.92	326	5.2	8.8
Norfolk sand	0.15	0.31	48	5.8	2.4
Norfolk sand	0.27	0.40	68	5.2	3.2
Norfolk sandy loam	0.21	0.08	260	5.1	0.2
Ooster silt loam	1.02	1.31	78	4.7	6.3
Oswego silty clay loam	3.90	1.67	234	6.4	21.0
Panoche loam	0.58	0.57	102	7.5	20.2
Park Grass 3A	2.50	5.60	45	7.0	—
Park Grass 3A	1.20	2.40	50	6.8	—
Park Grass 3A	1.80	3.60	50	6.9	—

Soil properties and adsorption data *(continued)*

Soil	K_d (mL/g)	f_{oc} (%)	K_{oc} (mL/g)	pH	CEC (meq/100 g)
Park Grass 3D	3.00	5.30	57	5.1	—
Park Grass 3D	1.90	3.10	61	5.1	—
Park Grass 3D	1.50	2.10	71	5.3	—
Park Grass 11/2D	1.40	1.80	78	3.8	—
Park Grass 11/2D	2.90	3.30	88	3.6	—
Park Grass 11/2D	1.19	10.90	109	3.7	—
Park Grass 13A	1.00	2.10	48	6.6	—
Park Grass 13A	1.80	3.50	51	6.5	—
Park Grass 13A	3.10	6.00	52	6.8	—
Park Grass 13D	1.00	2.80	36	4.7	—
Park Grass 13D	0.70	1.80	39	4.8	—
Park Grass 13D	2.00	4.50	44	4.8	—
Park Grass 14A	2.80	5.50	51	6.9	—
Park Grass 14A	1.90	3.30	58	6.9	—
Park Grass 14A	1.50	1.80	83	6.9	—
Pineda fine sand	0.55	1.22	45	7.1	—
Porterville adobe clay	1.27	1.90	67	7.3	37.6
Porterville adobe clay loam	0.75	0.75	100	6.4	18.1
Putnam sandy loam	2.20	1.09	202	5.3	12.3
Ramona sandy loam	0.97	1.78	54	5.9	15.2
Ramona sandy loam	0.62	1.38	45	6.7	10.4
Ramona sandy loam	0.29	0.51	57	6.6	9.5
Red Bay silty clay loam	0.92	1.20	77	6.0	7.7
Rhinebeck silty clay loam	1.97	3.13	63	6.7	—
Rothamsted Farm	0.73	1.51	48	5.1	—
Rincon loam	3.83	1.78	215	5.8	25.2
Riviera fine sand	0.59	0.94	63	6.2	—
Ruston loamy sand	0.59	0.60	98	5.3	6.0
Ruston sand	0.47	0.51	92	6.5	3.3
Ruston sandy loam	0.62	1.05	59	5.1	3.4
Ruston sandy loam	0.56	1.04	54	—	—
Salix loam	3.50	1.21	289	6.3	17.9
San Joaquin clay loam	1.10	1.21	91	7.4	20.8
San Joaquin loam	0.92	1.44	64	5.1	11.2
San Joaquin loam	0.94	1.15	82	6.6	16.7
San Joaquin loam	0.96	1.15	83	5.6	14.0
San Joaquin sandy clay loam	0.95	0.75	127	6.1	16.2
San Joaquin sandy loam	0.41	0.69	59	7.2	8.6
Sarpy loam	2.00	0.75	267	7.1	14.3
Sassafras sandy loam	0.28	0.53	53	6.4	4.4
Sassafras sandy loam	0.54	0.70	77	5.4	7.4
Sharkey clay	5.87	2.25	261	6.2	40.2
Sharkey clay	7.00	1.44	486	5.0	28.2
Sharkey clay	1.63	1.02	160	6.6	11.5

Soil properties and adsorption data *(continued)*

Soil	K_d (mL/g)	f_{oc} (%)	K_{oc} (mL/g)	pH	CEC (meq/100 g)
Shelby loam	5.10	2.07	246	4.3	20.1
Soil #1 (0–30 cm)	3.61	1.46	247	6.0	—
Soil #1 (30–50 cm)	4.50	0.93	484	6.1	—
Soil #1 (50–91 cm)	5.12	1.24	413	6.1	—
Soil #1 (91–135 cm)	2.58	0.70	369	6.4	—
Soil #1 (135–165 cm)	1.94	0.17	1,139	6.5	—
Soil #1 (>165 cm)	2.49	0.07	3,559	6.7	—
Soil #2 (0–27 cm)	5.42	1.50	361	5.9	—
Soil #2 (27–53 cm)	2.70	0.16	1,685	6.3	—
Soil #2 (53–61 cm)	2.37	0.14	1,690	6.4	—
Soil #2 (61–81 cm)	3.13	0.28	1,117	6.4	—
Soil #2 (81–157 cm)	4.66	0.50	931	6.0	—
Soil #2 (>157 cm)	8.53	1.30	656	5.9	—
Sorrento fine sandy loam	1.15	1.09	106	6.9	16.1
Staten peaty loam	0.21	14.00	151	6.9	72.1
Sterling clay loam	0.95	0.94	101	7.7	22.5
Summit sandy clay	7.90	2.82	280	4.8	35.1
Superstition stoney loamy sand	0.31	0.23	135	7.2	5.6
Thurlow clay loam	1.10	1.25	88	7.7	21.6
Tifton loamy sand	0.21	0.56	37	4.9	2.4
Toledo silty clay	3.06	2.80	109	5.5	29.8
Tripp loam	0.88	0.86	103	7.6	14.7
Truckton sandy loam	0.49	0.25	198	7.0	4.4
Union silt loam	3.80	1.04	365	5.4	6.8
Valois silt loam	1.97	1.64	120	5.9	—
Vertroz	2.88	3.25	89	6.7	—
Vista loamy sand	0.66	0.86	77	6.5	10.0
Vista sandy loam	1.10	0.86	128	6.4	12.6
Vista sandy loam	0.61	0.75	81	6.2	14.2
Vista sandy loam	0.60	0.69	100	6.8	13.2
Wabash clay	6.00	1.27	472	5.7	40.3
Wabasso sand	0.80	0.61	131	6.6	—
Waverley silt loam	3.10	1.15	270	6.4	12.8
Wehadkee silt loam	1.25	1.11	113	5.6	10.2
Wehadkee silt loam	2.70	1.09	248	5.6	10.2
Wooster silt loam	3.33	1.32	252	4.7	6.8
Wyman clay loam	2.52	1.78	142	6.4	32.0
Wyman loam	1.78	1.21	147	6.6	15.4
Yolo clay loam	2.37	2.07	114	6.8	31.0
Yolo clay loam	1.15	1.27	91	7.2	27.8
Yolo clay loam	4.95	2.13	232	6.1	31.0
Yolo loam	1.12	1.61	70	5.8	22.1
Yolo loam	1.43	1.78	80	7.2	27.4
Yolo loam	2.05	1.73	118	6.6	17.8
Yolo loam	2.53	1.03	246	7.3	24.0

Soil properties and adsorption data (continued)

Soil	K_d (mL/g)	f_{oc} (%)	K_{oc} (mL/g)	pH	CEC (meq/100 g)
Yolo sandy loam high fan phase	11.20	2.59	43	6.9	20.4
Yolo silt loam	1.09	2.30	47	7.0	21.2
Yolo sandy loam	0.90	1.32	68	7.0	17.0
Yolo sandy loam	0.96	1.32	73	6.9	17.1

Source: Harris and Sheets, 1965; Nearpass, 1965; Talbert and Fletchall, 1965; Harris, 1966; Day et al., 1968; Williams, 1968; Lord et al., 1980; Briggs, 1981; Brown and Flagg, 1981; Burkhard and Guth, 1981; Gamerdinger et al., 1991; Reddy et al., 1992; Sukop and Cogger, 1992.

Environmental Fate

Soil. The reported half-life in soil is 75 days (Alva and Singh, 1991). Under laboratory conditions, the half-lives of simazine in a Hatzenbühl soil (pH 4.8) and Neuhofen soil (pH 6.5) at 22°C were 45 and 100 days, respectively (Burkhard and Guth, 1981).

The half-lives for simazine in soil incubated in the laboratory under aerobic conditions ranged from 27 to 231 days (Zimdahl et al., 1970; Beynon et al., 1972; Walker, 1976, 1976a). In field soils, the disappearance half-lives were lower and ranged from 11 to 91 days (Roadhouse and Birk, 1961; Clay, 1973; Joshi and Datta, 1975; Marriage et al., 1975).

Groundwater. According to the U.S. EPA (1986) simazine has a high potential to leach to *groundwater.*

Plant. Simazine is metabolized by plants to the herbicidally inactive 6-hydroxysimazine which is further degraded via dealkylation of the side chains and hydrolysis of the amino group releasing carbon dioxide (Castelfranco et al., 1961; Humburg et al., 1989).

Photolytic. Pelizzetti et al. (1990) studied the aqueous photocatalytic degradation of simazine and other *s*-triazines (ppb level) using simulated sunlight (λ >340 nm) and titanium dioxide as a photocatalyst. Simazine rapidly degraded forming cyanuric acid, nitrates and other intermediate compounds similar to those found for atrazine. Mineralization of cyanuric acid to carbon dioxide was not observed (Pelizzetti et al., 1990). In aqueous solutions, simazine is converted exclusively to hydroxysimazine by UV light (λ = 253.7 nm). The UV irradiation of methanolic solutions of simazine afforded simetone (2-methoxy-4,6-bis(ethylamino-*s*-triazine). Photodegradation of simazine in methyl alcohol did not occur when irradiated at wavelengths >300 nm (Pape and Zabik, 1970).

Chemical/Physical. Emits toxic fumes of nitrogen oxides and chlorine when heated to decomposition (Sax and Lewis, 1987). In the presence of hydroxy or perhydroxy radicals generated from Fenton's reagent, simazine undergoes dealkylation to give 2-chloro-4,6-diamino-*s*-triazine as the major product (Kaufman and Kearney, 1970).

Symptoms of Exposure: May cause acute and moderate dermatitis; irritant to eyes, skin and mucous membranes.

Formulation Types: Emulsifiable concentrate; granules (4%); suspension concentrate; water-dispersible granules (90%); wettable powder (80%).

Toxicity: LC_{50} (96-hour) for bluegill sunfish 90 mg/L, guppies 49 mg/L, rainbow trout and crucian carp >100 mg/L (Hartley and Kidd, 1987); LC_{50} (48-hour) for coho salmon 6.60 mg/L, bluegill sunfish 130 ppm and rainbow trout 85 ppm (Verschueren, 1983); acute oral LD_{50} of the technical product for rats >5,000 mg/kg (Ashton and Monaco, 1991), 971 mg/kg (RTECS, 1985).

Uses: Selective preemergence systemic herbicide used to control many broad-leaved weeds and annual grasses in deep-rooted fruit and vegetable crops.

STRYCHNINE

Synonyms: Certox; Dolco mouse cereal; Kwik-kil; Mole death; Mole-nots; Mouse-rid; Mouse-tox; Pied piper mouse seed; RCRA waste number P108; Rodex; Sanaseed; **Strychnidin-10-one**; Strychnos; UN 1692.

Designations: CAS Registry Number: 57-24-9; DOT: 1692; mf: $C_{21}H_{22}N_2O_2$; fw: 334.42; RTECS: WL2275000.

Properties: Colorless to white, odorless crystals. Very bitter taste. Mp: 286–288°C with decomposition; bp: 270°C at 5 mmHg; ρ: 1.36 at 20/4°C; pK_a at 20°C: pK_1 = 6.0, pK_2 = 11.7; log K_{oc}: 2.45 (calculated); log K_{ow}: 1.93; S_o (g/L): soluble in alcohol (6.67), amyl alcohol (4.55), benzene (5.56), chloroform (200), glycerol (3.13), methanol (3.85), toluene (5); S_w: 143 mg/L at room temperature, 0.02 wt % at 20°C (pH = 9.5).

Environmental Fate
Chemical/Physical. Reacts with acids forming water-soluble salts (Worthing and Hance, 1991). Emits toxic nitrogen oxides when heated to decomposition (Lewis, 1990).

Exposure Limits: NIOSH REL: TWA 0.15 mg/m³, IDLH 3 mg/m³; OSHA PEL: TWA 0.15 mg/m³; ACGIH TLV: TWA 0.15 mg/m³.

Symptoms of Exposure: Stiff neck, muscle fasiculation, restlessness, apprehension, acuity of perception, reflex excitability, cyanosis, tetanic convulsions, opisthotonos.

Formulation Types: Bait.

Toxicity: Acute oral LD_{50} for rats 16 mg/kg (RTECS, 1985).

Uses: Destroying rodents and predatory animals.

SULFOMETURON-METHYL

Synonyms: Benzoic acid *o*-((3-(4,6-dimethyl-2-pyrimidinyl)ureido)sulfonyl) methyl ester; DPX 5648; **2-(((((4,6-Dimethyl-2-pyrimidinyl)amino)carbonyl)amino)sulfonyl)benzoic acid methyl ester**; Oust.

Designations: CAS Registry Number: 74222-97-2; mf: $C_{15}H_{16}N_4O_5S$; fw: 364.40; RTECS: DG9096550.

Properties: Colorless to white solid. Mp: 203–205°C; ρ: 1.48; fl p: nonflammable; pK_a: 5.2; H-$t_{1/2}$: 18 days at pH 5; log K_{oc}: 1.85–2.09; log K_{ow}: 1.18 (pH 5), -0.50 (pH 7); S_o (mg/kg at 25°C): acetone (2,380), acetonitrile (1,530), ethanol (137), ethyl ether (32), xylene (37); S_w at 25°C: 8–10 mg/L at pH 5, 70–300 mg/L at pH 7; vp: 5.5×10^{-16} mmHg at 25°C.

Soil properties and adsorption data

Soil	K_d (mL/g)	f_{oc} (%)	K_{oc} (mL/g)	pH	CEC (meq/100 g)
Fallsington sandy loam	0.97	0.81	120	5.6	4.8
Flanagan silt loam	2.85	2.33	122	6.5	23.4
Myakka sand	1.00	1.41	71	6.3	3.9

Source: Harvey et al., 1985.

Environmental Fate

Soil. In unsterilized soils, 58% of ^{14}C-labeled sulfometuron-methyl degraded after 24 weeks. Metabolites identified were 2,3-dihydro-3-oxobenzisosulfonazole (saccharin), methyl-2-(aminosulfonyl)benzoate, 2-aminosulfonyl benzoic acid, 2-(((aminocarbonyl)amino)sulfonyl)benzoate and [^{14}C]carbon dioxide. The rate of degradation in aerobic soils was primarily dependent upon pH and soil type (Anderson and Dulka, 1985). The reported half-life in soil was approximately 4 weeks (Hartley and Kidd, 1987).

Chemical/Physical. Sulfometuron-methyl is stable in water at pH values of 7 to 9 but is rapidly hydrolyzed at pH 5.0 forming methyl 2-(aminosulfonyl)benzoate and saccharin. When sulfometuron-methyl in an aqueous solution was exposed to UV light ($\lambda = 300$–400 nm), it degraded to the intermediate methyl benzoate which then mineralized to carbon dioxide (Harvey et al., 1985). The hydrolysis half-lives of metsulfuron-methyl at pH 5 and 25 and 45°C were 33 and 2.1 days, respectively. At pH 7 and 45°C, the hydrolysis half-life is 33 days (Beyer et al., 1988).

Symptoms of Exposure: May irritate eyes, nose, throat and skin.

Formulation Types: Water-dispersible granules (75%); flowable powder (60%).

Toxicity: LC_{50} (96-hour) for rainbow trout and bluegill sunfish >12.5 mg/L (Worthing and Hance, 1991); acute oral LD_{50} for rats >5,000 mg/kg (Ashton and Monaco, 1991).

Uses: Preemergence and postemergence herbicide used to control many annual and perennial grasses and broad-leaved weeds in noncropland and industrial turf areas.

SULFOTEPP

Synonyms: ASP 47; Bay E-393; Bayer E 393; Bis-*O,O*-diethylphosphorothionic anhydride; Bladafum; Bladafume; Bladafun; Dithio; Dithione; Dithiophos; Dithiophosphoric acid tetraethyl ester; Dithiotep; E 393; ENT 16273; Ethyl thiopyrophosphate; Lethalaire G 57; Pirofos; Plant dithio aerosol; Plantfume 103 smoke generator; Pyrophosphorodithioic acid tetraethyl ester; Pyrophosphorodithioic acid *O,O,O,O*-tetraethyl dithionopyrophosphate; RCRA waste number P109; Sulfatep; Sulfotep; TEDP; TEDTP; Tetraethyl dithionopyrophosphate; Tetraethyl dithiopyrophosphate; *O,O,O,O*-Tetraethyl dithiopyrophosphate; Thiodiphosphoric acid tetraethyl ester; **Thiophosphoric acid tetraethyl ester**; Thiopyrophosphoric acid tetraethyl ester; Thiotepp; UN 1704.

$$CH_3CH_2O \underset{CH_3CH_2O}{\overset{S}{\diagdown}} P - O - P \underset{OCH_2CH_3}{\overset{S}{\diagup}} OCH_2CH_3$$

Designations: CAS Registry Number: 3689-24-5; DOT: 1704; mf: $C_8H_{20}O_5P_2S_2$; fw: 322.30; RTECS: XN4375000.

Properties: Pale yellow liquid with a garlic-like odor. Mp: <25°C; bp: 136–139°C at 2 mmHg; ρ: 1.196 at 25/4°C; K_H: 2.9 × 10^{-6} atm · m^3/mol at 20°C (approximate — calculated from water solubility and vapor pressure); log K_{oc}: 2.66, 2.82; log K_{ow}: 3.02 (calculated); S_o: miscible with most organic solvents; S_w: 0.0025 wt % at 20°C; vap d: 13.17 g/L at 25°C, 11.13 (air = 1); vp: 1.7 × 10^{-4} mmHg at 20°C.

Soil properties and adsorption data

Soil	K_d (mL/g)	f_{oc} (%)	K_{oc} (mL/g)	pH	CEC (meq/100 g)
Coarse sand (Jutland, Denmark)	0.96	0.21	457	5.3	2.2
Sandy loam (Jutland, Denmark)	1.00	0.15	667	6.4	7.3

Source: Kjeldsen et al., 1990.

Exposure Limits: NIOSH REL: TWA 0.2 mg/m^3, IDLH 10 mg/m^3; ACGIH TLV: TWA 0.2 mg/m^3.

Environmental Fate

Soil. Cleavage of the molecule yields diethyl phosphate, monoethyl phosphate and phosphoric acid (Hartley and Kidd, 1987).

Chemical/Physical. Emits toxic oxides of sulfur and phosphorus oxides when heated to decomposition (Lewis, 1990).

Formulation Types: Fumigant.

Toxicity: LC_{50} (96-hour) for fathead minnow 178 µg/L, bluegill sunfish 1.6 µg/L and rainbow trout 18 µg/L (Verschueren, 1983); acute oral LD_{50} for rats 5–10 mg/kg (Hartley and Kidd, 1987), 5 mg/kg (RTECS, 1985).

Use: Insecticide.

SULPROFOS

Synonyms: Bayntn-9306; Bolstar; *O*-Ethyl *O*-(4-(methylmercapto)phenyl) *S-n*-propylphosphorothionothiolate; *O*-Ethyl *O*-(4-methylthio)phenyl) phosphorodithioic acid *S*-propyl ester; **O-Ethyl O-(4-methylthio)phenyl) S-propyl phosphorodithioate**; Helothion.

$$CH_3CH_2CH_2S - P \underset{OCH_2CH_3}{\overset{\underset{\displaystyle\|}{S}}{\diagup}} O - \bigcirc - SCH_3$$

Designations: CAS Registry Number: 35400-43-2; mf: $C_{12}H_{19}O_2PS_3$; fw: 322.45; RTECS: TE4165000.

Properties: Colorless to tan, oily liquid with a phosphorous-like odor. Mp: <25°C; bp: 155–158°C at 0.1 mmHg; ρ: 1.20 at 20/4°C; H-$t_{1/2}$: 52 hours (pH 11.5), 41 days (pH 2.2); K_H: 8.6 × 10^{-7} atm · m^3/mol at 20–25°C (approximate — calculated from water solubility and vapor pressure); log K_{oc}: 4.10–4.73; log K_{ow}: 5.48; P-$t_{1/2}$: 2 days (thin films exposed to sunlight); S_o (g/kg at 29°C): cyclohexanone (120), 2-propanol (400–600), toluene (>1,200); S_w: 310 µg/L at 25°C; vp: 6.30 × 10^{-7} mmHg at 20°C.

Soil properties and adsorption data

Soil	K_d (mL/g)	f_{oc} (%)	K_{oc} (mL/g)	pH
Clayey loam	155	0.29	53,448	6.0
Loamy sand	204	1.62	12,593	6.6
Silt loam	347	2.90	11,966	7.9

Source: U.S. Department of Agriculture, 1990.

Environmental Fate

Biological. From the first-order biotic and abiotic rate constants of sulprofos in estuarine water and sediment/water systems, the estimated biodegradation half-lives were 19.5–61.6 and 3.5–19 days, respectively (Walker et al., 1988).

Photolytic. When sulprofos was exposed to sunlight as deposits on cotton foliage, glass surfaces and in aqueous solution, the insecticide degraded rapidly (half-life <2 days). Irradiation of sulprofos in aqueous solution using UV light (λ >290 nm) was also very rapid (half-life <2 hours). The major degradative pathway involved oxidation of the methylthio sulfur to the corresponding sulfoxide and sulfones, hydrolysis of the *O*-phenyl ester and oxidative desulfuration of the P=S group. Products included the corresponding sulfone and sulfoxide from the parent compound, an *O*-analog sulfone, a phenol, a phenol sulfone and a phenol sulfoxide and four unidentified compounds (Ivie and Bull, 1976).

Chemical/Physical. Emits toxic oxides of sulfur and phosphorus when heated to decomposition (Lewis, 1990).

Exposure Limits: NIOSH PEL: TWA 1 mg/m^3; ACGIH TLV: TWA 1 mg/m^3.

Formulation Types: Emulsifiable concentrate.

Toxicity: LC$_{50}$ (96-hour) for rainbow trout 23 mg/L, bluegill sunfish 11 mg/L and carp 5.2 mg/L; acute oral LD$_{50}$ for rats 304 mg/kg (Hartley and Kidd, 1987), 65 mg/kg (RTECS, 1985).

Use: Insecticide.

2,4,5-T

Synonyms: Amine 2,4,5-T for rice; BCF-bushkiller; Brush-off 445 low volatile brush killer; Brush-rhap; Brushtox; Dacamine; Debroussaillant concentre; Debroussaillant super concentre; Decamine 4T; Dedweed brush killer; Dedweed LV-6 brush-kil and T-5 brush-kil; Dinoxol; Envert-T; Estercide T-2 and T-245; Esteron; Esterone 245; Esteron 245 BE; Esteron brush killer; Farmco fence rider; Fence rider; Forron; Forst U 46; Fortex; Fruitone A; Inverton 245; Line rider; NA 2765; Phortox; RCRA waste number U232; Reddon; Reddox; Spontox; Super weedone; Tippon; Tormona; Transamine; Tributon; **(2,4,5-Trichlorophenoxy)acetic acid**; Trinoxol; Trioxon; Trioxone; U 46; Veon; Veon 245; Verton 2T; Viskorhap low volatile ester; Weddar; Weedone; Weedone 2,4,5-T.

Designations: CAS Registry Number: 93-76-5; DOT: 2765; mf: $C_8H_5Cl_3O_3$; fw: 255.48; RTECS: AJ8400000.

Properties: White to pale brown, odorless crystals. Mp: 157–158°C (pure), 150–151°C (technical); ρ: 1.80 at 20/20°C; pK_a: 2.80–2.88; K_H: 4.87×10^{-8} atm · m³/mol at 20°C (approximate — calculated from water solubility and vapor pressure); log K_{oc}: 1.72, 2.27; log K_{ow}: 0.60–3.40; P-$t_{1/2}$: 15 days (near-surface waters under a midsummer sun); S_o (g/L at 20–25°C): 95% ethanol (548.2), ethyl ether (234.3), heptane (0.40), methanol (496.0), toluene (7.32), xylene (6.08), benzene, ethylbenzene, xylene and many other organic solvents; S_w: 220 mg/L at 20°C; vp: 3.75×10^{-5} mmHg at 20°C.

Soil properties and adsorption data

Soil	K_d (mL/g)	f_{oc} (%)	K_{oc} (mL/g)	pH
Ephrata sandy loam	0.31	0.80	38.8	7.5
Glendale silty clay loam	0.49	0.53	92.4	8.5
Ordinance sandy loam	2.40	3.66	65.6	6.6
Glendale silty clay loam	0.49	0.53	92.4	8.5
Ordinance sandy loam	2.40	3.66	65.6	6.6
Palouse silt loam	3.00	2.43	123.5	6.5
Webster sandy clay loam	6.20	3.34	185.6	—

Source: O'Connor and Anderson, 1974; Nkedi-Kizza et al., 1983.

Environmental Fate

Biological. 2,4,5-T degraded in anaerobic sludge by reductive dechlorination to 2,4,5-trichlorophenol, 3,4-dichlorophenol and 4-chlorophenol (Mikesell and Boyd, 1985). An anaerobic methanogenic consortium, growing on 3-chlorobenzoate, metabolized 2,4,5-T to (2,5-dichlorophenoxy)acetic acid at a rate of 1.02×10^{-7} M/hr. The half-life was reported to be 2 days at 37°C (Suflita et al., 1984). Under aerobic conditions, 2,4,5-T degraded to 2,4,5-trichlorophenol and 3,5-dichlorocatechol which may further degrade to 4-chlorocatechol or *cis,cis*-2,4-dichloromuconic acid, 2-chloro-4-carboxymethylenebut-2-enolide, chlorosuccinic acid and succinic acid (Byast and Hance, 1975). The cometabolic oxidation of 2,4,5-T by *Brevibacterium* sp. yielded a product tentatively identified as 3,5-dichlorocatechol (Horvath, 1970). The cometabolism of this compound by *Achromobacter* sp. gave 3,5-dichloro-2-hydroxymuconic semialdehyde (Horvath, 1970a). Rosenberg and Alexander (1980) reported that 2,4,5-trichlorophenol, the principal degradation product of 2,4,5-T by microbes, was further metabolized to 3,5-dichlorocatechol, 4-chlorocatechol, succinate, *cis,cis*-2,4-dichloromuconate, 2-chloro-4-(carboxymethylene)but-2-enolide and chlorosuccinate.

Soil. 2,4,5-Trichlorophenol and 2,4,5-trichloroanisole were the primary degradation products formed when 2,4,5-T was incubated in soil at 25°C under aerobic conditions. The half-life under these conditions was 14 days (McCall et al., 1981a). When 2,4,5-T (10 µg), in unsterilized tropical clay and silty clay soils, was incubated for 4 months, 5–35% degradation yields were observed (Rosenberg and Alexander, 1980). Hydrolyzes in soil to 2,4,5-trichlorophenol (Somasundaram et al., 1989, 1991) and 2,4,5-trichloroanisole (Somasundaram et al., 1989). The rate of 2,4,5-T degradation in soil remained unchanged in a soil pretreated with its hydrolysis metabolite (2,4,5-trichlorophenol) (Somasundaram et al., 1989).

The half-lives of 2,4,5-T in soil incubated in the laboratory under aerobic conditions ranged from 14 to 64 days with an average of 33 days (Altom and Stritzke, 1973; Foster and McKercher, 1973; Yoshida and Castro, 1975). In field soils, the disappearance half-lives were lower and ranged from 8 to 54 days with an average of 16 days (Radosevich and Winterlin, 1977; Stewart and Gaul, 1977).

Groundwater. According to the U.S. EPA (1986) 2,4,5-T has a high potential to leach to groundwater.

Photolytic. When 2,4,5-T (10^{-4} M) in oxygenated water containing titanium dioxide (2 g/L) suspension was irradiated by sunlight ($\lambda \geq 340$ nm), 2,4,5-trichlorophenol, 2,4,5-trichlorophenyl formate and nine chlorinated aromatic hydrocarbons formed as major intermediates. Complete mineralization yielded hydrochloric acid, carbon dioxide and water (Barbeni et al., 1987).

Crosby and Wong (1973) studied the photolysis of 2,4,5-T in aqueous solutions (100 mg/L) under alkaline conditions (pH 8) using both outdoor sunlight and indoor irradiation ($\lambda = 300-450$ nm). 2,4,5-Trichlorophenol and 2-hydroxy-4,5-dichlorophenoxyacetic acid formed as major products. Minor photodecomposition products included 4,6-dichlororesorcinol, 4-chlororesorcinol, 2,5-dichlorophenol and a dark polymeric substance. The rate of photolysis increased 11-fold in the presence of sensitizers (acetone or riboflavin) (Crosby and Wong, 1973). The rate of photolysis of 2,4,5-T was also higher in natural waters containing fulvic acids when compared to distilled water. The major photoproduct found in the humic acid-induced reaction was 2,4,5-trichlorophenol. In addition, the presence of ferric ions and/or hydrogen peroxides may contribute to the sunlight-induced photolysis of 2,4,5-T in acidic, weakly absorbing natural waters (Skurlatov et al., 1983). At 40° latitude in the summer, the direct photolysis half-life was calculated to be 15 days.

Chemical/Physical. Carbon dioxide, chloride, dichloromaleic, oxalic and glycolic acids were reported as ozonation products of 2,4,5-T in water at pH 8 (Struif et al., 1978). Reacts with alkali metals, amines and alkalies forming water-soluble salts (Worthing and Hance, 1991).

When 2,4,5-T was heated at 900°C, carbon monoxide, carbon dioxide, chlorine, hydrochloric acid and oxygen were produced (Kennedy et al., 1972, 1972a).

2,3,5-T will not hydrolyze to any reasonable extent (Kollig, 1993). The hydrolysis product is 2,4,5-trichlorophenol.

Exposure Limits: NIOSH REL: TWA 10 mg/m^3, IDLH 250 mg/m^3; OSHA PEL: TWA 10 mg/m^3; ACGIH TLV: TWA 10 mg/m^3.

Symptoms of Exposure: Skin irritation. May also cause eye, nose and throat irritation.

Formulation Types: Emulsifiable concentrate; soluble concentrate.

Toxicity: LC$_{50}$ (96-hour) for rainbow trout 350 mg/L and carp 355 mg/L (Hartley and Kidd, 1987); acute oral LD$_{50}$ for rats 500 mg/kg (Hartley and Kidd, 1987), 300 mg/kg (RTECS, 1985).

Uses: Plant hormone; defoliant; herbicide used to control undesirable brush and woody plants.

TEBUTHIURON

Synonyms: Brulan; Bushwacker; 1-(5-*t*-Butyl-1,3,4-thiadiazol-2-yl)-1,3-dimethylurea; *N*-**(5-(1,1-Dimethylethyl)-1,3,4-thiadiazol-2-yl)-*N,N*′-dimethylurea**; E 103; EI 103; EL 103; Graslan; Perflan; Perfmid; Preflan; Prefmid; Spike; Tebulan; Tiurolan.

$$(CH_3)_3C \underset{N-\!\!-N}{\overset{S}{\diagdown\!\!\diagup}} \overset{\overset{\displaystyle CH_3}{|}}{N}CONHCH_3$$

Designations: CAS Registry Number: 34014-18-1; mf: $C_9H_{16}N_4OS$; fw: 228.32; RTECS: YS4250000.

Properties: Colorless crystals with a slight pungent odor. Mp: 161.5–164°C (decomposes); fl p: nonflammable; H-$t_{1/2}$: >64 days at pH 3–9; K_d: 0.11 (sand), 1.82 (clay loam); K_H: 2.5 × 10^{-10} atm · m³/mol at 20–25°C (approximate — calculated from water solubility and vapor pressure); log K_{oc}: 2.79; log K_{ow}: 1.79; S_o (g/L at 25°C): acetone (70), acetonitrile (60), benzene (3.7), chloroform (250), hexane (6.1), methanol (170), methyl cellosolve (60); S_w: 2.3–2.5 g/L at 25°C; vp: 2 × 10^{-6} mmHg at 20°C.

Environmental Fate

Soil. In microbially active soils, tebuthiuron is degraded via demethylation. The average half-life in soil 12–15 months in areas receiving 60 inches of rainfall a year (Humburg et al., 1989).

When tebuthiuron was applied to a rangeland at a rate of 0.84 kg/ha, 38% remained after 21 months (Emmerich et al., 1984).

Groundwater. According to the U.S. EPA (1986) tebuthiuron has a high potential to leach to groundwater.

Plant. Degrades in plants via *N*-demethylation and hydroxylation of the *t*-butyl sidechain (Hartley and Kidd, 1987; Humburg et al., 1989).

Formulation Types: Dry flowable powder (85%), granules; pellets (20 and 40%); wettable powder.

Toxicity: LC_{50} (96-hour) for bluegill sunfish 112 mg/L, rainbow trout 144 mg/L, goldfish and fathead minnow >160 mg/L (Hartley and Kidd, 1987); acute oral LD_{50} for rats 644 mg/kg (Ashton and Monaco, 1991).

Uses: Nonselective herbicide used to control herbaceous and woody plants on noncrop land.

TERBACIL

Synonyms: 3-*tert*-Butyl-5-chloro-6-methyluracil; 5-Chloro-3-*tert*-butyl-6-methyluracil; **5-Chloro-3-(1,1-dimethylethyl)-6-methyl-2,4(1*H*,3*H*)-pyrimidinedione**; Compound 732; Du Pont 732; Du Pont herbicide 732; Experimental herbicide 732; Sinbar; Turbacil.

$$\text{CH}_3 \quad \text{N} \quad \text{O}$$
$$\text{Cl} \quad \| \quad \text{C(CH}_3)_3$$
$$\text{O}$$

Designations: CAS Registry Number: 5902-51-2; mf: $C_9H_{13}ClN_2O_2$; fw: 216.70; RTECS: YQ9360000.

Properties: Colorless to white crystalline solid. Mp: 175–177°C; bp: sublimes below mp; ρ: 1.34 at 25/25°C; fl p: nonflammable; K_H: 1.8×10^{-10} atm · m³/mol at 20–25°C (approximate — calculated from water solubility and vapor pressure); log K_{oc}: ≈ 1.74; log K_{ow}: 1.89, 1.90; S_o (g/kg at 25°C): *n*-butyl acetate (88), cyclohexanone (180), *N,N*-dimethylformamide (252), methyl isobutyl ketone (121), xylene (61); S_w: 710 mg/L at 25°C; vp: 4.5×10^{-7} mmHg at 20°C.

Soil properties and adsorption data

Soil	K_d (mL/g)	f_{oc} (%)	K_{oc} (mL/g)	pH	CEC (meq/100 g)
Cecil sandy loam	0.15	0.40	38	5.8	—
Cecil sandy loam	0.38	0.90	42	5.6	6.8
Eustis fine sand	0.12	0.50	21	5.6	5.2
Glendale sandy clay loam	0.38	0.56	76	7.4	—
Keyport silt loam	1.70	1.21	140	5.4	—
Webster silty clay loam	2.46	3.87	64	7.3	54.7

Source: Rhodes et al., 1970; Rao and Davidson, 1979; U.S. Department of Agriculture, 1990.

Environmental Fate

Soil. The half-life of terbacil on surface soil was approximately 5 to 6 months (Gardiner et al., 1969). The half-lives for terbacil in soil incubated in the laboratory under aerobic conditions ranged from 35 to 58 days (Gardiner et al., 1969; Zimdahl et al., 1970). In field soils, the half-lives for terbacil ranged from 75 to 408 days (Gardiner et al., 1969; Khan, 1977). The mineralization half-lives for terbacil in soil ranged from 82 days to 3.5 years (Gardiner et al., 1969; Wolf and Martin, 1974).

Groundwater. According to the U.S. EPA (1986) terbacil has a high potential to leach to groundwater.

Plant. The major degradation products of [2-^{14}C]terbacil identified in alfalfa using a mass spectrometer were (% of applied amount): 3-*tert*-butyl-5-chloro-6-hydroxymethyluracil (11.9) and 6-chloro-2,3-dihydro-7-(hydroxymethyl)-3,3-dimethyl-5*H*-oxazolo[3,2-

a]pyrimidin-5-one (41.2). Two additional compounds tentatively identified by TLC were 3-*tert*-butyl-6-hydroxymethyluracil and 6-chloro-2,3-dihydro-7-methyl-3,3-dimethyl-5*H*-oxazolo[3,2-*a*]pyrimidin-5-one (Rhodes, 1977).

Photolytic. Acher et al. (1981) studied the dye-sensitized photolysis of terbacil in aerated aqueous solutions over a wide pH range. After a 2-hour exposure to sunlight, terbacil in aqueous solution (pH range 3.0–9.2) in the presence of methylene blue (3 ppm) or riboflavin (10 ppm), decomposed to 3-*tert*-5-butyl-5-acetyl-5-hydroxyhydantoin. Deacylation was observed under alkaline conditions (pH 8.0 or 9.2) affording 3-*tert*-5-hydroxyhydantoin. In neutral or acidic conditions (pH 6.8 or 3.0) containing riboflavin, a mono-*N*-dealkylated terbacil dimer and an unidentified water-soluble product formed. Product formation, the relative amounts of products formed and the rate of photolysis were found to be all dependent upon pH, sensitizer, temperature and time (Acher et al., 1981).

Chemical/Physical. Stable in water (Worthing and Hance, 1991).

Symptoms of Exposure: May irritate eyes, nose, throat and skin.

Formulation Types: Wettable powder (80%).

Toxicity: LC_{50} (48-hour) for pumpkinseed sunfish 86 mg/L and fiddler crab >1,000 mg/L (Worthing and Hance, 1991); acute oral LD_{50} for nonfasted and fasted rats >5,000 and <7,500 mg/kg, respectively (Ashton and Monaco, 1991).

Uses: Herbicide used to control many annual broad-leaved weeds, some perennial weeds and grasses in alfalfa, sugarcane, mints and certain fruit and nut trees.

TERBUFOS

Synonyms: AC 92100; *S*-((*tert*-Butylthio)methyl) *O,O*-diethyl phosphorodithioate; Counter; Counter 15G soil insecticide; Counter 15G soil insecticide-nematocide; *S*-(((1,1-Dimethylethyl)thio)methyl) *O,O*-diethyl phosphorodithioate; Phosphorodithioic acid *S*-((*tert*-butylthio)methyl) *O,O*-diethyl ester; **Phosphorodithioic acid *S*-(((1,1-dimethylethyl)thio)methyl) *O,O*-diethyl ester.**

$$(CH_3)_3CSCH_2S - P \overset{\overset{\displaystyle S}{\|}}{\underset{OCH_2CH_3}{\diagdown}} {\diagup} OCH_2CH_3$$

Designations: CAS Registry Number: 13071-79-9; mf: $C_9H_{21}O_2PS_3$; fw: 288.43; RTECS: TD7200000.

Properties: Technical product (85–88%) is clear, colorless to pale yellow liquid with a mercaptan-like odor. Mp: –29.2°C; bp: 69°C at 0.01 mmHg (decomposes >120°C); ρ: 1.105 at 24/4°C; fl p: 88°C (open cup); K_H: 2.2 × 10⁻⁵ atm · m³/mol at 20–27°C (approximate — calculated from water solubility and vapor pressure); log K_{oc}: 2.46–3.03; log K_{ow}: 2.22–4.70; S_o: soluble in many organic solvents; S_w: 5.5 mg/L at 20°C (Bowman and Sans, 1982), 4.5 mg/L at 27°C; vap d: 11.79 g/L at 25°C, 9.99 (air = 1); vp: 2.63 × 10⁻⁴ mmHg at 20°C.

Soil properties and adsorption data

Soil	K_d (mL/g)	f_{oc} (%)	K_{oc} (mL/g)	pH	CEC (meq/100 g)
Clarion soil	8.30	2.64	314	5.00	21.02
Harps soil	21.33	3.80	561	7.30	37.84
Peat soil	52.72	18.36	287	6.98	77.34
Sarpy fine sandy loam	3.34	0.51	655	7.30	5.71
Thurman loamy fine sand	11.40	1.07	1,065	6.83	6.10

Source: Felsot and Dahm, 1979.

Environmental Fate

Soil. Oxidized in soil to its primary and secondary oxidation products, terbufos sulfoxide and terbufos sulfone, respectively (Bowman and Sans, 1982; Chapman et al., 1982; Wei, 1990). Both metabolites were formed due to micobial activity and chemical oxidation (Chapman et al., 1982). Incubation of terbufos (5 μg/g) in a loamy sand containing *Nitrosomonas* sp. and *Nitrobacter* sp. gave terbufos sulfoxide and terbufos sulfone as the primary products. After 2 weeks, the sulfoxide increased the bacterial population >55% and the sulfone increased the fungal population at least 66% (Tu, 1980). The half-life in soil is 9–27 days (Worthing and Hance, 1991).

Chemical/Physical. Terbufos and its degradation products terbufos sulfoxide and terbufos sulfone followed first-order disappearance in natural, sterilized natural and distilled

403

water at 20°C. In natural and distilled water, the sulfoxide and sulfone had half-lives of 18–40 and 280–250 days, respectively. Terbufos disappeared more rapidly (half-life 3 days) (Bowman and Sans, 1982). The hydrolysis half-lives of terbufos in a sterile 1% ethanol/water solution at 25°C and pH values of 4.5, 6.0, 7.0 and 8.0, were 0.29, 0.28, 0.28 and 0.32 weeks, respectively (Chapman and Cole, 1982).

Formulation Types: Granules.

Toxicity: LC_{50} (96-hour) for rainbow trout 10 mg/L and bluegill sunfish 4 mg/L (Hartley and Kidd, 1987); acute oral LD_{50} for rats 1,600 μg/kg (RTECS, 1985).

Use: Soil insecticide.

TERBUTRYN

Synonyms: 2-*tert*-Butylamino-4-ethylamino-6-methylmercapto-*s*-triazine; 2-*tert*-Butylamino-4-ethylamino-6-methylthio-*s*-triazine; Clarosan; **N-(1,1-Dimethylethyl)-N'-ethyl-6-(methylthio)-1,3,5-triazine-2,4-diamine**; GS 14260; HS 14260; Igran; Igran 50; Igran 80W; 2-Methylthio-4-ethylamino-6-*tert*-butylamino-*s*-triazine; Prebane; Shortstop; Shortstop E; Terbutrex B; Terbutryne.

$$CH_3S \quad\diagdown N \diagdown\quad NHC(CH_3)_3$$

$$NHCH_2CH_3$$

Designations: CAS Registry Number: 886-50-0; mf: $C_{10}H_{19}N_5S$; fw: 241.36; RTECS: XY4725000.

Properties: Colorless crystals or white powder. Mp: 104–105°C; bp: 154–160°C at 0.06 mmHg; ρ: 1.115 at 20/4°C; fl p; nonflammable; pK_a: 4.07; K_H: 1.2×10^{-7} atm · m^3/mol at 20°C (approximate — calculated from water solubility and vapor pressure); log K_{oc}: 3.21–4.07; log K_{ow}: 3.38 (Liu and Qian, 1995); S_o (g/L at 20°C): acetone (280), hexane (9), methanol (280), methylene chloride (300), toluene (45); S_w: 25 mg/L at 20°C; vp: 9.6 × 10^{-6} mmHg at 20°C.

Soil properties and adsorption data

Soil	K_d (mL/g)	f_{oc} (%)	K_{oc} (mL/g)	pH
Boyce loam	5.10	1.27	402	8.0
Boyce silt loam	13.70	0.41	3,341	9.6
Chehalis sandy loam	21.50	0.98	2,194	5.2
Clayey loam	183.00	1.57	11,656	—
Deschutes sandy loam	6.10	0.51	1,196	5.9
Gila silty clay	3.90	0.63	619	8.0
Loamy sand	2.84	0.17	1,632	7.4
Metolius sandy loam	4.80	0.81	593	7.1
Powder silt loam	7.20	1.67	431	7.7
Quincy sandy loam	1.70	0.28	607	8.0
Powder silt loam	7.20	1.67	431	7.7
Quincy sandy loam	1.70	0.28	607	8.0
Sand	49.30	1.80	2,742	5.1
Silt loam	14.10	0.87	1,621	6.7
Woodburn silt loam	14.70	1.32	1,113	5.2

Source: Colbert et al., 1975; U.S. Department of Agriculture, 1990.

Environmental Fate

Plant. In plants, the methylthio group is oxidized to hydroxy derivatives and by dealkylation of the side chains. Residual activity in soil is limited to approximately 3–10 weeks (Hartley and Kidd, 1987).

Symptoms of Exposure: Eye and skin irritant.

Formulation Types: Suspension concentrate; wettable powder; granules.

Toxicity: LC_{50} (96-hour) for rainbow trout 3 mg/L, bluegill sunfish 4 mg/L, carp 4 mg/L and perch 4 mg/L (Hartley and Kidd, 1987); acute oral LD_{50} for rats 2,500 mg/kg (Hartley and Kidd, 1987), 2,045 mg/kg (RTECS, 1985).

Uses: Selective herbicide for control of annual broad-leaved and grass weeds in wheat.

TETRAETHYL PYROPHOSPHATE

Synonyms: Bis-*O,O*-diethylphosphoric anhydride; Bladan; **Diphosphoric acid tetraethyl ester**; ENT 18771; Ethyl pyrophosphate; Fosvex; Grisol; Hept; Hexamite; Killax; Kilmite 40; Lethalaire G 52; Lirohex; Mortopal; NA 2783; Nifos; Nifos T; Nifost; Pyrophosphoric acid tetraethyl ester; RCRA waste number P111; TEP; TEPP; Tetraethyl diphosphate; Tetraethyl pyrofosfaat; Tetrastigmine; Tetron; Tetron-100; Vapotone.

$$CH_3CH_2O\underset{CH_3CH_2O}{\overset{O}{\underset{\diagup}{\overset{\parallel}{P}}}}-O-\underset{OCH_2CH_3}{\overset{O}{\underset{\diagdown}{\overset{\parallel}{P}}}}OCH_2CH_3$$

Designations: CAS Registry Number: 107-49-3; DOT: 2784; mf: $C_8H_{10}O_7P_2$; fw: 290.20; RTECS: UX6825000.

Properties: Colorless to amber liquid with a fruity odor. Hygroscopic. Mp: 0°C; bp: 135–138°C at 1 mmHg; ρ: 1.185 at 20/4°C; H-$t_{1/2}$: 6.8–7.5 hours at 25°C and pH 7; S_o: miscible with acetone, benzene, carbon tetrachloride, chloroform, ethanol, ethylbenzene, ethylene glycol, kerosene, methanol, methyl ethyl ketone, petroleum ether, propanol, toluene, xylene and many other organic solvents; S_w: miscible; vap d: 11.86 g/L at 25°C, 10.02 (air = 1); vp: 1.55×10^{-4} mmHg at 20°C.

Environmental Fate

Chemical/Physical. Tetraethyl pyrophosphate is quickly hydrolyzed by water. The hydrolysis half-lives at 25 and 38°C are 6.8 and 3.3 hours, respectively (Sittig, 1985).

Decomposes at 170–213°C releasing large amounts of ethylene (Hartley and Kidd, 1987; Keith and Walters, 1992).

Exposure Limits: NIOSH REL: TWA 0.05 mg/m³, IDLH 5 mg/m³; OSHA PEL: 0.05 mg/m³; ACGIH TLV: TWA 0.004 ppm.

Symptoms of Exposure: Eye pain, impaired vision, tears, headache, chest pain, cyanosis, anorexia, nausea, vomiting, diarrhea, local sweating, weakness, twitch, paralysis, Cheyne-Stokes respiratory, convulsions, low blood pressure.

Formulation Types: Aerosol; emulsifiable concentrate.

Toxicity: LC_{50} (96-hour) for fathead minnow 1.90 mg/L and bluegill sunfish 1.10 mg/L (Verschueren, 1983); acute oral LD_{50} for rats 1.12 mg/kg (Hartley and Kidd, 1987), 500 μg/kg (RTECS, 1985).

Uses: Insecticide for mites and aphids; rodenticide.

THIABENDAZOLE

Synonyms: Apl-Luster; Arbotect; Bovizole; Comfuval; Eprofil; Equizole; Lombristop; Mertec; Mertect; Mertect 160; Metasol TK-100; Mintesol; Mintezol; Minzolum; MK 360; Mycozol; Nemapan; Omnizole; Polival; Storite; TBDZ; TBZ; Tecto; Tecto 60; Tecto RPH; Thiaben; Thiabendazol; Thiabenzazole; Thiabenzol; 2-(Thiazol-4-yl)benzimidazole; 2-(1,3-Thiazol-4-yl)benzimidazole; 2-(4-Thiazolyl)benzimidazole; **2-(4-Thiazolyl)-1*H*-benzimidazole**; Thibenzol; Thibenzole; Thibenzole ATT; Top Form Wormer.

Designations: CAS Registry Number: 148–79–8; mf: $C_{10}H_7N_3S$; fw: 201.25; RTECS: DE0700000.

Properties: Colorless to pale white, odorless powder. Mp: 304–305°C; bp: sublimes at 310°C; log K_{oc}: 2.71 at pH 5–12 (calculated); log K_{ow}: 2.69 at pH 5–12 (calculated); S_o (g/L at 25°C): acetone (4.2), benzene (0.23), chloroform (80), *N,N*-dimethylformamide (39), dimethyl sulfoxide (80), ethanol (7.9), ethyl acetate (2.1), methanol (9.3); S_w: <50 mg/L at pH 5–12, ≈ 250 mg/L at pH 3–5, 39.9 g/L at pH 2.2.

Environmental Fate

Photolytic. When thin films of thiabendazole on glass plates were exposed to sunlight for 128 days, benzimidazole-2-carboxamide and benzimidazole formed as photolysis products. Both compounds were also formed when aqueous methanolic solutions of thiabendazole were subjected to UV light for 1 hour (Zbozinek, 1984).

Formulation Types: Smoke tablets; suspension concentrate; wettable powder.

Toxicity: Acute oral LD_{50} for rats 3,810 mg/kg (Hartley and Kidd, 1987), 3,100 mg/kg (RTECS, 1985).

Uses: Systemic fungicide used for diseases of fruits and vegetables and for control of Dutch elm disease.

THIAMETURON-METHYL

Synonyms: DPX M6316; Harmony; **3-(((((4-Methoxy-6-methyl-1,3,5-triazin-2-yl)amino)carbonyl)amino)sulfonyl)-2-thiophenecarboxylic acid methyl ester.**

Designations: CAS Registry Number: 79277-27-3; mf: $C_{12}H_{13}N_5O_6S_2$; fw: 387.40.

Properties: Colorless to white crystalline solid. Mp: 186°C; ρ: 1.49; fl p: nonflammable; pK_a: 4.0; log K_{oc}: $\approx$ 1.65; log K_{ow}: –1.60; S_o (mg/L at 25°C): acetone (11.9), acetonitrile (7.3), ethanol (0.9), ethyl acetate (2.6), hexane (<0.1), methanol (2.6), methylene chloride (27.5), xylene (0.2); S_w at 25°C: 24 mg/L at pH 4, 260 mg/L at pH 5, 2.4 g/L at pH 6; vp: 2.03×10^{-6} mmHg at 20°C.

Environmental Fate
Chemical/Physical. May hydrolyze in aqueous solutions forming methyl alcohol and 3-(((((4-methoxy-6-methyl-1,3,5-triazin-2-yl)amino)carbonyl)amino)sulfonyl)-2-thiophenecarboxylic acid.

Symptoms of Exposure: May irritate eyes, nose, throat and skin.

Formulation Types: Water-dispersible granules (75%).

Toxicity: LC_{50} (96-hour) for both bluegill sunfish and rainbow trout >100 mg/L; LC_{50} (48-hour) for *Daphnia magna* >1,000 mg/L (Humburg et al., 1989); acute oral LD_{50} for rats >5,000 mg/kg (Hartley and Kidd, 1987).

Uses: Postemergence herbicide used to control wild garlic and many broad-leaved weeds in barley and spring wheat.

THIDIAZURON

Synonyms: Defolit; Dropp; 1-Phenyl-3-(1,2,3-thiadiazol-5-yl)urea; *N*-**Phenyl-*N′*-(1,2,3-thiadiazol-5-yl)urea**; SN 49537.

NHCONH
S
N
N

Designations: CAS Registry Number: 51707-55-2; mf: $C_9H_8N_4OS$; fw: 220.20; RTECS: YU1395000.

Properties: Colorless crystals. Mp: 213°C (decomposes); pK_a: 8.86; K_d: 2.2–21; K_H: 3.2 × 10^{-13} atm · m^3/mol at 25°C (approximate — calculated from water solubility and vapor pressure); log K_{oc}: 2.86 (calculated); log K_{ow}: 1.77 at pH 7.3; S_o (g/L at 23°C): acetone (8), benzene (0.035), chloroform (0.013), cyclohexanone (21.5), *N,N*-dimethylformamide (>500), dimethyl sulfoxide (>500), ethyl acetate (0.8), hexane (0.006), methanol (4.5); S_w: 20–31 mg/L at 23–25°C; vp: 2.2 × 10^{-11} mmHg at 25°C.

Environmental Fate

Soil. The reported half-life in soil is approximately 26–144 days (Hartley and Kidd, 1987).

Photolytic. Rapidly converted to the photoisomer, 1-phenyl-3-(1,2,5-thiadiazol-3-yl)urea (Worthing and Hance, 1991). When thidiazuron adsorbed by soil was exposed to UV light (λ <290 nm), 1-phenyl-3-(1,2,5-thiadiazol-3-yl)urea formed as the major product (half-life = 0.5 hours). In addition, several unidentified polar products formed (Klehr et al., 1983).

Formulation Types: Wettable powder.

Toxicity: LC_{50} (96-hour) for bluegill sunfish, channel catfish and rainbow trout >1,000 mg/L (Hartley and Kidd, 1987); acute oral LD_{50} for rats >5,000 mg/kg (Hartley and Kidd, 1987), 5,350 mg/kg (RTECS, 1985).

Uses: Plant growth regulator used to defoliate cotton to facilitate harvesting.

THIODICARB

Synonyms: Bismethomyl thioether; CGA 45156; Dicarbosulf; **Dimethyl** N,N'-**(thio-bis((methylimino)carbonyloxy))bis(ethanimidothioate)**; Larvin; Lepicron; N,N'-(Thio-bis(methylimino)carbonyloxy)bisethanimidothioic acid dimethyl ester; UC 51762.

$$CH_3C = NOCN - S - NCON = CCH_3$$

with structure showing O double bonds above the carbonyls, and SCH₃, CH₃, CH₃, SCH₃ substituents below.

Designations: CAS Registry Number: 59669-26-0; mf: $C_{10}H_{18}N_4O_4S_3$; fw: 354.47; RTECS: KJ4301050.

Properties: Colorless to light tan crystals. Mp: 168–172°C; ρ: 1.4 at 20/4°C; H-$t_{1/2}$: ≈ 9 days at pH 3; K_H: 4.3×10^{-7} atm · m³/mol at 20–25°C (approximate — calculated from water solubility and vapor pressure); log K_{oc}: 1.81–3.07; log K_{ow}: 1.2–1.6; S_o (g/kg at 25°C): acetone (8), methanol (5), methylene chloride (150), xylene (3); S_w: 35 mg/L at 25°C; vp: 3.23×10^{-5} mmHg at 20°C.

Soil properties and adsorption data

Soil	K_d (mL/g)	f_{oc} (%)	K_{oc} (mL/g)	pH
Clay	14.00	1.21	1,160	—
Loam	1.34	0.75	178	8.1
Sand	0.58	0.46	125	5.8
Sand	0.16	0.25	489	—
Sandy loam	1.22	0.58	210	7.8
Sandy loam	1.34	0.40	335	—
Silt loam	4.47	1.21	371	—

Source: U.S. Department of Agriculture, 1990.

Environmental Fate
Soil. Under aerobic and anaerobic soil conditions, thiodicarb degrades to methomyl and methomyl oxime (Hartley and Kidd, 1987). The reported half-life in various soils is 3–8 days (Hartley and Kidd, 1987).

Formulation Types: Water-soluble granules; wettable powder; suspension concentrate; granular bait; dustable powder.

Toxicity: LC_{50} (96-hour) for rainbow trout 2.55 mg/L and bluegill sunfish 1.21 mg/L (Hartley and Kidd, 1987); acute oral LD_{50} for rats 66 mg/kg (in water) (Hartley and Kidd, 1987), 160 mg/kg (RTECS, 1985).

Use: Insecticide.

411

THIRAM

Synonyms: Aatack; Accelerator thiuram; Aceto TETD; Arasan; Arasan 70; Arasan 75; Arasan-M; Arasan 42-S; Arasan-SF; Arasan-SF-X; Aules; Bis(dimethylamino)carbono-thioyl disulfide; Bis(dimethylthiocarbamoyl) disulfide; Bis(dimethylthiocarbamyl) disulfide; Chipco thiram 75; Cyuram DS; α,α'-Dithiobis(dimethylthio) formamide; *N,N'*-(Dithiodicarbonothioyl)bis(*N*-methylmethanamine); Ekagom TB; ENT 987; Falitram; Fermide; Fernacol; Fernasan; Fernasan A; Fernide; Flo pro T seed protectant; Hermal; Hermat TMT; Heryl; Hexathir; Kregasan; Mercuram; Methyl thiram; Methyl thiuram disulfide; Methyl tuads; NA 2771; Nobecutan; Nomersan; Normersan; NSC 1771; Panoram 75; Polyram ultra; Pomarsol; Pomersol forte; Pomasol; Puralin; RCRA waste number U244; Rezifilm; Royal TMTD; Sadoplon; Spotrete; Spotrete-F; SQ 1489; Tersan; Tersan 75; Tetramethyldiurane sulphite; Tetramethylthiuram bisulfide; Tetramethylthiuram bisulphide; Tetramethylenethiuram disulfide; **Tetramethylthioperoxydicarbonic diamide**; Tetramethylthiocarbamoyl disulfide; Tetramethylthioperoxydicarbonic diamide; Tetramethylthiuram disulfide; Tetramethylthiuram disulphide; *N,N*-Tetramethylthiuram disulfide; *N,N,N',N'*-Tetramethylthiuram disulfide; Tetramethylthiurane disulphide; Tetramethylthiurum disulfide; Tetramethylthiurum disulphide; Tetrapom; Tetrasipton; Tetrathiuram disulfide; Tetrathiuram disulphide; Thillate; Thimer; Thiosan; Thiotex; Thiotox; Thiram 75; Thiramad; Thiram B; Thirasan; Thiulix; Thiurad; Thiuram; Thiuram D; Thiuramin; Thiuram M; Thiuram M rubber accelerator; Thiuramyl; Thylate; Tirampa; Tiuramyl; TMTD; TMTDS; Trametan; Tridipam; Tripomol; TTD; Tuads; Tuex; Tulisan; USAF B-30; USAF EK-2089; USAF P-5; Vancida TM-95; Vancide TM; Vulcafor TMTD; Vulkacit MTIC.

$$\underset{(CH_3)_2NC}{\overset{\overset{\displaystyle S}{\|}}{}} - SS - \underset{CN(CH_3)_2}{\overset{\overset{\displaystyle S}{\|}}{}}$$

Designations: CAS Registry Number: 137-26-8; DOT: 2771; mf: $C_6H_{12}N_2S_4$; fw: 269.35; RTECS: JO1400000.

Properties: Colorless to white to cream-colored crystalline solid. Mp: 155.6°C; bp: 310–315°C at 15 mmHg; ρ: 1.29 at 20/4°C; fl p: 88.9°C; H-$t_{1/2}$: 5.3 days at 25°C and pH 7; S_o: soluble in carbon tetrachloride, chloroform, methylene chloride, acetone (1.2 wt %), alcohol (<0.2 wt %), benzene (2.5 wt %), ethyl ether (<0.2 wt %), methyl ethyl ketone, toluene, xylene but only slightly soluble in carbon disulfide; S_w: 30 ppm.

Soil properties and adsorption data

Soil	K_d (mL/g)	f_{oc} (%)	K_{oc} (mL/g)	pH
Alluvial-Rio Nacimiento	11.96	0.91	1,314	8.0
Brown Clay-Almanzora Alto	12.00	1.17	1,026	8.5

Soil properties and adsorption data *(continued)*

Soil	K_d (mL/g)	f_{oc} (%)	K_{oc} (mL/g)	pH
Brown Lime-Almanzora Bajo	12.93	1.48	874	8.9
Brown Lime-Los Velez	13.73	2.06	667	8.1
Desert-Campo de Tabernas	8.08	0.33	2,448	7.9
Rendzine-Andarax	9.15	0.65	1,408	8.1
Saline-Campo de Dalías	11.17	1.65	677	8.2
Volcanic-Campo de Nijar	4.81	0.37	1,300	8.7

Source: Valverde-García et al., 1988.

Environmental Fate

Biological. In both soils and water, chemical and biological mediated reactions can transform thiram to compounds containing the mercaptan group (Alexander, 1981). Odeyemi and Alexander (1977) isolated three strains of *Rhizobium* sp. that degraded thiram. One of these strains, *Rhizobium meliloti*, metabolized thiram to yield dimethylamine (DMA) and carbon disulfide which formed spontaneously from dimethyldithiocarbamate (DMDT). The conversion of DMDT to DMA and carbon disulfide occurred via enzymatic and nonenzymatic mechanisms (Odeyemi and Alexander, 1977).

When thiram (100 ppm) was inoculated with activated sludge (30 ppm) at 25°C and pH 7.0 for two weeks, 30% degraded. Metabolites included methionine, elemental sulfur, formaldehyde, dimethyldithiocarbamate-α-aminobutyric acid and the corresponding keto acid (Kawasaki, 1980).

To a non-autoclaved alluvial sandy loam (pH 7.3) fortified and inoculated with the bacterium *Pseudomonas aeruginosa*, 40 and 86% degradation were observed after 4 and 24 days, respectively. In autoclaved soil, thiram degradation was not affected. Degradation yields were found to be inversely proportional to concentration. At thiram concentrations of 100 and 2,500 ppm, degradation yields after 24 days of incubation were approximately 85 and 50%, respectively (Shirkot and Gupta, 1985).

Soil. Decomposes in soils to carbon disulfide and dimethylamine (Sisler and Cox, 1954; Kaars Sijpesteijn et al., 1977). When a spodosol (pH 3.8) pretreated with thiram was incubated for 24 days at 30°C and relative humidity of 60–90%, dimethylamine formed as the major product. Minor degradative products included nitrite ions (nitration reduction) and dimethylnitrosamine (Ayanaba et al., 1973).

Plant. Major plant metabolites are ethylene thiourea, thiram monosulfide, ethylene thiram disulfide and sulfur (Hartley and Kidd, 1987).

Photolytic. In methanol, thiram absorbed UV light at wavelengths >290 nm (Gore et al., 1971).

Chemical/Physical. Emits toxic sulfur oxides when heated to decomposition (Lewis, 1990).

Though no products were reported, the calculated hydrolysis half-life at 25°C and pH 7 is 5.3 days (Ellington et al., 1988).

Exposure Limits: NIOSH REL: TWA 0.5 mg/m^3, IDLH 100 mg/m^3; OSHA PEL: 0.5 mg/m^3; ACGIH TLV: TWA 5 mg/m^3.

Symptoms of Exposure: Irritates mucous membranes, dermatitis; with ethanol consumption: flush, erythema, pruritus, urticaria, headache, nausea, vomiting, diarrhea, weakness, dizziness, difficulty in breathing. May cause allergic reaction in contact with skin.

Formulation Types: Suspension concentrate; wettable powder; dry seed treatment; dustable powder; water dispersible granules; water suspension.

Toxicity: LC_{50} (96-hour) for rainbow trout 0.13 mg/L, bluegill sunfish 0.23 mg/L and carp 4.0 mg/L (Hartley and Kidd, 1987); acute oral LD_{50} for rats 865 mg/kg (Hartley and Kidd, 1987), 560 mg/kg (RTECS, 1985).

Uses: Seed disinfectant; fungicide.

TOXAPHENE

Synonyms: Agricide maggot killer (F); Alltex; Alltox; Attac 4-2; Attac 4-4; Attac 6; Attac 6-3; Attac 8; Camphechlor; Camphochlor; Camphoclor; Chemphene M5055; Chlorinated camphene; Chloro-camphene; Clor chem T-590; Compound 3956; Crestoxo; Crestoxo 90; ENT 9735; Estonox; Fasco terpene; Geniphene; Gyphene; Hercules 3956; Hercules toxaphene; Huilex; Kamfochlor; M 5055; Melipax; Motox; NA 2761; NCI-C00259; Octachlorocamphene; PCC; Penphene; Phenacide; Phenatox; Phenphane; Polychlorcamphene; Polychlorinated camphenes; Polychlorocamphene; RCRA waste number P123; Strobane-T; Strobane T-90; Synthetic 3956; Texadust; Toxakil; Toxon 63; Toxyphen; Vertac 90%; Vertac toxaphene 90.

Designations: CAS Registry Number: 8001-35-2; DOT: 2761; mf: $C_{10}H_{10}Cl_8$; fw: 413.82; RTECS: XW5250000.

Properties: Yellow, waxy solid with a chlorine/terpene-like odor. Mp: 65–90°C; bp: >120°C (decomposes); ρ: 1.65 at 25/4°C; fl p: 28.9°C (10% xylene solution); lel: 1.1% (in solvent); uel: 6.4% (in solvent); H-$t_{1/2}$: 10 years at 25°C and pH 7; K_H: 6.3 × 10^{-2} atm · m³/mol; log K_{oc}: 3.18 (calculated); log K_{ow}: 3.23–5.50; S_o: soluble in chloroform, 95% ethanol (5–10 mg/mL at 19°C), hexane, mineral oil, petroleum oils, toluene, and xylene; S_w: 550 μg/L at 20°C; vp: 1.6 × 10^{-5} mmHg at 25°C.

Environmental Fate

Soil. Under reduced soil conditions, about 50% of the C-Cl bonds were cleaved (dechlorinated) by Fe^{2+} porphyrins forming two major toxicants having molecular formulas C10H10Cl8 (Toxicant A) and $C_{10}H_{11}Cl_7$ (Toxicant B). Toxicant A reacted with reduced hematin yielding two reductive dechlorination products ($C_{10}H_{11}Cl_7$), two dehydrodechlorination products ($C_{10}H_9Cl_7$) and two other products ($C_{10}H_{10}Cl_6$). Similarly, products formed from the reaction of Toxicant B with reduced hematin included two reductive dechlorination products ($C_{10}H_{12}Cl_6$), one dehydrochlorination product ($C_{10}H_{10}Cl_6$) and two products having the molecular formula $C_{10}H_{11}Cl_5$ (Khalifa et al., 1976). The reported dissipation rate of toxaphene from soil is 0.010/day (Seiber et al., 1979).

Photolytic. Dehydrochlorination will occur after prolonged exposure to sunlight releasing hydrochloric acid (U.S. Department of Health and Human Services, 1989). Two compounds isolated from toxaphene, 2-*exo*,3-*exo*,5,5,6-*endo*,8,9,10,10-nonachloroborane and 2-*exo*,3-*exo*,5,5,6-*endo*,8,10,10-octachloroborane were irradiated with UV light (λ >290 nm) in a neutral aqueous solution and on a silica gel surface. Both compounds underwent reductive dechlorination, dehydrochlorination and/or oxidation to yield numerous products including bicyclo[2.1.1]hexane derivatives (Parlar, 1988).

Chemical/Physical. Saleh and Casida (1978) demonstrated that Toxicant B (2,2,5-*endo*,6-*exo*,8,9,10-heptachlorobornane), the most active component of toxaphene, underwent reductive dechlorination at the geminal dichloro position yielding 2-*endo*,5-*endo*,6-*exo*,8,9,10-hexachlorobornane and 2-*exo*,5-*endo*,6-*exo*,8,9,10-hexachlorobornane in various chemical, photochemical and metabolic systems.

Toxaphene will slowly undergo hydrolysis resulting in the loss of chlorine atoms and the formation of hydrochloric acid (Kollig, 1993). The hydrolysis rate constant for toxaphene at pH 7 and 25°C was determined to be 8×10^{-6}/hour, resulting in a half-life of 9.9 years (Ellington et al., 1987).

Emits toxic chloride fumes when heated to decomposition (Lewis, 1990).

Exposure Limits: NIOSH REL: IDLH 200 mg/m^3; OSHA PEL: TWA 0.5 mg/m^3; ACGIH TLV: TWA 0.5 mg/m^3, STEL 1 mg/m^3.

Symptoms of Exposure: Nausea, confusion, agitation, tremors, convulsions, unconsciousness, dry red skin.

Formulation Types: Dustable powder; emulsifiable concentrate; wettable powder.

Toxicity: LC$_{50}$ for young rainbow trout 0.2 mg/L and young pike 0.1 mg/L (Hartley and Kidd, 1987); acute oral LD$_{50}$ for rats 40–90 mg/kg (Hartley and Kidd, 1987), 55 mg/kg (RTECS, 1985).

Uses: Pesticide used primarily on cotton, lettuce, tomatoes, corn, peanuts, wheat and soybean.

TRIADIMEFON

Synonyms: Amiral; Bay 6681 F; Bayleton; Maymeb-6647; **1-(4-Chlorophenoxy)-3,3-dimethyl-1-(1*H*-1,2,4-triazol-1-yl)-2-butanone**; 1-(4-Chlorophenoxy)-3,3-(1,2,4-triazol-1-yl)butan-2-one; MEB 6447.

Designations: CAS Registry Number: 43121-43-3; mf: $C_{14}H_{16}ClN_3O_2$; fw: 293.75; RTECS: EL7100000.

Properties: Colorless crystals. Mp: 82.3°C; ρ: 1.22 at 20/4°C; K_H: 1.1×10^{-9} atm · m³/mol at 20°C (approximate — calculated from water solubility and vapor pressure); log K_{oc}: 2.28–2.73; log K_{ow}: 3.18; So at 20°C: cyclohexanone (0.6–1.2 g/kg), hexane (10–20 g/L), methylene chloride (>200 g/L), 2-propanol (200–400 kg/kg), toluene (>200 g/L); S_w: 260 mg/L at 20°C; vp: 7.5×10^{-7} mmHg at 20°C.

Soil properties and adsorption data

Soil	K_d (mL/g)	f_{oc} (%)	K_{oc} (mL/g)	pH
Sandy loam	9.3	1.74	534	5.5
Sand	5.9	2.15	275	6.9
Sand	4.1	2.15	191	6.9
Sandy loam	5.3	1.74	305	5.5
Silt loam	3.5	1.22	287	6.7
Silt loam	2.4	1.22	197	6.7

Source: U.S. Department of Agriculture, 1990.

Environmental Fate

Soil. In a culture study, the microorganism *Aspergillus niger* degraded 32% of triadimefon to triadimenol after 5 days (Clark et al., 1978).

Plant. In soils and plants, triadimefon degrades to triadimenol (Clark et al., 1978; Rouchaud et al., 1981). In barley plants, triadimefon was metabolized to triadimenol and *p*-chlorophenol (Rouchaud et al., 1981; Rouchaud, 1982). In the grains and straw of ripe barley, the concentrations (μg/kg) of triadimefon, triadimenol and *p*-chlorophenol were 0.12, 0.26, 0.13 and 0.70, 1.71, 0.49, respectively (Rouchaud et al., 1981). When applied to tomato plants, triadimefon completed disappeared after one week following a single spraying. Residues remaining after repeated spraying were always <0.1 ppm. This suggests that triadimefon has a high degradation rate (Cabras et al., 1985).

417

Photolytic. When triadimefon was subjected to UV light for one week, *p*-chlorophenol, 4-chlorophenyl methyl carbamate and a 1,2,4-triazole formed as products (Clark et al., 1978).

Formulation Types: Paste; emulsifiable concentrate; wettable powder.

Toxicity: LC_{50} (96-hour) for rainbow trout 14 mg/L, bluegill sunfish 11 mg/L and golden orfe 13.8 mg/L; LC_{50} (48-hour) for carp 7.6 mg/L (Hartley and Kidd, 1987); acute oral LD_{50} for male and female rats 568 and 313 mg/kg, respectively (Hartley and Kidd, 1987), 400 mg/kg (RTECS, 1985).

Uses: Systemic fungicide used to control mildews and rusts that attack coffee, cereals, stone fruit, grapes and ornamentals.

TRIALLATE

Synonyms: Avadex BW; Buckle; CP 23426; *N*-Diisopropylthiocarbamic acid *S*-2,3,3-trichloro-2-propenyl ester; Diisopropyltrichloroallylthiocarbamate; Dipthal; Far-Go; *S*-2,3,3-Trichloroallyl diisopropylthiocarbamate; *S*-2,3,3-Trichloroallyl *N,N*-diisopropylthiocarbamate; *S*-2,3,3-**(Trichloro-2-propenyl) bis(1-methylethyl)carbamothioate.**

$$[(CH_3)_2CH]_2NCOSCH_2CCl=CCl_2$$

Designations: CAS Registry Number: 2303-17-5; mf: $C_{10}H_{16}Cl_3NOS$; fw: 304.70; RTECS: EZ8575000.

Properties: Colorless crystals or amber oil. Mp: 29–30°C; bp: 117°C at 0.3 mmHg (decomposes >200°C); ρ: 1.273 at 25/15.6°C; fl p: 90°C, 95°C (open cup); K_d: 5–35; K_H: 1×10^{-5} atm · m³/mol at 20–25°C (approximate — calculated from water solubility and vapor pressure); log K_{oc}: 3.31 (calculated); log K_{ow}: 4.29; S_o: soluble in acetone, benzene, ethanol, ethyl acetate, ethyl ether, heptane, hexane, methanol and many other organic solvents; S_w: 4 mg/L at 25°C; vp: 1.2×10^{-4} mmHg at 20°C.

Soil properties and adsorption data

Soil	K_d (mL/g)	f_{oc} (%)	K_{oc} (mL/g)	pH
Drummer silty clay loam	28.40	—	—	—
Dupo silt loam	18.50	—	—	—
Hanford sandy loam	12.07	0.43	2,807	6.05
Hanford sandy loam	13.10	0.43	3,046	6.05
Hanford sandy loam	14.49	0.43	3,370	6.05
Lintonia sandy loam	5.98	—	—	—
Spinks sandy loam	22.50	—	—	—
Tujunga loamy sand	8.90	0.33	2,697	6.30
Tujunga loamy sand	9.43	0.33	2,858	6.30
Tujunga loamy sand	11.27	0.33	3,415	6.30

Source: Singh et al., 1990; U.S. Department of Agriculture, 1990.

Environmental Fate

Soil. In an agricultural soil, $^{14}CO_2$ was the only biodegradation identified; however, bound residue and traces of benzene and water-soluble radioactivity were also detected in large amounts (Anderson and Domsch, 1980). In soil, triallate degrades via hydrolytic cleavage with the formation of dialkylamine, carbon dioxide and mercaptan moieties. The mercaptan compounds are further degraded to the corresponding alcohol (Hartley and Kidd, 1987). The reported half-life in soil is 100 days (Jury et al., 1987).

Formulation Types: Emulsifiable concentrate (4 lb/gal); granules (10%).

Toxicity: LC_{50} (96-hour) for rainbow trout 1.2 mg/L and bluegill sunfish 1.3 mg/L (Hartley and Kidd, 1987); acute oral LD_{50} of technical triallate for rats 1,100 mg/kg (Ashton and Monaco, 1991), 1,471 mg/kg (RTECS, 1985).

Uses: Herbicide used to control wild oats in lentils, barley, peas and winter wheat.

TRICHLORFON

Synonyms: Aerol 1; Agroforotox; Anthion; Bay 15922; Bayer 15922; Bayer L 13/59; Bilarcil; Bovinox; Britten; Briton; Cekufon; Chlorak; Chlorfos; Chlorofos; Chlorophos; Chlorophose; Chlorophthalm; Chloroxyphos; Ciclosom; Combot; Combot equine; Danex; DEP; Depthon; DETF; Dimethoxy-2,2,2-trichloro-1-hydroxyethylphosphine oxide; *O,O*-Dimethyl (1-hydroxy-2,2,2-trichloro)ethyl phosphate; Dimethyl-1-hydroxy-2,2,2-trichloroethyl phosphonate; *O,O*-Dimethyl-(1-hydroxy-2,2,2-trichloro)ethyl phosphonate; *O,O*-Dimethyl-1-oxy-2,2,2-trichloroethyl phosphonate; Dimethyltrichlorohydroxyethyl phosphonate; Dimethyl-2,2,2-trichloro-1-hydroxyethyl phosphonate; *O,O*-Dimethyl-2,2,2-trichloro-1-hydroxyethyl phosphonate; Dimetox; Dipterax; Dipterex; Dipterex 50; Diptevur; Ditrifon; Dylox; Dylox-metasystox-R; Dyrex; Dyvon; ENT 19763; Equino-acid; Equino-aid; Flibol E; Forotox; Foschlor; Foschlor 25; Foschlor R 50; 1-Hydroxy-2,2,2-trichloroethylphosphonic acid dimethyl ester; Hypodermacid; Leivasom; Loisol; Masoten; Mazoten; Methyl chlorophos; Metifonate; Metrifonate; Metriphonate; NA 2783; NCI-C54831; Neguvon; Neguvon A; Phoschlor; Phoschlor R50; Polfoschlor; Proxol; Ricifon; Ritsifon; Satox 20WSC; Soldep; Sotipox; Trichlorofon; 2,2,2-Trichloro-1-hydroxyethylphosphonate dimethyl ester; **2,2,2-Trichloro-1-hydroxyethylphosphonic acid dimethyl ester**; Trichlorophon; Trichlorphene; Trichlorphon; Trichlorphon FN; Trinex; Tugon; Tugon fly bait; Tugon stable spray; Vermicide Bayer 2349; Volfartol; Votexit; WEC 50; Wotexit.

$$CH_3O \diagdown \underset{CH_3O \diagup}{\overset{O}{\underset{\|}{P}}} - \overset{OH}{\underset{|}{C}}HCCl_3$$

Designations: CAS Registry Number: 52-68-6; mf: $C_4H_8ClO_4P$; fw: 257.45; RTECS: TA0700000.

Properties: Colorless to pale yellow crystals with an ethereal-like odor. Mp: 83–84°C; bp: 100°C at 0.1 mmHg; ρ: 1.73 at 20/4°C; K_H: 1.7×10^{-11} atm · m³/mol at 25°C; log K_{oc}: 0.99–1.58; log K_{ow}: 0.43, 0.43-0.76; S_o (g/kg at 25°C): benzene (152), chloroform (750), ethyl ether (170), hexane (0.80), pentane and heptane (≈ 0.80); S_w: 120 g/L at 20°C, 154 g/L at 25°C; vp: 7.8×10^{-6} mmHg at 20°C.

Soil properties and adsorption data

Soil	K_d (mL/g)	f_{oc} (%)	K_{oc} (mL/g)	pH
Sandy loam	9.25	0.81	31	6.4
Sandy loam	0.08	0.81	10	6.4
Silt loam	0.40	1.04	38	5.5

Soil properties and adsorption data *(continued)*

Soil	K_d (mL/g)	f_{oc} (%)	K_{oc} (mL/g)	pH
Silt loam	0.35	1.04	34	5.5
Silt loam	0.51	2.67	19	5.4
Silt loam	0.58	2.67	22	5.4

Source: U.S. Department of Agriculture, 1990.

Environmental Fate

Soil. Trichlorfon degraded in soil to dichlorvos (alkaline conditions) and desmethyl dichlorvos (Mattson et al., 1955).

Plant. In cotton leaves, the metabolites identified included dichlorvos, phosphoric acid, *O*-demethyl dichlorvos, *O*-demethyl trichlorfon, methyl phosphate and dimethyl phosphate (Bull and Ridgway, 1969). Chloral hydrate and trichloroethanol were reported as possible breakdown products of trichlorfon in plants (Anderson et al., 1966).

Pieper and Richmond (1976) studied the persistence of trichlorfon in various foliage following an application rate of 1.13 kg/ha. Concentrations of the insecticide found at day 0 and 14 were 81.7 ppm and 7 ppb for willow foliage, 12.6 ppm and 670 ppb for Douglas fir foliage and 113 and 2.1 ppm for grass.

Chemical/Physical. At 100°C, trichlorfon decomposes to chloral. Decomposed by hot water at pH <5 forming dichlorvos (Worthing and Hance, 1991).

Symptoms of Exposure: Irritates eyes.

Formulation Types: Water-soluble powder; suspension concentrate; wettable powder; granules; granular bait; coating agent.

Toxicity: LC_{50} (96-hour) for rainbow trout 1.4 mg/L bluegill sunfish 0.26 mg/L; LC_{50} (48-hour) for carp 6.2 mg/L and goldfish >10 mg/L (Hartley and Kidd, 1987); acute oral LD_{50} for rats 560 mg/kg (Hartley and Kidd, 1987).

Uses: Insecticide used to control flies and roaches.

2,3,6-TRICHLOROBENZOIC ACID

Synonyms: Benzabar; Benzac; Benzac-1281; Fen-All; HC 1281; NCI-C60242; T-2; TBA; 2,3,6-TBA; TCB; 2,3,5-TCB; TCBA; 2,3,6-TCBA; Tribac; Trichlorobenzoic acid; Tryben; Trysben; Trysben 200; Zobar.

Designations: CAS Registry Number: 50-31-7; mf: $C_7H_3Cl_3O_2$; fw: 225.47; RTECS: DH7700000.

Properties: Colorless to buff crystals. Mp: 125–126°C (pure), 80–100°C (technical); bp: decomposes; pK_a: <7; K_H: 9.3×10^{-7} atm · m^3/mol at 20–22°C (approximate — calculated from water solubility and vapor pressure); log K_{oc}: 1.50 (calculated); log K_{ow}: 0.31 (calculated); S_o (g/L): acetone (607), benzene (238), chloroform (237), ethanol (637), methanol (717), xylene (210); S_w: 7.7 g/L at 22°C; vp: 2.4×10^{-2} mmHg at 20°C.

Environmental Fate

Photolytic. Plimmer (1970) postulated that the irradiation of 2,3,6-trichlorobenzoic acid in methanol will undergo dechlorination forming 2,3-, 2,6- and 3,6-dichlorobenzoic acids. Further irradiation may yield 2- and 3-chlorobenzoic acids which may ultimately form benzoic acid (Plimmer, 1970).

Chemical/Physical. Reacts with alkalies and amines forming water-soluble salts.

Formulation Types: Solution.

Toxicity: Acute oral LD_{50} for rats 1,500 mg/kg (Hartley and Kidd, 1987), 650 mg/kg (RTECS, 1985).

Uses: Systemic growth-regulator herbicide used for postemergence control of broad-leaved perennial and annual weeds in cereals and grass seed crops.

TRICLOPYR

Synonyms: Dowco 233; Garlon; **3,5,6-((Trichloro-2-pyridyl)oxy)acetic acid.**

Designations: CAS Registry Number: 55335-06-3; mf: $C_7H_4Cl_3NO_3$; fw: 256.5; RTECS: AJ9000000.

Properties: Colorless solid or crystals. Mp: 148–150°C; pK_a: 2.68; S_o (g/kg): acetone (989), chloroform (27.3), hexane (410), *n*-octanol (307); S_w: 440 mg/L at 25°C; vp: 1.26 × 10^{-6} mmHg at 25°C.

Environmental Fate

Soil. In soil, triclopyr degraded to 3,5,6-trichloro-2-pyridinol and 2-methoxy-3,5,6-trichloropyridine (Norris et al., 1987). The major route of dissipation from soil is likely due to microbial degradation (Newton et al., 1990).

Fifty-four days after triclopyr was applied to soil at a rate of 5.6 kg/ha, the majority (65%) remained in the top 10 cm of soil (Lee et al., 1986a).

Ashton and Monaco (1991) reported triclopyr had an average half-life of 46 days and is influenced by soil type and climatic conditions.

Plant. Lewer and Owen (1989) found 3,5,6-trichloro-2-pyridinol as the major metabolite in plants. Cultured soybean cells metabolized triclopyr to dimethyl triclopyr-aspartate and dimethyl triclopyr-glutamate which can be rehydrolyzed to form the parent compound.

At an application rate of 3.4 kg/ha, the dissipation half-life was 3.7 days. At an application rate of 3.3 kg/ka, the dissipation half-lives of triclopyr in brown, browse and litter samples were 73.5, 202.3 and 31 days, respectively (Norris et al., 1987).

Photolytic. Based on laboratory results, the estimated photolytic half-life of triclopyr in clear water is 1–3 days (McCall and Gavit, 1986). Worthing and Hance (1991) reported a photolysis half-life of <12 hours.

Chemical/Physical: Forms water-soluble salts with alkalies.

Toxicity: LC_{50} (96-hour) for rainbow trout 117 mg/L, bluegill sunfish 148 mg/L (Worthing and Hance, 1991); acute oral LD_{50} for rats 713 mg/kg (RTECS, 1985), rabbits 550 mg/kg, guinea pigs 310 mg/kg; acute percutaneous LD_{50} for rabbits >2,000 mg/kg (Worthing and Hance, 1991).

Use: Selective postemergence herbicide to control many woody and broad-leaved weeds.

TRIFLURALIN

Synonyms: Agreflan; Agriflan 24; Crisalin; Digermin; 2,6-Dinitro-*N*,*N*-dipropyl-4-trifluoromethylaniline; **2,6-Dinitro-*N*,*N*-dipropyl-4-(trifluoromethyl)benzenamine**; 2,6-Dinitro-*N*,*N*-di-*n*-propyl-α,α,α-trifluoro-*p*-toluidine; 4-(Di-*n*-propylamino)-3,5-dinitro-1-trifluoromethylbenzene; *N*,*N*-Di-*n*-propyl-2,6-dinitro-4-trifluoromethylaniline; *N*,*N*-Dipropyl-4-trifluoromethyl-2,6-dinitroaniline; Elancolan; L 36352; Lilly 36352; NCI-C00442; Nitran; Olitref; Trefanocide; Treficon; Treflam; Treflan; Treflanocide elancolan; Trifluoralin; α,α,α-Trifluoro-2,6-dinitro-*N*,*N*-dipropyl-*p*-toluidine; Trifluraline; Triflurex; Trikepin; Trim.

Designations: CAS Registry Number: 1582-09-8; DOT: 1609; mf: $C_{13}H_{16}F_3N_3O_4$; fw: 335.29; RTECS: XU9275000.

Properties: Yellow to orange crystals. Mp: 46–47°C; bp: 139–140°C at 4.2 mmHg; ρ: 1.294 at 25/4°C; fl p: nonflammable; K_H: 4.84×10^{-5} atm · m³/mol at 23°C; log K_{oc}: 2.94–4.49; log K_{ow}: 5.07, 5.28, 5.34; S_o (g/L at 25°C): acetone (>500), acetonitrile (>500), chloroform (>500), *N*,*N*-dimethylformamide (820), 1,4-dioxane (830), hexane (>500), methanol (20), methyl cellosolve (440), methyl ethyl ketone (880), xylene (810); S_w: 4 ppm at 27°C; vp: 1.1×10^{-4} mmHg at 25°C.

Soil properties and adsorption data

Soil	K_d (mL/g)	f_{oc} (%)	K_{oc} (mL/g)	pH
Cecil sandy loam	0.46	0.90	51	5.6
Catlin	58.40	2.01	3,000	6.2
Commerce	36.50	0.68	5,360	6.7
Eustis fine sand	0.24	0.56	43	5.6
Gilat	58.70	0.73	8,041	—
Glendale sandy clay loam	1.60	0.50	178	7.4
Hickory Hill silt	999.00	3.27	30,550	—
Kinneret-A	1,354.00	4.55	29,758	—
Kinneret-G	699.00	2.55	27,412	—
Malkiya	427.00	3.38	12,633	—
Mivtahim	39.20	0.26	15,076	—
Kanuma high clay	5.30	1.35	397	5.7
Neve Yaar	139.00	1.64	8,476	—
Plano silt loam	300.09	2.32	129	6.3
Tracy	52.00	1.12	4,650	6.2

Soil properties and adsorption data *(continued)*

Soil	K_d (mL/g)	f_{oc} (%)	K_{oc} (mL/g)	pH
Tsukuba clay loam	56.60	4.24	1,335	6.5
Webster silty clay loam	2.93	3.87	76	7.3

Source: Harvey, 1974; Brown and Flagg, 1981; McCall et al., 1981; Gerstl and Mingelgrin, 1984; Kanazawa, 1989.

Environmental Fate

Biological. Laanio et al. (1973) incubated $^{14}CF_3$-trifluralin with *Paecilomyces, Fusarium oxysporum,* or *Aspergillus fumigatus* and reported that <1% was converted to $^{14}CO_2$. From the first-order biotic and abiotic rate constants of trifluralin in estuarine water and sediment/water systems, the estimated biodegradation half-lives were 2.5–9.7 and 2.4–7.1 days, respectively (Walker et al., 1988).

Soil. Anaerobic degradation in a Crowley silt loam yielded α,α,α-trifluoro-N^4,N^4-dipropyl-5-nitrotoluene-3,4-diamine and α,α,α-trifluoro-N^4,N^4-dipropyltoluene-3,4,5-triamine (Parr and Smith, 1973). Probst and Tepe (1969) reported that trifluralin degradation in many soils was probably by chemical reduction of the nitro groups into amino groups. Reported degradation products in aerobic soils include α,α,α-trifluoro-2,6-dinitro-N-propyl-p-toluidine, α,α,α-trifluoro-2,6-dinitro-p-toluidine, α,α,α-trifluoro-5-nitrotoluene-3,4-diamine and α,α,α-trifluoro-N,N-dipropyl-5-nitrotoluene-3,4-diamine. Anaerobic degradation products identified include α,α,α-trifluoro-N,N-dipropyl-5-nitrotoluene-3,4-diamine, α,α,α-trifluoro-N,N-dipropyltoluene-3,4,5-triamine, α,α,α-trifluorotoluene-3,4,5-triamine and α,α,α-trifluoro-N-propyltoluene-3,4,5-triamine. α,α,α-Trifluoro-5-nitro-N-propyltoluene-3,4-diamine was identified in both aerobic and anaerobic soils (Probst et al., 1967). The following compounds were reported as major soil metabolites: α,α,α-trifluoro-2,6-dinitro-N-propyl-p-toluidine, α,α,α-trifluoro-2,6-dinitro-p-toluidine, α,α,α-trifluoro-5-nitrotoluene-3,4-diamine, α,α,α-trifluorotoluene-3,4,5-triamine, 2-ethyl-7-nitro-1-propyl-5-(trifluoromethyl)benzimidazole, 2-ethyl-7-nitro-5-(trifluoromethyl)benzimidazole, 7-nitro-1-propyl-5-(trifluoromethyl)benzimidazole, 4-(dipropylamino)-3,5-dinitrobenzoic acid, 2,2′-azoxybis(α,α,α-trifluoro-6-nitro-N-propyl-p-toluidine), 2,2′-azobis(α,α,α-trifluoro-6-nitro-N-propyl-p-toluidine), 2,6-dinitro-N,N-dipropyl-4-(trifluoromethyl)-m-anisidine and α,α,α-trifluoro-2′,6′-dinitro-N-propyl-p-propionotoluidine (Koskinen et al., 1984, 1985).

Golab et al. (1979) studied the degradation of trifluralin in soil over a 3-year period. They found that the herbicide undergoes N-dealkylation, reduction of nitro substituents, followed by the formation cyclized products. Of the 28 transformation products identified, none exceeded 3% of the applied amount. These compounds were: α,α,α-trifluoro-2,6-dinitro-N-propyl-p-toluidine, α,α,α-trifluoro-2,6-dinitro-p-toluidine, α,α,α-trifluoro-5-nitro-N^4,N^4-dipropyltoluene-3,4-diamine, α,α,α-trifluoro-5-nitro-N^4-propyltoluene-3,4-diamine, α,α,α-trifluoro-5-nitrotoluene-3,4-diamine, α,α,α-trifluoro-N^4,N^4-dipropyltoluene-3,4,5-triamine, α,α,α-trifluoro-N^4-propyltoluene-3,4,5-triamine, α,α,α-trifluorotoluene-3,4,5-triamine, α,α,α-trifluoro-2′-hydroxyamino-6′-nitro-N-propyl-p-propionotoluidide, 2-ethyl-7-nitro-1-propyl-5-(trifluoromethyl)benzimidazole 3-oxide, 2-ethyl-7-nitro-5-(trifluoromethyl)benzimidazole 3-oxide, 2-ethyl-7-nitro-1-propyl-5-(trifluoromethyl)benzimidazole, 7-amino-2-ethyl-1-propyl-5-(trifluoromethyl)benzimidazole, 2-ethyl-7-nitro-5-(trifluoromethyl)benzimidazole, 7-amino-2-ethyl-5-(trifluoromethyl)benzimidazole, 7-nitro-1-propyl-5-(trifluoromethyl)benzimidazole, 7-nitro-5-(trifluoromethyl)benzimidazole, 7-amino-5-(trifluoromethyl)benzimidazole, α,α,α-trifluoro-2,6-dinitro-*p*-cresol, 4-(dipropyl-amino)-3,5-dinitrobenzoic acid, 2,2'-azoxybis(α,α,α-trifluoro-6-nitro-*N*-propyl-*p*-toluidine), *N*-propyl-2,2'-azoxybis(α,α,α-trifluoro-6-nitro-*p*-toluidine), 2,2'-azoxy-bis(α,α,α-trifluoro-6-nitro-*p*-toluidine), 2,2'-azobis(α,α,α-trifluoro-6-nitro-*N*-propyl-*p*-toluidine), α,α,α-trifluoro-4,6-dinitro-5-(dipropylamino)-*o*-cresol, α,α,α-trifluoro-2-hydroxyamino-6-nitro-*N*,*N*-dipropyl-*p*-toluidine, α,α,α-trifluoro-2',6'-dinitro-*N*-propyl-*p*-propionotoluidide and α,α,α-trifluoro-2,6-dinitro-*N*-(propan-2-ol)-*N*-propyl-*p*-toluidine (Golab et al., 1979).

Zayed et al. (1983) studied the degradation of trifluralin by the microbes *Aspergillus carneus, Fusarium oxysporum* and *Trichoderma viride*. Following an inoculation and incubation period of 10 days in the dark at 25°C, the following metabolites were identified: α,α,α-trifluoro-2,6-dinitro-*N*-propyl-*p*-toluidine, α,α,α-trifluoro-2,6-dinitro-*p*-toluidine, 2-amino-6-nitro-α,α,α-trifluoro-*p*-toluidine and 2,6-dinitro-4-trifluoromethyl phenol. The reported half-life in soil is 132 days (Jury et al., 1987).

The half-lives for trifluralin in soil incubated in the laboratory under aerobic and anaerobic conditions ranged from 33 to 375 days (Probst et al., 1967; Parr and Smith, 1973; Kearney et al., 1976; Zimdahl and Gwynn, 1977) to 4 to 70 days, respectively (Probst et al., 1967; Parr and Smith, 1973). In field soils, the disappearance half-lives ranged from approximately 15 to 69 days with an average half-life of 46 days (Savage and Barrentine, 1969; Smith, 1972; Menge and Tamez, 1974; Duseja and Holmes, 1978).

Groundwater. According to the U.S. EPA (1986) trifluralin has a high potential to leach to groundwater.

Plant. Trifluralin was absorbed by carrot roots in greenhouse soils pretreated with the herbicide (0.75 lb/acre). The major metabolite formed was α,α,α-trifluoro-2,6-dinitro-*N*-(*n*-propyltoluene)-*p*-toluidine (Golab et al., 1967). Two metabolites of trifluralin that were reported in goosegrass (*Eleucine indica*) were 3-methoxy-2,6-dinitro-*N*,*N'*-dipropyl-4-(trifluoromethyl)benzenamine and *N*-(2,6-dinitro-4-(trifluoromethyl)phenyl)-*N*-propylpropanamide (Duke et al., 1991).

Photolytic. Irradiation of trifluralin in hexane by laboratory light produced α,α,α-trifluoro-2,6-dinitro-*N*-propyl-*p*-toluidine and α,α,α-trifluoro-2,6-dinitro-*p*-toluidine. The sunlight irradiation of trifluralin in water yielded α,α,α-trifluoro-N^4,N^4-dipropyl-5-nitro-toluene-3,4-diamine, α,α,α-trifluoro-N^4,N^4-dipropyltoluene-3,4,5-triamine, 2-ethyl-7-nitro-5-trifluoromethylbenzimidazole, 2,3-dihydroxy-2-ethyl-7-nitro-1-propyl-5-trifluo-romethylbenzimidazoline and 2-ethyl-7-nitro-5-trifluoromethylbenzimidazole. 2-Amino-6-nitro-α,α,α-trifluoro-*p*-toluidine and 2-ethyl-5-nitro-7-trifluoromethyl-benzimidazole also were reported as major products under acidic and basic conditions, respectively (Crosby and Leitis, 1973). In a later study, Leitis and Crosby (1974) reported that trifluralin in aqueous solutions was very unstable to sunlight, especially in the presence of methanol. The photodecomposition of trifluralin involved oxidative *N*-dealkylation, nitro reduction and reductive cyclization. The principal photodecomposition products of trifluralin were 2-amino-6-nitro-α,α,α-trifluoro-*p*-toluidine, 2-ethyl-7-nitro-5-trifluoromethylbenzimida-zole 3-oxide, 2,3-dihydroxy-2-ethyl-7-nitro-1-propyl-5-trifluoromethylbenzimidazole, two azoxybenzenes. Under alkaline conditions, the principal photodecomposition product was 2-ethyl-7-nitro-5-trifluoro-methylbenzimidazole (Leitis and Crosby, 1974).

When trifluralin was released in the atmosphere on a sunny day, it was rapidly converted to the photochemical 2,6-dinitro-*N*-propyl-α,α,α-trifluoro-*p*-toluidine. The estimated half-life was 20 minutes (Woodrow et al., 1978). The vapor-phase photolysis of trifluralin was studied in the laboratory using a photoreactor which simulated sunlight conditions (Soder-quist et al., 1975). Vapor-phase photoproducts of trifluralin were identified as 2,6-dinitro-

N-propyl-α,α,α-trifluoro-*p*-toluidine, 2,6-dinitro-α,α,α-trifluoro-*p*-toluidine, 2-ethyl-7-nitro-1-propyl-5-trifluoromethylbenzimidazole, 2-ethyl-7-nitro-5-trifluoromethylbenzimidazole and four benzimidazole precursors reported by Leitis and Crosby (1974). Similar photo-products were also identified in air above both bare-surface treated soil and soil incorporated fields (Soderquist et al., 1975).

Sullivan et al. (1980) studied the UV photolysis of trifluralin in anaerobic benzene solutions. Products identified included the three azoxybenzene derivatives *N*-propyl-2,2′-azoxybis(α,α,α-trifluoro-6-nitro-*p*-toluidine), 2,2′-azoxybis(α,α,α-trifluoro-6-nitro-*N*-propyl-*p*-toluidine) and 2,2′-azoxybis(α,α,α-trifluoro-6-nitro-*p*-toluidine) and two azobenzene derivatives *N*-propyl-2,2′-azobis(α,α,α-trifluoro-6-nitro-*p*-toluidine) and 2,2′-azobis(α,α,α-trifluoro-6-nitro-*N*-propyl-*p*-toluidine) (Sullivan et al., 1980). ^{14}C-Labeled trifluralin on a silica plate was exposed to summer sunlight for 7.5 hours. Although 52% of trifluralin was recovered, no photodegradation products were identified. In soil, no significant photodegradation of trifluralin was observed after 9 hours of irradiation (Plimmer, 1978). When trifluralin on glass plates was irradiated for 4–6 hours, it photodegraded to 2,3-dihydroxy-2-ethyl-7-nitro-1-propyl-5-trifluoromethylbenzimidazoline and 2-ethyl-7-nitro-5-trifluoromethylbenzimidazole-3-oxide (Wright and Warren, 1965).

Chemical/Physical. Releases carbon monoxide, carbon dioxide and ammonia when heated to 900°C (Kennedy, 1972, 1972a). Incineration may also release hydrofluoric acid in the off-gases (Sittig, 1985).

Symptoms of Exposure: Irritates eyes and may cause skin sensitization in some individuals.

Formulation Types: Emulsifiable concentrate (4 lb/gal); granules (5 and 10%).

Toxicity: LC_{50} (96-hour) for young rainbow trout 10–40 µg/L and young bluegill sunfish 20–90 µg/L (Hartley and Kidd, 1987); LC_{50} (48-hour) for bluegill sunfish 19 ppb and rainbow trout 11 ppb (Verschueren, 1983); acute oral LD_{50} for mice 5,000 mg/kg (RTECS, 1985).

Uses: Preemergence herbicide for controlling many grasses and broad-leaved weeds.

WARFARIN

Synonyms: 3-(Acetonylbenzyl)-4-hydroxycoumarin; 3-(α-Acetonylbenzyl)-4-hydroxy-coumarin; Athrombine-K; Athrombin-K; Brumolin; Compound 42; Co-rax; Coumadin; Coumafen; Coumafene; Cov-r-tox; Dcon; Dethmor; Dethnel; Eastern states duocide; Fasco fascrat powder; 1-(4′-Hydroxy-3′-coumarinyl)-1-phenyl-3-butanone; 4-Hydroxy-3-(3-oxo-1-phenylbutyl)-2H-1-benzopyran-2-one; 4-Hydroxy-3-(1-phenyl-3-oxobutyl)cou-marin; Kumader; Kumadu; Kypfarin; Liquatox; Marfrin; Martin's marfrin; Maveran; Mouse-pak; 3-(1′-Phenyl-2′-acetylethyl)-4-hydroxycoumarin; 3-α-Phenyl-β-acetylethyl-4-hydroxycoumarin; Prothromadin; Rat-a-way; Rat-b-gon; Rat-gard; Rat-kill; Rat & mice bait; Rat-mix; Rat-o-cide #2; Ratola; Ratorex; Ratox; Ratoxin; Ratron; Ratron G; Rats-no-more; Rat-trol; Rattunal; Rax; RCRA waste number P001; Rodafarin; Rodeth; Rodex; Rodex blox; Rosex; Rough & ready mouse mix; Solfarin; Spraytrol brand rodentrol; Temus W; Toxhid; Twin light rat away; Vampirinip II; Vampirin III; Waran; W.A.R.F. 42; Warfarat; Warfarin plus; Warfarin Q; Warf compound 42; Warficide.

Designations: CAS Registry Number: 81-81-2; DOT: 3027; mf: $C_{19}H_{16}O_4$; fw: 308.33; RTECS: GN4550000.

Properties: Colorless to white, odorless, tasteless crystals. Mp: 161°C; bp: decomposes; H-$t_{1/2}$: 16 years at 25°C and pH 7; log K_{oc}: 2.96 (calculated); log K_{ow}: 3.20 (calculated); S_o: soluble in acetone, benzene, 1,4-dioxane, ethanol, ethylbenzene, ethyl ether, hexane, methyl ethyl ketone, pentane, toluene, xylene and many other common organic solvents but is only moderately soluble in methanol, isopropanol and some oils; S_w: 17 mg/L at 20°C.

Exposure Limits: NIOSH REL: TWA 0.1 mg/m³, IDLH 100 mg/m³; OSHA PEL: 0.1 mg/m³; ACGIH TLV: TWA 0.1 mg/m³.

Environmental Fate

Photolytic. Warfarin may undergo direct photolysis since the pesticide showed an absorption maximum of 330 nm (Gore et al., 1971).

Chemical/Physical. The hydrolysis half-lives at 68.0°C and pH values of 3.09, 7.11 and 10.18 were calculated to be 12.9, 57.4 and 23.9 days, respectively. At 25°C and pH 7, the half-life was estimated to be 16 years (Ellington et al., 1986).

Symptoms of Exposure: Hematuria, back pain; epistaxis, bleeding lips, mucous membrane hemorrhage; abdominal pain, vomiting, fecal blood; petechial rash; abnormal hematology.

Formulation Types: Granular bait; tracking powder; gel; bait concentrate.

Toxicity: EC_{50} (24-hour) for *Daphnia magna* 88.8 mg/L (Lilius et al., 1995); acute oral LD_{50} for rats 186 mg/kg (Hartley and Kidd, 1987), 3 mg/kg (RTECS, 1985).

Uses: Pesticide and rodenticide.

REFERENCES

Abdel-Wahab, A.M., R.J. Kuhr, and J.E. Casida. "Fate of C^{14}-Carbonyl-Labeled Aryl Methylcarbamate Insecticide Chemicals in and on Bean Plants," *J. Agric. Food Chem.*, 14(3):290–298 (1966).

Abernathy, J.R. and L.M. Wax. "Bentazon Mobility and Adsorption in Twelve Illinois Soils," *Weed Sci.*, 21:224–227 (1973).

Abou-Assaf, N. and J.R. Coats. "Degradation of [^{14}C]Isofenphos in Soil in the Laboratory Under Different Soil pH's, Temperatures, and Moistures," *J. Environ. Sci. Health*, B22(3):285–301 (1987).

Abou-Assaf, N., J.R. Coats, M.E. Gray, and J.J. Tollefson. "Degradation of Isofenphos in Cornfields with Conservation Tillage Practices," *J. Environ. Sci. Health*, B21(6):425–446 (1986).

Acher, A.J. and E. Dunkelblum. "Identification of Sensitized Photooxidation Products of Bromacil in Water," *J. Agric. Food Chem.*, 27(6):1164–1167 (1979).

Acher, A.J. and S. Saltzman. "Dye-Sensitized Photooxidation of Bromacil in Water," *J. Environ. Qual.*, 9(2):190–194 (1980).

Acher, A.J., S. Saltzman, N. Brates, and E. Dunkelblum. "Photosensitized Decomposition of Terbacil in Aqueous Solutions," *J. Agric. Food Chem.*, 29(4):707–711 (1981).

Adams, C.D. and S.J. Randtke. "Ozonation Byproducts of Atrazine in Synthetic and Natural Waters," *Environ. Sci. Technol.*, 26(11):2218–2227 (1992).

Addison, J.B., P.J. Silk, and I. Unger. "The Photochemical Reactions of Carbamates II. The Solution Photochemistry of Matacil (4-Dimethylamino-*m*-tolyl-*N*-meth-yl carbamate) and Landrin (3,4,5-Trimethylphenyl-*N*-methyl carbamate)," *Bull. Environ. Contam. Toxicol.*, 11(3):250–255 (1974).

Adewuyi, Y.G. and G.R. Carmichael. "Kinetics of Hydrolysis and Oxidation of Carbon Disulfide by Hydrogen Peroxide in Alkaline Medium and Application to Carbonyl Sulfide," *Environ. Sci. Technol.*, 21(2):170–177 (1987).

Adhya, T.K., Sudhakar-Barik, and N. Sethunathan. "Fate of Fenitrothion, Methyl Parathion, and Parathion in Anoxic Sulfur-Containing Soil Systems," *Pestic. Biochem. Physiol.*, 16(1):14–20 (1981).

Adhya, T.K., Sudhakar-Barik, and N. Sethunathan. "Stability of Commercial Formulation of Fenitrothion, Methyl Parathion, and Parathion in Anaerobic Soils," *J. Agric. Food Chem.*, 29(1):90–93 (1981a).

Ahel, M. and W. Giger. "Aqueous Solubility of Alkylphenols and Alkylphenol Polyethoxylates," *Environ. Sci. Technol.*, 26(8):1461–1470 (1993).

Ahmad, A., D.D. Walgenbach, and G.R. Sutter. "Degradation Rates of Technical Carbofuran and a Granular Formulation in Four Soils with Known Insecticide Use History," *Bull. Environ. Contam. Toxicol.*, 23(4/5):572–574 (1979).

Ahmed, M.K. and J.E. Casida. "Metabolism of Some Organophosphorus Insecticides by Microorganisms," *J. Econ. Entomol.*, 51(1):59–63 (1958).

Akesson, N.B. and W.E. Yates. "Problems Relating to Application of Agricultural Chemicals and Resulting Drift Residues," *Ann. Rev. Entomol.*, 9:285–318 (1964).

Akiyoshi, M., T. Deguchi, and I. Sanemasa. "The Vapor Saturation Method for Preparing Aqueous Solutions of Solid Aromatic Hydrocarbons," *Bull. Chem. Soc. Jpn.*, 60(11):3935–3939 (1987).

Alexander, M. "Biodegradation of Chemicals of Environmental Concern," *Science (Washington, DC)*, 211(4478):132–138 (1981).

Alley, E.G., B.R. Layton, and J.P. Minyard, Jr. "Identification of the Photo-products of the Insecticides Mirex and Kepone," *J. Agric. Food Chem.*, 22(3):442–445 (1974).

Altom, J.D. and J.F. Stritzke. "Degradation of Dicamba, Picloram, and Four Phenoxy Herbicides in Soils," *Weed Sci.*, 21:556–560 (1973).

Altshuller, A.P. "Measurements of the Products of Atmospheric Photochemical Reactions in Laboratory Studies and in Ambient Air-Relationships between Ozone and Other Products," *Atmos. Environ.*, 17(12):2383–2427 (1983).

Alva, A.K. and M. Singh. "Sorption-Desorption of Herbicides in Soils as Influenced by Electrolyte Cations and Ionic Strength," *J. Environ. Sci. Health*, B26(2):147–163 (1991).

Aly, O.M. and M.A. El-Dib. "Studies on the Persistence of Some Carbamate Insecticides in the Aquatic Environment-I. Hydrolysis of Sevin, Baygon, Pyrolam and Dimetilan in Waters," *Water Res.*, 5(12):1191–1205 (1971).

Aly, O.M. and S.D. Faust. "Studies on the Fate of 2,4-D and Ester Derivatives in Natural Surface Waters," *J. Agric. Food Chem.*, 12(6):541–546 (1964).

Ambrosi, D., P.C. Kearney, and J.A. Macchia. "Persistence and Metabolism of Oxadiazon in Soils," *J. Agric. Food Chem.*, 25(4):868–872 (1977).

Ambrosi, D., P.C. Kearney, and J.A. Macchia. "Persistence and Metabolism of Phosalone in Soil," *J. Agric. Food Chem.*, 25(2):342–347 (1977a).

Anderson, R.J., C.A. Anderson, and T.J. Olson. "A Gas-Liquid Chromatographic Method for the Determination of Trichlorfon in Plant and Animal Tissues," *J. Agric. Food Chem.*, 14(5):508–512 (1966).

Anderson, R.L. and D.L. DeFoe. "Toxicity and Bioaccumulation of Endrin and Methoxychlor in Aquatic Invertebrates and Fish," *Environ. Pollut. (Series A)*, 22(2):111–121 (1980).

Anderson, J.P.E. and K.H. Domsch. "Microbial Degradation of the Thiolcarbamate Herbicide, Diallate, in Soils and by Pure Cultures by Soil Microorganisms," *Arch. Environ. Contam. Toxicol.*, 4(1):1–7 (1976).

Anderson, J.P.E. and K.H. Domsch. "Relationship between Herbicide Concentration and the Rates of Enzymatic Degradation of ^{14}C-Diallate and ^{14}C-Triallate in Soil," *Arch. Environ. Contam. Toxicol.*, 9(3):259–268 (1980).

Anderson, J.J. and J.J. Dulka. "Environmental Fate of Sulfometuron Methyl in Aerobic Soils," *J. Agric. Food Chem.*, 33(4):596–602 (1985).

Anderson, J.F., G.R. Stephenson, and C.T. Corke. "Atrazine and Cyanazine Activity in Ontario and Manitoba Soils," *Can. J. Plant Sci.*, 60:773–781 (1980a).

Andrawes, N.R., W.P. Bagley, and R.A. Herrett. "Fate and Carryover Properties of Temik Aldicarb Pesticide [2-Methyl-2-(methylthio)propionaldehyde *O*-(methylcarbamoyl)oxime] in Soil," *J. Agric. Food Chem.*, 19(4):727–730 (1971).

Andrawes, N.R., R.R. Romine, and W.P. Bagley. "Metabolism and Residues of Temik Aldicarb Pesticide in Cotton Foliage and Under Field Conditions," *J. Agric. Food Chem.*, 21(3):379–386 (1973).

Andrews, L.J. and R.M. Keefer. "Cation Complexes of Compounds Containing Carbon-Carbon Double Bonds. VI. The Argentation of Substituted Benzenes," *J. Am. Chem. Soc.*, 72(7):3113–3116 (1950).

Andrews, L.J. and R.M. Keefer. "Cation Complexes of Compounds Containing Carbon-Carbon Double Bonds. VII. Further Studies on the Argentation of Substituted Benzenes," *J. Am. Chem. Soc.*, 72(11):5034–5037 (1950a).

Angemar, Y., M. Rebhun, and M. Horowitz. "Adsorption, Phytotoxicity, and Leaching of Bromacil in Some Israeli Soils," *J. Environ. Qual.*, 13(2):321–326 (1984).

Archer, T.E. "Malathion Residues on Ladino Clover Seed Screenings Exposed to Ultraviolet Irradiation," *Bull. Environ. Contam. Toxicol.*, 6(2):142–143 (1971).

Archer, T.E. "The Effect of Ultraviolet Radiation Filtered through Pyrex Glass Upon Residues of Dicofol [Kelthane; 1,1-Bis(*p*-chlorophenyl)-2,2,2-trichloroethanol on Apple Pomance," *Bull. Environ. Contam. Toxicol.*, 12(2):202–203 (1974).

Archer, T.E., I.K. Nazer, and D.G. Crosby. "Photodecomposition of Endosulfan and Related Products in Thin Films by Ultraviolet Light Irradiation," *J. Agric. Food Chem.*, 20(5):954–956 (1972).

Archer, T.E., J.D. Stokes, and R.S. Bringhurst. "Fate of Carbofuran and Its Metabolites on Strawberries in the Environment," *J. Agric. Food Chem.*, 25(3):536–541 (1977).

Argauer, R.J. and R.E. Webb. "Rapid Fluorometric Evaluation of the Deposition and Persistence of Carbaryl in the Presence of an Adjuvant on Bean and Tomato Leaves," *J. Agric. Food Chem.*, 20:732–734 (1972).

Arienzo, M., M. Sánchez-Camazano, T. Cristano Herrero, and M.J. Sánchez-Martín. "Effect of Organic Cosolvents on Adsorption of Organophosphorus Pesticides by Soils," *Environ. Sci. Technol.*, 27(8):1409–1417 (1993).

Armstrong, D.E., G. Chesters, and R.R. Harris. "Atrazine Hydrolysis in Soil," *Soil Sci. Soc. Am. Proc.*, 31:61–66 (1967).

Arnold, S.M., W.J. Hickey, and R.F. Harris. "Degradation of Atrazine by Fenton's Reagent: Condition Optimization and Product Quantification," *Environ. Sci. Technol.*, 29(8):2083–2089 (1995).

Arunachalam, K.D. and M. Lakshmanan. "Microbial Uptake and Accumulation of (^{14}C Carbofuran) 1,3-Dihydro-2,2-Dimethyl-7 Benzofuranylmethyl Carbamate in Twenty Fungal Strains Isolated by Miniecosystem Studies," *Bull. Environ. Contam. Toxicol.*, 41(1):127–134 (1988).

Ashton, F.M. and T.J. Monaco. *Weed Science: Principals and Practices* (New York: John Wiley & Sons, Inc., 1991), 466 p.

Atkinson, R. "Structure-Activity Relationship for the Estimation of Rate Constants for the Gas-Phase Reaction of OH Radicals with Organic Compounds," *Int. J. Chem. Kinetics*, 17:799–828 (1987).

Atkinson, R. and W.P.L. Carter. "Kinetics and Mechanisms of Gas-Phase Ozone with Organic Compounds Under Atmospheric Conditions," *Chem. Rev.*, 85:69–201 (1985).

Attaway, H.H., N.D. Camper, and M.J.B. Paynter. "Anaerobic Microbial Degradation of Diuron by Pond Sediment," *Pestic. Biochem. Physiol.*, 17:96–101 (1982).

Attaway, H.H., M.J.B. Paynter, and N.D. Camper. "Degradation of Selected Phen-ylurea Herbicides by Anaerobic Pond Sediment," *J. Environ. Sci. Health*, B17:683–700 (1982a).

Ayanaba, A., W. Verstraete, and M. Alexander. "Formation of Dimethylnitrosamine, a Carcinogen and Mutagen, in Soils Treated with Nitrogen Compounds," *Soil Sci. Soc. Am. Proc.*, 37:565–568 (1973).

Bachmann, A., W.P. Wijnen, W. de Bruin, J.L.M. Huntjens, W. Roelofsen, and A.J.B. Zehnder. "Biodegradation of *Alpha*- and *Beta*-Hexachlorocyclohexane in a Soil Slurry Under Different Redox Conditions," *Appl. Environ. Microbiol.*, 54(1):143–149 (1988).

Bailey, A.M. and M.D. Coffey. "Biodegradation of Metalaxyl in Avocado Soils," *Phytopathology*, 75:135–137 (1985).

Baker, M.D. and C.I. Mayfield. "Microbial and Non-biological Decomposition of Chlorophenols and Phenols in Soil," *Water, Air and Soil Pollut.*, 13:411–424 (1980).

Baker, M.D., C.I. Mayfield, and W.E. Inniss. "Degradation of Chlorophenol in Soil, Sediment and Water at Low Temperature," *Water Res.*, 14:765–771 (1980).

Banerjee, S., S.H. Yalkowsky, and S.C. Valvani. "Water Solubility and Octanol/Water Partition Coefficients of Organics. Limitations of the Solubility-Partition Coefficient Correlation," *Environ. Sci. Technol.*, 14(10):1227–1229 (1980).

Banks, S. and J.R. Tyrell. "Kinetic and Mechanism of Alkaline and Acidic Hydrolysis of Aldicarb," *J. Agric. Food Chem.*, 32:1233–1232 (1984).

Banks, S. and J.R. Tyrell. "Copper(II)-Promoted Aqueous Decomposition of Aldicarb," *J. Org. Chem.*, 50(24):4938–4943 (1985).

Barbash, J.E. and M. Reinhard. "Abiotic Dehalogenation of 1,2-Dichloroethane and 1,2-Dibromoethane in Aqueous Solution Containing Hydrogen Sulfide," *Environ. Sci. Technol.*, 23(11):1349–1358 (1989).

Barbeni, M., E. Pramauro, and E. Pelizzetti. "Photodegradation of Pentachlorophenol Catalyzed by Semiconductor Particles," *Chemosphere*, 14(2):195–208 (1985).

Barbeni, M., E. Pramauro, E. Pelizzetti, M. Vincenti, E. Borgarello, and N. Serpone. "Sunlight Photodegradation of 2,4,5-Trichlorophenoxyacetic Acid and 2,4,5-Trichlorophenol on TiO$_2$. Identification of Intermediates and Degradation Pathway," *Chemosphere*, 16(6):1165–1179 (1987).

Barik, S., D.M. Munnecke, and J.S. Fletcher. "Bacterial Degradation of Three Di-thioate Pesticides," *Agric. Wastes*, 10:81–94 (1984).

Barnes, C.J., T.L. Lavy, and R.E. Talbert. "Leaching, Dissipation, and Efficacy of Metolachlor Applied by Chemigation or Conventional Methods," *J. Environ. Qual.*, 21(2):232–236 (1992).

Bartha, R. "Biochemical Transformations of Anilide Herbicides in Soil," *J. Agric. Food Chem.*, 16(4):602–604 (1968).

Bartha, R. "Fate of Herbicide-Derived Chloroanilines in Soil," *J. Agric. Food Chem.*, 19(2):385–387 (1971).

Bartha, R., H.A.B. Linke, and D. Pramer. "Pesticide Transformations: Production of Chloroazobenzenes from Chloroanilines," *Science (Washington, DC)*, 161(3841):582–583 (1968).

Bartha, R. and D. Pramer. "Pesticide Transformation to Aniline and Azo Compounds in Soil," *Science (Washington, DC)*, 156(3782):1617–1618 (1967).

Bartl, P. and F. Korte. "Photochemisches und Thermisches Verhalten des Herbizids Sencor (4-Amino-6-*tert*-butyl-3-(methylthio)-1,2,4-triazin-5(4*H*)-on) als Festkörper und orf Oberflächen," *Chemosphere*, 4(3):173–176 (1975).

Bartley, W.J., N.R. Andrawes, E.L. Chancey, W.P. Bagley, and H.W. Spurr. "The Metabolism of Temik Aldicarb Pesticide [2-Methyl-2-methylthio)propionalde-hyde *O*-(methylcarbamoyl)oxime] in the Cotton Plant," *J. Agric. Food Chem.*, 18(3):446–453 (1970).

Bartsch, E. "Diazinon. II. Residues in Plants, Soil and Water," *Residue Rev.*, 51:37–68 (1974).

Battersby, N.S. and V. Wilson. "Evaluation of a Serum Bottle Technique for Assessing the Anaerobic Biodegradability of Organic Chemicals Under Methanogenic Conditions," *Chemosphere*, 17(12):2441–2460 (1988).

Battersby, N.S. and V. Wilson. "Survey of the Anaerobic Biodegradation Potential of Organic Compounds in Digesting Sludge," *Appl. Environ. Microbiol.*, 55:433–439 (1989).

Baude, F.J., H.L. Pease, and R.F. Holt. "Fate of Benomyl on Field Soil and Turf," *J. Agric. Food Chem.*, 22(3):413–418 (1974).

Baur, J.R. and R.W. Bovey. "Ultraviolet and Volatility Loss of Herbicides," *Arch. Environ. Contam.*, 2(3):275–288 (1974).

Baxter, R.M. "Reductive Dechlorination of Certain Chlorinated Organic Compounds by Reduced Hematin Compared with Their Behaviour in the Environment," *Chemosphere*, 21(4/5):451–458 (1990).

Baxter, J.N. and F.J. Johnston. "Photolysis Studies of the Chloroacetic Acids Using Light of 2537 Å," *Radiation Res.*, 33(2):303–310 (1968).

Beard, J.E. and G.W. Ware. "Fate of Endosulfan on Plants and Glass," *J. Agric. Food Chem.*, 17(2):216–220 (1969).

Beck, E.W., L.H. Dawsey, D.W. Wodham, and D.B. Leuck. "Dimethoate Residues on Soybean, Corn, and Grass Forage," *J. Econ. Entomol.*, 59:78–82 (1966).

Beck, J. and K.E. Hansen. "The Degradation of Quintozene, Pentachlorobenzene, Hexachlorobenzene, and Pentachloroaniline in Soil," *Pestic. Sci.*, 5(1):41–48 (1974).

Beeman, R.W. and F. Matsumura. "Metabolism of *cis*- and *trans*-Chlordane by a Soil Microorganism," *J. Agric. Food Chem.*, 29(1):84–89 (1981).

Belafdaĺ, O., M. Bergon, and J.P. Calmon. "Mechanism of Hydantoin Ring Opening in Prodione in Aqueous Media," *Pestic. Sci.*, 17:335–342 (1986).

Belal, M.H. and E.A.A. Gomaa. "Determination of Dimethoate Residue in Some Vegetables and Cotton Plants," *Bull. Environ. Contam. Toxicol.*, 22(6):726–730 (1979).

Beland, F.A., S.O. Farwell, A.E. Robocker, and R.D. Geer. "Electrochemical Reduction and Anaerobic Degradation of Lindane," *J. Agric. Food Chem.*, 24(4):753–756 (1976).

Belasco, I.J. and W.P. Langsdorf. "Synthesis of C^{14}-Labeled Siduron and Its Fate in Soil," *J. Agric. Food Chem.*, 17(5):1004–1007 (1969).

Belasco, I.J. and H.L. Pease. "Investigation of Diuron- and Linuron-Treated Soils for 3,3′,4,4′-Tetrachloroazobenzene," *J. Agric. Food Chem.*, 17(6):1414–1417 (1969).

Belay, N. and L. Daniels. "Production of Ethane, Ethylene, and Acetylene from Halogenated Hydrocarbons by Methanogenic Bacteria," *Appl. Environ. Microbiol.*, 53(7):1604–1610 (1987).

Bell, G.H. "The Action of Monocarboxylic Acids on Candida tropicalis Growing on Hydrocarbon Sunstrates," Antonie van Leeuwenhoek, 37:385–400 (1971).

Bell, G.R. "Photochemical Degradation of 2,4-Dichlorophenoxyacetic Acid and Structurally Related Compounds," *Botan. Gaz.*, 118:133–136 (1956).

Bell, R.P. and A.O. McDougall. "Hydration Equilibria of Some Aldehydes and Ketones," *Trans. Faraday Soc.*, 56:1281–1285 (1960).

Bender, M.E. "The Toxicity of the Hydrolysis and Breakdown Products of Malathion to the Fathead Minnow (*Pimephales Promelas*, Rafinesque)," *Water Res.*, 3(8):571–582 (1969).

Benezet, H.I. and F. Matsumura. "Isomerization of γ-BHC to α-BHC in the Environment," *Nature (London)*, 243(5408):480–481 (1973).

Bennett, G.M. and W.G. Philip. "The Influence of Structure on the Solubilities of Ethers. Part I. Aliphatic Ethers," *J. Chem. Soc. (London)*, 131:1930–1937 (1928).

Benoit-Guyod, J.L., D.G. Crosby, and J.B. Bowers. "Degradation of MCPA by Ozone and Light," *Water Res.*, 20(1):67–72 (1986).

Benson, W.R. "Photolysis of Solid and Dissolved Dieldrin," *J. Agric. Food Chem.*, 19(1):66–72 (1971).

Benson, W.R., P. Lombado, I.J. Egry, R.D. Ross, Jr., R.P. Barron, D.W. Mastbrook, and E.A. Hansen. "Chlordane Photoalteration Products: Their Preparation and Identification," *J. Agric. Food Chem.*, 19(5):857–862 (1971).

Best, J.A. and J.B. Weber. "Disappearance of *s*-Triazines as Affected by Soil pH Using a Balance-Sheet Approach," *Weed Sci.*, 22(4):364–373 (1974).

Beyer, E.M., Jr., M.J. Duffy, J.V. Hay, and D.D. Schlueter. "Sulfonylureas" in *Herbicides: Chemistry, Degradation, and Mode of Action*, Kearney, P.C. and D.D. Kaufman, Eds., (New York: Marcel Dekker, Inc. 1988), pp. 117–189.

Beynon, K.I., G. Stoydin, and A.N. Wright. "A Comparison of the Breakdown of the Triazine Herbicides Cyanazine, Atrazine, and Simazine in Soil and in Maize," *Pestic. Biochem. Physiol.*, 2:153–161 (1972).

Bharathi, C. "Toxicity of Insecticides and Effects on the Behavior of the Blood Clam *Anadaria granosa*," *Water, Air and Soil Pollut.*, 75(1–2):87–91 (1994).

Biggar, J.W., D.R. Nielson, and W.R. Tillotson. "Movement of DBCP in Laboratory Soil Columns and Field Soils to Groundwater," *Environ. Geol.*, 5(3):127–131 (1984).

Bilkert, J.N. and P.S.C. Rao. "Sorption and Leaching of Three Nonfumigant Nematocides in Soils," *J. Environ. Sci. Health*, B29(1):1–26 (1985).

Billington, J.W., G.-L. Huang, F. Szeto, W.Y. Shiu, and D. Mackay. "Preparation of Aqueous Solutions of Sparingly Soluble Organic Substances: I. Single Component Systems," *Environ. Toxicol. Chem.*, 7:117–124 (1988).

Blazquez, C.H., A.D. Vidyarthi, T.D. Sheehan, M.J. Bennett, and G.T. McGrew. "Effect of Pinolene (β-Pinene Polymer) on Carbaryl Foliar Residues," *J. Agric. Food Chem.*, 18(4):681–684 (1970).

Boerner, H. "Untersuchungen über den Abbau von Afalon [*N*-(3,4-Dichlorophen-yl)-*N'*-methoxy-*N'*-methylharnstoff] und Aresin [*N*-(4-Chlorophenyl)-*N'*-methoxy-*N'*-methylharnstoff] im Boden," *Z. Pflanzenkr. Pflanzenschutz.*, 72:516–531 (1965).

Boerner, H. "Der Abbau von Harnstoffherbiziden im Bodeno," *Z. Pflanzenkr. Pflanzens-chutz.*, 74:135–143 (1967).

Boesten, J.J.T.I., L.J.T. van der Pas, M. Leistra, J.H. Smelt, and N.W.H. Houx. "Transformation of ^{14}C-Labeled 1,2-Dichloropropane in Water-Saturated Subsoil Materials," *Chemosphere*, 24(8):993–1011 (1992).

Boethling, R.S. and M. Alexander. "Effect of Concentration of Organic Chemicals on Their Biodegradation by Natural Microbial Communities," *Appl. Environ. Microbiol.*, 37(6):1211–1216 (1979).

Bollag, J.-M. and S.-Y. Liu. "Degradation of Sevin by Soil Microorganisms," *Soil Biol. Biochem.*, 3:337–345 (1971).

Bollag, J.-M. and S.-Y. Liu. "Hydroxylations of Carbaryl by Soil Fungi," *Nature (London)*, 236(5343):177–178 (1972).

Bollag, J.-M. and S.-Y. Liu. "Biological Transformation Processes of Pesticides" in *Pesticides in the Soil Environment: Processes, Impacts, and Modeling*, SSSA Book Series: 2, Cheng, H.H., Ed., (Madison, WI: Soil Science of America, Inc., 1990), pp. 169–211.

Booze, T.F. and F.W. Oehme. "Metaldehyde Toxicity: A Review," *Vet. Human Toxicol.*, 27:11–19 (1985).

Borello, R., C. Minero, E. Pramauro, E. Pelizzetti, N. Serpone, and H. Hidaka. "Photocatalytic Degradation of DDT Mediated in Aqueous Semiconductor Slurries by Simulated Sunlight," *Environ. Toxicol. Chem.*, 8(11):997–1002 (1989).

Borsetti, A.P. and J.A. Roach. "Identification of Kepone Alteration Products in Soil and Mullet," *Bull. Environ. Contam. Toxicol.*, 20(2):241–247 (1978).

Bouchard, D.C., T.L. Lavy, and D.B. Marx. "Fate of Metribuzin, Metolachlor, and Fluometuron in Soil," *Weed Sci.*, 30:629–632 (1982).

Bove, J.L. and P. Dalven. "Pyrolysis of Phthalic Acid Esters: Their Fate," *Sci. Tot. Environ.*, 36:313–318 (1984).

Bovey, R.W., C.C. Bowler, and M.G. Merkle. "The Persistence and Movement of Picloram in Texas and Puerto Rican Soils," *Pest. Monitor. J.*, 3:177–181 (1969).

Bowman, B.T. "Mobility and Persistence of Metolachlor and Aldicarb in Field Lysimeters," *J. Environ. Qual.*, 17(4):689–694 (1988).

Bowman, B.T. and W.W. Sans. "Adsorption of Fenitrothion, Methyl Parathion, Amino-parathion and Paraoxon by Na^+, Ca^{2+}, and Fe^{3+} Montmorillonite Suspensions," *Soil Sci. Soc. Am. J.*, 41:514–519 (1977).

Bowman, B.T. and W.W. Sans. "Adsorption, Desorption, Soil Mobility, Aqueous Persistence and Octanol-Water Partitioning Coefficients of Terbufos, Terbufos Sulfoxide and Terbufos Sulfone," *J. Environ. Sci. Health*, B17(5):447–462 (1982).

Bowman, M.C., M.S. Schechter, and R.L. Carter. "Behavior of Chlorinated Insecticides in a Broad Spectrum of Soil Types," *J. Agric. Food Chem.*, 13(4):360–365 (1965).

Boyle, T.P., E.F. Robinson-Wilson, J.D. Petty, and W. Weber. "Degradation of Pentachlorophenol in Simulated Lenthic Environment," *Bull. Environ. Contam. Toxicol.*, 24(2):177–184 (1980).

Braverman, M.P., T.L. Lavy, and C.J. Barnes. "The Degradation and Bioactivity of Metolachlor in the Soil," *Weed Sci.*, 34:479–484 (1986).

Breaux, E.J., J.E. Patanella, and E.F. Sanders. "Chloroacetanilide Herbicide Selectivity: Analysis of Glutathione and Homoglutathione in Tolerant, Susceptible and Safened Seedlings," *J. Agric. Food Chem.*, 35(4):474–478 (1987).

Brett, C.H. and T.G. Bowery. "Insecticide Residues on Vegetables," *J. Econ. Entomol.*, 51:818–821 (1958).

Briggs, G.G. "Theoretical and Experimental Relationships between Soil Adsorption, Octanol-Water Partition Coefficients, Water Solubilities, Bioconcentration Factors, and the Parachor," *J. Agric. Food Chem.*, 29(5):1050–1059 (1981).

Briggs, G.G. and J.E. Dawson. "Hydrolysis of 2,6-Dichlorobenzonitrile in Soils," *J. Agric. Food Chem.*, 18(1):97–99 (1970).

Broadhurst, N.A., M.L. Montgomery, and V.H. Freed. "Metabolism of 2-Methoxy-3,6-dichlorobenzoic acid (Dicamba) by Wheat and Bluegrass Plants," *J. Agric. Food Chem.*, 14(6):585–588 (1966).

Bromilow, R.H., R.J. Baker, M.A.H. Freeman, and K. Görög. "The Degradation of Aldicarb and Oxamyl in Soil," *Pestic. Sci.*, 11(4):371–378 (1980).

Bromilow, R.H. and M. Leistra. "Measured and Simulated Behaviour of Aldicarb and Its Oxidation Products in Fallow Soils," *Pestic. Sci.*, 11(4):389–395 (1980).

Bro-Rasmussen, F., E. Noddegaard, and K. Voldum-Clausen. "Degradation of Diazinon in Soil," *J. Sci. Food Agric.*, 19:278–282 (1968).

Brown, D.S. and E.W. Flagg. "Empirical Prediction of Organic Pollutant Sorption in Natural Sediments," *J. Environ. Qual.*, 10(3):382–386 (1981).

Brown, M.A., M.T. Petreas, H.S. Okamoto, T.M. Mischke, and R.D. Stephens. "Monitoring of Malathion and Its Impurities and Environmental Transformation Products on Surfaces and in Air Following an Aerial Application," *Environ. Sci. Technol.*, 27(2):388–397 (1993).

Buck, N.A., B.J. Estesen, and G.W. Ware. "Dislodgable Insecticide Residues on Cotton Foliage: Fenvalerate, Permethrin, Sulprofos, Chlorpyrifos, Methyl Parathion, EPN, Oxamyl, and Profenofos," *Bull. Environ. Contam. Toxicol.*, 24(2):283–288 (1980).

Bull, D.L. "Metabolism of UC-21149 [2-Methyl-2-(methylthio)propionaldehyde *O*-(methylcarbamoyl)oxime] in Cotton Plants and Soil in the Field," *J. Econ. Entomol.*, 61:1598–1602 (1968).

Bull, D.L. "Fate and Efficacy of Acephate after Application to Plants and Insects," *J. Agric. Food Chem.*, 27(2):268–272 (1979).

Bull, D.L. and R.L. Ridgway. "Metabolism of Trichlorfon in Animals and Plants," *J. Agric. Food Chem.*, 17(4):837–841 (1969).

Bumpus, J.A. and S.D. Aust. "Biodegradation of DDT [1,1,1-Trichloro-2,2-bis(4-chlorophenyl)ethane] by the White Rot Fungus *Phanerochaete chrysosporium*," *Appl. Environ. Microbiol.*, 53(9):2001–2008 (1987).

Bumpus, J.A., M. Tien, D. Wright, and S.D. Aust. "Oxidation of Persistent Environmental Pollutants by a White Rot Fungus," *Science (Washington, DC)*, 228:1434–1436 (1985).

Burchfield, H.P. "Comparative Stabilities of Dyrene, 1-Fluoro-2,4-dinitrobenzene, Dichlone, and Captan in a Silt Loam Soil," *Contrib. S. Boyce Thompson Inst.*, 20:205–215 (1959).

Burczyk, L., K. Walczyk, and R. Burczyk. "Kinetics of Reaction of Water Addition to the Acrolein Double-Bond in Dilute Aqueous Solution," *Przem. Chem.*, 47(10):625–627 (1968).

Burge, W.D. "Anaerobic Decomposition of DDT in Soil. Acceleration by Volatile Components of Alfalfa," *J. Agric. Food Chem.*, 19(2):375–378 (1971).

Burkhard, N. and J.A. Guth. "Photodegradation of Atrazine, Atraton, and Ametryne in Aqueous Solution with Acetone as a Photosensitizer," *Pestic. Sci.*, 7:65 (1976).

Burkhard, N. and J.A. Guth. "Photolysis of Organophosphorus Insecticides on Soil Surfaces," *Pestic. Sci.*, 10(4):313–319 (1979).

Burkhard, N. and J.A. Guth. "Chemical Hydrolysis of 2-Chloro-4,6-bis(alkylamino)-1,3,5-triazine Herbicides and Their Breakdown in Soil Under the Influence of Adsorption," *Pestic. Sci.*, 12(1):45–52 (1981).

Burlinson, N.E., L.A. Lee, and D.H. Rosenblatt. "Kinetics and Products of Hydrolysis of 1,2-Dibromo-3-chloropropane," *Environ. Sci. Technol.*, 16(9):627–632 (1982).

Burns, R.G. "Loss of Phosdrin and Phorate Insecticides from a Range of Soil Types," *Bull. Environ. Contam. Toxicol.*, 6(4):316–321 (1971).

Burnside, O.C. and T.L. Lavy. "Dissipation of Dicamba," *Weeds*, 14:211–214 (1966).

Burris, D.R. and W.G. MacIntyre. "A Thermodynamic Study of Solutions of Liquid Hydrocarbon Mixtures in Water," *Geochim. Cosmochim. Acta*, 50(7):1545–1549 (1986).

Burton, W.B. and G.E. Pollard. "Rate of Photochemical Isomerization of Endrin in Sunlight," *Bull. Environ. Contam. Toxicol.*, 12(1):113–116 (1974).

Businelli, M., M. Patumi, and C. Marucchini. "Identification and Determination of Some Metalaxyl Degradation Products in Lettuce and Sunflower," *J. Agric. Food Chem.*, 32(3):644–647 (1984).

Bussacchini, V., B. Pouyet, and P. Meallier. "Photodegradation des Molecules Phytosanitaires. VI — Photodegradation du Phenmediphame," *Chemosphere*, 14(2):155–161 (1985).

Butler, J.A.V., D.W. Thomson, and W.H. Maclennan. "The Free Energy of the Normal Aliphatic Alcohols in Aqueous Solution. Part I. The Partial Vapour Pressures of Aqueous Solutions of Methyl, *n*-Propyl, and *n*-Butyl Alcohols. Part II. The Solubilities of Some Normal Aliphatic Alcohols in Water. Part III. The Theory of Binary Solutions, and its Application to Aqueous-Alcoholic Solutions," *J. Chem. Soc. (London)*, 136:674–686 (1933).

Byast, T.H. and R.J. Hance. "Degradation of 2,4,5-T by South Vietnamese Soils Incubated in the Laboratory," *Bull. Environ. Contam. Toxicol.*, 14(1):71–76 (1975).

Cabras, P., F. Cabitza, M. Meloni, and F.M. Pirisi. "Behavior of Some Pesticide Residues on Greenhouse Tomatoes. 2. Fungicides, Acaricides, and Insecticides," *J. Agric. Food Chem.*, 33(5):935–937 (1985).

Calderbank, A. and P. Slade. "Diquat and Paraquat" in *Herbicides: Chemistry, Degradation, and Mode of Action*, Kearney, P.C. and D.D. Kaufman, Eds., (New York: Marcel Dekker, Inc. 1976), pp. 503–540.

Calvert, J.G., J.A. Kerr, K.L. Demerjian, and R.D. McQuigg. "Photolysis of Formaldehyde as a Hydrogen Atom Source in the Lower Atmosphere," *Science (Washington, DC)*, 175(4023):751–752 (1972).

Calvert, J.G. and J.N. Pitts, Jr. *Photochemistry* (New York: John Wiley & Sons, Inc., 1966), 899 p.

Camazano, M.S. and M-J.S. Martin. "Interaction of Phosmet with Montmorillonite," *Soil Sci.*, 129(2):115–118 (1980).

Camper, N.D. "Effects of Pesticide Degradation Products on Soil Microflora" in *Pesticide Transformation Products. Fate and Significance in the Environment*, ACS Symposium Series 429, Somasundaram, L. and J.R. Coats, Eds., (New York: American Chemical Society, 1991), pp. 205–216.

Capel, P.D., W. Giger, P. Reichert, and O. Wanner. "Accidental Input of Pesticides into the Rhine River," *Environ. Sci. Technol.*, 22(9):992–997 (1988).

Capriel, P., A. Haisch, and S.U. Khan. "Distribution and Nature of Bound (Nonextractable) Residues of Atrazine in a Mineral Soil Nine Years after the Herbicide Application," *J. Agric. Food Chem.*, 33(4):567–569 (1985).

Caro, J.H., H.P. Freeman, D.W. Glotfelty, B.C. Turner, and W.M. Edwards. "Dissipation of Soil-Incorporated Carbofuran in the Field," *J. Agric. Food Chem.*, 21:1010–1015 (1973).

Caro, J.H., H.P. Freeman, and B.C. Turner. "Persistence in Soil and Losses in Runoff in Soil-Incorporated Carbaryl in a Small Watershed," *J. Agric. Food Chem.*, 22(5):860–863 (1974).

Casida, J.E., P.E. Gatterdam, L.W. Getzin, Jr., and R.K. Chapman. "Residual Properties of the Systemic Insecticide *O,O*-Dimethyl 1-Carbomethoxy-1-propen-2-yl Phosphate," *J. Agric. Food Chem.*, 4(3):236–243 (1956).

Casida, J.E., R.A. Gray, and H. Tilles. "Thiocarbamate Sulfoxides: Potent, Selective, and Biodegradable Herbicides," *Science (Washington, DC)*, 184(4136):573–574 (1974).

Castelfranco, P., C.L. Foy, and D.B. Deutsch. "Non-Enzymatic Detoxification of 2-Chloro-4,6-bis(ethylamino)-*s*-triazine by Extracts of Zea mays," *Weeds*, 9:580–591 (1961).

Castro, C.E. "The Rapid Oxidation of Iron(II) Porphyrins by Alkyl Halides. A Possible Mode of Intoxication of Organisms by Alkyl Halides," *J. Am. Chem. Soc.*, 86(2):2310–2311 (1964).

Castro, C.E. and N.O. Belser. "Hydrolysis of *cis*- and *trans*-1,3-Dichloropropene in Wet Soil," *J. Agric. Food Chem.*, 14(1):69–70 (1966).

Castro, C.E. and N.O. Belser. "Biodehalogenation. Reductive Dehalogenation of the Biocides Ethylene Dibromide, 1,2-Dibromo-3-chloropropane, and 2,3-Dibromobutane in Soil," *Environ. Sci. Technol.*, 2(10):779–783 (1968).

Castro, C.E. and N.O. Belser. "Photohydrolysis of Methyl Bromide and Chloropicrin," *J. Agric. Food Chem.*, 29(5):1005–1009 (1981).

Castro, C.E., R.S. Wade, and N.O. Belser. "Biodehalogenation. The Metabolism of Chloropicrin by *Pseudomonas* sp.," *J. Agric. Food Chem.*, 31(6):1184–1187 (1983).

Castro, T.F. and T. Yoshida. "Degradation of Organochlorine Insecticides in Flooded Soils in the Philippines," *J. Agric. Food Chem.*, 19(6):1168–1170 (1971).

Cayley, G.R. and G.A. Hide. "Uptake of Iprodione and Control of Diseases on Potato Stems," *Pestic. Sci.*, 11(1):15–19 (1980).

Cerniglia, C.E. and S.A. Crow. "Metabolism of Aromatic Hydrocarbons by Yeasts," *Arch. Microbiol.*, 129:9–13 (1981).

Ceron, J.J., C. Gutiérrez-Panizo, A. Barba, and M.A. Cámara. "Endosulfan Isomers and Metabolite Residue Degradation in Carnation (*Dianthus caryophyllus*) By-product Under Different Environmental Conditions," *J. Environ. Sci. Health,* B30(2):221–232 (1995).

Cessna, A.J. and D.C.G. Muir. "Photochemical Transformations" in *Environmental Chemistry of Herbicides*. Volume II, Grover, R. and A.J. Cessna, Eds. (Boca Raton, FL: CRC Press, Inc., 1991), pp. 199–263.

Chacko, C.I., J.L. Lockwood, and M. Zabik. "Chlorinated Hydrocarbon Pesticides: Degradation by Microbes," *Science (Washington, DC)*, 154(3751):893–895 (1966).

Chaigneau, M., G. LeMoan, and L. Giry. "Décomposition Pyrogénée du Bromurede Méthyle en L'absence D'oxygene et à 550°C," *Comp. Rend. Ser. C.*, 263:259–261 (1966).

Chapman, R.A. and P.C. Chapman. "Persistence of Granular and EC Formulations of Chlorpyrifos in a Mineral and Organic Soil Incubated in Open and Closed Containers," *J. Environ. Sci. Health*, B21(6):447–456 (1986).

Chapman, R.A. and M. Cole. "Observations on the Influence of Water and Soil pH on the Persistence of Insecticides," *J. Environ. Sci. Health*, B17(5):487–504 (1982).

Chapman, R.A. and C.R. Harris. "Persistence of Four Pyrethroid Insecticides in a Mineral and an Organic Soil," *J. Environ. Sci. Health*, B15(1):39–46 (1980).

Chapman, R.A., C.R. Harris, H.J. Svec, and J.R. Robinson. "Persistence and Mobility of Granular Insecticides in an Organic Soil Following Furrow Application for Onion Maggot Control," *J. Environ. Sci. Health*, B19(3):259–270 (1984).

Chapman, R.A., C.M. Tu, C.R. Harris, and C. Cole. "Persistence of Five Pyrethroid Insecticides in Sterile and Natural, Mineral and Organic Soil," *Bull. Environ. Contam. Toxicol.*, 26(4):513–519 (1981).

Chapman, R.A., C.M. Tu, C.R. Harris, and D. Dubois. "Biochemical and Chemical Transformations of Terbufos, Terbufos Sulfoxide, and Terbufos Sulfone in Natural and Sterile, Mineral and Organic Soil," *J. Econ. Entomol.*, 75:955–960 (1982).

Chaube, H.S., U.S. Sin, and P.K. Sinha. "Translocation and Persistence of Metalaxyl in Pigeon Pea," *Indian J. Agric. Sci.*, 54:777–779 (1984).

Chen, P.R., W.P. Tucker, and W.C. Dauterman. "Structure of Biologically Produced Malathion Monoacid," *J. Agric. Food Chem.*, 17(1):86–90 (1969).

Chesters, G., G.V. Simsiman, J. Levy, B.J. Alhajjar, R.N. Fathulla, and J.M. Harkin. "Environmental Fate of Alachlor and Metolachlor" in *Reviews of Environmental Contamination and Toxicology*, (New York: Springer-Verlag, Inc., 1989), pp. 1–74.

Chiba, M. and D.F. Veres. "Fate of Benomyl and Its Degradation Compound Methyl 2-Benzimidazolecarbamate on Apple Foliage," *J. Agric. Food Chem.*, 29(3):588–590 (1981).

Chiou, C.T., V.H. Freed, D.W. Schmedding, and R.L. Kohnert. "Partition Coefficients and Bioaccumulation of Selected Organic Chemicals," *Environ. Sci. Technol.*, 11(5):475–478 (1977).

Chiou, C.T., R.L. Malcolm, T.I. Brinton, and D.E. Kile. "Water Solubility Enhancement of Some Organic Pollutants and Pesticides by Dissolved Humic and Fulvic Acids," *Environ. Sci. Technol.*, 20(5):502–508 (1986).

Chiou, C.T. and M. Manes. "Application of the Flory-Higgins Theory to the Solubility of Solids in Glyceryl Trioleate," *J. Chem. Soc., Faraday Trans. 1*, 82(1):243–246 (1986).

Chiou, C.T., T.D. Shoup, and P.E. Porter. "Mechanistic Roles of Soil Humus and Minerals in the Sorption of Nonionic Organic Compounds from Aqueous and Organic Solutions," *Org. Geochem.*, 8(1):9–14 (1985).

Chiron, S., J. Abian, M. Ferrer, F. Sanchez-Baeza, A. Messeguer, and D. Barceló. "Comparative Photodegradation Rates of Alachlor and Bentazone in Natural Water and Determination of Breakdown Products," *Environ. Toxicol. Chem.*, 14(8):1287–1298 (1995).

Chisaka, H. and P.C. Kearney. "Metabolism of Propanil in Soils," *J. Agric. Food Chem.*, 18(5):854–858 (1970).

Chopra, N.M. and A.M. Mahfouz. "Metabolism of Endosulfan I, Endosulfan II, and Endosulfan Sulfate in Tobacco Leaf," *J. Agric. Food Chem.*, 25(1):32–36 (1977).

Chou, S.H. "Fate of Acylanilides in Soils and Polybrominated Biphenyls (PBBs) in Soils and Plants," Dissertation, Michigan State University, East Lansing, MI (1977).

Choudhry, G.G. and O. Hutzinger. "Acetone-Sensitized and Nonsensitized Photolyses of Tetra-, Penta-, and Hexachlorobenzenes in Acetonitrile-Water Mixtures: Photoisomerization and Formation of Several Products Including Polychlorobiphenyls," *Environ. Sci. Technol.*, 18(4):235–241 (1984).

Chu, W. and C.T. Jafvert. "Photodechlorination of Polychlorobenzene Congeners in Surfactant Micelle Solutions," *Environ. Sci. Technol.*, 28(13):2415–2422 (1994).

Clapés, P., J. Soley, M. Vicente, J. Rivera, J. Caixach, and F. Ventura. "Degradation of MCPA by Photochemical Methods," *Chemosphere*, 15(4):395–401 (1986).

Clapp, D.W., D.V. Naylor, and G.C. Lewis. "The Fate of Disulfoton in Portneuf Silt Loam Soil," *J. Environ. Qual.*, 5(2):207–210 (1976).

Clark, T., D.R. Clifford, A.H.B. Deas, P. Gendle, and D.A.M. Watkins. "Photolysis, Metabolism, and Other Factors Influencing the Performance of Triadimefon as a Powdery Mildew Fungicide," *Pestic. Sci.*, 9:497–506 (1978).

Clay, D.V. "The Persistence and Penetration of Large Doses of Simazine in Uncropped Soil," *Weed Res.*, 13:42–50 (1973).

Clemons, G.P. and H.D. Sisler. "Formation of a Fungitoxic Derivative from Benlate," *Phytopathology*, 59(5):705–706 (1969).

Coates, M., D.W. Connell, and D.M. Barron. "Aqueous Solubility and Octan-1-ol to Water Partition Coefficients of Aliphatic Hydrocarbons," *Environ. Sci. Technol.*, 19(7):628–632 (1985).

Coats, J.R. and N.L. O'Donnell-Jeffrey. "Toxicity of Four Synthetic Pyrethroid Insecticides to Rainbow Trout," *Bull. Environ. Contam. Toxicol.*, 23(1/2):250–255 (1979).

Colbert, F.O., V.V. Volk, and A.P. Appleby. "Sorption of Atrazine, Terbutryn, and GS-14254 on Natural and Lime-Amended Soils," *Weed Sci.*, 23(5):390–394 (1975).

Connors, T.F., J.D. Stuart, and J.B. Cope. "Chromatographic and Mutagenic Analyses of 1,2-Dichloropropane and 1,3-Dichloropropylene and Their Degradation Products," *Bull. Environ. Contam. Toxicol.*, 44(2):288–293 (1990).

Cook, A.M. and R. Hütter. "Ametryne and Prometryne as Sulfur Sources for Bacteria," *Appl. Environ. Microbiol.*, 43(4):781–786 (1982).

Coppedge, J.R., D.A. Lindquist, D.L. Bull, and H.W. Dorough. "Fate of 2-Methyl-2-(methylthio)propionaldehyde O-(Methylcarbamoyl) oxime (Temik) in Cotton Plants and.Soil," *J. Agric. Food Chem.*, 15(5):902–910 (1967).

Corbin, F.T. and R.P. Upchurch. "Influence of pH on Detoxification of Herbicides in Soils," *Weeds*, 15:370–377 (1967).

Corby, T.C. and P.H. Elworthy. "The Solubility of Some Compounds in Hexadecylpoly-oxyethylene Monoethers, Polyethylene Glycols, Water and Hexane," *J. Pharm. Pharmac.*, 23:39S-48S (1971).

Corwin, D.L. and W.J. Farmer. "Nonsingle-Valued Adsorption-Desorption of Bromacil and Diquat by Freshwater Sediments," *Environ. Sci. Technol.*, 18(7):507–514 (1984).

Coupland, D. and D.V. Peabody. "Adsorption, Translocations, and Exudation of Glyphosate, Fosamine, and Amitrole in Field Horsetail (*Equisetum arvense*)," *Weed Sci.*, 29:556–560 (1981).

Cowart, R.P., F.L. Bonner, and E.A. Epps, Jr. "Rate of Hydrolysis of Seven Organophosphorus Pesticides," *Bull. Environ. Contam. Toxicol.*, 6(3):231–234 (1971).

Cremlyn, R.J. *Agrochemicals — Preparation and Mode of Action* (New York: John Wiley & Sons, Inc., 1991), 396 p.

Cripe, C.R., W.W. Walker, P.H. Pritchard, and A.W. Bourquin. "A Shake-Flask Test for Estimation of Biodegradability of Toxic Organic Substances in the Aquatic Environment," *Ecotoxicol. Environ. Safety*, 14(3):239–251 (1987).

Crosby, D.G. "The Non-Metabolic Decomposition of Pesticides," *Ann. NY Acad. Sci.*, 160:82–90 (1969).

Crosby, D.G. and N. Hamadmad. "The Photoreduction of Pentachlorobenzenes," *J. Agric. Food Chem.*, 19(6):1171–1174 (1971).

Crosby, D.G. and E. Leitis. "Photodecomposition of Chlorobenzoic Acids," *J. Agric. Food Chem.*, 17(5):1033–1035 (1969).

Crosby, D.G. and E. Leitis. "The Photodecomposition of Trifluralin in Water," *Bull. Environ. Contam. Toxicol.*, 10(4):237–241 (1973).

Crosby, D.G., E. Leitis and W.L. Winterlin. "Photodecomposition of Carbamate Insecticides," *J. Agric. Food Chem.*, 13(3):204–207 (1965).

Crosby, D.G. and K.W. Moilanen. "Vapor-Phase Photodecomposition of Aldrin and Dieldrin," *Arch. Environ. Contam.*, 2(1):62–74 (1974).

Crosby, D.G. and C.S. Tang. "Photodecomposition of 3-(p-Chlorophenyl)-1-dimethylurea (Monuron)," *J. Agric. Food Chem.*, 17(5):1041–1044 (1969).

Crosby, D.G. and H.O. Tutass. "Photodecomposition of 2,4-Dichlorophenoxyacetic Acid," *J. Agric. Food Chem.*, 14(6):596–599 (1966).

Crosby, D.G. and A.S. Wong. "Photodecomposition of 2,4,5-Trichlorophenoxyacetic Acid (2,4,5-T) in Water," *J. Agric. Food Chem.*, 21(6):1052–1054 (1973).

Cserjsei, A.J. and E.L. Johnson. "Methylation of Pentachlorophenol by *Trichoderma virgatum*," *Can. J. Microbiol.*, 18:45–49 (1972).

Cullimore, D.R. "Interaction between Herbicides and Soil Microorganisms," *Residue Rev.*, 35:65–80 (1971).

Cupitt, L.T. "Fate of Toxic and Hazardous Materials in the Air Environment," Office of Research and Development, U.S. EPA Report-600/3-80–084 (1980), 28 p.

Dao, T.H. and T.L. Lavy. "Atrazine Adsorption on Soil as Influenced by Temperature, Moisture Content, and Electrolyte Concentration," *Weed Sci.*, 26(3):303–308 (1978).

Dauterman, W.C., G.B. Viado, J.E. Casida, and R.D. O'Brien. "Persistence of Dimethoate and. Metabolites Following Foliar Application to Plants," *J. Agric. Food Chem.*, 8(2):115–119 (1960).

David, W.A.L. and W.N. Aldridge. "The Insecticidal Material in Leaves and Plants Growing in Soil Treated with Parathion," *Ann. Appl. Biol.*, 45:332–346 (1957).

Davis, J.T. and W.S. Hardcastle. "Biological Assay of Herbicides for Fish Toxicity," *Weeds*, 7:397–404 (1959).

Davis, W.W., M.E. Krahl, and G.H.A. Clowes. "Solubility of Carcinogenic and Related Hydrocarbons in Water," *J. Am. Chem. Soc.*, 64(1):108–110 (1942).

Davis, A.C. and R.J. Kuhr. "Dissipation of Chlorpyrifos from Muck Soil and Onions," *J. Econ. Entomol.*, 69:665–666 (1976).

Davis, W.W. and T.V. Parke, Jr. "A Nephelometric Method for Determination of Solubilities of Extremely Low Order," *J. Am. Chem. Soc.*, 64(1):101–107 (1942).

Day, K.E. "Pesticide Transformation Products in Surface Waters" in P*esticide Transformation Products, Fate and Significance in the Environment*, ACS Symposium Series 459, Somasundaram, L. and J.R. Coats, Eds. (Washington, DC: American Chemical Society, 1991), pp. 217–241.

Day, B.E., L.S. Jordan, and V.L. Jolliffe. "The Influence of Soil Characteristics on the Adsorption and Phytotoxicity of Simazine," *Weed Sci.*, 16(2):209–213 (1968).

Debona, A.C. and L.J. Audus. "Effects of Herbicides on Soil Nitrification," *Weed Res.*, 10:250–263 (1970).

Deeley, G.M., M. Reinhard, and S.M. Stearns. "Transformation and Sorption of 1,2-Dibromo-3-Chloropropane in Subsurface Samples Collected at Fresno, California," *J. Environ. Qual.*, 20(4):547–556 (1991).

DeLaune, R.D., R.P. Gambrell, and K.S. Reddy. "Fate of Pentachlorophenol in Estuarine Sediment," *Environ. Pollut. Series B*, 6:297–308 (1983).

Dennis, W.H., Jr. and W.J. Cooper. "Catalytic Dechlorination of Organochlorine Compounds. II. Heptachlor and Chlordane," *Bull. Environ. Contam. Toxicol.*, 16(4):425–430 (1976).

Dennis, W.H., Jr., E.P. Meier, W.F. Randall, A.B. Rosencrance, and D.H. Rosenblatt. "Degradation of Diazinon by Sodium Hypochlorite. Chemistry and Aquatic Toxicity," *Environ. Sci. Technol.*, 13(5):594–598 (1979).

Deno, N.C. and H.E. Berkheimer. "Activity Coefficients as a Function of Structure and Media," *J. Chem. Eng. Data*, 5(1):1–5 (1960).

Derbyshire, M.K., J.S. Karns, P.C. Kearney, and J.O. Nelson. "Purification and Characterization of an *N*-Methylcarbamate Pesticide Hydrolyzing Enzyme," *J. Agric. Food Chem.*, 35(6):871–877 (1987).

Devine, M.D. and W.H.V. Born. "Absorption, Translocation, and Foliar Activity of Clopyralid and Chlorsulfuron in Canada Thistle (*Cirsium arevense*) and Perennial Sowthistle (*Sonchus arvensis*)," *Weed Sci.*, 33:524–530 (1985).

Dewey, J.E. and B.L. Parker. "Increase in Toxicity to *Drosophila melanogaster* of Phorate-Treated Soils," *J. Econ. Entomol.*, 58:491–497 (1965).

Dick, W.A., R.O. Ankumah, G. McClung, and N. Abou-Assaf. "Enhanced Degradation of *S*-Ethyl *N,N*-Dipropylcarbamothioate in Soil and by an Isolated Soil Microorganism" in *Enhanced Biodegradation of Pesticides in the Environment*, ACS Symposium Series 426, Racke, K.D. and J.R. Coats, Eds. (Washington, DC: American Chemical Society, 1990), pp. 98–112.

Dickhut, R.M., A.W. Andren, and D.E. Armstrong. "Aqueous Solubilities of Six Polychlorinated Biphenyl Congeners at Four Temperatures," *Environ. Sci. Technol.*, 20(8):807–810 (1986).

DiGeronimo, M.J. and A.D. Antoine. "Metabolism of Acetonitrile and Propionitrile by *Nocardia rhodochrous* LL 100-21," *Appl. Environ. Microbiol.*, 31(6):900–906 (1976).

Dilling, W.L. "Interphase Transfer Processes. II. Evaporation Rates of Chloro Methanes, Ethanes, Ethylenes, Propanes, and Propylenes from Dilute Aqueous Solutions. Comparisons with Theoretical Predictions," *Environ. Sci. Technol.*, 11(4):405–409 (1977).

Dilling, .W.L., L.C. Lickly, T.D. Lickly, P.G. Murphy, and R.L. McKellar. "Organic Photochemistry. 19. Quantum Yields for *O,O*-Diethyl *O*-(3,5,6-Trichlor-*o*-2-pyridinyl) Phosphorothioate (Chlorpyrifos) and 3,5,6-Trichloro-2-pyridinol in Dilute Aqueous Solutions and Their Environmental Phototransformation Rates," *Environ. Sci. Technol.*, 18(7):540–543 (1984).

Dobson, R.C., G.O. Throneberry, and T.E. Belling. "Residues of Established Alfalfa Treated with Granulated Phorate (Thimet) and Their Effect on Cattle Fed the Hay," *J. Econ. Entomol.*, 53:306–310 (1960).

Dodge R.H., C.E. Cerniglia, and D.T. Gibson. "Fungal Metabolism of Biphenyl," *Biochem. J.*, 178:223–230 (1979).

Dogger, J.R. and T.G. Bowery. "A Study of Residues of Some Commonly Used Insecticides on Alfalfa," *J. Econ. Entomol.*, 51:392–394 (1958).

Dorough, H.W., N.M. Randolph, and G.H. Wimbish. "Residual Nature of Certain Organophosphorus Insecticides in Grain Sorghum and Coastal Bermuda Grass," *Bull. Environ. Contam. Toxicol.*, 1(1):46–58 (1966).

Dorough, H.W., R.F. Skrentny, and B.C. Pass. "Residues in Alfalfa and Soils following Treatment with Technical Chlordane and High Purity Chlordane (HCS 3260) for Alfalfa Weevil Control," *J. Agric. Food Chem.*, 20(1):42–47 (1972).

Doucette, W.J. and A.W. Andren. "Aqueous Solubility of Biphenyl, Furan, and Dioxin Congeners," *Chemosphere*, 17(2):243–252 (1988).

Doyle, R.C., D.D. Kaufman, and G.W. Burt. "Effect of Dairy Manure and Sewage Sludge on ^{14}C-Pesticide Degradation in Soil," *J. Agric. Food Chem.*, 26:987–989 (1978).

Draper, W.M. and D.G. Crosby. "Solar Photooxidation of Pesticides in Dilute Hydrogen Peroxide," *J. Agric. Food Chem.*, 32(2):231–237 (1984).

Dreher, R.M. and B. Podratzki. "Development of an Enzyme Immunoassay for Endosulfan and Its Degradation Products," *J. Agric. Food Chem.*, 36(5):1072–1075 (1988).

Drinking Water Health Advisory. *Pesticides.* (Chelsea, MI: Lewis Publishers, Inc., 1989), 819 p.

Duff, W.G. and R.E. Menzer. "Persistence, Mobility, and Degradation of ^{14}C-Dimethoate in Soils," *Environ. Entomol.*, 2(3):309–318 (1973).

Duke, S.O. "Glyphosate" in *Herbicides: Chemistry, Degradation, and Mode of Action*, Kearney, P.C. and D.D. Kaufman, Eds., (New York: Marcel Dekker, Inc. 1988), pp. 1–70.

Duke, S.O., T.B. Moorman, and C.T. Bryson. "Phytotoxicity of Pesticide Degradation Products" in *Pesticide Transformation Products, Fate and Significance in the Environment*, ACS Symposium Series 459, Somasundaram, L. and J.R. Coats, Eds. (Washington, DC: American Chemical Society, 1991), pp. 188–204.

Dulfer, W.J., M.W.C. Bakker, and H.A.J. Govers. "Micellar Solubility and Micelle/Water Partitioning of Polychlorinated Biphenyls in Solutions of Sodium Dodecyl Sulfate," *Environ. Sci. Technol.*, 29(4):985–992 (1995).

Dunnivant, F.M. and A.W. Elzerman. "Aqueous Solubility and Henry's Law Constant for PCB Congeners for Evaluation of Quantitative Structure-Property Relationships (QSPRs)," *Chemosphere*, 7(3):525–531 (1988).

Dureja, P.S.W. and S.K. Mukerjee. "Photodecomposition of Isofenphos [O-Ethyl-O-(2-isopropoxycarbonyl)phenyl]isopropylphosphoramidothioate," *Toxicol. Environ. Chem.*, 19(3–4):187–192 (1989).

Dureja, P.S.W. and S. Walia. "Photodecomposition of Pendimethalin," *Pestic. Sci.*, 25(2):105–114 (1989).

Duseja. D.R. and E.E. Holmes. "Field Persistence and Movement of Trifluralin in Two Soil Types," *Soil Sci.*, 125:41–48 (1978).

Dzantor, E.K. and A.S. Felsot. Effects of Conditioning, Cross-Conditioning, and Microbial Growth on Development of Enhanced Biodegradation of Insecticides in Soil," *J. Environ. Sci. Health*, B24(6):569–597 (1989).

Eckert, J.W. "Fungistatic and Phytotoxic Properties of Some Derivatives of Nitrobenzene," *Phytopathology*, 52:642–650 (1962).

Eganhouse, R.P. and J.A. Calder. "The Solubility of Medium Weight Aromatic Hydrocarbons and the Effect of Hydrocarbon Co-Solutes and Salinity," *Geochim. Cosmochim. Acta*, 40(5):555–561 (1976).

Eichelberger, J.W. and J.J. Lichtenberg. "Persistence of Pesticides in River Water," *Environ. Sci. Technol.*, 5(6):541–544 (1971).

El-Dib, M.A. and O.S. Aly. "Persistence of Some Phenylamide Pesticides in the Aquatic Environment-I. Hydrolysis," *Water Res.*, 10(12):1047–1050 (1976).

Ellington, J.J., F.E. Stancil, and W.D. Payne. "Measurement of Hydrolysis Rate Constants for Evaluation of Hazardous Waste Land Disposal: Volume 1. Data on 32 Chemicals," Office of Research and Development, U.S. EPA Report-600/3-86-043 (1986), 122 p.

Ellington, J.J., F.E. Stancil, Jr., W.D. Payne, and C. Trusty. "Measurement of Hydrolysis Rate Constants for Evaluation of Hazardous Waste Land Disposal. Volume 2. Data on 54 Chemicals," Office of Research and Development, U.S. EPA Report-600/3-87/019 (1987), 152 p.

Ellington, J.J., F.E. Stancil, W.D. Payne, and C.D. Trusty. "Measurement of Hydrolysis Rate Constants for Evaluation of Hazardous Waste Land Disposal: Volume 3. Data on 70 Chemicals," Office of Research and Development, U.S. EPA Report-600/3-88-028 (1988), 233 p.

Elliott, S. "Effect of Hydrogen Peroxide on the Alkaline Hydrolysis of Carbon Disulfide," *Environ. Sci. Technol.*, 24(2):264–267 (1990).

Ellis, P.A. and N.D. Camper. "Aerobic Degradation of Diuron by Aquatic Microorganisms," *J. Environ. Sci. Health*, B17(3):277–289 (1982).

El-Refai, A. and T.L. Hopkins. "Parathion Absorption, Translocation, and Conversion to Paraoxon in Bean Plants," *J. Agric. Food Chem.*, 14(6):588–592 (1966).

Elsner, E., D. Bieniek, W. Klein, and F. Korte. "Verteilung und Umwandlumg von Aldrin, Heptachlor, und Lindan in der Grunalge *Chlorella pyrenoidosa*," *Chemosphere*, 1:247–250 (1972).

Emmerich, W.E., J.D. Helmer, K.G. Renard, and L.J. Lane. "Fate and Effectiveness of Tebuthiuron to a Rangeland Watershed," *J. Environ. Qual.*, 13(3):382–386 (1984).

Enfield C.G. and S.R. Yates. "Organic Chemical Transport in Groundwater" in *Pesticides in the Soil Environment: Processes, Impacts, and Modeling, Soil Science of Society of America*, Cheng, H.H., Ed., (Madison, WI: SSSA, 1990), pp. 271–302.

Engelhardt, G., L. Oehlmann, K. Wagner, P.R. Wallnöfer, and M. Wiedermann. "Degradation of the Insecticide Azinphosmethyl in Soil and by Isolated Soil Bacteria," *J. Agric. Food Chem.*, 32(1):102–108 (1984).

Engelhardt, G. and P.R. Wallnöfer. "Microbial Transformation of Benzazimide, a Microbial Degradation Product of the Insecticide Azinphos-Methyl," *Chemosphere*, 12(7/8):955–960 (1983).

Engelhardt, G., P.R. Wallnöfer, and O. Hutzinger. "The Microbial Metabolism of Di-*n*-Butyl Phthalate and Related Dialkyl Phthalates," *Bull. Environ. Contam. Toxicol.*, 13(3):342–347 (1975).

Engelhardt, G., P.R. Wallnöfer, and R. Plapp. "Identification of *N,O*-Dimethylhydroxylamine as a Microbial Degradation Product of the Herbicide, Linuron," *Appl. Microbiol.*, 23:664–666 (1972).

Ernst, W. "Determination of the Bioconcentration Potential of Marine Organisms. A Steady State Approach," *Chemosphere*, 6(11):731–740 (1977).

Erstfeld, K.M., M.S. Simmons, and Y.H. Atallah. "Sorption and Desorption Characteristics of Chlordane onto Sediments," *J. Environ. Sci. Health*, B31(1):43–58 (1996).

Esser, H.O., G. Dupuis, E. Ebert, G. Marco, and C. Vogel. "*s*-Triazines" in *Herbicides: Chemistry, Degradation, and Mode of Action*. Volume 1, Kearney, P.C. and D.D. Kaufman, Eds., (New York: Marcel Dekker, Inc. 1975), pp. 129–208.

Estesen, B.J., N.A. Buck, and G.W. Ware. Dislodgable Insecticide Residues on Foliage: Carbaryl, Cypermethrin, and Methamidophos," *Bull. Environ. Contam. Toxicol.*, 28:490–493 (1982).

Fang, C.H. "Studies on the Degradation and Dissipation of Herbicide Alachlor on Soil Thin Layers," *J. Chin. Agric. Chem. Soc.*, 17:47–53 (1979).

Fang, C.H. "Studies on the Degradation and Dissipation of Herbicide Alachlor in Different Soils," *J. Chin. Agric. Chem. Soc.*, 21:25–29 (1983).

Farmer, W.J. and Y. Aochi. "Picloram Sorption by Soils," *Soil Sci. Soc. Am. Proc.*, 38:418–423 (1974).

Fathepure, B.Z., J.M. Tiedje, and S.A. Boyd. "Reductive Dechlorination of Hexachlorobenzene to Tri- and Dichlorobenzenes in Anaerobic Sewage Sludge," *Appl. Environ. Microbiol.*, 54(2):327–330 (1988).

Felsot, A. and P.A. Dahm. "Sorption of Organophosphorus and Carbamate Insecticides by Soil," *J. Agric. Food Chem.*, 27(3):557–563 (1979).

Felsot, A., J.V. Maddox, and W. Bruce. "Dissipation of Pesticide from Soils with Histories of Furadan Use," *Bull. Environ. Contam. Toxicol.*, 26(6):781–788 (1981).

Felsot, A.S. and W.L. Pedersen. "Pesticidal Activity of Degradation Products Insecticides by Soil" in *Pesticide Transformation Products, Fate and Significance in the Environment*, ACS Symposium Series 459, Somasundaram, L. and J.R. Coats, Eds. (Washington, DC: American Chemical Society, 1991), pp. 172–187.

Felsot, A. and J. Wilson. "Adsorption of Carbofuran and Movement on Soil Thin Layers," *Bull. Environ. Contam. Toxicol.*, 24(5):778–782 (1980).

Fendinger, N.J. and D.W. Glotfelty. "A Laboratory Method for the Experimental Determination of Air-Water Henry's Law Constants for Several Pesticides," *Environ. Sci. Technol.*, 22(11):1289–1293 (1988).

Feng, J.C. "Persistence, Mobility and Degradation of Hexazinone in Forest Silt Loam Soils," *J. Environ. Sci. Health*, B22(2):221–233 (1987).

Ferreira, G.A.L. and J.N. Sieber. "Volatilization and Exudation Losses of Three *N*-Methylcarbamate Insecticides Applied Systemically to Rice," *J. Agric. Food Chem.*, 29(1):93–99 (1981).

Ferris, I.G. and E.P. Lichtenstein. "Interactions between Agricultural Chemicals and Soil Microflora and Their Effects on the Degradation of [^{14}C]Parathion in a Cranberry Soil," *J. Agric. Food Chem.*, 28(5):1011–1019 (1980).

Feung, C.-S., R.H. Hamilton, and R.O. Mumma. "Metabolism of 2,4–Dichlorophenoxy-acetic Acid. V. Identification of Metabolites in Soybean Callus Tissue Cultures," *J. Agric. Food Chem.*, 21(4):637–640 (1973).

Feung, C.S, R.H. Hamilton, and F.H. Witham. "Metabolism of 2,4-Dichlorophenoxyacetic Acid by Soybean Cotyledon Callus Tissue Cultures," *J. Agric. Food Chem.*, 19(3):475–479 (1971).

Feung, C.S., R.H. Hamilton, F.H. Witham, and R.O. Mumma. "The Relative Amounts and Identification of Some 2,4-Dichlorophenoxyacetic Acid Metabolites Isolated from Soybean Cotyledon Callus Cultures," *Plant Physiol.*, 50:80–86 (1972).

Fleeker, J. and R. Steen. "Hydroxylation of 2,4-D in Several Weed Species," Weed Sci., 19:507 (1971).

Fogel, S., R. Lancione, A. Sewall, and R.S. Boething. "Enhanced Biodegradation of Methoxychlor in Soil Under Enhanced Environmental Conditions," *Appl. Environ. Microbiol.*, 44:113–120 (1982).

Forrest, M., K.A. Lord, and N. Walker. "The Influence of Soil Treatments on the Bacterial Degradation of Diazinon and other Organophosphorus Insecticides," *Environ. Pollut. Series A*, 24:93–104 (1981).

Foster, R.K. and R.B. McKercher. "Laboratory Incubation Studies of Chlorophenoxyacetic Acids in Chernozemic Soils," *Soil Biol. Biochem.*, 5:333–337 (1973).

Francis, A.J., R.J. Spanggord, and G.I. Ouchi. "Degradation of Lindane by *Escherichia coli*," *Appl. Environ. Microbiol.*, 29(4):567–568 (1975).

Frank, R., H.E. Braun, M. Van Hove Holdrimet, G.J. Sirons, and B.D. Ripley. "Agriculture and Water Quality in the Canadian Great Lakes Basin: V. Pesticide Use in 11 Agricultural Watersheds and Presence in Stream Water, 1975–1977," *J. Environ. Qual.*, 11(3):497–505 (1982).

Frank, P.A. and R.J. Demint. "Gas Chromatographic Analysis of Dalapon in Water," *Environ. Sci. Technol.*, 3(1):69–71 (1969).

Frankel, L.S., K.S. McCallum, and L. Collier. "Formation of Bis(chloromethyl) Ether from Formaldehyde and Hydrogen Chloride," *Environ. Sci. Technol.*, 8(4):356–359 (1974).

Franks, F. "Solute-Water Interactions and the Solubility Behaviour of Long-Chain Paraffin Hydrocarbons," *Nature (London)*, 210(5031):87–88 (1966).

Fredrickson, D.R. and P.J. Shea. "Effect of Soil pH on Degradation, Movement, and Plant Uptake of Chlorsulfuron," *Weed Sci.*, 34:328–332 (1986).

Freed, V.H., C.T. Chiou, and R. Haque. "Chemodynamics: Transport and Behavior of Chemicals in the Environment — A Problem in Environmental Health," *Environ. Health Perspect.*, 20:55–70 (1977).

Freed, V.H., C.T. Chiou, and D.W. Schmedding. "Degradation of Selected Organophosphate Pesticides in Water and Soil," *J. Agric. Food Chem.*, 27(4):706–708 (1979).

Freed, V.H., D. Schmedding, R. Kohnert, and R. Haque. "Physical Chemical Properties of Several Organophosphates: Some Implication in Environmental and Biological Behavior," *Pestic. Biochem. Physiol.*, 10:203–211 (1979a).

Freeman, P.K. and K.D. McCarthy. "Photochemistry of Oxime Carbamates. 1. Phototransformations of Aldicarb," *J. Agric. Food Chem.*, 32(4):873–877 (1984).

Freitag, D., L. Ballhorn, H. Geyer, and F. Korte. "Environmental Hazard Profile of Organic Chemicals," *Chemosphere*, 14(10):1589–1616 (1985).

Fries, G.F. "Degradation of Chlorinated Hydrocarbons Under Anaerobic Conditions," in *Fate of Organic Pesticides in the Aquatic Environment*, Advances in Chemistry Series, R.F. Gould, Ed. (Washington, DC: American Chemical Society, 1972), pp. 256–270.

Fries, G.R., G.S. Marrow, and C.H. Gordon. "Metabolism of *o,p'*- and *p,p'*-DDT by Rumen Microorganisms," *J. Agric. Food Chem.*, 17(4):860–862 (1969).

Friesen, K.J., L.P. Sarna, and G.R.B. Webster. "Aqueous Solubility of Polychlorinated Dibenzo-*p*-Dioxins Determined by High Pressure Liquid Chromatography," *Chemosphere*, 14(9):1267–1274 (1985).

Friesen, K.J., J. Vilk, and D.C.G. Muir. "Aqueous Solubilities of Selected 2,3,7,8-Substituted Polychlorinated Dibenzofurans (PCDFs)," *Chemosphere*, 20(1/2):27–32 (1980).

Fu, J.-K. and R.G. Luthy. "Aromatic Compound Solubility in Solvent/Water Mixtures," *J. Environ. Eng.*, 112(2):328–345 (1986).

Fuchs, A. and F.W. de Vries. "Bacterial Breakdown of Benomyl. II. Mixed Cultures," Antonie van Leeuwenhoek, 44(3/4):293–309 (1978).

Fuhremann, T.W. and E.P. Lichtenstein. "A Comparative Study of the Persistence, Movement, and Metabolism of Six Carbon-14 Insecticides in Soils and Plants," *J. Agric. Food Chem.*, 28(2):446–452 (1980).

Fukui, S., T. Hirayama, H. Shindo, and M. Nohara. "Photochemical Reaction of Biphenyl (BP) and *o*-Phenylphenol (OPP) with Nitrogen Monoxide (1)," *Chemosphere*, 9(12):771–775 (1980).

Fukuto, T.R. and R.L. Metcalf. "Metabolism of Insecticides in Plants and Animals," *Ann. NY Acad Sci.*, 160:97–111 (1969).

Funderburk, H.H., Jr. and G.A. Bozarth. "Review of the Metabolism and Decomposition of Diquat and Paraquat," *J. Agric. Food Chem.*, 15(4):563–567 (1967).

Gäb, S., H. Parlar, S. Nitz, K. Hustert, and F. Korte. "Photochemischer Abbau von Aldrin, Dieldrin und Photodieldrin als Festkorper im Sauerstoffstrom," *Chemosphere*, 3(5):183–186 (1974).

Gäb, S., V. Saravanja, and F. Korte. "Irradiation Studies of Aldrin and Chlordene Adsorbed on a Silica Gel Surface," *Bull. Environ. Contam. Toxicol.*, 13(3):301–306 (1975).

Gälli, R. and P.L. McCarty. "Biotransformation of 1,1,1-Trichloroethane, Trichloromethane, and Tetrachloromethane by a *Clostridium* sp.," *Appl. Environ. Microbiol.*, 55(4):837–844 (1989).

Gamar, Y. and J.K. Gaunt. "Bacterial Metabolism of 4-Chloro-2-methylphenoxyacetate, Formation of Glyphosate by Side-Chain Cleavage," *Biochem. J.*, 122:527–531 (1971).

Gambrell, R.P., B.A. Taylor, K.S. Reddy, and W.H. Patrick, Jr. "Fate of Selected Toxic Compounds Under Controlled Redox Potential and pH Conditions in Soil and Sediment-Water Systems," U.S. EPA Report-600/3-83-018 (1984).

Gamerdinger, A.P., A.T. Lemley, and R.J. Wagenet. "Nonequilibrium Sorption and Degradation of Three 2-Chloro-*s*-Triazine Herbicides in Soil-Water Systems," *J. Environ. Qual.*, 20(4):815–822 (1991).

Gannon, N. and G.C. Decker. "The Conversion of Heptachlor to Its Epoxide on Plants," *J. Econ. Entomol.*, 51(1):3–7 (1958).

Gardiner, J.A., R.C. Rhodes, J.B. Adams, Jr., and E.J. Soboczenski. "Synthesis and Studies with 2-^{14}C-Labeled Bromacil and Terbacil," *J. Agric. Food Chem.*, 17(5):980–986 (1969).

Gardner, A.M., J.N. Damico, E.A. Hansen, E. Lustig and R.W. Storherr. "Previously Unreported Homolog of Malathion Found as Residue on Crops," *J. Agric. Food Chem.*, 17(6):1181–1185 (1969).

Garner, W.Y. and R.E. Menzer. "Metabolism of *N*-Hydroxymethyl Dimethoate and *N*-Desmethyl Dimethoate in Bean Plants," *Pestic. Biochem. Physiol.*, 25:218–232 (1986).

Gaston, L.A., M.A. Locke, and R.M. Zablotowicz. "Sorption of Bentazon in Conventional and No-till Dundee Soil," *J. Environ. Qual.*, 25(1):120–126 (1996).

Gaughan, L.C. and J.E. Casida. "Degradation of *trans*- and *cis*-Permethrin on Cotton and Bean Plants," *J. Agric. Food Chem.*, 26(3):525–528 (1978).

Gaynor, J.D. and V.V. Volk. "Surfactant Effects on Picloram Adsorption by Soils," *Weed Sci.*, 24(6):549–552 (1976).

Gear, J.R., J.G. Michel, and R. Grover. "Photochemical Degradation of Picloram," *Pestic. Sci.*, 13(2):189–194 (1982).

Geissbühler, H., C. Haselbach, and H. Aebi. "The Fate of *N'*-(4-Chlorophenoxy)phenyl-*N,N*-dimethylurea. I. Adsorption and Leaching in Different Soils," *Weed Res.*, 3(2):140–153 (1963).

Geissbühler, H., C. Haselbach, H. Aebi, and L. Ebner. "The Fate of *N'*-(4-Chlorophenoxy)phenyl-*N,N*-dimethylurea. III. Breakdown in Soils and Plants," *Weed Res.*, 3(4):277–297 (1963a).

Geissbühler, H., H. Martin, and G. Voss. "The Substituted Ureas" in *Herbicides: Chemistry, Degradation, and Mode of Action*, Kearney, P.C. and D.D. Kaufman, Eds., (New York: Marcel Dekker, Inc., 1975), pp. 209–292.

Geller, A. "Studies on the Degradation of Atrazine by Bacterial Communities Enriched from Various Biotypes," *Arch. Environ. Contam. Toxicol.*, 9:289–305 (1980).

Georgacakis, E. and M.A.Q. Khan. "Toxicity of the Photoisomers of Cyclodiene Insecticides to Freshwater Animals," *Nature (London)*, 233(5315):120–121 (1971).

Gerstl, Z. and L. Kliger. "Fractionation of the Organic Matter in Soils and Sediments and Their Contribution to the Sorption of Pesticides," *J. Environ. Sci. Health*, B25(6):729–741 (1990).

Gerstl, Z. and U. Mingelgrin. "Sorption of Organic Substances by Soils and Sediments," *J. Environ. Sci. Health*, B19(3):297–312 (1984).

Gerstl, Z. and B. Yaron. "Behavior of Bromacil and Napropamide in Soils: I. Adsorption and Degradation," *Soil Sci. Soc. Am. J.*, 47(3):474–478 (1983).

Gerstl, Z. and B. Yaron. "Behavior of Bromacil and Napropamide in Soils: II. Distribution after Application from a Point Source," *Soil Sci. Soc. Am. J.*, 47(3):478–483 (1983a).

Getzin, L.W. "Metabolism of Diazinon and Zinophos in Soils," *J. Econ. Entomol.*, 60(2):505–508 (1967).

Getzin, L.W. "Persistence of Diazinon and Zinophos in Soil: Effects of Autoclaving, Temperature, Moisture, and Acidity," *J. Econ. Entomol.*, 61(6):1560–1565 (1968).

Getzin, L.W. "Persistence and Degradation of Carbofuran in Soil," *Environ. Entomol.*, 2:461–467 (1973).

Getzin, L.W. "Degradation of Chlorpyrifos in Soil: Influence of Autoclaving, Soil Moisture, and Temperature," *J. Econ. Entomol.*, 74(2):158–162 (1981).

Getzin, L.W. "Dissipation of Chlorpyrifos from Dry Soil Surfaces," *J. Econ. Entomol.*, 74(6):707–713 (1981a).

Getzin, L.W. "Factors Influencing the Persistence and Effectiveness of Chlorpyrifos in Soil," *J. Econ. Entomol.*, 78:412–418 (1985).

Getzin, L.W. and R.K. Chapman. "The Fate of Phorate in Soils," *J. Econ. Entomol.*, 53(1):47–51 (1960).

Getzin, L.W. and I. Rosefield. "Organophosphorus Insecticide Degradation by Heat-Labile Substances in Soil," *J. Agric. Food Chem.*, 16(4):598–601 (1968).

Getzin, L.W. and J.C.H. Shanks. "Persistence, Degradation of Bioactivity of Phorate and Its Oxidative Analog in Soil," *J. Econ. Entomol.*, 63(1):52–58 (1970).

Giardina, M.C., M.T. Giardi, and G. Filacchioni. "4-Amino-2-chloro-1,3,5–triazine: A New Metabolite of Atrazine by a Soil Bacterium," *Agric. Biol. Chem.*, 44:2067–2072 (1980).

Giardina, M.C., M.T. Giardi, and G. Filacchioni. "Atrazine Metabolism by Nocardia: Elucidation of Initial Pathway and Synthesis of Potential Metabolites," *Agric. Biol. Chem.*, 46:1439–1445 (1982).

Gibson, W.P. and R.G. Burns. "The Breakdown of Malathion in Soil and Soil Components," *Microbial Ecol.*, 3:219–230 (1977).

Gilbert, M. "Fate of Chlorothalonil in Apple Foliage and Fruit," *J. Agric. Food Chem.*, 24(5):1004–1007 (1976).

Gilderhus, P.A. "Effects of Diquat on Bluegills and Their Food Organisms," *Progr. Fish-Cult.*, 29(2):67–74 (1967).

Gile, J.D. and J.W. Gillet. "Fate of Selected Fungicides in a Terrestrial Laboratory Ecosystem," *J. Agric. Food Chem.*, 27:1159–1164 (1979).

Ginnings, P.M. and R. Baum. "Aqueous Solubilities of the Isomeric Pentanols," *J. Am. Chem. Soc.*, 59(6):1111–1113 (1937).

Ginnings, P.M. and D. Coltrane. "Aqueous Solubility of 2,2,3-Trimethylpentanol-3," *J. Am. Chem. Soc.*, 61(2):525 (1939).

Ginnings, P.M. and M. Hauser. "Aqueous Solubilities of Some Isomeric Heptanols," *J. Am. Chem. Soc.*, 60(11):2581–2582 (1938).

Ginnings, P.M., E. Herring, and D. Coltrane. "Aqueous Solubilities of Some Unsaturated Alcohols," *J. Am. Chem. Soc.*, 61(4):807–808 (1939).

Ginnings, P.M., D. Plonk, and E. Carter. "Aqueous Solubilities of Some Aliphatic Ketones," *J. Am. Chem. Soc.*, 62(8):1923–1924 (1940).

Ginnings, P.M. and R. Webb. "Aqueous Solubilities of Some Isomeric Hexanols," *J. Am. Chem. Soc.*, 60(8):1388–1389 (1938).

Given, C.J. and F.E. Dierberg. "Effect of pH on the Rate of Aldicarb Hydrolysis," *Bull. Environ. Contam. Toxicol.*, 34(5):627–633 (1985).

Glass, B.L. "Relation between the Degradation of DDT and the Iron Redox System in Soils," *J. Agric. Food Chem.*, 20:324–327 (1972).

Glass, R.L. "Adsorption of Glyphosate by Soils and Clay Minerals," *J. Agric. Food Chem.*, 35(4):497–500 (1987).

Glooschenko, V., M. Holdrinet, J.N.A. Lott, and R. Frank. "Bioconcentration of Chlordane by the Green Alga Scenedesmus quadricauda," *Bull. Environ. Contam. Toxicol.*, 21(4/5):515–520 (1979).

Glotfelty, D.E., M.S. Majewski, and J.N. Seiber. "Distribution of Several Organophosphorus Insecticides and Their Oxygen Analogues in a Foggy Atmosphere," *Environ. Sci. Technol.*, 24(3):353–357 (1990).

Glotfelty, D.E., A.W. Taylor, B.C. Turner, and W.H. Zoller. "Volatilization of Surface-Applied Pesticides from Fallow Soil," *J. Agric. Food Chem.*, 32:638–643 (1984).

Gohre, K. and G.C. Miller. "Photooxidation of Thioether Pesticides on Soil Surfaces," *J. Agric. Food Chem.*, 34(4):709–713 (1986).

Golab, T., W.A. Althaus, and H.L. Wooten. "Fate of [14]Trifluralin in Soil," *J. Agric. Food Chem.*, 27(1):163–179 (1979).

Golab, T., R.J. Herberg, J.V. Gramlich, A.P. Raun, and G.W. Probst. "Fate of Benefin in Soils, Plants, Artificial Rumen Fluid, and the Ruminant Animal," *J. Agric. Food Chem.*, 18:838–844 (1970).

Golab, T., R.J. Herberg, S.J. Parka, and J.B. Tepe. "The Metabolism of Carbon-14 Diphenamid in Strawberry Plants," *J. Agric. Food Chem.*, 14(6):593–596 (1966).

Golab, T., R.J. Herberg, S.J. Parka, and J.B. Tepe. "Metabolism of Carbon-14 Trifluralin in Carrots," *J. Agric. Food Chem.*, 15(4):638–641 (1967).

Golovleva, L.A., A.B. Polyakova, R.N. Pertsova, and Z.I. Finkelstein. "The Fate of Methoxychlor in Soils and Transformation by Soil Microorganisms," *J. Environ. Sci. Health*, B19:523–538 (1984).

Gomaa, H.M., I.H. Suffet, and S.D. Faust. "Kinetics of Hydrolysis of Diazinon and Diazoxon," *Residue Rev.*, 29:171–190 (1969).

Gomez, J., C. Bruneau, N. Soyer, and A. Brault. "Identification of Thermal Degradation Products from Diuron and Iprodione," *J. Agric. Food Chem.*, 30(1):180–182 (1982).

González, V., J.H. Ayala, and A.M. Afonso. "Degradation of Carbaryl in Natural Waters: Enhanced Hydrolysis Rate in Micellar Solution," *Bull. Environ. Contam. Toxicol.*, 42(2):171–178 (1992).

Gorder, G.W., P.A. Dahm, and J.J. Tollefson. "Carbofuran Persistence in Cornfield Soils," *J. Econ. Entomol.*, 75(4):637–642 (1982).

Gore, R.C., R.W. Hannah, S.C. Pattacini, and T.J. Porro. "Infrared and Ultraviolet Spectra of Seventy-Six Pesticides," *J. Assoc. Off. Anal. Chem.*, 54(5):1040–1082 (1971).

Goring, C.A.I. "Control of Nitrification by 2-Chloro-6-(trichloromethyl) pyridine," *Soil Sci.*, 93:211–218 (1962).

Goring, C.A.I. and J.W. Hamaker. "Degradation and Movement of Picloram in Soil and Water," *Down to Earth*, 20(4):3–5 (1971).

Gormley, O.R. and R.F. Spalding. "Sources and Concentrations of Nitrate-Nitrogen in the Groundwater of the Central Platte Region, Nebraska," *Groundwater*, 17(3):291–301 (1979).

Goswami, K.P. and R.E. Green. "Microbial Degradation of the Herbicide Atrazine and Its 2-Hydroxy Analog in Submerged Soils," *Environ. Sci. Technol.*, 5(5):426–429 (1971).

Gowda, T.K.S. and N. Sethunathan. "Persistence of Endrin in Indian Rice Soils Under Flooded Conditions," *J. Agric. Food Chem.*, 24(4):750–753 (1976).

Gowda, T.K.S. and N. Sethunathan. "Endrin Decomposition in Soils as Influenced by Aerobic and Anaerobic Conditions," *Soil Sci.*, 125(1):5–9 (1977).

Graham, R.E., K.R. Burson, C.F. Hammer, L.B. Hansen, and C.T. Kenner. "Photochemical Decomposition of Heptachlor Epoxide," *J. Agric. Food Chem.*, 21(5):824–834 (1973).

Graham-Bryce, I.J. "Adsorption of Disulfoton by Soil," *J. Sci. Food Agric.*, 18:72–77 (1967).

Grayson, B.T. "Hydrolysis of Cyanazine and Related Diaminochloro-1,3,5-triazines; Hydrolysis in Sulfuric Acid Solutions," *Pestic. Sci.*, 11(5):493–505 (1980).

Green, R.E. and S.R. Obien. "Herbicide Equilibrium in Soils in Relation to Soil Water Content," *Weed Sci.*, 17(4):514–519 (1969).

Greenhalgh, R. and A. Belanger. "Persistence and Uptake of Carbofuran in a Humic Mesisol and the Effects of Drying and Storing Soil Samples on Residue Levels," *J. Agric. Food Chem.*, 29:231–235 (1981).

Greve, P.A. and S.L. Wit. "Endosulfan in the Rhine River," *J. Water Pollut. Control Fed.*, 43:2338–2348 (1971).

Gross, P.M. and J.H. Saylor. "The Solubilities of Certain Slightly Soluble Organic Compounds in Water," *J. Am. Chem. Soc.*, 53(5):1744–1751 (1931).

Gross, P.M., J.H. Saylor, and M.A. Gorman. "Solubility Studies. IV. The Solubilities of Certain Slightly Soluble Organic Compounds in Water," *J. Am. Chem. Soc.*, 55(2):650–652 (1933).

Grover, R. "Adsorption of Picloram by Soil Colloids and Various Other Adsorbants," *Weed Sci.*, 19(4):417–418 (1971).

Grover, R. "Adsorption and Desorption of Urea Herbicides in Soils," *Can. J. Soil Sci.*, 55:127–135 (1975).

Grover, R. "Mobility of Dicamba, Picloram, and 2,4-D in Soil Columns," *Weed Sci.*, 25(2):159–162 (1977).

Grover, R. and R.J. Hance. "Adsorption of Some Herbicides by Soil and Roots," *Can. J. Plant Sci.*, 49:378–380 (1969).

Grover, R. and A.E. Smith. "Adsorption Studies with the Acid and Dimethylamine Forms of 2,4-D and Dicamba," *Can. J. Soil Sci.*, 54:179–186 (1974).

Grunwell, J.R. and R.H. Erickson. "Photolysis of Parathion (O,O-Diethyl-O-(4-nitrophenyl)thiophosphate). New Products," *J. Agric. Food Chem.*, 21(5):929–931 (1973).

Grzenda, A.R., H.P. Nicholson, and W.S. Cox. "Persistence of Four Herbicides in Pond Water," *J. Am. Water Works Assoc.*, 58:326–332 (1966).

Guenzi, W.D. and W.E. Beard. "Anaerobic Biodegradation of DDT to DDD in Soil," *Science (Washington, DC)*, 156(3778):1116–1117 (1967).

Guenzi, W.D. and W.E. Beard. "DDT Degradation in Soil as Related to Temperature," *J. Environ. Qual.*, 5:391–394 (1976).

Guenzi, W.D. and W.E. Beard. "Picloram Degradation in Soils as Influenced by Soil Water Content and Temperature," *J. Environ. Qual.*, 5:185–192 (1976a).

Guenzi, W.D., W.E. Beard, and F.G. Viets, Jr. "Influence of Soil Treatment on Persistence of Six Chlorinated Hydrocarbon Insecticides in the Field," *Soil Sci. Soc. Am. Proc.*, 35:910–913 (1971).

Gysin, H. and A. Margot. "Chemistry and Toxicological Properties of O,O-Diethyl-O-(2-isopropyl-4-methyl-6-pyrimidinyl) Phosphorothioate (Diazinon)," *J. Agric. Food Chem.*, 6(12):900–903 (1958).

Haderlein, S.B. and R.P. Schwarzenbach. "Adsorption of Substituted Nitrobenzenes and Nitrophenols to Mineral Surfaces," *Environ. Sci. Technol.*, 27(2):316–326 (1993).

Hafkenscheid, T.L. and E. Tomlinson. "Estimation of Aqueous Solubilities of Organic Non-Electrolytes Using Liquid Chromatographic Retention Data," *J. Chromatogr.*, 218:409–425 (1981).

Hafkenscheid, T.L. and E. Tomlinson. "Isocratic Chromatographic Retention Data for Estimating Aqueous Solubilities of Acidic, Basic and Neutral Drugs," *Int. J. Pharm.*, 17:1–21 (1983).

Hagin, R.D., D.L. Linscott, and J.E. Dawson. "2,4-D Metabolism in Resistant Grasses," *J. Agric. Food Chem.*, 18(5):848–850 (1970).

Hall, R.C., C.S. Giam, and M.G. Merkle. "The Photolytic Degradation of Picloram," *Weed Res.*, 8(4):292–297 (1968).

Hall, J.K. and N.L. Hartwig. "Atrazine Mobility in Two Soils Under Conventional Tillage," *J. Environ. Qual.*, 7(1):63–68 (1978).

Hamilton, R.H., J. Hurter, J.K. Hall, and C.D. Ercegovich. "Metabolism of 2,4-Dichlorophenoxyacetic Acid and 2,4,5-Trichlorophenoxyacetic Acid on Bean Plants," *J. Agric. Food Chem.*, 19(3):480–483 (1971).

Hance, R.J. "The Adsorption of Urea and Some of Its Derivatives by a Variety of Soils," *Weed Res.*, 5(1):98–107 (1965).

Hance, R.J. "The Speed of Attainments of Sorption Equilibria in Some Systems Involving Herbicides," *Weed Res.*, 7(1):29–36 (1967).

Hance, R.J. "Decomposition of Herbicides in the Soil by Nonbiological Chemical Processes," *J. Sci. Food Agric.*, 18:544–547 (1967a).

Hance, R.J. "Soil Organic Matter and the Adsorption and Decomposition of the Herbicides Atrazine and Linuron," *Soil Biol. Biochem.*, 6:39–42 (1974).

Haniff, M.I., R.H. Zienius, C.H. Langford, and D.S. Gamble. "The Solution Phase Complexing of Atrazine by Fulvic Acid: Equilibria at 25°C," *J. Environ. Sci. Health*, B20:215–262 (1985).

Hansen, J.L. and M.H. Spiegel. "Hydrolysis Studies of Aldicarb, Aldicarb Sulfoxide and Aldicarb Sulfone," *Environ. Toxicol. Chem.*, 2(2):147–153 (1983).

Hanst, P.L. and B.W. Gay, Jr. "Photochemical Reactions Among Formaldehyde, Chlorine, and Nitrogen Dioxide in Air," *Environ. Sci. Technol.*, 11(12):1105–1109 (1977).

Hapeman-Somich, C.J. "Mineralization of Pesticide Degradation Products" in *Pesticide Transformation Products, Fate and Significance in the Environment*, ACS Symposium Series 459, Somasundaram, L. and J.R. Coats, Eds. (Washington, DC: American Chemical Society, 1991), pp. 133–147.

Haque, R. and R. Sexton. "Kinetic and Equilibrium Study of the Adsorption of 2,4-Dichlorophenoxy Acetic Acid on Some Surfaces," *J. Coll. Interface Sci.*, 27(4):818–827 (1968).

Hargrove, R.S. and M.G. Merkle. "The Loss of Alachlor from Soil," *Weed Sci.*, 19(6):652–654 (1971).

Harris, C.I. "Adsorption, Movement, and Phytotoxicity of Monuron and *s*-Triazine Herbicides in Soils," *Weeds*, 14:6–10 (1966).

Harris, C.I. "Fate of 2-Chloro-s-triazine Herbicides in Soil," *J. Agric. Food Chem.*, 15(1):157–162 (1967).

Harris, C.I. "Movement of Pesticides in Soil," *J. Agric. Food Chem.*, 17(1):80–82 (1969).

Harris, C.I., R.A. Chapman, C. Harris, and C.M. Tu. "Biodegradation of Pesticides in Soil: Rapid Induction of Carbamate Degrading Factors After Carbofuran Treatment," *J. Environ. Sci. Health*, B19(1):1–11 (1984).

Harris, C.I., D.D. Kaufman, T.J. Sheets, R.G. Nash, and P.C. Kearney. "Behavior and Fate of *s*-Triazines in Soil," *Adv. Pest Control Res.*, 8:1–55 (1968).

Harris, C.I. and E.P. Lichtenstein. "Factors Affecting the Volatilization of Insecticidal Residues from Soils," *J. Econ. Entomol.*, 54:1038–1045 (1961).

Harris, C.I. and T.J. Sheets. "Influence of Soil Properties on Adsorption and Phytotoxicity of CIPC, Diuron, and Simazine," *Weeds*, 13:215–219 (1965).

Harris, C.I., E.A. Woolson, and B.E. Hummer. "Dissipation of Herbicides at Three Soil Depths," *Weed Sci.*, 17(1):27–31 (1969).

Harrison, R.B., D.C. Holmes, J. Roburn, and J.O'G. Tatton. "The Fate of Some Organochlorine Pesticides on Leaves," *J. Sci. Food Agric.*, 18:10–15 (1967).

Hartley, D.M. and J.B. Johnston. "Use of the Freshwater Clam *Corbicula manilensis* as a Monitor for Organochlorine Pesticides," *Bull. Environ. Contam. Toxicol.*, 31(1):33–40 (1983).

Hartley, D. and H. Kidd, Eds. *The Agrochemicals Handbook*, 2nd ed. (England: Royal Society of Chemistry, 1987).

Harvey, R.G. "Soil Adsorption and Volatility of Dinitroaniline Herbicides," *Weed Sci.*, 22(2):120–124 (1974).

Harvey, J., Jr., J.J. Dulka, and J.J. Anderson. "Properties of Sulfometuron Methyl Affecting Its Environmental Fate: Aqueous Hydrolysis and Photolysis, Mobility and Adsorption on Soils, and Bioaccumulation Potential," *J. Agric. Food Chem.*, 33(4):590–596 (1985).

Harvey, J., Jr. and J.C.-Y. Han. "Decomposition of Oxamyl in Soil and Water," *J. Agric. Food Chem.*, 26(3):536–541 (1978).

Harvey, J., Jr. and H.L. Pease. "Decomposition of Methomyl in Soil," *J. Agric. Food Chem.*, 21(5):784–786 (1973).

Hashimoto, Y., K. Tokura, H. Kishi, and W.M.J. Strachan. "Prediction of Sea-water Solubility of Aromatic Compounds," *Chemosphere*, 13(8):881–888 (1984).

Hashimoto, Y., K. Tokura, K. Ozaki, and W.M.J. Strachan. "A Comparison of Water Solubilities by the Flask and Micro-Column Methods," *Chemosphere*, 11(10):991–1001 (1982).

Hatzios, K.K. "Biotransformations of Herbicides in Higher Plants" in *Environmental Chemistry of Herbicides*. Volume II, Grover, R. and A.J. Cessna, Eds. (Boca Raton, FL: CRC Press, Inc., 1991), pp. 141–185.

Hazardous Substances Data Bank. National Library of Medicine, Toxicology Information Program (1989).

Helling, C.S. "Pesticide Mobility in Soils II. Applications of Soil Thin-Layer Chromatography," *Soil Sci. Soc. Am. Proc.*, 35:737–743 (1971).

Helweg, A. "Microbial Breakdown of the Fungicide Benomyl," *Soil Biol. Biochem.*, 4:377–378 (1972).

Henderson, G.L. and D.G. Crosby. "The Photodecomposition of Dieldrin Residues in Water," *Bull. Environ. Contam. Toxicol.*, 3(3):131–134 (1968).

Henderson, C., Q.H. Pickering, and C.M. Tarzwell. "Relative Toxicity of Ten Chlorinated Hydrocarbon Insecticides to Four Species of Fish," *Trans. Am. Fish Soc.*, 88(1):23–32 (1959).

Heritage, A.D. and I.C. MacRae. "Degradation of Lindane by Cell-free Preparations of Clostridium sphenoides," *Appl. Environ. Microbiol.*, 34(2):222–224 (1977).

Heritage, A.D. and I.C. MacRae. "Identification of Intermediates Formed During the Degradation of Hexachlorocyclohexanes by Clostridium sphenoides," *Appl. Environ. Microbiol.*, 33(6):1295–1297 (1977a).

Herrett, R.A. and W.P. Bagley. "The Metabolism and Translocation of 3-Amino-1,2,4-triazole by Canada Thistle," *J. Agric. Food Chem.*, 12(1):17–20 (1964).

Herrmann, M., D. Kotzias, and F. Korte. "Photochemical Behavior of Chlorsulfuron in Water and In Adsorbed Phase," *Chemosphere*, 14(1):3–8 (1985).

Hill, D.W. and P.L. McCarty. "Anaerobic Degradation of Selected Chlorinated Hydrocarbon Pesticides," *J. Water Pollut. Control Fed.*, 39:1259–1277 (1967).

Hill, G.D., J.W. McGahen, H.M. Baker, D.W. Finnerty, and C.W. Bingeman. "The Fate of Substituted Urea Herbicides in Agricultural Soils," *Agron. J.*, 47:93–104 (1955).

Hiltbold, A.E. and G.A. Buchanan. "Influence of Soil pH on Persistence of Atrazine in the Field," *Weed Sci.*, 25(6):515–520 (1977).

Hinckley, D.A., T.F. Bidleman, W.T. Foreman, and J.R. Tuschall. "Determination of Vapor Pressures for Nonpolar and Semipolar Organic Compounds from Gas Chromatographic Retention Data," *J. Chem. Eng. Data*, 35(3):232–237 (1990).

Hine, J., H.W. Haworth, and O.B. Ramsey. "Polar Effects on Rates and Equilibria. VI. The Effect of Solvent on the Transmission of Polar Effects," *J. Am. Chem. Soc.*, 85(10):1473–1475 (1963).

Hine, J. and R.D. Weimar, Jr. "Carbon Basicity," *J. Am. Chem. Soc.*, 87(15):3387–3396 (1965).

Hirsch, M. and O. Hutzinger. "Naturally Occurring Proteins from Pond Water Sensitize Hexachlorobenzene Photolysis," *Environ. Sci. Technol.*, 23(10):1306–1307 (1989).

Hoagland, R.E. "Effects of Glyphosate on Metabolism of Phenolic Compounds. VI. Effects of Glyphosine and Glyphosate Metabolites on Phenylalanine Ammonia-Lyase Activity, Growth, and Protein, Chlorophyll, and Anthocuanin Levels in Soybean (Glycine mas) Seedlings Phytotoxicity," *Weed Sci.*, 28:393–400 (1980).

Hoffman, J. and D. Eichelsdoerfer. "Effect of Ozone on Chlorinated-Hydrocarbon-Group Insecticides in Water," *Vom. Wasser*, 38:197–206 (1971).

Hoffman, I., E.V. Parups, and R.B. Carson. "Analysis for Maleic Hydrazide," *J. Agric. Food Chem.*, 10(6):453–455 (1962).

Hollifield, H.C. "Rapid Nephelometric Estimate of Water Solubility of Highly Insoluble Organic Chemicals of Environmental Interest," *Bull. Environ. Contam. Toxicol.*, 23(4/5):579–586 (1979).

Holmstead, R.L, J.E. Casida, L.O. Ruzo, and D.G. Fullmer. "Pyrethroid Photodecomposition: Permethrin," *J. Agric. Food Chem.*, 26(3):590–595 (1978).

Holt, R.F. "Determination of Hexazinone and Metabolite Residues Using Nitrogen-Selective Gas Chromatography," *J. Agric. Food Chem.*, 29(1):165–172 (1981).

Hommelen, J.R. "The Elimination of Errors Due to Evaporation of the Solute in the Determination of Surface Tensions," *J. Colloid Sci.*, 34:385–400 (1959).

Honeycutt, R., H. Ballantine, H. LeBaron, D. Paulson, and V. Siem. "Degradation of High Concentrations of a Phosphorothioic Ester by Hydrolase" in *Treatment and Disposal of Pesticide Wastes*, R.F. Krueger and J.N. Seiber, Eds., (Washington, DC: ACE, 1984).

Horvath, R.S. "Microbial Cometabolism of 2,4,5-Trichlorophenoxyacetic Acid," *Bull. Environ. Contam. Toxicol.*, 5(6):537–541 (1970).

Horvath; R.S. "Co-metabolism of Methyl- and Chloro-substituted Catechols by an *Achromobacter* sp. Possessing a New meta-Cleavage Oxygenase," *Biochem. J.*, 119:871–876 (1970a).

Howard, P.H., R.S. Boethling, W.F. Jarvis, W.M. Meylan, and E.M. Michalenko. *Handbook of Environmental Degradation Rates* (Chelsea, MI: Lewis Publishers, Inc., 1991).

Hua, Z., Z. Manping, X. Zongfeng, and G.K.-C. Low. "Titanium Dioxide Mediated Photocatalytic Degradation of Monocrotophos," *Water Res.*, 29(12):2681–2688 (1995).

Huddleston, E.W. and G.G. Gyrisco. "Residues of Phosdrin on Alfalfa and Its Effectiveness on the Insect Complex," *J. Econ. Entomol.*, 54:209–210 (1961).

Humburg, N.E., S.R. Colby, R.G. Lym, E.R. Hill, W.J. McAvoy, L.M. Kitchen, and R. Prasad, Eds. *Herbicide Handbook of the Weed Science Society of America*, 6th ed. (Champaign, IL: Weed Science Society of America, 1989), 301 p.

Hustert, K. and P.N. Moza. "Photokatalytischer Abbau von Phthalaten an Titandioxid in Wässriger Phase," *Chemosphere*, 17(9):1751–1754 (1988).

Hwang, H.-M., R.E. Hodson, and R.F. Lee. "Degradation of Phenol and Chlorophenols by Sunlight and Microbes in Estuarine Water," *Environ. Sci. Technol.*, 20(10):1002–1007 (1986).

Ibrahim, F.B., J.M. Gilbert, R.T. Evans, and J.C. Cavagnol. "Decomposition of Di-Syston (*O,O*-Diethyl *S*-[2-(ethylthio)ethyl] Phosphorodithioate) on Fertilizers by Infrared, Gas-Liquid Chromatography, and Thin-Layer Chromatography," *J. Agric. Food Chem.*, 17(2):300–305 (1969).

Ide, A., Y. Ueno, K. Takaichi, and H. Watanabe. "Photochemical Reaction of 2,3-Dichloro-1,4-naphthoquinone (Dichlone) in the Presence of Oxygen," *Agric. Biol. Chem.*, 43(7):1387–1394 (1979).

Iizuka, H. and T. Masuda. "Residual Fate of Chlorphenamidine in Rice Plant and Paddy Soil," *Bull. Environ. Contam. Toxicol.*, 22(6):745–749 (1979).

Indira, K., S.C. Vyas, R.K. Verma, and A. Jain. "Translocation and Persistence of Acylon in Pigeon Pea," *Indian Phytopathol.*, 34:492–493 (1981).

Ishihara, H. "Photodimerization of Thymine and Ruacil on Filter Paper," *Photochem. Photobiol.*, 2:455–460 (1963).

Ismail, B.S. and H.J. Lee. "Persistence of Metsulfuronmethyl in Two Soils," *J. Environ. Sci. Health*, B30(4):485–497 (1995).

Ivie, G.W. and D.L. Bull. "Photodegradation of *O*-Ethyl *O*-[4-(Methylthio)phenyl] *S*-Propyl Phosphorodithioate (BAY NTN 9306)," *J. Agric. Food Chem.*, 24(5):1053–1057 (1976).

Ivie, G.W., D.L. Bull, and J.A. Veech. "Fate of Diflubenzuron in Water," *J. Agric. Food Chem.*, 28:330–337 (1980).

Ivie, G.W. and J.E. Casida. "Enhancement of Photoalteration of Cyclodiene Insecticide Chemical Residues by Rotenone," *Science (Washington, DC)*, 167(3925):1620–1622 (1970).

Ivie, G.W. and J.E. Casida. "Photosensitizers for the Accelerated Degradation of Chlorinated Cyclodienes and Other Insecticide Chemicals Exposed to Sunlight on Bean Leaves," *J. Agric. Food Chem.*, 19(3):410–416 (1971).

Ivie, G.W. and J.E. Casida. "Sensitized Photodecomposition and Photosensitizer Activity of Pesticide Chemicals Exposed to Sunlight on Silica Gel Chromatoplates," *J. Agric. Food Chem.*, 19(3):405–409 (1971a).

Ivie, G.W., J.R. Knox, S. Khalifa, I. Yamamoto, and J.E. Casida. "Novel Photoproducts of Heptachlor Epoxide, *trans*-Chlordane, and *trans*-Nonachlor," *Bull. Environ. Contam. Toxicol.*, 7(6):376–383 (1972).

Iwata, Y., G.E. Carman, T.M. Dinoff, and F.A. Gunther. "Metaldehyde Residues on and in Citrus Fruits after a Soil Broadcast of a Granular Formulation and after a Spray Application to Citrus Trees," *J. Agric. Food Chem.*, 30:606–608 (1982).

Jamet, P. and M.A. Piedallu. "Mouvement du Carbofuran dans Different Types de Sols," Phytiatrie-Phytopharmacie, 24:279–296 (1975).

Janko, Z., H. Stefaniak, and E. Czerwinska. "Microbiological Decomposition of Diuron [*N,N*-Dimethyl-*N*'-(3,4-dichlorophenyl)urea]," in Chem. Abstr. 78:25271n, *Pr. Inst. Przem. Org.*, 2:241–257 (1970).

Janssen, D.B., D. Jager, and B. Wilholt. "Degradation of *n*-Haloalkanes and α,ω-Dihaloalkanes by Wild Type and Mutants of *Acinetobacter* sp. Strain GJ70," *Appl. Environ. Microbiol.*, 53(3):561–566 (1987).

Jensen, S., R. Göthe, and M.-O. Kindstedt. "Bis(*p*-chlorophenyl)acetonitrile (DDN), a New DDT Derivative formed in Anaerobic Digested Sewage Sludge and Lake Sediment," *Nature (London)*, 240(5381):421–422 (1972).

Jensen-Korte, U., C. Anderson, and M. Spiteller. "Photodegradation of Pesticides in the Presence of Humic Substances," *Sci. Tot. Environ.*, 62:335–340 (1987).

Jenson, K.I.N. and E.R. Kimball. "Tolerance and Residues of Hexazinone in Lowbush Blueberries," *Can. J. Plant Sci.*, 65:223–227 (1985).

Jessee, J.A., R.E. Benoit, A.C. Hendricks, G.C. Allen, and J.L. Neal. "Anaerobic Degradation of Cyanuric Acid, Cysteine, and Atrazine by a Facultative Anaerobic Bacterium," *Appl. Environ. Microbiol.*, 45:97–102 (1983).

Johansen, C.A., W.E. Westlake, L.I. Butler, and R.E. Bry. "Residual Action and Toxicity of Methoxychlor and Parathion to the Cherry Fruit Fly," *J. Econ. Entomol.*, 47:746–749 (1954).

Johnsen, R.E. "DDT Metabolism in Microbial Systems," *Residue Rev.*, 61:1–28 (1976).

Johnson, D.P. and H.A. Stansbury. "Adaption of Sevin Insecticide (Carbaryl) Residue Method to Various Crops," *J. Agric. Food Chem.*, 13:235–238 (1965).

Johnson-Logan, L.R., R.E. Broshears, and S.J. Klaine. "Partitioning Behavior and the Mobility of Chlordane in Groundwater," *Environ. Sci. Technol.*, 26(11):2234–2239 (1992).

Joiner, R.L. and K.P. Baetcke. "Parathion: Persistence on Cotton and Identification of Its Photoalteration Products," *J. Agric. Food Chem.*, 21(3):391–396 (1973).

Joiner, R.L., H.W. Chambers, and K.P. Baetcke. "Toxicity of Parathion and Several of Its Photoalteration Products to Boll Weevils," *Bull. Environ. Contam. Toxicol.*, 6(3):220–224 (1971).

Jones, A.S. "Metabolism of Aldicarb by Five Soil Fungi," *J. Agric. Food Chem.*, 24(1):115–117 (1976).

Jones, R.L., J.L. Hansen, R.R. Romine, and T.E. Marquardt. "Unsaturated Zone Studies of the Degradation and Movement of Aldicarb and Aldoxycarb Residues," *Environ. Toxicol. Chem.*, 5:361–372 (1986).

Jones, T.W., W.M. Kemp, J.C. Stevenson, and J.C. Means. "Degradation of Atrazine in Estuarine Water/Sediment Systems and Soils," *J. Environ. Qual.*, 11(4):632–638 (1982).

Joshi, O.P. and N.P. Datta. "Persistence and Movement of Simazine in Soil," *J. Indian Soil Sci.*, 23:263–265 (1975).

Junk, G.A., R.F. Spalding, and J.J. Richard. "Areal, Vertical, and Temporal Differences in Ground Water Chemistry: II. Organic Constituents," *J. Environ. Qual.*, 9(3):479–483 (1980).

Jury, W.A., H. Elabd, and M. Resketo. "Field Study of Napropamide Movement Through Unsaturated Soil," *Water Resour. Res.*, 22(5):749–755 (1986).

Jury, W.A., D.D. Focht, and W.J. Farmer. "Evaluation of Pesticide Pollution Potential from Standard Indices of Soil-Chemical Adsorption and Biodegradation," *J. Environ. Qual.*, 16(4):422–428 (1987).

Kaars Sijpesteijn, A., H.M. Dekhuijzen, and J.W. Vonk. "Biological Conversion of Fungicides in Plants and Microorganisms" in *Antifungal Compounds*, (New York: Marcel Dekker, Inc., 1977), pp. 91–147.

Kale, S.P., N.B.K. Murthy, and K. Raghu. "Effect of Carbofuran, Carbaryl, and Their Metabolites on the Growth of *Rhizobium* sp. and *Azotobacter chroococcum*," *Bull. Environ. Contam. Toxicol.*, 42(5):769–772 (1989).

Kalkat, G.S., R.H. Davidson, and C.L. Brass. "Effect of Controlled Temperatures and Humidity on the Residual Life of Certain Insecticides," *J. Econ. Entomol.*, 54:1186–1190 (1961).

Kanazawa, J. "Biodegradability of Pesticides in Water by Microbes in Activated Sludge," *Environ. Monitor. Assess.*, 9:57–70 (1987).

Kanazawa, J. "Bioconcentration Ratio of Diazinon by Freshwater Fish and Snail," *Bull. Environ. Contam. Toxicol.*, 20(5):613–617 (1978).

Kanazawa, J. "Relationship between the Soil Sorption Constants for Pesticides and Their Physiochemical Properties," *Environ. Toxicol. Chem.*, 8(6):477–484 (1989).

Kaneda, Y., K. Nakamura, H. Nakahara, and M. Iwaida. "Degradation of Organochlorine Pesticides with Calcium Hypochlorite," in Chem. Abstr. 82:94190e, *I. Eisei Kagaku*, 20:296–299 (1974).

Kansouh, A.S.H. and T.L. Hopkins. "Diazinon Absorption, Translocation, and Metabolism in Bean Plants," *J. Agric. Food Chem.*, 16(3):446–450 (1968).

Kapoor, I.P., R.L. Metcalf, R.F. Nystrom, and G.K. Sangha. "Comparative Metabolism of Methoxychlor, Methiochlor, and DDT in Mouse, Insects, and in a Model Ecosystem," *J. Agric. Food Chem.*, 18(6):1145–1152 (1970).

Karickhoff, S.W., D.S. Brown, and T.A. Scott. "Sorption of Hydrophobic Pollutants on Natural Sediments," *Water Res.*, 13(3):241–248 (1979).

Karinen, J.F., J.G. Lamberton, N.E. Stewart, and L.C. Terriere. "Persistence of Carbaryl in the Marine Estuarine Environment. Chemical and Biological Stability in Aquarium Systems," *J. Agric. Food Chem.*, 15(1):148–156 (1967).

Katan, J., T.W. Fuhremann, and E.P. Lichtenstein. "Binding of [^{14}C] Parathion in Soil: A Reassessment by Pesticide Persistence," *Science (Washington, DC)*, 193:891–894 (1976).

Katz, M. "Acute Toxicity of Some Organic Insecticides to Three Species of Salmonids and to Threespine Stickleback," *Trans. Am. Fish Soc.*, 90(3):264–268 (1961).

Kaufman, D.D. "Degradation of Carbamate Herbicides in Soil," *J. Agric. Food Chem.*, 15(4):582–591 (1967).

Kaufman, D.D. and P.C. Kearney. "Microbial Degradation of *s*-Triazine Herbicides," *Residue Rev.*, 32:235–265 (1970).

Kaufman, D.D., J.R. Plimmer, P.C. Kearney, J. Blake, and F.S. Guardia. "Chemical Versus Microbial Decomposition of Amitrole in Soil," *Weed Sci.*, 16(1):266–272 (1968).

Kaufman, D.D., B.A. Russell, C.S. Helling, and A.J. Kayser. "Movement of Cypermethrin, Decamethrin, Permethrin, and Their Degradation Products in Soil," *J. Agric. Food Chem.*, 29(2):239–245 (1981).

Kauwatsuka, S. and M. Igarashi. "Degradation of PCP in Soils. II. The Relationship between the Degradation of PCP and the Properties of Soils, and Identification of the Degradation Products of PCP," *Soil Sci. Plant Nutr.*, 21:405–411 (1975).

Kawasaki, M. "Experiences with the Test Scheme Under the Chemical Control Law of Japan: An Approach to Structure-Activity Correlations," *Ecotoxicol. Environ. Safety*, 4:444–454 (1980).

Kay, B.D. and D.E. Elrick. "Adsorption and Movement of Lindane in Soils," *Soil Sci.*, 104(5):314–322 (1967).

Kazano, H., P.C. Kearney, and D.D. Kaufman. "Metabolism of Methylcarbamate Insecticides in Soils," *J. Agric. Food Chem.*, 20(5):975–979 (1972).

Kearney, P.C. and D.D. Kaufman. *Herbicides: Chemistry, Degradation and Mode of Action* (New York: Marcel Dekker, Inc., 1976), 1036 p.

Kearney, P.C., J.R. Plimmer, W.B. Wheeler, and A. Konston. "Persistence and Metabolism of Dinitroaniline Herbicides in Soils," *Pest. Biochem. Physiol.*, 6:229–238 (1976).

Kearney, P.C., E.A. Woolson, J.R. Plimmer, and A.R. Isensee. "Decontamination of Pesticide Residues in Soils," *Residue Rev.*, 29:137–149 (1969).

Keith, L.H. and D.B. Walters. *The National Toxicology Program's Chemical Data Compendium — Volume II. Chemical and Physical Properties* (Chelsea, MI: Lewis Publishers, Inc., 1992), 1642 p.

Kenaga, E.E. "Toxicological and Residue Data Useful in the Environmental Safety Evaluation of Dalapon," *Residue Rev.*, 53:109–151 (1974).

Kennedy, M.V., B.J. Stojanovic, and F.L. Shuman, Jr. "Chemical and Thermal Aspects of Pesticide Disposal," *J. Environ. Qual.*, 1(1):63–65 (1972).

Kennedy, M.V., B.J. Stojanovic, and F.L. Shuman, Jr. "Analysis of Decomposition Products of Pesticides," *J. Agric. Food Chem.*, 20(2):341–343 (1972a).

Khalifa, S., R.L. Holmstead, and J.E. Casida. "Toxaphene Degradation by Iron(II) Protoporphyrin Systems," *J. Agric. Food Chem.*, 24(2):277–282 (1976).

Khalil, M.A.K. and R.A. Rasmussen. "Global Sources, Lifetimes and Mass Balances of Carbonyl Sulfide (OCS) and Carbon Disulfide (CS2) in the Earth's Atmosphere," *Atmos. Environ.*, 18(9):1805–1813 (1984).

Khan, S.U. "Determination of Terbacil in Soil by Gas-Liquid Chromatography with 63Ni Electron Capture Detection," *Bull. Environ. Contam. Toxicol.*, 18(1):83–88 (1977).

Khan, S.U. "Kinetics of Hydrolysis of Atrazine in Aqueous Fulvic Acid Solution," *Pestic. Sci.*, 9(1):39–43 (1978).

Khan, S.U. "Distribution and Characteristics of Bound Residues of Prometryn in an Organic Soil," *J. Agric. Food Chem.*, 30(1):175–179 (1982).

Khan, S. and V. Bansal. "Thermodynamics of Adsorption of Methyl-2-(dimethylamino)-*N*-[{(methylamino)carbonyl}oxy]-2-oxoethanimidothioate on Saturated Cu- and Zn-Montmorillonites," *J. Coll. Interface Sci.*, 78(2):554–558 (1980).

Khan, S.U. and D.S. Gamble. "Ultraviolet Irradiation on an Aqueous Solution of Prometryn in the Presence of Humic Materials," *J. Agric. Food Chem.*, 31(5):1099–1104 (1983).

Khan, S.U. and H.A. Hamilton. "Extractable and Bound (Nonextractable) Residues of Prometryn and Its Metabolites in an Organic Soil," *J. Agric. Food Chem.*, 28(1):126–132 (1980).

Khan, S.U. and P.B. Marriage. "Residues of Atrazine and Its Metabolites in an Orchard Soil and Their Uptake by Oat Plants," *J. Agric. Food Chem.*, 25(6):1408–1413 (1977).

Khan, S.U. and M. Schnitzer. "UV Irradiation of Atrazine in Aqueous Fulvic Acid Solution," *J. Environ. Sci. Health*, B13(3):299–310 (1978).

Kiigemgi, U. and L.C. Terriere. "The Persistence of Zinophos and Dyfonate in Soil," *Bull. Environ. Contam. Toxicol.*, 6:355–361 (1971).

Kim, Y.H., J.E. Woodrow, and J.N. Seiber. "Evaluation of a Gas Chromatographic Method for Calculating Vapor Pressures with Organophosphorus Pesticides," *J. Chromatogr.*, 314:37–53 (1984).

King, P.H. and P.L. McCarty. "A Chromatographic Model for Predicting Pesticide Migration in Soils," *Soil Sci.*, 106(4):248–261 (1968).

Kishi, H., N. Kogure, and Y. Hashimoto. "Contribution of Soil Constituents in Adsorption Coefficient of Aromatic Compounds, Halogenated Alicyclic and Aromatic Compounds to Soil," *Chemosphere*, 21(7):867–876 (1990).

Kjeldsen, P., J. Kjølholt, B. Schultz, T.H. Christensen, and J.C. Tjell. "Sorption and Degradation of Chlorophenols, Nitrophenols and Organophosphorus Pesticides in the Subsoil Under Landfills," *J. Contam. Hydrol.*, 6(2):165–184 (1990).

Klehr, M., J. Iwan, and J. Riemann. "An Experimental Approach to the Photolysis of Pesticides Adsorbed on Soil: Thidiazuron," *Pestic. Sci.*, 14(4):359–366 (1983).

Klein, W., J. Kohli, I. Weisgerber, and F. Korte. "Fate of Aldrin-^{14}C in Potatoes and Soil Under Outdoor Conditions," *J. Agric. Food Chem.*, 21(2):152–156 (1973).

Klevens, H.B. "Solubilization of Polycyclic Hydrocarbons," *J. Phys. Colloid Chem.*, 54(2):283–298 (1950).

Knoevenagel, K. and R. Himmelreich. "Degradation of Compounds Containing Carbon Atoms by Photo-Oxidation in the Presence of Water," *Arch. Environ. Contam. Toxicol.*, 4:324–333 (1976).

Knowlton, M.F. and J.N. Huckins. "Fate of Radiolabeled Sodium Pentachlorophenate in Littoral Microcosms," *Bull. Environ. Contam. Toxicol.*, 30(2):206–213 (1983).

Knuesli, E., D. Berrer, G. Dupuis, and H. Esser. "*s*-Triazines" in *Degradation of Herbicides*, P.C. Kearney and D.D. Kaufman, Eds., (New York: Marcel Dekker, 1969), pp. 51–59.

Knuutinen, J., H. Palm, H. Hakala, J. Haimi, V. Huhta, and J. Salminen. "Polychlorinated Phenols and Their Metabolites in Soil and Earthworms of Sawmill Environment," *Chemosphere*, 20(6):609–623 (1990).

Ko, W.H. and J.D. Farley. "Conversion of Pentachloronitrobenzene to Pentachloroaniline in Soil and the Effect of these Compounds on Soil Microorganisms," *Phytopathology*, 59:64–67 (1969).

Kobayashi, H. and B.E. Rittman. "Microbial Removal of Hazardous Organic Compounds," *Environ. Sci. Technol.*, 16(3):170A–183A (1982).

Kobrinsky, P.C. and M.R. Martin. "High-Energy Methyl Radicals; the Photolysis of Methyl Bromide at 1850 Å," *J. Chem. Phys.*, 48(12):5728–5729 (1968).

Kolbe, A., A. Bernasch, M. Stock, H.R. Schütte, and W. Dedek. "Persistence of the Insecticide Dimethoate in Three Different Soils Under Laboratory Conditions," *Bull. Environ. Contam. Toxicol.*, 46(4):492–498 (1991).

Kole, R.K., J. Saha, S. Pal, S. Chaudhuri, and A. Chowdhury. "Bacterial Degradation of the Herbicide Pendimethalin and Activity Evaluation of Its Metabolites," *Bull. Environ. Contam. Toxicol.*, 52(5):779–786 (1994).

Kollig, H.P. "Environmental Fate Constants for Organic Chemicals Under Consideration for EPA's Hazardous Waste Identification Projects," Office of Research and Development, U.S. EPA Report-600/R-93/132 (1993), 172 p.

Konrad, J.G., G. Chesters, and D.E. Armstrong. "Soil Degradation of Diazinon, a Phosphorothioate Insecticide," *Agron. J.*, 59:591–594 (1967).

Konrad, J.G., G. Chesters, and D.E. Armstrong. "Soil Degradation of Malathion, a Phosphorothioate Insecticide," *Soil Sci. Soc. Am. Proc.*, 33:259–262 (1969).

Kopczynski, S.L., A.P. Altshuller, and F.D. Sutterfield. "Photochemical Reactivities of Aldehyde-Nitrogen Oxide Systems," *Environ. Sci. Technol.*, 8(10):909–918 (1974).

Korte, F., G. Ludwig, and J. Vogel. "Umwandlung von Aldrin-[^{14}C] und Dieldrin-[^{14}C] durch Mikroorganismen, Leberhomogenate, und Moskito-Larven," *Liebigs Ann. Chem.*, 656:135–140 (1962).

Koskinen, W.C., H.R. Leffler, J.E. Oliver, P.C. Kearney, and C.G. McWhorter. "Effect of Trifluralin Soil Metabolites on Cotton Boll Components and Fiber and Seed Properties," *J. Agric. Food Chem.*, 33(5):958–961 (1985).

Koskinen, W.C., J.E. Oliver, P.C. Kearney, and C.G. McWhorter. "Effect of Trifluralin Soil Metabolites on Cotton Growth and Yield," *J. Agric. Food Chem.*, 32(6):1246–1248 (1984).

Koskinen, W.C., K.E. Sellung, J.M. Baker, B.L. Barber, and R.H. Dowdy. "Ultrasonic Decomposition of Atrazine and Alachlor in Water," *J. Environ. Sci. Health*, B29(3):581–590 (1994).

Kotronarou, A., G. Mills, and M.R. Hoffmann. "Decomposition of Parathion in Aqueous Solution by Ultrasonic Irradiation," *Environ. Sci. Technol.*, 26(7):1460–1462 (1992).

Kozak, J., J.B. Weber, and T.J. Sheets. "Adsorption of Prometryn and Metolachlor by Selected Soil Organic Matter Fractions," *Soil Sci.*, 136(2):94–101 (1983).

Krause, A., W.G. Hancock, R.D. Minard, A.J. Freyer, R.C. Honeycutt, H.M.L. Baron, D.L. Paulson, S.-Y. Liu, and J.-M. Bollag. "Microbial Transformation of the Herbicide Metolachlor by a Soil Actinomycete," *J. Agric. Food Chem.*, 33(4):584–589 (1985).

Krause, A.A. and H.D. Niemczyk. "Gas-Liquid Chromatographic Analysis of Chlorthal-Dimethyl Herbicide and Its Degradates in Turfgrass Thatch and Soil using a Solid-Phase Extraction Technique," *J. Environ. Sci. Health*, B25(5):587–606 (1990).

Kriegman-King, M.R. and M. Reinhard. "Transformation of Carbon Tetrachloride in the Presence of Sulfide, Biotite, and Vermiculite," *Environ. Sci. Technol.*, 26(11):2198–2206 (1992).

Krumzdorov, A.M. "Transformation of 2-Methoxy-3,6-dichlorobenzoic Acid (Dicamba) in Corn Plants," *Agrokhimya*, 7:128–133 (1974).

Kuhr, R.J. "Metabolism of Methylcarbamate Insecticide Chemicals in Plants," *J. Agric. Food Chem. Suppl.*, 44–49 (1968).

Kuhr, R.J., A.C. Davis, and J.B. Bourke. "Dissipation of Guthion, Sevin, Polyram, Phygon, and Systox from Apple Orchard Soil," *Bull. Environ. Contam. Toxicol.*, 11(3):224–230 (1974).

Kulshrestha, G. and S.B. Singh. "Influence of Soil Moisture and Microbial Activity on Pendimethalin Degradation," *Bull. Environ. Contam. Toxicol.*, 48(2):269–274 (1992).

Kwok, E.S.C., R. Atkinson, and J. Arey. "Gas-Phase Atmospheric Chemistry of Selected Thiocarbamates," *Environ. Sci. Technol.*, 26(9):1798–1807 (1992).

Laanio, T.L., G. Dupuis, and H.O. Esser. "Fate of ^{14}C-Labeled Diazinon in Rice, Paddy Soil, and Pea Plants," *J. Agric. Food Chem.*, 20(6):1213–1219 (1972).

Laanio, T.L., P.C. Kearney, and D.D. Kaufman. "Microbial Metabolism of Dinitramine," *Pestic. Biochem. Physiol.*, 3:271–277 (1973).

Lacorte, S., S.B. Lartiges, P. Garrigues, and D. Barceló. "Degradation of Organophosphorus Pesticides and Their Transformation Products in Estuarine Waters," *Environ. Sci. Technol.*, 29(2):431–438 (1995).

LaFleur, K.S. "Metribuzin Movement in Soil Columns: Observation and Prediction," *Soil Sci.*, 129(2):107–114 (1980).

Lafrance, P., L. Marineau, L. Perreault, and J.-P. Villeneuve. "Effect of Natural Dissolved Organic Matter Found in Groundwater on Soil Adsorption and Transport of Pentachlorophenol," *Environ. Sci. Technol.*, 28(13):2314–2320 (1994).

Lamoureux, G.L., R.H. Shimabukuro, H.R. Swanson, and D.S. Frear. "Metabolism of 2-Chloro-4-ethylamino-6-isopropylamino-*s*-triazine (Atrazine) in Excised Sorghum Leaf Sections," *J. Agric. Food Chem.*, 18(1):81–86 (1970).

Lamoureux, G.L., L.E. Stafford, and F.S. Tanaka. "Metabolism of 2-Chloro-*N*-isopropylacetanilide (Propachlor) in the Leaves of Corn, Sorghum, Sugarcane, and Barley," *J. Agric. Food Chem.*, 19(2):346–350 (1971).

Lamparski, L.L., R.H. Stehl, and R.L. Johnson. "Photolysis of Pentachlorophenol-Treated Wood. Chlorinated Dibenzo-*p*-dioxin Formation," *Environ. Sci. Technol.*, 14(2):196–200 (1980).

Lange, A.H., B. Fischer, W. Humphrey, W. Seyman, and K. Baghott. "Herbicide Residues in California Agricultural Soils," *Calif. Agric.*, 22:2–4 (1968).

Langlois, B.E. "Reductive Dechlorination of DDT by *Escherichia coli*," *J. Dairy Sci.*, 50:1168–1170 (1967).

Laplanche, A., M. Bouvet, F. Venien, G. Martin, and A. Chabrolles. "Modelling Parathion Changes in the Natural Environment Laboratory Experiment," *Water Res.*, 15:543–545 (1981).

Laplanche, A., G. Martin, and F. Tonnard. "Ozonation Schemes of Organophosphorus Pesticides. Application in Drinking Water Treatment," *Ozone: Sci. Engrg.*, 6:207–219 (1984).

Lau, Y.L., D.L.S. Liu, G.J. Pacepavicius, and R.J. Maguire. "Volatilization of Metolachlor from Water," *J. Environ. Sci. Health*, B30(5):605–620 (1995).

Leafe, E.L. "Metabolism and Selectivity of Plant-Growth Regulator Herbicides," *Nature (London)*, 193(4814):485–486 (1962).

Leavitt, D.D. and M.A. Abraham. "Acid-Catalyzed Oxidation of 2,4-Dichlorophenoxyacetic Acid by Ammonium Nitrate in Aqueous Solution," *Environ. Sci. Technol.*, 24(4):566–571 (1990).

LeBaron, H.M., J.E. McFarland, B.J. Simoneaux, and E. Ebert. "Metolachlor" in *Herbicides: Chemistry, Degradation, and Mode of Action*. Volume 3, Kearney, P.C. and D.D. Kaufman, Eds., (New York: Marcel Dekker, Inc., 1988), pp. 336–383.

Lee, A. "EPTC (S-Ethyl N,N-dipropylthiocarbamate)-Degrading Microorganisms Isolated from a Soil Previously Exposed to EPTC," *Soil Biol. Biochem.*, 16:529–531 (1984).

Lee, J.K. "Degradation of the Herbicide, Alachlor, by Soil Microorganisms. III. Degradation Under an Upland Soil Condition," *J. Korean Agric. Chem. Soc.*, 29:182–189 (1986).

Lee, D.-Y., W.J. Farmer, and Y. Aochi. "Sorption of Napropamide on Clay and Soil in the Presence of Dissolved Organic Matter," *J. Environ. Qual.*, 19(3):567–573 (1990).

Lee, C.-C., R.E. Green, and W.J. Apt. "Transformation and Adsorption of Fenamiphos, F. Sulfoxide and F. Sulfone in Molokai Soil and Simulated Movement with Irrigations," *J. Contam. Hydrol.*, 1:211–225 (1986).

Lee, C.H., P.C. Oloffs, and S.Y. Szeto. "Persistence, Degradation, and Movement of Triclopyr and Its Ethylene Glycol Butyl Ester in a Forest Soil," *J. Agric. Food Chem.*, 34(6):1075–1079 (1986a).

Lee, P.W., S.M. Stearns, H. Hernandez, W.R. Powell, and M.V. Naidu. "Fate of Dicrotophos in the Soil Environment," *J. Agric. Food Chem.*, 37(4):1169–1174 (1989).

Lefsrud, C. and J.C. Hall. "Basis for Sensitivity Differences Among Crabgrass, Oat, and Wheat to Fenoxaprop-Ethyl," *Pestic. Biochem. Physiol.*, 34:218–227 (1989).

Leger, D.A. and V.N. Mallet. "New Degradation Products and a Pathway for the Degradation of Aminocarb [4-(Dimethylamino)-3-methylphenyl N-Methylcarbamate] in Purified Water," *J. Agric. Food Chem.*, 36(1):185–189 (1988).

Leigh, W. "Degradation of Selected Chlorinated Hydrocarbon Insecticides," *J. Water Pollut. Control Fed.*, 41(11):R450–R460 (1969).

Leinster, P., R. Perry, and R.J. Young. "Ethylene Bromide in Urban Air," *Atmos. Environ.*, 12:2382–2398 (1978).

Leistra, M. "Distribution of 1,3-Dichloropropene in Soil," *J. Agric. Food Chem.*, 18(6):1124–1126 (1970).

Leistra, M., J.H. Smelt, and R. Zandvoort. "Persistence and Mobility of Bromacil in Orchard Soils," *Weed Res.*, 15:243–247 (1975).

Leitis, E. and D.G. Crosby. "Photodecomposition of Trifluralin," *J. Agric. Food Chem.*, 22(5):842–848 (1974).

Leland, H.V., W.N. Bruce, and N.F. Shimp. "Chlorinated Hydrocarbon Insecticides in Sediments in Southern Lake Michigan," *Environ. Sci. Technol.*, 7(9):833–838 (1973).

Lemley, A.T., R.J. Wagenet, and W.Z. Zhong. "Sorption and Degradation of Aldicarb and Its Oxidation Products in a Soil-Water Flow System as a Function of pH and Temperature," *J. Environ. Qual.*, 17(3):408–414 (1988).

Leoni, V., C.B. Hollick, D.E. D'Alessandro, R.J. Collinson, and S. Merolli. "The Soil Degradation of Chlorpyrifos and the Significance of Its Presence in the Superficial Water in Italy," *Agrochimica*, 25:414–426 (1981).

Leuck, D.B. and M.C. Bowman. "Imidan Insecticide Residues (Imidan and Imidoxon): Their Persistence in Corn, Grass, and Soybeans," *J. Econ. Entomol.*, 61:705–707 (1968).

Leuck, D.B. and M.C. Bowman. "Residues of Phorate and Five of Its Metabolites: Their Persistence in Forage Corn and Grass," *J. Econ. Entomol.*, 63:1838–1842 (1970).

Leuck, D.B., R.L. Jones, and M.C. Bowman. "Chlorpyrifosmethyl Insecticide Residues: Their Analysis and Persistence in Coastal Bermuda Grass and Forage Corn," *J. Econ. Entomol.*, 69:287–290 (1975).

Lewer, P. and W.J. Owen. "Amino Acid Conjugation of Triclopyr by Soybean Cell Suspension Cultures," *Pestic. Biochem. Physiol.*, 33:249–256 (1989).

Lewis Publishers. 1989. *Drinking Water Health Advisory — Pesticides* (Chelsea, MI: Lewis Publishers, Inc.), 819 p.

Lewis, R.J., Sr. *Rapid Guide to Hazardous Chemicals in the Workplace* (New York: Van Nostrand Reinhold, 1990), 286 p.

Leyder, F. and P. Boulanger. "Ultraviolet Absorption, Aqueous Solubility and Octanol-Water Partition for Several Phthalates," *Bull. Environ. Contam. Toxicol.*, 30(2):152–157 (1983).

Li, A. and A.W. Andren. "Solubility of Polychlorinated Biphenyls in Water/Alcohol Mixtures. 1. Experimental Data," *Environ. Sci. Technol.*, 28(1):47–52 (1994).

Li, C.F. and R.L. Bradley. "Degradation of Chlorinated Hydrocarbon Pesticides in Milk and Butter Oil by Ultraviolet Energy," *J. Dairy Sci.,* 52:27–30 (1969).

Li, J., A.J. Dallas, D.I. Eikens, P.W. Carr, D.L. Bergmann, M.J. Hait, and C.A. Eckert. "Measurement of Large Infinite Dilution Activity Coefficients of Non-electrolytes in Water by Inert Gas Stripping and Gas Chromatography," *Anal. Chem.*, 65(22):3212–3218 (1993).

Li, A. and W.J. Doucette. "The Effect of Cosolutes on the Aqueous Solubilities and Octanol/Water Partition Coefficients of Selected Polychlorinated Biphenyl Congeners," *Environ. Toxicol. Chem.,* 12:2031–2035 (1993).

Li, G.-C. and G.T. Felbeck, Jr. "Atrazine Hydrolysis as Catalyzed by Humic Acids," *Soil Sci.*, 114(3):201–208 (1972).

Li, C.Y. and E.E. Nelson. "Persistence of Benomyl and Captan and Their Effects on Microbial Activity in Field Soils," *Bull. Environ. Contam. Toxicol.*, 34(4):533–540 (1985).

Liang, T.T. and E.P. Lichtenstein. "Effect of Light, Temperature, and pH on the Degradation of Azinphosmethyl," *J. Econ. Entomol.*, 65:315–321 (1972).

Lichtenstein, E.P. "'Bound' Residues in Soils and Transfer of Soil Residues in Crops," Residue Rev., 76:147–153 (1980).

Lichtenstein, E.P., L.J. DePew, E.L. Eshbaugh, and J.P. Sleesman. "Persistence of DDT, Aldrin, and Lindane in some Midwestern Soils," *J. Econ. Entomol.*, 53:136–142 (1960).

Lichtenstein, E.P., T.W. Fuhremann, and K.R. Schulz. "Effect of Sterilizing Agents on Persistence of Parathion and Diazinon in Soils and Water," *J. Agric. Food Chem.*, 16(5):870–873 (1968).

Lichtenstein, E.P., T.W. Fuhremann, and K.R. Schulz. "Persistence and Vertical Distribution of DDT, Lindane, and Aldrin Residues, 10 and 15 Years after a Single Soil Application," *J. Agric. Food Chem.*, 19(4):718–721 (1971).

Lichtenstein, E.P., T.W. Fuhremann, and K.R. Schulz. "Translocation and Metabolism of [14C]Phorate as Affected by Percolating Water in a Model Soil-Plant Ecosystem," *J. Agric. Food Chem.*, 22(6):991–996 (1974).

Lichtenstein, E.P., J. Katan, and B.N. Anderegg. "Binding of "Persistent" and "Nonpersistent" 14C-Labeled Insecticides in an Agricultural Soil," *J. Agric. Food Chem.*, 25(1):43–47 (1977).

Lichtenstein, E.P., T.T. Liang, and M.K. Koeppe. "Effects of Fertilizers, Captafol, and Atrazine on the Fate and Translocation of [14C]Fonofos and [14C]Parathion in a Soil-Plant Microcosm," *J. Agric. Food Chem.*, 30(5):871–878 (1982).

Lichtenstein, E.P. and K.R. Schulz. "Breakdown of Lindane and Aldrin in Soils," *J. Econ. Entomol.*, 52:118–124 (1959).

Lichtenstein, E.P. and K.R. Schulz. "Persistence of Some Chlorinated Hydrocarbon Insecticides as Influenced by Soil Types, Rate of Application, and Temperature," *J. Econ. Entomol.*, 52:124–131 (1959a).

Lichtenstein, E.P. and K.R. Schulz. "The Effects and Moisture and Microorganisms on the Persistence and Metabolism of Some Organophosphorus Insecticides in Soils, with Special Emphasis on Parathion," *J. Econ. Entomol.*, 57:618–627 (1964).

Lieberman, M.T. and M. Alexander. "Microbial and Nonenzymatic Steps in the Decomposition of Dichlorvos (2,2-Dichlorovinyl *O,O*-Dimethyl Phosphate)," *J. Agric. Food Chem.*, 31(2):265–267 (1983).

Lilius, H., T. Hästbacka, and B. Isomaa. "A Comparison of the Toxicity of 30 Reference Chemicals to *Daphnia magna* and *Daphnia pulex*," *Water Res.*, 14(12):2085–2088 (1995).

Lin, S., M.T. Lukasewycz, R.J. Liukkonen, and R.M. Carlson. "Facile Incorporation of Bromine into Aromatic Systems Under Conditions of Water Chlorination," *Environ. Sci. Technol.*, 18(12):985–986 (1984).

Linke, H.A.B. and R. Bartha. "Transformation Products of the Herbicide Propanil in Soil: A Balance Study," in *Agric. Indust. Microbiol. Abstr., Bacteriol. Proc. 70*, (Washington, DC: American Society of Microbiology, 1970), p. 9.

Linscott, D.L. and R.D. Hagin. "Additions to the Aliphatic Moiety of Chlorophenoxy Compounds," *Weed Sci.*, 18:197–198 (1970).

Liu, S.-Y. and J.M. Bollag. "Metabolism of Carbaryl by a Soil Fungus," *J. Agric. Food Chem.*, 19(3):487–490 (1971).

Liu, S.-Y. and J.M. Bollag. "Carbaryl Decomposition to 1-Naphthyl Carbamate by *Aspergillus terreus*," *Pestic. Biochem. Physiol.*, 1:366–372 (1971a).

Liu, L.C., H. Cibes-Viadé, and F.K.S. Koo. "Adsorption of Ametryne and Diuron by Soils," *Weed Sci.*, 18(4):470–474 (1970).

Liu, J. and C. Qian. "Hydrophobic Coefficients of *s*-Triazine and Phenylurea Herbicides," *Chemosphere*, 31(8):3951–3959 (1995).

Liu, D., W.M.J. Strachan, K. Thomson, and K. Kwasniewska. "Determination of the Biodegradability of Organic Compounds," *Environ. Sci. Technol.*, 15(7):788–793 (1981).

Locke, M.A. "Sorption-Desorption Kinetics of Alachlor in Surface Soil from Two Soybean Tillage Systems," *J. Environ. Qual.*, 21(4):558–566 (1992).

Loekke, H. "Residues in Carrots Treated with Linuron," *Pestic. Sci.*, 5:749–757 (1974).

Løkke, H. "Sorption of Selected Organic Pollutants in Danish Soils," *Ecotoxicol. Environ. Safety*, 8(5):395–409 (1984).

Lombardo, P., I.H. Pomerantz, and I.J. Egry. "Identification of Photoaldrin Chlorohydrin as a Photoalteration Product of Dieldrin," *J. Agric. Food Chem.*, 20(6):1278–1279 (1972).

Lopez, C.E. and J.I. Kirkwood. "Isolation of Microorganisms from a Texas Soil Capable of Degrading Urea Derivative Herbicides," *Soil Sci. Soc. Am. Proc.*, 38:309–312 (1974).

Lopez-Gonzales, J.De D. and C. Valenzuela-Calahorro. "Associated Decomposition of DDT to DDE in the Diffusion of DDT on Homoionic Clays," *J. Agric. Food Chem.*, 18(3):520–523 (1970).

Lord, K.A., G.G. Briggs, M.C. Neale, and R. Manlove. "Uptake of Pesticides from Water and Soil by Earthworms," *Pestic. Sci.*, 11(4):401–408 (1980).

Lowder, S.W. and J.B. Weber. "Atrazine Efficacy and Longevity as Affected by Tillage, Liming, and Fertilizer Type," *Weed Sci.*, 30:273–280 (1982).

Lu, P.-Y., R.L. Metcalf, A.S. Hirwe, and J.W. Williams. "Evaluation of Environ-mental Distribution and Fate of Hexachlorocyclopentadiene, Chlordene, Heptachlor, and Heptachlor Epoxide in a Model Ecosystem," *J. Agric. Food Chem.*, 23(5):967–973 (1975).

Lucier, G.W. and R.E. Menzer. "Metabolism of Dimethoate in Bean Plants in Relation to Its Mode of Application," *J. Agric. Food Chem.*, 16(6):936–945 (1968).

Lucier, G.W. and R.E. Menzer. "Nature of Oxidative Metabolites of Dimethoate Formed in Rats, Liver Microsomes, and Bean Plants," *J. Agric. Food Chem.*, 18(4):698–704 (1970).

Luckwill, L.C. and C.P. Lloyd-Jones. "2,4-Dichlorophenoxyacetic Acid in Leaves of Red and Black Currant," *Ann. of Appl. Biol.*, 48:613 (1960).

Lukens, R.J. and H.D. Sisler. "2-Thiazolidinethione-4-carboxylic Acid from the Reaction of Captan with Cysteine," *Science (Washington, DC)*, 127(3299):650 (1958).

Lund-Høie, K. and H.O. Friestad. "Photodegradation of the Herbicide Glyphosate in Water," *Bull. Environ. Contam. Toxicol.*, 36:723–729 (1986).

Lutz, J.F., G.E. Byers, and T.J. Sheets. "The Persistence and Movement of Picloram and 2,4,5-T in Soils," *J. Environ. Qual.*, 4:485–488 (1973).

Lyman, W.J., W.F. Reehl, and D.H. Rosenblatt. *Handbook of Chemical Property Estimation Methods: Environmental Behavior of Organic Compounds* (New York: McGraw-Hill, Inc., 1982).

Mabey, W. and T. Mill. "Critical Review of Hydrolysis of Organic Compounds in Water Under Environmental Conditions," *J. Phys. Chem. Ref. Data*, 7(2):383–415 (1978).

Macalady, D.L. and N.L. Wolfe. "New Perspectives on the Hydrolytic Degradation of the Organophosphorothioate Insecticide Chlorpyrifos," *J. Agric. Food Chem.*, 31(6):1139–1147 (1983).

Macalady, D.L., P.G. Tratnyek, and T.J. Grundy. "Abiotic Reduction Reactions of Anthropogenic Organic Chemicals in Anaerobic Systems: A Critical Review," J. Contam. Hydrol., 1(1):1–28 (1986).

Macalady, D.L. and N.L. Wolfe. "Effects of Sediment Sorption and Abiotic Hydrolyses. 1. Organophosphorothioate Esters," *J. Agric. Food Chem.*, 33(2):167–173 (1985).

Macek, K.J. and W.A. McAllister. "Insecticide Susceptibility of Some Common Fish Family Representatives," *Trans. Am. Fish Soc.*, 99(1):20–27 (1970).

Mackay, D. and W.-Y. Shiu. "Aqueous Solubility of Polynuclear Aromatic Hydrocarbons," *J. Chem. Eng. Data*, 22(4):399–402 (1977).

MacNamara, G. and S.J. Toth. "Adsorption of Linuron and Malathion by Soils and Clay Minerals," *Soil Sci.*, 109(4):234–240 (1970).

MacRae, I.C. "Microbial Metabolism of Pesticides and Structurally Related Compounds," *Rev. Environ. Contam. Toxicol.*, 109:2–87 (1989).

MacRae, I.C. and M. Alexander. "Microbial Degradation of Selected Herbicides in Soil," *J. Agric. Food Chem.*, 13(1):72–75 (1965).

MacRae, I.C., K. Raghu, and E.M. Bautista. "Anaerobic Degradation of the Insecticide Lindane by Clostridium sp.," *Nature (London)*, 221(5183):859–860 (1969).

MacRae, I.C., K. Raghu, and T.F. Castro. "Persistence and Biodegradation of Four Common Isomers of Benzene Hexachloride in Submerged Soils," *J. Agric. Food Chem.*, 15(5):911–914 (1967).

MacRae, I.C., Y. Yamaya, and T. Yoshida. "Persistence of Hexachlorocyclohexane Isomers in Soil Suspensions," *Soil Biol. Biochem.*, 16:285–286 (1984).

Madhun, Y.A. and V.H. Freed. "Degradation of the Herbicides Bromacil, Diuron and Chlortoluron in Soil," *Chemosphere*, 16(5):1003–1011 (1987).

Madhun, Y.A., V.H. Freed, J.L. Young, and S.C. Fang. "Sorption of Bromacil, Chlortoluron, and Diuron by Soils," *Soil Sci. Soc. Am. J.*, 50:1467–1471 (1986).

Majka, J.T. and T.L. Lavy. "Adsorption, Mobility, and Degradation of Cyanazine and Diuron in Soils," *Weed Sci.*, 25(5):401–406 (1977).

Malik, S.K. and P.R. Yadav. "Persistence of Aldicarb in Soil and Its Translocation into Cotton Seeds," *Indian J. Agric. Sci.*, 49:745–748 (1979).

Mann, R.K., W.W. Witt, and C.E. Rieck. "Fosamine Absorption and Translocation in Multiflora Rose (Rosa Multiflora)," *Weed Sci.*, 34:830–833 (1986).

Mansour, M., E. Feicht, and P. Méallier. "Improvement of the Photostability of Selected Substances in Aqueous Medium," *Toxicol. Environ. Chem.*, 20–21:139–147 (1989).

Mansour, M., S. Thaller, and F. Korte. "Action of Sunlight on Parathion," *Bull. Environ. Contam. Toxicol.*, 30(3):358–364 (1983).

Marriage, P.B., W.J. Saidak, and F.G. Von Stryk. "Residues of Atrazine, Simazine, Linuron, and Diuron after Repeated Annual Applications in a Peach Orchard," *Weed Res.*, 15:373–3379 (1975).

Martens, R. "Degradation of [8,9-^{14}C] Endosulfan by Soil Microorganisms," *Appl. Environ. Microbiol.*, 31(6):853–858 (1976).

Martens, R. "Degradation of Endosulfan-8,9-^{14}C in Soil Under Different Conditions," *Bull. Environ. Contam. Toxicol.*, 17(4):438–446 (1977).

Martin, G., A. Laplanche, J. Morvan, Y. Wei, and C. LeCloirec. "Action of Ozone on Organo-Nitrogen Products," in *Proceedings Symposium on Ozonization: Environmental Impact and Benefit* (Paris, France: International Ozone Association, 1983), pp. 379–393.

Mascolo, G., A. Lopez, R. Földényi, R. Passino, and G. Tiravanti. "Prometryne Oxidation by Sodium Hypochlorite in Aqueous Solution: Kinetics and Mechanism," *Environ. Sci. Technol.*, 29(12):2987–2991 (1995).

Massini, P. "Movement of 2,6-Dichlorobenzonitrile in Soils and in Plants in Relation to Its Physical Properties," *Weed Res.*, 1(2):142–146 (1961).

Matheson, L.J. and P.G. Tratnyek. "Reductive Dehalogenation of Chlorinated Methanes by Iron Metal," *Environ. Sci. Technol.*, 28(12):2045–2053 (1994).

Mathur, S.P. H.A. Mailton, R. Greenhalgh, K.A. Macmillan, and S.U. Khan. "Effect of Microorganisms and Persistence of Field-Applied Carbofuran and Dyfonate in a Humic Meisol," *Can. J. Soil Sci.*, 56:89–96 (1976).

Mathur, S.P. and J.G. Saha. "Microbial Degradation of Lindane-C^{14} in a Flooded Sandy Loam Soil," *Soil Sci.*, 120(4):301–307 (1975).

Mathur, S.P. and J.G. Saha. "Degradation of Lindane-^{14}C in a Mineral Soil and in an Organic Soil," *Bull. Environ. Contam. Toxicol.*, 17(4):424–430 (1977).

Matsumura, F. and G.M. Boush. "Malathion Degradation by *Trichoderma viride* and a *Pseudomonas* Species," *Science (Washington, DC)*, 153(3741):1278–1280 (1966).

Matsumura, F. and G.M. Boush. "Degradation of Insecticides by a Soil Fungus," *Trichoderma viride*," *J. Econ. Entomol.*, 61:610–612 (1968).

Matsumura, F., G.M. Boush, and A. Tai. "Breakdown of Dieldrin in the Soil by a Microorganism," *Nature (London)*, 219(5157):965–967 (1968).

Matsumura, F., K.C. Patil, and G.M. Boush. "DDT Metabolized by Microorganisms from Lake Michigan," *Nature (London)*, 230(5292):324–325 (1971).

Matsunaka, S. "Activation and Inactivation of Herbicides by Higher Plants," *Residue Rev.*, 25:45–58 (1969).

Matsuo, H. and J.E. Casida. "Photodegradation of two Dinitrophenolic Pesticide Chemicals, Dinobuton and Dinoseb, Applied to Bean Leaves," *Bull. Environ. Contam. Toxicol.*, 5(1):72–78 (1970).

Mattson, A.M., J.T. Spillane, and G.W. Pearce. "Dimethyl 2,2-Dichlorovinyl Phosphate (DDVP), an Organophosphorus Compound Highly Toxic to Insects," *J. Agric. Food Chem.*, 3(4):319–321 (1955).

Maule, A., S. Plyte, and A.V. Quirk. "Dehalogenation of Organochlorine Insecticides by Mixed Anaerobic Microbial Populations," *Pestic. Biochem. Physiol.*, 27:229–236 (1987).

Mazzochi, P.H. and M.P. Rao. "Photolysis of 3-(*p*-Chlorophenyl)-1,1-dimethylurea (Monuron) and 3-Phenyl-1,1-dimethylurea (Fenuron)," *J. Agric. Food Chem.*, 20(5):957–959 (1972).

McAuliffe, C. "Solubility in Water of C1-C9 Hydrocarbons," *Nature (London)*, 200(4911):1092–1093 (1963).

McAuliffe, C. "Solubility in Water of Paraffin, Cycloparaffin, Olefin, Acetylene, Cycloolefin, and Aromatic Compounds," *J. Phys. Chem.*, 70(4):1267–1275 (1966).

McBride, K.E., J.W. Kenny, and D.M. Stalker. "Metabolism of the Herbicide Bromoxynil by *Klebsiella pneumoniae* supus. ozaenae," *Appl. Environ. Microbiol.*, 52(2):325–330 (1986).

McCall, P.J., R.L. Swann, D.A. Laskowski, S.A. Vrona, S.M. Unger, and H.J. Dishburger. "Prediction of Chemical Mobility in Soil from Sorption Coefficients," in *Aquatic Toxicology and Hazard Assessment*, Fourth Conference, ASTM STP 737, Branson, D.R. and K.L. Dickson, Eds. (Philadelphia, PA: American Society for Testing and Materials, 1981), pp. 49–58.

McCall, P.J., S.A. Vrona, and S.S. Kelley. "Fate of Uniformly Carbon-14 Ring Labeled 2,4,5-Trichlorophenoxyacetic Acid and 2,4-Dichlorophenoxy Acid," *J. Agric. Food Chem.*, 29(1):100–107 (1981a).

McConnell, J.S. and L.R. Hossner. "pH-Dependent Adsorption Isotherms of Glyphosate," *J. Agric. Food Chem.*, 33(6):1075–1078 (1985).

McGahen, L.L. and J.M. Tiedje. "Metabolism of Two New Acylanilide Herbicides, Antor Herbicide (H-22234) and Dual (Metolachlor) by the Soil Fungus *Chaetomium globosum*," *J. Agric. Food Chem.*, 26(2):414–419 (1978).

McGlamery, M.D. and F.W. Slife. "The Adsorption and Desorption of Atrazine as Affected by pH, Temperature, and Concentration," *Weeds*, 14:237–239 (1966).

McGowan, J.C., P.N. Atkinson, and L.H. Ruddle. "The Physical Toxicity of Chemicals. V. Interaction Terms for Solubilities and Partition Coefficients," *J. Appl. Chem.*, 16:99–102 (1966).

McGuire, R.R., M.J. Zabik, R.D. Schuetz, and R.D. Flotard. "Photochemistry of Bioactive Compounds. Photochemical Reactions of Heptachlor: Kinetics and Mechanisms," *J. Agric. Food Chem.*, 20(4):856–861 (1972).

McMahon, P.B., F.H. Chapelle, and M.J. Jagucki. "Atrazine Mineralization Potential of Alluvial-Aquifer Sediments Under Aerobic Conditions," *Environ. Sci. Technol.*, 26(8):1556–1559 (1992).

Means, J.C., S.G. Wood, J.J. Hassett, and W.L. Banwart. "Sorption of Polynuclear Aromatic Hydrocarbons by Sediments and Soils," *Environ. Sci. Technol.*, 14(2):1524–1528 (1980).

Medley, D.R. and E.L. Stover. "Effects of Ozone on the Biodegradability of Biorefractory Pollutants," *J. Water Pollut. Control Fed.*, 55(5):489–494 (1983).

Meikle, R.W., N.H. Kurihara, and D.H. DeVries. "The Photocomposition Rates in Dilute Aqueous Solution and on a Surface, and the Volatilization Rate from a Surface," *Arch. Environ. Contam. Toxicol.*, 12(2):189–193 (1983).

Meikle, R.W., E.A. Williams, and C.T. Redemann. "Metabolism of Tordon Herbicide (4-Amino-3,5,6-trichloropicolinic Acid) in Cotton and Decomposition in Soil," *J. Agric. Food Chem.*, 14(4):384–387 (1966).

Meikle, R.W. and C.R. Youngson. "The Hydrolysis Rate of Chlorpyrifos, O,O-Diethyl O-(3,5,6-trichloro-2-pyridyl) Phosphorothioate, and Its Dimethyl Analog, Chloropyrifos-Methyl in Dilute Aqueous Solution," *Arch. Environ. Contam. Toxicol.*, 7(1):13–22 (1978).

Meikle, R.W., C.R. Youngson, R.T. Hedlund, C.A.I. Goring, and W.W. Addington. "Decomposition of Picloram by Soil Microorganisms: A Proposed Reaction Sequence," *Weed Sci.*, 22:263–268 (1974).

Meikle, R.W., C.R. Youngson, R.T. Hedlund, C.A.I. Goring, J.W. Hamaker, and W.W. Addington. "Measurement and Prediction of Picloram Disappearance rates from Soil," *Weed Sci.*, 21:549–555 (1973).

Melcer, H. and W.K. Bedford. "Removal of Pentachlorophenol in Municipal Activated Sludge Systems," *J. Water Pollut. Control Fed.*, 60(5):622–6626 (1988).

Melnikov, N.J. "Chemistry of Pesticides," *Residue Rev.*, 36:1–11 (1971).

Mendel, J.L. and M.S. Walton. "Conversion of *p,p'*-DDT to *p,p'*-DDD by Intestinal Flora of the Rat," *Science (Washington, DC)*, 151:1527–1528 (1966).

Menges, R.M. and S. Tamez. "Movement and Persistence of Bensulfide and Trifluralin in Irrigated Soil," *Weed Sci.*, 22:67–71 (1974).

Menn, J.J. and G.G. Still. "Metabolism of Insecticides and Herbicides in Higher Plants," *CRC Crit. Rev. Toxicol.*, 5:1–21 (1977).

Menzer, R.E. and L.P. Dittman. "Residues in Spinach Grown in Disulfoton and Phorate-Treated Soil," *J. Econ. Entomol.*, 61(1):225–229 (1968).

Menzer, R.E., E.L. Fontanilla, and L.P. Dittman. "Degradation of Disulfoton and Phorate in Soil Influenced by Environmental Factors and Soil Type," *Bull. Environ. Contam. Toxicol.*, 5(1):1–5 (1970).

Menzies, D.R., D.J. Pree, and R.W. Fisher. "Correlation of Spray Coverage Ratings and Phosmet Residues with Mortality of Oriental Fruit Moth Larvae," *J. Econ. Entomol.*, 72:721–724 (1979).

Merkle, M.G., R.W. Bovey, and F.S. Davis. "Factors Affecting the Persistence of Picloram in Soil," *Agron. J.*, 59:413–415 (1976).

Mersie, W. and C.L. Foy. "Adsorption, Desorption, and Mobility of Chlorsulfuron in Soils," *J. Agric. Food Chem.*, 34(1):89–92 (1986).

Metcalf, R.L. "A Century of DDT," *J. Agric. Food Chem.*, 21(4):511–519 (1973).

Metcalf, R.L., T.R. Fukuto, C. Collins, K. Borck, S. Abd El-Aziz, R. Munoz, and C.C. Cassil. "Metabolism of 2,2-Dimethyl-2,3-dihydrobenzofuranyl-7 *N*-Methylcarbamate (Furadan) in Plants, Insects, and Mammals," *J. Agric. Food Chem.*, 16(2):300–311 (1968).

Metcalf, R.L., T.R. Fukuto, C. Collins, K. Borck, J. Burk, H.T. Reynolds, and M.F. Osman. "Metabolism of 2-Methyl-(2-methylthio)propionaldehyde O-(Methylcarbamoyl)oxime in Plant and Insect," *J. Agric. Food Chem.*, 14(6):579–584 (1966).

Metcalf, R.L., T.R. Fukuto, and R.B. March. "Plant Metabolism of Dithiosystox and Thimet," *J. Econ. Entomol.*, 50:338–345 (1957).

Metcalf, R.L., G.K. Sangha, and I.P. Kapoor. "Model Ecosystem for the Evaluation of Pesticide Biodegradability and Ecological Magnification," *Environ. Sci. Technol.*, 5(8):709–713 (1971).

Metcalf, R.L., I.P. Kapoor, P.-Y. Lu, C.K. Schuth, and P. Sherman. "Model Ecosystem Studies of the Environmental Fate of Six Organochlorine Pesticides," *Environ. Health Perspect.*, (June 1973), pp. 35–44.

Meyers, N.L., J.L. Ahlrichs, and J.L. White. "Adsorption of Insecticides on Pond Sediments and Watershed Soils," *Proc. Indiana Acad. Sci.*, 79:432–437 (1970).

Mick, D.L. and P.A. Dahm. "Metabolism of Parathion by Two Species of *Rhizobium*," *J. Econ. Entomol.*, 63:1155–1159 (1970).

Mikesell, M.D. and S.A. Boyd. "Reductive Dechlorination of the Pesticides 2,4-D, 2,4,5-T, and Pentachlorophenol in Anaerobic Sludges," *J. Environ. Qual.*, 14(3):337–340 (1985).

Milano, J.C., A. Guibourg, and J.L. Vernet. "Non Biological Evolution, in Water, of Some Three- and Four-Carbon Atoms Organohalogenated Compounds: Hydrolysis and Photolysis," *Water Res.*, 22(12):1553–1562 (1988).

Miles, C.J. "Degradation Products of Sulfur-Containing Pesticides in Soil and Water" in *Pesticide Transformation Products. Fate and Significance in the Environment*, ACS Symposium Series 459, Somasundaram, L. and J.R. Coats, Eds. (Washington, DC: American Chemical Society, 1991), pp. 61–74.

Miles, C.J. "Degradation of Aldicarb, Aldicarb Sulfoxide, and Aldicarb Sulfone in Chlorinated Water," *Environ. Sci. Technol.*, 25(10):1774–1779 (1991a).

Miles, C.J. and J.J. Delfino. "Fate of Aldicarb, Aldicarb Sulfoxide, and Aldicarb Sulfone in Floridan Groundwater," *J. Agric. Food Chem.*, 33(3):455–460 (1985).

Miles, J.R.W., C.R. Harris, and C.M. Tu. "Influence of Temperature on the Persistence of Chlorpyrifos and Chlorfenvinphos in Sterile and Natural Mineral and Organic Soils," *J. Environ. Sci. Health*, B18(6):705–712 (1983).

Miles, J.R.W., C.R. Harris, and C.M. Tu. "Influence of Moisture on the Persistence of Chlorpyrifos and Chlorfenvinphos in Sterile and Natural Mineral and Organic Soils," *J. Environ. Sci. Health*, B19(2):237–243 (1984).

Miles, J.R.W. and P. Moy. "Degradation of Endosulfan and Its Metabolites by a Mixed Culture of Soil Microorganisms," *Bull. Environ. Contam. Toxicol.*, 23(1/2):13–19 (1979).

Miles, C.J. and W.C. Oshiro. "Degradation of Methomyl in Chlorinated Water," *Environ. Toxicol. Chem.*, 9(5):535–540 (1990).

Miles, C.J. and S. Takashima. "Fate of Malathion and *O,O,S*-Trimethyl Phosphorothioate By-Product in Hawaiian Soil and Water," *Arch. Environ. Contam. Toxicol.*, 20(3):325–329 (1991).

Miles, C.J., M.L. Trehy, and R.A. Yost. "Degradation of *N*-Methylcarbamate and Carbamoyl Oxime Pesticides in Chlorinated Water," *Bull. Environ. Contam. Toxicol.*, 41(6):838–843 (1988).

Miles, J.R.W., C.M. Tu, and C.R. Harris. "Metabolism of Heptachlor and Its Degradation Products by Soil Microorganisms," *J. Econ. Entomol.*, 62(6):1334–1339 (1969).

Miles, J.R.W., C.M. Tu, and C.R. Harris. "Degradation of Heptachlor Epoxide and Heptachlor by a Mixed Culture of Soil Microorganisms," *J. Econ. Entomol.*, 64:839–841 (1971).

Miles, J.R.W., C.M. Tu, and C.R. Harris. "Persistence of Eight Organophosphorus Insecticides in Sterile and Nonsterile Mineral and Organic Soils," *Bull. Environ. Contam. Toxicol.*, 23(3):312–318 (1979).

Miller, M.M., S. Ghodbane, S.P. Wasik, Y.B. Tewari, and D.E. Martire. "Aqueous Solubilities, Octanol/Water Partition Coefficients, and Entropies of Melting of Chlorinated Benzenes and Biphenyls," *J. Chem. Eng. Data*, 29(2):184–190 (1984).

Miller, D.E., F.R. Shaw, and C.T. Smith. "The Comparative Residual Life of Two Formulations of Imidan on Alfalfa," *J. Econ. Entomol.*, 62:720–721 (1969).

Miller, G.C. and R.G. Zepp. "Photoreactivity of Aquatic Pollutants Sorbed on Suspended Sediments," *Environ. Sci. Technol.*, 13(7):860–863 (1979).

Mills, A.C. and J.W. Biggar. "Solubility-Temperature Effect on the Adsorption of Gamma- and Beta-BHC from Aqueous and Hexane Solutions by Soil Materials," *Soil Sci. Soc. Am. Proc.*, 33:210–216 (1969).

Mills, G. and M.R. Hoffman. "Photocatalytic Degradation of Pentachlorophenol on TiO_2 Particles: Identification of Intermediates and Mechanism of Reaction," *Environ. Sci. Technol.*, 27(8):1681–1689 (1993).

Mills, M.S. and E.M. Thurman. "Reduction of Nonpoint Source Contamination of Surface Water and Groundwater by Starch Encapsulation of Herbicides," *Environ. Sci. Technol.*, 28(1):73–79 (1994).

Minero, C., E. Pramauro, E. Pelizzetti, M. Dolci, and A. Marchesini. "Photo-sensitized Transformations of Atrazine Under Simulated Sunlight in Aqueous Humic Acid Solution," *Chemosphere*, 24(11):1597–1606 (1992).

Mingelgrin, U. and Z. Gerstl. "Re-Evaluation of Partitioning as a Mechanism of Nonionic Chemicals Adsorption in Soils," *J. Environ. Qual.*, 12(1):1–11 (1983).

Mingelgrin, U., S. Saltzman, and B. Yaron. "A Possible Model for the Surface-Induced Hydrolysis of Organophosphorus Pesticides on Kaolinite Clays," *Soil Sci. Soc. Am. J.*, 41:519–523 (1977).

Mitchell, S. "A Method for Determining the Solubility of Sparingly Soluble Substances," *J. Chem. Soc. (London)*, 1333–1337 (1926).

Miyazaki, S., G.M. Boush, and F. Matsumura. "Metabolism of ^{14}C-Chlorobenzilate and ^{14}C-Chloropropylate by *Rhodotorula gracilis*," *Appl. Microbiol.*, 18(6):972–976 (1969).

Miyazaki, S., G.M. Boush, and F. Matsumura. "Microbial Degradation of Chlorobenzilate (Ethyl 4,4'-Dichlorobenzilate) and Chloropropylate (Isopropyl 4,4'-Dichlorobenzilate)," *J. Agric. Food Chem.*, 18(1):87–91 (1970).

Miyazaki, S., H.C. Sikka, and R.S. Lynch. "Metabolism of Dichlobenil by Microorganisms in the Aquatic Environment," *J. Agric. Food Chem.*, 23(3):365–368 (1975).

Moilanen, K.W. and D.G. Crosby. "Photodecomposition of 3'4'-Dichloropropionanilide (Propanil)," *J. Agric. Food Chem.*, 20(5):950–953 (1972).

Moilanen, K.W. and D.G. Crosby. "The Photodecomposition of Bromacil," *Arch. Environ. Contam. Toxicol.*, 2(1):3–8 (1974).

Moilanen, K.W., D.G. Crosby, J.R. Humphrey, and J.W. Giles. "Vapor-Phase Photodecomposition of *Chloropicrin (Trichloronitromethane)*," *Tetrahedron*, 34(22):3345–3349 (1978).

Montgomery, J.H. *Groundwater Chemicals Desk Reference* (Boca Raton, FL: CRC Press, Inc., 1996), 1345 p.

Montgomery, M.L. and V.H. Freed. "Metabolism of Triazine Herbicides by Plants," *J. Agric. Food Chem.*, 12(1):11–14 (1964).

Moore, J.W. and S. Ramamoorthy. *Organic Chemicals in Natural Waters — Applied Monitoring and Impact Assessment* (New York: Springer-Verlag, Inc., 1984), 289 p.

Moreale, A. and R. Van Bladel. "Behavior of 2,4-D in Belgian Soils," *J. Environ. Qual.*, 9(4):627–633 (1980).

Morrison, R.T. and R.N. Boyd. *Organic Chemistry* (Boston, MA: Allyn & Bacon, Inc., 1971), 1258 p.

Mortland, M.M. and K.V. Raman. "Catalytic Hydrolysis of Some Organic Phosphate Pesticides by Copper(II)," *J. Agric. Food Chem.*, 15(1):163–167 (1967).

Mosher, D.R. and A.M. Kadoum. "Effects of Four Lights on Malathion Residues on Glass Beads, Sorghum Grain, and Wheat Grain," *J. Econ. Entomol.*, 65:847–850 (1972).

Moshier, L.T. and D. Penner. "Factors Influencing Microbial Degradation of $^{14}CO_2$ in Soil," *Weed Sci.*, 26:686–691 (1978).

Mosier, A.R. and W.D. Guenzi. "Picloram Photolytic Decomposition," *J. Agric. Food Chem.*, 21(5):835–837 (1973).

Mosier, A.R., W.D. Guenzi, and L.L. Miller. "Photochemical Decomposition of DDT by a Free-Radical Mechanism," *Science (Washington, DC)*, 164(3883):1083–1085 (1969).

Mostafa, I.Y., M.R.E. Bahig, I.M.I. Fakhr, and Y. Adam. "Metabolism of Organophosphorus Pesticides. XIV. Malathion Breakdown by Soil Fungi," *Z. Natur-forsch. B.*, 27:1115–1116 (1972).

Moyer, J.R., R.J. Hance, and C.E. McKone. "The Effect of Adsorption of Adsorbents on the Rate of Degradation of Herbicides Incubated with Soil," *Soil Biol. Biochem.*, 4:307–311 (1972).

Muir, D.C.G. "Dissipation and Transformations in Water and Sediment" in *Environmental Chemistry of Herbicides*. Volume II, Grover, R. and A.J. Cessna, Eds. (Boca Raton, FL: CRC Press, Inc., 1991), pp. 1–87.

Muir, D.C.G. and B.E. Baker. "Detection of Triazine Herbicides and Their Degradation Products in Tile-Drain Water from Fields Under Corn (Maize) Production," *J. Agric. Food Chem.*, 21(1):122–125 (1976).

Muir, D.C.G. and B.E. Baker. "The Disappearance and Movement of Three Triazine Herbicides and Several of Their Degradation Products in Soil Under Field Conditions," *Weed Res.*, 18:111–120 (1978).

Mulla, M.S., L.S. Mian, and J.A. Kawecki. "Distribution, Transport and Fate of the Insecticides Malathion and Parathion in the Environment," *Residue Rev.*, 81:116–125 (1981).

Muller, M.M., C. Rosenberg, H. Siltanen, and T. Wartiovaara. "Fate of Glyphosate and Its Influence on Nitrogen Cycling in Two Finnish Agricultural Soils," *Bull. Environ. Contam. Toxicol.*, 27:724–730 (1981).

Munakata K. and M. Kuwahara. "Photochemical Degradation Products of Pentachlorophenol," *Residue Rev.*, 25:13–23 (1969).

Munnecke, D.M. and D.P.H. Hsieh. "Pathways of Microbial Metabolism of Parathion," *Appl. Environ. Microbiol.*, 31(1):63–69 (1976).

Munshi, H.B., K.V.S. Rama Rao, and R.M. Iyer. "Characterization of Products of Ozonolysis of Acrylonitrile in Liquid Phase," *Atmos. Environ.*, 23(9):1945–1948 (1989).

Munshi, H.B., K.V.S. Rama Rao, and R.M. Iyer. "Rate Constants of the Reactions of Ozone with Nitriles, Acrylates and Terpenes in Gas Phase," *Atmos. Environ.*, 23(9):1971–1976 (1989a).

Murray, D.S., P.W. Santelman, and J.M. Davidson. "Comparative Adsorption, Desorption, and Mobility of Dipropetryn and Prometryn in Soil," *J. Agric. Food Chem.*, 23(3):578–582 (1975).

Murthy, N.B.K. and D.D. Kaufman. "Degradation of Pentachloronitrobenzene (PCNB) in Anaerobic Soils," *J. Agric. Food Chem.*, 26:1151–1156 (1978).

Murthy, N.B.K., D.D. Kaufman, and G.F. Fries. "Degradation of Pentachlorophenol (PCP) in Aerobic and Anaerobic Soil," *J. Environ. Sci. Health*, B14(1):1–14 (1979).

Musoke, G.M.S., D.J. Roberts, and M. Cooke. "Heterogeneous Hydrodechlorination of Chlordan," *Bull. Environ. Contam. Toxicol.*, 28(4):467–472 (1982).

Nair, D.R. and J.L. Schnoor. "Effect of Two Electron Acceptors on Atrazine Mineralization Rates in Soil," *Environ. Sci. Technol.*, 26(11):2298–2300 (1992).

Nakajima, S., N. Nakato, and T. Tani. "Microbial Transformation of 2,4-D and Its Analog," *Chem. Pharm. Bull.*, 21:671–673 (1973).

Namdeo, K.N. "Persistence of Dalapon in Grassland Soil," *Plant and Soil*, 31:445–448 (1972).

Nash, R.G. "Comparative Volatilization and Dissipation Rates of Several Pesticides from Soil," *J. Agric. Food. Chem.*, 31(2):210–217 (1983).

Nash, R.G. "Dissipation from Soil," in *Environmental Chemistry of Herbicides.* Volume I, R. Grover, Ed. (Boca Raton, FL: CRC Press, Inc. 1988), pp. 131–169.

Nash, R.G. and E.A. Woolson. "Persistence of Chlorinated Insecticides in Soils," *Science (Washington, DC)*, 157(3791):924–927 (1967).

Natarajan, G.S. and K.A. Venkatachalam. "Solubilities of Some Olefins in Aqueous Solutions," *J. Chem. Eng. Data*, 17(3):328–329 (1972).

Nazer, I.K. and H.A. Mosoud. "Residues of Dicofol on Cucumber Grown Under Plastic Covers in Jordan," *J. Environ. Sci. Health*, B21(5):387–399 (1986).

Nearpass, D.C. "Effects of Soil Acidity on the Adsorption, Penetration, and Persistence of Simazine," *Weeds*, 13:341–346 (1965).

Nearpass, D.C. "Hydrolysis of Propazine by the Surface Acidity of Organic Matter," *Soil Sci. Soc. Am. Proc.*, 36:606–610 (1972).

Neary, D.G., P.B. Bush, and J.E. Douglass. "Off-Site Movement of Hexazinone in Stormflow and Baseflow from Forest Watersheds," *Weed Sci.*, 31:543–551 (1983).

Nelson, I.M. "Biologically Induced Hydrolysis of Parathion in Soil: Isolation of Hydrolyzing Bacteria," *Soil Biol. Biochem.*, 14:219–222 (1982).

Nesbitt, H.J. and J.R. Watson. "Degradation of the Herbicide 2,4-D in River Water. II. The Role of Suspended Sediment. Nutrients and Water Temperature," *Water Res.*, 14:1689–1694 (1980).

Neudorf, S. and M.A.Q. Khan. "Pick Up and Metabolism of DDT, Dieldrin, and Photodieldrin by a Fresh Water Alga (*Ankistrodesmus amalloides*) and a Microcrustacean (*Daphnia pulex*)," *Bull. Environ. Contam. Toxicol.*, 13(4):443–450 (1975).

Newland, L.W., G. Chesters, and G.B. Lee. "Degradation of γ-BHC in Simulated Lake Impoundments as Affected by Aeration," *J. Water Pollut. Control Fed.*, 41(5):R174–R188 (1969).

Newton, M., K.M. Howard, B.R. Kelpsas, R. Danhaus, C.M. Lottman, and S. Dubelman. "Fate of Glyphosate in an Oregon Forest Ecosystem," *J. Agric. Food Chem.*, 32(5):1144–1151 (1984).

Newton, M., F. Roberts, A. Allen, B. Kelpas, D. White, and P. Boyd. "Deposition and Dissipation of Three Herbicides in Foliage, Litter, and Soil of Brushfields of Southwest Oregon," *J. Agric. Food Chem.*, 38:574–583 (1990).

Ngabe, B., T.F. Bidleman, and R.L. Falconer. "Base Hydrolysis of α- and β-Hexachlorocyclohexanes," *Environ. Sci. Technol.*, 27(9):1930–1933 (1993).

Nigg, H.N., J.A. Reinert, and J.H. Stamper. "Disappearance of Acephate, Methamidophos, and Malathion from Citrus Foliage," *Bull. Environ. Contam. Toxicol.*, 26(2):267–272 (1981).

Niimi, A.J. "Solubility of Organic Chemicals in Octanol, Triolein and Cod Liver Oil and Relationships between Solubility and Partition Coefficients," *Water Res.*, 25(12):1515–1521 (1991).

Nkedi-Kizza, P., P.S.C. Rao, and J.W. Johnson. "Adsorption of Diuron and 2,4,5-T on Soil Particle-Size Separates," *J. Environ. Qual.*, 12(2):195–197 (1983).

Normura, N.S. and H.W. Hilton. "The Adsorption and Degradation of Glyphosate in Five Hawaiian Sugar Cane Soils," *Weed Res.*, 17:113–121 (1977).

Norris, L.A., M.L. Montgomery, and L.E. Warren. "Triclopyr Persistence in Western Oregon Hill Pastures," *Bull. Environ. Contam. Toxicol.*, 39:134–141 (1987).

Novick, N.J., R. Mukherjee, and M. Alexander. "Metabolism of Alachlor and Propachlor in Suspensions of Pretreated Soils and in Samples from Ground Water Aquifers," *J. Agric. Food Chem.*, 34(4):721–725 (1986).

Obien, S.R. and R.E. Green. "Degradation of Atrazine in Four Hawaiian Soils," *Weed Sci.*, 17(4):509–514 (1969).

O'Connell, K.M., E.J. Breaux, and R.T. Fraley. "Different Rates of Metabolism of Two Chloroacetanilide Herbicides in Pioneer 3220 Corn," *Plant Physiol.*, 86:359–363 (1988).

O'Connor, G.A. and J.U. Anderson. "Soil Factors Affecting the Adsorption of 2,4,5-T," *Soil Sci. Soc. Am. Proc.*, 38:433–436 (1974).

Odeyemi, O. and M. Alexander. "Resistance of Rhizobium Strains to Phygon, Spergon, and Thiram," *Appl. Environ. Microbiol.*, 33(4):784–790 (1977).

Opperhuizen, A., F.A.P.C. Gobas, J.M.D. Van der Steen, and O. Hutzinger. "Aqueous Solubility of Polychlorinated Biphenyls Related to Molecular Structure," *Environ. Sci. Technol.*, 22(6):638–646 (1988).

Osborne, A.D., J.N. Pitts, Jr., and E.F. Darley. "On the Stability of Acrolein Toward Photooxidation in the Near Ultra-Violet," *Int. J. Air and Water Pollut.*, 6:1–3 (1962)

Otto, S., P. Beutel, N. Decker, and R. Huber. "Investigations in the Degradation of Bentazon in Plant and Soil," *Adv. Pestic. Sci.*, 3:551–556 (1978).

Ou, L.-T., K.S.V. Edvarsson, and P.S.C. Rao. "Aerobic and Anaerobic Degradation of Aldicarb in Soils," *J. Agric. Food Chem.*, 33(1):72–78 (1985).

Ou, L.-T., D.H. Gancarz, W.B. Wheeler, P.S.C. Rao, and J.M. Davidson. "Influence of Soil Temperature and Soil Moisture on Degradation and Metabolism of Carbofuran in Soils," *J. Environ. Qual.*, 11(2):293–298 (1982).

Ou, L.-T. and P.S.C. Rao. "Degradation and Metabolism of Oxamyl and Phenamiphos in Soils," *J. Environ. Sci. Health*, B21(1):25–40 (1986).

Ou, L.-T., D.F. Rothwell, W.B. Wheeler, and J.M. Davidson. "The Effects of High 2,4-D Concentrations on Degradation and Carbon Dioxide Evolutions in Soils," *J. Environ. Qual.*, 7:241–246 (1978).

Owen, W.J. and B. Donzel. "Oxidative Degradation of Chlortoluron, Propiconazole, and Metalaxyl in Suspension Cultures of Various Crop Plants," *Pestic. Biochem. Physiol.*, 26:75–89 (1986).

Owens, J.W., S.P. Wasik, and H. DeVoe. "Aqueous Solubilities and Enthalpies of Solution of *n*-Alkylbenzenes," *J. Chem. Eng. Data*, 31(1):47–51 (1986).

Pape, B.E. and M.J. Zabik. "Photochemistry of Bioactive Compounds. Photochemistry of Selected 2-Chloro- and 2-Methylthio-4,6-di(alkylamino)-*s*-Triazine Herbicides," *J. Agric. Food Chem.*, 18(2):202–207 (1970).

Pape, B.E. and M.J. Zabik. "Photochemistry of Bioactive Compounds. Solution-Phase Photochemistry of Symmetrical Triazines," *J. Agric. Food Chem.*, 20(2):316–320 (1972).

Pardue, J.R., E.A. Hansen, R.P. Barron, and J.-Y.T. Chen. "Diazinon Residues on Field-Sprayed Kale. Hydroxydiazinon-A New Alteration Product of Diazinon," *J. Agric. Food Chem.*, 18(3):405–408 (1970).

Paris, D.F. and D.L. Lewis. "Chemical and Microbial Degradation of Ten Selected Pesticides in Aquatic Systems," *Residue Rev.*, 45:95–124 (1973).

Paris, D.F. and D.L. Lewis. "Accumulation of Methoxychlor by Microorganisms Isolated from Aqueous Systems," *Bull. Environ. Contam. Toxicol.*, 15(1):24–32 (1976).

Paris, D.F., D.L. Lewis, and N.L. Wolfe. "Rates of Degradation of Malathion by Bacteria Isolated from Aquatic System," *Environ. Sci. Technol.*, 9(2):135–138 (1975).

Paris, D.F., W.C. Steen, G.L. Baughham, and J.T. Barnett, Jr. "Second-Order Model to Predict Microbial Degradation of Organic Compounds in Natural Waters," *Appl. Environ. Microbiol.*, 41(3):603–609 (1981).

Parlar, H. "Photoinduced Reactions of Two Toxaphene Compounds in Aqueous Medium and Adsorbed on Silica Gel," *Chemosphere*, 17(11):2141–2150 (1988).

Parochetti, I.U. and J.G.W. Dec. "Photodecomposition of Eleven Dinitroaniline Herbicides," *Weed Sci.*, 26:153–156 (1978).

Parr, J.F. and S. Smith. "Degradation of Trifluralin Under Laboratory Conditions and Soil Anaerobiosis," *Soil Sci.*, 115(1):55–63 (1973).

Parr, J.F. and S. Smith. "Degradation of DDT in an Everglades Muck as Affected by Lime, Ferrous Ion, and Anaerobiosis," *Soil Sci.*, 118(1):45–52 (1974).

Patil, K.C. and F. Matsumura. "Degradation of Endrin, Aldrin, and DDT by Soil Microorganisms," *Appl. Microbiol.*, 19:879–881 (1970).

Patil, K.C., F. Matsumura, and G.M. Boush. "Metabolic Transformation of DDT, Dieldrin, Aldrin, and Endrin by Marine Microorganisms," *Environ. Sci. Technol.*, 6(7):629–632 (1972).

Patumi, M., C. Marucchini, M. Businelli, and F. Tafuri. "Effetto di Ripetuti Trattamenti con Atrazina Sulla sua Persistenza in Terreni Destinati alla Monosuccessione Di Mais," *Agrochimica*, 25(2):162–167 (1981).

Peek, D.C. and A.P. Appleby. "Phytotoxicity, Adsorption, and Mobility of Metribuzin and Its Ethylthio Analog as Influenced by Soil Properties," *Weed Sci.*, 37(3):419–423 (1989).

Pelizzetti, E., V. Maurino, C. Minero, V. Carlin, E. Pramauro, O. Zerbinati, and M.L. Tosato. "Photocatalytic Degradation of Atrazine and Other *s*-Triazine Herbicides," *Environ. Sci. Technol.*, 24(10):1559–1565 (1990).

Pelizzetti, E., C. Minero, V. Carlin, M. Vincenti, E. Pramauro, and M. Dolci. "Identification of Photocatalytic Degradation Pathways of 2-Cl-*s*-Triazine Herbicides and Detection of Their Decomposition Intermediates," *Chemosphere*, 24(7):891–910 (1992).

Pereira, W.E. and C.E. Rostad. "Occurrence, Distributions, and Transport of Herbicides and Their Degradation Products in the Lower Mississippi River and Its Tributaries," *Environ. Sci. Technol.*, 24(9):1400–1406 (1990).

Pereira, W.E., C.E. Rostad, C.T. Chiou, T.I. Brinton, L.B. Barber, II, D.K. Demcheck, and C.R. Demas. "Contamination of Estuarine Water, Biota, and Sediment by Halogenated Organic Compounds: A Field Study," *Environ. Sci. Technol.*, 22(7):772–778 (1988).

Peris-Cardells, E., J. Terol, A.R. Mauri, M. de la Guardia, and E. Pramauro. "Continuous Flow Photocatalytic Degradation of Carbaryl in Aqueous Media," *J. Environ. Sci. Health*, B28(4):431–445 (1993).

Perscheid, M., H. Schlueter, and K. Ballschmiter. "Aerober Abbau von Endosulfan Durch Bodenmikroorganismen," *Zeitschrift Naturforsch*, 28:761–763 (1973).

Peterson, C.A. and L.V. Edgington. "Quantitative Estimation of the Fungicide Benomyl Using a Bioautograph Technique," *J. Agric. Food Chem.*, 17(4):898–899 (1969).

Peterson, P.J. and B.A. Swisher. "Absorption, Translocation, and Metabolism of ^{14}C-Chlorsulfuron in Canada Thistle (Cirsium arvense)," *Weed Sci.*, 33:7–11 (1985).

Petrier, C., M. Micolle, G. Merlin, J.-L. Luche, and G. Reverdy. "Characteristics of Pentachlorophenate Degradation in Aqueous Solution by Means of Ultrasound," *Environ. Sci. Technol.*, 26(8):1639–1642 (1992).

Pettigrew, C.A., M.J.B. Paynter, and N.D. Camper. "Anaerobic Microbial Degradation of the Herbicide Propanil," *Soil Biol. Biochem.*, 17:815–818 (1985).

Peyton, T.O., R.V. Steel, and W.R. Mabey. "Carbon Disulfide, Carbonyl Sulfide: Literature Review and Environmental Assessment," U.S. EPA Report PB-257-947/2 (1976), 64 p.

Pfaender, F.K. and M. Alexander. "Extensive Microbial Degradation of DDT *in vitro* and DDT Metabolism by Natural Communities," *J. Agric. Food Chem.*, 20(4):842–846 (1972).

Pfaender, F.K. and M. Alexander. "Effect of Nutrient Additions on the Apparent Cometabolism of DDT," *J. Agric. Food Chem.*, 21(3):397–399 (1973).

Phillips, D.D., G.E. Pollard, and S.B. Soloway. "Thermal Isomerization of Endrin and Its Behavior in Gas Chromatography," *J. Agric. Food Chem.*, 10(3):217–221 (1962).

Piccolo, A., G. Celano, M. Arienzo, and A. Mirabella. "Adsorption and Desorption of Glyphosate in Some European Soils," *J. Environ. Sci. Health*, B29(6):1105–1115 (1994).

Pickering, Q.H., C. Henderson, and A.E. Lemke. "The Toxicity of Organic Phosphorus Insecticides to Different Species of Warmwater Fishes," *Trans. Am. Fish Soc.*, 91(2):175–184 (1962).

Pieper, G.R. and C.E. Richmond. "Residues of Trichlorfon and Lauroyl Trichlorfon in Douglas Fir, Willow, Grass, Aspen Foliage, and in Creek Water after Aerial Application," *Bull. Environ. Contam. Toxicol.*, 15(2):250–256 (1976).

Pierce, R.H., Jr., C.E. Olney, and G.T. Felbeck, Jr. "*p,p'*-DDT Adsorption to Suspended Particulate Matter in Sea Water," *Geochim. Cosmochim. Acta*, 38(7):1061–1073 (1974).

Pignatello, J.J. "Ethylene Dibromide Mineralization in Soils Under Aerobic Conditions," *Appl. Environ. Microbiol.*, 51(3):588–592 (1986).

Pignatello, J.J. "Microbial Degradation of 1,2-Dibromoethane in Shallow Aquifer Materials," *J. Environ. Qual.*, 16(4):307–312 (1987).

Pillai, C.G.P., J.D. Weete, and D.E. Davis. "Metabolism of Atrazine by *Spartina alterniflora*. 1. Chloroform-Soluble Metabolites," *J. Agric. Food Chem.*, 25(4):852–855 (1977).

Plimmer, J.R. "The Photochemistry of Halogenated Herbicides," *Residue Rev.*, 33:47–74 (1970).

Plimmer, J.R. "Photolysis of TCDD and Trifluralin on Silica and Soil," *Bull. Environ. Contam. Toxicol.*, 20(1):87–92 (1978).

Plimmer, J.R. and B.E. Hummer. "Photolysis of Amiben (3-Amino-2,5-dichlorobenzoic Acid) and Its Methyl Ester," *J. Agric. Food Chem.*, 17(1):83–85 (1969).

Plimmer, J.R., P.C. Kearney, H. Chisaka, J.B. Yount, and U.I. Klingebiel. "1,3-Bis(3,4-dichlorophenyl)triazene from Propanil in Soils," *J. Agric. Food Chem.*, 18(5):859–861 (1970a).

Plimmer, J.R., P.C. Kearney, D.D. Kaufman, and F.S. Guardia. "Amitrole Decomposition by Free Radical-Generating Systems and by Soils," *J. Agric. Food Chem.*, 15(6):996–999 (1967).

Plimmer, J.R., P.C. Kearney, and D.W. Von Endt. "Mechanism of Conversion of DDT to DDD by Aerobacter aerogenes," *J. Agric. Food Chem.*, 16(4):594–597 (1968).

Plimmer, J.R. and U.J. Klingebiel. "Photolysis of Hexachlorobenzene," *J. Agric. Food Chem.*, 24(4):721–723 (1976).

Plimmer, J.R., U.J. Klingebiel, and B.E. Hummer. "Photooxidation of DDT and DDE," *Science (Washington, DC)*, 167(3914):67–69 (1970).

Plust, S.J., J.R. Loehe, F.J. Feher, J.H. Benedict, and H.F. Herbrandson. "Kinetics and Mechanism of Hydrolysis of Chloro-1,3,5-triazines. Atrazine," *J. Org. Chem.*, 46:3661–3664 (1981).

Pogány, E., P.R. Wallnöfer, W. Ziegler, and W. Mücke. "Metabolism of *o*-Nitroaniline and Di-*n*-butyl Phthalate in Cell Suspension Cultures of Tomatoes," *Chemosphere*, 21(4/5):557–562 (1990).

Pollero, R. and S.C. dePollero. "Degradation of DDT by a Soil Amoeba," *Bull. Environ. Contam. Toxicol.*, 19(3):345–350 (1978).

Polles, S.G. and S.B. Vinson. "Effect of Droplet Size on Persistence of Malathion and Comparison of Toxicity of ULV and EC Malathion to Tobacco Budworm Larvae," *J. Econ. Entomol.*, 62:89–94 (1969).

Pree, D.J., K.P. Butler, E.R. Kimball, and D.K.R. Steward. "Persistence of Foliar Residues of Dimethoate and Azinphosmethyl and Their Toxicity to the Apple Maggot," *J. Econ. Entomol.*, 69:473–478 (1976).

Price, L.C. "Aqueous Solubility of Petroleum as Applied to its Origin and Primary Migration," *Am. Assoc. Pet. Geol. Bull.*, 60(2):213–244 (1976).

Probst, G.W., T. Golab, R.J. Herberg, F.J. Holzer, S.J. Parka, C. van der Schans, and J.B. Tepe. "Fate of Trifluralin in Soils and Plants," *J. Agric. Food Chem.*, 15(4):592–599 (1967).

Probst, G.W. and J.B. Tepe. "Trifluralin and Related Compounds" in *Degradation of Herbicides*, Kearney, P.C. and D.D. Kaufman, Eds., (New York: Marcel Dekker, Inc., 1969), pp. 255–282.

Putnam, T.B., D.D. Bills, and L.M. Libbey. "Identification of Endosulfan Based on the Products of Laboratory Photolysis," *Bull. Environ. Contam. Toxicol.*, 13(6):662–665 (1975).

Que Hee, S.S. and R.G. Sutherland. *The Phenoxyalkanoic Herbicides. Vol. 1. Chemistry, Analysis, and Environmental Pollution* (Boca Raton, FL: CRC Press, Inc, 1981), 321 p.

Que Hee, S.S., R.G. Sutherland, K.S. McKinlay, and J.G. Sara. "Factors Affecting the Volatility of DDT, Dieldrin, and Dimethylamine Salt of (2,4-Dichlorophenoxy)acetic Acid (2,4-D) from Leaf and Glass Surfaces," *Bull. Environ. Contam. Toxicol.*, 13(3):284–290 (1975).

Quirke, J.M.E., A.S.M. Marei, and G. Eglinton. "The Degradation of DDT and Its Degradative Products by Reduced Iron (III) Porphyrins and Ammonia," *Chemosphere*, 8(3):151–155 (1979).

Quistad, G.B. and K.M. Mulholland. "Photodegradation of Dienochlor [Bis(pentachloro-2,4-cyclopentadien-1-yl)] by Sunlight," *J. Agric. Food Chem.*, 31(3):621–624 (1983).

Racke, K.D. and J.R. Coats. "Enhanced Degradation of Isofenphos by Soil Microorganisms," *J. Agric. Food Chem.*, 35(1):94–99 (1987).

Racke, K.D. and J.R. Coats. "Enhanced Degradation and the Comparative Fate of Carbamate Insecticides in Soil," *J. Agric. Food Chem.*, 36(5):1067–1072 (1988).

Racke, K.D., J.R. Coats, and K.R. Titus. "Degradation of Chlorpyrifos and Its Hydrolysis Product, 3,5,6-Trichloro-2-pyridinol, in Soil," *J. Environ. Sci. Health*, B23(6):527–539 (1988).

Radosevich, S.R. and W.L. Winterlin. "Persistence of 2,4-D and 2,4,5-T in Chaparral Vegetation and Soil," *Weed Sci.*, 25:423–425 (1977).

Raghu, K. and I.C. MacRae. "Biodegradation of the Gamma Isomer of Benzene Hexachloride in Submerged Soils," *Science (Washington, DC)*, 154(3746):263–264 (1966).

Raha, P. and A.K. Das. "Photodegradation of Carbofuran," *Chemosphere*, 21(1/2):99–106 (1990).

Rajagopal, B.S., G.P. Brahmaprakash, B.R. Reddy, U.D. Singh, and N. Sethunathan. "Effect and Persistence of Selected Carbamate Pesticides in Soil," *Residue Rev.*, 93:1–199 (1984).

Rajagopal, B.S., K. Chendrayan, B.R. Reddy, and N. Sethunathan. "Persistence of Carbaryl in Flooded Soils and Its Degradation by Soil Enrichment Cultures," *Plant and Soil*, 73(1):35–45 (1973).

Rajagopal, B.S., S. Panda, and N. Sethunathan. "Accelerated Degradation of Carbaryl and Carbofuran in a Flooded Soil Pretreated with Hydrolysis Products, 1-Naphthol and Carbofuran Phenol," *Bull. Environ. Contam. Toxicol.*, 36(6):827–832 (1986).

Rajagopal, B.S., V.R. Rao, G. Nagendrappa, and N. Sethunathan. "Metabolism of Carbaryl and Carbofuran by Soil-Enrichment and Bacterial Cultures," *Can. J. Microbiol.*, 30(12):1458–1466 (1984a).

Rajaram, K.P. and N. Sethunathan. "Effect of Organic Sources on the Degradation of Parathion in Flooded Alluvial Soil," *Soil Sci.*, 119(4):296–300 (1975).

Ralls, J.W., D.R. Gilmore, and A. Cortes. "Fate of Radioactive *O,O*-Diethyl *O*-(2-Isopropyl-4-methylpyrimidin-6-yl) Phosphorothioate on Field-Grown Experimental Crops," *J. Agric. Food Chem.*, 14(4):387–293 (1966).

Ralston, A.W. and C.W. Hoerr. "The Solubilities of the Normal Saturated Fatty Acids," *J. Org. Chem.*, 7:546–555 (1942).

Ramakrishna, C., T.K.S. Gowda, and N. Sethunathan. "Effect of Benomyl and Its Hydrolysis Products, MBC and AB, on Nitrification in a Flooded Soil," *Bull. Environ. Contam. Toxicol.*, 21(3):328–333 (1979).

Ramanand, K., S. Panda, M. Sharmila, T.K. Adhya, and N. Sethunathan. "Development and Acclimatization of Carbofuran-Degrading Soil Enrichment Cultures at Different Temperatures," *J. Agric. Food Chem.*, 36(1):200–205 (1988).

Ramanand, K., M. Sharmila, and N. Sethunathan. "Mineralization of Carbofuran by a Soil Bacterium," *Appl. Environ. Microbiol.*, 54(8):2129–2133 (1988a).

Randall, T.L. and P.V. Knopp. "Detoxification of Specific Organic Substances by Wet Oxidation," *J. Water Pollut. Control Fed.*, 52(8):2117–2130 (1980).

Rao, P.S.C. and J.M. Davidson. "Adsorption and Movement of Selected Pesticides at High Concentrations in Soils," *Water Res.*, 13(4):375–380 (1979).

Rao, P.S.C. and J.M. Davidson. "Retention and Transformation of Selected Pesticides and Phosphorus in Soil-Water Systems: A Critical Review," Office of Research and Development, U.S. EPA Report-600/3–82-060 (1982), 321 p.

Rao, P.S.C. and N. Sethunathan. "Effect of Ferrous Sulfate on Parathion Degradation in Flooded Soil," *J. Environ. Sci. Health*, B14(3):335–351 (1979).

Reddy, B.R. and N. Sethunathan. "Mineralization of Parathion in the Rice Rhizophere," *Appl. Environ. Microbiol.*, 45(3):826–829 (1983).

Reddy, B.R. and N. Sethunathan. "Salinity and Persistence of Parathion in Flooded Soils," *Soil Biol. Biochem.*, 17(2):235–239 (1985).

Reddy, K.N., M. Singh, and A.K. Alva. "Sorption and Leaching of Bromacil and Simazine in Florida Flatwoods Soils," *Bull. Environ. Contam. Toxicol.*, 48(5):662–670 (1992).

Redemann, C.T., R.W. Meikle, P. Hamilton, V.S. Banks, and C.R. Youngson. "The Fate of 4-Amino-3,5,6-Trichloropicolinic Acid in Spring Wheat and Soil," *Bull. Environ. Contam. Toxicol.*, 3(2):80–96 (1968).

Reduker, S., C.G. Uchrin, and G. Winnett. "Characteristics of the Sorption of Chlorothalonil and Azinphos-Methyl to a Soil from a Commercial Cranberry Bog," *Bull. Environ. Contam. Toxicol.*, 41(5):633–641 (1988).

"Registry of Toxic Effects of Chemical Substances," U.S. Department of Health and Human Services, National Institute for Occupational Safety and Health (1985), 2050 p.

Rehberg, C.E. and M.B. Dixon. "*n*-Alkyl Lactates and Their Acetates," *J. Am. Chem. Soc.*, 72(5):1918–1922 (1950).

Reinert, K.H. M.L. Hinman, and J.H. Rodgers. "Fate of Endothall During the Pay Mayse Lake, Texas Aquatic Plant Management Program," *Arch. Environ. Contam. Toxicol.*, 17:195–199 (1988).

Reinert, K.H. and J.H. Rodgers, Jr. "Influence of Sediment Types on the Sorption of Endothall," *Bull. Environ. Contam. Toxicol.*, 32(5):557–564 (1984).

Reinert, K.H. and J.H. Rodgers, Jr. "Fate and Persistence of Aquatic Herbicides," *Rev. Environ. Contam. Toxicol.*, 98:61–98 (1987).

Rejtö, M., S. Saltzman, A.J. Acher, and L. Muszkat. "Identification of Sensitized Photo-oxidation Products of *s*-Triazine Herbicides in Water," *J. Agric. Food Chem.*, 31(1):138–142 (1983).

Rejtö, M., S. Saltzman, and A.J. Acher. "Photodecomposition of Propachlor," *J. Agric. Food Chem.*, 32(2):226–230 (1984).

Renberg, L., M. Tarkpea, and E. Lindén. "The Use of the Bivalve Mytilus edulis as a Test Organism for Bioconcentration Studies," *Ecotoxicol. Environ. Safety*, 9(2):171–178 (1982).

Reuber, M.D. "Carcinogenicity of Lindane," *Environ. Res.*, 19:460–491 (1979).

Rhodes, R.C. "Metabolism of [2-^{14}C]Terbacil in Alfalfa," *J. Agric. Food Chem.*, 25(5):1066–1068 (1977).

Rhodes, R.C. "Studies with ^{14}C-Labeled Hexazinone in Water and Bluegill Sunfish," *J. Agric. Food Chem.*, 28(2):306–310 (1980).

Rhodes, R.C. "Soil Studies with ^{14}C-Labeled Hexazinone," *J. Agric. Food Chem.*, 28(2):311–315 (1980a).

Rhodes, R.C., I.J. Belasco, and H.L. Pease. "Determination and Mobility of Agrochemicals on Soils," *J. Agric. Food Chem.*, 18(3):524–528 (1970).

Rhodes, R.C. and J.D. Long. "Run-off and Mobility Studies on Benomyl in Soils and Turf," *Bull. Environ. Contam. Toxicol.*, 12(4):385–393 (1974).

Rice, C.P., H.C. Sikka, and R.S. Lynch. "Persistence of Dichlobenil in a Farm Pond," *J. Agric. Food Chem.*, 22:533–535 (1974).

Richard, Y. and L. Bréner. "Removal of Pesticides from Drinking Water by Ozone," in *Handbook of Ozone Technology and Applications, Volume II. Ozone for Drinking Water Treatment*, Rice, A.G. and A. Netzer, Eds. (Montvale, MA: Butterworth Publishers, 1984), pp. 77–97.

Rippen, G., M. Ilgenstein, and W. Klöpffer. "Screening of the Adsorption Behavior of New Chemicals: Natural Soils and Model Adsorbents," *Ecotoxicol. Environ. Safety*, 6(3):236–245 (1982).

Roadhouse, F.E.B. and L.A. Birk. "Penetration of and Persistence in Soil of the Herbicide 2-Chloro-4,6-bis(ethylamino)-*s*-triazine (Simazine)," *Can. J. Plant Sci.*, 41:252–259 (1961).

Robb, I.D. "Determination of the Aqueous Solubility of Fatty Acids and Alcohols," *Aust. J. Chem.*, 18:2281–2285 (1966).

Roberts, J.R., A.S.W. DeFrietas, and M.A.J. Bidney. "Influence of Lipid Pool Size on Bioaccumulation of the Insecticide Chlordane by Northern Redhorse Suckers (Maxostoma macrolepidotum)," *J. Fish. Res. Board Can.*, 34(1):89–97 (1977).

Roberts, H.A. and B.J. Wilson. "Adsorption of Chlorpropham by Different Soils," *Weed Res.*, 5(4):348–350 (1965).

Robinson, J., A. Richardson, B. Bush, and K.E. Elgar. "A Photo-Isomerization Product of Dieldrin," *Bull. Environ. Contam. Toxicol.*, 1(4):127–132 (1966).

Roburn, J. "Effect of Sunlight and Ultraviolet Radiation on Chlorinated Pesticide Residues," *Chem. Ind.*, pp. 1555–1556 (1963).

Rosen, J.D. and W.F. Carey. "Preparation of the Photoisomers of Aldrin and Dieldrin," *J. Agric. Food Chem.*, 16(3):536–537 (1968).

Rosen, J.D., R.F. Strusz, and C.C. Still. "Photolysis of Phenylurea Herbicides," *J. Agric. Food Chem.*, 17(2):206–207 (1969).

Rosen, J.D. and D.J. Sutherland. "The Nature and Toxicity of the Photoconversion Products of Aldrin," *Bull. Environ. Contam. Toxicol.*, 2(1):1–9 (1967).

Rosen, J.D., J. Sutherland, and G.R. Lipton. "The Photochemical Isomerization of Dieldrin and Endrin and Effects on Toxicity," *Bull. Environ. Contam. Toxicol.*, 1(4):133–140 (1966).

Rosenberg, A. and M. Alexander. "Microbial Cleavage of Various Organophosphorus Insecticides," *Appl. Environ. Microbiol.*, 37(5):886–891 (1979).

Rosenfield, C. and W. Van Valkenburg. "Decomposition of (O,O-Dimethyl-O-2,4,5-trichlorophenyl) Phosphorothioate (Ronnel) Adsorbed on Bentonite and other Clays," *J. Agric. Food Chem.*, 13(1):68–72 (1972).

Ross, R.D. and D.G. Crosby. "The Photooxidation of Aldrin in Water," *Chemosphere*, 4(5):277–282 (1975).

Ross R.D. and D.G. Crosby. "Photooxidant Activity in Natural Waters," *Environ. Toxicol. Chem.*, 4(6):773–778 (1985).

Ross, L.J., S. Powell, J.E. Fleck, and B. Buechler. "Dissipation of Bentazon in Flooded Rice Fields," *J. Environ. Qual.*, 18(1):105–109 (1989).

Roth, W. and E. Knusli. "Beitrag zur Kenntnis der Resistenz-Phanomene Einzelner Pflanzen Gegeenuber dem Phtotoxischen Wirkstoff Simazin," *Experientia*, 17:312–313 (1961).

Rothman, A.M. "Low Vapor Pressure Determination by the Radiotracer Transpiration Method," *Anal. Chem.*, 28(6):1225–1228 (1980).

Rouchaud, J., C. Moons, and J.A. Meyer. "The Products of Metabolism of [^{14}C]Triadimefon in the Grain and in the Straw of Ripe Barley," *Bull. Environ. Contam. Toxicol.*, 27(4):543–550 (1981).

Rouchaud, J., C. Moons, and J.A. Meyer. "Metabolism of ^{14}C-Triadimefon in Barley Shoots," *Pestic. Sci.*, 13(2):169–176 (1982).

Rouchaud, J., P. Roucourt, and A. Vanachter. "Hydrolytic Biodegradation of Chlorothalonil in the Soil and Cabbage Crops," *Toxicol. Environ. Chem.*, 17:59–68 (1988).

Roy, D.N., S.K. Konar, D.A. Charles, J.C. Feng, R. Prasad, and R.A. Campbell. "Determination of Persistence, Movement, and Degradation of Hexazinone in Selected Canadian Boreal Forest Soils," *J. Agric. Food Chem.*, 37(2):443–447 (1989).

Roy, W.R. and I.G. Krapac. "Adsorption and Desorption of Atrazine and Deethylatrazine by Low Organic Carbon Geologic Materials," *J. Environ. Qual.*, 23:549–556 (1994).

Rubin, H.E., R.V. Subba-Rao, and M. Alexander. "Rates of Mineralization of Trace Concentrations of Aromatic Compounds in Lake Water and Sewage Samples," *Appl. Environ. Microbiol.*, 43:1133–1138 (1982).

Rueppel, M.L., B.B. Brightwell, J. Schaefer, and J.T. Marvel. "Metabolism and Degradation of Glyphosate in Soil and Water," *J. Agric. Food Chem.*, 25(3):517–528 (1977).

Ruzo, L.O. and J.E. Casida. "Photochemistry of Thiocarbamate Herbicides: Oxidative and Free Radical Processes of Thiobencarb and Diallate," *J. Agric. Food Chem.*, 33(2):272–276 (1985).

Ruzo, L.O., J.K. Lee, and M.J. Zabik. "Solution-Phase Photodecomposition of Several Substituted Diphenyl Ether Herbicides," *J. Agric. Food Chem.*, 28(6):1289–1292 (1980).

Russell, J.D., M. Cruz, and J.L. White. "Mode of Chemical Degradation of *s*-Triazines by Montmorillonite," *Science (Washington, DC)*, 160(3834):1340–1342 (1968).

Sacher, R.M.G., G.F. Ludvik, and J.M. Deming. "Bioactivity and Persistence of Some Parathion Formulations in Soil," *J. Econ. Entomol.*, 65:329–332 (1971).

Sahoo, A., S.K. Sahu, M. Sharmila, and N. Sethunathan. "Persistence of Carbamate Insecticides, Carbosulfan and Carbofuran in Soils as Influenced by Temperature and Microbial Activity," *Bull. Environ. Contam. Toxicol.*, 44(6):948–954 (1990).

Sahu, S.K., K.K. Patnaik, and N. Sethunathan. "Dehydrochlorination of δ-Isomer of Hexachlorocyclohexane by a Soil Bacterium, *Pseudomonas* sp.," *Bull. Environ. Contam. Toxicol.*, 48(2):265–268 (1992).

Saini, M.L. and H.W. Dorough. "Persistence of Malathion and Methyl Parathion when Applied as Ultra-Low-Volume and Emulsifiable Concentrate Sprays," *J. Econ. Entomol.*, 63:405–408 (1970).

Saleh, M.A. and J.E. Casida. "Reductive Dechlorination of the Toxaphene Component 2,2,5-*endo*,6-*exo*,8,9,10-Heptachlorobornane in Various Chemical, Photochemical, and·Metabolic Systems," *J. Agric. Food Chem.*, 26(3):583–590 (1978).

Sanborn, J.R., B.M. Francis, and R.L. Metcalf. "The Degradation of Selected Pesticides in Soil: A Review of the Published Literature," Office of Research and Development, U.S. EPA Report-600/9-77-022 (1977), 616 p.

Sanborn, J.R. and C. Yu. "The Fate of Dieldrin in a Model Ecosystem," *Bull. Environ. Contam. Toxicol.*, 10(6):340–346 (1973).

Sánchez-Camazano, M. and M.J. and Sánchez-Martin. "Montmorillonite-Catalyzed Hydrolysis of Phosmet," *Soil Sci.*, 136(2):89–93 (1983).

Sánchez-Martin, M.J. and M. Sánchez-Camazano. "Aspects of the Adsorption of Azinphosmethyl by Smectities," *J. Agric. Food Sci.*, 32(4):720–725 (1984).

Sanderman, W., H. Stockmann, and R. Casten. "Über die Pyrolyse des Pentachlorophenols," *Chem. Ber.*, 90:690–692 (1957).

Sanders, H.O. "Toxicities of Some Herbicides to Six Species of Freshwater Crustaceans," *J. Water Pollut. Control Fed.*, 42(8):1544–1550 (1970).

Sanders, H.O. and O.B. Cope. "The Relative Toxicities of Several Pesticides to Naiads of Three Species of Stoneflies," *Limnol. Oceanogr.*, 13(1):112–117 (1968).

Sanemasa, I., Y. Miyazaki, S. Arakawa, M. Kumamaru, and T. Deguchi. "The Solubility of Benzene-Hydrocarbon Binary Mixtures in Water," *Bull. Chem. Soc. Jpn.*, 60(2):517–523 (1987).

Saravanja-Bozanic, V., S. Gäb, K. Hustert, and F. Korte. "Reacktionen von Aldrin, Chlorden, und 2,2-Dichlorobiphenyl mit O(^{3}P)," *Chemosphere*, 6(1):21–26 (1977).

Savage, K.E. and W.L. Barrentine. "Trifluralin Persistence as Affected by Depth of Soil Incorporation," *Weed Sci.*, 17:349–352 (1969).

Savage, K.E. and T.N. Jordan. "Persistence of Three Dinitroaniline Herbicides on the Soil Surface," *Weed Sci.*, 28:105–110 (1980).

Sax, N.I. *Dangerous Properties of Industrial Materials* (New York: Van Nostrand Reinhold Co., 1984), 3124 p.

Sax, N.I. and R.J. Lewis, Sr., Eds. *Hazardous Chemicals Desk Reference* (New York: Van Nostrand Reinhold Co., 1987), 1084 p.

Schellenberg, K., C. Leuenberger, and R.P. Schwarzenbach. "Sorption of Chlorinated Phenols by Natural Sediments and Aquifer Materials," *Environ. Sci. Technol.*, 18(9):652–657 (1984).

Scheunert, I., Z. Qiao, and F. Korte. "Comparative Studies of the Fate of Atrazine-^{14}C and Pentachlorophenol-^{14}C in Various Laboratory and Outdoor Soil-Plant Systems," *J. Environ. Sci. Health*, B21(6):457–485 (1986).

Scheunert, I., D. Vockel, J. Schmitzer, and F. Korte. "Biomineralization Rates of ^{14}C-Labeled Organic Chemicals in Aerobic and Anaerobic Suspended Soil," *Chemosphere*, 16(5):1031–1041 (1987).

Schimmel, S.C., R.L. Garnas, J.M. Patrick, Jr., and J.C. Moore. "Acute Toxicity, Bioconcentration, and Persistence of AC 222,705, Benthiocarb, Chlorpyrifos, Fenvalerate, Methyl Parathion, and Permethrin in the Estuarine Environment," *J. Agric. Food Chem.*, 31(1):104–113 (1983).

Schimmel, S.C., J.M. Patrick, Jr., and J. Forester. "Heptachlor: Uptake, Depuration, Retention, and Metabolism by Spot (*Leiostomus xanthurus*)," *J. Toxicol. Environ. Health*, 2(1):169–178 (1976).

Schimmel, S.C., J.M. Patrick, Jr., and J. Forester. "Toxicity and Bioconcentration of BHC and Lindane in Selected Estuarine Animals," *Arch. Environ. Contam. Toxicol.*, 6:355–363 (1977).

Schliebe, K.A., O.C. Burnside, and T.L. Lavy. "Dissipation of Amiben," *Weeds*, 13:321–325 (1965).

Schoen, S.R. and W.L. Winterlin. "The Effects of Various Soil Factors and Amendments on the Degradation of Pesticide Mixtures," *J. Environ. Sci. Health*, B22(3):347–377 (1987).

Schulz, K.R. and E.P. Lichtenstein. "Field Studies on the Persistence and Movement of Dyfonate in Soil," *J. Econ. Entomol.*, 64:283–287 (1971).

Schulz, K.R., E.P. Lichtenstein, T.T. Liang, and T.W. Fuhremann. "Persistence and Degradation of Azinphosmethyl in Soils," *J. Econ. Entomol.*, 63:432–438 (1970).

Schumacher, H.G., H. Parlar, W. Klein, and F. Korte. "Beitrage zur Okologischen Chemie. LV. Photochemische Reaktion von Endosulfan," *Chemosphere*, 3(2):65–70 (1974).

Schwarz, F.P. "Determination of Temperature Dependence of Solubilities of Polycyclic Aromatic Hydrocarbons in Aqueous Solutions by a Fluorescence Method," *J. Chem. Eng. Data*, 22(3):273–277 (1977).

Schwarz, F.P. and S.P. Wasik. "A Fluorescence Method for the Measurement of the Partition Coefficients of Naphthalene, 1-Methylnaphthalene, and 1-Ethylnaphthalene in Water," *J. Chem. Eng. Data*, 22(3):270–273 (1977).

Schwarzenbach, R.P., R. Stierli, B.R. Folsom, and J. Zeyer. "Compound Properties Relevant for Assessing the Environmental Partitioning of Nitrophenols," *Environ. Sci. Technol.*, 22(1):83–92 (1988).

Scifres, D.J. and T.J. Allen. "Dissipation of Dicamba from Grassland Soils of Texas," *Weed Sci.*, 21:393–396 (1973).

Sears, M.K., C. Bowhey, H. Braun, and G.R. Stephenson. "Dislodgable Residues and Persistence of Diazinon, Chlorpyrifos and Isofenphos Following Their Application to Turfgrass," *Pestic. Sci.*, 20:223–231 (1987).

Seiber, J.N., M.P. Catahan, and C.R. Barril. "Loss of Carbofuran from Rice Paddy Water: Chemical and Physical Factors," *J. Environ. Sci. Health*, B13(2):131–148 (1978).

Seiber, J.N., S.C. Madden, M.M. McChesney, and W.L. Winterlin. "Toxaphene Dissipation from Treated Cotton Field Measurements: Component Residual Behavior on Leaves and in Air, Soil, and Sediments Determined by Gas Chromatography," *J. Agric. Food Chem.*, 27:284–290 (1979).

Sethunathan, N. "Organic Matter and Parathion Degradation in Flooded Soil," *Soil Biol. Biochem.*, 5(5):641–644 (1973).

Sethunathan, N. "Degradation of Parathion in Flooded Acid Soils," *J. Agric. Food Chem.*, 21(4):602–604 (1973a).

Sethunathan, N. and I.C. MacRae. "Persistence and Biodegradation of Diazinon in Submerged Soils," *J. Agric. Food Chem.*, 17(2):221–225 (1969).

Sethunathan, N. and M.D. Pathak. "Development of a Diazinon-Degrading Bacterium in Paddy Water after Repeated Applications of Diazinon," *Can. J. Microbiol.*, 17(5):699–702 (1971).

Sethunathan, N. and M.D. Pathak. "Increased Biological Hydrolysis of Diazinon after Repeated Application in Rice Paddies," *J. Agric. Food Chem.*, 20(3):586–589 (1972).

Sethunathan, N., R. Siddaramappa, K.P. Rajaram, S. Barik, and P.A. Wahid. "Parathion: Residues in Soil and Water," *Residue Rev.*, 68:91–122 (1977).

Sethunathan, N. and T. Yoshida. "Fate of Diazinon in Submerged Soil. Accumulation of Hydrolysis Product," *J. Agric. Food Chem.*, 17(6):1192–1195 (1969).

Sethunathan, N. and T. Yoshida. "A *Flavobacterium* sp. that Degrades Diazinon and Parathion," *Can. J. Microbiol.*, 19(5):873–875 (1973).

Sethunathan, N. and T. Yoshida. "Degradation of Chlorinated Hydrocarbons by *Clostridium* sp. Isolated from Lindane-Amended, Flooded Soil," *Plant and Soil*, 38:663–666 (1973a).

Sharma, S.R., R.P. Singh, and S.R. Ahmed. "Effect of Different Leachates on the Movement of Some Phosphorus Containing Pesticides in Soils Using Thin Layer Chromatography," *Ecotoxicol. Environ. Safety*, 11(2):229–240 (1986).

Sharmila, M., K. Ramanand, and N. Sethunathan. "Effect of Yeast Extract on the Degradation of Organophosphorus Insecticides by Soil Enrichment and Bacterial Cultures," *Can. J. Microbiol.*, 35(12):1105–1110 (1989).

Sharom, M.S. and L.V. Edgington. "Mobility and Dissipation of Metalaxyl in Tobacco Soils," *Can. J. Plant Sci.*, 66:761–771 (1986).

Sharom, M.S., J.R.W. Miles, J.W. Harris, and F.L. McEwen. "Persistence of 12 Insecticides in Water," *Water Res.*, 14(8):1089–1093 (1980).

Sharom, M.S. and K.R. Solomon. "Adsorption-Desorption, Degradation, and Distribution of Permethrin in Aqueous Systems," *J. Agric. Food Chem.*, 29(6):1122–1125 (1981).

Shelton, D.R., S.A. Boyd, and J.M. Tiedje. "Anaerobic Biodegradation of Phthalic Acid Esters in Sludge," *Environ. Sci. Technol.*, 18(2):93–97 (1984).

Sherblom, P.M., P.M. Gschwend, and R.P. Eganhouse. "Aqueous Solubilities, Vapor Pressures, and 1-Octanol-Water Partition Coefficients for C9-C14 Linear Alkylbenzenes," *J. Chem. Eng. Data*, 37(4):394–399 (1992).

Sherman, J.C., T.A. Nevin, and J.A. Lasater. "Hydrogen Sulfide Production from Ethion by Bacteria in Lagoonal Sediments," *Bull. Environ. Contam. Toxicol.*, 12(3):359–365 (1974).

Shevchenko, M.A., P.N. Taran, and P.V. Marchenko. "Technology of Water Treatment and Demineralization," *Soviet J. Water Chem. Technol.*, 4(4):53–71 (1982).

Shimabukuro, R.H. "Significance of Atrazine Dealkylation in Root and Shoot of Pea Plants," *J. Agric. Food Chem.*, 15(4):557–562 (1967).

Shin, Y.-O., J.J. Chodan, and A.R. Wolcott. "Adsorption of DDT by Soils, Soil Fractions, and Biological Materials," *J. Agric. Food Chem.*, 18(6):1129–1133 (1970).

Shirkot, C.K. and K.G. Gupta. "Accelerated Tetramethylthiuram Disulfide (TMTD) Degradation in Soil by Inoculation with TMTD-Utilizing Bacteria," *Bull. Environ. Contam. Toxicol.*, 35(3):354–361 (1985).

Siddaramappa, R., K.P. Rajaram, and N. Sethunathan. "Degradation of Parathion by Bacteria Isolated from Flooded Soil," *Appl. Microbiol.*, 26(6):846–849 (1973).

Sikka, H.C. and D.E. Davis. "Dissipation of Atrazine from Soil by Corn, Sorghum, and Johnson Grass," *Weeds*, 14:289–293 (1966).

Sikka, H.C. and C.P. Rice. "Persistence of Endothall in Aquatic Environment as Determined by Gas-Liquid Chromatography," *J. Agric. Food Chem.*, 21(5):842–845 (1973).

Sikka, H.C. and J. Saxena. "Metabolism of Endothall by Aquatic Microorganisms," *J. Agric. Food Chem.*, 21(3):402–406 (1973).

Simsiman, G.V. and G. Chesters. "Persistence of Diquat in the Aquatic Environment," *Water Res.*, 10(2):105–112 (1976).

Simsiman, G.V., T.C. Daniel, and G. Chesters. "Diquat and Endothall: Their Fates in the Environment," *Residue Rev.*, 62:131–174 (1976).

Singh, R.P. and M. Chiba. "Solubility of Benomyl in Water at Different pHs and Its Conversion to Methyl 2-Benzimidazolecarbamate, 3-Butyl-2,4-dioxo[1,2-*a*]-*s*-triazinobenzimidazole, and 1-(2-Benzimidazolyl)-3-*n*-butylurea," *J. Agric. Food Chem.*, 33(1):63–67 (1985).

Singh, S.B. and G. Kulshrestha. "Microbial Degradation of Pendimethalin," *J. Environ. Sci. Health*, B26(3):309–321 (1991).

Singh, A.K. and P.K. Seth. "Degradation of Malathion by Microorganisms Isolated from Industrial Effluents," *Bull. Environ. Contam. Toxicol.*, 43(1):28–35 (1989).

Singh, G., W.F. Spencer, M.M. Cliath, and M. Th. van Genuchten. "Sorption Behavior of *s*-Triazine and Thiocarbamate Herbicides on Soils," *J. Environ. Qual.*, 19(3):520–525 (1990).

Singh, N., P.A. Wahid, M.V.R. Murty, and N. Sethunathan. "Sorption-Desorption of Methyl Parathion, Fenitrothion, and Carbofuran in Soils," *J. Environ. Sci. Health*, B25(6):713–728 (1990).

Singmaster, J.A., III. "Environmental Behavior of Hydrophobic Pollutants in Aqueous Solutions," PhD Thesis, University of California, Davis (1975).

Sirons, G.J., R. Frank, and R.M. Dell. "Picloram Residues in Sprayed MacDonald-Carter Freeway Right-of-Way," *Bull. Environ. Contam. Toxicol.*, 18(5):526–533 (1977).

Sirons, G.J., R. Frank, and T. Sawyer. "Residues of Atrazine, Cyanazine, and Their Phytotoxic Metabolites in a Clay Loam Soil," *J. Agric. Food Chem.*, 21(6):1016–1019 (1973).

Sisler, H.D. and C.E. Cox. "Effects of Tetramethyl Thiuram Disulfide on Metabolism of Fusarium roseum," *Am. J. Bot.*, 41:338 (1954).

Sittig, M. *Handbook of Toxic and Hazardous Chemicals and Carcinogens* (Park Ridge, NJ: Noyes Publications, 1985), 950 p.

Skipper, H.D., C.M. Gilmour, and W.R. Furtick. "Microbial Versus Chemical Degradation of Atrazine in Soils," *Soil Sci. Soc. Am. Proc.*, 31:653–656 (1967).

Skipper, H.D. and V.V. Volk. "Biological and Chemical Degradation of Atrazine in Three Oregon Soils," *Weed Sci.*, 20(4):344–347 (1972).

Skipper, H.D., V.V. Volk, M.M. Mortland, and K.V. Raman. "Hydrolysis of Atrazine of Soil Colloids," *Weed Sci.*, 26:46–51 (1978).

Skurlatov, Y.I., R.G. Zepp, and G.L. Baughman. "Photolysis Rates of (2,4,5-Trichlorophenoxy)acetic Acid and 4-Amino-3,5,6-trichloropicolinic Acid in Natural Waters," *J. Agric. Food Chem.*, 31(5):1065–1071 (1983).

Slade, P. and A.E. Smith. "Photochemical Degradation of Diquat," *Nature (London)*, 213(5069):919–920 (1967).

Smelt, J.H., S.J.H. Crum, W. Teunissen, and M. Leistra. "Accelerated Transformation of Aldicarb, Oxamyl and Ethoprophos After Repeated Soil Treatments," *Crop Protection*, 6:295–303 (1987).

Smelt, J.H., M. Leistra, N.W.H. Houx, and A. Dekker. "Conversion Rates of Aldicarb and Its Oxidation Products in Soils. III. Aldicarb," *Pestic. Sci.*, 9(4):293–300 (1978).

Smith, A.E. "Persistence of Trifluralin in Small Field Plots as Analysis by a Rapid Gas Chromatograph Method," *J. Agric. Food Chem.*, 20:829–831 (1972).

Smith, A.E. "Degradation of Dicamba in Prairie Soils," *Weed Res.*, 13:373–378 (1973).

Smith, A.E. "Transformation of Dicamba in Regina Heavy Clay," *J. Agric. Food Chem.*, 21(4):708–710 (1973).

Smith, A.E. "Breakdown of the Herbicide Dicamba and Its Degradation Product 3,6-Dichlorosalicylic Acid in Prairie Soils," *J. Agric. Food Chem.*, 22(4):601–605 (1974).

Smith, A.E. "Degradation of the Herbicide Diclofop-Methyl in Prairie Soils," *J. Agric. Food Chem.*, 25(4):893–898 (1977).

Smith, A.E. "Degradation, Adsorption, and Volatility of Diallate and Triallate in Prairie Soils," *Weed Res.*, 18:275–279 (1978).

Smith, A.E. "Transformation of [^{14}C]Diclofop-Methyl in Small Field Plots," *J. Agric. Food Chem.*, 27(6):1145–1148 (1979).

Smith, A.E. "An Analytical Procedure for Bromoxynil and Its Octanoate in Soils: Persistence Studies with Bromoxynil Octanoate in Combination with other Herbicides in Soils," *Pestic. Sci.*, 11:341–346 (1980).

Smith, A.E. "Identification of 2,4-Dichloroanisole and 2,4-Dichlorophenol as Soil Degradation Products of Ring-Labeled [^{14}C]2,4-D," *Bull. Environ. Contam. Toxicol.*, 34(2):150–157 (1985).

Smith, A.E. "Transformations in Soil," in *Environmental Chemistry of Herbicides*. Volume I, R. Grover, Ed. (Boca Raton, FL: CRC Press, Inc. 1988), pp. 171–200.

Smith, A.E. and D.R. Cullimore. "Microbiological Degradation of the Herbicide Dicamba in Moist Soils at Different Temperatures," *Weed Res.*, 15:59–63 (1975).

Smith, L.R. and J. Dragun. "Degradation of Volatile Chlorinated Aliphatic Priority Pollutants in Groundwater," *Environ. Int.*, 19(4):291–298 (1984).

Smith, A.E. and G.S. Edmond. "Persistence of Linuron in Saskatchewan Soils," *Can. J. Soil Sci.*, 55:145–148 (1975).

Smith, A.E. and A. Fitzpatrick. "The Loss of Five Thiolcarbamate Herbicides in Nonsterile Soils and Their Stability in Acidic and Basic Solutions," *J. Agric. Food Chem.*, 18(4):720–722 (1970).

Smith, A.E. and J. Grove. "Photochemical Degradation of Diquat in Dilute Aqueous Solution and on Silica Gel," *J. Agric. Food Chem.*, 17(3):609–613 (1969).

Smith, C.A., Y. Iwata, and F.A. Gunther. "Conversion and Disappearance of Methidathion on Thin Layers of Dry Soil," *J. Agric. Food Chem.*, 26(4):959–962 (1978).

Smith, J.G., S.-F. Lee, and A. Netzer. "Model Studies in Aqueous Chlorination: The Chlorination of Phenols in Dilute Aqueous Solutions," *Water Res.*, 10(11):985–990 (1976).

Smith, S., T.E. Reagan, J.L. Flynn, and G.H. Willis. "Azinphosmethyl and Fenvalerate Runoff Loss from a Sugarcane-Insect IPM System," *J. Environ. Qual.*, 12(4):534–537 (1983).

Smith, R.V. and J.P. Rosazza. "Microbial Models of Mammalian Metabolism. Aromatic Hydroxylation," *Arch. Biochem. Biophys.*, 161(2):551–558 (1974).

Smith, C.T., F. Shaw, R. Lavigne, J. Archibald, H. Fenner, and D. Stern. "Residues of Malathion on Alfalfa and in Milk and Meat," *J. Econ. Entomol.*, 53:495–496 (1960).

Smith, A.E. and A. Walker. "A Quantitative Study of Asulam Persistence in Soil," *Pestic. Sci.*, 8:449–456 (1977).

Snider, E.H. and F.C. Alley. "Kinetics of the Chlorination of Biphenyl Under Conditions of Waste Treatment Processes," *Environ. Sci. Technol.*, 13(10):1244–1248 (1979).

Sobotka, H. and J. Kahn. "Determination of Solubility of Sparingly Soluble Liquids in Water," *J. Am. Chem. Soc.*, 53(8):2935–2938 (1931).

Soderquist, C.J., J.B. Bowers, and D.G. Crosby. "Dissipation of Molinate in a Rice Field," *J. Agric. Food Chem.*, 25(4):940–945 (1977).

Soderquist, C.J. and D.G. Crosby. "Dissipation of 4-Chloro-2-methylphenoxyacetic Acid (MCPA) in a Rice Field," *Pestic. Sci.*, 6(1):17–33 (1975).

Soderquist, C.J., D.G. Crosby, K.W. Moilanen, J.N. Seiber, and J.E. Woodrow. "Occurrence of Trifluralin and Its Photoproducts in Air," *J. Agric. Food Chem.*, 23(2):304–309 (1975).

Solon, J.M. and J.H. Nair. "The Effect of a Sublethal Concentration of LAS on the Acute Toxicity of Various Phosphate Pesticides to the Fathead Minnow Pimephales promelas Rafinesque," *Bull. Environ. Contam. Toxicol.*, 5(5):408–413 (1970).

Somasundaram, L. and J.R. Coats. "Interactions between Pesticides and Their Major Degradation Products," in *Pesticide Transformation Products. Fate and Significance in the Environment*, ACS Symposium Series 459, Somasundaram, L. and J.R. Coats, Eds., (Washington, DC: American Chemical Society, 1991), pp. 162–171.

Somasundaram, L., J.R. Coats, and K.D. Racke. "Degradation of Pesticides in Soil as Influenced by the Presence of Hydrolysis Metabolites," *J. Environ. Sci. Health*, B24(5):457–478 (1989).

Somasundaram, L., J.R. Coats, K.D. Racke, and V.M. Shanbhag. "Mobility of Pesticides and Their Metabolites in Soil," *Environ. Toxicol. Chem.*, 10(2):185–194 (1991).

Somich, C.J., P.C. Kearney, M.T. Muldoon, and S. Elsasser. "Enhanced Soil Degradation of Alachlor by Treatment with Ultraviolet Light and Ozone," *J. Agric. Food Chem.*, 36(6):1322–1326 (1988).

Sonobe, H., L.R. Kamps, E.P. Mazzola, and J.A.G. Roach. "Isolation and Identification of a New Conjugated Carbofuran Metabolite in Carrots: Angelic Acid Ester of 3-Hydroxycarbofuran," *J. Agric. Food Chem.*, 29(6):1125–1129 (1981).

Spencer, W.F., J.D. Adams, T.D. Shoup, and R.C. Spear. "Conversion of Parathion to Paraoxon on Soil Dusts and Clay Minerals as Affected by Ozone and UV Light," *J. Agric. Food Chem.*, 28(2):366–371 (1980).

Spencer, W.F., J.D. Adams, T.D. Shoup, and R.C. Spear. "Conversion of Parathion to Paraoxon on Soil Dusts As Related to Atmospheric Oxidants at Three California Locations," *J. Agric. Food Chem.*, 28(6):1295–1300 (1980a).

Spencer, W.F., M.M. Cliath, D.R. Davis, R.C. Spear, and W.J. Pependorf. "Persistence of Parathion and Its Oxidation to Paraoxon on the Soil Surface as Related to Worker Reentry into Treated Crops," *Bull. Environ. Contam. Toxicol.*, 14(3):265–272 (1975).

Splittstoesser, W.E. and H.J. Hopen. "Metabolism of Siduron by Barley and Crabgrass," *Weed Sci.*, 16:305–308 (1968).

Sprankle, P., W.F. Meggitt, and D. Penner. "Absorption Action and Translocation of Glyphosate," *Weed Sci.*, 23:235–240 (1975).

Sprankle, P., W.F. Meggitt, and D. Penner. "Absorption, Mobility, and Microbial Degradation of Glyphosate in Soil," *Weed Sci.*, 23:229–234 (1975a).

Staiff, D.C., S.W. Comer, J.F. Armstrong, and H.R.D. Wolfe. "Persistence of Azinphosmethyl in Soil," *Bull. Environ. Contam. Toxicol.*, 13(3):362–368 (1975).

Stalker, D.M., K.E. McBride, and L.D. Malyj. "Herbicide Resistance in Transgenic Plants Expressing a Bacterial Detoxification Gene," *Science (Washington, DC)*, 242(4877):419–423 (1988).

Stanley, J.G. and J.G. Trial. "Disappearance Constants of Carbaryl from Streams Contaminated by Forest Spraying," *Bull. Environ. Contam. Toxicol.*, 25(5):771–776 (1980).

Stearns, R.S., H. Oppenheimer, E. Simon, and W.D. Harkins. "Solubilization by Solutions of Long-Chain Colloidal Electrolytes," *J. Chem. Phys.*, 15(7):496–507 (1947).

Steen, W.C. "Project Summary. Microbial Transformation Rate Constants of Structurally Diverse Manmade Chemicals," U.S. EPA Report 600/S3-91/016 (1991), 2 p.

Steenson, T.I. and N. Walker. "The Pathway of Breakdown of 2,4-Dichloro- and 4-Chloro-2-methylphenoxyacetic Acid by Bacteria," *J. Gen. Microbiol.*, 16:146–155 (1957).

Steinwandter, H. "Experiments on Lindane Metabolism in Plants III. Formation of β-HCH," *Bull. Environ. Contam. Toxicol.*, 20(4):535–536 (1978).

Steinwandter, H. and H. Schluter. "Experiments on Lindane Metabolism in Plants IV. A Kinetic Investigation," *Bull. Environ. Contam. Toxicol.*, 20(2):174–179 (1978).

Steller, W.A. and W.W. Brand. "Analysis of Dimethoate-Treated Grapes for the *N*-Hydroxymethyl and De-*N*-Methyl Metabolites and for Their Sugar Adducts," *J. Agric. Food Chem.*, 22(3):445–449 (1974).

Stephenson, R. and J. Stuart. "Mutual Binary Solubilities: Water-Alcohols and Water Esters," *J. Chem. Eng. Data*, 31(1):56–70 (1986).

Stephenson, R., J. Stuart, and M. Tabak. "Mutual Solubility of Water and Aliphatic Alcohols," *J. Chem. Eng. Data*, 29(3):287–290 (1984).

Stepp, T.D., N.D. Camper, and M.J.B. Paynter. "Anaerobic Microbial Degradation of Selected 2,4-Dihalogenated Aromatic Compounds," *Pestic. Biochem. Physiol.*, 23:256–260 (1985).

Stewart, D.K.R. and K.G. Cairns. "Endosulfan Persistence in Soil and Uptake by Potato Tubers," *J. Agric. Food Chem.*, 22(6):984–986 (1974).

Stewart, D.K.R. and D. Chisholm. "Long-term Persistence of BHC, DDT, and Chlordane in a Sandy Loam Soil," *Can. J. Soil Sci.*, 51:379–383 (1971).

Stewart, D.K.R. and S.O. Gaul. "Persistence of 2,4-D and 2,4,5-T and Dicamba in a Dykeland Soil," *Bull. Environ. Contam. Toxicol.*, 18(2):210–218 (1977).

Still, G.G. "Metabolism of 3,4-Dichloropropionanilide in Plants: The Metabolic Fate of the 3,4-Dichloroaniline Moiety," *Science (Washington, DC)*, 159(3818):992–993 (1968).

Still, G.G. and R.A. Herrett. "Methylcarbamates, Carbanilates, and Acylanilides" in *Herbicides: Chemistry, Degradation, and Mode of Action*. Volume 2, Kearney, P.C. and D.D. Kaufman, Eds., (New York: Marcel Dekker, Inc., 1976), pp. 609–664.

Stojanovic, B.J., M.V. Kennedy, and F.L. Shuman, Jr. "Mild Thermal Degradation of Pesticides," *J. Environ. Qual.*, 1:397–401 (1972).

Stott, D.E., J.P. Martin, D.D. Focht, and K. Haider. "Biodegradation, Stabilization in Humus, and Incorporation into Soil Biomass of 2,4-D and Chlorocatechol Carbons," *Soil Sci. Soc. Am. J.*, 47:66–70 (1983).

Struif, B., L. Weil, and K.-E. Quentin. "Verhalten Herbizider Phenoxyalkan-carbonäuren bei der Wasseraufbereitung mit Ozon," *Zeit. fur Wasser und Abwasser-Forschung*, 3/4:118–127 (1978).

Su, F., J.G. Calvert, and J.H. Shaw. "Mechanism of the Photooxidation of Gaseous Formaldehyde," *J. Phys. Chem.*, 83(25):3185–3191 (1979).

Subba-Rao, R.V. and M. Alexander. "Effect of DDT Metabolites on Soil Respiration and on an Aquatic Alga," *Bull. Environ. Contam. Toxicol.*, 25(2):215–220 (1980).

Subba-Rao, R.V., H.E. Rubin, and M. Alexander. "Kinetics and Extent of Mineralization of Organic Chemicals at Trace Levels in Freshwater and Sewage," *Appl. Environ. Microbiol.*, 43:1139–1150 (1982).

Sud, R.K., A.K. Sud, and K.G. Gupta. "Degradation of Sevin (1-Naphthyl *N*-methylcar-bamate) by *Achromobacter* sp.," *Arch. Microbiol.*, 87:353–358 (1972).

Sudhakar-Barik and N. Sethunathan. "Biological Hydrolysis of Parathion in Natural Ecosystems," *J. Environ. Qual.*, 7(3):346–348 (1978).

Sudhakar-Barik, R. Siddaramappa, and N. Sethunathan. "Metabolism of Nitrophenols by Bacteria Isolated from Parathion-Amended Flooded Soil," Antonie van Leeuwenhoek, 42(4):461–470 (1976).

Sudhakar-Barik, R. Siddaramappa, P.A. Wahid, and N. Sethunathan. "Conversion of *p*-Nitrophenol to 4-Nitrocatechol by a *Pseudomonas* sp.," Antonie van Leeuwenhoek, 44(2):171–176 (1978a).

Sudhakar-Barik, P.A. Wahid, C. Ramakrishna, and N. Sethunathan. "A Change in the Degradation Pathway of Parathion after Repeated Applications to Flooded Soil," *J. Agric. Food Chem.*, 27(6):1391–1392 (1979).

Suett, D.L. "Persistence and Degradation of Chlorfenvinphos, Diazinon, Fonofos, and Phorate in Soils and Their Uptake by Carrots," *Pestic. Sci.*, 2:105–112 (1971).

Suffet, I.H., S.D. Faust, and W.F. Carey. "Gas-Liquid Chromatographic Separation of Some Organophosphate Pesticides, Their Hydrolysis Products, and Oxons," *Environ. Sci. Technol.*, 1(8):639–643 (1967).

Suflita, J.M., J. Stout, and J.M. Tiedje. "Dechlorination of (2,4,5-Trichlorophenoxy)acetic acid by Anaerobic Microorganisms," *J. Agric. Food Chem.*, 32(2):218–221 (1984).

Sukop, M. and C.G. Cogger. "Adsorption of Carbofuran. Metalaxyl, and Simazine: K_{oc} Evaluation and Relation to Soil Transport," *J. Environ. Sci. Health*, B27(5):565–590 (1992).

Sullivan, R.G., H.W. Knoche, and J.C. Markle. "Photolysis of Trifluralin: Characterization of Azobenzene and Azoxybenzene Photodegradation Products," *J. Agric. Food Chem.*, 28(4):746–755 (1980).

Sun, Y. and J.J. Pignatello. "Photochemical Reactions Involved in the Total Mineralization of 2,4-D by $Fe^{3+}/H_2O_2/UV$," *Environ. Sci. Technol.*, 27(2):304–310 (1993).

Surber, E.W. and Q.H. Pickering. "Acute Toxicity of Endothal, Diquat, Hyamine, Dalapon, and Silvex to Fish," *Progr. Fish-Cult.*, 24(4):164–171 (1962).

Sutton, C. and J.A. Calder. "Solubility of Higher-Molecular-Weight *n*-Paraffins in Distilled Water and Seawater," *Environ. Sci. Technol.*, 8(7):654–657 (1974).

Sutton, C. and J.A. Calder. "Solubility of Alkylbenzenes in Distilled Water and Seawater at 25°C," *J. Chem. Eng. Data*, 20(3):320–322 (1975).

Suzuki, M., Y. Yamato, and T. Watanabe. "Residue in Soil, Organochlorine Insecticide Residues in Field Soils of the Kita Kyusha District — Japan 1970–1974," *Pest. Monitor. J.*, 11:88–93 (1977).

Swann, R.L., D.A. Laskowski, P.J. McCall, K. Vander Kuy, and H.J. Dishburger. "A Rapid Method for the Estimation of the Environmental Parameters Octanol/Water Partition Coefficient, Soil Sorption Constant, Water to Air Ratio, and Water Solubility," *Residue Rev.*, 85:17–28 (1983).

Swoboda, A.R. and G.W. Thomas. "Movement of Parathion in Soil Columns," *J. Agric. Food Chem.*, 16(6):923–927 (1968).

Szalkowski, M.B. and D.E. Stallard. "Effect of pH on the Hydrolysis of Chlorothalonil," *J. Agric. Food Chem.*, 25(1):208–210 (1977).

Szeto, S.Y. and M.J. Brown. "Gas-Liquid Chromatographic Methods for the Determination of Disulfoton, Phorate, Oxydemetonmethyl and Their Toxic Metabolites in Asparagus Tissue and Soil," *J. Agric. Food Chem.*, 30(6):1082–1086 (1982).

Szeto, S.Y., R.S. Vernon, and M.J. Brown. "Degradation of Disulfoton in Soil and Its Translocation into Asparagus," *J. Agric. Food Chem.*, 31(2):217–220 (1983).

Szeto, S.Y., A.T.S. Wilkonson, and M.J. Brown. A Gas Chromatographic Method for the Determination of Bendiocarb in Soil and Corn: Application to the Analysis of Residues in Corn," *J. Agric. Food Chem.*, 32(1):78–80 (1984).

Tabak, H.H., S.A. Quave, C.I. Mashni, and E.F. Barth. "Biodegradability Studies with Organic Priority Pollutant Compounds," *J. Water Pollut. Control Fed.*, 53(10):1503–1518 (1981).

Takase, I, H, Tsuda, and Y. Yoshimoto. "Fate of Disyston Active Ingredient in Soil," *Pflanzen-Nachr*, 25:43–63 (1972).

Talbert, R.E. and O.H. Fletchall. "The Adsorption of Some *s*-Triazines in Soils," *Weeds*, 13:46–52 (1965).

Talbert, R.E., R.L. Runyan, and H.R. Baker. "Behavior of Amiben and Dinoben Derivatives in Arkansas Soils," *Weed Sci.*, 18(1):10–15 (1970).

Talekar, N.S., L.-T. Sun, E.-M. Lee, and J.-S. Chen. "Persistence of Some Insecticides in Subtropical Soil," *J. Agric. Food Chem.*, 25(2):348–352 (1977).

Tanaka, F.S., B.L. Hoffer, and R.G. Wien. "Detection of Halogenated Biphenyls from Sunlight Photolysis of Chlorinated Herbicides in Aqueous Solution," *Pestic. Sci.*, 16:265–270 (1985).

Tanaka, F.S., R.G. Wien, and B.L. Hoffer. "Biphenyl Formation in the Photolysis of 3-(4-Chlorophenyl)-1,1-dimethylurea in Aqueous Solution," *J. Agric. Food Chem.*, 29(6):1153–1158 (1981).

Tanaka, F.S., R.G. Wien, and B.L. Hoffer. "Investigation of the Mechanism and Pathway of Biphenyl Formation in the Photolysis of Monuron," *J. Agric. Food Chem.*, 30(5):957–963 (1982).

Tanaka, F.S., R.G. Wien, and B.L. Hoffer. "Biphenyl Formation in the Photolysis of Aqueous Herbicide Solutions," *Ind. Eng. Chem. Prod. Res. Dev.*, 23(1):1–5 (1984).

Tanaka, F.S., R.G. Wien, and E.R. Mansager. "Effect of Nonionic Surfactants on the Photochemistry of 3-(4-Chlorophenyl)-1,1-dimethylurea in Aqueous Solution," *J. Agric. Food Chem.*, 27(4):774–779 (1979).

Tanaka, F.S., R.G. Wien, and E.R. Mansager. "Survey for Surfactant Effects on the Photodegradation of Herbicides in Aqueous Media," *J. Agric. Food Chem.*, 29(2):227–230 (1981).

Tanaka, F.S., R.G. Wien, and E.R. Mansager. "Photolytic Demethylation of Monuron and Demethylmonuron in Aqueous Solution," *Pestic. Sci.*, 13(3):287–294 (1982a).

Tanaka, F.S., R.G. Wien, and R.G. Zaylskie. "Photolysis of 3-(4-Chlorophenyl)-1,1-dimethylurea in Dilute Aqueous Solution," *J. Agric. Food Chem.*, 25(5):1068–1072 (1977).

Tewari, Y.B., M.M. Miller, S.P. Wasik, and D.E. Martire. "Aqueous Solubility and Oc-tanol/Water Partition Coefficient of Organic Compounds at 25.0°C," *J. Chem. Eng. Data*, 27(4):451–454 (1982).

Thirunarayanan, K., R.L. Zimdahl, and D.E. Smika. "Chlorsulfuron Adsorption and Deg-radation in Soil," *Weed Sci.*, 33:558–563 (1985).

Thom, N.S. and A.R. Agg. "The Breakdown of Synthetic Organic Compounds in Biolog-ical Processes," *Proc. R. Soc. London*, B189(1096):347–357 (1975).

Thomas, E.W., B.C. Loughman, and R.G. Powell. "Metabolic Fate of Some Chlorinated Phenoxyacetic Acids in the Stem Tissue of *Avena sativa*," *Nature*, 204(4955):286 (1964).

Tiedje, J.M. and M.L. Hagedorn. "Degradation of Alachlor by a Soil Fungus, *Chaetomium globosum*," *J. Agric. Food Chem.*, 23(1):77–81 (1975).

Timms, P. and I.C. MacRae. "Conversion of Fensulfothion by *Klebsiella pneumoniae* to Fensulfothion Sulfide by Selected Microbes," *Aust. J. Biol. Sci.*, 35:661–668 (1982).

Timms, P. and I.C. MacRae. "Reduction of Fensulfothion and Accumulation of the Product, Fensulfothion Sulfide, by Selected Microbes," *Bull. Environ. Contam. Toxicol.*, 31(1):112–115 (1983).

Tirey, D.A., B. Dellinger, W. Rubey, and P. Taylor. "Thermal Degradation Characteristics of Environmentally Sensitive Pesticides Products," Office of Research and Develop-ment, U.S. EPA Report-600/R-93-102 (1993), 54 p.

Tomkiewicz, M.A., A. Groen, and M. Cocivera. "Electron Paramagnetic Resonance Spec-tra of Semiquinone Intermediates Observed during the Photooxidation of Phenol in Water," *J. Am. Chem. Soc.*, 93(25):7102–7103 (1971).

Treece, R.E. and G.W. Ware. "Lindane Residues on Alfalfa and in Milk," *J. Econ. Entomol.*, 58:218–219 (1965).

Trehy, M.L., R.A. Yost, and J.J. McCreary. "Determination of Aldicarb, Aldicarb Oxime, and Aldicarb Nitrile in Water by Gas Chromatography/Mass Spectrometry," *Anal. Chem.*, 56(8):1281–1285 (1984).

Tsao, R. and M. Eto. "Photoreactions of the Herbicide Naproanilide and the Effect of Some Photosensitizers," *J. Environ. Sci. Health*, B25(5):569–585 (1990).

Tu, C.M. "Utilization and Degradation of Lindane by Soil Microorganisms," *Arch. Mi-crobiol.*, 108(3):259–263 (1976).

Tu, C.M. "Influence of Pesticides and Some of the Oxidized Analogues on Microbial Populations, Nitrification and Respiration Activities in Soil," *Bull. Environ. Contam. Toxicol.*, 24(1):13–19 (1980).

Tuazon, E.C., R. Atkinson, A.M. Winer, and J.N. Pitts, Jr. "A Study of the Atmospheric Reactions of 1,3-Dichloropropene and Other Selected Organochlorine Compounds," *Arch. Environ. Contam. Toxicol.*, 13(6):691–700 (1984).

U.S. Department of Agriculture. Agricultural Research Service Pesticide Properties Data-base. Systems Research Laboratory, Beltsville, MD (1990).

U.S. Department of Health and Human Services. Hazardous Substances Data Bank. Na-tional Library of Medicine, Toxnet File (Bethedsa, MD: National Institute of Health, 1989).

U.S. EPA. 1986. "Pesticides in Ground Water: Background Document," Office of Ground-Water Protection, Washington, DC (1986).

Usorol, N.J. and R.J. Hance. "The Effect of Temperature and Water Contents on the Rate of Decomposition of the Herbicide Linuron," *Weed Sci.*, 16:19–21 (1974).

Uyeta, M., S. Taue, K. Chikasawa, and M. Mazaki. "Photoformation of Polychlorinated Biphenyls from Chlorinated Benzenes," *Nature (London)*, 264(5586):583–584 (1976).

Vadas, G.G., W.G. MacIntyre, and D.R. Burris. "Aqueous Solubility of Liquid Hydrocarbon Mixtures Containing Dissolved Solid Components," *Environ. Toxicol. Chem.*, 10:633–639 (1991).

Vaidyanathaswamy, R., P. Dureja, and S.K. Mukerjee. "Photoinduced Reaction: Part III — Photodegradation of *p,p'*-Dichlorodiphenyl-2,2,2-trichloroethanol (Dicofol), an Important Miticide," *Indain. J. Chem.*, 20B:866–869 (1981).

Valverde-García, A., E. González-Pradas, M. Villafranca-Sánchez, F. del Rey-Bueno, and A. García-Rodriguez. "Adsorption of Thiram and Dimethoate on Almeria Soils," *Soil Sci. Soc. Am. J.*, 52(6):1571–1574 (1988).

Veierov, D., A. Fenigstein, V. Melamed-Madjar, and M. Klein. "Effects of Concentration and Application Method on Decay and Residual Activity of Foliar Chlorpyrifos," *J. Econ. Entomol.*, 81(2):621–627 (1988).

Venkateswarlu, K., K. Chendrayan, and N. Sethunathan. "Persistence and Biodegradation of Carbaryl in Soils," *J. Environ. Sci. Health*, B15(4):421–429 (1980).

Venkateswarlu, K., T.K.S. Gowda, and N. Sethunathan. "Persistence and Biodegradation of Carbofuran in Flooded Soils," *J. Agric. Food Chem.*, 25(3):533–536 (1977).

Venkateswarlu, K. and N. Sethunathan. "Metabolism of Carbofuran in Rice Straw-Amended and Unamended Rice Soils," *J. Environ. Qual.*, 8(3):365–368 (1979).

Venkateswarlu, K. and N. Sethunathan. "Degradation of Carbofuran by *Azospirillium lipoferum* and *Streptomyces* spp. Isolated from Flooded Alluvial Soil," *Bull. Environ. Contam. Toxicol.*, 33(5):556–560 (1984).

Verschueren, K. *Handbook of Environmental Data on Organic Chemicals* (New York: Van Nostrand Reinhold Co., 1983), 1310 p.

Vesala, A. "Thermodynamics of Transfer of Nonelectrolytes from Light to Heavy Water. I. Linear Free Energy Correlations of Free Energy of Transfer with Solubility and Heat of Melting of a Nonelectrolyte," *Acta Chem. Scand.*, A28:839–845 (1974).

Voerman, S. and A.F.H. Besemer. "Persistence of Dieldrin, Lindane, and DDT in a Light Sandy Soil and Their Uptake by Grass," *Bull. Environ. Contam. Toxicol.*, 13(3):501–505 (1975).

Vogel, T.M. and M. Reinhard. "Reaction Products and Rates of Disappearance of Simple Bromoalkanes, 1,2-Dibromopropane, and 1,2-Dibromoethane in Water," *Environ. Sci. Technol.*, 20(10):992–997 (1986).

Vontor, T., J. Socha, and M. Vecera. "Kinetics and Mechanism of Hydrolysis of 1-Naphthyl, *N*-Methyl Carbamate and *N,N*-Dimethyl Carbamates," *Collect. Czech. Chem. Commun.*, 37:2183–2196 (1972).

Wahid, P.A., C. Ramakrishna, and N. Sethunathan. "Instantaneous Degradation of Parathion in Anaerobic Soils," *J. Environ. Qual.*, 9(1):127–130 (1980).

Wahid, P.A. and N. Sethunathan. "Sorption-Desorption of Parathion in Soils," *J. Agric. Food Chem.*, 26(1):101–105 (1978).

Wahid, P.A. and N. Sethunathan. "Involvement of Hydrogen Sulfide in the Degradation of Parathion in Flooded Acid Sulphate Soil," *Nature (London)*, 282(5737):401–402 (1979).

Waites, R.E. and C.H. Van Middelem. "Residue Studies of DDT and Malathion on Turnip Tops, Collards, Snap Beans, and Lettuce," *J. Econ. Entomol.*, 51:306–308 (1958).

Walker, A. "Simulation of Herbicide Persistence in Soil. I. Simazine and Prometryne," *Pestic. Sci.*, 7:41–49 (1976).

Walker, A. "Simulation of Herbicide Persistence in Soil. II. Simazine and Linuron in Long-Term Experiments," *Pestic. Sci.*, 7:50–58 (1976a).

Walker, A. "Further Observations on the Enhanced Degradation of Iprodione and Vinclozolin in Soil," *Pestic. Sci.*, 21:219–231 (1987).

Walker, C.R. "Toxicological Effects of Herbicides on the Fish Environment," *Water Sewer. Works*, 111(3):113–116 (1964).

Walker, W.W. "Chemical and Microbiological Degradation of Malathion and Parathion in an Estuarine Environment," *J. Environ. Qual.*, 5(2):210–216 (1976).

Walker, A. and P.A. Brown. "Measurement and Prediction of Chlorsulfuron Persistence in Soil," *Bull. Environ. Contam. Toxicol.*, 30(3):365–372 (1983).

Walker, A. and P.A. Brown. "The Relative Persistence in Soil of Five Acetanilide Herbicides," *Bull. Environ. Contam. Toxicol.*, 34(2):143–149 (1985).

Walker, A., P.A. Brown, and A.R. Entwistle. "Enhanced Degradation of Iprodione and Vinclozolin in Soil," *Pestic. Sci.*, 17:183–193 (1986).

Walker, A. and D.V. Crawford. "Diffusion Coefficients for Two Triazine Herbicides in Six Soils," *Weed Res.*, 10(2):126–132 (1970).

Walker, W.W. and B.J. Stojanovic. "Microbial Versus Chemical Degradation of Malathion in Soil," *J. Environ. Qual.*, 2:229–232 (1973).

Walker, W.W., C.R. Cripe, P.H. Pritchard, and A.W. Bourquin. "Biological and Abiotic Degradation of Xenobiotic Compounds in in vitro Estuarine Water and Sediment/Water Systems," *Chemosphere*, 17(12):2255–2270 (1988).

Walker, W.W. and B.J. Stojanovic. "Malathion Degradation by an *Arthrobacter* sp.," *J. Environ. Qual.*, 3(1):4–10 (1974).

Wallnöefer, P.R., S. Safe, and O. Hutzinger. "Microbial Demethylation and Debutylation of Four Phenylurea Herbicides," *Pestic. Biochem. Physiol.*, 3:253–258 (1973).

Walsh, P.R. and R.A. Hites. "Dicofol Solubility and Hydrolysis in Water," *Bull. Environ. Contam. Toxicol.*, 22(3):305–311 (1979).

Wang, C.H. and F.E. Broadbent. "Kinetic Losses of PCNB and DCNA in Three California Soils," *Soil Sci. Soc. Am. Proc.*, 36:742–745 (1972).

Wang, Y.-S., E.L. Madsen, and M. Alexander. "Microbial Degradation by Mineralization or Cometabolism Determined by Chemical Concentration and Environ-ment," *J. Agric. Food Chem.*, 33(3):495–499 (1985).

Wang, Y.-S., R.V. Subba-Rao, and M. Alexander. "Effect of Substrate Concentration and Organic and Inorganic Compounds on the Occurrence and Rate of Mineralization and Cometabolism," *Appl. Environ. Microbiol.*, 47:1195–1200 (1984).

Wanner, O., T. Egli, T. Fleischmann, K. Lanz, P. Reichert, and R.P. Schwarzenbach. "Behavior of the Insecticides Disulfoton and Thiometon in the Rhine River: A Chemodynamic Study," *Environ. Sci. Technol.*, 23(10):1232–1242 (1989).

Ward, T.M. and J.B. Weber. "Aqueous Solubility of Alkylamino-*s*-triazines as a Function of pH and Molecular Structure," *J. Agric. Food Chem.*, 16(6):959–961 (1968).

Ware, G.W., B.J. Estesen, and N.A. Buck. "Dislodgable Insecticide Residues on Cotton," *Bull. Environ. Contam. Toxicol.*, 20(1):24–27 (1978).

Watanabe, I. "Pentachlorophenol-Decomposing and PCP-Tolerant Bacteria in Field Soil Treated with PCP," *Soil Biol. Biochem.*, 9:99–103 (1977).

Watanabe, I. "Pentachlorophenol (PCP) Decomposing Activity of Field Soils Treated Annually with PCP," *Soil Biol. Biochem.*, 10:71–753 (1978).

Wauchope, R.D. Pesticide Properties Database, U.S. Department of Agriculture, Washington, DC (1988).

Wauchope, R.D. and R. Haque. "Effects of pH, Light and Temperature on Carbaryl in Aqueous Media," *Bull. Environ. Contam. Toxicol.*, 9(5):257–261 (1973).

Way, M.J. and N.E.A. Scopes. "Studies on the Persistence and Effects on Soil Fauna and Some Soil-Applied Systemic Insecticides," *Ann. Appl. Biol.*, 62:199–214 (1968).

Weber, J.B. and J.A. Best. "Activity and Movement of 13 Soil-Applied Herbicides as Influenced by Soil Reaction," *Proc. S. Weed Sci. Soc.*, 25:403–413 (1972).

Weber, J.B. and H.D. Coble. "Microbial Decomposition of Diquat Adsorbed on Montmorillonite and Kaolinite Clays," *J. Agric. Food Chem.*, 16(3):475–478 (1968).

Webster, G.R.B., K.J. Friesen, L.P. Sarna, and D.C.G. Muir. "Environmental Fate Modelling of Chlorodioxins: Determination of Physical Constants," *Chemosphere*, 14(6/7):609–622 (1985).

Wedemeyer, G.A. "Dechlorination of DDT by Aerobacter aerogenes," *Science (Washington, DC)*, 152(3722):647 (1966).

Wei, L.Y. "Gas Chromatographic and Mass Spectrometric Characterization of Terbufos Sulfoxide and Its Hydration Product," *J. Environ. Sci. Health*, B25(5):607–613 (1990).

Weil, L., G. Dure, and K.E. Quentin. "Solubility in Water of Insecticide Chlorinated Hydrocarbons and Polychlorinated Biphenyls in View of Water Pollution," *Z. Wasser Forsch.*, 7(6):169–175 (1974).

Weiss, U.M., I. Scheunert, W. Klein, and F. Korte. "Fate of Pentachlorophenol-^{14}C in Soil Under Controlled Conditions," *J. Agric. Food Chem.*, 30(6):1191–1194 (1982).

Werner, R.A. "Distribution and Toxicity of Root Absorbed ^{14}C-Orthene and Its Metabolites in Loblolly Pine Seedlings," *J. Econ. Entomol.*, 67(5):588–591 (1974).

Wheeler, W.B., D.F. Rothwell, and D.H. Hubbell. "Persistence and Microbial Effects of Acrol and Chlorobenzilate to Two Florida Soils," *J. Environ. Qual.*, 2:115–118 (1973).

Wheeler, H.G., F.F. Smith, A.H. Yeomans, and E. Fields. "Persistence of Low-Volume and Standard Formulations of Malathion on Lima Bean Foliage," *J. Econ. Entomol.*, 60:400–402 (1967).

Whitehouse, B.G. "The Effects of Temperature and Salinity on the Aqueous Solubility of Polynuclear Aromatic Hydrocarbons," *Mar. Chem.*, 14:319–332 (1984).

Wierzbicki, T. and O. Wojcik. "Preliminary Trials on the Decomposition of Acro-lein, Allyl Alcohol, and Glycerol by Activated Sludge," in *Chem. Abstr.*, 68:1589z, Zesz. Nauk. Politech Slaska Inz. Sanit., 8:173–185 (1968).

Wildung, R.E., G. Chesters, and D.E. Armstrong. "Chloramben (Amiben) Degradation in Soil," *Weed Res.*, 8(3):213–225 (1968).

Williams, J.D.H. "Adsorption and Desorption of Simazine by Some Rothamsted Soils," *Weed Res.*, 8(4):327–335 (1968).

Williams, P.P. "Metabolism of Synthetic Organic Pesticides by Anaerobic Microorganisms," *Residue Rev.*, 66:63–135 (1977).

Williams, I.H., M.J. Brown, and P. Whitehead. "Persistence of Carbofuran Residues in Some British Columbia Soils," *Bull. Environ. Contam. Toxicol.*, 15(2):242–243 (1976).

Williams, J.H. and D.J. Eagle. "Persistence of Dichlorbenil in a Sandy Soil and Effects of Residues on Plant Growth," *Weed Res.*, 19:315–319 (1979).

Willis, G.H. and L.L. McDowell. "Pesticide Persistence on Foliage," *Rev. Environ. Contam. Toxicol.*, 100:23–73 (1987).

Wilson, R.G. and H.H. Cheng. "Fate of 2,4-D in a Naff Silt Loam Soil," *J. Environ. Qual.*, 7:281–286 (1978).

Wilson, J.T., C.G. Enfield, and W.J. Dunlap. "Transport and Fate of Selected Organic Pollutants in a Sandy Soil," *J. Environ. Qual.*, 10(4):501–506 (1981).

Windholz, M., S. Budavari, R.F. Blumetti, and E.S. Otterbein, Eds., *The Merck Index*, 10th ed. (Rahway, NJ: Merck & Co., 1983), 1463 p.

Wink, O. and U. Luley. "Enantioselective Transformation of the Herbicides Diclofop-Methyl and Fenoxaprop-Ethyl in Soil," *Pestic. Sci.*, 22(1):31–40 (1988).

Winkelmann, D.A. and S.J. Klaine. "Atrazine Metabolite Behavior in Soil-Core Microcosms" in *Pesticide Transformation Products. Fate and Significance in the Environment*, ACS Symposium Series 459, Somasundaram, L. and J.R. Coats, Eds., (Washington, DC: American Chemical Society, 1991), pp. 75–92.

Witkontòn, S. and C.D. Ercegovich. "Degradation of N'-(4-Chloro-o-tolyl)-N,N-dimethylformamidine in Six Different Fruit," *J. Agric. Food Chem.*, 20(3):569–573 (1972).

Wolf, D.C. and J.P. Martin. "Microbial Degradation of Bromacil-2-^{14}C and Terbacil-2-^{14}C," Soil Sci. Soc. Am. Proc., 38:921–925 (1974).

Wolfe, N.L. "Abiotic Transformations of Pesticides in Natural Waters and Sediments," in *Fate of Pesticides and Chemicals in the Environment*, Schnoor, J.L., Ed., (New York: John Wiley & Sons, Inc., 1992), pp. 93–104.

Wolfe, N.L., U. Mingelgrin, and G.C. Miller. "Abiotic Transformations in Water, Sediments, and Soil" in *Pesticides in the Soil Environment: Processes, Impacts, and Modeling*, SSSA Book Series: 2, Cheng, H.H., Ed., (Madison, WI: Soil Science Society of America, Inc., 1990), pp. 103–168.

Wolfe, N.L., D.C. Staff, J.F. Armstrong, and S.W. Comer. "Persistence of Parathion in Soil," *Bull. Environ. Contam. Toxicol.*, 10(1):1–9 (1973).

Wolfe, N.L., W.C. Steen, and L.A. Burns. "Phthalate Ester Hydrolysis: Linear Free Energy Relationships," *Chemosphere*, 9(7/8):403–408 (1980).

Wolfe, N.L., R.G. Zepp, G.L. Baughman, R.C. Fincher, and J.A. Gordon. "Chemical and Photochemical Transformations of Selected Pesticides in Aquatic Systems," U.S. EPA Report 600/3-76-067 (1976), 141 p.

Wolfe, N.L., R.G. Zepp, J.C. Doster, and H.C. Hollis. "Captan Hydrolysis," *J. Agric. Food Chem.*, 24(5):1041–1045 (1976).

Wolfe, N.L., R.G. Zepp, J.A. Gordon, G.L. Baughman, and D.M. Cline. "Kinetics of Chemical Degradation of Malathion in Water," *Environ. Sci. Technol.*, 11(1):88–93 (1977).

Wolfe, N.L., R.G. Zepp, and D.F. Paris. "Use of Structure-Reactivity Relationships to Estimate Hydrolytic Persistence of Carbamate Pesticides," *Water Res.*, 12(8):561–563 (1978).

Wolfe, N.L., R.G. Zepp, and D.F. Paris. "Carbaryl, Propham, and Chlorpropham: A Comparison of the Rates of Hydrolysis and Photolysis with the Rate of Biolysis," *Water Res.*, 12(8):565–571 (1978a).

Wolfe, N.L., R.G. Zepp, D.F. Paris, G.L. Baughman, and R.C. Hollis. "Methoxy-chlor and DDT Degradation in Water: Rates and Products," *Environ. Sci. Technol.*, 11(12):1077–1081 (1977a).

Wong, A.S. and D.G. Crosby. "Photodecomposition of Pentachlorophenol in Water," *J. Agric. Food Chem.*, 29(1):125–130 (1981).

Wood, A.L., D.C. Bouchard, M.L. Brusseau, and P.S.C. Rao. "Cosolvent Effects on Sorption and Mobility of Organic Contaminants in Soils," *Chemosphere*, 21(4/5):575–587 (1990).

Woodrow, J.E., D.G. Crosby, T. Mast, K.W. Moilanen, and J.N. Seiber. "Rates of Transformation of Trifluralin and Parathion Vapors in Air," *J. Agric. Food Chem.*, 26(6):1312–1316 (1978).

Woodrow, J.E., D.G. Crosby, and J.N. Seiber. "Vapor-Phase Photochemistry of Pesticides," *Residue Rev.*, 85:111–125 (1983).

Worthing, C.R. and R.J. Hance, Eds. *The Pesticide Manual — A World Compendium*, 9th ed. (Great Britain: British Crop Protection Council, 1991), 1,141 p.

Wright, C.G., R.B. Leidy, and H.E. Dupree, Jr. "Insecticides in the Ambient Air of Rooms Following Their Application for Control of Pests," *Bull. Environ. Contam. Toxicol.*, 26(4):548–553 (1981).

Wright, W.L. and G.F. Warren. "Photochemical Decomposition of Trifluralin," *Weeds*, 13:329–331 (1965).

Yalkowsky, S.H., R.J. Orr, and S.C. Valvani. "Solubility and Partitioning. 3. The Solubility of Halobenzenes in Water," *Indust. Eng. Chem. Fundam.*, 18(4):351–353 (1979).

Yan, H., C. Ye, and C. Yin. "Kinetics of Phthalate Ester Biodegradation by Chlorella pyrenoidosa," *Environ. Toxicol. Chem.*, 14(6):931–938 (1995).

Yaron, B., B. Heuer, and Y. Birk. "Kinetics of Azinphosmethyl Losses in the Soil Environment," *J. Agric. Food Chem.*, 22(3):439–441 (1974).

Yih, R.Y., D.H. McRae, and H.F. Wilson. "Metabolism of 3′,4′-Dichloropropionanilide: 3,4-Dichloroaniline-Lignin Complex in Rice Plants," *Science (Washington, DC)*, 161(3839):376–377 (1968).

Yoshida, T. and T.F. Castro. "Degradation of 2,4-D, 2,4,5-T, and Picloram in Two Philippine Soils," *Soil Sci. Plant Nutr.*, 21:397–404 (1975).

Young, W.F., H. Horth, R. Crane, T. Ogden, and M. Arnott. "Taste and Threshold Concentrations of Potential Potable Water Contaminants," *Water Res.*, 30(2):331–340 (1996).

Young, J.C. and S.U. Khan. "Kinetics of Nitrosation of the Herbicide Glyphosate," *J. Environ. Sci. Health*, B13:59–72 (1978).

Youngson, C.R., C.A.I. Goring, R.W. Meikle, H.H. Scott, and J.D. Griffith. "Factors Influencing the Decomposition of Tordon Herbicide in Soils," *Down to Earth*, 23:3–11 (1967).

Yu, C.-C., G.M. Booth, D.J. Hansen, and J.R. Larson. "Fate of Alachlor and Propachlor in a Model Ecosystem," *J. Agric. Food Chem.*, 23(4):877–879 (1975).

Yu, C.-C., D.J. Hansen, and G.M. Booth. "Fate of Dicamba in a Model Eco-system," *Bull. Environ. Contam. Toxicol.*, 13(3):280–283 (1975).

Yu, C.-C. and J.R. Sanborn. "The Fate of Parathion in a Model Ecosystem," *Bull. Environ. Contam. Toxicol.*, 13(5):543–550 (1975).

Yuen, Q.H. and H.W. Hilton. "The Adsorption of Monuron and Diuron by Hawaiian Sugarcane Soils," *J. Agric. Food Chem.*, 10(5):386–392 (1962).

Yule, W.N., M. Chiba, and H.V. Morley. "Fate of Insecticide Residues. Decomposition of Lindane in Soil," *J. Agric. Food Chem.*, 15(6):1000–1004 (1967).

Zabik, M.J., R.D. Schuetz, W.L. Burton, and B.E. Pape. "Photochemistry of Bio-reactive Compounds. Studies of a Major Product of Endrin," *J. Agric. Food Chem.*, 19(2):308–313 (1971).

Zayed, S.M.A.D., I.Y. Mostafa, M.M. Farghaly, H.S.H. Attaby, Y.M. Adam, and F.M. Maḥdy. "Microbial Degradation of Trifluralin by Aspergillus carneus, Fusarium oxysporum and Trichoderma viride," *J. Environ. Sci. Health*, B18(2):253–267 (1983).

Zbozinek, J.V. "Environmental Transformations of DPA, SOPP, Benomyl, and TBZ," *Residue Rev.*, 92:113–155 (1984).

Zepp, R.G., G.L. Baughman, and P.F. Schlotzhauer. "Comparison of Photochemical Behavior of Various Humic Substances in Water. I. Sunlight Induced Reactions of Aquatic Pollutants Photosensitized by Humic Substances," *Chemosphere*, 10:109–117 (1981).

Zepp, R.G., N.L. Wolfe, L.V. Azarraga, R.H. Cox, and C.W. Pape. "Photochemical Transformation of the DDT and Methoxychlor Degradation Products, DDE and DMDE, by Sunlight," *Arch. Environ. Contam. Toxicol.*, 6:305–314 (1977).

Zepp, R.G., N.L. Wolfe, J.A. Gordon, and R.C. Fincher. "Light-Induced Transformations of Methoxychlor in Aquatic Systems," *J. Agric. Food Chem.*, 24(4):727–733 (1976).

Zhong, W.Z., A.T. Lemley, and R.J. Wagenet. *Evaluation of Pesticides in Ground Water*, ACS Symposium Series 315, Garner, W.J., R.C. Honeycutt, and H.N. Nigg, Eds., (Washington, DC: American Chemical Society, 1986), pp. 61–67.

Zimdahl, R.L. and S.K. Clark. "Degradation of Three Acetanilide Herbicides in Soil," *Weed Sci.*, 30:545–548 (1982).

Zimdahl, R.L., V.H. Freed, M.L. Montgomery, and W.R. Furtick. "The Degradation of Triazine and Uracil Herbicides in Soil," *Weed Res.*, 10:18–26 (1970).

Zimdahl, R.L. and S.M. Gwynn. "Soil Degradation of Three Dinitroanilines," *Weed Sci.*, 25:247–251 (1977).

Zoro, J.A., J.M. Hunter, and G.C. Ware. "Degradation of *p,p'*-DDT in Reducing Environments," *Nature (London)*, 247(5438):235–237 (1974).

Appendix A. Conversion Factors between Various Concentration Units

To obtain	From	Compute
g/L of solution	molality	$g/L = 1000dm(MW_1)/[1000 + m(MW_1)]$
g/L of solution	molarity	$g/L = M(MW_1)$
g/L of solution	mole fraction	$g/L = 1000dn(MW_1)/[n(MW_1) + (1 - n)MW_2]$
g/L of solution	wt %	$g/L = 10d(\text{wt \%})$
molality	molarity	$m = 1000M/[1000d - M(MW_1)]$
molality	mole fraction	$m = 1000n/[MW_2 - n(MW_2)]$
molality	g/L of solution	$m = 1000G/[MW_1(1000d - G)]$
molality	wt % of solute	$m = 1000(\text{wt \%})/[MW_1(100 - \text{wt \%})]$
molarity	g/L of solution	$M = G/MW_1$
molarity	molality	$M = 1000dm/[1000 + m(MW_1)]$
molarity	mole fraction	$M = 1000dn/[n(MW_1) + (1 - n)MW_2]$
molarity	wt % of solute	$M = 10d(\text{wt \%})/MW_1$
mole fraction	g/L of solution	$n = G(MW_2)/[G(MW_2 - MW_1) + 1000d(MW_1)]$
mole fraction	molality	$n = m(MW_2)/[m(MW_2) + 1000]$
mole fraction	molarity	$n = M(MW_2)/[M(MW_2 - MW_1)+1000d]$
mole fraction	wt % of solute	$n = [(\text{wt \%})/MW1]/[(\text{wt \%}/MW_1) + (100 - \text{wt \%})MW_2]$
wt % of solute	g/L of solution	$\text{wt \%} = G/10d$
wt % of solute	molality	$\text{wt \%} = 100m(MW_1)/[1000 + m(MW_1)]$
wt % of solute	molarity	$\text{wt \%} = M(MW_1)/10d$
wt % of solute	mole fraction	$\text{wt \%} = 100n(MW_1)/[n(MW_1) + (1 - n)MW_2]$

d = density of solution (g/L); G = g of solute/L of solution; m = molality; M = mass of solute; MW_1 = formula weight of solute; MW_2 = formula weight of solvent; n = mole fraction; wt % = weight percent of solute

Appendix B. Conversion Factors

To convert	Into	Multiply by
A		
acre-feet	feet^3	4.356×10^4
acre-feet	gallons (U.S.)	3.529×10^5
acre-feet	inches^3	7.527×10^7
acre-feet	liters	1.233×10^6
acre-feet	meters^3	1,233
acre-feet	yards^3	1,613
acre-feet/day	feet^3/second	5.042×10^{-1}
acre-feet/day	gallons (U.S.)/minute	226.3
acre-feet/day	liters^3/second	14.28
acre-feet/day	meters^3/day	1,234
acre-feet/day	meters^3/second	1.428×10^{-2}
acres	feet^2	43,560
acres	hectares	4.047×10^{-1}
acres	inches^2	6.273×10^6
acres	kilometers^2	4.047×10^{-3}
acres	meters^2	4,047
acres	miles^2	1.563×10^{-3}
angstroms	inches	3.937×10^{-6}
angstroms	meters	10^{-10}
angstroms	microns	10^{-4}
atmospheres	bars	1.01325
atmospheres	inches of Hg	29.92126
atmospheres	millibars	1013.25
atmospheres	millimeters of Hg	760
atmospheres	millimeters of water	1.033227×10^4
atmospheres	pascals	1.01325×10^5
atmospheres	pounds/inch^2	14.70
atmospheres	torrs	760
B		
barrels (petroleum)	liters	159.0
bars	atmospheres	9.869×10^{-1}
bars	pounds/inch^2	14.50
British thermal unit	joules	1.055×10^{-3}
British thermal unit	kilowatt-hour	2.928×10^{-4}
British thermal unit	watts	2.931×10^{-1}
C		
calories	joules	4.187
Celsius (°C)	Fahrenheit (°F)	$(1.8 \times °C) + 32$

To convert	Into	Multiply by
Celsius (°C)	Kelvin (°K)	°C + 273.15
centimeters	feet	3.291×10^{-2}
centimeters	inches	3.937×10^{-1}
centimeters	miles	6.214×10^{-6}
centimeters	millimeters	10
centimeters	mils	393.7
centimeters	yards	1.094×10^{-2}
centimeters^2	feet^2	1.076×10^{-3}
centimeters^2	inches^2	1.55×10^{-1}
centimeters^2	meters^2	1×10^{-4}
centimeters^2	yards^2	1.196×10^{-4}
centimeters^3	fluid ounces	3.381×10^{-2}
centimeters^3	feet^3	3.5314×10^{-5}
centimeters^3	inches^3	6.102×10^{-2}
centimeters^3	liters	1×10^{-3}
centimeters^3	ounces (U.S., fluid)	3.381×10^{-2}
centimeters/second	feet/day	2,835
centimeters/second	feet/minute	1.968
centimeters/second	feet/second	3.281×10^{-2}
centimeters/second	kilometers/hour	3.6×10^{-2}
centimeters/second	liters/meter/second	9.985
centimeters/second	meters/minute	6×10^{-1}
centimeters/second	miles/hour	2.237×10^{-2}
centimeters/second	miles/minute	3.728×10^{-4}

D

Darcy	centimeters/second	9.66×10^{-4}
Darcy	feet/second	3.173×10^{-5}
Darcy	liters/meter/second	8.58×10^{-3}
days	seconds	86,400.0
dynes/centimeter^2	atmospheres	9.869×10^{-7}
dynes/centimeter^2	inches of water at 4°C	4.015×10^{-4}

F

Fahrenheit (°F)	Celsius (°C)	5(°F − 32)/9
Fahrenheit (°F)	Kelvin (°K)	5(°F + 459.67)/9
Fahrenheit (°F)	Rankine (°R)	°F + 459.67
feet	centimeters	30.48
feet	inches	12
feet	kilometers	3.048×10^{-4}
feet	meters	3.048×10^{-1}
feet	miles	1.894×10^{-4}
feet	millimeters	304.8
feet	yards	3.33×10^{-1}
feet/day	feet/second	1.157×10^{-5}

To convert	Into	Multiply by
feet/day	meters/second	3.528×10^{-6}
feet^2	acres	2.296×10^{-5}
feet^2	hectares	9.29×10^{-9}
feet^2	inches	2144
feet^2	kilometers	29.29×10^{-8}
feet^2	meters	29.29×10^{-2}
feet^2	miles^2	3.587×10^{-8}
feet^2	yards^2	1.111×10^{-1}
feet^2/day	meters^2/day	9.290×10^{-2}
feet^2/sec	meters^2/day	8,027
feet^3	acre-feet	2.296×10^{-5}
feet^3	gallons (U.S.)	7.481
feet^3	inches^3	1,728
feet^3	liters^2	8.32
feet^3	meters^3	2.832×10^{-2}
feet^3	yards^3	3.704×10^{-2}
feet^3/foot/day	gallons/foot/day	7.48052
feet^3/foot/day	liters/meter/day	92.903
feet^3/foot/day	meters^3/meter/day	9.29×10^{-2}
feet^3/foot/day	feet^3/foot^2/minute	6.944×10^{-4}
feet^3/foot/day	gallons/foot^2/day	7.4805
feet^3/foot/day	inches^3/inch^2/hour	5×10^{-1}
feet^3/foot/day	liters/meter^2/day	304.8
feet^3/foot/day	meters^3/meter^2/day	3.048×10^{-1}
feet^3/foot/day	millimeters^3/inch^2/hour	5×10^{-1}
feet^3/foot^2/day	millimeters^3/millimeter^2/hour	25.4
feet/mile	meters/kilometer	1.894×10^{-1}
feet/second	centimeters/second	5.080×10^{-1}
feet/second	feet/day	86,400
feet/second	feet/hour	3,600
feet/second	gallons (U.S.)/foot^2/day	5.737×10^{5}
feet/second	kilometers/hour	1.097
feet/second	meters/second	3.048×10^{-1}
feet/second	miles/hour	6.818×10^{-1}
feet/year	centimeters/second	9.665×10^{-7}
feet^3/second	acre-feet/day	1.983
feet^3/second	feet^3/minute	60.0
feet^3/second	gallons (U.S.)/minute	448.8
feet^3/second	liters/second	28.32
feet^3/second	meters^3/second	2.832×10^{-2}
feet^3/second	meters^3/day	2,447

G

gallons (U.K.)	gallons (U.S.)	1.200
gallons (U.S.)	acre-feet	3.068×10^{-6}
gallons (U.S.)	feet^3	1.337×10^{-1}

To convert	Into	Multiply by
gallons (U.S.)	fluid ounces	128.0
gallons (U.S.)	liters	3.785
gallons (U.S.)	meters^3	3.785×10^{-3}
gallons (U.S.)	yards^3	4.951×10^{-3}
gallons/day	acre-feet/year	1.12×10^{-3}
gallons/foot/day	feet^3/foot/day	1.3368×10^{-1}
gallons/foot/day	liters/meter/day	
gallons/foot/day	meters^3/meter/day	1.242×10^{-2}
gallons (U.S.)/foot^2/day	centimeters/second	4.717×10^{-5}
gallons (U.S.)/foot^2/day	Darcy	5.494×10^{-2}
gallons (U.S.)/foot^2/day	feet/day	1.3368×10^{-1}
gallons (U.S.)/foot^2/day	feet/second	1.547×10^{-6}
gallons (U.S.)/foot^2/day	gallons/foot^2/minute	6.944×10^{-4}
gallons (U.S.)/foot^2/day	liters/meter^2/day	40.7458
gallons (U.S.)/foot^2/day	meters/day	4.07458×10^{-2}
gallons (U.S.)/foot^2/day	meters/minute	2.83×10^{-5}
gallons (U.S.)/foot^2/day	meters/second	4.716×10^{-7}
gallons (U.S.)/foot^2/minute	meters/day	58.67
gallons (U.S.)/foot^2/minute	meters/second	6.791×10^{-2}
gallons (U.S.)/minute	acre-feet/day	4.419×10^{-3}
gallons (U.S.)/minute	feet^3/second	2.228×10^{-3}
gallons (U.S.)/minute	feet^3/hour	8.0208
gallons (U.S.)/minute	liters/second	6.309×10^{-2}
gallons (U.S.)/minute	meters^3/day	5.30
gallons (U.S.)/minute	meters^3/second	6.309×10^{-5}
grams/centimeter^3	kilograms/meter	31,000
grams/centimeter^3	pounds/feet^3	62.428
grams/centimeter^3	pounds/gallon (U.S.)	8.345
grams	kilograms	1×10^{-3}
grams	ounces (avoirdupois)	3.527×10^{-2}
grams	pounds	2.2046×10^{-3}
grams/liter	grains/gallon (U.S.)	58.4178
grams/liter	grams/centimeter^3	1×10^{-3}
grams/liter	kilograms/meter^3	1
grams/liter	pounds/feet^3	6.24×10^{-2}
grams/liter	pounds/inch^3	3.61×10^{-5}
grams/liter	pounds/gallon (U.S.)	8.35×10^{-3}
grams/meter^3	grains/feet^3	4.37×10^{-1}
grams/meter^3	milligrams/liter	1.0
grams/meter^3	pounds/gallon (U.S.)	8.345×10^{-5}
grams/meter^3	pounds/inch^3	7.433×10^{-3}

H

hectares	acres	2.471
hectares	feet^2	1.076×10^{5}
hectares	inches^2	1.55×10^{7}

To convert	Into	Multiply by
hectares	kilometers^2	1×10^{-2}
hectares	meters^2	10,000
hectares	miles^2	3.861×10^{-3}
hectares	yards^2	11,959.90
horsepower	kilowatts/hour	7.457×10^{-1}

I

inches	centimeters	2.540
inches	feet	8.333×10^{-1}
inches	kilometers	2.54×10^{-5}
inches	meters	2.54×10^{-2}
inches	miles	1.578×10^{-5}
inches	millimeters	25.4
inches	yards	2.778×10^{-2}
inches of Hg	pascals	3,386
inches^2	acres	1.594×10^{-8}
inches^2	centimeters	26.4516
inches^2	feet	26.944×10^{-3}
inches^2	hectares	6.452×10^{-8}
inches^2	kilometers^2	6.452×10^{-10}
inches^2	meters^2	6.452×10^{-4}
inches^2	millimeters^2	645.16
inches^3	acre-feet	1.329×10^{-8}
inches^3	centimeters^3	16.39
inches^3	feet^3	5.787×10^{-4}
inches^3	gallons (U.S.)	4.329×10^{-3}
inches^3	liters	1.639×10^{-2}
inches^3	meters^3	1.639×10^{-5}
inches^3	milliliters	16.387
inches^3	yards^3	2.143×10^{-5}

K

Kelvin (°K)	Celsius	°K − 273.15
Kelvin (°K)	Fahrenheit	1.8(°K) − 459.67
kilograms	grams	1,000
kilograms	milligrams	1×10^6
kilograms	ounces (avoirdupois)	35.28
kilograms	pounds	2.205
kilograms	tons (long)	9.842×10^{-4}
kilograms	tons (metric)	1×10^{-3}
kilograms/meter^3	grams/centimeter^3	1×10^{-3}
kilograms/meter^3	grams/liter	1.0
kilograms/meter^3	pounds/inch^3	3.613×10^{-5}
kilograms/meter^3	pounds/feet^3	0.0624
kilometers	feet	3,281

To convert	Into	Multiply by
kilometers	inches	39,370
kilometers	meters	1,000
kilometers	miles	6.214×10^{-1}
kilometers	millimeters	1×10^{6}
kilometers^2	acres	247.1
kilometers^2	hectares	100
kilometers^2	inches^2	1.55×10^{9}
kilometers^2	meters^2	1×10^{6}
kilometers^2	miles^2	0.3861
kilometers/hour	feet/day	78,740
kilometers/hour	feet/second	9.113×10^{-1}
kilometers/hour	meters/second	2.778×10^{-1}
kilometers/hour	miles/hour	6.214×10^{-1}

L

liters	acre-feet	8.106×10^{-7}
liters	feet^3	3.531×10^{-2}
liters	fluid ounces	33.814
liters	gallons (U.S.)	2.642×10^{-1}
liters	inches^3	61.02
liters	meters^3	10^{-3}
liters	yards^3	1.308×10^{-3}
liters/second	acre-feet/day	7.005×10^{-2}
liters/second	feet^3/second	3.531×10^{-2}
liters/second	gallons (U.S.)/minute	15.85
liters/second	meters^3/day	86.4

M

meters	feet	3.28084
meters	inches	39.3701
meters	kilometers	1×10^{-3}
meters	miles	6.214×10^{-4}
meters	millimeters	1,000
meters	yards	1.0936
meters^2	acres	2.471×10^{-4}
meters^2	feet^2	10.76
meters^2	hectares	1×10^{-4}
meters^2	inches^2	1.550
meters^2	kilometers	1×10^{-6}
meters^2	miles^2	3.861×10^{-7}
meters^2	yards^2	1.196
meters^3	acre-feet	8.106×10^{-4}
meters^3	feet^3	35.31
meters^3	gallons (U.S.)	264.2

To convert	Into	Multiply by
meters^3	inches^3	6.102×10^4
meters^3	liters	1,000
meters^3	yards^3	1.308
meters/second	feet/minute	196.8
meters/second	feet/day	283,447
meters/second	kilometers/hour	3.6
meters/second	feet/second	3.281
meters/second	miles/hour	2.237
meters^3/day	acre-feet/day	6.051×10^6
meters^3/day	feet^3/second	3.051×10^6
meters^3/day	gallons (U.S.)/minute	1.369×10^9
meters^3/day	liters/second	8.64×10^7
miles	feet	5,280
miles	kilometers	1.609
miles	meters	1,609
miles	millimeters	1.609×10^6
miles^2	acres	640
miles^2	feet^2	2.778×10^7
miles^2	hectares	259
miles^2	inches^2	4.014×10^9
miles^2	kilometers^2	2.590
miles^2	meters^2	2.59×10^6
miles/hour	feet/day	1.267×10^5
miles/hour	kilometers/hour	1.609
miles/hour	meters/second	4.47×10^{-1}
millibars	pascals	100.0
milligrams/liter	grams/meter^3	1.0
milligrams/liter	parts/million	1.0
milligrams/liter	pounds/feet^3	6.2428×10^{-5}
milliliters	centimeters^3	1.0
milliliters	liters	1×10^{-3}
millimeters	centimeters	1×10^{-1}
millimeters	feet	3.281×10^{-3}
millimeters	inches	0.03937
millimeters	kilometers	1.0×10^{-6}
millimeters	meters	1.0×10^{-3}
millimeters	miles	6.214×10^{-7}
millimeters^2	centimeters^2	1×10^{-2}
millimeters^2	inches^2	1.55×10^{-3}
millimeters^3	centimeters^3	1×10^{-3}
millimeters^3	inches^3	6.102×10^{-5}
millimeters^3	liters	10^{-6}
millimeters of Hg	atmospheres	1.316×10^{-3}
millimeters of Hg	pascals	133.3224
millimeters of Hg	torrs	1.0

To convert	Into	Multiply by

O

ounces (avoirdupois)	grams	28.35
ounces (avoirdupois)	kilograms	2.8355×10^{-2}
ounces (avoirdupois)	pounds	6.25×10^{-2}

P

pascals	atmospheres	9.869×10^{-6}
pascals	millimeters of Hg	7.501×10^{-3}
pounds (avoirdupois)	kilograms	4.535×10^{-1}
pounds (avoirdupois)	ounces (avoirdupois)	16
pounds/centimeter^3	pounds/inch^3	3.61×10^{-2}
pounds/inch^2	pascals	6,895
pounds/inch^2	kilograms/centimeter^2	7.31×10^{-2}
pounds/feet^3	grams/centimeter^3	1.6×10^{-2}
pounds/feet^3	grams/liter	27,680
pounds/feet^3	pounds/inch^3	5.787×10^{-4}
pounds/feet^3	pounds/gallon (U.S.)	1.337×10^{-1}
pounds/foot^2	pascals	47.88
pounds/inch^3	pounds/centimeter^3	27.68
pounds/inch^3	pounds/feet^3	1,728
pounds/inch^3	pounds/gallon (U.S.)	231
pounds/gallon (U.S.)	grams/centimeter^3	1.198×10^{-1}
pounds/gallon (U.S.)	grams/liter	119.8
pounds/gallon (U.S.)	pounds/feet^3	7.481
pounds/gallon (U.S.)	pounds/inch^3	4.329×10^{-3}

Q

quarts	liters	9.464×10^{-1}

T

torrs	atmospheres	1.316×10^{-3}
torrs	millimeters of Hg	1.0
torrs	pascals	133.322

W

watts	BTU/hour	3.4129

Y

yards	centimeters	91.44
yards	fathoms	5×10^{-1}
yards	feet	3.0
yards	inches	36.0

To convert	Into	Multiply by
yards	meters	9.144×10^{-1}
yards	miles	5.682×10^{-4}
yards	feet	327
yards	gallons (U.S.)	202
yards	inches^3	4.666×10^4
yards	liters	764.6
yards	meters^3	7.646×10^{-1}
yards^2	meters^2	8.361×10^{-1}
yards^3	meters^3	7.646×10^{-1}
years (normal calendar)	hours	8,760
years (normal calendar)	minutes	5.256×10^5
years (normal calendar)	seconds	3.1536×10^7
years (normal calendar)	weeks	52.1428

Appendix C. U.S. EPA Approved Test Methods[a]

The numbers that follow each compound name below are the U.S. EPA approved test methods based on information obtained from the "Guide to Environmental Analytical Methods" (Schenectady, NY: Genium Publishing Corp., 1992) and the U.S. EPA's Sampling and Analysis Methods Database. The database information is available in hard copy from Keith, L.H. *Compilation of EPA's Sampling and Analysis Methods* (Chelsea, MI: Lewis Publishers, Inc., 1991), 803 p.

Chemical Name	Test Methods
Acrolein	8030, 8240
Acrylonitrile	8030, 8240
Alachlor	525
Aldrin	508, 525, 608, 625, SM-6410, SM-6630, 8080, 8270
Anilazine	8270
Atrazine	525
Azinphos-methyl	8140, 8270
α-BHC	508, 608, 625, SM-6410, SM-6630, 8080, 8270
β-BHC	508, 608, 625, SM-6410, SM-6630, 8080, 8270
δ-BHC	508, 608, 625, SM-6410, SM-6630, 8080, 8270
Bis(2-chloroethyl)ether	625, SM-6410, 8270
Bis(2-chloroisopropyl)ether	625, SM-6410, 8270
Bromoxynil	8270
Carbaryl	8270
Carbofuran	8270
Carbon disulfide	8240
Carbon tetrachloride	502.1, 502.2, 601, 524.1, 524.2, 624, SM-6210, SM-6230, 8010, 8240
Carbophenothion	8270
Chloramben	515.1
Chlordane (technical)	508, 608, SM-6630, 8080, 8270
Chlorobenzilate	508, 8270
Chlorothalonil	508
Chlorpyrifos	508, 8140
Crotoxyphos	8270
2,4-D	515.1, 8150
Dalapon	8150
p,p'-DDD	508, 608, SM-6630, 8080, 8270
p,p'-DDE	508, 608, SM-6630, 8080, 8270
p,p'-DDT	508, 608, SM-6630, 8080, 8270
Diallate	8270
Diazinon	8140
1,2-Dibromo-3-chloropropane	502.2, 504, 524.1, 524.2, SM-6210, 8240
Di-*n*-butyl phthalate	525, 625, SM-6410, 8060, 8270
Dicamba	515.1, 8150
Dichlone	8270

Chemical Name	Test Methods
1,2-Dichloropropane	502.1, 502.2, 524.1, 524.2, 601, 624, SM-6210, SM-6230, 8010, 8240
cis-1,3-Dichloropropylene	502.1, 502.2, 524.1, 601, 624, SM-6210, SM-6230, 8010, 8240
trans-1,3-Dichloropropylene	502.1, 502.2, 524.1, 601, 624, SM-6210, SM-6230, 8010, 8240
Dichlorvos	8140, 8270
Dicrotophos	8270
Dieldrin	508, 608, SM-6630, 8080, 8270
Dimethoate	8270
Dimethyl phthalate	525, 625, SM-6410, 8060, 8270
4,6-Dinitro-*o*-cresol	8040
Dinoseb	515.1, 8150, 8270
Disulfoton	8140, 8270
α-Endosulfan	508, 608, 625, SM-6410, SM-6630, 8080, 8270
β-Endosulfan	508, 608, 625, SM-6410, SM-6630, 8080, 8270
Endosulfan sulfate	508, 608, 625, SM-6410, SM-6630, 8080, 8270
Endrin	508, 525, 608, 625, SM-6410, SM-6630, 8080, 8270
Endrin aldehyde	508, 608, 625, SM-6410, SM-6630, 8080, 8270
EPN	8270
Ethion	8270
Ethoprop	8140
Ethylene dibromide	502.1, 502.2, 504, 524.1, 524.2, SM-6210, 8240
Fensulfothion	8140, 8270
Fenthion	8140, 8270
Heptachlor	508, 525, 608, 625, SM-6410, SM-6630, 8080, 8270
Heptachlor epoxide	508, 525, 608, 625, SM-6410, SM-6630, 8080, 8270
Hexachlorobenzene	508, 525, 625, SM-6410, 8120, 8270
Kepone	8270
Lindane	508, 525, 608, SM-6630, 8080, 8270
Malathion	8270
MCPA	8150
Methoxychlor	508, 525, SM-6630, 8080, 8270
Methyl bromide	502.1, 502.2, 524.1, 524.2, 601, 624, SM-6210, SM-6230, 8010, 8240
Mevinphos	8140, 8270
Monocrotophos	8270
Naled	8140, 8270
Parathion	8270
Pentachlorobenzene	8270
Pentachlorophenol	525, 625, SM-6410, 8040, 8270
Phorate	8140, 8270
Phosalone	8270
Phosmet	8270
Phosphamidon	8270
Picloram	515.1
Propachlor	508

Chemical Name	Test Methods
Propyzamide	8270
Ronnel	8140
Simazine	525
Strychnine	525
Sulprofos	8140
2,4,5-T	515.1, 8150, SM-6640
Terbufos	8270
Tetraethyl pyrophosphate	8270
Toxaphene	508, 525, 608, 625, SM-6630, 8080, 8270
Trifluralin	508, 8270

a) Methods beginning with the prefix "SM-" refer to standard methods found in the 17th edition of the "Standard Methods for the Examination of Water and Waste Water," published by the American Water Works Association in 1989.

Appendix D. CAS Index

51218-45-2	Metolachlor
51235-04-2	Hexazinone
51338-27-3	Diclofop-methyl
51630-58-1	Fenvalerate
51707-55-2	Thidiazuron
52315-07-8	Cypermethrin
52645-53-1	Permethrin
53780-34-0	Mefluidide
55335-06-3	Triclopyr
57837-19-1	Metalaxyl
59669-26-0	Thiodicarb
64902-72-3	Chlorsulfuron
66215-27-8	Cyromazine
66230-04-4	Esfenvalerate
66441-23-4	Fenoxaprop-ethyl
68359-37-5	Cyfluthrin
70124-77-5	Flucythrinate
74222-97-2	Sulfometuron-methyl
74223-64-6	Metsulfuron-methyl
76578-14-8	Quizalofop-ethyl
78587-05-0	Hexythiazox
79277-27-3	Thiameturon-methyl

XY9100000	Ametryn	YS6640000	Chlorsulfuron
XZ2990000	Metribuzin	YS8925000	Diuron
XZ3850000	Amitrole	YS9100000	Linuron
YQ9100000	Bromacil	YT1575000	Fluometuron
YQ9360000	Terbacil	YT7350000	Siduron
YS3810000	Neburon	YT9275000	ANTU
YS4250000	Tebuthiuron	YU1395000	Thidiazuron
YS6200000	Diflubenzuron	YV6010000	Monalide
YS6300000	Monuron	ZB3200000	Mancozeb

Appendix F. Empirical Formula Index

CCl_3NO_2	Chloropicrin
CCl_4	Carbon tetrachloride
CH_2O	Formaldehyde
CH_3Br	Methyl bromide
CS_2	Carbon disulfide
C_2H_3NS	Methyl isothiocyanate
$C_2H_4Br_2$	Ethylene dibromide
$C_2H_4N_4$	Amitrole
$C_2H_6ClO_3P$	Ethephon
$C_2H_8NO_2PS$	Methamidophos
$C_3H_3Cl_2NaO_2$	Dalapon-sodium
C_3H_3N	Acrylonitrile
$C_3H_4Cl_2$	*cis*-1,2-Dichloropropylene
	trans-1,2-Dichloropropylene
C_3H_4O	Acrolein
$C_3H_5Br_2Cl$	1,2-Dibromo-3-chloropropane
$C_3H_6Cl_2$	1,2-Dichloropropane
$C_3H_8NO_5P$	Glyphosate
$C_3H_{11}N_2O_4P$	Fosamine-ammonium
$C_4H_4N_2O_2$	Maleic hydrazide
$[C_4H_6N_2S_2Mn]_xZn_y$	Mancozeb
$C_4H_7Br_2Cl_2O_4P$	Naled
$C_4H_7Cl_2O_4P$	Dichlorvos
$C_4H_8ClO_4P$	Trichlorfon
$C_4H_8Cl_2O$	Bis(2-chloroethyl)ether
$C_4H_{10}NO_3PS$	Acephate
$C_5H_{10}N_2O_2S$	Methomyl
$C_5H_{10}N_2S_2$	Dazomet
$C_5H_{12}NO_3PS_2$	Dimethoate
$C_6Cl_5NO_2$	Pentachloronitrobenzene
C_6Cl_6	Hexachlorobenzene
$C_6H_3Cl_4N$	Nitrapyrin
C_6HCl_5	Pentachlorobenzene
C_6HCl_5O	Pentachlorophenol
$C_6H_3Cl_3N_2O_2$	Picloram
$C_6H_6Cl_6$	α-BHC
	β-BHC
	δ-BHC
	Lindane
$C_6H_{10}N_6$	Cyromazine
$C_6H_{11}N_2O_4PS_3$	Methidathion
$C_6H_{12}Cl_2O$	Bis(2-chloroisopropyl)ether
$C_6H_{12}N_2S_4$	Thiram
$C_6H_{15}O_4PS_2$	Oxydemeton-methyl
$C_6H_{18}AlO_9P_3$	Fosetyl-aluminum
$C_7H_3Br_2NO$	Bromoxynil

$C_7H_3Cl_2N$	Dichlobenil
$C_7H_3Cl_3O_2$	2,3,6-Trichlorobenzoic acid
$C_7H_4Cl_3NO_3$	Triclopyr
$C_7H_5Cl_2NO_2$	Chloramben
$C_7H_6N_2O_5$	4,6-Dinitro-*o*-cresol
$C_7H_{12}ClN_5$	Simazine
$C_7H_{13}N_3O_3S$	Oxamyl
$C_7H_{13}O_6P$	Mevinphos
$C_7H_{14}NO_5P$	Monocrotophos
$C_7H_{14}N_2O_2S$	Aldicarb
$C_7H_{17}O_2PS_3$	Phorate
$C_8Cl_4N_2$	Chlorothalonil
$C_8H_5Cl_3O_3$	2,4,5-T
$C_8H_6Cl_2O_3$	2,4-D
	Dicamba
$C_8H_8Cl_3O_3PS$	Ronnel
$C_8H_{10}N_2O_4S$	Asulam
$C_8H_{10}O_5$	Endothall
$C_8H_{10}O_7P_2$	Tetraethyl pyrophosphate
$C_8H_{12}ClNO$	Allidochlor
$C_8H_{14}ClN_5$	Atrazine
$C_8H_{14}N_4OS$	Metribuzin
$C_8H_{16}NO_5P$	Dicrotophos
$C_8H_{16}O_4$	Metaldehyde
$C_8H_{19}O_2PS_3$	Disulfoton
$C_8H_{19}O_2PS_2$	Ethoprop
$C_8H_{20}O_5P_2S_2$	Sulfotepp
$C_9H_5Cl_3N_4$	Anilazine
$C_9H_6Cl_6O_3S$	α-Endosulfan
	β-Endosulfan
$C_9H_6Cl_6O_4S$	Endosulfan sulfate
$C_9H_8Cl_3NO_2S$	Captan
$C_9H_8N_4OS$	Thidiazuron
$C_9H_9ClO_3$	MCPA
$C_9H_9Cl_2NO$	Propanil
$C_9H_{10}Cl_2N_2O$	Diuron
$C_9H_{10}Cl_2N_2O_2$	Linuron
$C_9H_{11}ClN_2O$	Monuron
$C_9H_{11}Cl_3NO_3PS$	Chlorpyrifos
$C_9H_{13}BrN_2O_2$	Bromacil
$C_9H_{13}ClN_2O_2$	Terbacil
$C_9H_{13}ClN_6$	Cyanazine
$C_9H_{16}ClN_5$	Propazine
$C_9H_{16}N_4OS_2$	Tebuthiuron
$C_9H_{17}NOS$	Molinate
$C_9H_{17}N_2S$	Ametryn
$C_9H_{18}FeN_3S_6$	Ferbam
$C_9H_{19}NOS$	EPTC
$C_9H_{21}O_2PS_3$	Terbufos

$C_9H_{22}O_4P_2S_4$	Ethion
$C_{10}Cl_{10}$	Dienochlor
$C_{10}Cl_{10}O$	Kepone
$C_{10}H_4Cl_2O_2$	Dichlone
$C_{10}H_5Cl_7$	Heptachlor
$C_{10}H_5Cl_7O$	Heptachlor epoxide
$C_{10}H_6Cl_4O_4$	Chlorthal-dimethyl
$C_{10}H_6Cl_8$	Chlordane
	cis-Chlordane
	trans-Chlordane
$C_{10}H_6N_2OS_2$	Quinomethionate
$C_{10}H_7N_3S$	Thiabendazole
$C_{10}H_{10}Cl_8$	Toxaphene
$C_{10}H_{10}O_4$	Dimethyl phthalate
$C_{10}H_{11}F_3N_2O$	Fluometuron
$C_{10}H_{12}ClNO_2$	Chlorpropham
$C_{10}H_{12}N_2O_3S$	Bentazone
$C_{10}H_{12}N_2O_5$	Dinoseb
$C_{10}H_{12}N_3O_3PS_2$	Azinphos-methyl
$C_{10}H_{13}ClN_2$	Chlordimeform
$C_{10}H_{13}NO_2$	Propham
$C_{10}H_{14}NO_5PS$	Parathion
$C_{10}H_{15}OPS_2$	Fonofos
$C_{10}H_{15}O_3PS_2$	Fenthion
$C_{10}H_{16}Cl_3NOS$	Triallate
$C_{10}H_{17}Cl_2NOS$	Diallate
$C_{10}H_{18}N_4O_4S_3$	Thiodicarb
$C_{10}H_{19}ClNO_5P$	Phosphamidon
$C_{10}H_{19}N_5O$	Prometon
$C_{10}H_{19}N_5S$	Prometryn
	Terbutryn
$C_{10}H_{19}O_6PS_2$	Malathion
$C_{10}H_{21}NOS$	Pebulate
$C_{11}H_{10}N_2S$	ANTU
$C_{11}H_{12}NO_4PS_2$	Phosmet
$C_{11}H_{13}F_3N_2O_3$	Mefluidide
$C_{11}H_{13}NO_4$	Bendiocarb
$C_{11}H_{14}ClNO$	Propachlor
$C_{11}H_{15}BrClO_3PS$	Profenofos
$C_{11}H_{15}NO_2S$	Ethiofencarb
	Methiocarb
$C_{11}H_{15}NO_3$	Propoxur
$C_{11}H_{16}ClO_2PS_3$	Carbophenothion
$C_{11}H_{16}N_2O_2$	Aminocarb
$C_{11}H_{17}O_4PS_2$	Fensulfothion
$C_{11}H_{21}NOS$	Cycloate
$C_{11}H_{21}N_5S$	Dipropetryn
$C_{11}H_{23}NOS$	Butylate
$C_{12}H_8Cl_6$	Aldrin

$C_{16}H_{17}NO$	Diphenamid
$C_{16}H_{22}O_4$	Di-*n*-butyl phthalate
$C_{17}H_{21}ClN_2O_2$	Hexythiazox
$C_{17}H_{21}NO_2$	Napropamide
$C_{18}H_{13}NO_3$	Naptalam
$C_{18}H_{16}ClNO_5$	Fenoxaprop-ethyl
$C_{18}H_{20}N_2O_4S$	Difenzoquat methyl sulfate
$C_{19}H_{16}O_4$	Warfarin
$C_{19}H_{17}ClN_2O_4$	Quizalofop-ethyl
$C_{21}H_{20}Cl_2O_3$	Permethrin
$C_{21}H_{21}N_2O_3PS$	Diazinon
$C_{21}H_{22}N_2O_2$	Strychnine
$C_{22}H_{18}Cl_2FNO_3$	Cyfluthrin
$C_{22}H_{19}Cl_2NO_3$	Cypermethrin
$C_{25}H_{22}ClNO_3$	Esfenvalerate
	Fenvalerate
$C_{26}H_{23}F_2NO_4$	Flucythrinate
$C_{60}H_{78}OSn_2$	Fenbutatin oxide

Appendix G. Typical Bulk Density Values for Selected Soils and Rocks

Soil/Rock Type	Bulk Density, B
Silt	1.38
Clay	1.49
Loess	1.45
Sand, dune	1.58
Sand, fine	1.55
Sand, medium	1.69
Sand, coarse	1.73
Gravel, fine	1.76
Gravel, medium	1.85
Gravel, coarse	1.93
Till, predominantly silt	1.78
Till, predominantly sand	1.88
Till, predominantly gravel	1.91
Glacial drift, predominantly silt	1.38
Glacial drift, predominantly sand	1.55
Glacial drift, predominantly gravel	1.60
Siltstone	1.61
Claystone	1.51
Sandstone, fine-grained	1.76
Sandstone, medium-grained	1.68
Limestone	1.94
Dolomite	2.02
Schist	1.76
Basalt	2.53
Shale	2.53
Gabbro, weathered	1.73
Granite, weathered	1.50

Morris and Johnson (1967)

Appendix H. Ranges of Porosity Values for Selected Soils and Rocks

Soil/Rock Type	Porosity, n
Peat	0.60–0.80
Silt	0.34–0.61
Clay	0.34–0.57
Loess	0.40–0.57
Sand, dune	0.35–0.51
Sand, fine	0.25–0.55
Sand, medium	0.28–0.49
Sand, coarse	0.30–0.46
Gravel, fine	0.25–0.40
Gravel, medium	0.24–0.44
Gravel, coarse	0.25–0.35
Sand + gravel	0.20–0.35
Till, predominantly silt	0.30–0.41
Till, predominantly sand	0.22–0.37
Till, predominantly gravel	0.22–0.30
Till, clay-loam	0.30–0.35
Glacial drift, predominantly silt	0.38–0.59
Glacial drift, predominantly sand	0.36–0.48
Glacial drift, predominantly gravel	0.35–0.42
Siltstone	0.20–0.41
Claystone	0.41–0.45
Sandstone, fine-grained	0.14–0.49
Sandstone, medium-grained	0.30–0.44
Volcanic, dense	0.01–0.10
Volanic, pumice	0.80–0.90
Volcanic, vesicular	0.10–0.50
Volcanic, tuff	0.10–0.40
Limestone	0.05–0.56
Dolomite	0.19–0.33
Schist	0.05–0.55
Basalt	0.03–0.35
Shale	0.01–0.10
Igneous, dense metamorphic and plutonic	0.01–0.05
Igneous, weathered metamorphic and plutonic	0.34–0.55

Davis and DeWiest (1966); Grisak et al. (1980); Morris and Johnson (1967)

Appendix I. Solubility Data of Miscellaneous Compounds

Compound	CAS Registry No.	Temp. °C	Log S mol/L	Reference
Acetal	105-57-7	25.0	−0.12	1
Acetophenone	98-86-2	25.0	−1.34	2
Allyl bromide	106-95-6	25.0	−1.50	1
Anisole	100-66-3	25.00	−1.85	3
		25.0	−1.89	2
Atrazine	1912-24-9	22	−3.80	4
2,3-Benzanthracene	92-24-0	25	−8.60	5
		27	−8.36	6
1,2-Benzofluorene	238-84-6	25	−6.68	7
2,3-Benzofluorene	243-17-4	25	−7.73	8
		25	−8.03	7
Benzonitrile	100-47-0	25	−1.38	9
Benzyl acetate	140-11-4	25.0	−1.77	10
Bibenzyl	103-29-7	20	−5.62	11
		25.0	−4.80	12
Bis(2-ethylhexyl)isophthalate	137-89-3	24	−7.55	13
Borneol	464-45-9	25	−2.32	14
4-Bromobiphenyl	92-66-0	25	−5.55	15
1-Bromobutane	109-65-9	25.0	−2.20	1
		30	−2.35	16
4-Bromo-1-butene	5162-44-7	25.0	−2.25	1
2-Bromochlorobenzene	694-80-4	25	−3.19	17
3-Bromochlorobenzene	108-37-2	25	−3.21	17
4-Bromochlorobenzene	106-39-8	25	−3.63	17
1-Bromo-2-chloroethane	107-04-0	30.00	−1.32	18
1-Bromo-3-chloropropane	109-70-6	25.0	−1.85	1
2-Bromoethylacetate	927-68-4	25.0	−0.67	1
1-Bromoheptane	629-04-9	25.0	−4.43	1
1-Bromohexane	111-25-1	25.0	−3.81	1
4-Bromoiodobenzene	589-87-7	25	−4.56	17
1-Bromooctane	111-83-1	25.0	−5.06	1
1-Bromopentane	110-53-2	25.0	−3.08	1
n-Butyl butyrate	109-21-7	25.0	−2.37	10
n-Butyl formate	592-84-7	25.0	−0.97	10
n-Butyl isobutyrate	97-87-0	25.0	−2.27	10
n-Butyl lactate	138-22-7	25	−0.53	19
4-sec-Butyl-2-nitrophenol	3555-18-8	20.0	−3.84	20
n-Butyl propionate	590-01-2	25.0	−1.85	10
Carbazole	86-74-8	20	−5.14	21
Carbon tetrabromide	558-13-4	30	−3.14	16
		30.00	−2.71	18
2-Chlorobiphenyl	2051-60-7	25	−4.57	22
4-Chlorobiphenyl	2051-62-9	25	−5.15	23

531

Compound	CAS Registry No.	Temp. °C	Log S mol/L	Reference
		25	−5.15	24
1-Chlorobutane	109-69-3	25.0	−2.03	1
2-Chlorodibenzo-*p*-dioxin	39227-540-8	25	−5.84	15
1-Chloroheptane	629-06-1	25.0	−4.00	1
2-Chloroiodobenzene	615-41-8	25	−3.54	17
3-Chloroiodobenzene	625-99-0	25	−3.55	17
4-Chloroiodobenzene	637-87-6	25	−4.03	17
2-Chloronitrobenzene	88-73-3	20	−2.55	25
3-Chloronitrobenzene	121-73-3	20	−2.76	25
4-Chloronitrobenzene	100-00-5	20	−2.85	26
		20	−2.54	25
4-Chloro-2-nitrophenol	89-64-5	20.0	−3.09	20
4-Chlorophenol	106-48-9	20	−0.99	26
6-Chloro-3-nitrotoluene	7147-89-9	20.0	−4.39	20
4-Chlorotoluene	106-43-4	20	−2.97	26
		20	−3.08	27
Chlorprothixene	113-59-7	20	−4.41	26
Chlorpyrifos	2921-88-2	20	−8.00	11
Cholanthrene	479-23-2	27	−7.86	6
Cinnamic acid	621-82-9	30.00	−2.39	18
Coronene	191-07-1	25	−9.48	8
		25	−9.33	5
Cycloheptanol	502-41-0	25.0	−0.89	10
Cycloheptatriene	544-25-2	25	−2.17	28
Cycloheptene	628-92-2	25	−3.16	28
1,4-Cyclohexadiene	628-41-1	25	−2.06	28
Cyclohexyl acetate	622-45-7	25.0	−1.71	10
Cyclohexyl formate	4351-54-6	25.0	−1.19	10
Cyclohexylmethanol	100-49-2	25.0	−1.14	10
Cyclohexyl propionate	6222-35-1	25.0	−2.31	10
Cyclooctane	292-64-8	25	−4.15	28
Cyclooctanol	696-71-9	25.0	−1.30	10
Cyclopentanol	96-41-3	25.0	0.07	10
Decachlorobiphenyl	2051-24-3	22.0	−10.38	29
		25	−11.89	30
		25	−10.83	22
Decanoic acid	334-48-5	20.0	−3.06	31
1-Decanol	112-30-1	20.0	−3.57	32
		25	−3.50	33
		29.6	−2.88	10
2-Decanone	693-54-9	25.0	−3.30	1
1-Decene	872-05-9	25	−4.39	34
n-Decylbenzene	104-72-3	25.0	−7.94	35
Dially phthalate	131-17-9	20	−3.13	36
Diazinon	333-41-5	25	−2.47	37
1,2,3,4-Dibenzanthracene	215-58-7	25	−8.24	8

Compound	CAS Registry No.	Temp. °C	Log S mol/L	Reference
Dibenzo-*p*-dioxin	262-12-4	25	−5.31	15
1,2-Dibromobenzene	583-53-9	25	−3.50	17
1,3-Dibromobenzene	108-36-1	35.0	−3.54	38
		25	−3.38	17
Dibromomethane	74-95-3	30	−1.16	16
1,3-Dibromopropane	109-64-8	30.00	−2.08	18
Dicapthon	2463-84-5	20	−4.68	39
3,5-Dichloroaniline	626-43-7	23	−2.37	40
2,2′-Dichlorobiphenyl	13029-08-8	25	−5.27	41
		25	−5.45	41
2,4-Dichlorobiphenyl	33284-50-3	25	−5.29	41
2,5-Dichlorobiphenyl	34883-39-1	25	−5.30	41
		25	−5.06	22
		25	−5.59	41
2,6-Dichlorobiphenyl	33146-45-1	22.0	−5.62	29
		25	−4.97	41
		25	−5.21	22
3,3′-Dichlorobiphenyl	2050-67-1	25	−5.80	41
3,4-Dichlorobiphenyl	2974-92-7	25	−7.45	41
3,5-Dichlorobiphenyl	31883-41-5	25	−6.37	42
4,4′-Dichlorobiphenyl	2050-08-2	25	−6.56	39
		25	−6.32	42
2,8-Dichlorodibenzofuran	5409-83-6	25	−7.21	15
2,3-Dichloronitrobenzene	3209-22-1	20	−3.49	25
2,5-Dichloronitrobenzene	89-61-2	20	−3.32	25
		25	−6.79	41
		25	−6.60	41
3,4-Dichloronitrobenzene	99-54-7	20	−3.20	25
Dicyclohexyl phthalate	84-61-7	24	−4.92	13
Di-*n*-decyl phthalate	2432-90-8	24	−6.18	13
Diethyl carbonate	105-58-8	25.0	−0.70	10
Diethyl oxalate	95-92-1	25.0	−0.59	10
1,2-Difluorobenzene	367-11-3	25	−2.00	17
1,3-Difluorobenzene	372-18-9	25	−2.00	17
1,4-Difluorobenzene	540-36-3	25	−1.97	17
1,2-Diiodobenzene	615-42-9	25	−4.24	17
1,4-Diiodobenzene	624-38-4	25	−5.25	17
		25.0	−5.50	2
Diiodomethane	75-11-6	30	−2.33	16
Diisobutyl phthalate	84-69-5	20	−4.14	36
		24	−4.65	13
Diisopropyl phthalate	605-45-8	20	−2.88	36
1,3-Dimethoxybenzene	151-10-1	25	−2.06	9
Dimethyl adipate	627-93-0	25.0	−0.77	10
9,10-Dimethylanthracene	781-43-1	25	−6.57	7
7,12-Dimethylbenz[*a*]anthracene	57-97-6	25	−7.02	43

Compound	CAS Registry No.	Temp. °C	Log S mol/L	Reference
9,10-Dimethylbenz[a]anthracene	58429-99-5	25	−6.62	5
		24	−6.67	13
		27	−6.78	6
4,4′-Dimethylbiphenyl	613-33-2	25	−6.02	15
2,2-Dimethyl-1-butanol	1185-33-7	25	−1.13	44
2,3-Dimethyl-2-butanol	594-60-5	25	−0.34	44
3,3-Dimethyl-2-butanol	464-07-3	25	−0.62	44
		25.0	−0.68	10
3,3-Dimethyl-2-butanone	75-97-8	25	−0.72	45
Dimethyl carbonate	616-38-6	25.0	0.03	10
5,6-Dimethylchrysene	3697-27-6	27	−7.01	6
2,6-Dimethylcyclohexanol	5337-72-4	25.0	−1.46	10
Dimethyl glutarate	1119-40-0	25.0	−0.43	10
2,6-Dimethyl-4-heptanol	108-82-7	25.0	−2.35	10
Dimethyl maleate	624-48-6	25.0	−0.27	10
1,3-Dimethylnaphthalene	575-41-7	25	−3.79	5
1,4-Dimethylnaphthalene	571-58-4	25	−4.29	5
1,5-Dimethylnaphthalene	571-61-9	25.0	−4.74	46
		25	−4.14	5
2,3-Dimethylnaphthalene	581-40-8	25.0	−4.89	46
		25	−4.66	5
2,6-Dimethylnaphthalene	581-42-0	25.0	−5.20	47
		25.0	−5.08	46
		25	−4.72	5
2,2-Dimethylpentane	590-35-2	25	−4.36	48
2,2-Dimethyl-3-pentanol	3970-62-5	25	−1.15	49
2,3-Dimethyl-2-pentanol	4911-70-0	25	−0.88	49
2,3-Dimethyl-3-pentanol	595-41-5	25	−0.85	49
2,4-Dimethyl-2-pentanol	625-06-9	25	−0.94	49
2,4-Dimethyl-3-pentanol	600-36-2	25	−1.22	49
		25	−1.24	50
2,4-Dimethyl-3-pentanone	565-80-0	25	−1.30	45
2,2-Dimethyl-1-propanol	75-84-3	24.4	−0.44	50
2,7-Dimethylquinoline	93-37-8	25	−1.94	48
Dimethyl sulfide	75-18-3	25	−0.45	51
2,5-Dinitrophenol	329-71-5	20.0	−2.68	20
Di-n-pentyl phthalate	131-18-0	20	−5.84	36
1,1-Diphenylethylene	530-48-3	25.0	−4.52	12
Diphenyl phosphate	838-85-7	24	−2.97	13
Diphenyl phthalate	84-62-8	24	−6.59	13
Diphenylamine	122-39-4	20	−3.55	21
Diphenylmethane	101-81-5	24	−4.75	13
		25.0	−5.08	52
Diphenylmethyl phosphate	115-89-9	24	−5.44	13
Di-n-propyl phthalate	131-16-8	20	−3.36	36
Ditridecyl phthalate	119-06-2	24	−6.19	13

Compound	CAS Registry No.	Temp. °C	Log S mol/L	Reference
		25	−0.40	53
Dodecanoic acid	143-07-7	20.0	−3.56	31
		25	−4.58	54
1-Dodecanol	112-53-8	25	−4.64	54
		29.5	−2.67	10
Eicosane	112-95-8	25.0	−8.18	55
Ethyl adipate	141-28-6	20	−1.68	57
		30	−1.68	18
2-Ethylanthracene	52251-71-5	25.0	−6.78	47
		25.3	−6.89	58
10-Ethylbenz[a]anthracene	3697-30-1	27	−6.81	59
Ethyl benzoate	93-89-0	25.0	−2.40	2
		25.0	−2.26	10
2-Ethyl-1-butanol	97-95-0	25.0	−1.06	10
Ethyl butyrate	105-54-4	25.0	−1.26	10
Ethyl chloroacetate	105-39-5	25.0	−0.80	10
Ethyl decanoate	110-38-3	20	−4.13	57
Ethyl heptanoate	106-30-9	20	−2.74	57
2-Ethyl-1,3-hexanediol	94-96-2	25.0	−2.81	1
Ethyl hexanoate	123-66-0	20	−2.36	57
		25.0	−2.26	10
2-Ethyl-1-hexanol	104-76-7	20.0	−2.17	32
		24.7	−2.07	50
Ethyl hydrocinnamate	2021-28-5	25.0	−3.01	12
Ethyl iodide	75-03-6	30	−1.59	16
Ethyl isobutyrate	97-62-1	25.0	−1.27	10
Ethyl isopentanoate	108-64-5	25.0	−1.87	10
Ethyl malonate	105-53-3	20	−0.89	57
1-Ethylnaphthalene	1127-26-0	25	−4.19	60
		25.0	−4.19	61
		25	−4.16	5
2-Ethylnaphthalene	939-27-5	25.0	−4.29	46
Ethyl octanoate	123-29-5	20	−3.79	57
3-Ethyl-3-pentanol	597-49-9	25	−0.84	49
Ethyl propionate	105-37-3	25.0	−0.83	1
		25.0	−0.69	10
Ethyl salicylate	119-61-6	25.0	−2.65	10
Ethyl succinate	123-25-1	20	−0.96	57
Ethyl succinate	123-25-1	25.0	−0.93	10
2-Ethylthiophene	872-55-9	25	−2.58	48
2-Ethyltoluene	611-14-3	25.0	−3.21	1
Ethyl trimethylacetate	3938-95-2	25.0	−1.38	10
Fluorobenzene	462-06-6	25.0	−1.87	2
m-Fluorobenzyl chloride	456-42-8	25.0	−2.54	1
		30.00	−1.79	18
o-Fluorobenzyl chloride	345-35-7	25.0	−2.54	1

Compound	CAS Registry No.	Temp. °C	Log S mol/L	Reference
4-Formyl-2-nitrophenol	3011-34-5	20.0	−2.95	20
2,2′,3,3′,4,4′,6-Heptachloro-biphenyl	52663-71-5	25	−8.26	22
2,2′,3,4,4′,5,5′-Heptachloro-biphenyl	35065-29-3	25	−8.71	42
1,2,3,4,6,7,8-Heptachlorodibenzo-p-dioxin	35822-46-9	20.0	−11.25	62
		26.0	−11.22	23
1,2,3,4,6,7,8-Heptachlorodibenzo-benzofuran	67462-39-4	22.7	−11.48	62
Heptadecanoic acid	506-12-7	20.0	−4.81	31
1,6-Heptadiene	3070-53-9	25	−3.34	28
1,6-Heptadiyne	2396-63-6	25	−1.75	28
Heptanoic acid	111-14-8	20.0	−1.73	31
		30	−1.85	63
1-Heptanol	111-70-6	20.0	−1.84	64
		25	−1.89	23
		25.0	−1.95	1
		25.4	−1.84	50
2-Heptanol	543-49-7	25.1	−1.48	50
3-Heptanol	589-82-2	20.0	−1.39	32
		25.0	−1.40	10
4-Heptanol	589-55-9	20.0	−1.39	32
		25.0	−1.42	10
1-Heptene	592-76-7	25	−3.55	34
		25.0	−3.73	1
n-Heptyl acetate	112-06-1	25.0	−2.88	10
n-Heptyl formate	112-23-2	25.0	−1.98	10
1-Heptyne	628-71-7	25	−3.01	28
2,2′,3,3′,4,4′-Hexachlorobiphenyl	38380-07-3	25	−9.01	41
		25	−9.11	22
		25	−8.87	41
		25	−8.41	42
2,2′,3,3′,4,5-Hexachlorobiphenyl	55215-18-4	25	−7.79	41
		25	−8.59	41
2,2′,3,3′,6,6-Hexachlorobiphenyl	38411-22-2	25	−7.90	30
		25	−7.78	22
		25	−8.35	42
2,2′,4,4′,5,5′-Hexachlorobiphenyl	35065-27-1	22.0	−8.50	29
		24	−8.58	39
		25	−7.63	15
		25	−8.62	41
2,2′,4,4′,6,6′-Hexachlorobiphenyl	33979-03-2	22.0	−8.52	29
		25	−8.20	41
		25.0	−8.04	65
		25	−8.10	23

Compound	CAS Registry No.	Temp. °C	Log S mol/L	Reference
		25	−8.04	24
		25	−8.95	22
		25	−8.56	41
		25	−8.46	42
1,2,3,4,7,8-Hexachlorodibenzo-*p*-dioxin	39227-28-6	20.0	−10.95	62
		26.0	−10.69	23
1,2,3,4,7,8-Hexachlorodibenzofuran	70658-26-9	22.7	−10.66	66
Hexacosane	630-01-3	25.0	−8.33	55
Hexadecane	544-76-3	25.0	−8.40	55
		25	−7.56	67
Hexadecanoic acid	57-10-3	20.0	−4.55	31
		25	−5.47	54
1-Hexadecanol	36653-82-4	25	−6.77	54
1,5-Hexadiene	592-42-7	25	−2.69	28
Hexanoic acid	142-62-1	20.0	−1.08	31
		30	−1.19	63
1-Hexanol	111-27-3	20.0	−1.22	32
		24.9	−1.22	50
		25	−1.28	23
		25.0	−1.38	1
2-Hexanol	623-93-7	20.0	−0.99	32
		24.9	−0.92	50
		25	−0.87	44
3-Hexanol	623-37-0	25	−0.80	44
3-Hexanone	589-38-8	25	−0.83	45
2-Hexene	592-43-8	25	−3.10	34
1-Hexen-3-ol	4798-44-1	25	−0.60	68
4-Hexen-3-ol	4798-58-7	25	−0.42	68
n-Hexyl acetate	142-92-7	25.0	−2.48	10
Hexylbenzene	1077-16-3	25.0	−5.20	1
		25.0	−5.25	70
n-Hexyl formate	629-33-4	25.0	−1.97	10
n-Hexyl propionate	2445-76-3	25.0	−1.94	10
Hexyne	693-02-7	25.0	−2.08	1
		25	−2.36	28
4-Hydroxybenzoic acid	99-96-7	20	−1.46	69
1-Iodoheptane	4282-40-0	25.0	−4.81	1
4-Isopropyltoluene	99-87-6	25.0	−3.76	71
Iodobenzene	591-50-4	25.00	−2.95	3
		25.0	−3.11	2
Isobutyl butyrate	97-85-8	25.0	−2.46	10
		25.0	−3.01	1
		30.00	−2.78	18
Isononyl acetate	40379-24-6	25.0	−2.99	10
Isopentyl acetate	123-92-2	25.0	−1.79	10

Compound	CAS Registry No.	Temp. °C	Log S mol/L	Reference
Isopentyl butyrate	106-27-4	25.0	−2.80	10
Isopentyl propionate	105-68-0	25.0	−2.37	10
Isopropyl butyrate	638-11-9	25.0	−0.68	10
Leptophos	21609-90-5	20	−7.94	39
Methane	74-82-8	25	−2.79	28
		25.0	−2.79	72
4-Methoxy-2-nitrophenol	1568-70-3	20.0	−2.84	20
9-Methylanthracene	779-02-2	25.0	−5.56	47
		25	−5.87	5
1-Methylbenz[a]anthracene	2498-77-3	27	−6.64	6
9-Methylbenz[a]anthracene	2381-16-0	24	−6.82	13
		27	−6.56	6
10-Methylbenz[a]anthracene	2381-15-9	24	−7.34	13
		27	−6.64	6
Methyl benzoate	93-58-3	25.0	−1.74	10
4-Methylbiphenyl	644-08-6	25	−4.62	15
		25	−5.15	65
2-Methyl-1-butanol	137-32-6	24.6	−0.52	50
		25	−0.47	53
2-Methyl-2-butanol	75-85-4	25	0.10	53
		25.2	0.08	50
3-Methyl-2-butanol	598-75-4	25	−0.20	53
		25.0	−0.20	50
3-Methyl-2-butanone	563-80-4	25	−0.15	45
2-Methyl-2-butene	513-35-9	25	−2.34	34
Methyl butyrate	623-42-7	25.0	−0.76	10
Methyl chloroacetate	96-34-4	25.0	−0.33	10
Methyl 2-chlorobutyrate	3153-37-5	26.0	−1.09	10
Methyl 2-chloropropionate	17639-93-9	25.0	−0.79	10
3-Methylcholanthrene	56-49-5	25	−7.92	43
		25	−7.97	5
5-Methylchrysene	3697-24-3	27	−5.69	6
6-Methylchrysene	1705-85-7	27	−6.57	6
2-Methylcyclohexanol	583-59-5	25.0	−0.83	10
3-Methylcyclohexanol	591-23-1	25.0	−0.93	10
4-Methylcyclohexanol	589-91-3	25.0	−0.97	10
1-Methylcyclohexene	591-47-9	25	−3.27	28
Methyl decanoate	110-42-9	25.0	−4.69	1
Methyl dichloroacetate	116-54-1	25.0	−1.64	10
6-Methyl-2,4-dinitrophenol	534-52-1	20.0	−3.00	20
Methyl enanthate	106-73-0	25.0	−2.20	10
1-Methylfluorene	1730-37-6	25	−4.96	8
Methyl hexanoate	106-70-7	25.0	−1.89	10
2-Methyl-2-hexanol	625-23-0	25	−1.08	49
3-Methyl-3-hexanol	597-96-6	25	−0.99	49

Compound	CAS Registry No.	Temp. °C	Log S mol/L	Reference
5-Methyl-2-hexanol	627-59-8	25.0	−1.38	10
Methyl isobutyrate	547-63-7	25.0	−0.75	10
Methyl *p*-methoxybenzoate	121-98-2	20	−2.41	69
1-Methylnaphthalene	90-12-0	25.0	−3.74	46
		25	−3.68	60
		25.0	−3.85	61
		25	−3.65	5
3-Methyl-2-nitrophenol	4920-77-8	20.0	−1.64	20
3-Methyl-4-nitrophenol	2581-34-2	20.0	−2.11	20
4-Methyl-2-nitrophenol	119-33-5	20.0	−2.55	20
5-Methyl-2-nitrophenol	700-838-9	20.0	−2.75	20
Methyl nonanoate	1731-84-6	25.0	−3.88	1
Methyl pentanoate	624-24-8	25.0	−6.05	48
2-Methyl-1-pentanol	105-30-6	25.2	−1.37	10
2-Methyl-2-pentanol	590-36-3	25	−0.50	44
2-Methyl-3-pentanol	565-67-3	25	−0.71	44
3-Methyl-1-pentanol	589-35-5	25.0	−1.04	10
3-Methyl-2-pentanol	565-60-6	25	−0.72	44
3-Methyl-3-pentanol	77-74-7	24.7	−0.52	50
		25	−0.38	44
4-Methyl-2-pentanol	108-11-2	25	−0.79	44
		25.0	−0.81	50
2-Methyl-3-Pentanone	565-69-5	25	−0.82	45
3-Methyl-2-pentanone	565-61-7	25	−0.68	45
4-Methyl-1-penten-3-ol	4798-45-2	25	−0.51	68
3-Methylphenol	108-39-4	25.0	−1.59	1
Methyl propionate	554-12-1	25.0	−0.17	10
Methyl salicylate	119-36-8	25.0	−2.10	10
Methyl trichloroacetate	598-99-2	25.0	−2.29	10
Methyl trimethylacetate	598-98-1	25.0	−1.74	10
Mirex	2385-85-5	24	−6.44	13
1-Naphthol	90-15-3	23	−2.23	40
2-Nitroanisole	91-23-6	30.00	−1.96	18
4-Nitroanisole	100-02-7	30.00	−2.41	18
3-Nitrophenol	554-84-7	20	−1.08	74
2,2′,3,3′,4,4′,5,5′,6-Nonachloro-biphenyl	40186-72-9	22.0	−9.77	29
		25	−10.26	30
2,2′,3,3′,4,5,5′,6,6′-Nonachloro-biphenyl	52663-77-1	25	−10.41	22
1,8-Nonadiyne	2396-65-8	25	−2.98	28
Nonanoic acid	112-05-0	20.0	−2.78	31
1-Nonanol	143-08-8	20.0	−3.03	32
		25.0	−3.13	1
		25.0	−2.68	10
2-Nonanol	628-99-9	25.0	−2.59	10

Compound	CAS Registry No.	Temp. °C	Log S mol/L	Reference
2-Nonanone	821-55-6	25.0	−2.58	1
1-Nonene	124-11-8	25.0	−5.05	1
4-Nonylphenol	104-40-5	20.5	−4.61	75
1-Nonyne	3452-09-3	25	−4.24	28
2,2′,3,3′,4,4′,5,5′-Octachloro-biphenyl	35694-08-7	22.0	−9.54	29
2,2′,3,3′,5,5′,6,6′-Octachloro-biphenyl	2136-99-4	25	−9.47	30
		25	−9.04	22
Octachlorodibenzo-p-dioxin	3268-87-9	20.0	−12.06	62
		20	−12.06	76
Octadecane	593-45-3	25.0	−8.08	55
Octadecanoic acid	57-11-4	20.0	−4.99	31
		25	−5.70	54
Octanoic acid	127-04-2	20.0	−2.33	31
		30	−2.46	63
1-Octanol	111-87-5	20.0	−2.43	32
		25.0	−2.39	77
		25	−2.49	23
		25	−2.35	78
		25.6	−2.36	50
2-Octanol	123-96-6	20.0	−2.07	32
		25.0	−2.07	10
		25	−2.01	14
2-Octanone	111-13-7	25.0	−2.05	1
Octanol	589-98-0	25.0	−1.97	10
2-Octylbenzene	777-22-0	25	−5.80	79
n-Octyl formate	112-32-3	25.0	−2.48	10
1-Octyne	629-05-0	25	−3.66	28
2,2′,4,5,5′-Pentachlorobiphenyl	37680-72-3	20	−7.91	11
2,2′,4,6,6′-Pentachlorobiphenyl	56558-16-8	25	−7.32	41
2,3,4,5,6-Pentachlorobiphenyl	18259-05-7	22.0	−7.38	29
		25	−7.91	41
		25	−7.77	22
		25	−7.68	41
		25	−7.91	42
1,2,3,4,7-Pentachloro-p-dioxin	39227-61-7	20.0	−9.47	62
		26.0	−9.33	23
2,3,4,7,8-Pentachlorodibenzofuran	57117-31-4	22.7	−9.16	66
Pentadecanoic acid	1002-84-2	20.0	−4.30	31
1-Pentadecanol	629-76-5	25	−6.35	54
Pentanoic acid	109-52-4	30	−0.60	63
1-Pentanol	71-41-0	25	−0.64	23
		25	−0.60	53
		25.0	−0.88	1
		25	−0.60	78
		25.4	−0.61	50

Compound	CAS Registry No.	Temp. °C	Log S mol/L	Reference
2-Pentanol	6032-29-7	25	−0.30	53
		25.1	−0.31	50
3-Pentanol	584-02-1	25	−0.23	53
		25.0	−0.20	50
3-Pentanone	96-22-0	25	−0.25	45
		25.0	−0.28	1
1-Penten-3-ol	616-25-1	25	−0.02	68
3-Penten-2-ol	1569-50-2	25	0.02	68
4-Penten-1-ol	821-09-0	25	−0.18	68
n-Pentyl acetate	628-63-7	25.0	−1.84	10
Pentylbenzene	528-68-1	25.0	−4.59	1
		25.0	−4.64	70
tert-Pentylbenzene	2049-95-8	25.0	−4.25	12
n-Pentyl butyrate	540-18-1	25.0	−2.80	10
1-Pentyne	627-19-0	25.0	−1.81	1
		25	−1.64	28
Perylene	198-55-0	25	−8.80	5
Phenetole	103-73-1	25.00	−2.33	3
		25	−2.35	9
2-Phenyldecane	4537-13-7	25.0	−7.59	35
3-Phenyldecane	4621-36-7	25.0	−7.42	35
4-Phenyldecane	4537-12-6	25.0	−7.44	35
5-Phenyldecane	4537-11-5	25.0	−7.46	35
2-Phenyldodecane	2719-61-1	25.0	−8.40	35
3-Phenyldodecane	2400-00-2	25.0	−8.15	35
4-Phenyldodecane	2719-64-4	25.0	−8.30	35
5-Phenyldodecane	2719-63-3	25.0	−8.30	35
6-Phenyldodecane	2719-62-2	25.0	−8.40	35
2-Phenylethanol	60-12-8	25.0	−0.67	10
Phenylbutazone	50-33-9	20	−4.08	26
4-Phenyl-2-nitrophenol	885-82-5	20.0	−4.41	20
1-Phenyl-1-propanol	93-54-9	25.0	−1.12	10
2-Phenyltetradecane	4534-59-2	25.0	−8.40	35
3-Phenyltetradecane	4534-58-1	25.0	−8.30	35
4-Phenyltetradecane	4534-57-0	25.0	−8.40	35
5-Phenyltetradecane	4534-56-9	25.0	−8.30	35
6-Phenyltetradecane	4534-55-8	25.0	−8.40	35
2-Phenyltridecane	4534-53-6	25.0	−8.40	35
3-Phenyltridecane	4534-52-5	25.0	−8.40	35
4-Phenyltridecane	4534-51-4	25.0	−8.40	35
5-Phenyltridecane	4534-50-3	25.0	−8.40	35
6-Phenyltridecane	4534-49-0	25.0	−8.40	35
2-Phenylundecane	4536-88-3	25.0	−8.10	35
3-Phenylundecane	4536-87-2	25.0	−7.92	35
4-Phenylundecane	4536-86-1	25.0	−8.05	35
5-Phenylundecane	4537-15-9	25.0	−8.00	35

Compound	CAS Registry No.	Temp. °C	Log S mol/L	Reference
6-Phenylundecane	4537-14-8	25.0	−7.96	35
Phosalone	2310-17-0	20	−5.23	39
Prometryn	7287-19-6	25	−4.68	37
Propene	115-07-1	25	−2.32	28
n-Propyl bromide	106-94-5	30	−1.73	16
n-Propyl butyrate	105-66-8	25.0	−1.88	10
n-Propyl ether	111-43-3	25	−1.61	73
n-Propyl propionate	106-36-5	25.0	−1.29	10
Propyne	115-07-1	25	−1.06	28
Quinoline	91-22-5	23	−1.28	40
trans-Stilbene	103-03-0	25.0	−5.93	12
Terbufos	13071-79-9	20	−4.72	80
m-Terphenyl	92-06-8	25.0	−5.18	81
o-Terphenyl	84-15-1	25.0	−5.27	81
p-Terphenyl	92-94-4	25.0	−7.11	81
2,2′,3,3′-Tetrachlorobiphenyl	38444-93-8	25	−7.27	41
		25	−7.01	42
2,2′,4,4′-Tetrachlorobiphenyl	2437-79-8	22.0	−6.73	29
2,2′,4′,5-Tetrachlorobiphenyl	41464-40-8	25	−7.25	22
2,2′,5,5′-Tetrachlorobiphenyl	35693-99-3	22.0	−7.28	29
		25	−6.43	41
2,2′,5,6′-Tetrachlorobiphenyl	41464-41-9	25	−6.79	41
2,2′,6,6′-Tetrachlorobiphenyl	15968-05-5	22.0	−8.03	29
		25	−7.39	41
2,3,4,5-Tetrachlorobiphenyl	33284-53-6	25	−7.32	41
		25.0	−7.33	65
		25	−7.33	24
		25	−7.14	22
		25	−7.18	41
		25	−7.26	42
2,4,4′,6-Tetrachlorobiphenyl	32598-12-2	25	−6.51	41
3,3′,4,4′-Tetrachlorobiphenyl	32598-13-3	22.0	−8.21	29
		25	−8.73	41
		25	−8.71	30
		25	−8.59	41
3,3′,5,5′-Tetrachlorobiphenyl	33284-52-5	25	−8.37	41
1,2,3,4-Tetrachloro-p-dioxin	30746-58-8	25	−8.84	15
1,2,3,7-Tetrachloro-p-dioxin	67028-18-6	20.0	−8.87	62
		26.0	−8.65	23
1,3,6,8-Tetrachlorodibenzo-p-dioxin	33423-92-6	20.0	−9.00	62
		20	−9.00	76
2,3,7,8-Tetrachlorodibenzofuran	51207-31-9	22.7	−8.86	66
Tetradecane	629-59-4	23	−8.78	56
		25.0	−7.96	55
		25	−7.46	67

Compound	CAS Registry No.	Temp. °C	Log S mol/L	Reference
Tetradecanoic acid	544-63-8	20.0	−4.06	31
		25	−5.43	54
1-Tetradecanol	112-72-1	25	−5.84	54
1,2,3,4-Tetrahydronaphthalene	119-64-2	20.0	−3.49	82
2,3,4,5-Tetrachloronitrobenzene	879-39-0	20	−4.55	25
2,3,5,6-Tetrachloronitrobenzene	117-18-0	20	−5.10	25
Thioanisole	110-68-5	25	−2.39	51
Thiophene	110-02-1	25	−1.45	48
Thiophenol	108-98-5	25	−2.12	51
p-Toluidine	106-49-0	20	−1.21	74
p-Tolunitrile	104-85-8	20	−1.89	9
1,2,4-Tribromobenzene	615-54-3	25	−4.50	17
Tri-n-butylamine	102-82-9	25.00	−3.12	3
2,2′,5-Trichlorobiphenyl	37680-65-2	25	−5.70	41
		25	−5.60	41
2,3′,5-Trichlorobiphenyl	38444-81-4	25	−6.01	41
2,4,4′-Trichlorobiphenyl	7012-37-5	22.0	−6.58	29
		25	−6.34	41
		25	−6.35	83
		25	−6.00	41
2,4,5-Trichlorobiphenyl	15862-07-4	25	−6.20	22
		25	−6.08	42
2,4,6-Trichlorobiphenyl	35693-92-6	25	−6.14	15
		25	−6.01	41
		25	−6.03	65
		25	−6.06	22
		25	−5.83	42
2,3,4-Trichloronitrobenzene	17700-09-3	20	−3.94	25
2,4,5-Trichloronitrobenzene	89-69-0	20	−3.89	25
n-Tridecanoic acid	638-53-9	20.0	−3.81	31
4-Trifluoromethyl-2-nitrophenol	400-99-7	20.0	−2.50	20
2,3,3-Trimethyl-2-butanol	594-83-2	40	−0.72	49
1,1,3-Trimethylcyclopentane	4516-69-2	25	−4.48	48
3,3,5-Trimethyl-1-hexanol	1484-87-3	25.0	−1.99	10
		20.0	−2.51	32
1,4,5-Trimethylnaphthalene	213-41-1	25	−4.91	5
2,2,3-Trimethyl-3-pentanol	7294-83-2	25	−1.28	84
Tri-n-propylamine	102-69-2	25.00	−2.28	3
Triphenylene	217-59-4	25.0	−6.74	81
		25	−6.73	7
		25	−6.73	85
		27	−6.78	6
Tris(1,3-dichloroisopropyl) phosphate	13674-87-8	24	−4.79	13
Tris(2-ethylhexyl) phosphate	126-72-7	24	−4.94	13

Compound	CAS Registry No.	Temp. °C	Log S mol/L	Reference
1-Undecanol	112-42-5	25.0	−2.63	10
Undecylbenzene	6742-54-7	25.0	−9.05	35
4-Vinyl-1-cyclohexene	100-40-3	25	−3.33	28

1) Tewari et al., 1982; 2) Andrews and Keefer, 1950; 3) Vesala, 1974; 4) Mills and Thurman, 1994; 5) Mackay and Shiu, 1977; 6) Davis et al., 1942; 7) Sutton and Calder, 1975; 8) Billington et al., 1988; 9) McGowan et al., 1966; 10) Stephenson and Stuart, 1986; 11) Swann et al., 1983; 12) Andrews and Keefer, 1950a; 13) Hollifield, 1979; 14) Mitchell, 1926; 15) Doucette and Andren, 1988; 16) Gross and Saylor, 1931; 17) Yalkowsky et al., 1979; 18) Gross et al., 1933; 19) Rehberg and Dixon, 1950; 20) Schwarzenbach et al., 1988; 21) Hashimoto et al., 1982; 22) Miller et al., 1984; 23) Li and Andren, 1994; 24) Li and Doucette, 1993; 25) Eckert, 1962; 26) Hafkenscheid and Tomlinson, 1983; 27) Hafkenscheid and Tomlinson, 1981; 28) McAuliffe, 1966; 29) Opperhuizen et al., 1988; 30) Dickhut et al., 1986; 31) Ralston and Hoerr, 1942; 32) Hommelen, 1959; 33) Stearns et al., 1947; 34) Natarajan and Venkatachalam, 1972; 35) Sherblom et al., 1992; 36) Leyder and Boulanger, 1983; 37) Somasundaram et al., 1991; 38) Hine et al., 1963; 39) Chiou et al., 1977; 40) Fu and Luthy, 1986; 41) Dunnivant and Elzerman, 1988; Weil et al., 1974; 42) Dulfer et al., 1995; 43) Means et al., 1980; 44) Ginnings and Webb, 1938; 45) Ginnings et al., 1940; 46) Eganhouse and Calder, 1976; 47) Vadas et al., 1991; 48) Price, 1976; 49) Ginnings and Hauser, 1938; 50) Stephenson et al., 1984; 51) Hine and Weimar, 1965; 52) Andrews and Keefer, 1949; 53) Ginnings and Baum, 1937; 54) Robb, 1966; 55) Sutton and Calder, 1974; 56) Coates et al., 1985; 57) Sobotka and Kahn, 1931; 58) Whitehouse, 1984; 59) Davis and Parke, 1942; 60) Schwarz and Wasik, 1977; 61) Schwarz, 1977; 62) Friesen et al., 1985; 63) Bell, 1971; 64) Hommelen, 1979; 65) Li et al., 1993; 66) Friesen et al., 1990; 67) Franks, 1966; 68) Ginnings et al., 1939; 69) Corby and Elworthy, 1971; 70) Owens et al., 1986; 71) Banerjee et al., 1980; 72) McAuliffe, 1963; 73) Bennett and Philip, 1928; 74) Hashimoto et al., 1984; 75) Ahel and Giger, 1993; 76) Webster et al., 1985; 77) Sanemassa et al., 1987; 78) Butler et al., 1933; 79) Deno and Berkheimer, 1960; 80) Bowman and Sans, 1982; 81) Akiyoshi et al., 1987; 82) Burris and MacIntyre, 1986; 83) Chiou et al., 1986; 84) Ginnings and Coltrane, 1939; 85) Klevens, 1950.

A 42, see Aspon
A 361, see Atrazine
A 363, see Aminocarb
A 1068, see Chlordane
A 1093, see Ametryn
A 2079, see Simazine
Aacaptan, see Captan
Aafertis, see Ferbam
Aahepta, see Heptachlor
Aalindan, see Lindane
AAT, see Parathion
Aatack, see Thiram
AATP, see Parathion
Aatrex, see Atrazine
Aatrex 4L, see Atrazine
Aatrex 4LC, see Atrazine
Aatrex nine-o, see Atrazine
Aatrex 80W, see Atrazine
AC 3422, see Ethion, Parathion
AC 3911, see Phorate
AC 12880, see Dimethoate
AC 18682, see Dimethoate
AC 84777, see Difenzoquat methyl sulfate
AC 92100, see Terbufos
AC 92553, see Pendimethalin
AC 222705, see Flucythrinate
Acar, see Chlorobenzilate
Acaraben, see Chlorobenzilate
Acaraben 4E, see Chlorobenzilate
Acarin, see Dicofol
Acarithion, see Carbophenothion
Acaron, see Chlordimeform
Accelerate, see Endothall
Accelerator thiuram, see Thiram
Acephate
Acephate-met, see Methamidophos
Acetaldehyde homopolymer, see Metaldehyde
Acetaldehyde polymer, see Metaldehyde
3-(Acetonylbenzyl)-4-hydroxycoumarin, see Warfarin
3-(α-Acetonylbenzyl)-4-hydroxycoumarin, see Warfarin
Aceto TETD, see Thiram
Acetylene dibromide, see Ethylene dibromide
Acetylphosphoramidothioic acid O,S-dimethyl ester, see Acephate
N-Acetylphosphoramidothioic acid O,S-dimethyl ester, see Acephate
ACP 322, see Naptalam
ACP-M-728, see Chloramben

Acquinite, see Chloropicrin
Acraldehyde, see Acrolein
Acritet, see Acrylonitrile
Acrolein
Acrylaldehyde, see Acrolein
Acrylic aldehyde, see Acrolein
Acrylon, see Acrylonitrile
Acrylonitrile
Acrylonitrile monomer, see Acrylonitrile
Acutox, see Pentachlorophenol
Aerol 1, see Trichlorfon
AF 101, see Diuron
Afalon, see Linuron
Afalon inuron, see Linuron
Aficide, see Lindane
Agreflan, see Trifluralin
Agricide maggot killer (F), see Toxaphene
Agriflan 24, see Trifluralin
Agrisol G-20, see Lindane
Agritan, see *p,p'*-DDT
Agritox, see MCPA
Agroceres, see Heptachlor
Agrocide, see Lindane
Agrocide 2, see Lindane
Agrocide 6G, see Lindane
Agrocide 7, see Lindane
Agrocide III, see Lindane
Agrocide WP, see Lindane
Agroforotox, see Trichlorfon
Agronexit, see Lindane
Agrosol S, see Captan
Agrotect, see 2,4-D
Agroxon, see MCPA
Agroxone, see 2,4-D, MCPA
Agrox 2-way and 3-way, see Captan
AI3-29158, see Permethrin
Akar, see Chlorobenzilate
Akar 50, see Chlorobenzilate
Akar 338, see Chlorobenzilate
Akarithion, see Carbophenothion
Aktikon, see Atrazine
Aktikon PK, see Atrazine
Aktinit A, see Atrazine
Aktinit PK, see Atrazine
Aktinit S, see Simazine
Alachlor
Alanap, see Naptalam

Alanap 10G AT, see Naptalam
Alanap L, see Naptalam
Alanape, see Naptalam
Alanex, see Alachlor
Albrass, see Propachlor
Aldecarb, see Aldicarb
Aldrec, see Aldrin
Aldrex, see Aldrin
Aldrex 30, see Aldrin
Aldrin
Aldrite, see Aldrin
Aldrosol, see Aldrin
Alfatox, see Diazinon
Algistat, see Dichlone
Alidochlor, see Allidochlor
Aliette, see Fosetyl-aluminum
Alkron, see Parathion
Alleron, see Parathion
Allidochlor
Alltex, see Toxaphene
Alltox, see Toxaphene
Ally, see Metsulfuron-methyl
Ally 20DF, see Metsulfuron-methyl
Allyl aldehyde, see Acrolein
Alochlor, see Alachlor
Altox, see Aldrin
Aluminum tris(*O*-ethyl phosphonate), see Fosetyl-aluminum
Alvit, see Dieldrin
Amatin, see Hexachlorobenzene
Amaze, see Isofenphos
Amazol, see Amitrole
Ambiben, see Chloramben
Ambush, see Aldicarb
Ambush, see Permethrin
Amchem 68-250, see Ethephon
Amdon grazon, see Picloram
Ameisenatod, see Lindane
Ameisenmittel merck, see Lindane
Amercide, see Captan
American Cyanamid 3422, see Parathion
American Cyanamid 3911, see Phorate
American Cyanamid 4049, see Malathion
American Cyanamid 12880, see Dimethoate
Amerol, see Amitrole
Ametryn
Ametryne, see Ametryn
Amiben, see Chloramben

Amiben DS, see Chloramben
Amibin, see Chloramben
Amidox, see 2,4-D
Amine 2,4,5-T for rice, see 2,4,5-T
Aminocarb
3-Amino-2,5-dichlorobenzoic acid, see Chloramben
4-Amino-6-(1,1-dimethylethyl)-3-(methylthio)-1,2,4-triazin-5-(4*H*)-one, see Metribuzin
Aminotriazole, see Amitrole
2-Aminotriazole, see Amitrole
3-Aminotriazole, see Amitrole
2-Amino-1,3,4-triazole, see Amitrole
3-Amino-1,2,4-triazole, see Amitrole
3-Amino-1*H*-1,2,4-triazole, see Amitrole
3-Amino-*s*-triazole, see Amitrole
Aminotriazole-spritzpulver, see Amitrole
Amino triazole weed killer 90, see Amitrole
4-Amino-3,5,6-trichloropicolinic acid, see Picloram
4-Amino-3,5,6-trichloropyridine-2-carboxylic acid, see Picloram
4-Amino-3,5,6-trichloro-2-pyridinecarboxylic acid, see Picloram
Amiral, see Triadimefon
Amitol, see Amitrole
Amitril, see Amitrole
Amitril T.L., see Amitrole
Amitrol, see Amitrole
Amitrol 90, see Amitrole
Amitrole
Amitrol T, see Amitrole
Amizol, see Amitrole
Amizol D, see Amitrole
Amizol F, see Amitrole
Ammo, see Cypermethrin
Ammonium ethyl carbamoylphosphonate solution, see Fosamine-ammonium
Ammonium salt, ammonium ethyl(aminocarbonyl)phosphonate, see Fosamine-ammonium
Amoxone, see 2,4-D
Amtrex, see Ametryn
An, see Acrylonitrile
Anicon Kombi, see MCPA
Anicon M, see MCPA
Anilazin, see Anilazine
Anilazine
Anofex, see *p,p'*-DDT
Anthion, see Trichlorfon
Anticarie, see Hexachlorobenzene
Antinonin, see 4,6-Dinitro-*o*-cresol
Antinonnon, see 4,6-Dinitro-*o*-cresol
ANTU
Anturat, see ANTU
Apadrin, see Monocrotophos
Apamidon, see Phosphamidon

Aparasin, see Lindane
Apavap, see Dichlorvos
Apavinphos, see Mevinphos
Aphalon, see Linuron
Aphamite, see Parathion
Aphtiria, see Lindane
Aplidal, see Lindane
Apl-Luster, see Thiabendazole
Appa, see Phosmet
Aprocarb, see Propoxur
Apron 2E, see Metalaxyl
Aquacide, see Diquat
Aqua-kleen, see 2,4-D
Aqualin, see Acrolein
Aqualine, see Acrolein
Aquathol, see Endothall
Aquazine, see Simazine
Aquazine 80W, see Simazine
Arasan, see Thiram
Arasan 42-S, see Thiram
Arasan 70, see Thiram
Arasan 75, see Thiram
Arasan-M, see Thiram
Arasan-SF, see Thiram
Arasan-SF-X, see Thiram
Arbitex, see Lindane
Arborol, see 4,6-Dinitro-*o*-cresol
Arbotect, see Thiabendazole
Ardap, see Cypermethrin
Aretit, see Dinoseb
Argezin, see Atrazine
Arilate, see Benomyl
Arkotine, see *p,p'*-DDT
Arprocarb, see Propoxur
Arthodibrom, see Naled
Arylam, see Carbaryl
Asilan, see Asulam
ASP 47, see Sulfotepp
ASP 51, see Aspon
Aspon
Aspon-chlordane, see Chlordane
Assure, see Quizalofop-methyl
Astrobot, see Dichlorvos
Asulam
Asulox, see Asulam
Asulox 40, see Asulam
Asulfox F, see Asulam
AT, see Amitrole
AT-90, see Amitrole

Banvel D, see Dicamba
Banvel herbicide, see Dicamba
Banvel II herbicide, see Dicamba
Banvel 4S, see Dicamba
Banvel 4WS, see Dicamba
Barricade, see Cypermethrin
Barrier 2G, see Dichlobenil
Barrier 50W, see Dichlobenil
Basagran, see Bentazone
Basaklor, see Heptachlor
Basamid, see Dazomet
Basamid G, see Dazomet
Basamid-granular, see Dazomet
Basamid P, see Dazomet
Basamid-puder, see Dazomet
Basanite, see Dinoseb
Basfapon F, see Dalapon-sodium
BAS 351-H, see Bentazone
Basudin, see Diazinon
Basudin 10 G, see Diazinon
Batazina, see Simazine
Batrilex, see Pentachloronitrobenzene
Bay 6681 F, see Triadimefon
Bay 9010, see Propoxur
Bay 9026, see Methiocarb
Bay 9027, see Azinphos-methyl
Bay 15922, see Trichlorfon
Bay 17147, see Azinphos-methyl
Bay 19149, see Dichlorvos
Bay 19639, see Disulfoton
Bay 21097, see Oxydemeton-methyl
Bay 25141, see Fensulfothion
Bay 29493, see Fenthion
Bay 30130, see Propanil
Bay 36205, see Quinomethionate
Bay 37344, see Methiocarb
Bay 39007, see Propoxur
Bay 44646, see Aminocarb
Bay 61597, see Metribuzin
Bay 68138, see Fenamiphos
Bay 70143, see Carbofuran
Bay 71628, see Methamidophos
Bay 92114, see Isofenphos
Baycid, see Fenthion
Bay dic 1468, see Metribuzin
Bay E-393, see Sulfotepp
Bay E-605, see Parathion
Bayer 4964, see Quinomethionate
Bayer 5080, see Aminocarb

Bayer 6159H, see Metribuzin
Bayer 6443H, see Metribuzin
Bayer 9007, see Fenthion
Bayer 9027, see Azinphos-methyl
Bayer 15922, see Trichlorfon
Bayer 17147, see Azinphos-methyl
Bayer 19639, see Disulfoton
Bayer 21097, see Oxydemeton-methyl
Bayer 24493, see Fenthion
Bayer 25141, see Fensulfothion
Bayer 36205, see Quinomethionate
Bayer 37344, see Methiocarb
Bayer 39007, see Propoxur
Bayer 44646, see Aminocarb
Bayer 71268, see Methamidophos
Bayer 94337, see Metribuzin
Bayer E 393, see Sulfotepp
Bayer L 13/59, see Trichlorfon
Bayer S 767, see Fensulfothion
Bayer S 1752, see Fenthion
Bay FCR 1272, see Cyfluthrin
Baygon, see Propoxur
Bay-Hox-1901, see Ethiofencarb
Bayleton, see Triadimefon
Bayntn-9306, see Sulprofos
Baysra-12869, see Isofenphos
Baytex, see Fenthion
Baythroid, see Cyfluthrin
Baythroid H, see Cyfluthrin
Bazinon, see Diazinon
Bazuden, see Diazinon
BBC, see Benomyl
BBC 12, see 1,2-Dibromo-3-chloropropane
BBX, see Lindane
BCF-bushkiller, see 2,4,5-T
BCIE, see Bis(2-chloroisopropyl)ether
BCMEE, see Bis(2-chloroisopropyl)ether
Bean seed protectant, see Captan
Beet-Kleen, see Chlorpropham, Propham
Beetomax, see Phenmedipham
Beetup, see Phenmedipham
Belmark, see Fenvalerate
Belt, see Chlordane
Bencarbate, see Bendiocarb
Bendex, see Fenbutatin oxide
Bendiocarb
Bendioxide, see Bentazone
Benefex, see Benfluralin
Benfluralin

Benfluraline, see Benfluralin
Benfos, see Dichlorvos
Benhex, see Lindane
Benlat, see Benomyl
Benlate, see Benomyl
Benlate 50, see Benomyl
Benlate 50W, see Benomyl
Benomyl
Benomyl 50W, see Benomyl
Bentanex, see Desmedipham
Bentazon, see Bentazone
Bentazone
Bentox 10, see Lindane
Benzabar, see 2,3,5-Trichlorobenzoic acid
Benzac, see 2,3,5-Trichlorobenzoic acid
Benzac-1281, see 2,3,5-Trichlorobenzoic acid
1,2-Benzenedicarboxylate, see Di-*n*-butyl phthalate
1,2-Benzenedicarboxylic acid dibutyl ester, see Di-*n*-butyl phthalate
o-Benzenedicarboxylic acid dibutyl ester, see Di-*n*-butyl phthalate
Benzene-*o*-dicarboxylic acid di-*n*-butyl ester, see Di-*n*-butyl phthalate
1,2-Benzenedicarboxylic acid dimethyl ester, see Dimethyl phthalate
Benzene hexachloride, see Lindane
Benzene-*cis*-hexachloride, see β-BHC
trans-α-Benzene hexachloride, see β-BHC
Benzene-γ-hexachloride, see Lindane
α-Benzene hexachloride, see α-BHC
β-Benzene hexachloride, see β-BHC
γ-Benzene hexachloride, see Lindane
δ-Benzene hexachloride, see δ-BHC
Benzene hexachloride-α-isomer, see α-BHC
Benzilan, see Chlorobenzilate
Benzinoform, see Carbon tetrachloride
Benz-o-chlor, see Chlorobenzilate
Benzoepin, see α-Endosulfan, see β-Endosulfan
Benzoic acid *o*-((3-(4,6-dimethyl-2-pyrimidinyl)ureido)sulfonyl) methyl ester, see
    Sulfometuron-methyl
Benzophosphate, see Phosalone
Benzotriazinedithiophosphoric acid dimethoxy ester, see Azinphos-methyl
S-((3-Benzoxazolinyl-6-chloro-2-oxo)methyl) *O,O*-diethyl phosphorodithioate, see
    Phosalone
Benzphos, see Phosalone
Beosit, see α-Endosulfan, β-Endosulfan
Bercema Fertam 50, see Ferbam
Bermat, see Chlordimeform
Betanal, see Phenmedipham
Betanal AM, see Desmedipham
Betanal E, see Phenmedipham
Betanex, see Desmedipham
Bethrodine, see Benfluralin

Betosip, see Phenmedipham

Bexol, see Lindane

Bexton, see Propachlor

Bexton 4L, see Propachlor

BFV, see Formaldehyde

BHC, see Lindane

α-BHC

β-BHC

δ-BHC

γ-BHC, see Lindane

BH 2,4-D, see 2,4-D

BH MCPA, see MCPA

BI 58, see Dimethoate

BI 58 EC, see Dimethoate

Bibenzene, see Biphenyl

Bibesol, see Dichlorvos

Bicep, see Metolachlor

Bidirl, see Dicrotophos

Bidrin, see Dicrotophos

Bifenox

Bifex, see Propoxur

Bilarcil, see Trichlorfon

Biloborb, see Monocrotophos

Bilobran, see Monocrotophos

Bio 5462, see α-Endosulfan, β-Endosulfan

Biocide, see Acrolein

Biphenyl

1,1′-Biphenyl, see Biphenyl

2,2-Bis(*p*-anisyl)-1,1,1-trichloroethane, see Methoxychlor

S-1,2-Bis(carbethoxy)ethyl-*O,O*-dimethyl dithiophosphate, see Malathion

Bis(2-chloroethyl)ether

Bis(β-chloroethyl)ether, see Bis(2-chloroethyl)ether

Bis(2-chloroisopropyl)ether

Bis(β-chloroisopropyl)ether, see Bis(2-chloroisopropyl)ether

Bis(2-chloro-1-methylethyl)ether, see Bis(2-chloroisopropyl)ether

1,1-Bis(4-chlorophenyl)-2,2-dichloroethane, see *p,p′*-DDD

1,1-Bis(*p*-chlorophenyl)-2,2-dichloroethane, see *p,p′*-DDD

2,2-Bis(4-chlorophenyl)-1,1-dichloroethane, see *p,p′*-DDD

2,2-Bis(*p*-chlorophenyl)-1,1-dichloroethane, see *p,p′*-DDD

2,2-Bis(4-chlorophenyl)-1,1-dichloroethene, see *p,p′*-DDE

2,2-Bis(*p*-chlorophenyl)-1,1-dichloroethene, see *p,p′*-DDE

1,1-Bis(4-chlorophenyl)-2,2-dichloroethylene, see *p,p′*-DDE

1,1-Bis(*p*-chlorophenyl)-2,2-dichloroethylene, see *p,p′*-DDE

1,1-Bis(*p*-chlorophenyl)-2,2,2-trichloroethane, see *p,p′*-DDT

2,2-Bis(4-chlorophenyl)-1,1,1-trichloroethane, see *p,p′*-DDT

2,2-Bis(*p*-chlorophenyl)-1,1,1-trichloroethane, see *p,p′*-DDT

α,α-Bis(*p*-chlorophenyl)-β,β,β-trichloroethane, see *p,p′*-DDT

1,1-Bis(chlorophenyl)-2,2,2-trichloroethanol, see Dicofol

1,1-Bis(4-chlorophenyl)-2,2,2-trichloroethanol, see Dicofol

1,1-Bis(*p*-chlorophenyl)-2,2,2-trichloroethanol, see Dicofol
Bis-*O,O*-diethylphosphoric anhydride, see Tetraethyl pyrophosphate
Bis-*O,O*-diethylphosphorothionic anhydride, see Sulfotepp
Bis(*S*-(dimethoxyphosphinothioyl)mercapto)methane, see Ethion
Bis(dimethylamino)carbonothioyl disulfide, see Thiram
Bis(dimethylthiocarbamoyl) disulfide, see Thiram
Bis(dimethylthiocarbamyl) disulfide, see Thiram
Bis-*O,O*-di-*n*-propylphosphorothionic anhydride, see Aspon
S-1,2-Bis(ethoxycarbonyl)ethyl-*O,O*-dimethyl phosphorodithioate, see Malathion
S-1,2-Bis(ethoxycarbonyl)ethyl-*O,O*-dimethyl thiophosphate, see Malathion
1,1-Bis(*p*-ethoxyphenyl)-2,2,2-trichloroethane, see Methoxychlor
2,4-Bis(ethylamino)-6-chloro-*s*-triazine, see Simazine
8014 Bis HC, see Dimethoate
2,4-Bis(isopropylamino)-6-chloro-*s*-triazine, see Propazine
2,4-Bis(isopropylamino)-6-methoxy-*s*-triazine, see Prometon
2,4-Bis(isopropylamino)-6-methylmercapto-*s*-triazine, see Prometryn
2,4-Bis(isopropylamino)-6-methylthio-1,3,5-triazine, see Prometryn
2,4-Bis(isopropylamino)-6-methylthio-*s*-triazine, see Prometryn
Bismethomyl thioether, see Thiodicarb
2,2-Bis(*p*-methoxyphenyl)-1,1,1-trichloroethane, see Methoxychlor
Bis(1-methylethyl)carbamothioic acid *S*-(2,3-dichloro-2-propenyl) ester, see Diallate
N,N'-Bis(1-methylethyl)-6-(methylthio)-1,3,5-triazine-2,4-diamine, see Prometryn
Bis(2-methylpropyl)carbamothioic acid *S*-ethyl ester, see Butylate
Bis(pentachlor-2,4-cyclopentadien-1-yl); see Dienochlor
Bis(pentachloroclopentadienyl); see Dienochlor
Bis(pentachloro-2,4-cyclopentadien-1-yl); see Dienochlor
Bis(tris(β,β-dimethylphenethyl)tin)oxide, see Fenbutatin oxide
Bis(tris(2-methyl-2-phenylpropyl)tin)oxide, see Fenbutatin oxide
Bitemol, see Simazine
Bitemol S 50, see Simazine
Bladafum, see Sulfotepp
Bladafume, see Sulfotepp
Bladafun, see Sulfotepp
Bladan, see Ethion, Parathion, Tetraethyl pyrophosphate
Bladan F, see Parathion
Bladex, see Cyanazine
Bladex 4L, see Cyanazine
Bladex 80WP, see Cyanazine
Bladex 90DF, see Cyanazine
Blattanex, see Propoxur
BNM, see Benomyl
BNP 30, see Dinoseb
Bolstar, see Sulprofos
Bonalan, see Benfluralin
Bordermaster, see MCPA
Borea, see Bromacil
Borolin, see Picloram
Bortrysan, see Anilazine
Bosan Supra, see *p,p'*-DDT

Botrilex, see Pentachloronitrobenzene
Bovidermol, see *p,p'*-DDT
Bovinox, see Trichlorfon
Bovizole, see Thiabendazole
Boygon, see Propoxur
Brassicol, see Pentachloronitrobenzene
Bravo, see Chlorothalonil
Bravo 6F, see Chlorothalonil
Bravo-W-75, see Chlorothalonil
Brevinyl, see Dichlorvos
Brevinyl E50, see Dichlorvos
Briton, see Trichlorfon
Britten, see Trichlorfon
Brittox, see Bromoxynil
Brodan, see Chlorpyrifos
Bromacil
Bromax, see Bromacil
Bromax 4G, see Bromacil
Bromax 4L, see Bromacil
Bromazil, see Bromacil
Bromchlophos, see Naled
Bromex, see Naled
Brominal, see Bromoxynil
Brominal M & Plus, see MCPA
Brominex, see Bromoxynil
Brominil, see Bromoxynil
5-Bromo-3-*sec*-butyl-6-methyluracil, see Bromacil
O-(4-Bromo-2-chlorophenyl) *O*-ethyl *S*-propyl phosphorothioate, see Profenofos
Bromoflor, see Ethephon
Bromofume, see Ethylene dibromide
Brom-o-gas, see Methyl bromide
Brom-o-gaz, see Methyl bromide
Bromomethane, see Methyl bromide
5-Bromo-6-methyl-3-(1-methylpropyl)-2,4(1*H*3*H*)-pyrimidinedione, see Bromacil
5-Bromo-6-methyl-3-(1-methylpropyl)uracil, see Bromacil
Bromoxynil
Bromoxynil octanoate
Bronate, see Bromoxynil octanoate
Bronco, see Alachlor
Broxynil, see Bromoxynil
Brulan, see Tebuthiuron
Brumolin, see Warfarin
Brush buster, see Dicamba
Brush-off 445 low volatile brush killer, see 2,4,5-T
Brush-rhap, see 2,4-D, 2,4,5-T
Brushtox, see 2,4,5-T
Brygou, see Propoxur
B-Selektonon, see 2,4-D
B-Selektonon M, see MCPA

Buckle, see Triallate
Buctril, see Bromoxynil, Bromoxynil octanoate
Buctril 20, see Bromoxynil
Buctril 21, see Bromoxynil
Buctril industrial, see Bromoxynil
Bud-nip, see Chlorpropham
Bullet, see Alachlor
Bunt-cure, see Hexachlorobenzene
Bunt-no-more, see Hexachlorobenzene
Burtolin, see Maleic hydrazide
Bushwacker, see Tebuthiuron
Butaphene, see Dinoseb
2-Butenoic acid 3-((dimethoxyphosphinyl)oxy)methyl ester, see Mevinphos
Butifos
Butilate, see Butylate
Butilchlorofos, see Bromoxynil
Butiphos, see Butifos
(1-((Butylamino)carbonyl)-1H-benzimidazol-2-yl)carbamic acid methyl ester, see
  Benomyl
2-*tert*-Butylamino-4-ethylamino-6-methylmercapto-*s*-triazine, see Terbutryn
2-*tert*-Butylamino-4-ethylamino-6-methylthio-*s*-triazine, see Terbutryn
Butylate
1-(Butylcarbamoyl)-2-benzimidazolecarbamic acid methyl ester, see Benomyl
3-*tert*-Butyl-5-chloro-6-methyluracil, see Terbacil
2-*tert*-Butyl-4-(2-2,4-dichloro-5-isopropyloxyphenyl)-1,3,4-oxadiazolin-5-one, see
  Oxadiazon
1-Butyl-3-(3,4-dichlorophenyl)-1-methylurea, see Neburon
N-Butyl-N'-(3,4-dichlorophenyl)-N-methylurea, see Neburon
2-*sec*-Butyl-2,4-dinitrophenol, see Dinoseb
N-Butyl-N-ethyl-2,6-dinitro-4-trifluoromethylaniline, see Benfluralin
N-Butyl-N-ethyl-2,6-dinitro-4-(trifluoromethyl)benzenamine, see Benfluralin
Butylethylthiocarbamic acid S-propyl ester, see Pebulate
N-Butyl-N-ethyl-α,α,α-trifluoro-2,6-dinitro-p-toluidine, see Benfluralin
Butyl phosphorotrithioate, see Butifos
Butyl phthalate, see Di-n-butyl phthalate
n-Butyl phthalate, see Di-n-butyl phthalate
1-(5-t-Butyl-1,3,4-thiadiazol-2-yl)-1,3-dimethylurea, see Tebuthiuron
S-(($tert$-Butylthio)methyl) O,O-diethyl phosphorodithioate, see Terbufos
BW-21-Z, see Permethrin
C 570, see Phosphamidon
C 709, see Dicrotophos
C 1414, see Monocrotophos
C 1983, see Chloroxuron
C 2059, see Fluometuron
C 8514, see Chlordimeform
Caldon, see Dinoseb
Caliper, see Simazine
Caliper 90, see Simazine
Calmathion, see Malathion

Campaprim A 1544, see Amitrole
Camphechlor, see Toxaphene
Camphochlor, see Toxaphene
Camphoclor, see Toxaphene
Camposan, see Ethephon
Candex, see Atrazine
Cannon, see Alachlor
Canogard, see Dichlorvos
Caparol, see Prometryn
Caparol 80W, see Prometryn
Capsine, see 4,6-Dinitro-*o*-cresol
Captaf, see Captan
Captaf 85W, see Captan
Captan
Captan 50W, see Captan
Captane, see Captan
Captan-streptomycin 7.5-0.1 potato seed piece protectant, see Captan
Captex, see Captan
Carbacryl, see Acrylonitrile
Carbamate, see Ferbam
Carbamic acid ethylenebis(dithio-, manganese zinc complex), see Mancozeb
Carbamine, see Carbaryl
Carbanilic acid isopropyl ester, see Propham
Carbanolate, see Aldicarb
Carbaryl
Carbatox, see Carbaryl
Carbatox 60, see Carbaryl
Carbatox 75, see Carbaryl
Carbax, see Dicofol
Carbethoxy malathion, see Malathion
Carbetovur, see Malathion
Carbetox, see Malathion
Carbicron, see Dicrotophos
Carbofos, see Malathion
Carbofuran
2-Carbomethoxy-1-methylvinyl dimethyl phosphate, see Mevinphos
α-Carbomethoxy-1-methylvinyl dimethyl phosphate, see Mevinphos
2-Carbomethoxy-1-propen-2-yl dimethyl phosphate, see Mevinphos
Carbon bisulfide, see Carbon disulfide
Carbon bisulphide, see Carbon disulfide
Carbon chloride, see Carbon tetrachloride
Carbon disulfide
Carbon disulphide, see Carbon disulfide
Carbon sulfide, see Carbon disulfide
Carbon sulphide, see Carbon disulfide
Carbona, see Carbon tetrachloride
Carbon tet, see Carbon tetrachloride
Carbon tetrachloride
Carbophenothion

Carbophos, see Malathion
Carbothialdin, see Dazomet
Carbothialdine, see Dazomet
5-Carboxanilido-2,3-dihydro-6-methyl-1,4-oxathiin, see Carboxin
Carboxin
Carboxine, see Carboxin
Carfene, see Azinphos-methyl
Carmazine, see Mancozeb
Carpolin, see Carbaryl
Carsoron, see Dichlobenil
Carylderm, see Carbaryl
Carzol, see Chlordimeform
Casaron, see Dichlobenil
Casoron, see Dichlobenil
Casoron 133, see Dichlobenil
Casoron G, see Dichlobenil
Casoron G-4, see Dichlobenil
Casoron G-10, see Dichlobenil
Casoron W-50, see Dichlobenil
CAT, see Simazine
CCN52, see Cypermethrin
CD-68, see Chlordane
CDAA, see Allidochlor
CDAAT, see Allidochlor
CDM, see Chlordimeform
CDT, see Simazine
Cekiuron, see Diuron
Cekubaryl, see Carbaryl
Cekudifol, see Dicofol
Cekufon, see Trichlorfon
Cekusan, see Dichlorvos
Cekusan, see Simazine
Cekuthoate, see Dimethoate
Cekuzina-S, see Simazine
Cekuzina-T, see Atrazine
Celanex, see Lindane
Celfume, see Methyl bromide
Celluflex DPB, see Di-*n*-butyl phthalate
Celmide, see Ethylene dibromide
Celthion, see Malathion
CEP, see Ethephon
2-CEPA, see Ethephon
Cepha, see Ethephon
Cephalon, see Linuron
Cepha 10LS, see Ethephon
Cerone, see Ethephon
Certox, see Strychnine
CGA 117, see Metalaxyl
CGA 15324, see Profenofos

CGA 24705, see Metolachlor
CGA 45156, see Thiodicarb
CGA 48988, see Metalaxyl
CGA 72662, see Cyromazine
Chemagro 90, see Propoxur
Chemagro 1776, see Butifos
Chemagro 25141, see Fensulfothion
Chemagro B1776, see Butifos
Chemathion, see Malathion
Chemathoate, see Dimethoate
Chemform, see Maleic hydrazide, Methoxychlor
Chem-Hoe, see Propham
Chemical 109, see ANTU
Chemox general, see Dinoseb
Chemox P.E., see Dinoseb
Chempenta, see Pentachlorophenol
Chem-phene M5055, see Toxaphene
Chemrat, see Pindone
Chem-Rice, see Propanil
Chemsect DNOC, see 4,6-Dinitro-*o*-cresol
Chemtol, see Pentachlorophenol
Chevron 9006, see Methamidophos
Chevron ortho 9006, see Methamidophos
Chevron RE 12420, see Acephate
Chinomethionat, see Quinomethionate
Chinomethionate, see Quinomethionate
Chipco 26019, see Iprodione
Chipco Buctril, see Bromoxynil
Chipco crab-kleen, see Bromoxynil
Chipco Ronstar G, see Oxadiazon
Chipco Ronstar 50WP, see Oxadiazon
Chipco thiram 75, see Thiram
Chipco turf herbicide 'D', see 2,4-D
Chipman 11974, see Phosalone
Chiptox, see MCPA
Chlorak, see Trichlorfon
Chloramben
Chloran, see Lindane
Chlorbenzilat, see Chlorobenzilate
Chlorbenzilate, see Chlorobenzilate
Chlordan, see Chlordane
α-Chlordan, see *trans*-Chlordane
γ-Chlordan, see Chlordane
cis-Chlordan, see *trans*-Chlordane
Chlordane
α-Chlordane, see *cis*-Chlordane, *trans*-Chlordane
β-Chlordane, see *cis*-Chlordane
γ-Chlordane, see *trans*-Chlordane
cis-**Chlordane**

α(*cis*)-Chlordane, see *trans*-Chlordane

***trans*-Chlordane**

Chlordecone, see Kepone

Chlordimeform

Chloresene, see Lindane

Chlorethephon, see Ethephon

Chlorex, see Bis(2-chloroethyl)ether

Chlorfenamidine, see Chlordimeform

Chlorfenidim, see Monuron

Chlorfos, see Trichlorfon

Chloridan, see Chlordane

Chlor-IFC, see Chlorpropham

Chlorinated camphene, see Toxaphene

Chlorindan, see Chlordane

Chlor-IPC, see Chlorpropham

Chlor kil, see Chlordane

Chloroalonil, see Chlorothalonil

(*o*-Chloroanilo)dichlorotriazine, see Anilazine

Chlorobenzilate

(±)-2-(4-(6-Chlorobenzoxazolyl)oxy)phenoxy)propionic acid, see Fenoxaprop-ethyl

(±)-2-(4-(6-Chlorobenzoxazol-2-yloxy)phenoxy)propionic acid, see Fenoxaprop-ethyl

(±)-2-(4-(6-Chloro-1,3-benzoxazol-2-yloxy)phenoxy)propionic acid, see Fenoxaprop-
   ethyl

Chlorobenzylate, see Chlorobenzilate

2-Chloro-4,6-bis(ethylamino)-1,3,5-triazine, see Simazine

2-Chloro-4,6-bis(ethylamino)-*s*-triazine, see Simazine

2-Chloro-4,6-bis(isopropylamino)-*s*-triazine, see Propazine

6-Chloro-*N,N'*-bis(1-methylethyl)-1,3,5-triazine-2,4-diamine, see Propazine

5-Chloro-3-*tert*-butyl-6-methyluracil, see Terbacil

Chloro-camphene, see Toxaphene

3-Chlorochlordene, see Heptachlor

1-Chloro-2-(β-chloroethoxy)ethane, see Bis(2-chloroethyl)ether

1-Chloro-2-(β-chloroisopropoxy)propane, see Bis(2-chloroisopropyl)ether

4-Chloro-α-(4-chlorophenyl)-α-(trichloromethyl)benzenemethanol, see Dicofol

4-Chloro-*o*-cresolxyacetic acid, see MCPA

2-Chloro-4-(1-cyano-1-methylethylamino)-6-ethylamino-1,3,5-triazine, see Cyanazine

Chlorodane, see Chlordane

2-Chloro-*N,N*-diallylacetamide, see Allidochlor

α-Chloro-*N,N*-diallylacetamide, see Allidochlor

1-Chloro-2,3-dibromopropane, see 1,2-Dibromo-3-chloropropane

3-Chloro-1,2-dibromopropane, see 1,2-Dibromo-3-chloropropane

2-Chloro-*N*-(4,6-dichloro-1,3,5-triazin-2-yl)aniline, see Anilazine

2-Chloro-3-(diethylamino)-1-methyl-3-oxo-1-propenyl dimethyl phosphate, see
   Phosphamidon

2-Chloro-2-diethylcarbamoyl-1-methylvinyl dimethylphosphate, see Phosphamidon

2-Chloro-2',6'-diethyl-*N*-(methoxymethyl)acetanilide, see Alachlor

2-Chloro-*N*-(2,6-diethylphenyl)-*N*-(methoxymethyl)acetamide, see Alachlor

6-Chloro-*N',N'*-diethyl-1,3,5-triazine-2,4-diamine, see Simazine

6-Chloro-N^2,N^4-diethyl-1,3,5-triazine-2,4-diyldiamine, see Simazine

3-(*p*-(*p*-Chlorophenoxy)phenyl)-1,1-dimethylurea, see Chloroxuron
N'-(4-(4-Chlorophenoxy)phenyl)-*N*,*N*-dimethylurea, see Chloroxuron
1-(4-Chlorophenoxy)-3,3-(1,2,4-triazol-1-yl)butan-2-one, see Triadimefon
N-(((4-Chlorophenyl)amino)carbonyl)-2,6-difluorobenzamide, see Diflubenzuron
(3-Chlorophenyl)carbamic acid 1-methylethyl ester, see Chlorpropham
trans-5-(4-Chlorophenyl)-*N*-cyclohexyl-4-methyl-2-oxo-3-thiazolidinecarboxamide, see
    Hexythiazox
1-(4-Chlorophenyl)-3-(2,6-difluorobenzoyl)urea, see Diflubenzuron
N-(4-Chlorophenyl)-2,2-dimethylpentanamide, see Monalide
1-(4-Chlorophenyl)-3,3-dimethylurea, see Monuron
1-(*p*-Chlorophenyl)-3,3-dimethylurea, see Monuron
3-(4-Chlorophenyl)-1,1-dimethylurea, see Monuron
3-(*p*-Chlorophenyl)-1,1-dimethylurea, see Monuron
N-(4-Chlorophenyl)-*N'*,*N'*-dimethylurea, see Monuron
N'-(4-Chlorophenyl)-*N*,*N*-dimethylurea, see Monuron
N'-(*p*-Chlorophenyl)-*N*,*N*-dimethylurea, see Monuron
N-(4-Chlorophenyl)-2,2-dimethylvaleramide, see Monalide
1-((*o*-Chlorophenyl)sulfonyl)-3-(4-methoxy-6-methyl-*s*-triazin-2-yl) urea, see
    Chlorsulfuron
S-((4-Chlorophenyl)thio)methyl *O*,*O*-diethyl phosphorodithioate, see Carbophenothion
S-((*p*-Chlorophenyl)thio)methyl *O*,*O*-diethyl phosphorodithioate, see Carbophenothion
S-(4-Chlorophenylthiomethyl)diethyl phosphorothiolothionate, see Carbophenothion
Chlorophos, see Trichlorfon
Chlorophose, see Trichlorfon
S-(2-Chloro-1-phthalimidoethyl) *O*,*O*-diethyl phosphorodithioate, see Dialifos
Chlorophthalm, see Trichlorfon
Chlor-o-pic, see Chloropicrin
Chloropicrin
Chloropropham, see Chlorpropham
2-Chloro-4-(2-propylamino)-6-ethylamino-*s*-triazine, see Atrazine
Chlorothal, see Chlorthal-dimethyl
4-Chloro-*o*-toloxyacetic acid, see MCPA
N'-(4-Chloro-*o*-tolyl)-*N*,*N*-dimethylformamidine, see Chlordimeform
2-Chloro-6-(trichloromethyl)pyridine, see Nitrapyrin
Chloroxifenidim, see Chloroxuron
Chloroxone, see 2,4-D
Chloroxuron
Chloroxyphos, see Trichlorfon
Chlorphenamidine, see Chlordimeform
Chlorpropham
Chlorpyrifos
Chlorpyrifos-ethyl, see Chlorpyrifos
Chlorsulfuron
Chlorthal-dimethyl
Chlorthiepin, see α-Endosulfan, β-Endosulfan
Chlortox, see Chlordane
Chlorvinphos, see Dichlorvos
Chwastox, see MCPA
CIBA 570, see Phosphamidon

Cortilan-neu, see Chlordane
Cotneon, see Azinphos-methyl
Cotnion methyl, see Azinphos-methyl
Cotofor, see Dipropetryn
Cotoran, see Fluometuron
Cotoran multi, see Metolachlor
Cotoran multi 50WP, see Fluometuron
Cottonex, see Fluometuron
Cotton-Pro, see Prometryn
Coumadin, see Warfarin
Coumafen, see Warfarin
Coumafene, see Warfarin
Counter, see Terbufos
Counter 15G soil insecticide, see Terbufos
Counter 15G soil insecticide-nematocide, see Terbufos
Cov-r-tox, see Warfarin
CP 6343, see Allidochlor
CP 15336, see Diallate
CP 23426, see Triallate
CP 31393, see Propachlor
CP 50144, see Alachlor
CPCA, see Dicofol
Crag 974, see Dazomet
Crag fungicide 974, see Dazomet
Crag nematocide, see Dazomet
Crag sevin, see Carbaryl
Crag 85W, see Dazomet
Crestoxo, see Toxaphene
Crestoxo 90, see Toxaphene
Crisalin, see Trifluralin
Crisatine, see Ametryn
Crisatrina, see Atrazine
Crisazine, see Atrazine
Crisodrin, see Monocrotophos
Crisulfan, see α-Endosulfan, β-Endosulfan
Crisuron, see Diuron
Crolean, see Acrolein
Croneton, see Ethiofencarb
Crop rider, see 2,4-D
Crotilin, see 2,4-D
Crotoxyphos
Cryptogil OL, see Pentachlorophenol
Crystal Propanil-4, see Propanil
Crysthion 2L, see Azinphos-methyl
Crysthyon, see Azinphos-methyl
Curacron, see Profenofos
Cyanazine
Cyanoethylene, see Acrylonitrile

Cyano(4-fluoro-3-phenoxyphenyl)methyl 3-(2,2-dichloroethenyl)-2,2-
    dimethylcyclopropanecarboxylate, see Cyfluthrin
α-Cyano-3-phenoxybenzyl-2-(4-chlorophenyl)-3-methylbutyrate, see Fenvalerate
(+)-α-Cyano-3-phenoxybenzyl 2,2-dimethyl-3-(2,2-dichlorovinyl)cyclopropane
    carboxylate, see Cypermethrin
(S-(R*,R*))-Cyano(3-phenoxyphenyl)methyl 4-chloro-α-(1-methylethyl)benzeneacetate,
    see Esfenvalerate
Cyano(3-phenoxyphenyl)methyl 3-(2,2-dichloroethenyl)-2,2-
    dimethylcyclopropanecarboxylate, see Cypermethrin
(RS)-Cyano-(3-phenoxyphenyl)methyl (S)-4-(difluoromethoxy)phenyl)-α-(1-
    methylethyl)benzeneacetate, see Flucythrinate
Cyanophos, see Dichlorvos
Cyazin, see Atrazine
Cybolt, see Flucythrinate
Cyclic S,S-(6-methyl-2,3-quinoxalinediyl) dithiocarbonate, see Quinomethionate
Cycloate
Cyclodan, see α-Endosulfan, β-Endosulfan
3-Cyclohexyl-6-(dimethylamino)-1-methyl-1,3,5-triazine-2,4(1H,3H)dione, see
    Hexazinone
3-Cyclohexyl-6-(dimethylamino)-1-methyl-s-triazine-2,4(1H,3H)dione, see Hexazinone
2-Cyclopropylamino-4,6-diamino-s-triazine, see Cyromazine
N-Cyclopropyl-1,3,5-triazine-2,4,6-triamine, see Cyromazine
Cyfluthrin
Cygon, see Dimethoate
Cygon 4E, see Dimethoate
Cygon insecticide, see Dimethoate
Cymbush, see Cypermethrin
Cynogan, see Bromacil
Cyodrin, see Crotoxyphos
Cyperkill, see Cypermethrin
Cypermethrin
Cypona, see Dichlorvos
Cypona E.C., see Crotoxyphos
Cyromazine
Cythion, see Malathion
Cytrol, see Amitrole
Cytrol Amitrole-T, see Amitrole
Cytrole, see Amitrole
Cyuram DS, see Thiram
2,4-D
D 50, see 2,4-D
D 735, see Carboxin
D 1410, see Oxamyl
D 1991, see Benomyl
D-90-A, see Monalide
DAC 893, see Chlorthal-dimethyl
DAC 2787, see Chlorothalonil
Dacamine, see 2,4-D, 2,4,5-T
2,4-D acid, see 2,4-D

Daconil, see Chlorothalonil
Daconil 2787, see Chlorothalonil
Daconil 2787 flowable fungicide, see Chlorothalonil
Dacosoil, see Chlorothalonil
Dacthal, see Chlorthal-dimethyl
Dacthalor, see Chlorthal-dimethyl
Dagadip, see Carbophenothion
Dailon, see Diuron
Dalapon, see Dalapon-sodium
Dalapon-sodium
Dalapon sodium salt, see Dalapon-sodium
Danex, see Trichlorfon
Danthion, see Parathion
Daphene, see Dimethoate
Dasanit, see Fensulfothion
DATC, see Diallate
Dawson 100, see Methyl bromide
Dazomet
Dazomet-powder BASF, see Dazomet
Dazzel, see Diazinon
DBCP, see 1,2-Dibromo-3-chloropropane
DBD, see Azinphos-methyl
DBE, see Ethylene dibromide
DBH, see Lindane
2,6-DBN, see Dichlobenil
DBP, see Di-*n*-butyl phthalate
DCB, see Dichlobenil
2,3-DCDT, see Diallate
DCEE, see Bis(2-chloroethyl)ether
DCMO, see Carboxin
DCMU, see Diuron
Dcon, see Warfarin
DCPA, see Chlorthal-dimethyl, Propanil
DDD, see *p,p'*-DDD
4,4'-DDD, see *p,p'*-DDD
***p,p'*-DDD**
DDE, see *p,p'*-DDE
4,4'-DDE, see *p,p'*-DDE
***p,p'*-DDE**
DDP, see Parathion
DDT, see *p,p'*-DDT
4,4'-DDT, see *p,p'*-DDT
***p,p'*-DDT**
DDT dehydrochloride, see *p,p'*-DDE
DDVF, see Dichlorvos
DDVP, see Dichlorvos
Debroussaillant 600, see 2,4-D
Debroussaillant concentre, see 2,4,5-T
Debroussaillant super concentre, see 2,4,5-T

Decabane, see Dichlobenil

1,2,3,5,6,7,8,9,10,10-Decachloro[5.2.1.0^{2,6}.0^{3,9}.0^{5,8}]decano-4-one, see Kepone

Decachlor; see Dienochlor

Decachlorobi-2,4-cyclopentadien-1-yl; see Dienochlor

1,1′,2,2′,3,3′,4,4′,5,5′-Decachlorobi-2,4-cyclopentadien-1-yl; see Dienochlor

Decachloroketone, see Kepone

Decachloro-1,3,4-metheno-2*H*-cyclobuta[*cd*]pentalen-2-one, see Kepone

Decachlorooctahydrokepone-2-one, see Kepone

Decachlorooctahydro-1,3,4-metheno-2*H*-cyclobuta[*cd*]pentalen-2-one, see Kepone

1,1a,3,3a,4,5,5a,5b,6-Decachlorooctahydro-1,3,4-metheno-2*H*-cyclobuta[*cd*]pentalen-2-one, see Kepone

Decachloropentacyclo[5.2.1.0^{2,6}.0^{3,9}.0^{5,8}]decan-3-one, see Kepone

Decachloropentacyclo[5.2.1.0^{2,6}.0^{4,10}.0^{5,9}]decan-3-one, see Kepone

Decachlorotetracyclodecanone, see Kepone

Decachlorotetrahydro-4,7-methanoindeneone, see Kepone

Decamine, see 2,4-D

Decamine 4T, see 2,4,5-T

Decofol, see Dicofol

Decrotox, see Crotoxyphos

De-cut, see Maleic hydrazide

Dedelo, see *p,p′*-DDT

Dedevap, see Dichlorvos

Dedweed, see 2,4-D, MCPA

Dedweed brush killer, see 2,4,5-T

Dedweed LV-6 brush-kil and T-5 brush-kil, see 2,4,5-T

Dedweed LV-69, see 2,4-D

Def, see Butifos

Def defoliant, see Butifos

De-Fend, see Dimethoate

Defolit, see Thidiazuron

Degrassan, see 4,6-Dinitro-*o*-cresol

Degreen, see Butifos

Deiquat, see Diquat

Dekrysil, see 4,6-Dinitro-*o*-cresol

Demeton-methyl sulfoxide, see Oxydemeton-methyl

Demeton-*O*-methyl sulfoxide, see Oxydemeton-methyl

Demeton-*S*-methyl sulfoxide, see Oxydemeton-methyl

Demos-L40, see Dimethoate

Denapon, see Carbaryl

Deoval, see *p,p′*-DDT

DEP, see Trichlorfon

Depthon, see Trichlorfon

Deriban, see Dichlorvos

Dermaphos, see Ronnel

Derribante, see Dichlorvos

Desapon, see Diazinon

Des-i-cate, see Endothall

Desmedipham

Desormone, see 2,4-D

Detal, see 4,6-Dinitro-*o*-cresol
DETF, see Trichlorfon
Dethmor, see Warfarin
Dethnel, see Warfarin
Detmol-extrakt, see Lindane
Detmol MA, see Malathion
Detmol MA 96%, see Malathion
Detmol U.A., see Chlorpyrifos
Detox, see *p,p′*-DDT
Detox 25, see Lindane
Detoxan, see *p,p′*-DDT
Devicarb, see Carbaryl
Devigon, see Dimethoate
Devikol, see Dichlorvos
Devoran, see Lindane
Devrinol, see Napropamide
Devrinol 2EC, see Napropamide
Devrinol 10G, see Napropamide
Devrinol 50WP, see Napropamide
Dextrone, see Diquat
Dfluron, see Diflubenzuron
Dialifor, see Dialifos
Dialifos
Diallate
Diallylchloroacetamide, see Allidochlor
N,N-Diallylchloroacetamide, see Allidochlor
N,N-Diallyl-2-chloroacetamide, see Allidochlor
N,N-Diallyl-α-chloroacetamide, see Allidochlor
Diamide, see Diphenamid
Dianate, see Dicamba
2,2-Di-*p*-anisyl-1,1,1-trichloroethane, see Methoxychlor
Dianon, see Diazinon
Diater, see Diuron
Diaterr-fos, see Diazinon
Diazatol, see Diazinon
Diazide, see Diazinon
Diazinon
Dibovan, see *p,p′*-DDT
Dibrom, see Naled
Dibromochloropropane, see 1,2-Dibromo-3-chloropropane
1,2-Dibromo-3-chloropropane
2,6-Dibromo-4-cyanophenol, see Bromoxynil
2,6-Dibromo-4-cyanophenyl octanoate, see Bromoxynil octanoate
1,2-Dibromo-2,2-dichloroethyldimethyl phosphate, see Naled
Dibromoethane, see Ethylene dibromide
1,2-Dibromoethane, see Ethylene dibromide
α,β-Dibromoethane, see Ethylene dibromide
sym-Dibromoethane, see Ethylene dibromide
3,5-Dibromo-4-hydroxybenzonitrile, see Bromoxynil

3,5-Dibromo-4-hydroxyphenylcyanide, see Bromoxynil
3,5-Dibromo-4-octanoyloxybenzonitrile, see Bromoxynil octanoate
Dibutyl-1,2-benzenedicarboxylate, see Di-*n*-butyl phthalate
Dibutyl phthalate, see Di-*n*-butyl phthalate
Di-*n*-butyl phthalate
DIC 1468, see Metribuzin
Dicamba
Dicambe, see Dicamba
Dicarbam, see Carbaryl
S-1,2-Dicarbethoxyethyl-*O,O*-dimethyl dithiophosphate, see Malathion
Dicarboethoxyethyl-*O,O*-dimethyl phosphorodithioate, see Malathion
Dicarbosulf, see Thiodicarb
Dichlobenil
Dichlone
Dichlor-fenidim, see Diuron
Dichlorman, see Dichlorvos
Dichloroallyl diisopropylthiocarbamate, see Diallate
S-Dichloroallyl diisopropylthiocarbamate, see Diallate
Dichloroallyl *N,N*-diisopropylthiolcarbamate, see Diallate
3,6-Dichloro-*o*-anisic acid, see Dicamba
4,4′-Dichlorobenzilate, see Chlorobenzilate
4,4′-Dichlorobenzilic acid ethyl ester, see Chlorobenzilate
2,6-Dichlorobenzonitrile, see Dichlobenil
1,1-Dichloro-2,2-bis(*p*-chlorophenyl)ethane, see *p,p′*-DDD
1,1-Dichloro-2,2-bis(*p*-chlorophenyl)ethylene, see *p,p′*-DDE
Dichlorochlordene, see Chlordane
2,4-Dichloro-6-(2-chloroanilo)-1,3,5-triazine, see Anilazine
2,4-Dichloro-6-*o*-chloroanilo-*s*-triazine, see Anilazine
4,6-Dichloro-*N*-(2-chlorophenyl)-1,3,5-triazin-2-amine, see Anilazine
1,1-Dichloro-2,2-di(4-chlorophenyl)ethane, see *p,p′*-DDD
1,1-Dichloro-2,2-di(*p*-chlorophenyl)ethane, see *p,p′*-DDD
Dichlorodiethyl ether, see Bis(2-chloroethyl)ether
2,2′-Dichlorodiethyl ether, see Bis(2-chloroethyl)ether
β,β′-Dichlorodiethyl ether, see Bis(2-chloroethyl)ether
Dichlorodiisopropyl ether, see Bis(2-chloroisopropyl)ether
S-2,3-Dichlorodiisopropylthiocarbamate, see Diallate
3,5-Dichloro-*N*-(1,1-dimethyl-2-propynyl)benzamide, see Propyzamide
3,5-Dichloro-*N*-(1,1-dimethylpropynyl)benzamide, see Propyzamide
Dichlorodiphenyldichloroethane, see *p,p′*-DDD
4,4′-Dichlorodiphenyldichloroethane, see *p,p′*-DDD
p,p′-Dichlorodiphenyldichloroethane, see *p,p′*-DDD
Dichlorodiphenyldichloroethylene, see *p,p′*-DDE
p,p′-Dichlorodiphenyldichloroethylene, see *p,p′*-DDE
Dichlorodiphenyltrichloroethane, see *p,p′*-DDT
4,4′-Dichlorodiphenyltrichloroethane, see *p,p′*-DDT
p,p′-Dichlorodiphenyltrichloroethane, see *p,p′*-DDT
3-(2,2-Dichloroethenyl)-2,2-dimethylcyclopropanecarboxylic acid (3-
    phenoxyphenyl)methyl ester, see Permethrin
2,2-Dichloroethenyl dimethyl phosphate, see Dichlorvos

1,1'-(Dichloroethenylidene)bis(4-chlorobenzene), see *p,p'*-DDE
2,2-Dichloroethenyl phosphoric acid dimethyl ester, see Dichlorvos
Dichloroether, see Bis(2-chloroethyl)ether
Dichloroethyl ether, see Bis(2-chloroethyl)ether
2,2'-Dichloroethyl ether, see Bis(2-chloroethyl)ether
α,α'-Dichloroethyl ether, see Bis(2-chloroethyl)ether
sym-Dichloroethyl ether, see Bis(2-chloroethyl)ether
Di(2-chloroethyl)ether, see Bis(2-chloroethyl)ether
Di(β-chloroethyl)ether, see Bis(2-chloroethyl)ether
1,1'-(2,2-Dichloroethylidene)bis(4-chlorobenzene), see *p,p'*-DDD
Dichloroethyl oxide, see Bis(2-chloroethyl)ether
Dichloroisopropyl ether, see Bis(2-chloroisopropyl)ether
2,2'-Dichloroisopropyl ether, see Bis(2-chloroisopropyl)ether
Dichlorokelthane, see Dicofol
2,5-Dichloro-6-methoxybenzoic acid, see Dicamba
3,6-Dichloro-2-methoxybenzoic acid, see Dicamba
3-(2,4-Dichloro-5-(1-methylethoxy)phenyl)-5-(1,1-dimethylethyl)-1,3,4-oxadiazol-
    2(3*H*)-one, see Diuron
2,3-Dichloro-1,4-naphthalenedione, see Dichlone
2,3-Dichloro-1,4-naphthaquinone, see Dichlone
Dichloronaphthoquinone, see Dichlone
2,3-Dichloronaphthoquinone, see Dichlone
2,3-Dichloronaphthoquinone-1,4, see Dichlone
2,3-Dichloro-1,4-naphthoquinone, see Dichlone
2,3-Dichloro-α-naphthoquinone, see Dichlone
Dichlorophenoxyacetic acid, see 2,4-D
(2,4-Dichlorophenoxy)acetic acid, see 2,4-D
2-(4-(2,4-Dichlorophenoxy)phenoxy)methyl propionoate, see Diclofop-methyl
2-(4-(2,4-Dichlorophenoxy)phenoxy)propanoic acid methyl ester, see Diclofopmethyl
3-(3,4-Dichlorophenyl)-1,1-dimethylurea, see Diuron
N'-(3,4-Dichlorophenyl)-*N,N*-dimethylurea, see Diuron
3-(3,4-Dichlorophenyl)-1-methoxymethylurea, see Linuron
3-(3,4-Dichlorophenyl)-1-methoxy-1-methylurea, see Linuron
N'-(3,4-Dichlorophenyl)-*N*-methoxy-*N*-methylurea, see Linuron
3-(3,5-Dichlorophenyl)-*N*-(1-methylethyl)-2,4-dioxo-1-imidazolidinecarboxamide, see
    Iprodione
N-(3,4-Dichlorophenyl)-*N'*-methyl-*N'*-methoxyurea, see Linuron
N-(3,4-Dichlorophenyl)propanamide, see Propanil
N-(3,4-Dichlorophenyl)propionamide, see Propanil
Di(*p*-chlorophenyl)trichloromethylcarbinol, see Dicofol
Dichlorophos, see Dichlorvos
1,2-Dichloropropane
α,β-Dichloropropane, see 1,2-Dichloropropane
cis-1,3-Dichloropropene, see *cis*-1,3-Dichloropropylene
(*E*)-1,3-Dichloropropene, see *trans*-1,3-Dichloropropylene
trans-1,3-Dichloropropene, see *trans*-1,3-Dichloropropylene
(*Z*)-1,3-Dichloropropene, see *cis*-1,3-Dichloropropylene
1,3-Dichloroprop-1-ene, see *cis*-1,3-Dichloropropylene, *trans*-1,3-Dichloropropylene
cis-1,3-Dichloro-1-propene, see *cis*-1,3-Dichloropropylene

(*E*)-1,3-Dichloro-1-propene, see *trans*-1,3-Dichloropropylene

trans-1,3-Dichloro-1-propene, see *trans*-1,3-Dichloropropylene

(*Z*)-1,3-Dichloro-1-propene, see *cis*-1,3-Dichloropropylene

Dichloropropionanilide, see Propanil

3′,4′-Dichloropropionanilide, see Propanil

2,2-Dichloropropionic acid sodium salt, see Dalapon-sodium

α,α-Dichloropropionic acid sodium salt, see Dalapon-sodium

cis-1,2-Dichloropropylene

cis-1,3-Dichloro-1-propylene, see *cis*-1,3-Dichloropropylene

trans-1,2-Dichloropropylene

trans-1,3-Dichloro-1-propylene, see *trans*-1,3-Dichloropropylene

4,4′-Dichloro-α-(trichloromethyl)benzhydrol, see Dicofol

2,2-Dichlorovinyl dimethyl phosphate, see Dichlorvos

2,2-Dichlorovinyl dimethyl phosphoric acid ester, see Dichlorvos

Dichlorovos, see Dichlorvos

Dichlorvos

Diclofop-methyl

Dicofol

Dicophane, see *p,p′*-DDT

Dicopur, see 2,4-D

Dicopur-M, see MCPA

Dicotex, see MCPA

Dicotox, see 2,4-D

Dicrotophos

1,3-Dicyanotetrachlorobenzene, see Chlorothalonil

Didigam, see *p,p′*-DDT

Didimac, see *p,p′*-DDT

Dieldrin

Dieldrite, see Dieldrin

Dieldrix, see Dieldrin

Dienochlor

Diethion, see Ethion

1,2-Di(ethoxycarbonyl)ethyl-*O,O*-dimethyl phosphorodithioate, see Malathion

S-1,2-Di(ethoxycarbonyl)ethyl dimethyl phosphorothiolothionate, see Malathion

(*E*)-3-(Diethylamino)-1-methyl-3-oxo-1-propenyl dimethyl phosphate, see Dicrotophos

O,O-Diethyl *S*-(6-chlorobenzoxazolinyl-3-methyl) dithiophosphate, see Phosalone

O,O-Diethyl *S*-((6-chloro-2-oxobenzoxazolin-3-yl)methyl) phosphorodithioate, see Phosalone

O,O-Diethyl *S*-(6-chloro-2-oxobenzoxazolin-3-yl)methyl phosphorothiolothionate, see Phosalone

O,O-Diethyl *p*-chlorophenylmercaptomethyl dithiophosphate, see Carbophenothion

O,O-Diethyl *S*-(4-chlorophenylthio)methyl dithiophosphate, see Carbophenothion

O,O-Diethyl *S*-(*p*-chlorophenylthio)methyl phosphorodithioate, see Carbophenothion

O,O-Diethyl *S*-(2-chloro-1-phthalimidoethyl) phosphorodithioate, see Dialifos

Diethyl (dimethoxyphosphinothioylthio) butanedioate, see Malathion

Diethyl (dimethoxyphosphinothioylthio) succinate, see Malathion

3-Diethyldithiophosphorymethyl-6-chlorobenzoxazolone-2, see Phosalone

O,O-Diethyl *S*-(2-eththioethyl) phosphorodithioate, see Disulfoton

O,O-Diethyl *S*-(2-ethylmercaptoethyl) dithiophosphate, see Disulfoton

O,O-Diethyl *S*-ethylmercaptomethyl dithiophosphonate, see Phorate
O,O-Diethyl *S*-(2-(ethylthio)ethyl) phosphorodithioate, see Disulfoton
O,O-Diethyl *S*-(2-ethylthioethyl) thiothionophosphate, see Disulfoton
O,O-Diethyl *S*-ethylthiomethyl dithiophosphonate, see Phorate
O,O-Diethyl *S*-ethylthiomethyl phosphorodithioate, see Phorate
Diethyl 4-(2-isopropyl-6-methylpyrimidinyl)phosphorothioate, see Diazinon
O,O-Diethyl *O*-(2-isopropyl-4-methyl-6-pyrimidinyl)phosphorothioate, see Diazinon
O,O-Diethyl *O*-(2-isopropyl-6-methyl-4-pyrimidinyl)phosphorothioate, see Diazinon
O,O-Diethyl *O*-(2-isopropyl-4-methyl-6-pyrimidinyl)thionophosphate, see Diazinon
O,O-Diethyl 2-isopropyl-4-methylpyrimidinyl-6-thiophosphate, see Diazinon
Diethyl mercaptosuccinate, *O,O*-dimethyl phosphorodithioate, see Malathion
Diethyl mercaptosuccinate, *O,O*-dimethyl thiophosphate, see Malathion
Diethyl mercaptosuccinic acid *O,O*-dimethyl phosphorodithioate, see Malathion
O,O-Diethyl *O*-6-methyl-2-isopropyl-4-pyrimidinyl phosphorothioate, see Diazinon
O,O-Diethyl *O*-(6-methyl-2-(1-methylethyl)-4-pyrimidinyl) phosphorothioate, see
    Diazinon
O,O-Diethyl *O*-4-methylsulphinylphenyl phosphorothioate, see Fensulfothion
O,O-Diethyl *O*-*p*-methylsulphinylphenyl phosphorothioate, see Fensulfothion
O,O-Diethyl O-p-methylsulphinylphenyl thiophosphate, see Fensulfothion
N,N-Diethyl-2-(1-naphthalenyloxy)propionamide, see Napropamide
(*RS*)-*N,N*-Diethyl-2-(1-naphthyloxy)propionamide, see Napropamide
Diethyl-4-nitrophenyl phosphorothionate, see Parathion
O,O-Diethyl-*O*-4-nitrophenyl phosphorothioate, see Parathion
O,O-Diethyl *O*-*p*-nitrophenyl phosphorothioate, see Parathion
Diethyl-*p*-nitrophenyl thionophosphate, see Parathion
O,O-Diethyl-*O*-4-nitrophenyl thionophosphate, see Parathion
O,O-Diethyl-*O*-*p*-nitrophenyl thionophosphate, see Parathion
Diethyl-*p*-nitrophenyl thiophosphate, see Parathion
O,O-Diethyl-*O*-*p*-nitrophenyl thiophosphate, see Parathion
Diethylparathion, see Parathion
O,O-Diethyl phosphorodithioate, *S*-ester with 6-chloro-3-(mercaptomethyl)-2-
    benzoxazolinone, see Phosalone
O,O-Diethyl-*O*-3,5,6-trichloro-2-pyridyl phosphorothioate, see Chlorpyrifos
Dif 4, see Diphenamid
Difenzoquat methyl sulfate
Diflubenzuron
4-(Difluoromethoxy)-α-(1-methylethyl)benzeneacetic acid cyano(3-
    phenoxyphenyl)methyl ester, see Flucythrinate
Difonate, see Fonofos
Digermin, see Trifluralin
2,3-Dihydro-5-carboxanilido-6-methyl-1,4-oxatiin, see Carboxin
9,10-Dihydro-8a,10-diazoniaphenanthrene dibromide, see Diquat
Dihydro-8a,10a-diazoniaphenanthrene-(1,1′-ethylene-2,2′-bipyridylium)dibromide, see
    Diquat
2,3-Dihydro-2,2-dimethyl-7-benzofuranol methylcarbamate, see Carbofuran
5,6-Dihydrodipyrido[1,2*a*:2,1*c*]pyrazinium dibromide, see Diquat
S-(2-Dihydro-5-methoxy-2-oxo-1,3,4-thiadiazol-3-methyl)dimethyl
    phosphorothiolothionate, see Methidathion

S-2-Dihydro-5-methoxy-2-oxo-1,3,4-thiadiazol-3-ylmethyl *O,O*-dimethyl
    phosphorodithionate, see Methidathion
2,3-Dihydro-6-methyl-1,4-oxathiin-5-carboxanilide, see Carboxin
5,6-Dihydro-2-methyl-1,4-oxathiin-3-carboxanilide, see Carboxin
5,6-Dihydro-2-methyl-*N*-phenyl-1,4-oxathiin-3-carboxamide, see Carboxin
S-(3,4-Dihydro-4-oxobenzo(α)(1,2,3)triazin-3-ylmethyl)-*O,O*-dimethyl
    phosphorodithioate, see Azinphos-methyl
S-(3,4-Dihydro-4-oxo-1,2,3-benzotriazin-3-ylmethyl)-*O,O*-dimethyl phosphorodithioate,
    see Azinphos-methyl
1,2-Dihydro-3,6-pyradizinedione, see Maleic hydrazide
1,2-Dihydroxypyridazine-3,6-dione, see Maleic hydrazide
1,2-Dihydro-3,6-pyridizinedione, see Maleic hydrazide
6,7-Dihydropyrido(1,2-α;2′,1′-*c*)pyrazinedium dibromide, see Diquat
Diisobutylthiocarbamic acid *S*-ethyl ester, see Butylate
Diisocarb, see Butylate
2,6-Diisopropylamino-4-methoxytriazine, see Prometon
N,N′-Diisopropyl-6-methoxy-1,3,5-triazine-2,4-diyldiamine, see Prometon
N,N′-Diisopropyl-6-methylthio-1,3,5-triazine-2,4-diyldiamine, see Prometryn
N-Diisopropylthiocarbamic acid S-2,3,3-trichloro-2-propenyl ester, see Triallate
Diisopropyltrichloroallylthiocarbamate, see Triallate
Dikotes, see MCPA
Dikotex, see MCPA
Dilene, see *p,p′*-DDD
Dimate 267, see Dimethoate
Dimaz, see Disulfoton
Dimecron, see Phosphamidon
Dimecron 100, see Phosphamidon
Dimetate, see Dimethoate
Dimethoate
Dimethoate-267, see Dimethoate
Dimethoat tecvhnisch 95%, see Dimethoate
Dimethogen, see Dimethoate
Dimethoxy-DDT, see Methoxychlor
Dimethoxy-DT, see Methoxychlor
p,p′-Dimethoxydiphenyltrichloroethane, see Methoxychlor
2,2-Di-(*p*-methoxyphenyl)-1,1,1-trichloroethane, see Methoxychlor
Di(*p*-methoxyphenyl)trichloromethyl methane, see Methoxychlor
((Dimethoxyphosphinothioyl)thio)butanedioic acid diethyl ester, see Malathion
N-Dimethoxyphosphinothioylthiomethyl)phthalimide, see Phosmet
3-((Dimethoxyphosphinyl)oxy)-2-butenoic acid methyl ester, see Mevinphos
3-((Dimethoxyphosphinyl)oxy)-2-butenoic acid 1-phenylethyl ester, see Crotoxyphos
3-(Dimethoxyphosphinyloxy)-*N,N*-dimethyl-*cis*-crotonamide, see Dicrotophos
3-(Dimethoxyphosphinyloxy)-*N*-methylisocrotonamide, see Monocrotophos
3-(Dimethoxyphosphinyloxy)-*N,N*-dimethylisocrotonamide, see Dicrotophos
Dimethoxy-2,2,2-trichloro-1-hydroxyethylphosphine oxide, see Trichlorfon
O,S-Dimethylacetylphosphoroamidothioate, see Acephate
4-Dimethylamine-*m*-cresyl methylcarbamate, see Aminocarb
4-Dimethylamino-3-cresyl methylcarbamate, see Aminocarb

2-(Dimethylamino)-*N*-(((methylamino)carbonyl)oxy)-2-oxoethanimidothioic acid methyl ester, see Oxamyl

3-(Dimethylamino)-1-methyl-3-oxo-1-propenyl dimethyl phosphate, see Dicrotophos

4-(Dimethylamino)-3-methylphenol methylcarbamate, see Aminocarb

2-Dimethylamino-1-(methylthio)glyoxal *O*-methylcarbamoylmonoxime, see Oxamyl

4-(Dimethylamino)-*m*-tolyl methylcarbamate, see Aminocarb

O,O-Dimethyl-*S*-(benzaziminomethyl) dithiophosphate, see Azinphos-methyl

Dimethyl-1,2-benzenedicarboxylate, see Dimethyl phthalate

Dimethylbenzeneorthodicarboxylate, see Dimethyl phthalate

2,2-Dimethylbenzo-1,3-dioxol-4-ol methylcarbamate, see Bendiocarb

2,2-Dimethyl-1,3-benzodioxol-4-ol methylcarbamate, see Bendiocarb

2,2-Dimethyl-1,3-benzodioxol-4-ol *N*-methylcarbamate, see Bendiocarb

O,O-Dimethyl-*S*-(1,2,3-benzotriazinyl-4-keto)methyl phosphorodithioate, see Azinphos-methyl

O,O-Dimethyl-*S*-1,2-bis(ethoxycarbonyl)ethyldithiophosphate, see Malathion

Dimethylcarbamodithioc acid iron complex, see Ferbam

Dimethylcarbamodithioc acid iron(3+) salt, see Ferbam

cis-2-Dimethylcarbamoyl-1-methylvinyl dimethylphosphate, see Dicrotophos

O,O-Dimethyl-*O*-(2-carbomethoxy-1-methylvinyl)phosphate, see Mevinphos

Dimethyl-1-carbomethoxy-1-propen-2-yl phosphate, see Mevinphos

O,O-Dimethyl 1-carbomethoxy-1-propen-2-yl phosphate, see Mevinphos

Dimethyl 2-chloro-2-diethylcarbamoyl-1-methylvinyl phosphate, see Phosphamidon

O,O-Dimethyl *O*-(2-chloro-2-(*N,N*-diethylcarbamoyl)-1-methylvinyl) phosphate, see Phosphamidon

N,N-Dimethyl-*N*′-(4-chlorophenyl)urea, see Monuron

N,N-Dimethyl-*N*′-(*p*-chlorophenyl)urea, see Monuron

2,2-Dimethyl-7-coumaranyl *N*-methylcarbamate, see Carbofuran

Dimethyl 1,2-dibromo-2,2-dichloroethyl phosphate, see Naled

O,O-Dimethyl-*O*-(1,2-dibromo-2,2-dichloroethyl)phosphate, see Naled

O,O-Dimethyl-*S*-(1,2-dicarbethoxyethyl)dithiophosphate, see Malathion

O,O-Dimethyl-*S*-(1,2-dicarbethoxyethyl)phosphorodithioate, see Malathion

O,O-Dimethyl-*S*-(1,2-dicarbethoxyethyl)thiothionophosphate, see Malathion

O,O-Dimethyl *O*-(2,2-dichloro-1,2-dibromoethyl)phosphate, see Naled

Dimethyl 2,2-dichloroethenyl phosphate, see Dichlorvos

1,1-Dimethyl-3-(3,4-dichlorophenyl)urea, see Diuron

Dimethyl dichlorovinyl phosphate, see Dichlorvos

Dimethyl 2,2-dichlorovinyl phosphate, see Dichlorvos

O,O-Dimethyl *O*-(2,2-dichlorovinyl)phosphate, see Dichlorvos

O,O-Dimethyl-*S*-1,2-di(ethoxycarbamyl)ethyl phosphorodithioate, see Malathion

2,2-Dimethyl-2,3-dihydro-7-benzofuranyl-*N*-methylcarbamate, see Carbofuran

O,O-Dimethyl-*S*-(3,4-dihydro-4-keto-1,2,3-benzotriazinyl-3-methyl) dithiophosphate, see Azinphos-methyl

2,2-Dimethyl-*N,N*-dimethylacetamide, see Diphenamid

O,O-Dimethyl *O*-(*N,N*-dimethylcarbamoyl-1-methylvinyl) phosphate, see Dicrotophos

O,O-Dimethyl-*O*-(1,4-dimethyl-3-oxo-4-azapent-1-enyl)phosphate, see Dicrotophos

N,N-Dimethyldiphenylacetamide, see Diphenamid

N,N-Dimethyl-2,2-diphenylacetamide, see Diphenamid

N,N-Dimethyl-α,α-diphenylacetamide, see Diphenamid

1,2-Dimethyl-3,5-diphenyl-1*H*-pyrazolium methyl sulfate, see Difenzoquat methyl sulfate

Dimethyldithiocarbamic acid iron salt, see Ferbam
Dimethyldithiocarbamic acid iron(3+) salt, see Ferbam
O,O-Dimethyldithiophosphate dimethylmercaptosuccinate, see Malathion
Dimethyldithiophosphoric acid *N*-methylbenzazimide ester, see Azinphos-methyl
O,O-Dimethyl dithiophosphorylacetic acid *N*-monomethylamide salt, see Dimethoate
O,S-Dimethyl ester amide of aminothioate, see Methamidophos
O,O-Dimethyl *S*-(2-eththionylethyl) phosphorothioate, see Oxydemeton-methyl
Dimethyl *S*-(2-eththionylethyl) thiophosphate, see Oxydemeton-methyl
N-(1,1-Dimethylethyl)-*N'*-ethyl-6-(methylthio)-1,3,5-triazine-2,4-diamine, see Terbutryn
O,O-Dimethyl *S*-ethylsulphinylethyl phosphorothioate, see Oxydemetonmethyl
O,O-Dimethyl *S*-2-(ethylsulfinyl)ethyl phosphorothioate, see Oxydemetonmethyl
O,O-Dimethyl *S*-2-(ethylsulfinyl)ethyl thiophosphate, see Oxydemetonmethyl
N-(5-(1,1-Dimethylethyl)-1,3,4-thiadiazol-2-yl)-*N,N'*-dimethylurea, see Tebuthiuron
S-((((1,1-Dimethylethyl)thio)methyl) *O,O*-diethyl phosphorodithioate, see Terbufos
Dimethylformocarbothialdine, see Dazomet
Dimethyl-1-hydroxy-2,2,2-trichloroethyl phosphonate, see Trichlorfon
O,O-Dimethyl-(1-hydroxy-2,2,2-trichloro)ethyl phosphate, see Trichlorfon
O,O-Dimethyl-(1-hydroxy-2,2,2-trichloro)ethyl phosphonate, see Trichlorfon
Dimethyl 2-methoxycarbonyl-1-methylvinyl phosphate, see Mevinphos
Dimethyl methoxycarbonylpropenyl phosphate, see Mevinphos
Dimethyl (1-methoxycarboxypropen-2-yl)phosphate, see Mevinphos
O,O-Dimethyl *S*-(5-methoxy-1,3,4-thiadiazolinyl-3-methyl) dithiophosphate, see
    Methidathion
O,O-Dimethyl *S*-(2-methoxy-1,3,4-thiadiazol-5(4*H*)-onyl-4-methyl)phosphorodithioate,
    see Methidathion
O,O-Dimethyl *S*-(2-(methylamino)-2-oxoethyl) phosphorodithioate, see Dimethoate
O,O-Dimethyl *S*-(*N*-methylcarbamoylmethyl) dithiophosphate, see Dimethoate
O,O-Dimethyl *S*-(*N*-methylcarbamoylmethyl) phosphorodithioate, see Dimethoate
O,O-Dimethyl-*O*-(2-*N*-methylcarbamoyl-1-methylvinyl) phosphate, see Monocrotophos
N,N-Dimethyl-α-methylcarbamoyloxyimino-α-(methylthio)acetamide, see Oxamyl
N',N'-Dimethyl-*N*-((methylcarbamoyl)oxy)-1-thiooxamimidic acid methyl ester, see
    Oxamyl
O,O-Dimethyl *S*-(*N*-methylcarbamylmethyl) thiothionophosphate, see Dimethoate
O,O-Dimethyl *O*-(1-methyl-2-carboxy-α-phenylethyl)vinyl phosphate, see Crotoxyphos
O,O-Dimethyl *O*-(1-methyl-2-carboxyvinyl)phosphate, see Mevinphos
N,N-Dimethyl-*N'*-(2-methyl-4-chlorophenyl)formamidine, see Chlordimeform
O,O-Dimethyl *O*-4-(methylmercapto)-3-methylphenyl phosphorothioate, see Fenthion
O,O-Dimethyl *O*-4-(methylmercapto)-3-methylphenyl thiophosphate, see Fenthion
(*E*)-Dimethyl 1-methyl-3-(methylamino)-3-oxo-1-propenyl phosphate, see
    Monocrotophos
Dimethyl-1-methyl-2-(methylcarbamoyl)vinyl phosphate, see Monocrotophos
Dimethyl (*E*)-1-methyl-2-(methylcarbamoyl)vinyl phosphate, see Monocrotophos
O,O-Dimethyl *O*-(3-methyl-4-methylmercaptophenyl) phosphorothioate, see Fenthion
O,O-Dimethyl *O*-(3-methyl-4-methylthiophenyl) phosphorothioate, see Fenthion
Dimethyl-*cis*-1-methyl-2-(1-phenylethoxycarbonyl)vinyl phosphate, see Crotoxyphos
O,O-Dimethyl *O*-(4-methylthio-3-methylphenyl) phosphorothioate, see Fenthion
3,5-Dimethyl-4-(methylthio)phenyl methylcarbamate, see Methiocarb
3,5-Dimethyl-4-methylthiophenyl *N*-methylcarbamate, see Methiocarb

O,O-Dimethyl *O*-(4-methylthio-*m*-tolyl) phosphorothioate, see Fenthion

O,O-Dimethyl-*S*-(*N*-monomethyl)carbamyl methyldithiophosphate, see Dimethoate

O,O-Dimethyl-*S*-(4-oxo-3*H*-1,2,3-benzotriazine-3-methyl) phosphorodithioate, see
  Azinphos-methyl

O,O-Dimethyl-*S*-(4-oxobenzotriazino-3-methyl) phosphorodithioate, see
  Azinphos-methyl

O,O-Dimethyl-*S*-(4-oxo-1,2,3-benzotriazino-3-methyl) thiothionophosphate, see
  Azinphos-methyl

O,O-Dimethyl *S*-((4-oxo-1,2,3-benzotriazin-3-yl)methyl) phosphorodithioate, see
  Azinphos-methyl

O,O-Dimethyl *S*-((4-oxo-1,2,3-benzotriazin-3(4*H*)-yl)methyl) phosphorodithioate, see
  Azinphos-methyl

2-(2,2-Dimethyl-1-oxopropyl)-1*H*-indene-1,3(2*H*)-dione, see Pindone

O,O-Dimethyl-1-oxy-2,2,2-trichloroethyl phosphonate, see Trichlorfon

N,N-Dimethyl-α-phenylbenzeneacetamide, see Diphenamid

N-(2,6-Dimethylphenyl)-*N*-methoxyacetyl)alanine methyl ester, see Metalaxyl

N-(2,6-Dimethylphenyl)-*N*-methoxyacetyl)-DL-alanine methyl ester, see Metalaxyl

Dimethyl phosphate ester with 3-hydroxy-*N,N*-dimethyl-*cis*-crotonamide, see Dicrotophos

Dimethyl phosphate ester with (*E*)-3-hydroxy-*N,N*-dimethylcrotonamide, see Dicrotophos

Dimethyl phosphate ester of 3-hydroxy-*N*-methyl-*cis*-crotonamide, see Monocrotophos

Dimethyl phosphate of chloro-*N,N*-diethyl-3-hydroxycrotonamide, see Phosphamidon

Dimethyl phosphate of 3-hydroxy-*N,N*-dimethyl-*cis*-crotonamide, see Dicrotophos

Dimethyl phosphate of α-methylbenzyl 3-hydroxy-*cis*-crotonate, see Crotoxyphos

Dimethyl phosphate of methyl-3-hydroxy-*cis*-crotonate, see Mevinphos

O,S-Dimethyl phosphoramidothioate, see Methamidophos

O,O-Dimethyl phosphorodithoate *S*-ester with *N*-(mercaptomethyl)phthalimide, see
  Phosmet

Dimethyl phthalate

O,O-Dimethyl *S*-phthalimidomethyl phosphorodithioate, see Phosmet

2-(((((4,6-Dimethyl-2-pyrimidinyl)amino)carbonyl)amino)sulfonyl)benzoic acid methyl
  ester, see Sulfometuron-methyl

Dimethyl 2,3,5,6-tetrachloro-1,4-benzenedicarboxylate, see Chlorthal-dimethyl

Dimethyl tetrachloroterephthalate, see Chlorthal-dimethyl

Dimethyl 2,3,5,6-tetrachloroterephthalate, see Chlorthal-dimethyl

Dimethyl-2*H*-1,3,5-tetrahydrothiadiazine-2-thione, see Dazomet

3,5-Dimethyl-1,2,3,5-tetrahydro-1,3,5-thiadiazinethione-2, see Dazomet

3,5-Dimethyl-1,3,5,2*H*-tetrahydrothiadiazine-2-thione, see Dazomet

3,5-Dimethyltetrahydro-1,3,5-2*H*-thiadiazine-2-thione, see Dazomet

3,5-Dimethyltetrahydro-1,3,5-thiadiazine-2-thione, see Dazomet

3,5-Dimethyltetrahydro-2*H*-1,3,5-thiadiazine-2-thione, see Dazomet

Dimethyl *N,N'*-(thiobis((methylimino)carbonyloxy))bis(ethanimidothioate), see
  Thiodicarb

3,5-Dimethyl-2-thionotetrahydro-1,3,5-thiadiazine, see Dazomet

Dimethyltrichlorohydroxyethyl phosphonate, see Trichlorfon

Dimethyl-2,2,2-trichloro-1-hydroxyethyl phosphonate, see Trichlorfon

O,O-Dimethyl-2,2,2-trichloro-1-hydroxyethyl phosphonate, see Trichlorfon

O,O-Dimethyl-*O*-2,4,5-trichlorophenyl phosphorothioate, see Ronnel

Dimethyl trichlorophenyl thiophosphate, see Ronnel

O,O-Dimethyl *O*-(2,4,5-trichlorophenyl)thiophosphate, see Ronnel
1,1-Dimethyl-3-(3-trifluoromethylphenyl)urea, see Fluometuron
N,N-Dimethyl-*N'*-(3-(trifluoromethyl)phenyl)urea, see Fluometuron
N-(2,4-Dimethyl-5-(((trifluoromethyl)sulfonyl)amino)phenyl)acetamide, see Mefluidide
1,1-Dimethyl-3-(α,α,α-trifluoro-*m*-tolyl)urea, see Fluometuron
Dimeton, see Dimethoate
Dimetox, see Trichlorfon
Dimevur, see Dimethoate
Dimid, see Diphenamid
Dimilin, see Diflubenzuron
Dimpylate, see Diazinon
Dinitro, see Dinoseb
Dinitro-3, see Dinoseb
Dinitrobutylphenol, see Dinoseb
2-Dinitro-6-*sec*-butylphenol, see Dinoseb
4,6-Dinitro-2-*sec*-butylphenol, see Dinoseb
4,6-Dinitro-*o*-*sec*-butylphenol, see Dinoseb
Dinitrocresol, see 4,6-Dinitro-*o*-cresol
Dinitro-*o*-cresol, see 4,6-Dinitro-*o*-cresol
2,4-Dinitro-*o*-cresol, see 4,6-Dinitro-*o*-cresol
3,5-Dinitro-*o*-cresol, see 4,6-Dinitro-*o*-cresol
Dinitrodendtroxal, see 4,6-Dinitro-*o*-cresol
2,6-Dinitro-*N,N*-dipropyl-4-trifluoromethylaniline, see Trifluralin
2,6-Dinitro-*N,N*-dipropyl-4-(trifluoromethyl)benzenamine, see Trifluralin
2,6-Dinitro-*N,N*-di-*n*-propyl-α,α,α-trifluoro-*p*-toluidine, see Trifluralin
3,5-Dinitro-2-hydroxytoluene, see 4,6-Dinitro-*o*-cresol
Dinitrol, see 4,6-Dinitro-*o*-cresol
Dinitromethyl cyclohexyltrienol, see 4,6-Dinitro-*o*-cresol
2,4-Dinitro-2-methylphenol, see 4,6-Dinitro-*o*-cresol
2,4-Dinitro-6-methylphenol, see 4,6-Dinitro-*o*-cresol
4,6-Dinitro-2-methylphenol, see 4,6-Dinitro-*o*-cresol
4,6-Dinitro-2-(1-methyl-*n*-propyl)phenol, see Dinoseb
Dinitrosol, see 4,6-Dinitro-*o*-cresol
Dinoc, see 4,6-Dinitro-*o*-cresol
Dinoseb
S-((1,3-Dioxo-2*H*-isoindol-2-yl)methyl) *O,O*-dimethyl phosphorodithioate, see Phosmet
Dinoxol, see 2,4-D, 2,4,5-T
Dinurania, see 4,6-Dinitro-*o*-cresol
Dion, see Diuron
Diphenamid
Diphenamide, see Diphenamid
Diphenyl, see Biphenyl
Diphenylamide, see Diphenamid
Diphenyltrichloroethane, see *p,p'*-DDT
Diphosphoric acid tetraethyl ester, see Tetraethyl pyrophosphate
Dipofene, see Diazinon
Dipram, see Propanil
Dipropetryn
Dipropetryne, see Dipropetryn

4-(Di-*n*-propylamino)-3,5-dinitro-1-trifluoromethylbenzene, see Trifluralin
Dipropylcarbamothioic acid *S*-ethyl ester, see EPTC
N,N-Di-*n*-propyl-2,6-dinitro-4-trifluoromethylaniline, see Trifluralin
N,N-Dipropylthiocarbamic acid *S*-ethyl ester, see EPTC
N,N-Dipropyl-4-trifluoromethyl-2,6-dinitroaniline, see Trifluralin
Dipterax, see Trichlorfon
Dipterex, see Trichlorfon
Dipterex 50, see Trichlorfon
Diptevur, see Trichlorfon
Dipthal, see Triallate
Diquat
Direx, see Diuron
Direx 4L, see Diuron
Direz, see Anilazine
Disulfaton, see Disulfoton
Disulfoton
Disyston, see Disulfoton
Disystox, see Disulfoton
Dithane M 45, see Mancozeb
Dithane S 60, see Mancozeb
Dithane SPC, see Mancozeb
Dithane ultra, see Mancozeb
Dithio, see Sulfotepp
α,α′-Dithiobis(dimethylthio) formamide, see Thiram
Dithiocarbonic anhydride, see Carbon disulfide
Dithiodemeton, see Disulfoton
N,N′-(Dithiodicarbonothioyl)bis(*N*-methylmethanamine), see Thiram
Dithione, see Sulfotepp
Dithiophos, see Sulfotepp
Dithiophosphoric acid tetraethyl ester, see Sulfotepp
Dithiosystox, see Disulfoton
Dithiotep, see Sulfotepp
Di(tri(2,2-dimethyl-2-phenylpropyl)tin)oxide, see Fenbutatin oxide
Ditrifon, see Trichlorfon
Diurex, see Diuron
Diurol, see Amitrole, Diuron
Diurol 5030, see Amitrole
Diuron
Diuron 4L, see Diuron
Diuron 80W, see Diuron
Divipan, see Dichlorvos
Dixon, see Phosphamidon
Dizinon, see Diazinon
DMA-4, see 2,4-D
DMDT, see Methoxychlor
4,4′-DMDT, see Methoxychlor
p,p′-DMDT, see Methoxychlor
DMP, see Dimethyl phthalate
DMSP, Fensulfothion

DMTD, see Methoxychlor
DMTP, see Fenthion, Methidathion
DMTT, see Dazomet
DMU, see Diuron
DN, see 4,6-Dinitro-*o*-cresol
DN 289, see Dinoseb
DNBP, see Dinoseb
DNC, see 4,6-Dinitro-*o*-cresol
DN-dry mix no. 2, see 4,6-Dinitro-*o*-cresol
DNOC, see 4,6-Dinitro-*o*-cresol
Dnosbp, see Dinoseb
DNSBP, see Dinoseb
DNTP, see Parathion
Dodat, see *p,p'*-DDT
Dolco mouse cereal, see Strychnine
Dol granule, see Lindane
Dolochlor, see Chloropicrin
Domatol, see Amitrole
Domatol 88, see Amitrole
Dormone, see 2,4-D
Dowcide 7, see Pentachlorophenol
Dowco 163, see Nitrapyrin
Dowco 179, see Chlorpyrifos
Dowco 233, see Triclopyr
Dow ET 14, see Ronnel
Dow ET 57, see Ronnel
Dowfume, see Methyl bromide
Dowfume 40, see Ethylene dibromide
Dowfume EDB, see Ethylene dibromide
Dowfume MC-2, see Methyl bromide
Dowfume MC-2 soil fumigant, see Methyl bromide
Dowfume MC-33, see Methyl bromide
Dowfume W-8, see Ethylene dibromide
Dowfume W-85, see Ethylene dibromide
Dowfume W-90, see Ethylene dibromide
Dowfume W-100, see Ethylene dibromide
Dow general, see Dinoseb
Dow general weed killer, see Dinoseb
Dowicide 7, see Pentachlorophenol
Dowicide EC-7, see Pentachlorophenol
Dowicide G, see Pentachlorophenol
Dowklor, see Chlordane
Dow MCP amine weed killer, see MCPA
Dow pentachlorophenol DP-2 antimicrobial, see Pentachlorophenol
Dowpon, see Dalapon-sodium
Dow selective weed killer, see Dinoseb
DPA, see Propanil
2,2-DPA, see Dalapon-sodium
DPP, see Parathion

DPX 1108, see Fosamine-ammonium
DPX 1410, see Oxamyl
DPX 3674, see Hexazinone
DPX 4189, see Chlorsulfuron
DPX 5648, see Sulfometuron-methyl
DPX M6316, see Thiameturon-methyl
Draza, see Methiocarb
Drexel parathion 8E, see Parathion
Drexel-super P, see Maleic hydrazide
Drinox, see Aldrin, Heptachlor
Drinox H-34, see Heptachlor
Dropp, see Thidiazuron
DTMC, see Dicofol
DU 112307, see Diflubenzuron
Dual, see Metolachlor
Dual 8E, see Metolachlor
Dual 25G, see Metolachlor
Duo-kill, see Crotoxyphos, Dichlorvos
Du Pont 326, see Linuron
Du Pont 732, see Terbacil
Du Pont 1179, see Methomyl
Du Pont 1991, see Benomyl
Du Pont herbicide 326, see Linuron
Du Pont herbicide 732, see Terbacil
Du Pont herbicide 976, see Bromacil
Du Pont herbicide 1318, see Siduron
Du Pont insecticide 1179, see Methomyl
Dusprex, see Dichlobenil
Duraphos, see Mevinphos
Duravos, see Crotoxyphos, Dichlorvos
Durotox, see Pentachlorophenol
Dursban, see Chlorpyrifos
Dursban F, see Chlorpyrifos
DW 3418, see Cyanazine
Dyanap, see Naptalam
Dycarb, see Bendiocarb
Dyclomec, see Dichlobenil
Dyclomec G2, see Dichlobenil
Dyclomec G4, see Dichlobenil
Dyfonate, see Fonofos
Dykol, see *p,p'*-DDT
Dylox, see Trichlorfon
Dylox-metasystox-R, see Trichlorfon
Dymid, see Diphenamid
Dynex, see Diuron
Dyphonate, see Fonofos
Dyrene, see Anilazine
Dyrene 50W, see Anilazine
Dyrex, see Trichlorfon

Emulsamine E-3, see 2,4-D
Endocel, see α-Endosulfan, β-Endosulfan
3,6-Endooxohexahydrophthalic acid, see Endothall
Endosol, see α-Endosulfan, β-Endosulfan
Endosulfan, see α-Endosulfan, β-Endosulfan
Endosulfan I, see a-Endosulfan
Endosulfan II, see β-Endosulfan
α-Endosulfan
β-Endosulfan
Endosulfan sulfate
Endosulphan, see α-Endosulfan, see β-Endosulfan
Endothal, see Endothall
Endothall
Endrex, see Endrin
Endrin
Endyl, see Carbophenothion
Enide, see Diphenamid
Enide 50, see Diphenamid
Enide 90, see Diphenamid
ENT 54, see Acrylonitrile
ENT 154, see 4,6-Dinitro-*o*-cresol
ENT 262, see Dimethyl phthalate
ENT 987, see Thiram
ENT 1122, see Dinoseb
ENT 1506, see *p,p'*-DDT
ENT 1716, see Methoxychlor
ENT 3776, see Dichlone
ENT 4225, see *p,p'*-DDD
ENT 4504, see Bis(2-chloroethyl)ether
ENT 4705, see Carbon tetrachloride
ENT 7796, see Lindane
ENT 8538, see 2,4-D
ENT 9232, see α-BHC
ENT 9233, see β-BHC
ENT 9234, see δ-BHC
ENT 9735, see Toxaphene
ENT 9932, see Chlordane
ENT 14689, see Ferbam
ENT 15108, see Parathion
ENT 15152, see Heptachlor
ENT 15349, see Ethylene dibromide
ENT 15406, see 1,2-Dichloropropane
ENT 15949, see Aldrin
ENT 16225, see Dieldrin
ENT 16273, see Sulfotepp
ENT 16391, see Kepone
ENT 16894, see Aspon
ENT 17034, see Malathion
ENT 17251, see Endrin

ENT 27396, see Methamidophos
ENT 27567, see Chlordimeform
ENT 27572, see Fenamiphos
ENT 27738, see Fenbutatin oxide
ENT 27822, see Acephate
ENT 29054, see Diflubenzuron
Entex, see Fenthion
Entomoxan, see Lindane
Envert 171, see 2,4-D
Envert DT, see 2,4-D
Envert T, see 2,4,5-T
EP 30, see Pentachlorophenol
EP 161E, see Methylisothiocyanate
EP 333, see Chlordimeform
EP 452, see Phenmedipham
EP 475, see Desmedipham
Epal, see Fosetyl-aluminum
EPN
EPN 300, see EPN
3,6-Epoxycyclohexane-1,2-dicarboxylic acid, see Endothall
3,6-*endo*-Epoxy-1,2-cyclohexanedicarboxylic acid, see Endothall
Epoxy heptachlor, see Heptachlor epoxide
Eprofil, see Thiabendazole
Eptam, see EPTC
EPTC
Equigard, see Dichlorvos
Equigel, see Dichlorvos
Equino-acid, see Trichlorfon
Equino-aid, see Trichlorfon
Equizole, see Thiabendazole
Erade, see Quinomethionate
Eradex, see Chlorpyrifos
Erazidon, see Quinomethionate
Erban, see Propanil
Erbanil, see Propanil
Escort, see Metsulfuron-methyl
Esfenvalerate
Essofungicide 406, see Captan
Estercide T-2 and T-245, see 2,4,5-T
Esteron, see 2,4-D, 2,4,5-T
Esteron 245 BE, see 2,4,5-T
Esteron 76 BE, see 2,4-D
Esteron 99, see 2,4-D
Esteron 99 concentrate, see 2,4-D
Esteron 44 weed killer, see 2,4-D
Esteron brush killer, see 2,4-D, 2,4,5-T
Esterone 4, see 2,4-D
Esterone 245, see 2,4,5-T
Estonate, see *p,p'*-DDT

Estone, see 2,4-D
Estonox, see Toxaphene
Estrosel, see Dichlorvos
Estrosol, see Dichlorvos
ET 14, see Ronnel
ET 57, see Ronnel
((1,2-Ethanediylbis(carbamodithioato))(2-))manganese mixture with ((1,2-
    ethanediylbis(carbamodithioato))(2-))zinc, see Mancozeb
Ethanox, see Ethion
Ethel, see Ethephon
Ethephon
Etheverse, see Ethephon
Ethiofencarb
Ethiol, see Ethion
Ethiolacar, see Malathion
Ethion
Ethiophencarp, see Ethiofencarb
Ethlon, see Parathion
Ethodan, see Ethion
Ethoprop
Ethoprophos, see Ethoprop
3-Ethoxycarbonylaminophenyl-*N*-phenylcarbamate, see Desmedipham
2-((Ethoxy((1-methylethyl)amino)phosphinothioyl)oxy)benzoic acid 1-methylethyl ester,
    see Isofenphos
Ethoxy-4-nitrophenoxy phenylphosphine sulfide, see EPN
Ethrel, see Ethephon
2-Ethylamino-4-isopropylamino-6-methylmercapto-*s*-triazine, see Ametryn
2-Ethylamino-4-isopropylamino-6-methylthio-1,3,5-triazine, see Ametryn
2-Ethylamino-4-isopropylamino-6-methylthio-*s*-triazine, see Ametryn
S-Ethyl azepane-1-carbothioate, see Molinate
S-Ethyl bis(2-methylpropyl)carbamothioate, see Butylate
(±)-Ethyl 2-(4-((6-chloro-2-benzoxazolyl)oxy)phenoxy) propanoate, see Fenoxapropethyl
Ethyl 4-chloro-α-(4-chlorophenyl)-α-hydroxybenzene acetate, see Chlorobenzilate
S-Ethyl cyclohexylethylcarbamothioate, see Cycloate
Ethyl-*p,p′*-dichlorobenzilate, see Chlorobenzilate
Ethyl-*p,p′*-dichlorodiphenyl glycollate, see Chlorobenzilate
Ethyl *N,N*-diisobutylthiocarbamate, see Butylate
S-Ethyl diisobutylthiocarbamate, see Butylate
S-Ethyl *N,N*-diisobutylthiocarbamate, see Butylate
Ethyl-*N,N*-diisobutylthiolcarbamate, see Butylate
S-Ethyl dipropylcarbamothioate, see EPTC
O-Ethyl *S,S*-dipropyl phosphorodithioate, see Ethoprop
S-Ethyl dipropylthiocarbamate, see EPTC
S-Ethyl di-*n*-propylthiocarbamate, see EPTC
S-Ethyl-*N,N*-di-*n*-propylthiolcarbamate, see EPTC
Ethylene aldehyde, see Acrolein
1,1-Ethylene-2,2-bipyridylium dibromide, see Diquat
1,1′-Ethylene-2,2′-bipyridylium dibromide, see Diquat
Ethylene bromide, see Ethylene dibromide

Ethylene bromide glycol dibromide, see Ethylene dibromide
Ethylene dibromide
1,2-Ethylene dibromide, see Ethylene dibromide
Ethyl-4,4-dichlorobenzilate, see Chlorobenzilate
Ethyl-4,4′-dichlorodiphenyl glycollate, see Chlorobenzilate
S-Ethyl *N*-ethyl *N*-cyclohexylthiolcarbamate, see Cycloate
O,O-Ethyl *S*-2-((ethylthio)ethyl) phosphorodithioate, see Disulfoton
S-Ethyl hexahydro-1*H*-azepine-1-carbothioate, see Molinate
S-Ethyl *N,N*-hexamethylenethiocarbamate, see Molinate
Ethyl-2-hydroxy-2,2-bis(4-chlorophenyl)acetate, see Chlorobenzilate
O-Ethyl *O*-(2-isopropoxycarbonyl)phenyl isopropylphosphoramidothioate, see Isofenphos
N-Ethyl-*N′*-isopropyl-6-methylthio-1,3,5-triazine-2,4-diyldiamine, see Ametryn
2-Ethylmercapotomethylphenyl-*N*-methylcarbamate, see Ethiofencarb
Ethyl methylene phosphorodithioate, see Ethion
N-Ethyl-*N′*-(1-methylethyl)-6-(methylthio)-1,3,5-triazine-2,4-diamine, see Ametryn
O-Ethyl *O*-(4-(methylmercapto)phenyl) *S-n*-propylphosphorothionothiolate, see Sulprofos
2-Ethyl-6-methyl-1-*N*-(2-methoxy-1-methylethyl)chloroacetanilide, see Metolachlor
Ethyl 3-methyl-4-(methylthio)phenyl (1-methylethyl)phosphoramidate, see Fenamiphos
O-Ethyl *O*-(4-methylthio)phenyl) phosphorodithioic acid *S*-propyl ester, see Sulprofos
O-Ethyl *O*-(4-methylthio)phenyl) *S*-propyl phosphorodithioate, see Sulprofos
Ethyl-4-methylthio-*m*-tolyl isopropyl phosphoramidate, see Fenamiphos
Ethyl *p*-nitrophenyl benzenethionophosphate, see EPN
Ethyl *p*-nitrophenyl benzenethiophosphonate, see EPN
Ethyl *p*-nitrophenyl ester, see EPN
Ethyl *p*-nitrophenyl phenylphosphonothioate, see EPN
O-Ethyl *O*-4-nitrophenyl phenylphosphonothioate, see EPN
O-Ethyl *O-p*-nitrophenyl phenylphosphonothioate, see EPN
Ethyl *p*-nitrophenyl thionobenzenephosphate, see EPN
Ethyl *p*-nitrophenyl thionobenzenephosphonate, see EPN
Ethyl parathion, see Parathion
S-Ethyl perhydroazepine-1-thiocarboxylate, see Molinate
Ethyl 3-phenylcarbamoyloxyphenylcarbamate, see Desmedipham
O-Ethyl *S*-phenyl ethyldithiophosphonate, see Fonofos
O-Ethyl *S*-phenyl ethylphosphonodithioate, see Fonofos
O-Ethyl phenyl *p*-nitrophenyl phenylphosphorothioate, see EPN
O-Ethyl phenyl *p*-nitrophenyl thiophosphonate, see EPN
Ethylphosphonodithioic acid *O*-ethyl *S*-phenyl ester, see Fonofos
N-(1-Ethylpropyl)-3,4-dimethyl-2,6-dinitrobenzenamine, see Pendimethalin
N-(1-Ethylpropyl)-2,6-dinitro-3,4-xylidine, see Pendimethalin
Ethyl pyrophosphate, see Tetraethyl pyrophosphate
S-2-(Ethylsulfinyl)ethyl *O,O*-dimethyl phosphorothioate, see Oxydemeton-methyl
2-Ethylthio-4,6-bis(isopropylamino)-*s*-triazine, see Dipropetryn
6-(Ethylthio)-*N,N′*-bis(1-methylethyl)-1,3,5-triazine-2,4-diamine, see Dipropetryn
S-2-(Ethylthio)ethyl *O,O*-diethyl ester of phosphorodithioic acid, see Disulfoton
2-((Ethylthio)methyl)phenol methylcarbamate, see Ethiofencarb
2-Ethylthiomethylphenyl methylcarbamate, see Ethiofencarb
2-Ethylthiomethylphenyl-*N*-methylcarbamate, see Ethiofencarb
Ethyl thiopyrophosphate, see Sulfotepp
α-Ethylthio-*o*-tolyl methylcarbamate, see Ethiofencarb

Etilon, see Parathion
Etiol, see Malathion
Etrolene, see Ronnel
Eurex, see Cycloate
Evik, see Ametryn
Evik 80W, see Ametryn
Exagama, see Lindane
Exmin, see Permethrin
Exotherm, see Chlorothalonil
Exotherm termil, see Chlorothalonil
Experimental herbicide 732, see Terbacil
Experimental insecticide 269, see Endrin
Experimental insecticide 3911, see Phorate
Experimental insecticide 4049, see Malathion
Experimental insecticide 7744, see Carbaryl
Experimental insecticide 12880, see Dimethoate
Extermathion, see Malathion
Extrar, see 4,6-Dinitro-*o*-cresol
E-Z-Off, see Butifos
F 735, see Carboxin
F 1991, see Benomyl
F 2966, see Mancozeb
FA, see Formaldehyde
Fair-2, see Maleic hydrazide
Fair-30, see Maleic hydrazide
Fairplus, see Maleic hydrazide
Fair PS, see Maleic hydrazide
Falitram, see Thiram
Famfos, see Phosphamidon
Fannoform, see Formaldehyde
Far-Go, see Triallate
Farmco, see 2,4-D
Farmco atrazine, see Atrazine
Farmco diuron, see Diuron
Farmco fence rider, see 2,4,5-T
Farmco propanil, see Propanil
Fartox, see Pentachloronitrobenzene
Fasciolin, see Carbon tetrachloride
Fasco fascrat powder, see Warfarin
Fasco terpene, see Toxaphene
Fasco Wy-hoe, see Chlorpropham
Fatal, see Chlorthal-dimethyl
FB/2, see Diquat
FDA 1541, see EPTC
FDN, see Diphenamid
Fecama, see Dichlorvos
Felan, see Molinate
Fen-All, see 2,3,5-Trichlorobenzoic acid

Fenam, see Diphenamid
Fenamin, see Atrazine
Fenamine, see Amitrole, Atrazine
Fenamiphos
Fenatrol, see Atrazine
Fenavar, see Amitrole
Fenbutatin oxide
Fence rider, see 2,4,5-T
Fenchchlorphos, see Ronnel
Fenchlorfos, see Ronnel
Fenchlorophos, see Ronnel
Fender, see Phenmedipham
Fenmedifam, see Phenmedipham
Fennosan B 100, see Dazomet
Fenoxaprop-ethyl
Fensulfothion
Fenthion
Fenvalerate
Ferbam
Ferbam 50, see Ferbam
Ferbame, see Ferbam
Ferbame, iron salt, see Ferbam
Ferbeck, see Ferbam
Ferberk, see Ferbam
Ferkethion, see Dimethoate
Fermate, see Ferbam
Fermate ferbam fungicide, see Ferbam
Fermide, see Thiram
Fermine, see Dimethyl phthalate
Fermocide, see Ferbam
Fernacol, see Thiram
Fernasan, see Thiram
Fernasan A, see Thiram
Fernesta, see 2,4-D
Fernide, see Thiram
Fernimine, see 2,4-D
Fernoxone, see 2,4-D
Ferradow, see Ferbam
Ferric dimethyldithiocarbamate, see Ferbam
Ferxone, see 2,4-D
Ficam, see Bendiocarb
Ficam D, see Bendiocarb
Ficam ULV, see Bendiocarb
Ficam W, see Bendiocarb
Finaven, see Difenzoquat methyl sulfate
Fip, see Dimethoate
Fisons NC 2964, see Methidathion
Flibol E, see Trichlorfon

Flit 406, see Captan
Flo pro T seed protectant, see Thiram
Flo pro V seed protectant, see Carboxin
Flordimex, see Ethephon
Florel, see Ethephon
Flucythrinate
Flukoids, see Carbon tetrachloride
Fluometuron
Fly-die, see Dichlorvos
Fly fighter, see Dichlorvos
FMC 1240, see Ethion
FMC 5462, see α-Endosulfan, β-Endosulfan
FMC 30980, see Cypermethrin
FMC 33297, see Permethrin
FMC 41665, see Permethrin
FMC 45497, see Cypermethrin
FMC 45806, see Cypermethrin
Folbex, see Chlorobenzilate
Folbex smoke-strips, see Chlorobenzilate
Folidol, see Parathion
Folidol E605, see Parathion
Folidol E & E 605, see Parathion
Folosan, see Pentachloronitrobenzene
Fomac 2, see Pentachloronitrobenzene
Fonofos
Fonophos, see Fonofos
Fore, see Mancozeb
Foredex 75, see 2,4-D
Forlin, see Lindane
Formal, see Malathion
Formaldehyde
Formalin, see Formaldehyde
Formalin 40, see Formaldehyde
Formalith, see Formaldehyde
Formic aldehyde, see Formaldehyde
Formol, see Formaldehyde
Formula 40, see 2,4-D
Forotox, see Trichlorfon
Forron, see 2,4,5-T
Forstan, see Quinomethionate
Forst U 46, see 2,4,5-T
Fortex, see 2,4,5-T
Forthion, see Malathion
Fortion NM, see Dimethoate
Fortrol, see Cyanazine
Forturf, see Chlorothalonil
Fosamine-ammonium
Foschlor, see Trichlorfon
Foschlor 25, see Trichlorfon

Foschlor R 50, see Trichlorfon
Fosdrin, see Mevinphos
Fosetyl AL, see Fosetyl-aluminum
Fosetyl-aluminum
Fos-fall 'A', see Butifos
Fosfamid, see Dimethoate
Fosfermo, see Parathion
Fosferno, see Parathion
Fosfex, see Parathion
Fosfive, see Parathion
Fosfono 50, see Ethion
Fosfothion, see Malathion
Fosfotion, see Malathion
Fosfotox, see Dimethoate
Fosfotox R, see Dimethoate
Fosfotox R 35, see Dimethoate
Fosova, see Parathion
Fostern, see Parathion
Fostion MM, see Dimethoate
Fostox, see Parathion
Fosvex, see Tetraethyl pyrophosphate
Fozalon, see Phosalone
Framed, see Simazine
Freon 10, see Carbon tetrachloride
Fruitone A, see 2,4,5-T
Frumin AL, see Disulfoton
Frumin G, see Disulfoton
Fuklasin, see Ferbam
Fuklasin ultra, see Ferbam
Fumagon, see 1,2-Dibromo-3-chloropropane
Fumazone, see 1,2-Dibromo-3-chloropropane
Fumazone 86, see 1,2-Dibromo-3-chloropropane
Fumazone 86E, see 1,2-Dibromo-3-chloropropane
Fumigant-1, see Methyl bromide
Fumigrain, see Acrylonitrile
Fumogas, see Ethylene dibromide
Fundal, see Chlordimeform
Fundal 500, see Chlordimeform
Fundasol, see Benomyl
Fundazol, see Benomyl
Fundex, see Chlordimeform
Fungicide 1991, see Benomyl
Fungiclor, see Pentachloronitrobenzene
Fungifen, see Pentachlorophenol
Fungol, see Pentachlorophenol
Fungus ban type II, see Captan
Furadan, see Carbofuran
Furloe, see Chlorpropham
Furloe 4EC, see Chlorpropham

FW 293, see Dicofol
FW 734, see Propanil
Fyde, see Formaldehyde
Fyfanon, see Malathion
G 25, see Chloropicrin
G 301, see Diazinon
G 338, see Chlorobenzilate
G 996, see Ethephon
G 23992, see Chlorobenzilate
G 24480, see Diazinon
G 27692, see Simazine
G 30027, see Atrazine
G 30028, see Propazine
G 31435, see Prometon
G 34161, see Prometryn
G 34162, see Ametryn
Galecron, see Chlordimeform
Gallogama, see Lindane
Gamacid, see Lindane
Gamaphex, see Lindane
Gamene, see Lindane
Gamiso, see Lindane
Gammahexa, see Lindane
Gammalin, see Lindane
Gammexene, see Lindane
Gammopaz, see Lindane
Gamonil, see Carbaryl
Gardentox, see Diazinon
Garlon, see Triclopyr
Garnitan, see Linuron
Garrathion, see Carbophenothion
Garvox, see Bendiocarb
GC 1189, see Kepone
GC 3944-3-4, see Pentachloronitrobenzene
Gearphos, see Parathion
Gebutox, see Dinoseb
Geigy 338, see Chlorobenzilate
Geigy 13005, see Methidathion
Geigy 24480, see Diazinon
Geigy 27692, see Simazine
Geigy 30027, see Atrazine
Geigy 30028, see Propazine
Geigy GS-13005, see Methidathion
Genate Plus 6.7EC, see Butylate
General chemicals 1189, see Kepone
Geniphene, see Toxaphene
Genithion, see Parathion
Genitox, see *p,p'*-DDT
Germain's, see Carbaryl

Gesafid, see *p,p'*-DDT
Gesafram, see Prometon
Gesafram 50, see Prometon
Gesagard, see Prometryn
Gesamil, see Propazine
Gesapax, see Ametryn
Gesapon, see *p,p'*-DDT
Gesaprim, see Atrazine
Gesaran, see Simazine
Gesarex, see *p,p'*-DDT
Gesarol, see *p,p'*-DDT
Gesatop, see Simazine
Gesatop 50, see Simazine
Gesfid, see Mevinphos
Gesoprim, see Atrazine
Gestid, see Mevinphos
Gexane, see Lindane
Glazd penta, see Pentachlorophenol
Glean, see Chlorsulfuron
Glean 20DF, see Chlorsulfuron
Glycol bromide, see Ethylene dibromide
Glycol dibromide, see Ethylene dibromide
Glycophen, see Iprodione
Glycophene, see Iprodione
Glyodex 3722, see Captan
Glyphosate
Goliath, see Phenmedipham
Gothnion, see Azinphos-methyl
GPKh, see Heptachlor
Gramevin, see Dalapon-sodium
Granox NM, see Hexachlorobenzene
Granox PFM, see Captan
Granurex, see Neburon
Granutox, see Phorate
Grascide, see Propanil
Graslan, see Tebuthiuron
Grazon, see Picloram
Green-daisen M, see Mancozeb
Griffex, see Atrazine
Grisol, see Tetraethyl pyrophosphate
Grundier arbezol, see Pentachlorophenol
GS 13005, see Methidathion
GS 14260, see Terbutryn
GS 16068, see Dipropetryn
Guesapon, see *p,p'*-DDT
Gusathion, see Azinphos-methyl
Gusathion 20, see Azinphos-methyl
Gusathion 25, see Azinphos-methyl
Gusathion K, see Azinphos-methyl

Gusathion M, see Azinphos-methyl
Gusathion methyl, see Azinphos-methyl
Gusto, see Phenmedipham
Gustafson captan 30-DD, see Captan
Guthion, see Azinphos-methyl
Gyphene, see Toxaphene
Gyron, see *p,p'*-DDT
H 34, see Heptachlor
H 133, see Dichlobenil
H 321, see Methiocarb
H 1313, see Dichlobenil
H 1318, see Siduron
H 1803, see Simazine
Halon 104, see Carbon tetrachloride
Halon 1001, see Methyl bromide
Hamidop, see Methamidophos
Harmony, see Thiameturon-methyl
Havero-extra, see *p,p'*-DDT
Hazodrin, see Monocrotophos
HC 1281, see 2,3,5-Trichlorobenzoic acid
HCB, see Hexachlorobenzene
HCCH, see Lindane
HCE, see Heptachlor epoxide
HCH, see Lindane
α-HCH, see α-BHC
β-HCH, see β-BHC
δ-HCH, see δ-BHC
γ-HCH, see Lindane
HCS 3260, see Chlordane
Heclotox, see Lindane
Hedapur M 52, see MCPA
Hederax M, see MCPA
Hedolit, see 4,6-Dinitro-*o*-cresol
Hedolite, see 4,6-Dinitro-*o*-cresol
Hedonal, see 2,4-D
Hedonal M, see MCPA
Hel-fire, see Dinoseb
Helothion, see Sulprofos
HEOD, see Dieldrin
Hept, see Tetraethyl pyrophosphate
Heptachlor
Heptachlorane, see Heptachlor
Heptachlor epoxide
3,4,5,6,7,8,8-Heptachlorodicyclopentadiene, see Heptachlor
3,4,5,6,7,8,8a-Heptachlorodicyclopentadiene, see Heptachlor
3,4,5,6,7,8,8a-Heptachloro-α-dicyclopentadiene, see Heptachlor
1,2,3,4,5,6,7,8,8-Heptachloro-2,3-epoxy-3a,4,7,7a-tetrahydro-4,7-methanoindene, see
    Heptachlor epoxide

1,4,5,6,7,8,8-Heptachloro-2,3-epoxy-2,3,3a,4,7,7a-hexahydro-4,7-methanoindene, see
    Heptachlor epoxide
2,3,4,5,6,7,7-Heptachloro-1a,1b,5,5a,6,6a-hexahydro-2,5-methano-2*H*-oxireno[*a*]-
    indene, see Heptachlor
1(3a),4,5,6,7,8,8-Heptachloro-3a(1),4,7,7a-tetrahydro-4,7-methanoindene, see Heptachlor
1,4,5,6,7,8,8-Heptachloro-3a,4,7,7a-tetrahydro-4,7-methanoindene, see Heptachlor
1,4,5,6,7,8,8a-Heptachloro-3a,4,7,7a-tetrahydro-4,7-methanoindene, see Heptachlor
1,4,5,6,7,8,8-Heptachloro-3a,4,7,7a-tetrahydro-4,7-*endo*-methanoindene, see Heptachlor
1,4,5,6,7,8,8-Heptachloro-3a,4,7,7a-tetrahydro-4,7-methanol-1*H*-indene, see Heptachlor
1,4,5,6,7,8,8-Heptachloro-3a,4,7,7a-tetrahydro-4,7-methyleneindene, see Heptachlor
1,4,5,6,7,10,10-Heptachloro-4,7,8,9-tetrahydro-4,7-methyleneindene, see Heptachlor
1,4,5,6,7,10,10-Heptachloro-4,7,8,9-tetrahydro-4,7-*endo*-methyleneindene, see
    Heptachlor
Heptadichlorocyclopentadiene, see Heptachlor
Heptagran, see Heptachlor
Heptagranox, see Heptachlor
Heptamak, see Heptachlor
Heptamul, see Heptachlor
Heptasol, see Heptachlor
Heptox, see Heptachlor
Herbadox, see Pendimethalin
Herbatox, see Diuron
Herbax, see Propanil
Herbax technical, see Propanil
Herbazin, see Simazine
Herbazin 50, see Simazine
Herbex, see Simazine
Herbicide 326, see Linuron
Herbicide 976, see Bromacil
Herbicide C 2059, see Fluometuron
Herbicide M, see MCPA
Herbidal, see 2,4-D
Herbidal total, see Amitrole
Herbizole, see Amitrole
Herboxy, see Simazine
Hercules 3956, see Toxaphene
Hercules 14503, see Dialifos
Hercules toxaphene, see Toxaphene
Herkal, see Dichlorvos
Herkol, see Dichlorvos
Hermal, see Thiram
Hermat TMT, see Thiram
Heryl, see Thiram
Hexa, see Lindane
Hexacap, see Captan
Hexa C.B., see Hexachlorobenzene
γ-Hexachlor, see Lindane
Hexachloran, see Lindane
α-Hexachloran, see α-BHC

γ-Hexachloran, see Lindane
Hexachlorane, see Lindane
α-Hexachlorane, see α-BHC
γ-Hexachlorane, see Lindane
α-Hexachlorcyclohexane, see α-BHC
Hexachlorobenzene
β-Hexachlorobenzene, see β-BHC
γ-Hexachlorobenzene, see Lindane
1,2,3,7,7-Hexachlorobicyclo[2.2.1]-2-heptene-5,6-bisoxymethylene sulfite, see
    α-Endosulfan, β-Endosulfan
α,β-1,2,3,7,7-Hexachlorobicyclo[2.2.1]-2-heptene-5,6-bisoxymethylene sulfite, see
    α-Endosulfan, β-Endosulfan
1,2,3,4,5,6-Hexachlorocyclohexane, see Lindane
α-Hexachlorocyclohexane, see α-BHC
β-Hexachlorocyclohexane, see β-BHC
δ-Hexachlorocyclohexane, see δ-BHC
γ-Hexachlorocyclohexane, see Lindane
β-1,2,3,4,5,6-Hexachlorocyclohexane, see β-BHC
γ-1,2,3,4,5,6-Hexachlorocyclohexane, see Lindane
1,2,3,4,5,6-Hexachloro-α-cyclohexane, see α-BHC
1,2,3,4,5,6-Hexachloro-β-cyclohexane, see β-BHC
1,2,3,4,5,6-Hexachloro-δ-cyclohexane, see δ-BHC
1,2,3,4,5,6-Hexachloro-γ-cyclohexane, see Lindane
α-1,2,3,4,5,6-Hexachlorocyclohexane, see α-BHC
δ-(aeeeee)-1,2,3,4,5,6-Hexachlorocyclohexane, see δ-BHC
δ-1,2,3,4,5,6-Hexachlorocyclohexane, see δ-BHC
1α,2α,3α,4β,5β,6β-Hexachlorocyclohexane, see δ-BHC
1α,2α,3β,4α,5α,6β-Hexachlorocyclohexane, see Lindane
1α,2α,3β,4α,5β,6β-Hexachlorocyclohexane, see α-BHC
1α,2β,3α,4β,5α,6β-Hexachlorocyclohexane, see β-BHC
1,2,3,4,5,6-Hexachloro-*trans*-cyclohexane, see β-BHC
Hexachloroepoxyoctahydro-*endo,endo*-dimethanonaphthalene, see Endrin
Hexachloroepoxyoctahydro-*endo,exo*-dimethanonaphthalene, see Dieldrin
1,2,3,4,10,10-Hexachloro-6,7-epoxy-1,4,4a,5,6,7,8,8a-octahydro-*endo,endo*-1,4:5,8-
    dimethanonaphthalene, see Endrin
1,2,3,4,10,10-Hexachloro-6,7-epoxy-1,4,4a,5,6,7,8,8a-octahydro-1,4-*endo-exo*-5,8-
    dimethanonaphthalene, see Dieldrin
Hexachlorohexahydro-*endo,exo*-dimethanonaphthalene, see Aldrin
1,2,3,4,10,10-Hexachloro-1,4,4a,5,8,8a-hexahydro-1,4:5,8-dimethanonaphthalene, see
    Aldrin
1,2,3,4,10,10-Hexachloro-1,4,4a,5,8,8a-hexahydro-1,4-*endo,exo*-5,8-
    dimethanonaphthalene, see Aldrin
1,2,3,4,10,10-Hexachloro-1,4,4a,5,8,8a-hexahydro-*exo*-1,4-*endo*-5,8-
    dimethanonaphthalene, see Aldrin
6,7,8,9,10,10-Hexachloro-1,5,5a,6,9,9a-hexahydro-3,3-dioxide, see Endosulfan sulfate
Hexachlorohexahydromethano-2,4,3-benzodioxathiepin-3-oxide, see α-Endosulfan,
    β-Endosulfan
(3α,5αβ,6α,9α,9αβ)-6,7,8,9,10,10-Hexachloro-1,5,5a,6,9,9a-hexahydro-6,9-methano-
    2,4,3-benzodioxathiepin-3-oxide, see α-Endosulfan, β-Endosulfan

1,4,5,6,7,7-Hexachloro-5-norborene-2,3-dimethanol cyclic sulfite, see α-Endosulfan,
β-Endosulfan
3,4,5,6,9,9-Hexachloro-1a,2,2a,3,6,6a,7,7a-octahydro-2,7:3,6-dimethanonaphth[2,3-
b]oxirene, see Dieldrin
Hexadrin, see Endrin
Hexaferb, see Ferbam
1,4,4a,5,8,8a-Hexahydro-1,4-*endo,exo*-5,8-dimethanonaphthalene, see Aldrin
5a,6,6a-Hexahydro-2,5-methano-2*H*-indeno[1,2-*b*]oxirene, see Heptachlor epoxide
Hexakis(β,β-dimethylphenethyl)distannoxane, see Fenbutatin oxide
Hexakis(2-methyl-2-phenylpropyl)distannoxane, see Fenbutatin oxide
Hexamite, see Tetraethyl pyrophosphate
Hexaplas M/B, see Di-*n*-butyl phthalate
Hexathir, see Thiram
Hexatox, see Lindane
Hexaverm, see Lindane
Hexavin, see Carbaryl
Hexazinone
Hexicide, see Lindane
Hexyclan, see Lindane
Hexythiazox
Hexylthiocarbam, see Cycloate
HGI, see Lindane
HHDN, see Aldrin
Hibrom, see Naled
Hifol, see Dicofol
Hifol 18.5 EC, see Dicofol
Higalnate, see Molinate
Hildan, see α-Endosulfan, β-Endosulfan
Hilthion, see Malathion
Hilthion 25WDP, see Malathion
HOCH, see Formaldehyde
Hoe 2671, see α-Endosulfan, β-Endosulfan
Hoe 2810, see Linuron
Hoe 23408, see Diclofop-methyl
Hoe 33171, see Fenoxaprop-ethyl
Hoe-Grass, see Diclofop-methyl
Hoelon, see Diclofop-methyl
Hoelon 3EC, see Diclofop-methyl
Hokmate, see Ferbam
Hooker HRS 16; see Dienochlor
Hooker HRS 1654; see Dienochlor
Horbadox, see Pendimethalin
Hormotuho, see MCPA
Hornotuho, see MCPA
Hortex, see Lindane
HOX 1901, see Ethiofencarb
HRS 16; see Dienochlor
HRS 16A; see Dienochlor
HRS 1654; see Dienochlor

Insecticide 4049, see Malathion
Insecticide-nematocide 1410, see Oxamyl
Insectophene, see α-Endosulfan, β-Endosulfan
Inverton 245, see 2,4,5-T
Invisi-gard, see Propoxur
Ipaner, see 2,4-D
IPC, see Propham
Iprodione
Iron flowable, see Ferbam
Iron tris(dimethyldithiocarbamate), see Ferbam
Iscobrome, see Methyl bromide
Iscobrome D, see Ethylene dibromide
Isocarb, see Propoxur
Isodrin epoxide, see Endrin
Isofenphos
β-Isomer, see β-BHC
γ-Isomer, see Lindane
Isomethylsystox sulfoxide, see Oxydemeton-methyl
Isophenphos, see Isofenphos
IsoPPC, see Propham
O-Isopropoxyphenyl methylcarbamate, see Propoxur
O-Isopropoxyphenyl *N*-methylcarbamate, see Propoxur
Isopropylamino-*o*-ethyl-(4-methylmercapto)-3-methylphenyl)phosphate, see Fenamiphos
3-Isopropyl-1*H*-2,1,3-benzothiadiazin-4(3*H*)-one-2,2-dioxide, see Bentazone
1-Isopropyl carbamoyl-3-(3,5-dichlorophenyl)hydantoin, see Iprodione
Isopropyl carbanilate, see Propham
Isopropyl carbanilic acid ester, see Propham
N-Isopropyl-2-chloroacetanilide, see Propachlor
N-Isopropyl-α-chloroacetanilide, see Propachlor
Isopropyl 3-chlorocarbanilate, see Chlorpropham
Isopropyl *m*-chlorocarbanilate, see Chlorpropham
Isopropyl 3-chlorophenylcarbamate, see Chlorpropham
Isopropyl-*N*-(3-chlorophenyl)carbamate, see Chlorpropham
Isopropyl-*N*-*m*-chlorophenylcarbamate, see Chlorpropham
O-Isopropyl *N*-(3-chlorophenyl)carbamate, see Chlorpropham
2,3-Isopropylidenedioxyphenyl methylcarbamate, see Bendiocarb
O-2-Isopropyl-4-methylpyrimidinyl-*O*,*O*-diethyl phosphorothioate, see Diazinon
Isopropylmethylpyrimidinyl diethyl thiophosphate, see Diazinon
Isopropyl phenylcarbamate, see Propham
O-Isopropyl *N*-phenylcarbamate, see Propham
Isopropyl *N*-phenylurethane, see Propham
Isopropyl salicylate *O*-ester with *O*-ethylisopropylphosphoramidothioate, see Isofenphos
Isothiocyanatomethane, see Methylisothiocyanate
Isothiocyanic acid methyl ester, see Methylisothiocyanate
Isotox, see Lindane
Itopaz, see Ethion
Ivalon, see Formaldehyde
Ivoran, see *p,p'*-DDT
Ixodex, see *p,p'*-DDT

Jacutin, see Lindane
Jalan, see Molinate
JF 5705F, see Cypermethrin
Jolt, see Ethoprop
Jonnix, see Asulam
Julin's carbon chloride, see Hexachlorobenzene
K III, see 4,6-Dinitro-*o*-cresol
K IV, see 4,6-Dinitro-*o*-cresol
Kafil super, see Cypermethrin
Kamfochlor, see Toxaphene
Kamposan, see Ethephon
Kaptan, see Captan
Karamate, see Mancozeb
Karbam black, see Ferbam
Karbaspray, see Carbaryl
Karbatox, see Carbaryl
Karbofos, see Malathion
Karbosep, see Carbaryl
Karlan, see Ronnel
Karmex, see Diuron, Monuron
Karmex diuron herbicide, see Diuron
Karmex DW, see Diuron
Karsan, see Formaldehyde
Kayafume, see Methyl bromide
Kayazinon, see Diazinon
Kayazol, see Diazinon
Keltane, see Dicofol
Kelthane, see Dicofol
p,p'-Kelthane, see Dicofol
Kelthane A, see Dicofol
Kelthane dust base, see Dicofol
Kelthanehanol, see Dicofol
Kemate, see Anilazine
Kemifam, see Phenmedipham
Kemolate, see Phosmet
Kepone
Kerb, see Propyzamide
Kerb 50W, see Propyzamide
Kestrel, see Permethrin
Killax, see Tetraethyl pyrophosphate
Kilmite 40, see Tetraethyl pyrophosphate
Kiloseb, see Dinoseb
Kilsem, see MCPA
Kleer-lot, see Amitrole
Kloben, see Neburon
4K-2M, see MCPA
KMH, see Maleic hydrazide
Kobu, see Pentachloronitrobenzene
Kobutol, see Pentachloronitrobenzene

Kokotine, see Lindane
Kolphos, see Parathion
Kopfume, see Ethylene dibromide
Kop-mite, see Chlorobenzilate
Kopsol, see *p,p'*-DDT
Kop-thiodan, see α-Endosulfan, β-Endosulfan
Kopthion, see Malathion
Korlan, see Ronnel
Korlane, see Ronnel
KP 2, see Pentachloronitrobenzene
K-pin, see Picloram
Krecalvin, see Dichlorvos
Kregasan, see Thiram
Krenite, see Fosamine-ammonium
Krenite brush control agent, see Fosamine-ammonium
Kresamone, see 4,6-Dinitro-*o*-cresol
Krezone, see MCPA
Krezotol 50, see 4,6-Dinitro-*o*-cresol
Krotiline, see 2,4-D
Krovar I, see Bromacil
Krovar II, see Bromacil
Krysid, see ANTU
Kumader, see Warfarin
Kumadu, see Warfarin
Kwell, see Lindane
Kwik-kil, see Strychnine
Kwit, see Ethion
Kypchlor, see Chlordane
Kypfarin, see Warfarin
Kypfos, see Malathion
Kypthion, see Parathion
L 11/6, see Phorate
L 395, see Dimethoate
L 34314, see Diphenamid
L 36352, see Trifluralin
Lanex, see Fluometuron
Lannate, see Methomyl
Lannate L, see Methomyl
Lariat, see Alachlor
Larvacide 100, see Chloropicrin
Larvin, see Thiodicarb
Lasso, see Alachlor
Lasso II, see Alachlor
Lasso EC, see Alachlor
Lauxtol, see Pentachlorophenol
Lauxtol A, see Pentachlorophenol
Lawn-keep, see 2,4-D
Lazo, see Alachlor
Lebaycid, see Fenthion

Le captane, see Captan
Legumex DB, see MCPA
Leivasom, see Trichlorfon
Lemonene, see Biphenyl
Lendine, see Lindane
Lentox, see Lindane
Lepicron, see Thiodicarb
Lethalaire G 52, see Tetraethyl pyrophosphate
Lethalaire G 54, see Parathion
Lethalaire G 57, see Sulfotepp
Lethox, see Carbophenothion
Leuna M, see MCPA
Lexone, see Metribuzin
Lexone DF, see Metribuzin
Lexone 4L, see Metribuzin
Leyspray, see MCPA
LFA 2043, see Iprodione
Lidenal, see Lindane
Lilly 34314, see Diphenamid
Lilly 36352, see Trifluralin
Lindafor, see Lindane
Lindagam, see Lindane
Lindagrain, see Lindane
Lindagranox, see Lindane
Lindan, see Dichlorvos
Lindane
α-Lindane, see α-BHC
β-Lindane, see β-BHC
δ-Lindane, see δ-BHC
γ-Lindane, see Lindane
Lindapoudre, see Lindane
Lindatox, see Lindane
Lindosep, see Lindane
Line rider, see 2,4,5-T
Linex 4L, see Linuron
Linormone, see MCPA
Linorox, see Linuron
Lintox, see Lindane
Linurex, see Linuron
Linuron
Linuron 4L, see Linuron
Lipan, see 4,6-Dinitro-*o*-cresol
Liquatox, see Warfarin
Lirobetarex, see Monuron
Liro CIPC, see Chlorpropham
Lirohex, see Tetraethyl pyrophosphate
Liroprem, see Pentachlorophenol
Lirothion, see Parathion
Loisol, see Trichlorfon

Lombristop, see Thiabendazole
Lorex, see Linuron
Lorexane, see Lindane
Lorox, see Linuron
Lorox DF, see Linuron
Lorox L, see Linuron
Lorox linuron weed killer, see Linuron
Lorsban, see Chlorpyrifos
LS 74783, see Fosetyl-aluminum
Lurgo, see Dimethoate
Lysoform, see Formaldehyde
M 40, see MCPA
M 74, see Disulfoton
M 140, see Chlordane
M 410, see Chlordane
M 5055, see Toxaphene
M&B 10064, see Bromoxynil
M&B 10731, see Bromoxynil octanoate
Maax, see Propazine
Macrondray, see 2,4-D
Mafu, see Dichlorvos
Mafu strip, scc Dichlorvos
Magnacide, see Acrolein
MAH, see Maleic hydrazide
Maintain 3, see Maleic hydrazide
Malacide, see Malathion
Malafor, see Malathion
Malagran, see Malathion
Malakill, see Malathion
Malamar, see Malathion
Malamar 50, see Malathion
Malaphele, see Malathion
Malaphos, see Malathion
Malasol, see Malathion
Malaspray, see Malathion
Malathion
Malathion E50, see Malathion
Malathion LV concentrate, see Malathion
Malathion ULV concentrate, see Malathion
Malathiozoo, see Malathion
Malathon, see Malathion
Malathyl LV concentrate & ULV concentrate, see Malathion
Malatol, see Malathion
Malatox, see Malathion
Maldison, see Malathion
Maleic acid hydrazide, see Maleic hydrazide
Maleic hydrazide
Maleic hydrazide 30%, see Maleic hydrazide
Malein 30, see Maleic hydrazide

N,N-Maleohydrazine, see Maleic hydrazide
Malipur, see Captan
Malix, see α-Endosulfan, β-Endosulfan
Malmed, see Malathion
Malphos, see Malathion
Maltox, see Malathion
Maltox MLT, see Malathion
Malzid, see Maleic hydrazide
Mancofol, see Mancozeb
Mancozeb
Maneb-zinc, see Mancozeb
Manoseb, see Mancozeb
Manzate 200, see Mancozeb
Manzeb, see Mancozeb
Manzin, see Mancozeb
Manzin 80, see Mancozeb
Maralate, see Methoxychlor
Marfrin, see Warfarin
Marlate, see Methoxychlor
Marlate 50, see Methoxychlor
Marmer, see Diuron
Martin's mar-frin, see Warfarin
Marvex, see Dichlorvos
Masoten, see Trichlorfon
Matacil, see Aminocarb
Mataven, see Difenzoquat methyl sulfate
Maveran, see Warfarin
Maymeb-6647, see Triadimefon
Mazide, see Maleic hydrazide
Mazoten, see Trichlorfon
MB, see Methyl bromide
MB 9057, see Asulam
MB 10064, see Bromoxynil
MBC, see Benomyl
M-B-C Fumigant, see Methyl bromide
MBR 12325, see Mefluidide
MBX, see Methyl bromide
MC-4379, see Bifenox
2M-4C, see MCPA
2M-4CH, see MCPA
MCP, see MCPA
MCPA
2,4-MCPA, see MCPA
MDBA, see Dicamba
ME 1700, see *p,p'*-DDD
MEB 6447, see Triadimefon
MEBR, see Methyl bromide
ME4 Brominal, see Bromoxynil
Mediben, see Dicamba

Mefluidide
Melipax, see Toxaphene
Mendrin, see Endrin
Meniphos, see Mevinphos
Menite, see Mevinphos
Mephanac, see MCPA
Mercaptodimethur, see Methiocarb
3-(Mercaptomethyl)-1,2,3-benzotriazin-4(3*H*)-one-*O,O*-dimethyl phosphorodithioate-*S*-
    ester, see Azinphos-methyl
Mercaptophos, see Fenthion
Mercaptosuccinic acid diethyl ester, see Malathion
Mercaptothion, see Malathion
Mercuram, see Thiram
Merex, see Kepone
Merkazin, see Prometryn
Merpan, see Captan
Mertec, see Thiabendazole
Mertect, see Thiabendazole
Mertect 160, see Thiabendazole
Mesomile, see Methomyl
Mesurol, see Methiocarb
Metacetaldehyde, see Metaldehyde
Metachlor, see Alachlor
Metafume, see Methyl bromide
Metaisosystox sulfoxide, see Oxydemeton-methyl
Metalaxil, see Metalaxyl
Metalaxyl
Metaldehyde
Metamidofos estrella, see Methamidophos
Metasol TK-100, see Thiabendazole
Metasystemox, see Oxydemeton-methyl
Metasystox-R, see Oxydemeton-methyl
Metaxon, see MCPA
Metelilachlor, see Metolachlor
Methachlor, see Alachlor
Methamidophos
Methanal, see Formaldehyde
Methanedithiol-*S,S*-diester with *O,O*-diethyl phosphorodithioate, see Ethion
Methane tetrachloride, see Carbon tetrachloride
6,9-Methano-2,4,3-benzodioxathiepin, see Endosulfan sulfate
Methidathion
Methidathion 50S, see Methidathion
Methiocarb
Methogas, see Methyl bromide
Methomyl
Methoxcide, see Methoxychlor
Methoxo, see Methoxychlor
Methoxone, see MCPA
2-Methoxy-4,6-bis(isopropylamino)-1,3,5-triazine, see Prometon

6-Methoxy-*N,N'*-bis(1-methylethyl)-1,3,5-triazine-2,4-diamine, see Prometon

3-Methoxycarbonylaminophenyl 3-methylcarbanilate, see Phenmedipham

3-((Methoxycarbonyl)amino)phenyl (3-methylphenyl)carbamate, see Phenmedipham

2-Methoxycarbonyl-1-methylvinyl dimethyl phosphate, see Mevinphos

cis-2-Methoxycarbonyl-1-methylvinyl dimethyl phosphate, see Mevinphos

1-Methoxycarbonyl-1-propen-2-yl dimethyl phosphate, see Mevinphos

Methoxychlor

4,4'-Methoxychlor, see Methoxychlor

p,p'-Methoxychlor, see Methoxychlor

Methoxy-DDT, see Methoxychlor

2-Methoxy-3,6-dichlorobenzoic acid, see Dicamba

Methoxydiuron, see Linuron

1-Methoxy-1-methyl-3-(3,4-dichlorophenyl)urea, see Linuron

N-(Methoxy(methylthio)phosphinoyl)acetamide, see Acephate

2-(((((4-Methoxy-6-methyl-1,3,5-triazin-2-yl)amino)carbonyl)amino)sulfonyl)-benzoic
    acid, see Metsulfuron-methyl

3-(((((4-Methoxy-6-methyl-1,3,5-triazin-2-yl)amino)carbonyl)amino)sulfonyl)-2-
    thiophenecarboxylic acid methyl ester, see Thiameturon-methyl

S-((5-Methoxy-2-oxo-1,3,4-thiadiazol-3(2*H*)-yl)methyl) *O,O*-dimethyl
    phosphorodithioate, see Methidathion

Methoxypropazine, see Prometon

Methyl aldehyde, see Formaldehyde

Methyl *N*-(4-aminobenzenesulfonyl)carbamate, see Asulam

N-(((Methylamino)carbonyl)oxy)ethanimidothioate, see Methomyl

N-(((Methylamino)carbonyl)oxy)ethanimidothioic acid methyl ester, see Methomyl

Methyl((4-aminophenyl)sulfonyl)carbamate, see Asulam

Methylazinphos, see Azinphos-methyl

N-Methylbenzazimide, dimethyldithiophosphoric acid ester, see Azinphos-methyl

1-Methylbenzyl-3-(dimethoxyphosphinyloxo)isocrotonate, see Crotoxyphos

α-Methyl benzyl-3-(dimethoxyphosphinyloxy)-*cis*-crotonate, see Crotoxyphos

α-Methylbenzyl 3-hydroxycrotonate dimethyl phosphate, see Crotoxyphos

Methyl bromide

Methyl (1-((butylamino)carbonyl)-1*H*-benzimidazol-2-yl)carbamate, see Benomyl

Methyl 1-(butylcarbamoyl)-2-benzimidazolylcarbamate, see Benomyl

Methylcarbamate-1-naphthalenol, see Carbaryl

Methylcarbamate-1-naphthol, see Carbaryl

Methyl carbamic acid 2,3-dihydro-2,2-dimethyl-7-benzofuranyl ester, see Carbofuran

Methyl carbamic acid 4-(methylthio)-3,5-xylyl ester, see Methiocarb

Methyl carbamic acid 1-naphthyl ester, see Carbaryl

S-Methylcarbamoylmethyl *O,O*-dimethyl phosphorodithioate, see Dimethoate

N-((Methylcarbamoyl)oxy)thioacetimidic acid methyl ester, see Methomyl

2-Methyl-4-chlorophenoxyacetic acid, see MCPA

Methyl chlorophos, see Trichlorfon

1-(2-Methylcychohexyl)-3-phenylurea, see Siduron

N-(2-Methylcyclohexyl)-*N'*-phenylurea, see Siduron

Methyl demeton-*O*-sulfoxide, see Oxydemeton-methyl

Methyl 5-(2,4-dichlorophenoxy)-2-nitrobenzoate, see Bifenox

Methyl 2-(4-(2,4-dichlorophenoxy)phenoxy)propanoate, see Diclofop-methyl

Methyl 3-(dimethoxyphosphinyloxy)crotonate, see Mevinphos

Methyl 2-(dimethylamino)-*N*-(((methylamino)carbonyl)oxy)-2-oxoethanimidothioate, see
 Oxamyl
Methyl-1-(dimethylcarbamoyl)-*N*-((methylcarbamoyl)oxy)thioformimidate, see Oxamyl
S-Methyl 1-(dimethylcarbamoyl)-*N*-((methylcarbamoyl)oxy)thioformimidate, see Oxamyl
Methyl-*N′,N′*-dimethyl-*N*-((methylcarbamoyl)oxy)-1-thiooxamimidate, see Oxamyl
Methyl *N*-(2,6-dimethylphenyl)-*N*-(methoxyacetyl)-DL-alaninate, see Metalaxyl
2-Methyl-4,6-dinitrophenol, see 4,6-Dinitro-*o*-cresol
6-Methyl-2,4-dinitrophenol, see 4,6-Dinitro-*o*-cresol
6-Methyl-1,3-dithiolo[4,5-*b*]quinoxalin-2-one, see Quinomethionate
S,S′-Methylene bis(*O,O*-diethyl phosphorodithioate), see Ethion
Methylene glycol, see Formaldehyde
Methylene oxide, see Formaldehyde
S,S′-Methylene *O,O,O′,O′*-tetraethyl phosphorodithioate, see Ethion
2-(1-Methylethoxy)phenol methylcarbamate, see Propoxur
3-(1-Methylethyl)-1*H*-2,1,3-benzothiadiazin-4(3*H*)-one-2,2-dioxide, see Bentazone
1-Methylethyl (3-chlorophenyl)carbamate, see Chlorpropham
1-Methylethyl-2-((ethoxy((1-methylethyl)amino)phosphinothioyl)oxy)benzoate, see
 Isofenphos
1-(Methylethyl)ethyl 3-methyl-4-(methylthio)phenyl phosphoramidate, see Fenamiphos
N-(1-Methylethyl)-*N*-phenylacetamide, see Propachlor
1-Methylethyl phenylcarbamate, see Propham
Methyl guthion, see Azinphos-methyl
Methyl *m*-hydroxycarbanilate *m*-methylcarbanilate, see Phenmedipham
N-Methyl-2-isopropoxyphenylcarbamate, see Propoxur
Methyl isothiocyanate
2-Methylmercapto-4,6-bis(isopropylamino)-*s*-triazine, see Prometryn
4-Methylmercapto-3,5-dimethylphenyl *N*-methylcarbamate, see Methiocarb
2-Methylmercapto-4-ethylamino-*s*-triazine, see Ametryn
4-Methylmercapto-3-methylphenyl dimethyl thiophosphate, see Fenthion
Methyl-*N*-(((methylamino)carbonyl)oxy)ethanimidothioate, see Methomyl
Methyl-*N*-((methylcarbamoyl)oxy)thioacetimidate, see Methomyl
S-Methyl *N*-((methylcarbamoyl)oxy)thioacetimidate, see Methomyl
Methyl *O*-(methylcarbamoyl)thioacethoydroxamate, see Methomyl
cis-1-Methyl-(2-methylcarbamoyl)vinyl phosphate, see Monocrotophos
Methyl 3-(3-methylcarbaniloyloxy)carbanilate, see Phenmedipham
2-Methyl-2-(methylthio)propanal *O*-((methylamino)carbonyl)oxime, see Aldicarb
2-Methyl-2-(methylthio)propionaldehyde *O*-(methylcarbamoyl)oxime, see Aldicarb
Methyl mustard oil, see Methylisothiocyanate
N-Methyl-1-naphthylcarbamate, see Carbaryl
N-Methyl-α-naphthylcarbamate, see Carbaryl
N-Methyl-α-naphthylurethan, see Carbaryl
6-Methyl-2-oxo-1,3-dithiolo[4,5-*b*]quinoxaline, see Quinomethionate
(Methylphenyl)carbamic acid 3-((methoxycarbonyl)amino)phenyl ester, see
 Phenmedipham
Methyl phthalate, see Dimethyl phthalate
2-(1-Methylpropyl)-4,6-dinitrophenol, see Dinoseb
6-Methyl-2,3-quinoxaline dithiocarbonate, see Quinomethionate
6-Methyl-2,3-quinoxalinedithiol cyclic carbonate, see Quinomethionate
6-Methyl-2,3-quinoxalinedithiol cyclic dithiocarbonate, see Quinomethionate

6-Methyl-2,3-quinoxalinedithiol cyclic *S,S*-dithiocarbonate, see Quinomethionate
6-Methylquinoxaline-2,3-dithiolcyclocarbonate, see Quinomethionate
S,S-(6-Methylquinoxaline-2,3-diyl) dithiocarbonate, see Quinomethionate
Methyl sulfanilylcarbamate, see Asulam
2-Methylthio-4,6-bis(isopropylamino)-*s*-triazine, see Prometryn
4-Methylthio-3,5-dimethylphenyl methylcarbamate, see Methiocarb
2-Methylthio-4-ethylamino-6-*tert*-butylamino-*s*-triazine, see Terbutryn
2-Methylthio-4-ethylamino-6-isopropylamino-*s*-triazine, see Ametryn
4-Methylthio-3,5-xylyl isomethylcarbamate, see Methiocarb
Methyl thiram, see Thiram
Methyl thiuramdisulfide, see Thiram
Methyl 3-(*m*-tolylcarbamoyloxy)phenylcarbamate, see Phenmedipham
Methyl tuads, see Thiram
Metifonate, see Trichlorfon
Metilmercaptofosoksid, see Oxydemeton-methyl
Metiltriazotion, see Azinphos-methyl
Metmercapturon, see Methiocarb
Metolachlor
Metox, see Methoxychlor
Metoxon, see Chlorpropham
Metribuzin
Metrifonate, see Trichlorfon
Metriphonate, see Trichlorfon
Metsulfuron-methyl
Meturon, see Fluometuron
Meturon 4L, see Fluometuron
Mevinphos
MH, see Maleic hydrazide
MH 30, see Maleic hydrazide
MH 36 Bayer, see Maleic hydrazide
MH 40, see Maleic hydrazide
MIC, see Methylisothiocyanate
Micofume, see Dazomet
Micro-check 12, see Captan
Microlysin, see Chloropicrin
Milbol, see Dicofol
Milbol 49, see Lindane
Miller's fumigrain, see Acrylonitrile
Milocep, see Metolachlor
Milogard, see Propazine
Milogard 4L, see Propazine
Milogard 80W, see Propazine
Mintesol, see Thiabendazole
Mintezol, see Thiabendazole
Minzolum, see Thiabendazole
Mipax, see Dimethyl phthalate
Miracle, see 2,4-D
MIT, see Methylisothiocyanate
Mitacil, see Aminocarb

Mitigan, see Dicofol
MITC, see Methylisothiocyanate
MK 360, see Thiabendazole
2M-4KH, see MCPA
ML 97, see Phosphamidon
MLT, see Malathion
Mobil V-C 9-104, see Ethoprop
Mocap, see Ethoprop
Modown, see Bifenox
Mole death, see Strychnine
Mole-nots, see Strychnine
Molinate
Molmate, see Molinate
Mon 0573, see Glyphosate
Monalide
Monitor, see Methamidophos
Monobromomethane, see Methyl bromide
Monocil 40, see Monocrotophos
Monocron, see Monocrotophos
Monocrotophos
N-Monomethylamide of *O,O*-dimethyldithiophosphorylacetic acid, see Dimethoate
Monosan, see 2,4-D
Monsanto penta, see Pentachlorophenol
Montrose Propanil, see Propanil
Monurex, see Monuron
Monuron
Monurox, see Monuron
Monuuron, see Monuron
Moosuran, see Pentachlorophenol
Mopari, see Dichlorvos
Morbicid, see Formaldehyde
Mor-cran, see Naptalam
Morestan, see Quinomethionate
Morestane, see Quinomethionate
Morton EP 161E, see Methylisothiocyanate
Mortopal, see Tetraethyl pyrophosphate
Moscardia, see Malathion
Motox, see Toxaphene
Mouse-pak, see Warfarin
Mouse-rid, see Strychnine
Mouse-tox, see Strychnine
Mowchem, see Mefluidide
Moxie, see Methoxychlor
Moxone, see 2,4-D
MPP, see Fenthion
MRC 910, see Iprodione
Mszycol, see Lindane
MTD, see Methamidophos

Multamat, see Bendiocarb
Multimet, see Bendiocarb
Murfos, see Parathion
Mustard oil, see Methylisothiocyanate
Mutoxin, see *p,p'*-DDT
Mycozol, see Thiabendazole
Mylon, see Dazomet
Mylone, see Dazomet
Mylone 85, see Dazomet
NA 521, see Dazomet
NA 1583, see Chloropicrin
NA 2757, see Carbaryl, Methiocarb
NA 2761, see Aldrin, Dichlone, Dicofol; Dieldrin, Endrin, Heptachlor, Kepone, Lindane, *p,p'*-DDD, Toxaphene
NA 2762, see Aldrin, Chlordane
NA 2763, see Diazinon
NA 2765, see 2,4-D, 2,4,5-T
NA 2767, see Diuron
NA 2769, see Dicamba, Dichlobenil
NA 2771, see Thiram
NA 2781, see Diquat
NA 2783, see Azinphos-methyl, Chlorpyrifos, Dichlorvos, Disulfoton, Ethion, Malathion, Mevinphos, Naled, Parathion, Tetraethyl pyrophosphate, Trichlorfon
NA 2790, see Fonofos
NA 9099, see Captan
NAC, see Carbaryl
Nalcon 243, see Dazomet
Naled
Nalkil, see Bromacil
Nanchor, see Ronnel
Nanker, see Ronnel
Nankor, see Ronnel
Naptalame, see Naptalam
1-Naphthalenol methylcarbamate, see Carbaryl
2-((1-Naphthalenylamino)carbonyl)benzoic acid, see Naptalam
1-Naphthalenylthiourea, see ANTU
1-Naphthol-*N*-methylcarbamate, see Carbaryl
2-(α-Napthoxy)-*N,N*-diethylpropionamide, see Napropamide
1-Naphthyl methylcarbamate, see Carbaryl
1-Naphthyl-*N*-methylcarbamate, see Carbaryl
α-Naphthyl-*N*-methylcarbamate, see Carbaryl
α-Naphthylphthalamic acid, see Naptalam
N-1-Naphthylphthalamic acid, see Naptalam
α-Naphthylthiocarbamide, see ANTU
1-(1-Naphthyl)-2-thiourea, see ANTU
α-Naphthylthiourea, see ANTU
N-1-Naphthylthiourea, see ANTU
Napropamide
Naptalam

Niagara 5006, see Dichlobenil
Niagara 5462, see α-Endosulfan, β-Endosulfan
Niagara 5996, see Dichlobenil
Niagara 9241, see Phosalone
Niagara 10242, see Carbofuran
Niagara 33297, see Permethrin
Nialate, see Ethion
Nicochloran, see Lindane
Nifos, see Tetraethyl pyrophosphate
Nifos T, see Tetraethyl pyrophosphate
Nifost, see Tetraethyl pyrophosphate
Niomil, see Bendiocarb
Nip-A-Thin, see Naptalam
Nipsan, see Diazinon
Niran, see Chlordane, Parathion
Niran E-4, see Parathion
Niticid, see Propachlor
Nitrador, see 4,6-Dinitro-*o*-cresol
Nitran, see Trifluralin
Nitrapyrin
Nitrile, see Acrylonitrile
Nitrochloroform, see Chloropicrin
Nitrofan, see 4,6-Dinitro-*o*-cresol
Nitropone P, see Dinoseb
Nitrostigmine, see Parathion
Nitrostygmine, see Parathion
Nitrotrichloromethane, see Chloropicrin
Niuif-100, see Parathion
Nobecutan, see Thiram
No bunt, see Hexachlorobenzene
No bunt 40, see Hexachlorobenzene
No bunt 80, see Hexachlorobenzene
No bunt liquid, see Hexachlorobenzene
Nogos, see Dichlorvos
Nogos G, see Dichlorvos
Nogos 50, see Dichlorvos
Nomersan, see Thiram
Nopcocide, see Chlorothalonil
Nopcocide N-96, see Chlorothalonil
Nopcocide N40D & N96, see Chlorothalonil
No-pest, see Dichlorvos
No-pest strip, see Dichlorvos
Nor-Am, see Diphenamid
Norex, see Chloroxuron
Normersan, see Thiram
Norosac, see Dichlobenil
Norosac 4G, see Dichlobenil
Norosac 10G, see Dichlobenil
Nourithion, see Parathion

Oleoparathion, see Parathion
Oleophosphothion, see Malathion
Olitref, see Trifluralin
Olpisan, see Pentachloronitrobenzene
Omnitox, see Lindane
Omnizole, see Thiabendazole
OMS 2, see Fenthion
OMS 14, see Dichlorvos
OMS 29, see Carbaryl
OMS 33, see Propoxur
OMS 37, see Fensulfothion
OMS 93, see Methiocarb
OMS 570, see α-Endosulfan, β-Endosulfan
OMS 771, see Aldicarb
OMS 971, see Chlorpyrifos
OMS 1804, see Diflubenzuron
Ontracic 800, see Prometon
Ontrack, see Prometon
Ontrack 8E, see Metolachlor
Ontrack-WE-2, see Prometon
OR 1191, see Phosphamidon
Ordram, see Molinate
Ordram 8E, see Molinate
Ordram 10G, see Molinate
Ordram 15G, see Molinate
Orga 414, see Amitrole
Ornamental weeder, see Chloramben
Orthene, see Acephate
Orthene 755, see Acephate
Ortho, see Diquat
Ortho 4355, see Naled
Ortho 9006, see Methamidophos
Ortho 12420, see Acephate
Orthocide, see Captan
Orthocide 7.5, see Captan
Orthocide 50, see Captan
Orthocide 406, see Captan
Orthodibrom, see Naled
Orthodibromo, see Naled
Ortho grass killer, see Propham
Orthoklor, see Chlordane
Orthomalathion, see Malathion
Orthophos, see Parathion
Ortho phosphate defoliant, see Butifos
Ortran, see Acephate
Ortril, see Acephate
OS 1987, see 1,2-Dibromo-3-chloropropane
OS 2046, see Mevinphos
Osocide, see Captan

Oust, see Sulfometuron-methyl

Outflanf-stockade, see Permethrin

Outflank, see Permethrin

Ovadziak, see Lindane

Owadziak, see Lindane

7-Oxabicyclo[2.2.1]heptane-2,3-dicarboxylic acid, see Endothall

Oxadiazon

Oxamyl

Oxomethane, see Formaldehyde

1,1′-Oxybis(2-chloroethane), see Bis(2-chloroethyl)ether

2,2′-Oxybis(1-chloropropane), see Bis(2-chloroisopropyl)ether

Oxy DBCP, see 1,2-Dibromo-3-chloropropane

Oxydemetonmethyl, see Oxydemeton-methyl

Oxydemeton-methyl

Oxymethylene, see Formaldehyde

Oxythioquinox, see Quinomethionate

Oxytril M, see Bromoxynil

P-974, see Phosalone

PA, see Naptalam

Pac, see Parathion

Pakhtaran, see Fluometuron

Palatinol C, see Di-*n*-butyl phthalate

Palatinol M, see Dimethyl phthalate

Panam, see Carbaryl

Pandar, see Monocrotophos

Panoram 75, see Thiram

Panoram D-31, see Dieldrin

Panthion, see Parathion

Pantozol 1, see Crotoxyphos

Parachlorocidum, see *p,p′*-DDT

Paradust, see Parathion

Paraflow, see Parathion

Paraform, see Formaldehyde

Paramar, see Parathion

Paramar 50, see Parathion

Paraphos, see Parathion

Paraspray, see Parathion

Parathene, see Parathion

Parathion

Parathionethyl, see Parathion

Parawet, see Parathion

Partner, see Bromoxynil

Pathclear, see Diquat

Pay-off, see Flucythrinate, Pendimethalin

Payze, see Cyanazine

PCC, see Toxaphene

PCNB, see Pentachloronitrobenzene

PCP, see Pentachlorophenol

PD 5, see Mevinphos

PDD 60401, see Diflubenzuron
Peach-Thin, see Naptalam
PEB1, see *p,p'*-DDT
PEBC, see Pebulate
Pebulate
Pedraczak, see Lindane
PEI 35, see Dimethoate
Penchlorol, see Pentachlorophenol
Pendimethalin
Pennamine, see 2,4-D
Pennamine D, see 2,4-D
Pennant, see Metolachlor
Pennant 5G, see Metolachlor
Pennout, see Endothall
Penoxalin, see Pendimethalin
Penoxaline, see Pendimethalin
Penoxyn, see Pendimethalin
Penphene, see Toxaphene
Penphos, see Parathion
Penta, see Pentachlorophenol
Pentac; see Dienochlor
Pentac WP;
Pentachlorin, see *p,p'*-DDT
Pentachlorobenzene
Pentachlorofenol, see Pentachlorophenol
Pentachloronitrobenzene
Pentachloronitrobenzol, see Pentachloronitrobenzene
Pentachlorophenate, see Pentachlorophenol
Pentachlorophenol
Pentachlorphenol, see Pentachlorophenol
2,3,4,5,6-Pentachlorophenol, see Pentachlorophenol
Pentachlorophenyl chloride, see Hexachlorobenzene
Pentacon, see Pentachlorophenol
Pentagen, see Pentachloronitrobenzene
Pentakil, see Pentachlorophenol
Pentasol, see Pentachlorophenol
Pentech, see *p,p'*-DDT
Penwar, see Pentachlorophenol
Peratox, see Pentachlorophenol
Perchlorobenzene, see Hexachlorobenzene
Perchloromethane, see Carbon tetrachloride
Perfecthion, see Dimethoate
Perfekthion, see Dimethoate
Perfektion, see Dimethoate
Perflan, see Tebuthiuron
Perfmid, see Tebuthiuron
Permacide, see Pentachlorophenol
Permaguard, see Pentachlorophenol
Permasan, see Pentachlorophenol

Permatox DP-2, see Pentachlorophenol
Permatox Penta, see Pentachlorophenol
Permethrin
Permite, see Pentachlorophenol
Pestmaster, see Ethylene dibromide, Methyl bromide
Pestmaster EDB-85, see Ethylene dibromide
Pestox plus, see Parathion
Pethion, see Parathion
Pflanzol, see Lindane
PH 60-40, see Diflubenzuron
Phasolon, see Phosalone
PHC, see Propoxur
Phenacide, see Toxaphene
Phenamiphos, see Fenamiphos
Phenatox, see Toxaphene
Phenmedipham
Phenotan, see Dinoseb
Phenox, see 2,4-D
Phenoxalin, see Pendimethalin
Phenoxybenzyl-2-(4-chlorophenyl)isovalerate, see Fenvalerate
Phenoxylene 50, see MCPA
Phenoxylene Plus, see MCPA
Phenoxylene Super, see MCPA
cis-2-(1-Phenylethoxy)carbonyl-1-methylvinyl dimethyl phosphate, see Crotoxyphos
(3-Phenoxyphenyl)methyl 3-(2,2-dichloroethenyl)-2,2-dimethylcyclopropane carboxylate,
   see Permethrin
Phenphane, see Toxaphene
3-(1'-Phenyl-2'-acetylethyl)-4-hydroxycoumarin, see Warfarin
3-α-Phenyl-β-acetylethyl-4-hydroxycoumarin, see Warfarin
Phenylbenzene, see Biphenyl
Phenylcarbamic acid 1-methylethyl ester, see Propham
N-Phenyl isopropylcarbamate, see Propham
Phenyl perchloryl, see Hexachlorobenzene
Phenylphosphonothioic acid *O*-ethyl *O*-*p*-nitrophenyl ester, see EPN
1-Phenyl-3-(1,2,3-thiadiazol-5-yl)urea, see Thidiazuron
N-Phenyl-*N'*-(1,2,3-thiadiazol-5-yl)urea, see Thidiazuron
Philips-duphar PH 60-40, see Diflubenzuron
Phorate
Phorate-10G, see Phorate
Phortox, see 2,4,5-T
Phosalon, see Phosalone
Phosalone
Phoschlor, see Trichlorfon
Phoschlor R50, see Trichlorfon
Phosdrin, see Mevinphos
cis-Phosdrin, see Mevinphos
Phosethyl, see Fosetyl-aluminum
Phosethyl AL, see Fosetyl-aluminum
Phosfene, see Mevinphos

Phoskil, see Parathion
Phosmet
Phosphamid, see Dimethoate
Phosphamide, see Dimethoate
Phosphamidon
Phosphemol, see Parathion
Phosphenol, see Parathion
N-(Phosphonomethyl)glycine, see Glyphosate
Phosphonothioic acid O,O-diethyl O-(3,5,6-trichloro-2-pyridinyl) ester, see EPN
Phosphoramidothioic acid O,S-dimethyl ester, see Methamidophos
Phosphoric acid 2-chloro-3-(dimethylamino)-1-methyl-3-oxo-1-propenyl dimethyl ester, see Phosphamidon
Phosphoric acid 1,2-dibromo-2,2-dichloroethyl dimethyl ester, see Naled
Phosphoric acid 2,2-dichloroethenyl dimethyl ester, see Dichlorvos
Phosphoric acid 2,2-dichlorovinyl dimethyl ester, see Dichlorvos
Phosphoric acid dimethyl (1-methyl-3-(methylamino)-3-oxo-1-propenyl) ester, see Monocrotophos
Phosphoric acid (1-methoxycarboxypropen-2-yl) dimethyl ester, see Mevinphos
Phosphorodithioic acid S-((tert-butylthio)methyl) O,O-diethyl ester, see Terbufos
Phosphorodithioic acid S-(2-chloro-1-(1,3-dihydro-1,3-dioxo-2H-isoindol-2-yl)ethyl) O,O-diethyl ester, see Dialifos
Phosphorodithioic acid S-((6-chloro-2-oxo-3(2H)-benzoxazolyl)methyl) O,O-diethyl ester, see Phosalone
Phosphorodithioic acid S-(((4-chlorophenyl)thio)methyl) O,O-diethyl ester, see Carbophenothion
Phosphorodithioic acid S-(2-chloro-1-phthalimidoethyl) O,O-diethyl ester, see Dialifos
Phosphorodithioic acid O,O-diethyl ester, S,S-diester with methanedithiol, see Ethion
Phosphorodithioic acid O,O-diethyl ester S-((ethylthio)methyl) ester, see Phorate
Phosphorodithioic acid S-((1,3-dihydro-1,3-dioxo-2H-isoindol-2-yl)methyl) O,O-dimethyl ester, see Phosmet
Phosphorodithioic acid O,O-dimethyl ester, ester with 2-mercapto-N-methylacetamide, see Dimethoate
Phosphorodithioic acid O,O-dimethyl ester, S-ester with 3-mercaptomethyl-1,2,3-benzotriazin-4(3H)-one, see Azinphos-methyl
Phosphorodithioic acid O,O-dimethyl ester, S-ester with 4-(mercaptomethyl)-2-methoxy-Δ^2-1,3,4-thiadiazolin-5-one, see Methidathion
Phosphorodithioic acid O,O-dimethyl ester with N-(mercaptomethyl)-phthalimide, see Phosmet
Phosphorodithioic acid S-(((1,1-dimethylethyl)thio)methyl) O,O-diethyl ester, see Terbufos
Phosphorodithioic acid O,O-dimethyl S-(2-(methylamino)-2-oxoethyl) ester, see Dimethoate
Phosphorodithioic acid O,O-dimethyl ((4-oxo-1,2,3-benzotriazin-3(4H)-yl)methyl) ester, see Azinphos-methyl
Phosphorodithioic acid S-ester of 6-chloro-3-mercaptomethylbenzoxazyol-2-one, see Phosalone
Phosphorodithioic acid O-ethyl S,S-dipropyl ester, see Ethoprop
Phosphorothioic acid O-(4-bromo-2-chlorophenyl)-O-ethyl-S-propyl ester, see Profenofos
Phosphorothioic acid O,O-diethyl O-(p-(methylsulfinyl)phenyl) ester, see Fensulfothion

Phosphorothioic acid *O,O*-diethyl *O*-(4-nitrophenyl) ester, see Parathion
Phosphorothioic acid *O,O*-dimethyl *O*-(3-methyl-4-(methylthio)phenyl) ester, see
    Fenthion
Phosphorothioic acid *O,O*-dimethyl *O*-(2,4,5-trichlorophenyl)ester, see Ronnel
Phosphorothionic acid *O,O*-diethyl *O*-(3,5,6-trichloro-2-pyridyl)ester, see Chlorpyrifos
Phosphostigmine, see Parathion
Phosphothioic acid *O,O*-dimethyl *S*-2-(ethylsulfinyl)ethyl ester, see Oxydemetonmethyl
Phosphothion, see Malathion
Phosphotox E, see Ethion
Phosvit, see Dichlorvos
Phozalon, see Phosalone
Phtalofos, see Phosmet
Phthalic acid dibutyl ester, see Di-*n*-butyl phthalate
Phthalic acid dimethyl ester, see Dimethyl phthalate
Phthalic acid methyl ester, see Dimethyl phthalate
Phthalimidomethyl-*O,O*-dimethylphosphorodithioate, see Phosmet
Phygon, see Dichlone
Phygon paste, see Dichlone
Phygon seed protectant, see Dichlone
Phygon XL, see Dichlone
Pic-clor, see Chloropicrin
Picfume, see Chloropicrin
Picloram
Picride, see Chloropicrin
Pied piper mouse seed, see Strychnine
Pielik, see 2,4-D
Pillardrin, see Monocrotophos
Pillaron, see Methamidophos
Pillarzo, see Alachlor
Pin, see EPN
Pindone
Pirofos, see Sulfotepp
Pistol, see Phenmedipham
Pivacin, see Pindone
Pival, see Pindone
2-Pivaloylindane-1,3-dione, see Pindone
2-Pivaloyl-1,3-indanedione, see Pindone
Pivalyl, see Pindone
Pivalyl indandione, see Pindone
2-Pivalyl-1,3-indandione, see Pindone
Pivalyl valone, see Pindone
PKhNB, see Pentachloronitrobenzene
Planotox, see 2,4-D
Plant dithio aerosol, see Sulfotepp
Plantdrin, see Monocrotophos
Plantfume 103 smoke generator, see Sulfotepp
Plantgard, see 2,4-D
Plantulin, see Propazine
PMP, see Phosmet

Polfoschlor, see Trichlorfon
Policar MZ, see Mancozeb
Policar S, see Mancozeb
Polisin, see Prometryn
Polival, see Thiabendazole
Polychlorcamphene, see Toxaphene
Polychlorinated camphenes, see Toxaphene
Polychlorocamphene, see Toxaphene
Polycizer DBP, see Di-*n*-butyl phthalate
Polycron, see Profenofos
Polyoxymethylene glycols, see Formaldehyde
Polyram ultra, see Thiram
Pomarsol, see Thiram
Pomasol, see Thiram
Pomersol forte, see Thiram
Potablan, see Monalide
Pounce, see Permethrin
PP 383, see Cypermethrin
PP 557, see Permethrin
PPzeidan, see *p,p′*-DDT
Pramitol, see Prometon
Prebane, see Terbutryn
Preeglone, see Diquat
Preflan, see Tebuthiuron
Prefmid, see Tebuthiuron
Premalin, see Linuron
Premalox, see Propham
Premazine, see Simazine
Premerge, see Dinoseb
Premerge 3, see Dinoseb
Prep, see Ethephon
Prevenol, see Chlorpropham
Prevenol 56, see Chlorpropham
Preventol, see Chlorpropham
Preventol 56, see Chlorpropham
Preview, see Metribuzin
Preweed, see Chlorpropham
Prezervit, see Dazomet
Priltox, see Pentachlorophenol
Primagram, see Metolachlor
Primatol, see Atrazine, Prometon, Prometryn
Primatol A, see Atrazine
Primatol 25E, see Prometon
Primatol P, see Propazine
Primatol Q, see Prometryn
Primatol S, see Simazine
Primaze, see Atrazine
Primextra, see Metolachlor
Princep, see Simazine

Princep 4G, see Simazine
Princep 4L, see Simazine
Princep 80W, see Simazine
Printop, see Simazine
Prioderm, see Malathion
Profam, see Propham
Profenofos
Profume, see Methyl bromide
Profume A, see Chloropicrin
Prokarbol, see 4,6-Dinitro-*o*-cresol
Prolate, see Phosmet
Prolex, see Propachlor
Promamide, see Propyzamide
Prometon
Prometone, see Prometon
Prometrex, see Prometryn
Prometrin, see Prometryn
Prometryn
Prometryne, see Prometryn
Promidione, see Iprodione
Pronamide, see Propyzamide
Propachlor
Propachlore, see Propachlor
Propanex, see Propanil
Propanid, see Propanil
Propanide, see Propanil
Propanil
Propasin, see Propazine
Propazin, see Propazine
Propazine
Propenal, see Acrolein
2-Propenal, see Acrolein
Prop-2-en-1-al, see Acrolein
Propenenitrile, see Acrylonitrile
2-Propenenitrile, see Acrylonitrile
2-Propen-1-one, see Acrolein
Propham
Prophame, see Propham
Prophos, see Ethoprop
Propionic acid 3,4-dichloroanilide, see Propanil
Prop-Job, see Propanil
Propotox M, see Propoxur
Propozur
Propoxure, see Propoxur
S-Propyl butylethylthiocarbamate, see Pebulate
Propylene chloride, see 1,2-Dichloropropane
Propylene dichloride, see 1,2-Dichloropropane
α,β-Propylene dichloride, see 1,2-Dichloropropane
Propylethylbutylthiocarbamate, see Pebulate

Propyl *N*-ethyl-*n*-butylthiocarbamate, see Pebulate
n-Propyl-*N*-ethyl-*N*-(*n*-butyl)thiocarbamate, see Pebulate
S-(*n*-Propyl)-*N*-ethyl-*N,N*-butylthiocarbamate, see Pebulate
Propylethylbutylthiolcarbamate, see Pebulate
Propyl ethyl-*n*-butylthiolcarbamate, see Pebulate
Propylthiopyrophosphate, see Aspon
Propyon, see Propoxur
Propyzamide
Prothromadin, see Warfarin
Protrum K, see Phenmedipham
Prowl, see Pendimethalin
Proxol, see Trichlorfon
Prozinex, see Propazine
PS, see Chloropicrin
Puralin, see Thiram
PX 104, see Di-*n*-butyl phthalate
Pyradex, see Diallate
Pyrinex, see Chlorpyrifos
Pyrophosphoric acid tetraethyl ester, see Tetraethyl pyrophosphate
Pyrophosphorodithioic acid *O,O,O,O*-tetraethyl dithionopyrophosphate, see Sulfotepp
Pyrophosphorodithioic acid tetraethyl ester, see Sulfotepp
6Q8, see Naptalam
QCB, see Pentachlorobenzene
Queletox, see Fenthion
Quellada, see Lindane
Quilan, see Benfluralin
Quinomethionate
Quinomethoate, see Quinomethionate
Quintar, see Dichlone
Quintar 540F, see Dichlone
Quintocene, see Pentachloronitrobenzene
Quintox, see Dieldrin
Quintozen, see Pentachloronitrobenzene
Quintozene, see Pentachloronitrobenzene
Quizalofop-ethyl
R 10, see Carbon tetrachloride
R 40B1, see Methyl bromide
R 1303, see Carbophenothion
R 1504, see Phosmet
R 1582, see Azinphos-methyl
R 1608, see EPTC
R 1910, see Butylate
R 2061, see Pebulate
R 2063, see Cycloate
R 2170, see Oxydemeton-methyl
R 4572, see Molinate
R 7465, see Napropamide
R 7475, see Napropamide
Racusan, see Dimethoate

Radapon, see Dalapon-sodium
Radazin, see Atrazine
Radizine, see Atrazine
Radocon, see Simazine
Radokor, see Simazine
Radox, see Allidochlor
Radoxone TL, see Amitrole
Rafex, see 4,6-Dinitro-*o*-cresol
Rafex 35, see 4,6-Dinitro-*o*-cresol
Ramizol, see Amitrole
Rampart, see Phorate
Ramrod, see Propachlor
Ramrod 65, see Propachlor
Randox, see Allidochlor
Randox T, see Allidochlor
Raphatox, see 4,6-Dinitro-*o*-cresol
Raphone, see MCPA
Rat-a-way, see Warfarin
Rat & mice bait, see Warfarin
Rat-b-gon, see Warfarin
Rat-gard, see Warfarin
Rat-kill, see Warfarin
Rat-mix, see Warfarin
Rat-o-cide #2, see Warfarin
Ratola, see Warfarin
Ratorex, see Warfarin
Ratox, see Warfarin
Ratoxin, see Warfarin
Ratron, see Warfarin
Ratron G, see Warfarin
Rats-no-more, see Warfarin
Rattrack, see ANTU
Rat-trol, see Warfarin
Rattunal, see Warfarin
Ravyon, see Carbaryl
Rax, see Warfarin
Razol dock killer, see MCPA
RB, see Parathion
RCRA waste number P001, see Warfarin
RCRA waste number P003, see Acrolein
RCRA waste number P004, see Aldrin
RCRA waste number P020, see Dinoseb
RCRA waste number P022, see Carbon disulfide
RCRA waste number P037, see Dieldrin
RCRA waste number P039, see Disulfoton
RCRA waste number P044, see Dimethoate
RCRA waste number P047, see 4,6-Dinitro-*o*-cresol
RCRA waste number P050, see α-Endosulfan, β-Endosulfan
RCRA waste number P051, see Endrin

RCRA waste number P059, see Heptachlor
RCRA waste number P066, see Methomyl
RCRA waste number P070, see Aldicarb
RCRA waste number P088, see Endothall
RCRA waste number P089, see Parathion
RCRA waste number P094, see Phorate
RCRA waste number P108, see Strychnine
RCRA waste number P109, see Sulfotepp
RCRA waste number P111, see Tetraethyl pyrophosphate
RCRA waste number P123, see Toxaphene
RCRA waste number U009, see Acrylonitrile
RCRA waste number U011, see Amitrole
RCRA waste number U025, see Bis(2-chloroethyl)ether
RCRA waste number U027, see Bis(2-chloroisopropyl)ether
RCRA waste number U029, see Methyl bromide
RCRA waste number U036, see Chlordane
RCRA waste number U038, see Chlorobenzilate
RCRA waste number U060, see *p,p'*-DDD
RCRA waste number U061, see *p,p'*-DDT
RCRA waste number U062, see Diallate
RCRA waste number U066, see 1,2-Dibromo-3-chloropropane
RCRA waste number U067, see Ethylene dibromide
RCRA waste number U069, see Di-*n*-butyl phthalate
RCRA waste number U083, see 1,2-Dichloropropane
RCRA waste number U102, see Dimethyl phthalate
RCRA waste number U122, see Formaldehyde
RCRA waste number U127, see Hexachlorobenzene
RCRA waste number U129, see Lindane
RCRA waste number U142, see Kepone
RCRA waste number U148, see Maleic hydrazide
RCRA waste number U183, see Pentachlorobenzene
RCRA waste number U185, see Pentachloronitrobenzene
RCRA waste number U192, see Propyzamide
RCRA waste number U211, see Carbon tetrachloride
RCRA waste number U232, see 2,4,5-T
RCRA waste number U240, see 2,4-D
RCRA waste number U242, see Pentachlorophenol
RCRA waste number U244, see Thiram
RCRA waste number U247, see Methoxychlor
RE 4355, see Naled
RE 12420, see Acephate
Rebelate, see Dimethoate
Reddon, see 2,4,5-T
Reddox, see 2,4,5-T
Reglon, see Diquat
Reglone, see Diquat
Regulox, see Maleic hydrazide
Regulox W, see Maleic hydrazide
Regulox 50 W, see Maleic hydrazide

Retard, see Maleic hydrazide
Rezifilm, see Thiram
RH-315, see Propyzamide
Rhodia, see 2,4-D
Rhodiachlor, see Heptachlor
Rhodiacide, see Ethion
Rhodia RP 11974, see Phosalone
Rhodiasol, see Parathion
Rhodiatox, see Parathion
Rhodiatrox, see Parathion
Rhodocide, see Ethion
Rhomenc, see MCPA
Rhomene, see MCPA
Rhonox, see MCPA
Rhothane, see *p,p'*-DDD
Rhothane D-3, see *p,p'*-DDD
Ricifon, see Trichlorfon
Ridomil, see Metalaxyl
Ridomil E, see Metalaxyl
Ripcord, see Cypermethrin
Ripenthal, see Endothall
Riselect, see Propanil
Ritsifon, see Trichlorfon
Rodafarin, see Warfarin
Rodeth, see Warfarin
Rodex, see Strychnine, Warfarin
Rodex blox, see Warfarin
Rodocid, see Ethion
Rogodial, see Dimethoate
Rogor 20L, see Dimethoate
Rogor 40, see Dimethoate
Rogor, see Dimethoate
Rogor L, see Dimethoate
Rogor P, see Dimethoate
Rogue, see Propanil
Rollfruct, see Ethephon
Ro-neet, see Cycloate
Ronit, see Cycloate
Ronnel
Ronstar, see Oxadiazon
Ronstar 25EC, see Oxadiazon
Ronstar 2G, see Oxadiazon
Ronstar 12L, see Oxadiazon
Rop 500 F, see Iprodione
Rosanil, see Propanil
Rosex, see Warfarin
Rosuran, see Monuron
Rotate, see Bendiocarb
Rothane, see *p,p'*-DDD

Rotox, see Methyl bromide
Rough & ready mouse mix, see Warfarin
Rovral, see Iprodione
Roxion, see Dimethoate
Roxion U.A., see Dimethoate
Royal MH-30, see Maleic hydrazide
Royal Slo-Gro, see Maleic hydrazide
Royal TMTD, see Thiram
RP 8167, see Ethion
RP 11974, see Phosalone
RP 16272, see Bromoxynil octanoate
RP 17623, see Oxadiazon
RP 26019, see Iprodione
RP 32545, see Fosetyl-aluminum
RS 141, see Chlordimeform
Rubitox, see Phosalone
Rukseam, see *p,p'*-DDT
Rylam, see Carbaryl
S 1, see Chloropicrin
S 276, see Disulfoton
S 1752, see Fenthion
S 3151, see Permethrin
S 5602, see Fenvalerate
S 10165, see Propanil
Sadofos, see Malathion
Sadophos, see Malathion
Sadoplon, see Thiram
Sakkimol, see Molinate
Salvo, see 2,4-D
Sanaseed, see Strychnine
Sancap 80W, see Dipropetryn
Sandolin, see 4,6-Dinitro-*o*-cresol
Sandolin A, see 4,6-Dinitro-*o*-cresol
Saniclor, see Pentachloronitrobenzene
Saniclor 30, see Pentachloronitrobenzene
Sanmarton, see Fenvalerate
Sanocide, see Hexachlorobenzene
Sanquinon, see Dichlone
Santobane, see *p,p'*-DDT
Santobrite, see Pentachlorophenol
Santophen, see Pentachlorophenol
Santophen 20, see Pentachlorophenol
Santox, see EPN
Sarclex, see Linuron
Sarolex, see Diazinon
Satecid, see Propachlor
Satox 20WSC, see Trichlorfon
SBP-1513, see Permethrin
Scarclex, see Linuron

Schering 35830, see Monalide
Schering 36268, see Chlordimeform
Schering 38107, see Desmedipham
Schering 38584, see Phenmedipham
SD 40, see Isofenphos
SD 897, see 1,2-Dibromo-3-chloropropane
SD 1750, see Dichlorvos
SD 3562, see Dicrotophos
SD 4294, see Crotoxyphos
SD 5532, see Chlordane
SD 9129, see Monocrotophos
SD 14114, see Fenbutatin oxide
SD 14999, see Methomyl
SD 15418, see Cyanazine
SD 43775, see Fenvalerate
Seedrin, see Aldrin
Seedrin liquid, see Aldrin
Seffein, see Carbaryl
Selecron, see Profenofos
Selektin, see Prometryn
Selephos, see Parathion
Selinon, see 4,6-Dinitro-*o*-cresol
Sencor, see Metribuzin
Sencor 4, see Metribuzin
Sencor DF, see Metribuzin
Sencoral, see Metribuzin
Sencorer, see Metribuzin
Sencorex, see Metribuzin
Sendran, see Propoxur
Seppic MMD, see MCPA
Septene, see Carbaryl
Sesagard, see Prometryn
Sevimol, see Carbaryl
Sevin, see Carbaryl
SF 60, see Malathion
Shamrox, see MCPA
Shell atrazine herbicide, see Atrazine
Shell SD 3562, see Dicrotophos
Shell SD 4294, see Crotoxyphos
Shell SD 5532, see Chlordane
Shell SD 9129, see Monocrotophos
Shell SD 14114, see Fenbutatin oxide
Shortstop, see Terbutryn
Shortstop E, see Terbutryn
Siduron
Silvanol, see Lindane
Sim-Trol, see Simazine
Simadex, see Simazine
Simanex, see Simazine

Simazin, see Simazine
Simazine
Simazine 80W, see Simazine
Simazol, see Amitrole
Sinbar, see Terbacil
Sinituho, see Pentachlorophenol
Sinoratox, see Dimethoate
Sinox, see 4,6-Dinitro-*o*-cresol
Sinox general, see Dinoseb
Sinuron, see Linuron
Siperin, see Cypermethrin
Siptox I, see Malathion
Sixty-three special E.C. insecticide, see Parathion
Slimicide, see Acrolein
Slo-Grow, see Maleic hydrazide
Smut-go, see Hexachlorobenzene
SN 4075, see Phenmedipham
SN 35830, see Monalide
SN 36268, see Chlordimeform
SN 38107, see Desmedipham
SN 38584, see Phenmedipham
SN 49537, see Thidiazuron
Snieciotox, see Hexachlorobenzene
SNP, see Parathion
Sodium dalapon, see Dalapon-sodium
Sodium 2,2-dichloropropanoate, see Dalapon-sodium
Sodium α,α-dichloropropionate, see Dalapon-sodium
Soilbrom-40, see Ethylene dibromide
Soilbrom-85, see Ethylene dibromide
Soilbrom-90, see Ethylene dibromide
Soilbrom-90EC, see Ethylene dibromide
Soilbrom-100, see Ethylene dibromide
Soilbrome-85, see Ethylene dibromide
Soilfume, see Ethylene dibromide
Sok, see Carbaryl
Soldep, see Trichlorfon
Soleptax, see Heptachlor
Solfarin, see Warfarin
Solo, see Naptalam
Solvanom, see Dimethyl phthalate
Solvarone, see Dimethyl phthalate
Solvirex, see Disulfoton
Somonil, see Methidathion
Soprathion, see Ethion
Soprathion, see Parathion
Sotipox, see Trichlorfon
Soviet technical herbicide 2M-4C, see MCPA
75 SP, see Acephate
Spanon, see Chlordimeform

Stopgerme-S, see Chlorpropham
Storite, see Thiabendazole
Strathion, see Parathion
Strazine, see Atrazine
Strel, see Propanil
Streunex, see Lindane
Strobane-T, see Toxaphene
Strobane T-90, see Toxaphene
Strychnidin-10-one, see Strychnine
Strychnine
Strychnos, see Strychnine
Stunt-Man, see Maleic hydrazide
Subdue, see Metalaxyl
Subdue 2E, see Metalaxyl
Subdue 5SP, see Metalaxyl
Subitex, see Dinoseb
Sucker-Stuff, see Maleic hydrazide
Sulfatep, see Sulfotepp
Sulfometuron-methyl
Sulfotep, see Sulfotepp
Sulfotepp
Sulphocarbonic anhydride, see Carbon disulfide
Sulphos, see Parathion
Sulprofos
Sumicidin, see Fenvalerate
Sumifly, see Fenvalerate
Sumipower, see Fenvalerate
Sumitox, see Malathion
Suncide, see Propoxur
Superaven, see Difenzoquat methyl sulfate
Super-De-Sprout, see Maleic hydrazide
Super D weedone, see 2,4-D, 2,4,5-T
Superlysoform, see Formaldehyde
Supernox, see Propanil
Super rodiatox, see Parathion
Super Sprout Stop, see Maleic hydrazide
Super Sucker-Stuff, see Maleic hydrazide
Super Sucker-Stuff HC, see Maleic hydrazide
Sup'r flo, see Diuron
Surcopur, see Propanil
Surpracide, see Methidathion
Surpur, see Propanil
Susvin, see Monocrotophos
Sutan, see Butylate
Sweep, see Chlorothalonil
Symazine, see Simazine
Synklor, see Chlordane
Synpran N, see Propanil
Synthetic 3956, see Toxaphene

Szklarniak, see Dichlorvos
2,4,5-T
T-2, see 2,3,5-Trichlorobenzoic acid
T-47, see Parathion
Tafazine, see Simazine
Tafazine 50-W, see Simazine
Tahmabon, see Methamidophos
Tak, see Malathion
Talcord, see Permethrin
Talodex, see Fenthion
Tamaron, see Methamidophos
Tap 85, see Lindane
Tap 9VP, see Dichlorvos
Taphazine, see Simazine
Task, see Dichlorvos
Task tabs, see Dichlorvos
Tat chlor 4, see Chlordane
Taterpex, see Chlorpropham
Tattoo, see Bendiocarb
TBA, see 2,3,5-Trichlorobenzoic acid
2,3,6-TBA, see 2,3,5-Trichlorobenzoic acid
TBDZ, see Thiabendazole
TBH, see α-BHC, β-BHC, δ-BHC, Lindane
TBZ, see Thiabendazole
TCB, see 2,3,5-Trichlorobenzoic acid
2,3,5-TCB, see 2,3,5-Trichlorobenzoic acid
TCBA, see 2,3,5-Trichlorobenzoic acid
2,3,6-TCBA, see 2,3,5-Trichlorobenzoic acid
TCIN, see Chlorothalonil
m-TCPN, see Chlorothalonil
TDE, see *p,p'*-DDD
4,4'-TDE, see *p,p'*-DDD
p,p'-TDE, see *p,p'*-DDD
Tebulan, see Tebuthiuron
Tebuthiuron
Tecto 60, see Thiabendazole
Tecto RPH, see Thiabendazole
Tecto, see Thiabendazole
TEDP, see Sulfotepp
TEDTP, see Sulfotepp
Telar, see Chlorsulfuron
Telvar Diuron Weed Killer, see Diuron
Telvar Monuron Weed Killer, see Monuron
Telvar, see Diuron, Monuron
Temic, see Aldicarb
Temik, see Aldicarb
Temik 10 G, see Aldicarb
Temik G10, see Aldicarb
Temus W, see Warfarin

3a,4,7,7a-Tetrahydro-*N*-(trichloromethanesulphenyl)phthalimide, see Captan
3a,4,7,7a-Tetrahydro-2-((trichloromethyl)thio)-1*H*-isoindole-1,3(2*H*)-dione, see Captan
1,2,3,6-Tetrahydro-*N*-trichloromethylthio)phthalimide, see Captan
Tetramethyldiurane sulphite, see Thiram
Tetramethylenethiuram disulfide, see Thiram
2,4,6,8-Tetramethyl-1,3,5,7-tetraoxacyclooctane, see Metaldehyde
Tetramethylthiocarbamoyl disulfide, see Thiram
Tetramethylthioperoxydicarbonic diamide, see Thiram
Tetramethylthiuram bisulfide, see Thiram
Tetramethylthiuram bisulphide, see Thiram
Tetramethylthiuram disulfide, see Thiram
N,N-Tetramethylthiuram disulfide, see Thiram
N,N,N′,N′-Tetramethylthiuram disulfide, see Thiram
Tetramethylthiuram disulphide, see Thiram
Tetramethylthiurane disulphide, see Thiram
Tetramethylthiurum disulfide, see Thiram
Tetrapom, see Thiram
Tetra-*n*-propyl dithionopyrophosphate, see Aspon
Tetrapropyl dithiopyrophosphate, see Aspon
Tetra-n-propyl dithiopyrophosphate, see Aspon
Tetrasipton, see Thiram
Tetrasol, see Carbon tetrachloride
Tetrastigmine, see Tetraethyl pyrophosphate
Tetrathiuram disulfide, see Thiram
Tetrathiuram disulphide, see Thiram
Tetravos, see Dichlorvos
Tetron, see Tetraethyl pyrophosphate
Tetron-100, see Tetraethyl pyrophosphate
Texadust, see Toxaphene
TH 6040, see Diflubenzuron
Thiaben, see Thiabendazole
Thiabendazol, see Thiabendazole
Thisbendazole
Thiabenzazole, see Thiabendazole
Thiabenzol, see Thiabendazole
Thiameturon-methyl
2-(1,3-Thiazol-4-yl)benzimidazole, see Thiabendazole
2-(4-Thiazolyl)benzimidazole, see Thiabendazole
2-(4-Thiazolyl)-1*H*-benzimidazole, see Thiabendazole
2-(Thiazol-4-yl)benzimidazole, see Thiabendazole
Thiazon, see Dazomet
Thiazone, see Dazomet
Thibenzol, see Thiabendazole
Thibenzole, see Thiabendazole
Thibenzole ATT, see Thiabendazole
Thidiazuron
Thifor, see α-Endosulfan, β-Endosulfan
Thillate, see Thiram
Thimer, see Thiram

Thimet, see Phorate

Thimul, see α-Endosulfan, β-Endosulfan

N,N'-(Thiobis(methylimino)carbonyloxy)bisethanimidothioic acid dimethyl ester, see
  Thiodicarb

Thiodan, see α-Endosulfan, β-Endosulfan

Thiodemeton, see Disulfoton

Thiodemetron, see Disulfoton

Thiodicarb

2-Thio-3,5-dimethyltetrahydro-1,3,5-thiadiazine, see Dazomet

Thiodiphosphoric acid tetraethyl ester, see Sulfotepp

Thiofor, see α-Endosulfan, β-Endosulfan

Thiofos, see Parathion

Thiomul, see α-Endosulfan, β-Endosulfan

Thionex, see α-Endosulfan, β-Endosulfan

Thiophos 3422, see Parathion

Thiophosphoric acid tetraethyl ester, see Sulfotepp

Thiopyrophosphoric acid tetraethyl ester, see Sulfotepp

Thiosan, see Thiram

Thiosulfan, see α-Endosulfan, β-Endosulfan

Thiotepp, see Sulfotepp

Thiotex, see Thiram

Thiotox, see Thiram

Thiozamyl, see Oxamyl

Thiram

Thiram 75, see Thiram

Thiram B, see Thiram

Thiramad, see Thiram

Thirasan, see Thiram

Thiulix, see Thiram

Thiurad, see Thiram

Thiuram, see Thiram

Thiuram D, see Thiram

Thiuram M, see Thiram

Thiuram M rubber accelerator, see Thiram

Thiuramin, see Thiram

Thiuramyl, see Thiram

Thompson-Hayward TH6040, see Diflubenzuron

Thompson's wood fix, see Pentachlorophenol

Thylate, see Thiram

Tiazon, see Dazomet

Tiguvon, see Fenthion

Tilcarex, see Pentachloronitrobenzene

Tillam, see Pebulate

Timet, see Phorate

Timmam-6-E, see Pebulate

Tionel, see α-Endosulfan, β-Endosulfan

Tiophos, see Parathion

Tiovel, see α-Endosulfan, β-Endosulfan

Tippon, see 2,4,5-T

Tirampa, see Thiram
Tiuramyl, see Thiram
Tiurolan, see Tebuthiuron
TL 314, see Acrylonitrile
TM-4049, see Malathion
TMTD, see Thiram
TMTDS, see Thiram
Tomathrel, see Ethephon
Top Form Wormer, see Thiabendazole
Topichlor 20, see Chlordane
Topiclor, see Chlordane
Topiclor 20, see Chlordane
Torax, see Dialifos
Tordon, see Picloram
Tordon 10K, see Picloram
Tordon 101 mixture, see Picloram
Tormona, see 2,4,5-T
Torque, see Fenbutatin oxide
Tox 47, see Parathion
Toxaphene
Toxakil, see Toxaphene
Toxan, see Carbaryl
Toxhid, see Warfarin
Toxichlor, see Chlordane
Toxon 63, see Toxaphene
Toxyphen, see Toxaphene
TPN, see Chlorothalonil
Trametan, see Thiram
Transamine, see 2,4-D, 2,4,5-T
Trapex, see Methylisothiocyanate
Trapexide, see Methylisothiocyanate
Trasan, see MCPA
Trefanocide, see Trifluralin
Treficon, see Trifluralin
Treflam, see Trifluralin
Treflan, see Trifluralin
Treflanocide elancolan, see Trifluralin
Tri-6, see Lindane
Triadimefon
Triallate
Triasyn, see Anilazine
Triazin, see Anilazine
Triazine, see Anilazine
Triazine A, see Atrazine
Triazine A 384, see Simazine
Triazine A 1294, see Atrazine
Triazolamine, see Amitrole
1*H*-1,2,4-Triazol-3-amine, see Amitrole
Tribac, see 2,3,5-Trichlorobenzoic acid

Tri-Ban, see Pindone
Tributon, see 2,4-D, 2,4,5-T
S,S,S-Tributyl phosphorotrithioate, see Butifos
S,S,S-Tributyl trithiophosphate, see Butifos
Tricarnam, see Carbaryl
Trichlorfon
S-2,3,3-Trichloroallyl diisopropylthiocarbamate, see Triallate
S-2,3,3-Trichloroallyl *N,N*-diisopropylthiocarbamate, see Triallate
1,1,1-Trichloro-2,2-bis(*p*-anisyl)ethane, see Methoxychlor
Trichlorobenzoic acid, see 2,3,5-Trichlorobenzoic acid
2,3,5-Trichlorobenzoic acid
Trichlorobis(4-chlorophenyl)ethane, see *p,p'*-DDT
Trichlorobis(*p*-chlorophenyl)ethane, see *p,p'*-DDT
1,1,1-Trichloro-2,2-bis(*p*-chlorophenyl)ethane, see *p,p'*-DDT
2,2,2-Trichloro-1,1-bis(*p*-chlorophenyl)ethanol, see Dicofol
1,1,1-Trichloro-2,2-bis(*p*-methoxyphenyl)ethane, see Methoxychlor
1,1,1-Trichloro-2,2-bis(*p*-methoxyphenol)ethanol, see Methoxychlor
1,1,1-Trichloro-2,2-di(4-chlorophenyl)ethane, see *p,p'*-DDT
1,1,1-Trichloro-2,2-di(*p*-chlorophenyl)ethane, see *p,p'*-DDT
2,2,2-Trichloro-1,1-di(4-chlorophenyl)ethanol, see Dicofol
1,1,1-Trichloro-2,2-di(4-methoxyphenyl)ethane, see Methoxychlor
1,1'-(2,2,2-Trichloroethylidene)bis(4-chlorobenzene), see *p,p'*-DDT
1,1'-(2,2,2-Trichloroethylidene)bis(4-methoxybenzene), see Methoxychlor
Trichlorofon, see Trichlorfon
2,2,2-Trichloro-1-hydroxyethylphosphonate dimethyl ester, see Trichlorfon
2,2,2-Trichloro-1-hydroxyethylphosphonic acid dimethyl ester, see Trichlorfon
Trichlorometafos, see Ronnel
N-Trichloromethylmercapto-4-cyclohexene-1,2-dicarboximide, see Captan
N-Trichloromethylthiocyclohex-4-ene-1,2-dicarboximide, see Captan
N-((Trichloromethylthio)-4-cyclohexene-1,2-dicarboximide, see Captan
N-((Trichloromethyl)thio)tetrahydrophthalamide, see Captan
Trichloromethylthio-1,2,5,6-tetrahydrophthalamide, see Captan
N-Trichloromethylthio-3a,4,7,7a-tetrahydrophthalamide, see Captan
Trichloronitromethane, see Chloropicrin
Trichlorophon, see Trichlorfon
(2,4,5-Trichlorophenoxy)acetic acid, see 2,4,5-T
S-2,3,3-(Trichloro-2-propenyl) bis(1-methylethyl)carbamothioate, see Triallate
3,5,6-((Trichloro-2-pyridyl)oxy)acetic acid, see Triclopyr
Trichlorphene, see Trichlorfon
Trichlorphon, see Trichlorfon
Trichlorphon FN, see Trichlorfon
Triclopyr
Tri-clor, see Chloropicrin
Tridipam, see Thiram
Triendothal, see Endothall
Trifina, see 4,6-Dinitro-*o*-cresol
Trifluoralin, see Trifluralin
α,α,α-Trifluoro-2,6-dinitro-*N,N*-dipropyl-*p*-toluidine, see Trifluralin
5'-(1,1,1-Trifluoromethanesulphonamido)acet-2',4'-xylidide, see Mefluidide

Trifluralin
Trifluraline, see Trifluralin
Triflurex, see Trifluralin
3-(*m*-Trifluoromethylphenyl)-1,1-dimethylurea, see Fluometuron
N-(3-Trifluoromethylphenyl)-*N'*,*N'*-dimethylurea, see Fluometuron
N-(*m*-Trifluoromethylphenyl)-*N'*,*N'*-dimethylurea, see Fluometuron
Trifocide, see 4,6-Dinitro-*o*-cresol
Trifungol, see Ferbam
Triherbicide CIPC, see Chlorpropham
Triherbide, see Propham
Triherbide-IPC, see Propham
Trikepin, see Trifluralin
Trim, see Trifluralin
Trimetion, see Dimethoate
Trinex, see Trichlorfon
Trinoxol, see 2,4-D, 2,4,5-T
Trioxon, see 2,4,5-T
Trioxone, see 2,4,5-T
Tri-PCNB, see Pentachloronitrobenzene
Tripomol, see Thiram
Tris(dimethylcarbamodithioato-*S*,*S'*)iron, see Ferbam
Tris(dimethyldithiocarbamato)iron, see Ferbam
Trithion, see Carbophenothion
Trithion miticide, see Carbophenothion
Tritisan, see Pentachloronitrobenzene
Triziman, see Mancozeb
Triziman D, see Mancozeb
Trolen, see Ronnel
Trolene, see Ronnel
Troysan 142, see Dazomet
Tryben, see 2,3,5-Trichlorobenzoic acid
Trysben, see 2,3,5-Trichlorobenzoic acid
Trysben 200, see 2,3,5-Trichlorobenzoic acid
TTD, see Thiram
Tuads, see Thiram
Tuberit, see Propham
Tuberite, see Propham
Tuex, see Thiram
Tugon, see Trichlorfon
Tugon fliegenkugel, see Propoxur
Tugon fly bait, see Trichlorfon
Tugon stable spray, see Trichlorfon
Tulisan, see Thiram
Tupersan, see Siduron
Turbacil, see Terbacil
Turcam, see Bendiocarb
Twin light rat away, see Warfarin
U 46, see 2,4-D, MCPA, 2,4,5-T
U 46 D, see 2,4-D

U 46DP, see 2,4-D
U 46 M-fluid, see MCPA
U 4513, see Diphenamid
U 5043, see 2,4-D
UC 974, see Dazomet
UC 7744, see Carbaryl
UC 21149, see Aldicarb
UC 51762, see Thiodicarb
UDVF, see Dichlorvos
Ultracide, see Methidathion
UN 1062, see Methyl bromide
UN 1092, see Acrolein
UN 1093, see Acrylonitrile
UN 1131, see Carbon disulfide
UN 1198, see Formaldehyde
UN 1580, see Chloropicrin
UN 1605, see Ethylene dibromide
UN 1692, see Strychnine
UN 1704, see Sulfotepp
UN 1846, see Carbon tetrachloride
UN 1916, see Bis(2-chloroethyl)ether
UN 2209, see Formaldehyde
UN 2472, see Pindone
UN 2477, see Methylisothiocyanate
UN 2490, see Bis(2-chloroisopropyl)ether
UN 2588, see Amitrole
UN 2729, see Hexachlorobenzene
UN 2872, see 1,2-Dibromo-3-chloropropane
Unden, see Propoxur
Unicrop CIPC, see Chlorpropham
Unicrop DNBP, see Dinoseb
Unidron, see Diuron
Unifos, see Dichlorvos
Unifos 50 EC, see Dichlorvos
Unifume, see Ethylene dibromide
Union Carbide 7744, see Carbaryl
Union Carbide 21149, see Aldicarb
Union Carbide UC 21149, see Aldicarb
Unipon, see Dalapon-sodium
Uniroyal, see Dichlone
Univerm, see Carbon tetrachloride
Uragan, see Bromacil
Uragon, see Bromacil
Urox, see Monuron
Urox B, see Bromacil
Urox B water soluble concentrate weed killer, see Bromacil
Urox D, see Diuron
Urox HX, see Bromacil
Urox HX granular weed killer, see Bromacil

Velsicol compound 'R', see Dicamba
Velsicol 53-CS-17, see Heptachlor epoxide
Velsicol 58-CS-11, see Dicamba
Velsicol heptachlor, see Heptachlor
Vendex, see Fenbutatin oxide
Ventox, see Acrylonitrile
Veon, see 2,4,5-T
Veon 245, see 2,4,5-T
Verdican, see Dichlorvos
Verdipor, see Dichlorvos
Verdone, see MCPA
Vergemaster, see 2,4-D
Vergfru foratox, see Phorate
Vermicide Bayer 2349, see Trichlorfon
Vermoestricid, see Carbon tetrachloride
Vertac, see Propanil
Vertac 90%, see Toxaphene
Vertac dinitro weed killer, see Dinoseb
Vertac general weed killer, see Dinoseb
Vertac selective weed killer, see Dinoseb
Vertac toxaphene 90, see Toxaphene
Verton, see 2,4-D
Verton D, see 2,4-D
Verton 2D, see 2,4-D
Verton 2T, see 2,4,5-T
Vertron 2D, see 2,4-D
Vesakontuho MCPA, see MCPA
Vetiol, see Malathion
Vetrazine, see Cyromazine
Vidon 638, see 2,4-D
Vinyl alcohol 2,2-dichlorodimethyl phosphate, see Dichlorvos
Vinyl cyanide, see Acrylonitrile
Vinylofos, see Dichlorvos
Vinylophos, see Dichlorvos
Viozene, see Ronnel
Virginia-Carolina VC 9-104, see Ethoprop
Visko-rhap, see 2,4-D
Visko-rhap drift herbicides, see 2,4-D
Visko-rhap low volatile ester, see 2,4,5-T
Visko-rhap low volatile 4L, see 2,4-D
Vistar, see Mefluidide
Vistar herbicide, see Mefluidide
Vitavax, see Carboxin
Viton, see Lindane
Vitrex, see Parathion
Volfartol, see Trichlorfon
Volfazol, see Crotoxyphos
Vondcaptan, see Captan
Vondalhyde, see Maleic hydrazide

Vondozeb, see Mancozeb
Vondrax, see Maleic hydrazide
Vonduron, see Diuron
Vorlex, see Methylisothiocyanate
Vorox, see Amitrole
Vorox AA, see Amitrole
Vorox SS, see Amitrole
Vortex, see Methylisothiocyanate
Votexit, see Trichlorfon
Vulcafor TMTD, see Thiram
Vulkacit MTIC, see Thiram
Vydate, see Oxamyl
Vydate L insecticide/nematocide, see Oxamyl
Vydate L oxamyl insecticide/nematocide, see Oxamyl
80W, see Diphenamid
W 6658, see Simazine
Waran, see Warfarin
W.A.R.F. 42, see Warfarin
Warfarat, see Warfarin
Warfarin
Warfarin plus, see Warfarin
Warfarin Q, see Warfarin
Warf compound 42, see Warfarin
Warficide, see Warfarin
WEC 50, see Trichlorfon
Weddar, see 2,4,5-T
Weddar-64, see 2,4-D
Weddatul, see 2,4-D
Weedar ADS, see Amitrole
Weedar AT, see Amitrole
Weedar, see 2,4-D, MCPA
Weedar MCPA concentrate, see MCPA
Weedazin, see Amitrole
Weedazin arginit, see Amitrole
Weedazol, see Amitrole
Weedazol GP2, see Amitrole
Weedazol super, see Amitrole
Weedazol T, see Amitrole
Weedazol TL, see Amitrole
Weed-b-gon, see 2,4-D
Weed Blast, see Bromacil
Weedex, see Simazine
Weedex A, see Atrazine
Weedex granulat, see Amitrole
Weedez wonder bar, see 2,4-D
Weedoclor, see Amitrole
Weedol, see Diquat
Weedone, see 2,4-D, MCPA, Pentachlorophenol, 2,4,5-T
Weedone 2,4,5-T, see 2,4,5-T

Weedone LV4, see 2,4-D
Weedone MCPA ester, see MCPA
Weedrhap, see 2,4-D, MCPA
Weed tox, see 2,4-D
Weedtrine-D, see Diquat
Weedtrol, see 2,4-D
Weeviltox, see Carbon disulfide
Wham EZ, see Propanil
Winterwash, see 4,6-Dinitro-*o*-cresol
Witicizer 300, see Di-*n*-butyl phthalate
Witophen P, see Pentachlorophenol
WL 18236, see Methomyl
WL 19805, see Cyanazine
WL 43467, see Cypermethrin
WL 43479, see Permethrin
WL 43775, see Fenvalerate
WN 12, see Methylisothiocyanate
Wonuk, see Atrazine
Wotexit, see Trichlorfon
X-all liquid, see Amitrole
Y 2, see Propham
Y 3, see Chlorpropham
Yalan, see Molinate
Yeh-Yan-Ku, see Difenzoquat methyl sulfate
Yulan, see Molinate
Zeapur, see Simazine
Zeazin, see Atrazine
Zeazine, see Atrazine
Zeidane, see *p,p'*-DDT
Zelan, see MCPA
Zerdane, see *p,p'*-DDT
Zimanat, see Mancozeb
Zimaneb, see Mancozeb
Zimman-dithane, see Mancozeb
Zinochlor, see Anilazine
Zithiol, see Malathion
Zobar, see 2,3,5-Trichlorobenzoic acid
Zolon, see Phosalone
Zolone, see Phosalone
Zolon PM, see Phosalone
Zytox, see Methyl bromide

Appendix K. Cumulative Index[a]

645

[a] I and II refers to Montgomery (1996) and this work, respectively.